ENERGY MATERIALS

ENERGY MATERIALS

FUNDAMENTALS TO APPLICATIONS

Edited by

S.J. DHOBLE
*Department of Physics, RTM Nagpur University,
Nagpur, Maharashtra, India*

N. THEJO KALYANI
*Department of Applied Physics, Laxminarayan Institute of Technology,
Nagpur, Maharashtra, India*

B. VENGADAESVARAN
*Higher Institution Centre of Excellence (HICoE),
UM Power Energy Dedicated Advanced Center (UMPEDAC), Level 4,
Wisma R&D, University of Malaya, Jalan Pantai Baharu, Kuala Lumpur, Malaysia*

ABDUL KARIEM AROF
*Centre for Ionics University of Malaya,
Department of Physics, Faculty of Science,
University of Malaya Kuala Lumpur, Malaysia*

ELSEVIER

Elsevier
Radarweg 29, PO Box 211, 1000 AE Amsterdam, Netherlands
The Boulevard, Langford Lane, Kidlington, Oxford OX5 1GB, United Kingdom
50 Hampshire Street, 5th Floor, Cambridge, MA 02139, United States

Copyright © 2021 Elsevier Ltd. All rights reserved.

No part of this publication may be reproduced or transmitted in any form or by any means, electronic or mechanical, including photocopying, recording, or any information storage and retrieval system, without permission in writing from the publisher. Details on how to seek permission, further information about the Publisher's permissions policies and our arrangements with organizations such as the Copyright Clearance Center and the Copyright Licensing Agency, can be found at our website: www.elsevier.com/permissions.

This book and the individual contributions contained in it are protected under copyright by the Publisher (other than as may be noted herein).

Notices

Knowledge and best practice in this field are constantly changing. As new research and experience broaden our understanding, changes in research methods, professional practices, or medical treatment may become necessary.

Practitioners and researchers must always rely on their own experience and knowledge in evaluating and using any information, methods, compounds, or experiments described herein. In using such information or methods they should be mindful of their own safety and the safety of others, including parties for whom they have a professional responsibility.

To the fullest extent of the law, neither the Publisher nor the authors, contributors, or editors, assume any liability for any injury and/or damage to persons or property as a matter of products liability, negligence or otherwise, or from any use or operation of any methods, products, instructions, or ideas contained in the material herein.

Library of Congress Cataloging-in-Publication Data
A catalog record for this book is available from the Library of Congress

British Library Cataloguing-in-Publication Data
A catalogue record for this book is available from the British Library

ISBN: 978-0-12-823710-6

For information on all Elsevier publications visit our website at
https://www.elsevier.com/books-and-journals

Publisher: Matthew Deans
Acquisitions Editor: Christina Gifford
Editorial Project Manager: Isabella Silva
Production Project Manager: Prem Kumar Kaliamoorthi
Cover Designer: Victoria Pearson

Typeset by TNQ Technologies

Contents

Contributors   ix
Preface   xi

I
Fundamentals and overarching topics

1. Energy materials: Fundamental physics and latest advances in relevant technology
R.S. Gedam, N. Thejo Kalyani and S.J. Dhoble

1.1 Introduction   3
1.2 Energy materials   4
1.3 Fundamental physics   8
1.4 Recent advances in technology   21
1.5 Potential applications   23
1.6 The future   23
1.7 Conclusions   24
References   25

2. Solar cell technology
R.M. Pujahari

2.1 Introduction   27
2.2 Solar cell design   27
2.3 Solar cell design   38
2.4 Thin-film solar cell technologies   43
References   60

3. Energy materials: synthesis and characterization techniques
N.S. Bajaj and R.A. Joshi

3.1 Introduction   61
3.2 Synthesis techniques   62
3.3 Characterization techniques   65
3.4 Failure parameters   79
3.5 Failure analysis/damage evaluation   79

3.6 Conclusions   80
References   80

4. An extensive study of the adhesion and antifogging of the transparent polydimethylsiloxane/Sylgard coating system
A. Syafiq, B. Vengadaesvaran, N.A. Rahim, A.K. Pandey, A.R. Bushroa, K. Ramesh and S. Ramesh

4.1 Introduction   83
4.2 Methodology   84
4.3 Results and discussions   88
4.4 Conclusion   100
Acknowledgment   100
References   101

II
Photovoltaic Materials and devices

5. Crystalline silicon solar cells
R.M. Pujahari

5.1 Introduction   107
5.2 Efficient solar cells   111
5.3 Silicon solar cells and fabrication processes   118
5.4 Large area crystalline silicon solar cells   121
5.5 2D and 3D screen-printed metallization effect on industrial crystalline silicon solar cells   125
References   137

6. Introduction to solar energy and its conversion into electrical energy by using dye-sensitized solar cells
N.H. Rased, B. Vengadaesvaran, S.R.S. Raihan and N.A. Rahim

6.1 Introduction   139
6.2 Solar cells   139

vi CONTENTS

6.3 Dye-sensitized solar cells 141
6.4 Working principle of DSSCs 141
6.5 Anatomy of DSSCs 142
6.6 Materials used as electrodes of DSSCs 143
6.7 Synthesis 158
6.8 Summary 174
Acknowledgment 174
References 174

7. Dye-sensitized solar cells

Prashant K. Baviskar and Babasaheb R. Sankapal

7.1 Introduction 179
7.2 Construction and working of DSSCs 180
7.3 Materials requirement for DSSCs 183
7.4 Solar cell performance 185
7.5 Performance of DSSCs 190
7.6 Advanced colorful approach 193
7.7 DSSC limitations, commercialization, and future prospects 194
7.8 Summary and scope 198
References 200

8. Potential of nanooxidic materials and structures of photoanodes for DSSCs

Markus Diantoro, Siti Wihdatul Himmah, Thathit Suprayogi, Ulwiyatus Sa'adah, Arif Hidayat, Nandang Mufti and Nasikhudin

8.1 Introduction 213
8.2 DSSCs 215
8.3 Nanooxidic photoanode materials for DSSCs (TiO_2, ZnO, and SnO_2) 218
8.4 1D construction for photoanode material (fibers and rods) 229
8.5 Performance of DSSCs 236
8.6 Summary 241
Acknowledgments 241
References 241

9. Perovskite solar cells

Amol Nande, Swati Raut and S.J. Dhoble

9.1 Introduction 249
9.2 Perovskite materials 250
9.3 Synthesis techniques 252

9.4 Perovskite solar cell 254
9.5 Fabrication approach and device anatomy 257
9.6 Requirement of each layer 259
9.7 Working mechanism of PVSCs or device operation 262
9.8 Characterization technique 263
9.9 Device evaluation 267
9.10 Key challenges 270
9.11 Future outlook 271
9.12 Conclusion 272
References 272

III

Electrochemical energy conversion and storage

10. Layered and spinel structures as lithium-intercalated compounds for cathode materials

Z.I. Radzi, B. Vengadaesvaran, S. Ramesh and N.A. Rahim

10.1 Introduction 285
10.2 Layered structures 287
10.3 Spinel structures 301
10.4 Summary 304
References 305

11. Prospects and challenges in the selection of polymer electrolytes in advanced lithium—air batteries

M.Z. Kufian and Z. Osman

11.1 Introduction 313
11.2 Lithium—air batteries 314
11.3 Types of LABs 314
11.4 Electrolytes 318
11.5 Polymer host 322
11.6 Type of salt 324
11.7 Anion trapping agent 326
11.8 Solvents 328
11.9 Summary 329
Acknowledgments 329
References 329

CONTENTS

12. Li-S ion batteries: a substitute for Li-ion storage batteries
Kalpana R. Nagde and S.J. Dhoble

12.1 Introduction 335
12.2 Energy storage materials 335
12.3 Batteries 338
12.4 Novel materials for batteries 342
12.5 Lithium-ion batteries 345
12.6 Prior state-of-the-art of Li-ion batteries 351
12.7 Shortcomings of Li-ion batteries 355
12.8 Lithium—sulfur batteries 357
12.9 Extensive comparison 359
12.10 Current research on Li—S batteries 360
12.11 Applications 363
12.12 Future prospective 365
12.13 Conclusions 366
References 366

13. Glasses and glass-ceramics as sealants in solid oxide fuel cell applications
P.S. Anjana, M.S. Salinigopal and N. Gopakumar

13.1 Introduction 373
13.2 Solid oxide fuel cell 375
13.3 Operation of the SOFC 375
13.4 Components of the SOFC 377
13.5 SOFC component requirements 378
13.6 Requirements of sealants 379
13.7 Methodology 380
13.8 Sealant materials currently used 381
13.9 Conclusions 397
Acknowledgments 398
References 398
Further reading 404

IV
Lighting and Light emitting diodes

14. Role of rare-earth ions for energy-saving LED lighting devices
Nutan S. Satpute and S.J. Dhoble

14.1 Introduction 407
14.2 Fundamentals 407
14.3 Applications of various phosphor materials 427
14.4 Phosphor for LED lighting 427

14.5 Prospects for energy-saving technology 435
14.6 Conclusion and future outlook 436
References 438

15. Synthesis and luminescence study of silicate-based phosphors for energy-saving light-emitting diodes
Nilesh Ugemuge, Yatish R. Parauha and S.J. Dhoble

15.1 Introduction 445
15.2 Approach for energy-saving and eco-friendly lighting 446
15.3 Synthesis methods for alkaline-earth silicate-based phosphor 450
15.4 Review of recently reported silicate-based phosphors 461
15.5 Applications of silicate-based phosphor 469
15.6 Conclusions 473
15.7 Future scope 474
Acknowledgments 474
References 475

16. Investigations of energy-efficient RE(TTA)$_3$dmphen complexes dispersed in polymer matrices for solid-state lighting applications
Akhilesh Ugale, N. Thejo Kalyani and S.J. Dhoble

16.1 Introduction 481
16.2 Solid-state lighting 482
16.3 Requisite of solid-state lighting 483
16.4 Eco-friendly and energy-efficient organic complexes 483
16.5 Role of rare earths in SSL 486
16.6 Experimental 488
16.7 Result and discussion 491
16.8 Conclusions 501
References 502

17. Synthesis and characterization of energy-efficient Mq$_2$ (M = Zn, Cd, Ca, and Sr) organometallic complexes for OLED display applications
Prajakta P. Varghe, N. Thejo Kalyani, P.G. Shende and S.J. Dhoble

17.1 Introduction 505
17.2 Role of quinoline in display technology 507
17.3 Quinoline complexes: prior state of the art 508

viii CONTENTS

17.4 Synthesis of green phosphor for displays 509
17.5 Experimental 509
17.6 Results and discussion 510
17.7 Superiority: OLED displays 517
17.8 Limitations 519
17.9 Applications 519
17.10 Conclusions 519
17.11 Future perspectives 520
References 520

V

Practical concerns and beyond

18. Spectral response and quantum efficiency evaluation of solar cells: a review

M.Z. Farah Khaleda, B. Vengadaesvaran and N.A. Rahim

18.1 Introduction 526
18.2 Factors affecting the performance of solar cells 527
18.3 Research methodology 533
18.4 Spectral response and measurement quantities 534
18.5 Recent advancements 544
18.6 Challenges and limitations 556
18.7 Conclusions, future recommendations, and publication trends 557
Acknowledgments 560
References 561

19. Energy materials: Applications and propelling opportunities

N. Thejo Kalyani and S.J. Dhoble

19.1 Introduction 567
19.2 Overview of novel energy materials for energy-related applications 567
19.3 Research area connections 577
19.4 Opportunities and daunting challenges 577
19.5 Conclusions 578
19.6 Future outlook 579
References 579

20. Sustainability, recycling, and lifetime issues of energy materials

N. Thejo Kalyani, S.J. Dhoble, B. Vengadaesvaran and Abdul Kariem Arof

20.1 Introduction 581
20.2 Sustainability, recycling, and lifetime issues 582
20.3 Energy-harvesting and -conversion materials 586
20.4 Energy-storing materials 592
20.5 Energy-saving materials 597
20.6 Possible solutions 597
20.7 Conclusions 598
References 599

Index 603

Contributors

P.S. Anjana Department of Physics, All Saints' College, University of Kerala, Trivandrum, Kerala, India

Abdul Kariem Arof Centre for Ionics University of Malaya, Department of Physics, Faculty of Science, University of Malaya Kuala Lumpur, Malaysia

N.S. Bajaj Department of Physics, Toshniwal Arts, Commerce and Science College, Sengaon, Maharashtra, India

Prashant K. Baviskar Department of Physics, SN Arts, DJ Malpani Commerce and BN Sarda Science College (Autonomous), Sangamner, Maharashtra, India

A.R. Bushroa AMMP Centre Level 8, Engineering Tower, Faculty of Engineering, University of Malaya, Kuala Lumpur, Malaysia

S.J. Dhoble Department of Physics, RTM Nagpur University, Nagpur, Maharashtra, India

Markus Diantoro Department of Physics, Faculty of Mathematics and Natural Science, Universitas Negeri Malang, Malang, East Java, Indonesia; Center of Advanced Materials for Renewable Energy (CAMRY), Universitas Negeri Malang, Malang, East Java, Indonesia

M.Z. Farah Khaleda Higher Institution Centre of Excellence (HICoE), UM Power Energy Dedicated Advanced Center (UMPEDAC), Level 4, Wisma R&D, University of Malaya, Jalan Pantai Baharu, Kuala Lumpur, Malaysia

R.S. Gedam Department of Physics, Visvesvaraya National Institute of Technology, Nagpur, Maharashtra, India

N. Gopakumar Department of Physics, Mahatma Gandhi College, Research Centre, University of Kerala, Trivandrum, Kerala, India

Arif Hidayat Department of Physics, Faculty of Mathematics and Natural Science, Universitas Negeri Malang, Malang, East Java, Indonesia; Center of Advanced Materials for Renewable Energy (CAMRY), Universitas Negeri Malang, Malang, East Java, Indonesia

Siti Wihdatul Himmah Department of Physics, Faculty of Mathematics and Natural Science, Universitas Negeri Malang, Malang, East Java, Indonesia

R.A. Joshi Department of Physics, Toshniwal Arts, Commerce and Science College, Sengaon, Maharashtra, India

N. Thejo Kalyani Department of Applied Physics, Laxminarayan Institute of Technology, Nagpur, Maharashtra, India

M.Z. Kufian Centre for Ionics University of Malaya (C.I.U.M), Faculty of Science, Universiti Malaya, Kuala Lumpur, Malaysia

Nandang Mufti Department of Physics, Faculty of Mathematics and Natural Science, Universitas Negeri Malang, Malang, East Java, Indonesia; Center of Advanced Materials for Renewable Energy (CAMRY), Universitas Negeri Malang, Malang, East Java, Indonesia

Kalpana R. Nagde Department of Physics, Institute of Science, Nagpur, Maharashtra, India

Amol Nande Guru Nanak College of Science, Ballarpur, Maharashtra, India

Nasikhudin Department of Physics, Faculty of Mathematics and Natural Science, Universitas Negeri Malang, Malang, East Java, Indonesia; Center of Advanced Materials for Renewable Energy (CAMRY), Universitas Negeri Malang, Malang, East Java, Indonesia

Z. Osman Centre for Ionics University of Malaya (C.I.U.M), Faculty of Science, Universiti Malaya, Kuala Lumpur, Malaysia; Department of Physics, Faculty of Science, Universiti Malaya, Kuala Lumpur, Malaysia

A.K. Pandey Research Centre for Nano-Materials and Energy Technology (RCNMET), School of Science and Technology, Sunway University, Petaling Jaya, Selangor, Malaysia

Yatish R. Parauha Department of Physics, RTM Nagpur University, Nagpur, Maharashtra, India

R.M. Pujahari Assistant Professor in Physics, Department of Applied Sciences, ABES Institute of Technology (ABESIT), Ghaziabad, Uttar Pradesh, India

Z.I. Radzi Higher Institution Centre of Excellence (HICoE), UM Power Energy Dedicated Advanced Center (UMPEDAC), Level 4, Wisma R&D, University of Malaya, Jalan Pantai Baharu, Kuala Lumpur, Malaysia

N.A. Rahim Higher Institution Centre of Excellence (HICoE), UM Power Energy Dedicated Advanced Center (UMPEDAC), Level 4, Wisma R&D, University of Malaya, Jalan Pantai Baharu, Kuala Lumpur, Malaysia; Renewable Energy Research Group, King Abdulaziz University, Jeddah, Makkah Province, Saudi Arabia

S.R.S. Raihan Higher Institution Centre of Excellence (HICoE), UM Power Energy Dedicated Advanced Center (UMPEDAC), Level 4, Wisma R&D, University of Malaya, Jalan Pantai Baharu, Kuala Lumpur, Malaysia

S. Ramesh Center for Ionics University of Malaya, Department of Physics, Faculty of Science, University of Malaya, Kuala Lumpur, Malaysia

K. Ramesh Center for Ionics University of Malaya, Department of Physics, University of Malaya, Kuala Lumpur, Malaysia

N.H. Rased Higher Institution Centre of Excellence (HICoE), UM Power Energy Dedicated Advanced Center (UMPEDAC), Level 4, Wisma R&D, University of Malaya, Jalan Pantai Baharu, Kuala Lumpur, Malaysia

Swati Raut Department of Physics, RTM Nagpur University, Nagpur, Maharashtra, India

Ulwiyatus Sa'adah Department of Physics, Faculty of Mathematics and Natural Science, Universitas Negeri Malang, Malang, East Java, Indonesia

M.S. Salinigopal Department of Physics, All Saints' College, University of Kerala, Trivandrum, Kerala, India; Department of Physics, Mahatma Gandhi College, Research Centre, University of Kerala, Trivandrum, Kerala, India

Babasaheb R. Sankapal Nano Materials and Device Laboratory, Department of Physics, Visvesvaraya National Institute of Technology, Nagpur, Maharashtra, India

Nutan S. Satpute Department of Physics, RTM Nagpur University, Nagpur, Maharashtra, India

P.G. Shende Department of Surface Coating Technology, Laxminarayan Institute of Technology, Nagpur, Maharashtra, India

Thathit Suprayogi Department of Physics, Faculty of Mathematics and Natural Science, Universitas Negeri Malang, Malang, East Java, Indonesia

A. Syafiq Higher Institution Centre of Excellence (HICoE), UM Power Energy Dedicated Advanced Center (UMPEDAC), Level 4, Wisma R&D, University of Malaya, Jalan Pantai Baharu, Kuala Lumpur, Malaysia

Akhilesh Ugale Department of Applied Physics, G.H. Raisoni Academy of Engineering and Technology, Nagpur, Maharashtra, India

Nilesh Ugemuge Department of Physics, Anand Niketan College, Warora, Chandrapur, India

Prajakta P. Varghe Department of Surface Coating Technology, Laxminarayan Institute of Technology, Nagpur, Maharashtra, India

B. Vengadaesvaran Higher Institution Centre of Excellence (HICoE), UM Power Energy Dedicated Advanced Center (UMPEDAC), Level 4, Wisma R&D, University of Malaya, Jalan Pantai Baharu, Kuala Lumpur, Malaysia

Preface

The energy crisis and environmental pollution are extremely serious issues faced by the world at large. Despite great progress being made designing new materials, it is necessary to improve the situation before using them for various practical applications. Hence, eco-friendly energy conversion, harvesting, and storage device architecture are the foremost challenges for researchers in the 21st century. To address these issues, this edited book entitled *Energy Materials: Fundamentals to Applications* emphasizes the need for cutting-edge research advances in the field of energy conversion, harvesting, storing, and energy-saving materials.

Scope and coverage

This book is divided into four key sections.

Section I deals with fundamentals and overarching topics (Chapters 1–4). In this section, Chapter 1 looks at the fundamentals and closely related topics of energy conversion, harvesting, energy storage materials, and recent advances in technology along with potential applications and future prospectives, while Chapter 2 illustrates various synthesis and characterization techniques pertaining to structural and surface evaluation, electrical characterization, optical microscopy, carrier lifetime measurement, simulation techniques, failure parameters, and damage evaluation. Chapter 3 delves into solar cell technology, basic solar cell design, cell parameters and requirements, and losses

and various analytical techniques to evaluate solar efficiency. An extensive study of the adhesion and antifogging of the transparent polydimethyl siloxane/Sylgard coating system is detailed in Chapter 4.

Section II looks at photovoltaic materials and devices (Chapters 5–9). Chapter 5 highlights $I-V$ characteristics, spectral response, various means to optimize solar cell efficiency, and fabrication technologies. Chapter 6 deals with dye-sensitized solar cells (DSSCs), generally considered to be the third generation of photovoltaic technologies. A basic understanding of each component of a DSSC and their requirements are discussed at length. Focus is centered on electrode materials of a DSSC. Chapter 7 focuses on the basic performance and advanced colorful approach of the efficient energy conversion and harvesting of DSSCs. Extensive efforts to overcome the disadvantages of TiO_2 nanoparticles are highlighted in Chapter 8. Chapter 9 deals with novel methods of synthesizing perovskite solar cells along with device anatomy, characterization techniques, fabrication approaches, and device evaluation.

Section III focuses on electrochemical energy conversion and storage (Chapters 10–13). Chapter 10 deals with intercalation compounds for cathode materials in lithium-ion batteries as energy storage devices, the working principle of lithium-ion batteries, cell design and configuration, and synthesis methods of cathode materials for lithium-ion batteries and their specifications. Based on previous studies of polymer electrolytes,

xi

Chapter 11 establishes the necessary requirements for the selection of polymer electrolytes in lithium—air applications. Chapter 12 deals with lithium—sulfur batteries, their prior state of the art, and the current prospective and future scope of these new-fangled batteries. Chapter 13 deals with solid oxide fuel cells (SOFCs), considered an alternative to fossil fuel combustion systems in power generation due to their efficient energy conversion with low emissions of pollutants and high operating temperature (800—1000°C). Their components, basic operation, sealant materials, requirements, and methodological approaches are discussed in detail.

Section IV details lighting and light-emitting diodes (LEDs) (Chapters 14—17). The role of rare-earth ions for energy-saving LED lighting devices is detailed in Chapter 14. An overview of rare-earth elements, phosphor, basic fundamentals of luminescence, and applications of these phosphors for energy-saving LED lighting is detailed at length. Chapter 15 deals with the synthesis and luminescence of silicate-based phosphors for energy-saving LEDs, considered for energy-saving and eco-friendly lighting. Chapter 16 makes an attempt to synthesize and analyze orange/red light-emitting rare-earth-doped polymer matrices in the solid-state as well as the solvated state. These complexes blended in polymers prove their potential as orange/red light-emissive materials for organic light-emitting diode (OLED) devices, lasers, displays, and solid-state lighting. Chapter 17 reports on the synthesis and characterization of energy-efficient green/bluish-green/blue light-emitting 8-hydroxyquinoline derivative metal complexes, which can be employed for the fabrication of OLED devices, displays, and solid-state lighting.

Section V covers the practical concerns and beyond (Chapters 18—20). Spectral response and quantum efficiency evaluation of solar cells, major factors affecting performance of solar cells, research methodology, recent advancements, challenges, and future recommendations are discussed in Chapter 18. Novel materials for energy-related applications, research area connections, and the opportunities and challenges of these versatile energy materials are discussed in Chapter 19. Lastly, Chapter 20 focuses on the practical concerns of these energy materials correlated to sustainability, recycling, and lifetime issues.

Thus a serious attempt has been made by the editors to convey pioneering methodologies and strategies approved in the research and development of the subject with high scientific and technological merit. It is hoped that this book will create immense interest among new researchers interested in working in the field of energy conversion, harvesting, energy storing, and energy-saving materials, and will also act as a reference to undergraduate and postgraduate courses related to material science in the national and international community.

SECTION I

Fundamentals and overarching topics

CHAPTER 1

Energy materials: Fundamental physics and latest advances in relevant technology

R.S. Gedam[1], N. Thejo Kalyani[2], S.J. Dhoble[3]

[1]Department of Physics, Visvesvaraya National Institute of Technology, Nagpur, Maharashtra, India; [2]Department of Applied Physics, Laxminarayan Institute of Technology, Nagpur, Maharashtra, India; [3]Department of Physics, RTM Nagpur University, Nagpur, Maharashtra, India

1.1 Introduction

Society use materials in different ways and life can become very difficult without them. Particularly, energy materials play a very important role at every stage of energy production, distribution, conversion, and utilization, depending on the properties of the material [1]. Intensification in understanding the properties and structures of materials helps us to search for new materials and make significant progress in older versions so as to enhance their efficiencies at affordable costs. Energy materials also hold the key for many advanced energy technologies, including photovoltaic solar cells, thermoelectrics, fuel cells, batteries, supercapacitors, light-emitting diodes (LEDs), organic light-emitting diodes (OLEDs), etc. Potential areas of these materials and relevant technologies are extensive, but are not limited to those mentioned here [2]. It is necessary to know about the areas where these technologies are falling behind and the ways to bridge the gap of conversion and storage efficiencies. Hence, now, is the right time to accelerate the rate at which energy-related materials should be developed and to create effective methods to upgrade their efficiencies at a larger scale. Various fast, scalable, computational methods need to be identified and novel designs in their architecture are essential to explore the world of these energy materials [3,4].

1.2 Energy materials

The increase in energy demand due to the growing world population has encouraged the discovery and development of economical, durable, environmentally friendly, and efficient energy materials. These materials are versatile; some have the ability to harness, and others have the ability to convert, store, and save. Based on their properties and the means of exploitation, these materials are classified as energy-harvesting, -converting, -storing, and -saving materials as depicted in Fig. 1.1.

1.2.1 Energy harvesting

The process that has the potential to capture energy from a system's environment and convert it into usable electric power is generally known as energy harvesting. Various sources such as mechanical load, vibrations, and temperature gradient are scavenged and converted into power; however, the power attained is small, ranging between nanowatts and milliwatts. This mechanism is very similar to renewable energy generation; the main difference is the scale of conversion (megawatts of power) [5], and depends on the size of the initial mass, frequency and amplitude of driving vibration, mass displacement, and damping. It offers a means of powering autonomous devices, such as vehicles, satellites, laptops, etc. in the absence of conventional power sources or rechargeable batteries. The various means of harvesting energy include the harnessing of mechanical, thermal, electromagnetic, and vibrational energy into electrical energy. However, efficiency is material specific and hence proper choice plays a key role.

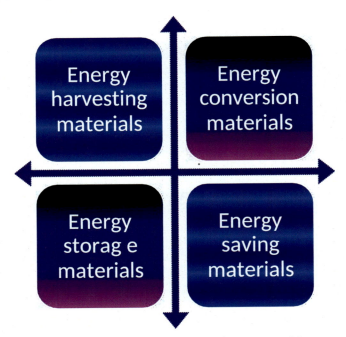

FIGURE 1.1 Broad classification of energy materials.

1.2.2 Energy-harvesting sources

Nature itself is bestowed with many natural ambient energy-harvesting resources that can be converted into electrical energy by various means. With the growing thirst for generating more and more electricity, which is considered a basic right today, researchers and scientists have come up with external energy-harvesting sources. Thus there are two kinds of energy-harvesting sources, namely ambient or external [6]. However, there is a basic difference between them: in the former case, sources are available within the environment without any external energy supply like solar, radio frequency, thermal, and vibration energy-harvesting sources, which exploit energy through an appropriate transducer [7], while the latter sources emit energy to the environment, which helps to monitor the surroundings. These ambient sources can be classified into four categories [8]: mechanical (wind, vibrations, and deformations), thermal (temperature gradient), radiant (sun, infrared, radio frequency), and chemical (chemical or biochemical reactions) as portrayed in Fig. 1.2. Interestingly, each source has the potential to harness different scales of power densities, ranging between 1 and 100,000 µW/cm^3. A comparison of the pros and cons these mechanical sources—piezoelectric (kinetic energy to electrical energy), electromagnetic (electromagnetic waves to electrical energy), and electrostatic (electrostatic force to electric energy)—can be found in Fig. 1.3. The devices employed for these types of conversions include piezoelectric oscillators (employing piezoelectric materials), electromagnetic converters (based on electromagnetic induction, governed by Faraday's law and Lenz's law), and electrostatic converters (using variable capacitor structures). Mechanical energy can be harvested from vibrations, subjecting a body to pressure, deformation, or strain.

Thermal energy harvesting is the process of capturing heat, which is either freely available in the environment or waste energy given off by engines, machines, and other sources, and putting it to use by means of a thermoelectric generator. On the other hand, electromagnetic

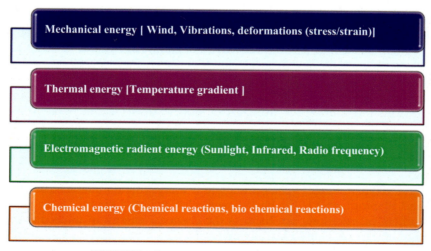

FIGURE 1.2 Various energy-harvesting sources.

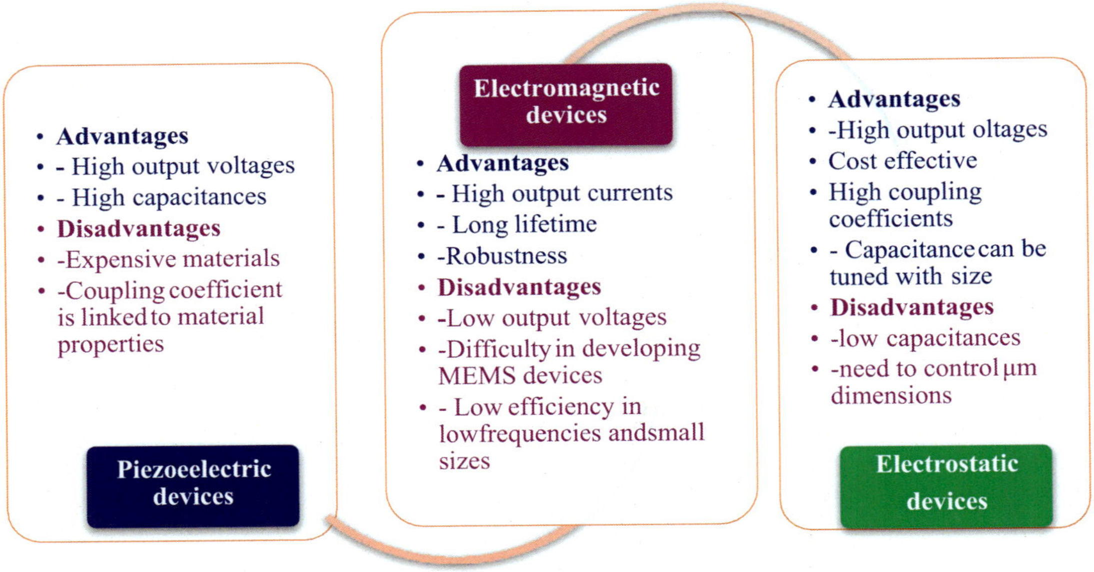

FIGURE 1.3 Comparison of mechanical energy-harvesting devices [8]. *MEMS*, Microelectromechanical systems.

energy (sunlight, infrared, radio frequencies) is converted into electrical energy by employing a photovoltaic, pyroelectric infrared detector and antenna as transducers, respectively. Solar energy can be harvested by solar-powered photovoltaic panels that convert the sun's rays into electricity by exciting electrons in the photovoltaic material with the help of photons of light from the sun by the photovoltaic effect. Electromagnetic energy harvesting generally uses microelectromechanical systems technology to harvest energy from low-frequency vibrations of less than 100 Hz [9] using Faraday's law of electromagnetic induction. NdFeB is one popular magnetic material employed with the conducting coil material aluminum that can generate a voltage of 2.34 mV. An array of cantilevers can be used to attain larger voltages. Radio frequency energy harvesters use the transmission of radio waves (3 kHz—300 GHz) to subsequently convert into direct current [10] by a single or multistage converter based on various requirements. The conversion efficiency mainly depends on antenna gain and distance from the source. Electromagnetic and piezoelectric technologies are widely used to harvest kinetic energy for versatile applications such as wearable electronic devices, mobile electronics, etc. A piezoelectric generator converts mechanical strain on an active material into electric charge, while an electromagnetic generator employs relative motion between a conductor and a magnetic flux to induce charge in the conductor [11], ranging between 10 and 100 mV. Lead zirconate titanate is a crystalline material that contains lead, zirconium, and titanium, and is the best piezoelectric material known. Even chemical reactions and biochemical reactions can harness the chemical energy of molecules and convert it into electrical energy, as shown by galvanic cells.

1.2.3 Energy conversion

Exploring effective energy conversion technologies is essential due to increasing demands for energy requirements. These materials play a key role in energy sustainability, energy conversion, pollution control, curtailing the carbon footprint, and also meeting the energy challenges of the modern era. Energy conversion efficiency (η) of any system is the ratio between the useful output and the input of an energy conversion machine in terms of energy. Globally, efforts are under way in this area.

1.2.4 Pressure to electrical energy

When certain crystals like quartz, Rochelle salt, tourmaline, etc. are stretched or compressed along a mechanical axis (an axis through opposite faces), an electric potential difference is produced along the electric axis (an axis passing through any set of opposite corners). During this process, kinetic energy in the form of vibrations is converted into electrical energy. The most popular crystal is quartz crystal—a six-sided prism with pyramid-shaped ends and three major axes: X-axis (electrical), Y-axis (mechanical), and Z-axis (optical). Piezoelectricity is exploited in a number of useful applications and scientific instrument techniques. Lead zirconate titanate is currently the most commonly used piezoelectric ceramic. Potassium niobate ($KNbO_3$) and lithium niobate ($LiNbO_3$) are other known piezoelectric materials.

1.2.5 Solar to electric energy

Solar energy is derived from natural sources that do not harm the environment. To increase performance to its theoretical limits, it is necessary to optimize light management, as discussed in the following sections.

1.2.6 Thermal to electric energy

Thermoelectric generators are generally used to convert the unwanted heat generated by nuclear power plants, thermal power plants, and many other such sources into useful electric energy. However, they have low efficiency (only about 5%—6%). Thus every conversion mechanism has its advantages as well as disadvantages, which dictate the efficiency of the harnessing mechanism.

1.2.7 Energy storage

Energy storage mechanisms have been well known for some time [12]. They play a noteworthy role in creating a more flexible and consistent grid system. At night or when dwellings are unoccupied, energy supply exceeds demand, and the surplus electricity generated can be stored in power storage devices, i.e., the energy captured today can be stored for future use. In that sense, batteries and capacitors play a key role [3]. Particularly, lithium-based energy storage batteries and super- or ultracapacitors are widely used currently. Smartphones and laptops, e.g., which are recharged once a day, are some examples of mechanisms of energy storage. A basic battery consists of two electrodes (anode: Li, Al, Zn, Fe, graphite and cathode:

MgO_2, HgO, Li_2O, PbO_2) and an electrolyte. The electrodes of a lithium-ion battery are made of lightweight lithium and carbon. As lithium is a highly reactive element, much energy can be stored in its atomic bonds, translating into a very high energy density.

The other potential device that stores electrical potential energy in an electric field is the electrostatic capacitor, designated as $E = \frac{1}{2}QV$, where Q and V are charge and voltage on the capacitor. Their anatomy includes two parallel plates separated by dielectric media (ceramic, mica, oxides of various metals, dry air, etc.). A supercapacitor is a high-capacity capacitor with a capacitance value much higher than other capacitors, but with lower voltage limits, which bridges the gap between electrolytic capacitors and rechargeable batteries [13]. It typically stores 10—100 times more energy per unit volume or mass than electrolytic capacitors, can accept and deliver charge much faster than batteries, and tolerates many more charge and discharge cycles than rechargeable batteries. Thus supercapacitors prevail as batteries in some applications. Thus, depending on the purpose and requirement, it is up to the user whether to use a conventional battery, lithium-based battery [14], capacitor, or supercapacitor. Apart from lithium-based batteries, other battery with lead, manganese, nickel also play role in energy storage.

1.2.8 Energy saving

Today's scenario of energy crises endorses the idea of handling electrical energy with care. Energy-saving materials and their devices play a crucial role in saving energy without any compromise to the lifetime of devices. Though many energy-saving devices have been explored, here we will concentrate on energy-efficient lighting sources. The history of artificial lighting sources started with tungsten filament bulbs but these subsequently lost their popularity as the new cold emission lighting fluorescent tube lights and compact fluorescent lamps came into existence in the mid-1990s [15]. However, the legacy left behind by these incandescent lamps remains forever. The fact that they employ mercury—a toxic substance that creates the problem of disposal—is repeatedly ignored but the advent of solid-state lighting sources with LEDs (organic/inorganic: LED/OLED) is a welcome alternative. These are promising energy-efficient lighting sources for displays and general lighting, has and they have proved to limit the usage of energy to a certain extent [16,17]. Versatile energy-harvesting, -converting, -storing, and -saving technologies are listed in Fig. 1.4 and their comparision is tabulated in Table 1.1.

1.3 Fundamental physics

To develop a novel and revolutionary technology, raw materials play a crucial role; so, it is essential to have up-to-date knowledge of application-specific materials, including their basic functional science, properties, and earlier/current performance before preliminary selection. Such materials can then be employed in developing a device by designing suitable anatomy and manufacturing techniques. The developed product/device should be assessed for its performance and environmental safety before commercialization. The basic five steps employed in such an evaluation are specified in Fig. 1.5. Some energy materials and their technologies are discussed in the following sections.

1.3 Fundamental physics

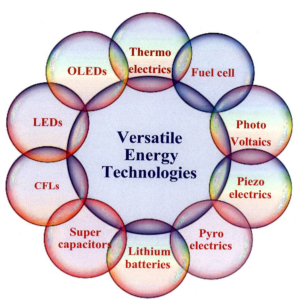

FIGURE 1.4 Versatile energy technologies. *CFLs*, Compact fluorescent lamps; *LEDs*, light-emitting diodes; *OLEDs*, organic light-emitting diodes.

TABLE 1.1 Comparison: basic energy technologies [18–25].

Energy-harvesting/-converting technologies						
Energy	Source	Device	Phenomenon	Advantages	Disadvantages	Applications
Solar energy	Sun	Solar panels	Photovoltaic effect	Renewable Free of cost Pollution free Low maintenance Can be stored	Most of the energy is wasted Low conversion efficiency High initial cost restricted to sunny days Occupies lot of space	Power generation Solar furnaces Solar water heaters Solar distillation Solar pumping Solar cooking Power source for automobiles, domestic applications, etc.

(Continued)

10   1. Energy materials: Fundamental physics and latest advances in relevant technology

TABLE 1.1   Comparison: basic energy technologies [18–25].—cont'd

Energy-harvesting/-converting technologies						
Energy	**Source**	**Device**	**Phenomenon**	**Advantages**	**Disadvantages**	**Applications**
Thermal energy	Temperature gradient	Thermoelectric generator	Seebeck effect	Reliable source of energy High scalability Low production cost Wasted heat energy is recycled	Low energy conversion efficiency rate Slow technology progression Requires constant heat source Lower coefficient of performance High output resistance	Power generation Heating Electronics cooling Powers lights and fans Space and military applications Automobiles Power plants
Fuel energy	Chemical reaction in Fuel	Fuel cell	Electrochemical reactions	Clean Efficient Reliable	High initial cost Storage problems Safety hazards	Auxiliary power portable power Stationary power
Kinetic energy (in the form of vibration)	Piezoelectrics (quartz/ tourmaline)	Piezoelectric oscillator	Piezoelectric effect	Wide operating temperature Low carbon footprint Green energy Materials can be reused unaffected by external electromagnetic field Low maintenance	Not used for truly static measures Can pick up stray voltages in connecting wires Crystal is prone to crack if overstressed May be affected by long use at high temperature noisy response at higher frequency	Pressure sensor Pressure gauge actuators Gas grills Blow torches butane lighters
Mechanical oscillations	Pyroelectrics (naturally polarized crystals/polar crystal symmetry class)	Pyroelectric generator	Pyroelectric effect	Low maintenance cost High-frequency response Pollution free	Requires high impedance cable Difficult to measure static motion Stable up to 1200°C	Radiometry Fire detectors Laser diagnostics Heat sensors Power generation Steers nuclear fusion

I. Fundamentals and overarching topics

1.3 Fundamental physics

11

TABLE 1.1 Comparison: basic energy technologies [18–25].—cont'd

Energy-harvesting/-converting technologies						
Energy	Source	Device	Phenomenon	Advantages	Disadvantages	Applications
Energy-storing technologies						
—	Chemical energy	Lithium batteries	Electrochemical reactions	High energy density Cheap Low self-discharge Low maintenance	Protection required Suffer from aging Costlier	Uninterrupted power supply Power source for mobile phones, laptops e-scooters, digital cameras
—	—	Supercapacitor	Electrochemical reactions	Backup power Longer lifetime Reduced weight Cheap Wide operating temperature	High self-discharge rate Requires series connection Energy stored is low Low energy density Cannot be used in air-conditioning and high-frequency circuits	Power source for voltage stabilizer Camera flashes Wind turbines Startup mechanism for automobiles Auto equipment
Energy-saving technologies						
—	Electric current through a tube containing Ar and Hg	CFLs		Cool emission Save energy	Contains Hg (toxic) Disposing problem	Residential lighting
—	Current through direct bandgap semiconductors	LEDs	Electroluminescence	Longer lifetime Energy efficient Eco-friendly Self-luminescence	Point sources Light pollution	Lamps Indicators Decoration Traffic signals Displays
—	Current through direct bandgap semiconductors	OLEDs	Electroluminescence	170-degree viewing angle Fast response time Energy efficient Eco-friendly High resolution Full emission	Poor lifetime Susceptible to water Luminance degradation Limited market Costly manufacturing process	Mobile screens flexible displays Solid-state lighting

CFLs, Compact fluorescent lamps; *LEDs*, light emitting diodes; *OLEDs*, organic light emitting diodes.

I. Fundamentals and overarching topics

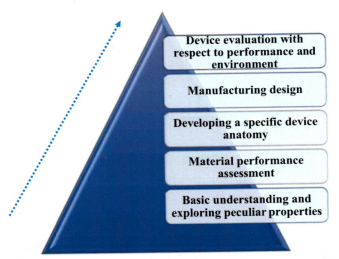

FIGURE 1.5 Steps involved in the development of raw materials into a suitable device.

1.3.1 Thermoelectrics

This is the collective application of the thermoelectric effect, which encompasses the Seebeck effect, Peltier effect, and Thomson effect, which are thermodynamically reversible. According to the Seebeck effect, temperature gradient can be converted into electricity by means of a thermoelectric generator [26]. Let us consider that the heat emitted from nuclear power plants, thermal power plants, automobiles, factories, incinerators, etc. is converted into electrical energy as interpreted in Fig. 1.6. In other words, the heat otherwise considered as waste is converted into electrical energy by utilizing the Seebeck effect.

On the other hand, the generated electrical energy can also create heating as well as cooling effects through the Peltier effect as shown in Fig. 1.7 i.e., when a suitable amount of electric current is passed through a thermocouple circuit, heat is created at one junction and absorbed at the other junction. The Thomson effect is an extension of the Peltier–Seebeck model, which is beyond the scope of this chapter.

Conversely, electricity can be employed to create a temperature difference flow directly by means of the Peltier effect. The Peltier effect has applications in cooling and heating and the Seebeck effect is used in thermoelectric generators for power generation.

1.3.2 Fuel cell

This is one of the greatest devices invented so far; it converts chemical potential energy into electrical energy, and was first envisaged by Sir William Robert Grove. He observed that when hydrogen and oxygen are mixed in the presence of an electrolyte, electricity and water are generated. However, the electricity generated is minimal. Subsequently, modifications were carried by researchers and scientists to enhance the efficiency of the fuel cell. An ideal fuel cell consists of an anode, electrolyte, and cathode. The role of electrodes is to collect

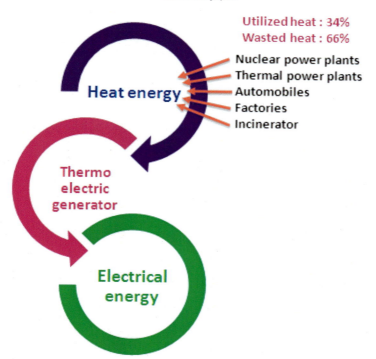

FIGURE 1.6 Mechanism: generation of electricity from waste heat.

the charge carriers, which generate electricity, while the electrolyte maintains the right proportion of ions to travel between the two electrodes [27]. The most commonly used anode gases include hydrogen, methanol solution in water, and methane, while cathode gases include pure or atmospheric oxygen. Different electrolyte materials employed so far include solid polymer membrane, potassium hydroxide, phosphorus, alkali carbonate, ceramic oxides, zinc, occasionally bacteria, etc. The electrolyte chosen depends on the type of fuel cell. With these versatile anode, cathode, and electrolyte materials, developments in this field led to different types of fuel cells, namely alkaline fuel cells, phosphoric acid fuel cells, direct methanol fuel cells, polymer electrolyte membrane fuel cells, molten carbonate fuel cells, solid oxide fuel cells, reversible fuel cells, proton exchange membrane fuel cells, microbial fuel cells, and zinc—air fuel cells (Fig. 1.8); so far the efficiency of fuel cells has reached in excess

FIGURE 1.7 Energy conversions: heat energy—electrical energy—heating and cooling.

I. Fundamentals and overarching topics

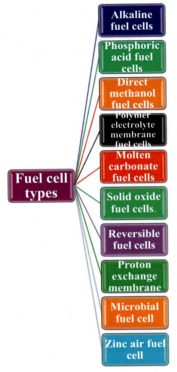

FIGURE 1.8 Types of fuel cells [28].

of 70% and the temperature of the heat generated varies from 75°C to 1000°C. For example, in the proton exchange membrane fuel cell, H_2 gas and O_2 gas are used as fuel; as a result of the reaction, electricity, water, and heat are generated [28].

1.3.3 Photovoltaics

A photovoltaic cell, popularly known as a solar cell, works on the principle of the photovoltaic effect—the phenomenon in which solar energy is converted into electrical energy [29]. This process of conversion is carried out in three steps as shown in Fig. 1.9.

Initially, when light of a suitable energy is incident on the material (energy greater than the energy gap of the absorbing material), covalent bonds break down and create electron—hole pairs. Electrons move toward the conduction band, leaving behind a hole in the valence band. In this process, the electrons move toward the depletion region, creating a barrier potential at the junction as the next step. Finally, these generated electrons and holes are collected at opposite electrodes, thereby generating electricity, which can be either (1) stored in batteries or passed through a storing device for future use or (2) converted into power for domestic applications and powering electronic gadgets [30].

FIGURE 1.9 Illustration: mechanism of energy conversion in a solar cell.

1.3.4 Batteries

A battery is a cell or combination of cells connected in series, generally used as a source of electrical energy. The battery was invented in the early 1800s, and they are still popular because they enhance the convenience of users, and are cheap and robust. However, they have limited lifetime [31]. Batteries can be broadly classified as primary and secondary. In the former case, the reaction occurs only once and after long usage the battery dies. Examples are dry cells, mercury cells, etc. They cannot be reused. In the latter case, batteries can be recharged and used again and again. The best examples of this class are lead storage batteries and nickel—cadmium cells. However, their capacity to store energy is poor. This has led to further developments, and the evolution of lithium—ion batteries is seen as an ideal solution to this problem [3,32]. Later, with the discovery of supercapacitors in the late 1950s, the weight of storing devices was reduced to a greater extent.

1.3.5 Supercapacitor

This is a high-capacity storage device with higher values of capacitance, and was created to bridge the gap between rechargeable batteries and capacitors. Like a normal capacitor, it has two plates separated by a suitable dielectric material. However, it differs from an ordinary capacitor by using plates of larger surface area and the separation between the plates is substantially reduced. The metal plates are coated with porous materials like activated charcoal so that they can store additional charges on them at a particular voltage. In an electrostatic capacitor, the plates of the capacitor are filled with dielectric medium such as mica, while in supercapacitors, insulators such as carbon, paper, and plastic are placed between the plates. When the capacitor is charged, an electric field is created between the plates of the capacitor because one plate has an accumulation of positive charges, and the other plate has an accumulation of negative charges due to polarization of the field.

In the case of a supercapacitor, both plates are soaked in an electrolyte, whereby opposite charges accumulate on the plates, creating an electric double layer, which is similar to two capacitors being placed side by side as shown in Fig. 1.10. Common applications of supercapacitors are found in wind turbines, electric vehicles, and flywheels. Table 1.2 portrays exhaustive comparision of various energy storage devices.

1.3.6 Compact fluorescent lamps

Compact fluorescent lamps (CFLs) were produced as a persuasive substitute for fluorescent tube lights. The inner glass wall of a fluorescent lamp is generally coated with thin phosphor and is filled with mercury vapor at low pressure, which emits ultraviolet (UV) light. When current is passed through it, the phosphor absorbs this UV light and converts it into visible light. CFLs are compact and have good lifetime. However, they are less popular due to the toxic mercury used in their operation.

1.3.7 Optoelectronic light-emitting devices

Recent years have witnessed a swift change in the pattern of general lighting sources and flat panel display devices due to the approval of energy efficient and environmentally friendly LEDs and OLEDs. In contrast to traditional lamps, which emit warm light, these devices emit cool light by the process of electroluminescence.

Hence, this technology has the prospect of reinstating incandescent and fluorescent lamps for a wide range of lighting. LEDs and OLEDs are semiconducting devices competent at emitting a fairly constricted bandwidth of visible or infrared or UV light when forward biased (p-type and n-type semiconductors are connected to the positive and negative terminals of the battery, respectively). A wide range of colors can be achieved from these devices in the visible region (orange, red, yellow, blue, green, and white) and in the UV and infrared light regions [36,37]. Displays made of these devices are slimline.

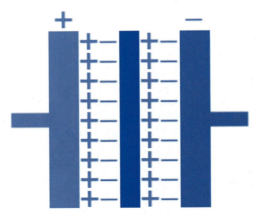

FIGURE 1.10 Demonstration: electric double layer in a supercapacitor.

1.3 Fundamental physics

TABLE 1.2 Exhaustive comparison: energy storage devices [33–35].

S. No.	Parameter	Batteries	Lithium–ion batteries	Electrostatic capacitor	Supercapacitor
1.	Minimum operating voltage (V)	1.2–1.4	3.6–3.7	6–800	2.3–2.75
2.	Maximum rating voltage (V)	13.8	4.1–4.3	–	–
3.	Charging time	1–5 h	10–60 min	Ps to ms	1 to 10 s
4.	Operating temperature (°C)	−20 to 100	−20 to 60	−20 to 65	−40 to 650
5.	Energy density (Wh/kg)	10–1000 8–600		<0.1 0.01–0.05	1–10 1–5
6.	Life (cycles)	150–1500	0.5–1 K	>1005 K	>100 K
7.	Self-discharge	High	Very low	High	Low
8.	Safety monitoring	Not required	Required	Not required	Required
9.	Voltage monitoring	Not required	Required	Not required	Required
10.	Capacitance/energy versus device volume	–	Highest	–	–
11.	Weight	1g–10 kg	–	1g–10 kg	1–2g

1.3.8 Various approaches for obtaining white light

Any light source used for domestic applications should have the potential of emitting bright and intense white light and this can be achieved by various mechanisms [38], and the most employed ones include (1) a combination of red, blue, and green phosphor in an appropriate ratio (60:30:10, respectively), and (2) a combination of yellow and blue phosphor as shown in Fig. 1.11.

1.3.9 Light-emitting diodes

LEDs and OLEDs work on the same principle of electroluminescence; both employ direct bandgap semiconductors because (1) they do not require a photon to generate electron–hole pairs, so the electrooptical devices is very efficient, and (2) the top of the valence band and bottom of the conduction band occur at the same values of momentum. In these material-based devices, electrons and holes recombine at the interface of the junction and create photons through recombination. Conversely, in indirect bandgap semiconductors, a photon cannot be emitted because the electron must pass through an intermediate state and transfer momentum to the crystal lattice [39]. Direct versus indirect bandgap semiconductors in an application perspective are shown in Fig. 1.12. However, there are certain differences such as the materials (inorganic/organic) employed, device anatomy, and fabrication process.

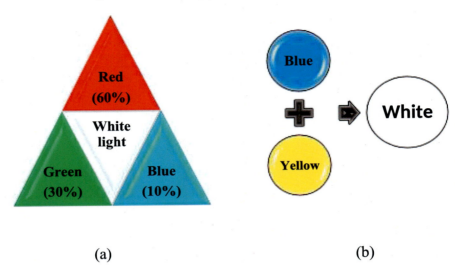

FIGURE 1.11 Possible means of creating white light.

An LED structure consists of an n-type layer grown on a p-type substrate by the process of diffusion or any suitable technique. Metal contacts are made at the outer edge of the p-layer so that more of the upper surface is left free for light to escape. For making cathode connections, metal film is coated at the bottom of the substrate. Transparent encapsulation is made to protect the LED device from moisture and other undesirable particles in the atmosphere. This film reflects as much light (visible, UV, and infrared) as possible to the surface of the device under forward bias. Forward bias is preferred because such biasing lowers the potential barrier across the junction, thereby offering a path to the majority charge carriers to diffuse across the junction (electrons from n to p and holes from p to n) and recombine and emit monochromatic light of a particular color, depending on the selection of semiconducting

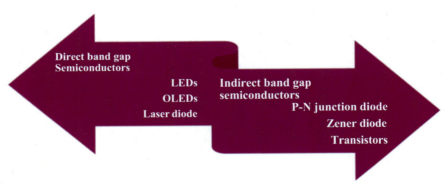

FIGURE 1.12 Direct versus indirect bandgap semiconductors: application perspective. *LED*, Light-emitting diode; *OLED*, organic light-emitting diode.

material. If E_g represents the semiconductor bandgap, then the light emitted by the diode can be calculated by the relation:

$$\lambda = \frac{hc}{E_g}$$

where h is Planck's constant, c is velocity of light, and λ is the wavelength of emitted radiation.

Thus based on the material's optical energy bandgap, light emission from the device can be tuned in the infrared, visible, and UV regions of the electromagnetic spectrum (Table 1.3).

1.3.9.1 Where do LEDs end and OLEDs begin?

LEDs are point source devices, generally made of inorganic rare earths; although they have been in use for some time, they have been used as indicators and have now emerged as solid-state lighting sources [41]. Due to their limited directionality in emission, a bundle or an array of LEDs is required, as in the case of traffic signal lights and LED lamp sources, for example. It has been established that blue LEDs particularly can lead to light pollution [42]. On the other hand, OLEDs utilize organic molecules as the emissive layer and these materials can be easily amalgamated in polymer matrices to spread them into larger display sheets; this makes them suitable for lighting panels and windows. They can be fabricated by various thermal deposition techniques, too, thus offering a uniform morphology on the total exposed substrate [43–45].

1.3.10 Organic light-emitting diodes

An OLED is a thin film optoelectronic device with organic materials (small molecules, dendrimers, or polymeric substances) sandwiched between two electrodes, the anode and cathode, all deposited on a substrate. The organic molecules are electrically conductive with conductivity levels ranging between insulators and conductors and hence they are considered as organic semiconductors. The highest occupied molecular orbitals and the lowest unoccupied molecular orbitals of organic semiconductors are equivalent to the valence and conduction bands of inorganic semiconductors. The anatomy of OLEDs can be single-layer, two-layer, triple-layer, or multilayer OLEDs. A single-layer OLED is made up of a single organic layer sandwiched between the cathode and the anode. This layer should have hole transport, electron transport, and emission capabilities. In this case, the injection of

TABLE 1.3 Light-emitting diode materials [40].

Semiconducting material	Energy gap (eV)	Emission wavelength (A^0)	Part of electromagnetic spectrum
GaAs	1.45	8500	Infrared region
GaAsP	1.9	6500	Visible region
AlGaP	3.4	5500	Visible region
AlN	5.9	2100	Ultraviolet region

both the carriers should be the same; otherwise, the device will result in low efficiency because the excess of electrons or holes cannot combine. However, in a two-layer OLED, one organic layer is explicitly chosen to transport holes and the other layer is chosen to transport electrons, namely hole transport layer (HTL) and electron transport layer (ETL), respectively. Recombination of the hole–electron pair takes place at the interface between the two layers, generating electroluminescence [46].

In the structure of a triple-layer OLED, the function of the individual organic layer is distinct and can therefore be optimized independently. Hence, the luminescence or recombination layer can be chosen to the desired emission color with high luminescence efficiency. Multilayer OLEDs consist of different layers, namely indium tin oxide glass substrate, hole injection layer, HTL, emissive layer, ETL, and anode. This architecture eliminates the charge carrier leakage as well as exciton quenching at the interface of the organic layer and the metal. The general structure of a multilayer OLED is sketched in Fig. 1.13. The materials to be used for different layers for OLEDs should meet certain requirements like high luminescence efficiency, slender spectra and accurate CIE coordinates, adequate conductivity, good temperature stability, good oxidative stability (water and oxygen), and good radical cation/anion stability; they should also not degrade during sublimation. Various materials generally used in different layers of OLED are shown in Table 1.4.

The performance of these devices is evaluated by quantum efficiency and power efficiency. Different metrics for quantum efficiency include internal and external quantum efficiency.

1.3.11 Internal quantum efficiency

$$\text{Internal quantume efficiency}(\eta_{\text{int}}) = \frac{\text{No. of photons generated internally}}{\text{No. of carriers}}$$

$$= \frac{\text{Radiative recombination rate}}{\text{Total recombination rate}} = \frac{R_r}{R_r + R_{nr}}$$

where R_r = radiative recombination and R_{nr} is nonradiative recombination. For better efficiency, $R_r \gg R_{nr}$.

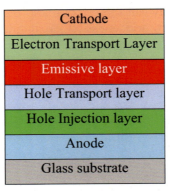

FIGURE 1.13 Structure of a multilayer organic light-emitting diode.

TABLE 1.4 Materials generally used in different layers of organic light-emitting diodes (OLEDs) [47,48].

Layer of OLED	Materials generally used in different layers of OLEDs
Anode	Indium tin oxide, indium zinc oxide, graphene
Hole transport layer	4,4'-Bis[N-(naphthyl)-N-phenylamino]biphenyl, N,N'-bis(naphthalen-1-yl)-n,N'-bis(phenyl) benzidine, di-[4-(N,N-ditolyl-amino)phenyl] cyclohexane
Hole injection layer	4,4',4''-Tris(3-methylphenylphenylamino)triphenylamine, phthalocyanine—copper complex, phthalocyanine—platinum complex
Electron transport layer	2,9-Dimethyl-4,7-diphenyl-1,10-phenanthroline, 3-(4-biphenyl)-4-phenyl-5-*tert*-butylphenyl-1,2,4-triazole, 2,9-dimethyl-4,7-diphenyl-1,10-phenanthroline
Emissive layer	Aluminum 8-hydroxyquinoline, rare earth complexes, rubrene
Cathode	Low work function, magnesium, lithium fluoride, silver

1.3.12 External quantum efficiency

External quantum efficiency is defined as the product of internal and optical quantum efficiency, given by:

$$\eta_{ext} = \eta_{int} \times \eta_{opt}$$

It further depends on various parameters such as device geometry, reflectance, transmission, and absorbance of different layers of the device [49].

1.3.13 Power efficiency

$$\text{Power efficiency}(\eta_P) = \frac{\text{Optical power output}}{\text{Input power}}$$

$$= \frac{\text{No. of photons} \times \text{Energy}}{\text{Input power}} = \frac{\rho_{Ph} \times E}{P_{in}} = \frac{\rho_{Ph} \times h\upsilon}{VI}$$

Fig. 1.14 illustrates different OLED devices and related information, while a comparison of general lighting sources is shown in Table 1.5.

1.4 Recent advances in technology

The advent of nanotechnology has ramped up developments in the field of material science due to the performance of materials for energy conversion, energy storage, and energy saving, which have increased many times. These new innovations have already portrayed a positive impact on the energy sector. Nanomaterials play a crucial role in thermoelectricity generation. Nanooptimized membranes, fuel additives, and electrodes are essential for the

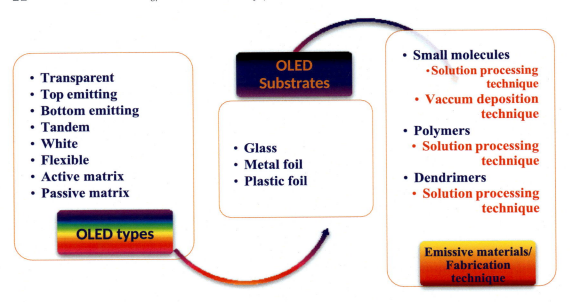

FIGURE 1.14 Organic light-emitting diode (OLED) types, substrates, materials, and fabrication technologies.

efficient conversion of fuels into electricity by combustion. The use of nanooptimized cells with polymeric dyes, double heterojunction structures, up/down conversion phosphors, perovskite materials, antireflection coatings, and the efficiency of solar cells can be enhanced many times. Nanostructured flexible electrodes can be used in lithium-ion and lithium-sulfur batteries for enhanced performance, whereas in supercapacitors, plates made of carbon

TABLE 1.5 Comparison: general lighting sources.

Parameter	Organic light-emitting diode	Light-emitting diode (LED)	Compact fluorescent lamp	Electric bulb
Color brilliance	High	High	High	Highest
Efficiency	High	Very high	High	Low
Heat loss	Low	Low	Low	High
Illumination	Large-area illumination	Point light source	Compact light source	Point source
Glare	Low	High	Medium	High
Body	Flexible	Rigid; flexible connection of many LEDs possible	Rigid and ductile	Glass
Lifetime	Short	Long	Moderate	Short

nanotubes, graphene, barium titanate, and aerogel lead to better storing capacity. Though we are now self-sufficient, a few years from now we may need more and more advanced devices that can effectively and efficiently convert, store, and save electricity and hence the global struggle could be to seek energy materials that would make the development of a nation sustainable.

1.5 Potential applications

The potential applications of energy materials are wide and varied. This book covers a small fraction of the gigantic field of energy materials and their technologies, which are depicted in Fig. 1.15. Detailed applications of these materials are discussed in Chapter 19.

1.6 The future

The new-fangled energy materials and relevant technologies are going to offer window of opportunities in the up-coming future. Auxiliary progress can be made in this field by employing (1) lightweight materials (e.g., titanium aluminates) and (2) anticorrosion layers (aluminide coatings) for turbine blades in power plants, which could augment efficiency. Using nanostructured anodes and cathodes can effectively accelerate the conversion efficiency of fuel cells. More effective sealants such as glass and glass ceramics, which enhance the performance of fuel cells, are currently hotspot research areas. Solid oxide fuel cells are now an emerging technology with an expectation of over 60% energy conversion efficiency and higher power density as compared to other fuel cells. The performance of solar cells can be improvised by many trial methods such as the development of dye-sensitized solar cells, perovskite solar cells, graphene-based solar cells, and the use of various anode materials for extracting additional charge carriers, up-conversion and down-conversion phosphors, and now nanooxidic and photoanode materials such as ZnO, TiO_2, and SnO_2, which are proving to be the prevailing investigatory materials. Mesopores, rods, fibers, and needles are the next-generation photoanode materials with the hope of exceeding efficiency beyond 50%.

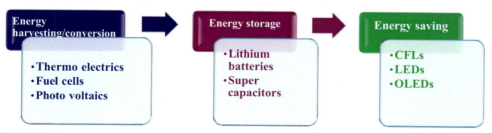

FIGURE 1.15 Applications of popular energy materials. *CFLs*, Compact fluorescent lamps; *LEDs*, light emitting diodes; *OLEDs*, organic light emitting diodes.

In the field of electrical energy storage, lithium—ion and lithium—sulfur batteries have superseded conventional batteries; similarly, ultra- or supercapacitors have superseded conventional capacitors. By exploiting nanoscience and technology, the capacity, performance, and safety of storage devices can be improvised. Subsequently, hydrogen seems to be a promising energy store for environmentally friendly energy supplies. Intercalation compounds for cathode materials in lithium—ion batteries as energy storage devices and the selection of polymer electrolytes such as solid polymer electrolytes and gel polymer electrolytes in advanced lithium—air batteries are the new-fangled upcoming advancements in the field of storage devices. With the discovery of smart storing materials, supercapacitors will be the capacitors of choice in the near future.

Energy efficient and environmentally friendly solid-state lighting is also an upcoming, highly competent, and viable alternative to existing lighting technologies. This technology has the potential to reduce the carbon footprint, without any intervention in the color rendering index. Hence, the current intention in the field of optoelectronics is to reinstate these conventional lighting sources with more power-efficient semiconducting light sources; this stems from their ability to trim down ecological damage to our environment. The fabrication of extremely consistent, efficient, and long-lived blue OLEDs is challenging, due to the complications in aligning the energy levels at the layer interfaces. Novel energy-efficient materials have to be synthesized. Their operating lifetime may be enhanced by using materials with higher glass transition temperature, adopting better encapsulation techniques, and optimizing device processes and structures. Once these hurdles are overcome, sold-state lighting can be embraced, which is currently proving to be of vital importance for the future of sustainable development, energy, and community health.

1.7 Conclusions

The dearth of feasible plentiful electrical energy in a society is likely to emphasize the subsistence of social asymmetry in the standard of living. This may lead to lack of progress, leading to a migration toward urban areas because of lack of prospects for the future. With a belief that sufficient generation of electrical energy could lead to an elevated degree of economic sustainability and lifestyle, many techniques to convert, store, and save electricity have emerged in cutting-edge research areas. This book caters to all these areas of research, and is divided into three sections, dealing with concise information about various means of the converting, harvesting, storing, and saving energy, new-fangled materials, and key mechanisms in the operation of respective devices. It also details with the practical work carried out by researches along with the fundamentals and phenomenological theories of novel energy materials. These include solar cells, lithium—ion/lithium—air/ lithium—sulfur batteries, fuel cells and their energy storage mechanisms, energy-saving and eco-friendly light-emissive devices like LEDs and OLEDs, solid-state lighting flat panel displays and their diversified applications, and sustainability issues. The book concludes with practical concerns such sustainability and lifetime and recycling issues of emerging energy materials, which are all addressed in a contemporary setting.

References

[1] P. Musilek, M. Prauzek, P. Kromer, J. Rodway, Barton, Intelligent energy management for environmental monitoring systems, Smart Sens. Netw. (2017) 67–94.

[2] F. Akhtar, M.H. Rehmani, Energy replenishment using renewable and traditional energy resources for sustainable wireless sensor networks: a review, Renew. Sustain. Energy Rev. 45 (2015) 769–784.

[3] M. Aneke, M. Wang, Energy storage technologies and real life applications—a state of the art review, Appl. Energy 179 (2016) 350–377.

[4] S. Sharma, V. Panwar, S.C. Yadav, Sachin, Lokesh, Different sources of energy harvesting: a survey, in: International Conference on Computing, Communication and Automation (ICCCA), Greater Noida, 2017, pp. 1370–1373.

[5] T.J. Kazmieriski, S. Bee, Energy Harvesting System, Springer, 2014.

[6] F.K. Shaikh, S. Zeadally, Energy harvesting in wireless sensor networks: a comprehensive review, Renew. Sustain. Energy Rev. 55 (2016) 1041–1054.

[7] P. Jiao, W. Borchani, H. Hasni, N. Lajnef, Enhancement of quasi-static strain energy harvesters using non-uniform cross-section post-buckled beams, Smart Mater. Struct. (2017) 085045.

[8] S. Boisseau, G. Despesse, B. Ahmed Seddik, Electrostatic Conversion for Vibration Energy Harvesting, Intech publications, 2012, pp. 91–134. Ch.5.

[9] A. Kumar, S.C. Anjankar, Electromagnetic energy harvester using MEMS, Proc. Comput. Sci. 79 (2016) 785–792.

[10] M. Imran, Energy Harvesting Mechanism, 2015. US Patent 8,948,870.

[11] M. Prauzek, J. Konecny, M. Borova, K. Janosova, J. Hlavica, P. Musilek, Energy harvesting sources, storage devices and system topologies for environmental wireless sensor networks: a review, Sensors 18 (8) (2018) 2446.

[12] A. Khaligh, P. Zeng, C. Zheng, Kinetic energy harvesting using piezoelectric and electromagnetic technology state of art, IEEE Trans. Ind. Electron. 57 (3) (2009) 850–860.

[13] M. Prauzek, J. Konecny, M. Borova, K. Janosova, J. Hlavica, P. Musilek, Review topologies for environmental wireless sensor networks: a energy harvesting sources, storage devices and system, Sensors 18 (8) (2018) 2446–2474.

[14] M. Habibzadeh, M. Hassanalieragh, A. Ishikawa, T. Soyata, G. Sharma, Hybrid solar-wind energy harvesting for embedded applications: supercapacitors-based system architectures and design tradeoffs, IEEE Circ. Syst. Mag. 17 (2017) 29–63.

[15] M. Asadi, B. Sayahpour, P. Abbasi, A.T. Ngo, K. Karis, J.R. Jokisaari, C. Liu, B. Narayanan, M. Gerard, P. Yasaei, et al., A lithium-oxygen battery with a long cycle life in an air-like atmosphere, Nature 555 (2018) 502–506.

[16] D. Chitnis, N.T. kalyani, S.J. Dhoble, Escalating opportunities in the field of lighting, Renew. Sustain. Energy Rev. 64 (2016) 727–748.

[17] N. Thejo kalyani, S.J. Dhoble, Novel approaches for energy efficient solid state lighting by RGB organic light emitting diodes – a review, Renew. Sustain. Energy Rev. 32 (2014) 448–467.

[18] P. Pawar, S.R. Munishwar, S. Gautam, R.S. Gedam, Physical, thermal, structural and optical properties of Dy3þdopedlithium alumino-borate glasses for bright W-LEDP, J. Lumin. 183 (2017) 79–88.

[19] G.D. Szarka, B.,H. stark, S.G. Burrow, Review of power conditioning for kinetic energy harvesting systems art, IEEE Trans. Ind. Electron. 27 (2) (2011) 803–815.

[20] https://en.wikipedia.org/wiki/Energy_harvesting.

[21] https://www.americanpiezo.com/blog/top-uses-of-piezoelectricity-in-everyday-applications/.

[22] https://www.cleanenergyreviews.info/blog/most-efficient-solar-panels.

[23] https://en.wikipedia.org/wiki/Supercapacitor.

[24] https://en.wikipedia.org/wiki/Compact_fluorescent_lamp.

[25] https://en.wikipedia.org/wiki/LED_lamp.

[26] N.Q. Nguyen, K.V. Pochiraju, Behavior of thermoelectric generators exposed to transient heat sources, Appl. Therm. Eng. 51 (2013) 1–9.

[27] L. Pauling, Oxidation-reduction reactions; electrolysis, in: General Chemistry, Dover Publications, Inc., New York, 1988, ISBN 978-0-486-65622-9, p. 539.

[28] L. Carrette, K.A. Friedrich, U. Stimming, Fuel cells: fundamentals to applications, Fuel Cells 1 (1) (2001) 1–39.

[29] N.A. Kelly, T.L. Gibson, Increasing the solar photovoltaic energy capture on sunny and cloudy days, Sol. Energy 85 (2011) 111–125.

[30] N.T. Kalyani, S.J. Dhoble, Empowering the future with organic solar cell devices, in: Nanomaterials for Green Energy, 2018, ISBN 9780128137314, pp. 325–350, https://doi.org/10.1016/B978-0-12-813731-4.00010-2.21. Ch 5.

[31] B. Munir, V. Dyo, On the impact of mobility on battery-less RF energy harvesting system performance, Sensors 18 (11) (2018) 3597.

[32] X. Tan, Q. Li, H. Wang, Advances and trends of energy storage technology in Microgrid, Int. J. Electr. Power Energy Syst. 44 (2013) 179–191.

[33] P. Simon, Y. Gogotsi, B. Dunn, Where do batteries end and supercapacitors begin? Science 343 (6176) (2014) 1210–1211.

[34] J. Xia, F. Chen, J. Li, N. Tao, Measurement of the quantum capacitance of graphene, Nat. Nanotechnol. 4 (2009) 505–509.

[35] M. Jayalakshmi, K. Balasubramaniam, Simple capacitors to super capacitors-An overview, Int. J. Electrochem. Sci. 3 (2008) 1196–1217.

[36] O. Ostroverkhova, Handbook of Organic Materials for Optical and (Opto)Electronic Devices: Properties and Applications, Elsevier, Technology & Engineering, 2013, pp. 1–832.

[37] L. Yu, J. Liu, S. Hu, R. He, W. Yang, H.bin Wu, J. Peng, R. Xia, Red, green, and blue light-emitting polyfluorenes containing a dibenzothiophene-S,S - dioxide unit and efficient high-color-rendering-index white-light-emitting diodes made therefrom, Adv. Funct. Mater. 23 (35) (2013) 4273–4385.

[38] A.M. Suhail, M.J. Khalifa, N.M. Saeed, O.A. Ibrahim, White light generation from CdS nanoparticles illuminated by UV-LED, Eur. Phys. J. Appl. Phys. 49 (2010) 30601–30606.

[39] I.G. Lezama, A. Arora, A. Ubaldini, C. Barreteau, E. Giannini, M. Potemski, A.F. Morpurgo, Indirect-to-Direct band gap crossover in few- layer MoTe2, Nano Lett. 15 (4) (2015) 2336–2342.

[40] N.T. Kalyani, S.J. Dhoble, Organic light emitting diodes: novel energy saving lighting technology-A review, Renew. Sustain. Energy Rev. 16 (2012) 2696–2723.

[41] N. Narendran, Y. Gu, Life of LED-based white light sources, J. Disp. Technol. 1 (1) (2005) 167.

[42] F. Falchi, P. Cinzano, C.D. Elvidge, A. Haim, Limiting the impact of light pollution on human health, environment and stellar visibility, J. Environ. Manag. 92 (10) (2011) 2714–2722.

[43] N.T. Kalyani, S.J. Dhoble, R.B. Pode, Fabrication of red organic light emitting diodes (OLEDs) using EuxY(1-x)(TTA)3Phen organic complexes for solid state lighting, Adv. Mater. Lett. 2 (1) (2011) 65–70.

[44] Doosan DND: OLED System Instruction Manual.

[45] N.T. Kalyani, S.J. Dhoble, Novel materials for fabrication and encapsulation of OLEDs, Renew. Sustain. Energy Rev. 44 (2015) 319–347.

[46] N.T. Kalyani, H. Swart, S.J. Dhoble, Organic light emitting diodes: the future lighting sources, in: Principles and Applications of Organic Light Emitting Diodes, Elsevier: Imprint Wood head Publishing Series in Electronic and Optical Materials, United Kingdom, 2017, ISBN 978-0-08-101213-0, pp. 141–170. Ch.6.

[47] S.I. Kim, K.W. Lee, B.B. Sahu, J.G. Han, Flexible OLED fabrication with ITO thin film on polymer substrate, Jpn. J. Appl. Phys. 54 (9) (2015), 090301-1-4.

[48] B. Geffroy, P. le Roy, C. Prat, Organic light-emitting diode (OLED) technology: materials, devices and display technologies 55 (2006) 572–582.

[49] K. Albrecht, K. Matsuoka, D. Yokoyama, Y. Sakai, A. Nakayama, K. Fujita, K. Yamamoto, Thermally activated delayed fluorescence OLEDs with fully solution processed organic layers exhibiting nearly 10% external quantum efficiency, Chem. Commun. 53 (2017) 2439–2442.

CHAPTER

2

Solar cell technology

R.M. Pujahari

Assistant Professor in Physics, Department of Applied Sciences, ABES Institute of Technology (ABESIT), Ghaziabad, Uttar Pradesh, India

2.1 Introduction

Solar cells and their applications are usually connected to the semiconductor material they are made from. These materials must have certain characteristics for absorbing sunlight. Most cells are built to withstand the sunlight that reaches Earth's surface, while others are adapted for use in space. Solar cells may consist of only one layer of light-absorbing material (single junction) or several physical configurations (multijunctions) may be used. According to their technologies, solar cells can be divided into first-, second-, and third-generation cells. First-generation cells, also known as conventional, traditional, or wafer-based cells, are made of crystalline silicon, the prevalent commercial photovoltaic (PV) technology that includes materials such as polysilicon and monocrystalline silicon. Solar cells of the second generation are thin-film cells that include polysilicon and monocrystalline silicon. Second-generation cells are thin-film solar cells, which include amorphous silicon (a-Si), cadmium telluride (CdTe), and copper indium gallium selenide (CIGS) cells, and are important commercially in PV power plants, built-in PVs, or small standalone power plants. A variety of thin-film technologies are also integrated into the third generation of solar cells. Described as modern PVs, most of them are currently in the research or development phase and have not yet been commercially deployed. Many use organic materials, particularly inorganic and organometallic compounds.

2.2 Solar cell design

Solar cell architecture is straightforward in harnessing solar power to optimum efficiency. Many factors that are addressed in this chapter are limiting factors for solar cell performance. Likewise, loss mechanisms are also responsible for significantly reducing cell efficiency. Solar cell design is very important for achieving high solar cell efficiency, which has been addressed in this context [1].

Energy Materials
https://doi.org/10.1016/B978-0-12-823710-6.00007-8

Copyright © 2021 Elsevier Ltd. All rights reserved.

2.2.1 Upper limits of cell parameters

2.2.1.1 Short-circuit current

Short-circuit current (I_{SC}) is the average current produced by a solar cell when its terminals are shorted, indicating that the solar cell lacks voltage. This photon incident on the energy cell, which is larger than the bandgap, gives rise to an electron flowing under ideal conditions in the outer circuit. To determine the average current of the short circuit, the sunlight photon flux must be known. The photon flux can be determined by comparing a solar spectrum's energy content at a given wavelength with a photon's energy at that wavelength. The energy of photons E and wavelength λ are related as:

$$E(\text{eV}) = \frac{hc}{\lambda} = \frac{1.24}{\lambda(\mu m)} \tag{2.1}$$

With the full current of the short circuit, we will presume that the material is not recombined. I_{SC} can be measured by integrating and multiplying the photon flux from the maximum photon energy level to the cutoff energy level (band energy) multiplying it by elementary charge $q(1.6 \times 10^{-19}\ \text{C})$ [2].

2.2.1.2 Open-circuit voltage

Open-circuit voltage (V_{OC}) is the maximum voltage that can be derived from a solar cell while its terminals remain open. Because of the light-generated current, the amount of forward bias of a p—n junction is open-circuit voltage (the light-induced I_L at this open-circuit voltage is equal to and opposite the forward bias diffusion current of a p—n junction diode).

If an energy photon exceeds the bandgap energy, it excites an electron from the valence band to the conductive band, which in effect increases potential energy by an amount equal to $E_g(qV)$. The upper limit of a solar cell's open-circuit voltage is defined by the material's band distance. For instance, Si's bandgap is 1.1 eV; hence, the maximum possible V_{OC} is 1.1 V. Open-circuit voltage that can be obtained from the solar cell when there is no current drawn from is termed:

$$V_{OC} = \frac{kT}{q} \ln\left(\frac{I_L}{I_0} + 1\right) \tag{2.2}$$

When current densities are very low, the value 1 in the foregoing expression when compared with I_L/I_0 and $I_L \approx I_{SC}$ can be neglected, therefore the expression of V_{OC} is:

$$V_{OC} = U_T \ln(I_{SC} / I_0) \tag{2.3}$$

The open-circuit voltage in this case is equal to the ratio of the short-circuit current and dark current [3].

2.2.1.3 Fill factor

The fill factor (FF) can be expressed as $I - V$ curve squareness and is primarily related to solar cell resistive losses. It is always a fact that a suitable R_S load resistor is needed for maximum power output, which corresponds to the V_M/I_M ratio. V_M and I_M are the highest voltage and maximum current, and the maximum power output attainable is P_m.

Consequently, the peak output ratio $(V_M I_M)$ to the variable $(V_{OC} I_{SC})$ is called the solar cell's FF:

$$FF = V_M I_M / V_{OC} I_{SC} \qquad (2.4)$$

2.2.1.4 *Efficiency*

The electric output of a cell generated photovoltaically, the luminous power electric output that falls on it, and their ratio are called the efficiency (η) of solar cells:

$$\eta = \frac{V_M I_M}{P_{light}} = \frac{FF \; V_{OC} I_{SC}}{P_{light}} \qquad (2.5)$$

For a solar cell's maximum bandgap, the output will be highest.

2.2.2 Losses in solar cells

The loss of photon energy, which means it cannot pull an electron out of the solar cell, results in loss of solar cells.

Material properties or potential causes for losses in solar cells may be limited by cell-processing capabilities. The different types of fundamental losses are:

- Loss of low-energy photons;
- Loss due to excess energy of photons;
- Voltage loss; and
- FF loss.

The losses due to technological reasons are:

- Loss by reflection;
- Loss due to incomplete absorption;
- Loss due to metal coverage; and
- Recombination losses.

2.2.3 Solar cell design

A solar cell's basic functions are: (1) light absorption, (2) separation of generated charge carriers, and (3) transport of separate charge carriers to external loads without resistive losses. Solar cells of a given material must be engineered for conversion efficiency optimization of light to electricity.

This in effect optimizes the parameters of solar cells such as I_{SC}, V_{OC}, and FF so that higher values of these parameters are obtained to increase the performance of cells. The efficiency in cells is:

$$\eta = \frac{I_{SC} V_{OC} \; FF}{P_{in}} \qquad (2.6)$$

2.2.4 Design of high short-circuit current

2.2.4.1 Requirements of high short-circuit current

Short-circuit current expression, from which the requirement for obtaining high I_{SC}, can be deduced as follows:

$$I_{SC} = qAG(L_n + W + L_p) \tag{2.7}$$

The short-circuit current is directly proportional to the area of the cell; the greater the area of the solar cell, the greater the current.

2.2.4.2 Choice of junction depth and its orientation

Generation rate

A p–n junction diode is illuminated by the solar radiation from the n-side of the semiconductor.

The generation rate $G(x)$ of the carriers decays exponentially along the depth of the material, which is expressed as:

$$G(x) = \int_0^\lambda [1 - R(\lambda)]\alpha(\lambda)\varphi_P e^{-\alpha(\lambda)x} d\lambda \tag{2.8}$$

where $R(\lambda)$ represents reflected solar radiation from the semiconductor, $\alpha(\lambda)$ is the recombination coefficient, and $\varphi_P(\lambda)$ is the photon flux in solar radiation per unit wavelength.

Collection probability: Carriers produced in the depletion region and within the diffusion length of the minority carriers have good chances of separation (or collection) and hence they contribute to I_{SC} (it will not combine) and will have 100% collection probability. Carriers that are generated near the surface may also be collected and contribute to I_{SC}.

A light-generated carrier will contribute to the current if it has a probability of collection. The light-generated current $I_L = I_{SC}$ can be written as:

$$I_L = qA \int_0^W G(x)C_P(x)dx \tag{2.9}$$

where W is the width of the solar cell.

Junction depth and its orientation: The rate of generation and collection probability should be high for large light-generated current.

2.2.4.3 Minimization of optical losses

Optical losses can be reduced by changing the design pattern as follows:

- By putting an antireflection coating (ARC) on the solar cell surface;
- By texturing the front surface;
- By using the minimum front metal contact coverage area to reduce the contact covering; and
- By making the solar cell thicker to increase absorption.

ARC: The theory of "optical quarter wavelength" is observed for the reduction of reflection from the solar cell surface. The barrier layer reflects the penetrating light beam between the antireflection medium and the silicon.

Here, energy conservation law and continuity principles are used and, upon reaching a more optically dense medium, the electromagnetic wave undergoes a phase change of π. The refractive index element and the thickness of the coating are equal to one-quarter of the wavelength and the wavelength light falling vertically is completely extinguished.

The thickness of antireflection layer plays the main role using the basic principle of interference. This principle can be written as a formula:

$$n \cdot d = \frac{\lambda}{4} \tag{2.10}$$

As per Fresnel's formula, the reflection factor is:

$$R = \frac{r_1^2 + r_2^2 + 2r_1r_2 \cos 2\theta}{1 + r_1^2r_2^2 \cos 2\theta} \tag{2.11}$$

With:

$$r_1 = \frac{n_0 - n_1}{n_0 + n_1} \tag{2.12}$$

$$r_2 = \frac{n_1 - n_2}{n_1 + n_2} \tag{2.13}$$

$$\theta = \frac{2\pi n_1 d_1}{\lambda} \tag{2.14}$$

Reflection is maximum at its minimum where $n_1d_1 = \lambda/4$:

$$R_{\min} = \left[\frac{n_1^2 - n_0n_2}{n_1^2 + n_0n_2}\right]^2 \tag{2.15}$$

$R_{\min}$ is equal to zero when:

$$n_1^2 = n_0n_2 \tag{2.16}$$

i.e., when:

$$n_1 = \sqrt{n_0n_2} \tag{2.17}$$

Full thickness reflection d_1 in relation to wavelength is calculated as the refractive index of silicon changes with respect to the wavelength. The minimum reflection occurs at $\lambda = 0.6$ μm, as per the reflection layer thickness and the foregoing relationship.

Surface texturing: Surface texturing is another technique for reducing reflection. It is accomplished by making the surface rough. Due to multiple reflections, the structured surface absorbs most of the light incident on it [4].

Absorber material thickness: Reasons for radiation loss are photon emission and providing energy greater than bandgap energy through the cell without being absorbed by it. To absorb all the radiations above the bandgap energy, the thickness of the absorbing material should be equal to the absorption length of the energy photon and the same as the bandgap energy. Therefore material thickness for zero transmission loss is written as:

$$d_{\text{trans}=0} = \frac{1}{\alpha(E_{\text{g}})} \tag{2.18}$$

where $\alpha(E_{\text{g}})$ is the energy E_{g} photon absorption coefficient. The absorption length is approximately 10 mm in silicon to absorb photons that have energy equal to the bandgap [5].

Light trapping: Around 10% of sunlight (AM1.5) in a 200 μm thick silicon solar cell, which is incident on the surface, is not absorbed. An efficient method for fitting the solar cell at the back surface with an anonymous surface structure is called reflection due to Lambert's law.

Because silicon has a high refractive index, radiation can evacuate through a narrow leakage cone.

The leakage cone angle Φ_{C} is calculated as:

$$\sin^2\Phi_{\text{C}} = \frac{1}{n^2} \tag{2.19}$$

Using the value of $n_{\text{Si}} = 3.5$, it is found that $\Phi_{\text{C}} = 17$ degrees. About 8.5% of the total radiation incident into this cone is lost due to reflection [6,7].

2.2.4.4 *Minimization of recombination*

Recombination of a generated carrier is possible at the surface, in the depletion region, and in the bulk of the semiconductor. All possible recombinations should be minimized to obtain a high illuminated current (I_{L}).

Surface passivation: Passivation at the surface is accomplished by depositing a dielectric layer on the front surface so that it passivates the unfinished bonds at the surface (by removing the energy states from the middle of the bandgap) [8,9]. Surface recombination effectiveness is expressed in terms of surface recombination velocity (SRV). It is expressed as the rate of recombination at the surface divided by the surface concentration of the excess carrier. Mathematically:

$$S_{\text{n}} = \frac{R_{\text{sur}}}{\Delta n_{\text{sur}}} \text{ and } S_{\text{p}} = \frac{R_{\text{sur}}}{\Delta p_{\text{sur}}} \tag{2.20}$$

where S_{n} and S_{p} are SRVs for electrons and holes, respectively. The lower the SRV, the better the passivation.

Bulk passivation: To produce low-cost solar cells, the substrates used in them cannot be of very high quality (as in float zone wafers). To keep the cost very low, the use of multicrystal-line silicon (mC-Si) wafers has become very common. Mc-Si wafers or in general a deposited thin-film active material (in thin-film solar cell technologies) may contain crystallographic defects, grain boundaries, metallic impurities, oxygen, and carbon.

2.2.5 Design of high open-circuit voltage

In a solar cell, the solar cell controls the voltage of the open circuit in the recombination process. Higher V_{OC} can be accomplished by allowing recombination in the solar cell to be tiny as possible.

The V_{OC} is specified by the voltage at which I_{SC} becomes equal and is in contrast to the current of forward-biased diffusion. The forward-biased current depends on the recombination of the surface and the size of the solar cell. If the current of reverse saturation I_0 increases, which is shown by:

$$I = I_0 \left(e^{qV/kT} - 1 \right) - I_L \tag{2.21}$$

then the current of forward-biased diffusion increases by an increase in the current of reverse saturation. Therefore the rise in recombination in the solar cell reduces the V_{OC} of the solar cell [10].

2.2.5.1 Requirements for high open-circuit voltage

The specific parameters controlling open-circuit voltage are:

- The number of minority carriers at the edge of the depletion region is created because of injection of minority carriers.
- Minimizing minority carrier injection means reducing the concentration of the minority equilibrium carrier. Minority carriers injected switch forward and vanish through recombination.
- The forward bias current increases during this process and V_{OC} reduces.
- Crystallographic defects in the bulk material and on the surface cause recombination of the carrier, which in fact reduces V_{OC}.
- The thickness of the solar cell should be small for lower recombination, which leads to high V_{OC}.

The V_{OC} depends on I_0, which is contributed by n-side, p-side, and depletion regions. The contribution to I_0 from the p-side, I_{0p}, in the presence of back surface field (BSF) can be written as:

$$I_{0p} = \frac{qD_n \, n_i^2}{L_n N_A} \tan h \, \frac{W_p}{L_n} \tag{2.22}$$

where W_p is the width of the p-side. This is the case when the width of the p-side is more than the minority electron diffusion length L_n. When p-side thickness is reduced such that $W_p \ll L_n$, then Eq. (2.22) modifies to the following form:

$$I_{0p} = \frac{qD_n n_i^2}{\tau_n N_A} \tag{2.23}$$

With an increase in p-side doping, the minority electron lifetime (τ_n) increases. It can be concluded that V_{OC} will be independent of p-side doping at a lower doping level [11].

2.2.6 Design of high fill factor

Fig. 2.1 displays solar cell series resistance.
The participating resistances are:

R_1: resistance related to contact due to metal—semiconductor on the total surface back;
R_2: semiconductor material (base) resistance;
R_3 : resistance due to an emitter between two grid fingers;
R_4: resistance due to metal—semiconductor contact on the grid finger;
R_5: resistance due to grid finger; and
R_6: resistance due to collection bus.

2.2.7 Analytical techniques

It is important to define parameters and their complex relationships in case of solar cells the measurements prerequirement of solar cells.

2.2.7.1 *Current—voltage characteristics*

Solar cell efficiency can be calculated within the range of AM1.5 according to international agreement. Radiation materializes in the laboratory by using a sun simulator [12,13].

Often for greater accuracy, observation readings from the test cell, the reference cell, and the intensity of radiation are calculated at the same time (Fig. 2.2).

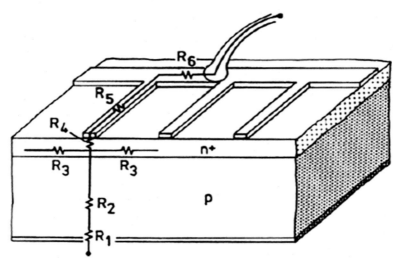

FIGURE 2.1   Solar cell series resistance.

I. Fundamentals and overarching topics

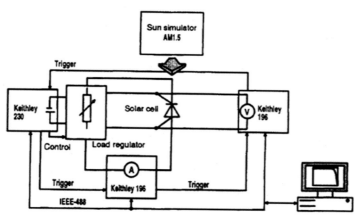

FIGURE 2.2 Block diagram of a sun simulator.

2.2.7.2 Measuring the current–voltage curve under illuminated conditions

The current–voltage (I–V) characteristic is calculated by using an electrical charge regulator to calculate the current from the solar cell from zero to point by point of a short-circuit current [14]. The computer system calculates the following parameters from observations about current–voltage:

- Open-circuit voltage V_{OC};
- Short-circuit current I_{SC};
- Current I_m and voltage V_m at maximum power point;
- Fill factor FF; and
- Efficiency η.

2.2.7.3 Measurement of dark current characteristics

Using the measurement systems already described, current–voltage characteristics using dark current characteristics for a solar cell as a regular diode can be determined. In a two-diode model, different variables can be allocated to the individual ranges of the characteristics.

The diode equation used here is:

$$I(V) = I_{01}\left[\exp\frac{V - IR_S}{n_1 V_T} - 1\right] + I_{02}\left[\exp\frac{V - IR_S}{n_1 V_T} - 1\right] + \frac{V - IR_S}{R_P} \tag{2.24}$$

- The beginning two terms in Eq. (2.24) have minimal impact in the range 0 to about 0.15 V, and the current in the dark condition is calibrated using shunt resistance R_P (the proportionality relation, which is linear, is defined as a curve in the logarithmically linear representation).

36
2. Solar cell technology

- In the adjacent area, the current at the dark condition can be designated by the second term of the two-diode models (from 0.2 to 0.4 V).
- The role term and its range is 0.4–0.6 V. If the current is based on $n = 1$, then it can be used to determine the saturation current I_0, which is responsible for open-circuit voltage.
- Series resistance of around 0.6 V has a significant impact on characteristics in the region.

Series resistance can be more accurately measured under both illumination and dark conditions. To measure dark current, it is appropriate to avoid a higher voltage (V_S) than the voltage at the open circuit (V_{OC}) to calculate the current counterpart as like as short-circuit current, voltage drop which is in addition to the original at the series resistance (R_S). Resistance to the sequence could be calculated by taking the difference between the two voltages:

$$V_S - V_{OC} = R_S\, I_{SC} \tag{2.25}$$

and thus:

$$R_S = \frac{V_S - V_{OC}}{I_{SC}} \tag{2.26}$$

The determination of dark current characteristics can also indicate the behavior of solar cells and also solar modules.

2.2.7.4 Spectral response solar cell

A front-illuminated solar cell's spectral response: Spectral response is simply recording the dependency of the collected charge carriers (solar current) at various wavelength ranges on the radiated photons [15].

To achieve the spectral response, the solar cell is irradiated by light from different spectral ranges.

Internal spectral response is often measured with respect to the wavelength and using the exact reflective conditions on the surface. The same conditions can then be measured for polished surfaces and the standard result is compared with the experimental data [16].

Figure also indicates the emitter's response at the region of the base as well as space charge with respect to the radiated light wavelength.

Spectral response of a solar cell illuminated from the back surface: For determination of surface recombination velocities at lower values as well; a separate methodology can be used as demonstrated in Fig. 2.10.

On the back surface of the finger grid there is a finger grid where the finger width is around 10 times the length of the centered load carrier. Thus heavy recombination is prevented under the contact fingers. Also, by applying a BSF metal coating for the reduction of recombination further improvement can be achieved [17].

The spectral response curve is illustrated in Fig. 2.3. With short wavelength light when absorbed near the surface and then at some distance from the p-n junction, charge carriers are produced, and the spectral behavior is strongly influenced by surface recombination velocity.

I. Fundamentals and overarching topics

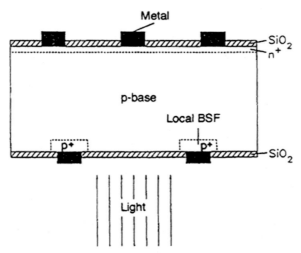

FIGURE 2.3 Rear side illuminated solar cell structure.

2.2.7.5 Technique for PCVD analysis

Sometimes it is very difficult for high-efficiency solar cells to calibrate the diffusion lengths when the values are more than the thickness. In the case of plasma chemical vapour deposition (PCVD) measurements, one can measure the diffusion length [18].

This principle has certain advantages compared to the static process, here:

- Absolute measurement of light intensity is not required.
- There must be no knowledge of surface reflection.
- The precise coefficient of absorption must be understood (Fig. 2.4).

The laser pulse exerts excitation for the specimen cell. To operate the apparatus, an Nd–YAg laser that emits 1064 nm of light with wavelength is used. Due to an extremely low absorption coefficient at this wavelength, it results in a continuous charge carrier at high level across the entire base.

2.2.7.6 The PCD method

To perform measurements, ohmic contacts need to be connected in the photocurrent decay (PCD) measuring technique. Therefore a thorough separation of all recombination values can be achieved using the PCD technique based on volume or surface. Ohmic contacts are not required in this case.

At a high-frequency resonant circuit, the silicon wafer is placed at a frequency of 13.56 MHz at about a 1 mm distance. Damping of the resonant circuit occurs due to electrical conductivity [19]. A high-frequency bridge is balanced in such a way that the bridge's differential current is zero.

In this way, two parameters are measured: (1) the saturation current of the emitter and (2) the velocity of the surface recombination [20].

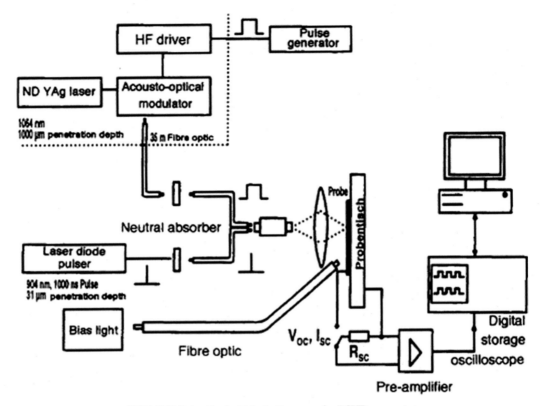

FIGURE 2.4 Typical block diagram of a PCVD apparatus.

2.3 Solar cell design

2.3.1 High-efficiency Si solar cells

It is always a challenging task to produce highly pure crystalline silicon. In the course of discussion, various technologies to manufacture solar cells have been enumerated as follows.

2.3.1.1 Development of commercial Si solar cells

Silicon is one of the most common materials on the upper layer of Earth. Silicon is found mainly as quartz and sand silicon dioxide (SiO_2). The fabrication process has been used for many decades. Using an arc furnace, process it is typically extracted from quartzite reduction with carbon.

First, pulverized quartz and carbon are mixed inside a graphite crucible. Then, using an arc, the mixture is melted at a very high temperature around 1800°C. The reduction cycle materializes as follows according to the formula:

$$SiO_2 + 2C \rightarrow Si + 2CO \tag{2.27}$$

Then, the liquid silicon produced at the crucible base (1415°C melting point) can be drawn off. Its purity is approximately 98%. This is called metallurgical-grade silicon, and a large amount is used in the iron and aluminum industries [21].

Because the energy that this cycle absorbs is very high at around 14 kWh/kg, production is based only on the availability of power.

2.3.1.2 *Refractioning processes*

To produce electronic-grade silicon, standardized processes were developed and certified by Siemens in the 1950s.

In a fluidized-bed reactor, metallic metallurgical silicon is exposed to hydrochloric gas in the process step. The chemical reaction (exothermic reaction) yields trichlorosilane and hydrogen:

$$Si + 3HCl \rightarrow SiHCl_3 + H_2 \tag{2.28}$$

2.3.2 Process flow of commercial Si cell technology

2.3.2.1 *Polycrystalline silicon material manufacturing*

Silicon having a high purity grade is produced in this process. According to chemical vapor deposition (CVD) theory, polycrystalline silicon is deposited in a reactor tank. The method is as follows: a U-shaped thin silicone rod is electrically heated to 1350°C. A mixture of (high purity) hydrogen and trichlorosilane is added to the reactor vessel. On the silicon's hot surface, trichlorosilane is reduced to silicon, which is placed on the top surface of the rod. The method is performed according to the formula:

$$4SiHCl_3 + 2H_2 \rightarrow 3Si + SiCl_4 + 8HCl \tag{2.29}$$

2.3.2.2 *Crystal pulling process*

There are primarily two processes: the crucible pulling process (also called the Czochralski process) and the float zone pulling process.

The Czochralski (CZ) process: In this process, fragmented mC-Si derived from polysilicon is put in a quartz crucible; it is then put in a graphite crucible and melted with heat of induction. The pulling process starts with the dipping of single silicon crystal seed.

Float zone pulling: In this process, the puller is situated within a boundary filled with a gas, which is inert in nature. On the polycrystalline rod is a melted form of polycrystalline silicon seed, which is produced by induction heating at the bottom end. A region of liquid silicon is pushed upward by the induction coil's vertical movement while spinning after melting [22].

It then solidifies into a single crystal when the silicon cools. Appropriate gaseous dopants (such as phosphine PH_3 or diborane B_2H_6) are doped into the inert gas. Additional crystal cleaning is also achieved by this method [23].

2.3.3 Process used for solar cell technologies

2.3.3.1 Silicon wafer manufacture

Largely, semiconductor devices such as solar cells require thin wafers in the range of 0.2—0.5 mm. The traditional wafer method used the so-called inner diameter saw, in which the diamond particles are lodged in the blade of the saw around a cavity. This is a cost-intensive process, which has the drawback of losing almost half of the material. A modern approach is propounded in the shape of a multiwire saw where a multikilometer long wire is passed through the crystal rod in several coils inside an abrasive suspension, while being wound from one coil to another.

2.3.3.2 Silicon wafers out of polycrystalline material

Later on, the block-casting method was used to reduce the expense of fabricating silicon wafers. Silicon is melted in a quartz and graphite crucible. A controlled cooling process produces polycrystalline blocks of silicon with large grain structures.

2.3.3.3 Sheet materials

So far, the focus has been on the making of so-called sheet material so that the sawing process can be avoided in the production of silicon. The two processes that are mostly used are:

The edge-defined film growth (EFG) process: The EFG process is based on the principle of an octagonal tube derived from the melting of silicon using appropriate templates of graphite. The edge length of the octagonal segments is slightly above 100 mm, giving a tube diameter of approximately 30 cm. The thickness of the shaped slices is a few tenths of a millimeter, set by the capillary graphite, as well as the prevailing temperature and pulling speed (in cm/min). This technique can be used to produce tubes around 4—5 m long [24].

Laser cutting can be used to separate individual wafers with dimensions 100×100 mm^2. As in block cast silicon, the content is multicrystalline. Multicrystalline solar cells have achieved efficiencies of 13%—15% in large-scale output.

The silicon sheet from powder (SSP) process: In the SSP process, silicon powder is put in a quartz form. First, the powder is sintered together. Second, the self-supporting foil goes through a zone-melting process [25]. A multicrystalline material is produced that has very large grains (millimeters to centimeters). Incoherent light materializes the melting process. In this way, silicon sheets are made with a thickness of around 400 μm. This enables the manufacture of solar cells in a laboratory with an efficiency of about 13%.

2.3.4 High-efficiency silicon solar cell technologies

The technologies to fabricate crystalline silicon solar cells can be graded as:

- p—n junction production technologies;
- Growing of SiO_2 layers technologies;
- Making of electrical contacts technologies;
- Reflection loss minimization technologies.

2.3.4.1 p–n Junction production technologies

The two important methods for producing p-n junctions are the techniques of diffusion and ion implantation. There is another process that is also used: the alloying technique, which was used in the early 1950s [26].

Diffusion technologies: The diffusion of solid substances in solid silicon obeys the second law of Fick, as follows:

$$\frac{\partial N(x,t)}{\partial t} = D\frac{\partial^2 N(x,t)}{\partial x^2} \tag{2.30}$$

where:

$N(x,t)$ is the concentration of the substances used for diffusion at a particular time t at point x
D is another parameter, specific to each material, called the diffusion coefficient

Emitter diffusion processes: For some time solar cell wafers have often been doped with boron, and therefore a p-conductive material (with a concentration of 10^{15}–10^{17} cm^{-3}) using a phosphorous n-doped emitter is fabricated [2,27].

Phosphorus is put into a gas diffusion furnace either as phosphine or as gaseous phosphorous oxychloride (POCl$_3$). The POCl$_3$ is blended as a carrier gas using nitrogen. Once the oxygen is added at about 800°C, the dopant gases react with the silicon sheet. First, according to the equation Si + O$_2$ → SiO$_2$, silicon dioxide is created on the surface, and second, phosphine changes into phosphorous pentoxide (P$_2$O$_5$):

$$2PH + 3O_2 \rightarrow P_2O_5 + H_2O \tag{2.31}$$

Using the diffusion technique, P$_2$O$_5$ is produced; it is then mixed with silicon dioxide, and liquid phosphorous glass is formed on the silicon surface, which is a source of diffusion.

The BSF diffusion processes: As it is understood, solar cells that are highly efficient need a BSF. Aluminum is used in industry for generating a BSF. The doping process is done using screen printing at around 800°C, and aluminum is induced into the surface by vacuum evaporation as an ink. The aluminum partly diffuses at this temperature and produces a p$^+$-doping material. The recrystallized layers also play the role of good getter sinks (Fig. 2.5).

2.3.4.2 Oxidation technologies

The process of oxidation without vapor is as per:

$$Si + O_2 \rightarrow SiO_2 \tag{2.32}$$

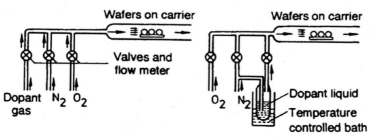

FIGURE 2.5 Diffusion process (open-tube process).

Oxygen is diffused into the SiO_2 layer. Therefore there is no saturation thickness in this process and growth rate slows with thickness. The thickness of the layer develops with respect to time at the periphery. Typically, the SiO_2 layer requires an approximately 45% silicon layer of its own thickness [3].

Wet oxidation occurs according to:

$$2Si + O_2 + 2H_2O \rightarrow 2SiO_2 + 2H_2 \tag{2.33}$$

In this case, the rate of growth is much higher than for dry oxidation because the hydrogen controls the reaction cycle. The effect of dry or wet oxidation on the SiO_2 layer thickness is governed by time and temperature of oxidation.

Adding chlorine is helpful during oxidation. The removal of metal traces by volatile metal chlorides and sodium atoms (alkali atoms) is also excluded and thus improves the passivating characteristics of the SiO_2 base. Chlorine is also mixed in the process, for example, with the mixing of trichloroethane (TCA) ($C_2H_3Cl_3$).

The oxidation reaction is:

$$C_2H_3Cl_3 + 2O_2 \rightarrow 3HCl + 2CO_2 \tag{2.34}$$

and:

$$4HCl + O_2 \rightarrow 2H_2O + 2Cl_2 \tag{2.35}$$

Addition of TCA to the supply of oxygen is essential; if this is not done, it may produce poisonous phosgene because TCA is now substituted with a trans-liquid crystal display (also containing chlorine) for environmental safety.

2.3.4.3 Auxiliary technologies

For the manufacture of mC-Si solar cells, auxiliary technologies such as etching, cleaning, and photolithography are essential [28].

Etching and cleaning techniques: To manufacture solid-state semiconductor devices, etching and cleaning of solar cell wafers is necessary. The silicon wafer surfaces must be as contaminant free as possible. There are ionic, molecular, or radioactive contaminants. Residues of the polishing, lapping, and photoresist methods are basically molecular.

Absorption of ions from the etching solution may lead to contamination; on the other hand, heavy metals like copper, silver, and gold have an atomic character.

RCA cleaning is one surface-cleaning process.

Silicon isotropic etching occurs in nitric acid solution and hydrofluoric acid. The hydrofluoric (HF) solution can then be buffered with ammonium fluoride (NH_4F) as needed by the etching rate.

Isotropic silicon etching occurs in hydrofluoric acid and nitric acid solution. The silicon is typically oxidized by the nitric acid to SiO_2, and dissolved with hydrofluoric acid.

Photolithography: Photolithography is a special technique used with silicon dioxide for masking. A thin photoresist film is spun in a yellow area because it is sensitive to light on a silicon wafer. A homogeneous film with a thickness of about 0.5—2 μm is formed due to viscosity and revolution speed values (around thousands of rpm). After cleaning the film, it is allowed to pass through masks, and short wavelength light (around 0.4 μm) is used for this purpose [15,16].

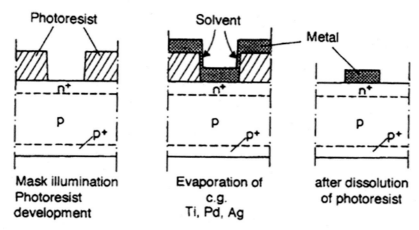

FIGURE 2.6 The principle of the lift-off process.

2.3.4.4 Solar cell metallization technique

There are several technical approaches for semiconductor contact making. In this section, we discuss the correct metallization techniques used in solar cell technology. First of all, we look at the structuring of the finger grid.

The structuring of the finger grid: There are various methods that can be implemented for fabricating the grid finger. The appropriate layer sequence is evaporated by a vacuum as a whole layer onto a photoresist having a definite structure. Usage of ultrasound at the same time during dissolution is beneficial. By using this method, structures that are only some micrometers long can be generated (Fig. 2.6).

The next step is screen printing by using metal paste, which is popular in industry as it is cost effective and can handle a maximum of 100 μm. In the laboratory, shadowing masks are used with metal evaporation for defining patterns on the cells.

2.3.4.5 Antireflection technologies

ARC: High-vacuum evaporation techniques and dense film methods are used to produce ARCs. Titanium dioxide (TiO$_2$) is vigorously used for a single layer of ARC. During the evaporation process, the refractive index is regulated within a fixed limit by selecting the rate of evaporation and adding minimal quantities of oxygen.

Manufacturing textured silicon surfaces: Silicon etching rate is monitored by using structures the chemical—physical effect in the direction of the crystal in an alkaline solution.

Therefore to dissolve an atom from the <100> crystal orientation, higher expenditures are required, selected vertically to the surface [17].

2.4 Thin-film solar cell technologies

Thin-film cells have emerged on the market as silicon-driven solar calculators. These are now available in very large modules, which are used for integrating buildings in advanced structures and vehicle charging systems.

2.4.1 Advantages of thin-film technologies

Thin-film high-efficiency modules have a proven advantage in generating more energy available per nameplate watt than regular c-Si modules.

2.4.1.1 Higher energy production in hot conditions

Under standard test conditions (STCs), the power for all PV modules, regardless of manufacturer or technology, is set. STCs do not set an ambient operating temperature, but describe the temperature of the PV module as 25°C [18]. Typically, module temperatures are above ambient temperature between 25 and 30°C.

The power output for all PV modules decreases as the module temperature reaches an STC of 25°C. Thin-film modules produce up to 5% more power under these conditions than average c-Si modules. Under hot climate conditions, this high-temperature advantage adds up to 3% more energy than c-Si modules over 1 year.

2.4.1.2 Spectral advantage in humid environments

PV technologies react differently to different luminous wavelengths. Water in the atmosphere creates various light wavelengths on hot days. Since CdTe modules are less vulnerable to wavelength reductions that are most affected by this type of high-atmospheric water content, thin-film modules produce up to 6% more annual energy per year under humid conditions.

2.4.1.3 Minimal power loss with better response to shading

The specific cell design of thin film ensures that the shaded portion is effective while the rest of the module continues to produce electricity. To protect them from damage, traditional c-Si modules turn off disproportionately large portions of the chip. A thin-film module will still produce 90% power in an environment with 10% shading.

2.4.2 Materials for thin-film technologies

Thin-film techniques cut down the volume of raw materials contained in a typical cell. Many active materials are sandwiched between two glass panes. Because panels of silicon solar cells use only a single glass plate, panels of thin films are almost double the weight of silicon solar cell panels made of crystalline material, although they have minimal ecological impact (determined from lifecycle studies).

For outdoor applications, CIGS, CdTe, and a-Si are three thin-film technologies that are often used [19].

2.4.2.1 Cadmium telluride

Among the prevailing materials for thin film is CdTe. This makes up more than half of the thin-film industry with around 5% of worldwide PV output. Similarly, CdTe is one of the thin-film materials that has an energy payback period, which is the lowest of all mass-produced PV technologies, generally around 8 months on designated locations.

2.4.2.2 Copper indium gallium selenide

A typical CIGS cell is made of copper, indium, gallium, and selenide. The abbreviation CIS is used in gallium-free types of semiconductor material [20]. The most widely used thin-film technologies are CdTe and a-Si, with laboratory performance values around 20%.

2.4.2.3 Gallium arsenide

In solar cells, gallium arsenide (GaAs) semiconductor material is also used as single-crystalline thin films. Though GaAs cells are very expensive, this high-performance single-junction solar cell holds the world record of 28.8% output. For spacecraft solar panels, GaAs are more widely used in multijunction solar cells, because the power-to-weight ratio decreases the costs (InGaP/(In)GaAs/Ge) (Fig. 2.7).

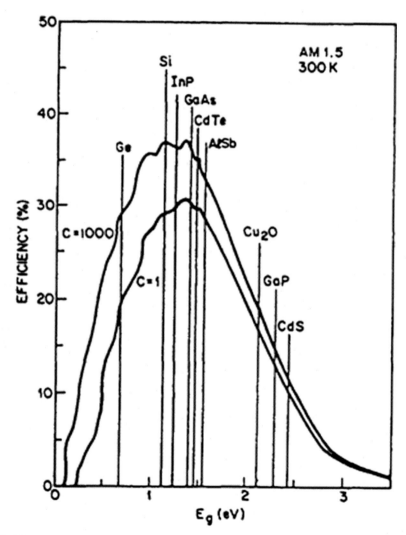

FIGURE 2.7 Maximum solar cell efficiency of radiation: 1 sun/1000 sun (300 K) versus energy gap.

2.4.3 Methods of thin-film deposition

Thin-film deposition is a critical step in the manufacture of cells using thin-film solar cell technologies (Fig. 2.8).

2.4.3.1 Physical vapor deposition

Physical vapor deposition (PVD), also known as physical vapor transport, describes a variety of vacuum deposition methods that can be used to make thin films and coatings. PVD is distinguished by a process in which the material moves from a condensed phase to a vapor phase, and then back to a condensed thin-film phase. PVD is used in the development of articles requiring thin films for mechanical, electrical, chemical, or electronic functions [21] (Fig. 2.9).

Common PVD processes are:

- Evaporation (thermal and electron beam evaporation);
- Sputtering (DC sputtering, DC sputtering of magnetrons, and RF sputtering); and
- Molecular beam epitaxy.

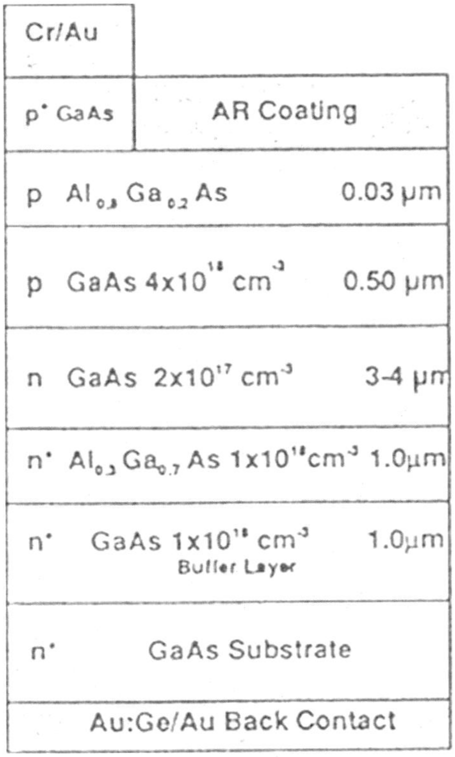

FIGURE 2.8 The structure of high-efficiency gallium arsenide solar cells.

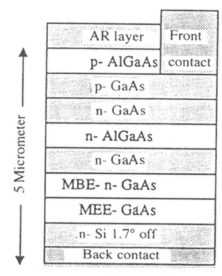

FIGURE 2.9 The structure gallium arsenide solar cells on a silicon substrate.

2.4.3.2 Chemical vapor deposition

CVD is a type of vacuum deposition used in high-quality, high-performance solids manufacturing. This process is also used within the semiconductor industry for the development of thin films. The wafer (substrate) in typical CVD is exposed to one or more volatile precursors that react and/or decompose to create the desired deposit on the substrata surface. Sometimes, volatile by-products are produced that are extracted through the reaction gas-flow chamber.

CVD is researched in a wide variety of formats. Such processes typically differ in the mechanism by which chemical reactions are triggered [22].

The operating conditions are:

- Atmospheric pressure CVD (APCVD): CVD at atmospheric pressure.
- Low-pressure CVD(LPCVD): CVD under subatmospheric pressure.
- Ultrahigh CVD vacuum: low pressure CVD, usually operates at 10^{-6} Pa ($\approx 10^{-8}$ torr).
- CVD supported by aerosols: CVD where the precursors are transferred to the substrate using a liquid/gas aerosol that can be produced ultrasonically.
- Direct liquid injection CVD: CVD where the precursors are in liquid form (liquid or solid dissolved in a suitable solvent).
- Hot wall CVD: CVD in which the chamber is heated by an external power source and the substrate is heated by radiation from the heated chamber walls.
- Cold wall CVD: CVD in which only the substratum is directly heated either through induction or by transmitting current through the substrate surface itself or a heater in contact with the substrate. The chamber walls stay at room temperature.

Various plasma methods:

- Microwave plasma CVD.
- Plasma-enhanced CVD (PECVD).

- Remote plasma-enhanced CVD.
- Low-energy plasma-enhanced CVD.
- Atomic layer CVD.
- Combustion CVD.
- Hot filament CVD.

2.4.3.3 Evaporation

Evaporation is a popular method for thin-film deposition. The source material will evaporate within a vacuum. The vacuum allows vapor particles to pass straight to the target area (substrate) where they condense back to a solid state. Evaporation is used in microfabrication to render items of macro size, such as metallic plastic film.

Evaporation involves two simple processes: the liquid evaporates from hot sources and then condenses on the substrate. This is analogous to the traditional process where water (liquid) surfaces on the upper surface (top) of the boiling pot. In this case, atmosphere (gaseous) and source of heat are different [23] (Fig. 2.10).

2.4.3.4 Sputtering

Sputter deposition is a method for sputtering PVD into the thin-film deposition. This includes expelling deposited material on the "top" just in case of a source to a "substrate" which usually happens in a silicon wafer. Reemission by ion or atom bombardment of the deposited material during the deposition process is done by resputtering.

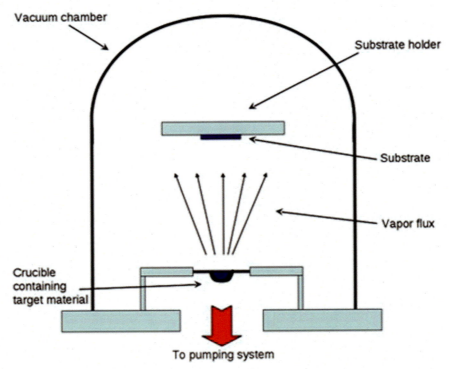

FIGURE 2.10 Schematic diagram showing thermal evaporation.

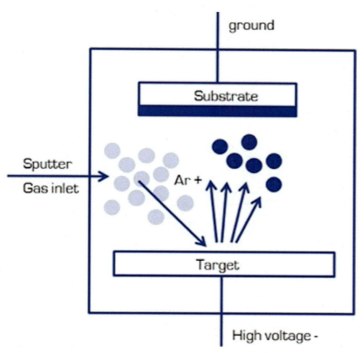

FIGURE 2.11 Schematic diagram showing sputtering.

Atoms that are expelled by sputtering from the target have an energy distribution range that is very wide, typically as much as equal to multiples of tens of eV (100,000 K). The presence of various criteria controlling sputter deposition makes it a difficult process but also allows experts to track the film's growth and microstructure to a large extent [24] (Fig. 2.11).

The sputtered atoms are charged neutrally, and therefore the magnetic trap is unaffected. Charge accumulation on isolating by using radio frequency sputtering where the anode-cathode bias is varied at a high rate.

2.4.3.5 Ion-beam sputtering

Ion-beam sputtering (IBS) is a process where the destination is external to the ion source. As in the ionization gauge for hot filaments, a source can operate without any magnetic field. Inside a Kaufman source, ions are formed by electron collisions that are confined similar to a magnetron by a magnetic field. Then, the generated electric field from the grid accelerates them to a target.

2.4.3.6 Reactive sputtering

Material responds to the sputtered particles in reactive sputtering until they coat the substrate. Therefore the film that has been deposited is different from the target material. The particles undergo a chemical reaction in a reactive gas such as oxygen or nitrogen, which is added to the chamber for sputtering,; oxide and nitride films are also made using reactive sputtering [15].

2.4.3.7 Ion-assisted deposition

In ion-assisted deposition (IAD), a secondary ion beam running at a lower intensity than the sputter gun is transmitted to the substrate. The secondary beam is normally supplied by a Kaufman source similar to that used in IBS. IAD may be used to deposit carbon on a substrate material in a diamond-like design.

2.4.3.8 High-target utilization sputtering

Remote generation of a high-density plasma can cause sputtering as well. The plasma is generated in a side chamber, which opens into the main process chamber, which contains the target and the coating substrate.

2.4.3.9 Gas-flow sputtering

The sputtering of gas flow utilizes the hollow cathode effect, the same effect that hollow cathode lamps use. In gas flowing, a working gas like argon is performed by a metal opening which is subjected to negative electrical potential.

2.4.3.10 LPCVD and APCVD

LPCVD is a process where heat is used to make the solid substrate a precondition for gas. This surface reaction forms the material for the solid phase. Low pressure is used to decrease any unintended reactions in the gas phase and also improves substrate-wide uniformity [16].

A cold- or hot-walled quartz tube reactor will materialize the LPCVD process. The drawbacks of LPCVD are that it operates at higher temperatures, which hinders the substrate types, and other types of materials may be present in the samples (Fig. 2.12).

APCVD systems are used to deposit a coating of material on wafers or other types of substrates usually many micrometers thick. These are used for the growth of Si epitaxial films, compound semiconductors, SiO_2, ARCs, and transparent conductive oxide (TCO) coatings. APCVD is also used as an article surface finishing process (Fig. 2.13).

2.4.3.11 PECVD

PECVD is a CVD process that deposits thin films on a substrate from a gaseous (vapor) state to a solid state. The process involves chemical reactions that occur after a plasma of the reacting gases has been formed.

In the case of low-energy PECVD reactors, the low-energy plasma having high density is used for epitaxial deposition at high levels (Fig. 2.14).

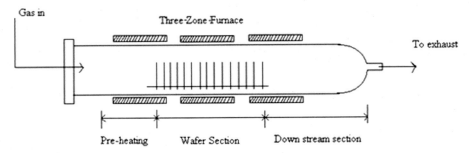

FIGURE 2.12 Schematic diagram of low-pressure chemical vapor deposition setup.

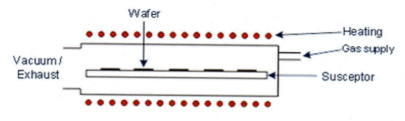

FIGURE 2.13 Illustration of a horizontal atmospheric pressure chemical vapor deposition reactor.

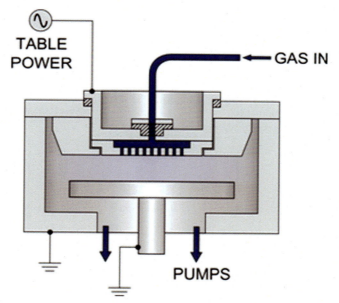

FIGURE 2.14 Schematic diagram of plasma-enhanced chemical vapor deposition setup.

2.4.3.12 Hot wire CVD

Hot wire CVD (HWCVD) is a technique of deposition involving heat decomposition of precursor gases to form radicals on the surface of resistively heated the substrate [17] (Fig. 2.15).

2.4.3.13 Closed space sublimation

Closed space sublimation is a method of making thin-film solar cells with CdTe, although it is used for other materials such as antimony triselenide. This is a process of PVD that keeps the source material and substrate surface (to be coated) close to each other.

Henceforth, this technique is quite useful for large-scale production (Fig. 2.16).

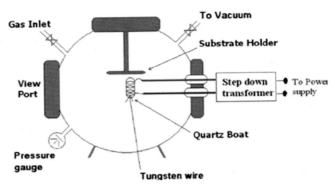

FIGURE 2.15   Schematic diagram of hot wire chemical vapor deposition setup.

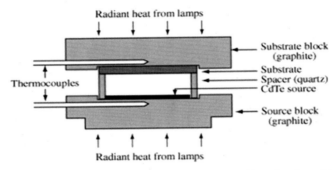

FIGURE 2.16   Schematic drawing of a closed space sublimation deposition tool.

2.4.3.14 Ion-beam-assisted deposition

Ion-beam-assisted deposition is a material-engineering technique that combines ion implantation with simultaneous sputtering or another PVD technique. This technique can independently regulate parameters such as energy from ions, temperature, and the rate of arrival of atomic species during deposition (Fig. 2.17).

2.4.4 Common features of thin-film technologies

2.4.4.1 Use of transparent conductive oxide and light trapping

Doped metal oxides used in optoelectronic devices such as flat panel displays and PVs (including inorganic devices, organic devices, and dye-sensitized solar cells) are TCOs. Generally, these films consist of amorphous or polycrystalline microstructures.

The mobility of electrons is generally higher than the mobility of holes, making it difficult to find acceptors to create a large hole population in wide bandgap oxides.

Appropriate p-type TCOs are being studied, although they lag behind n-type TCOs. The TCOs have a lower concentration of carriers compared to metals, which affects their resonance related to plasma in the near-infrared and shortwave infrared ranges.

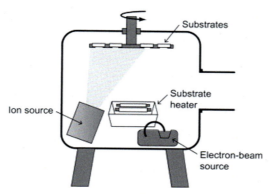

FIGURE 2.17 Schematic drawing of an ion-beam-assisted deposition tool.

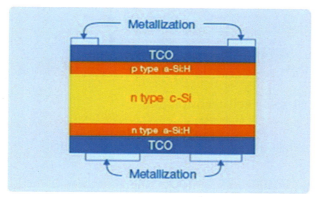

FIGURE 2.18 Schematic demonstration of a transparent conductive oxide (TCO) at the front and back of a solar cell.

For an alternative source, doped binary compounds such as zinc oxide doped with aluminum (AZO) and cadmium oxide doped with indium were suggested for this. AZO consists of two popular and inexpensive materials, aluminum and zinc, whereas indium-doped cadmium oxide uses only low concentrations of indium. There are a variety of transition-metal choices for indium oxide, in particular molybdenum (Fig. 2.18).

2.4.4.2 Possible solar cell structures

Multijunction solar cells are solar cells with multiple p−n junctions made of different semiconductor materials. In response to different wavelengths of light, the p−n junction of each material will produce electrical current. The use of many semiconducting materials allows for the absorption of a wider range of wavelengths, enhancing the cell's absorption of sunlight to convert to electrical energy.

Techniques of existing tandem manufacturing designs were used to enhance the performance. In fact, unlike conventional crystalline silicon, a-Si can be used to reduce the cost of thin-film solar cells to generate cells with a lightweight and versatile capacity of about 10% (Fig. 2.19).

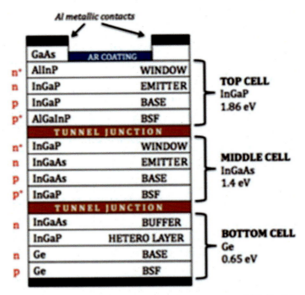

FIGURE 2.19 Possible multijunction solar cell structure. *BSF*, Back surface field.

2.4.5 Amorphous Si solar cell technologies

2.4.5.1 Silicon

Three important module designs based on silicon dominate:

- A-Si-based cells;
- Amorphous/microcrystalline (micromorph) tandem cells; and
- Polycrystalline silicon thin-film on glass.

2.4.5.2 Amorphous silicon

A-Si is the most advanced thin-film technology to date, consisting of noncrystalline allotropic silicon. Thin-film silicon is an option to traditional crystalline silicon wafer (or bulk). Although CdTe and CIS thin-film cells based on chalcogenide have been produced with great success in the laboratory, industry also has a keen interest in silicon-based thin-film cells [18,19].

This kind of thin-film cell is mostly produced using a technique called PECVD.

Nevertheless, during the first 6 months of service, an a-Si cell experiences a substantial 10%–30% decrease in efficiency. This is called the Staebler–Wronski effect—a common loss of electrical efficiency due to variation in photoconductivity and dark conductivity resulting from continuous exposure to sunlight (Fig. 2.20).

2.4.5.3 A-Si/µc-Si tandem cell

To create a multijunction solar cell, the a-Si layer may be mixed with layers of other allotropic silicon forms. A mixture of only two layers (two p–n junctions) is called a tandem cell. By piling these layers together, light radiation of a larger range is absorbed and the cell's efficiency increases [20]. A microcrystalline silicon (µc-Si) layer is amalgamated with a-Si, forming a tandem membrane to form a micromorphous silicon.

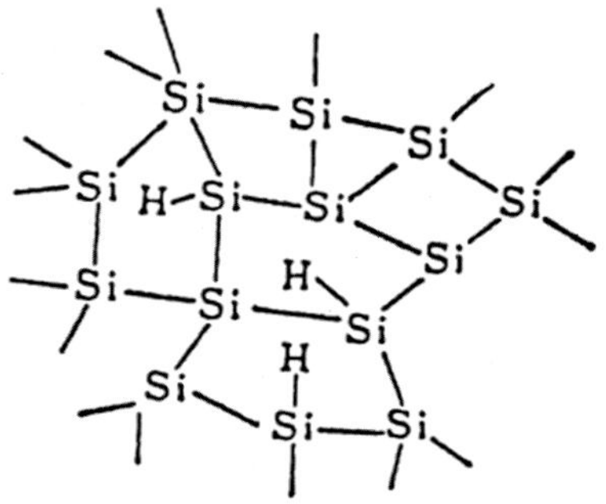

FIGURE 2.20 The atomic structure of amorphous silicon.

2.4.5.4 A-Si/pc-Si tandem cell

In addition, a-Si can be fused into a tandem cell with protocrystalline silicon (pc-Si). For high open-circuit voltage, pc-Si with a small volume fraction of nanocrystalline silicon is suitable. Such forms of silicon are also deformed by hanging and twisted bonds, resulting in deep defects (energy levels in the bandgap) as well as valence bands and conduction bands (band tails).

2.4.6 Chalcopyrite (CIGS) solar cell technology

A solar cell called a CIGS cell is a solar thin-film cell used to turn sunlight into electricity. It is formed by depositing on glass or plastic a thin layer of copper, indium, gallium, and selenium, along with electrodes at the front and back for collecting current.

Among popular thin-film PV technologies, CIGS, CdTe, and a-Si are widely used. CIGS layers are much thinner and therefore more versatile in comparison to these materials, allowing versatile substrates to be deposited [21].

2.4.7 Thin-film crystalline Si solar cell technologies

A solar thin-film cell is a second-generation solar cell made from PV material such as glass, plastic, or metal on which single or multiple thin layers or thin films on a substrate are deposited. Many technologies use thin-film solar cells commercially as in CIGS, CdTe, and amorphous thin-film silicon.

2.4.7.1 Polycrystalline silicon on glass

An attempt to merge the benefits of thin-film polycrystalline silicon on glass applications and bulk silicon applications has materialized. The modules are created using PECVD by depositing an ARC and doped silicon on textured glass substrates.

Crystalline silicon on glass is is known for its toughness and durability, where polycrystalline silicon is 1–2 μm thick; thin film techniques and its use also leads to cost savings as compared to photovoltaic bulk [22].

2.4.8 Microcrystalline Si thin-film technology

μc-Si or hydrogenated microcrystalline Si (μc-Si:H) is produced in almost the same manner as hydrogenated amorphous silicon (a-Si:H). Normally, tools based on PECVD techniques are

56 2. Solar cell technology

used. The properties of μc-Si:H are very different from those of a-Si:H. The material properties of μc-Si:H are as follows:

- It has a complex microstructure having a mixture of crystalline and amorphous phases.
- It has a smaller bandgap around 1.1 eV compared to 1.7−1.8 eV of a-Si:H, thus a wider spectrum is absorbed in it.
- It is an indirect bandgap material.
- Light-induced degradation or Staebler−Wronski degradation in contrast to a-Si: H is much smaller.

2.4.8.1 Microstructure of films and their deposition

μc-Si:H is a complex material. Its microstructure comprises the crystalline phase of Si conglomerated in the a-Si phase. The size of the crystalline phase is in the order of tens of nanometers. The conglomerates are about a micrometer in size [23].

The thin-film μc-Si:H is suitable for particular applications, such as thin-film transistors, due to its higher room temperature conductivity and carrier mobility (compared to those of a-Si:H or hydrogenated polymorphous silicon), which are converted into faster modules. Also, the bandgap and absorption coefficient are a-Si:H specific, and as a result, thin solar cells are more infrared absorbent and more stable against sunlight. A-Si:H/μc-Si:H tandem solar cells (or micromorphic solar cells) are currently being developed with a stabilized efficiency of approximately 12% [24].

2.4.8.2 μc-Si:H deposition conditions

We carried out an analysis in a PECVD reactor for the development of μc-Si:H thin films. High pressure, moderated strength, and high H_2 dilution are the key parameters for the production of μc-Si:H PECVD deposited in a sequence of μc-Si:H at a radio frequency of 13.56 MHz from a mixture of SiH_4, H_2, and Ar.

Table 2.1 indicates the deposition conditions for thin films of μc-Si:H; these films were deposited at a substratum temperature of $T_s = 200°C$ for 30 min. Ar dilution was used for the growth of the μc-Si:H films, and the effect of radio frequency power density variance was observed.

2.4.9 Thin-film polycrystalline Si solar cell technology

Polysilicon is manufactured from metallurgical-grade silicon using the Siemens process, which is a chemical purification technique. This process involves distilling, at high

TABLE 2.1 Deposition conditions of hydrogenated microcrystalline silicon films.

Sample	Pressure (mTorr)	Power (W)	Power density (mW/cm²)	SiH₄ flow (sccm)	H₂ flow (sccm)	Ar flow
1	1500	25	83	10	41	Yes
2	1500	30	103	10	41	Yes
3	1500	35	121	10	41	Yes

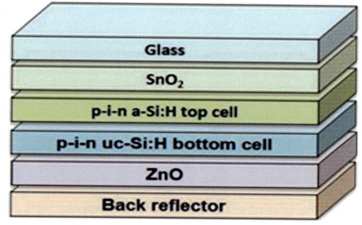

FIGURE 2.21 Typical structure of a micromorph solar cell.

temperatures, volatile silicon compounds and their decomposition into silicon. The current alternative method of refining uses a fluidized reactor bed (Fig. 2.21).

Currently, intrinsic and doped polysilicon in large-scale electronics is used as active and/or doped layers in thin-film transistors. While it may be deposited in some processing steps, LPCVD, PECVD, or a-Si solid-phase crystallization also require relatively high temperatures of at least 300°C [25,26].

2.4.9.1 Polycrystalline silicon deposition

A desire to create digital displays on flexible screens is motivated by pc-Si deposition on plastic substrates. A relatively new technique, called laser crystallization, has been developed, which is used for crystallizing an a-Si plastic substrate on a precursor material without damaging or melting the plastic. Relatively small, high-intensity, ultraviolet laser pulses are used to heat the deposited a-Si material above the silicon melting point without melting the whole substrate [27].

However, for device homogeneity, a crystal grain size smaller than the device feature size is required to create devices on polysilicon over large areas. Another process for producing polysilicon at low temperatures is metal-induced crystallization, in which an a-Si thin film can be crystallized at temperatures of 150°C or below.

A-Si and the polysilicon systems are utilized in the same phase, which is simply hybrid manufacturing. In the case where a minute pixel size is needed, —active layer process such as a complete polysilicon in projection displays is also used [2].

2.4.10 Large-grain thin-film crystalline Si solar cells on foreign substrate

Surface management of mC-Si solar cells by texture is one of the critical issues of mass production of high-performance, wide-area, low-cost industrial cells [3,28].

The direction-dependent anisotropic alkaline texturization solution is standard for crystalline silicon solar cells. This anisotropic etching often results in high surface roughness due to different etch rates of the solution for different crystal orientations in mC-Si solar cells. There are other texturization processes available for mC-S with their own advantages and disadvantages. The very common acid etchant with HF, nitric (HNO_3), and acetic (CH_3COOH) acids gives a rapid isotropic etching of the surface of silicon. After this texturing, the silicon coating is smoother regardless of the orientations of the grain and thus the reflectivity of the film is increased. Excessive control is needed to etch the solution during the etching process and the cost of chemicals limits its use to specific purposes only [29]. An isotropic etching process that contains an $HF-HNO_3$-deionized (DI) water-etching step followed by an $HF-HNO_3$-etching step provides a good choice for texturing.

A low-cost chemical texture process developed for industrial mC-Si solar cells with sodium hydroxide-sodium hypochlorite (NaOH-NaOCl) offers a simple and low-cost alternative development. This optimized etching solution has no effect on mC-Si grain boundaries, and therefore has excellent isotropic etching features to improve parameters such as V_{OC} and FF [30,31].

In addition, the evolution of chlorine during the silicon-etching process eliminates the additional step of chlorine neutralization that is typically needed for processes involving NaOH.

The starting material for the experiment was Deutsche Solar's 125 mm $\times$ 125 mm square boron-doped p-type mC-Si wafers of base resistivity 0.5–3.0 Ωcm from SOLSIX. MC-Si texturing is done using the solution NaOH-NaOCl. The wafers are thoroughly rinsed in DI water and dried after the final etching phase [11]. Using atomic force microscopy, the wafer samples are then analyzed.

The wafers are then doped with phosphorus (P) as the source of dopant, using optimized diffusion conditions with phosphorous oxychloride ($POCl_3$) [12]. Thermal silicon dioxide is produced after removal of the phosphosilicate glass coating from the wafer surface with diluted HF. This serves as a passivation layer and lowers the sheet resistance to a uniform, required range for manufacturing solar cells.

After removal of thermal oxide and etching of the base, a silicon nitride ARC layer is formed by the process of plasma-enhanced deposition of chemical vapors. Full solar cells are then processed with screen-printed metallization and proper firing of the contacts. The complete process is shown in the flow chart in Fig. 2.22.

2.4.11 Thin-film epitaxial Si solar cells

Because of the reduced material and energy consumption, film crystal silicon solar cells produce energy at much lower costs than PVs based on Si wafers. Silicon films can be epitaxially grown on crystalline seed layers affixed to cheap foreign substrates.

Cells with absorber layers of 2–10 μm may show efficiencies above 15% through the integration of light-trapping arrangements [32]. These efficiencies would involve the growth of high-quality absorber layers on substrates confined to low temperatures in comparison

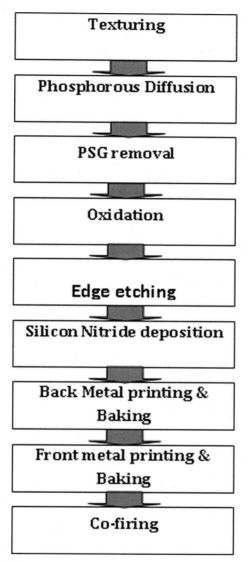

FIGURE 2.22 Solar cell fabrication process. *PSG*, Phosphosilicate glass.

with those used in standard c-Si epitaxies. Hot wire chemical vapor deposition (HWCVD) is a versatile approach to low-temperature Si epitaxy and has been demonstrated to generate c-Si layers with regulated dopant concentrations and carrier mobilities that are nearly equal to those of bulk crystalline content. The reactor's simple design and high deposition rates (nearly 300 nm/min are shown) make HWCVD an economical route for the development of c-Si films.

Cells made of Si wafer and seed-on-glass have heavily doped substrates.

References

[1] R.A. Arndt, J.F. Allision, J.G. Haynos, A. Meulenberg, Solar cell materials and their basic parameters, in: Proceedings of 11th IEEE International Specialist Conference, New York, 1975, p. 40.

[2] M.A. Green, Solar Cells, Operating Principles, Technology and System Applications, Prentice-Hall, Sydney, 1982.

[3] A.M. Jaffry, Solar Cells: An Introduction to Crystalline Photovoltaic Technology, Kluwer Academic Publishers, Dordrecht, 1997, p. 137.

[4] D.H. Macdonald, A. Cuevas, M.J. Kerr, C. Samundselt, d. Ruby, S. Winderbaum, A. Leo, Texturing industrial multicrystalline silicon solar cells, Sol. Energy 76 (2004) 277–283.

[5] Y. Nishimoto, T. Ishihara, K. Namba, Investigation of acidic texturization for multicrystalline silicon solar cells, J. Electrochem. Soc. 146 (1999) 457–461.

[6] J. Lindmayer, J.F. Allison, An improved silicon solar cell, COSMAT Tech. Rev. 3 (1973) 1–22.

[7] U. Gangopadhyay, S.K. Dhungel, K. Kim, U. Manna, P.K. Basu, H.J. Kim, B. Karungaran, K.S. Leo, J.S. Yow, J. Yi, Novel low cost chemical texturing for very large area industrial multicrystalline silicon solar cells, Semicond. Sci. Technol. 20 (2005) 938–946.

[8] E. Vazsonyi, Z. Vertesy, A. Toth, J. Szlufcik, Anisotropic etching of silicon in a two component alkaline solution, J. Micromech. Microeng. 13 (2003) 165–169.

[9] U. Gangopadhyay, S.K. Dhungel, P.K. Basu, S.K. Dutta, H. Saha, J. Yi, Comparative study of different approaches of multicrystalline silicon texturing for silicon solar cell fabrication, Sol. Energy Mater. Sol. Cell. 91 (2007) 285–289.

[10] A. Ebony, Y.H. Cho, M. Hilalii, A. Rohatgi, D. Ruby, Sol. Energy Mater. Sol. Cell. 74 (2002) 51.

[11] M. Bohm, E. Urbanski, A.E. Delahoy, Z. Kiss, Sol. Cell. 20 (1987) 155.

[12] P.K. Basu, B.C. Chakravarty, S.N. Singh, P. Dutta, R. Kesavan, Sol. Energy Mater. Sol. Cell. 43 (1996) 15.

[13] A. Goetberger, J. Knobloch, B. Voss, Crystalline Silicon Solar Cells, John Wiley & Sons Ltd., England, 1998.

[14] C.S. Solanki, Solar Photovoltaics, Fundamentals, Technologies and Applications, PHI, India, 2009.

[15] W. Schottky, Zeitschrift f. Physik 113 (1939) 367.

[16] W. Schockley, Bell System Tech. 28 (1949) 435.

[17] D.K. Schroder, D.L. Meier, IEEE TED 31 (1984) 637.

[18] C.Y. Chang, S.M. Sze, Solid State Electron. 13 (1970) 727.

[19] F.A. Padovani, R. Stratton, Solid State Electron. 9 (1966) 695.

[20] A. Goetzberger (Ed.), Proceedings of 15th IEEE PV Spec. Conference, Kissimmee, Florida, USA, 1981, p. 867.

[21] P. Campbell, S.R. Wenham, M.A. Green, IEEE TED 37 (1990) 331.

[22] P. Campbell, S.R. Wenham, M.A. Green, IEEE TED 35 (1988) 713.

[23] T. Uematsu, M. Ida, K. Hane, S. Kokunai, T. Saitoh, IEEE TED 37 (1990) 344.

[24] M.A. Green (Ed.), Proceedings of 10th EC PV Solar Energy Conference, Lissabon, Portugal, 1991, p. 250.

[25] S. Sterck, J. Knobloch, W. Wettling, Progr. Photovolt. 2 (1994) 19.

[26] J. Knobloch, S.W. Glunz, D. Biro, W. Warta, E. Schaffer, W. Wettling (Eds.), Proceedings of 25th IEEE PV Spec. Conference, Washington DC, USA, 1996, p. 405.

[27] R.A. Arndt, J.F. Allision, J.G. Haynos, A. Meulenberg (Eds.), Proceedings of 11th IEEE International Specialist Conference, New York, 1975, p. 40.

[28] Y. Nishimoto, T. Ishihara, K. Namba, J. Electrochem. Soc. 146 (1999) 457–461.

[29] U. Gangopadhyay, S.K. Dhungel, K. Kim, U. Manna, P.K. Basu, H.J. Kim, B. Karungaran, K.S. Leo, J.S. Yow, J. Yi, Semisconduct. Sci. Technol. 20 (2005) 938–946.

[30] E. Vazsonyi, Z. Vertesy, A. Toth, J. Szlufcik, J. Micromech. Microeng. 13 (2003) 165–169.

[31] U. Gangopadhyay, S.K. Dhungel, P.K. Basu, S.K. Dutta, H. Saha, J. Yi, Sol. Energy Mater. Sol. Cell. 91 (2007) 285–289.

[32] P.K. Basu, R.M. Pujahari, H. Kaur, D. Singh, D. Varandani, B.R. Mehta, Sol. Energy 84 (2010) 1658–1665.

CHAPTER 3

Energy materials: synthesis and characterization techniques

N.S. Bajaj, R.A. Joshi

Department of Physics, Toshniwal Arts, Commerce and Science College, Sengaon, Maharashtra, India

3.1 Introduction

In the era of 21st-century energy production, conversion and application are the most important topics. Due to the vast consumption of fossil-fuel and natural resources, environmental pollution is worsening. Hence, use of renewable and clean energy sources that can be substituted for fossil fuels is in high demand. This will help to enable the sustainable development of our economy and society. These abilities will allay worldwide concern and increase research interest into energy storage materials and their production because they can provide versatile, clean, and efficient use of energy.

If we look at the era of energy storage or production from the past to the present, it seems to be quite simple and natural. Initially, coal, oil, and natural gas were the main energy sources. These have been naturally collecting and storing solar energy for billions of years. But after the discovery of electric motors and generators in the 1870s, electrical energy has become the most significant secondary energy source and the primary form of consumed energy. In earlier times, electricity was generated from burning fuels but with evolution and invention we can now generate electricity from solar power, hydropower, wind power, nuclear power, tidal power, and biopower systems. Hence, it has become essential for every part of our lives, from lighting to cooking to entertainment, transportation, and communication.

The rate of electrical energy consumption has dramatically increased and has diversified due to the rapid development of modern industries and the increase in the global population. Thus energy storage has become even more complex and important. Also, desirable and high-performance energy storage techniques are needed to enable the efficient, versatile, and environmentally friendly use of energy, including electricity.

To overcome the issues of the current era, many researchers have developed numerous techniques based on materials science and provided materials and methods to society that

Energy Materials
https://doi.org/10.1016/B978-0-12-823710-6.00019-4

Copyright © 2021 Elsevier Ltd. All rights reserved.

can store natural and artificial energy. These materials and methods are termed energy materials and energy devices, respectively. In this chapter, we will focus on the methods of synthesis of these materials, as well as characterization techniques, to analyze their structural, chemical, and physical properties. However, it is been noticed that the top-down process is less approachable than the bottom-up process. Because in the top-down process there is limited control over particle size and morphology, the production of reliable and reproducible materials for technological applications requires strict control over their characteristics, including chemical homogeneity, low impurity levels, small particle size, narrow distribution, and reduced agglomeration. To meet all these requirements, in this chapter we will focus on the bottom-up synthesis process.

3.2 Synthesis techniques

The synthesis part of energy materials has always fascinated researchers from the field because these materials can be improved and improvised. So far, many methods have been reported regarding modification of the synthesis of these kinds of materials that can store and modify energy. However, it has always been noted by researchers that the surface area, morphology, and size of prepared materials define their function and application. All these materials characteristics are dependent on the method used or applied for their synthesis, such as the chemical process involved and the steps followed.

Generally, the methods of material synthesis are of two types: one is top down and the other is bottom up. In the top-down process, larger (macroscopic) initial structures were used that could be externally controlled in processing by using mechanical, thermal, optical, or chemical processes, for example, mechanochemical milling of micron-scale or bulk-scale materials, where larger particles are broken down into smaller ones. However, the bottom-up process involves contraction of material components (up to the atomic level) with a further self-assembly process leading to the formation of nanostructures. In this process of self-assembly, the physical forces are those operating at a lower scale, which are used to combine basic units into larger stable structures such as the synthesis of quantum dots and formation of nanoparticles.

However, it is been noticed that the top-down process is less approachable compared to the bottom-up process because in the top-down process there is limited control over particle size and morphology. Also, the process lends itself to wide particle size distributions and, in some cases, it can be an intensive time-consuming process [1—8]. Moreover, the production of reliable and reproducible materials for technological applications requires strict control over their characteristics, which includes chemical homogeneity, low impurity levels, small particle size, narrow distribution, and reduced agglomeration. To meet all these requirements, in this chapter we will focus on the bottom-up synthesis process and discuss them in detail.

3.2.1 Solid-state diffusion method

In solid-state diffusion, the ingredients or precursors react through a self-diffusion process. This is a kind of top-down approach for the synthesis of materials. However, it is expected to achieve the intermixing of precursors at the atomic level assisted by high temperature without melting the constituents. Reaction time and temperature bear a sort of reciprocal relation. The process operates at high temperatures ($>1600°C$) and has a long reaction time (2−3 days); however, many researchers frequently use this technique. It can also be seen that due to insufficient mixing and the low reactivity of raw materials, several intermediate phases easily exist in the products developed through the solid-state diffusion method. Hence, the possibility of obtaining a single-phase material becomes difficult and also restricts the use of doping concentration. Many researchers have suggested repeated grinding and calcinations to eliminate these intermediate compounds, but this will take longer and increase energy. Moreover, control over particle size and morphology is difficult [9].

3.2.2 Coprecipitation method

The coprecipitation technique is becoming increasingly important to distribute materials and precursors used in a reaction to produce a required material. The aim in coprecipitation is to prepare multicomponent materials through the formation of intermediate precipitates, usually hydrous oxides or oxalates, so that an intimate mixture of components is formed during precipitation and chemical homogeneity is maintained on calcination. In the typical process of coprecipitation, aqueous metal salts are mixed at sufficient temperatures with a base, which acts as a precipitating agent. This method is widely employed for synthesis [10,11]. However, in a few cases, the process is performed under an inert atmosphere [12].

The advantage of the coprecipitation method is that it gives a crystalline size in the small range compared to other synthesis processes depending on the precipitating agent selected during the reaction. Also, the crystallite size and morphology of the material prepared using this method can be controlled through the use of capping agents. However, continuous washing, drying, and calcination to achieve a pure phase of phosphor are major disadvantages of the coprecipitation method.

3.2.3 Sol-gel method

The sol-gel method is one of the most used wet chemical techniques for synthesis materials by using the following chemical processes: hydrolysis and gelation, followed by drying and finally thermal treatment. Generally, the sol is defined as colloidal particles suspended in a liquid from which a gel can be prepared. However, the gel is an interconnected rigid network having submicrometer pores and a polymeric chain with an average length in the order of microns. In sol-gel processing, alkoxides are used as a reactive precursor and hydrolyzed with water. By adding appropriate reagents, a homogeneous gel is obtained from the mixture of alkoxides through the processes of hydrolysis and gelation. After gelation, the precipitate is subsequently washed, dried, and then sintered at an elevated temperature to obtain crystalline phase material [13,14].

The main advantage of this method is that it has good chemical homogeneity due to the mixing of components at the colloidal level. Also, it requires a lower reaction temperature than the solid-state reaction/diffusion method. On the other hand, the disadvantage associated with the method is that alkoxides are relatively expensive compared to precursors for the sol-gel processing of colloids. Also, the high purity and high cost of chemicals required for gel formation must also be considered.

3.2.4 Combustion method

The combustion method is yet another wet-chemical method that does not require further calcinations and repeated heating. This method was accidentally discovered in 1988 in the lab of Prof. K.C. Patil in India [15,16]. It is an exothermic reaction and occurs with the evolution of heat and light. Such a high temperature leads to formations of crystallized phosphor materials. In recent years, combustion synthesis has been investigated to produce homogeneous, crystalline materials, compared to time-consuming techniques such as solid-state reaction and sol-gel processing [17−19]. The method is a promising technique due to its ability to produce fine-sized particles without high-temperature annealing and extra steps such as grinding or milling.

In combustion synthesis, a starting aqueous precursor of corresponding nitrates (oxidizers) and a suitable organic fuel (hydrazine-based compounds or urea or glycine as reducer) predetermined in a stoichiometric ratio are induced to boil. The mixture is ignited and a self-sustaining fast combustion reaction resulting from the appropriate combination of oxidizers and reducers produces fine crystalline materials. In this process, energy is released from the highly exothermic ($\Delta H < -170$ kJ/mol) reaction between the nitrates and the fuel. The only disadvantage associated with the method is the evolution of large amounts of toxic gases (NH_3, H_2O, CO_2). Also, it is very difficult to maintain the fuel/oxidizer ratio to preserve the enthalpy of the reaction.

3.2.5 Spray pyrolysis

Recently, spray pyrolysis has attracted attention because it is easy to produce a fine size and spherical morphology of materials. In addition, it has some advantages such as inexpensive precursor materials and high production rate. To prepare materials by spray pyrolysis, liquid precursors as a starting solution are required; usually, these materials are inexpensive source materials such as nitrate, acetate, and chloride. The solution is prepared by dissolving precursors in water, alcohol, or desired precursor solvent. In the spray pyrolysis process, the droplets, which are atomized from a starting solution, are supplied to a series of furnaces. The aerosol droplets then experience evaporation of the solvent, diffusion of solute, drying and precipitation, and reaction between precursor and surrounding gas. The precipitate undergoes pyrolysis or sintering inside the furnace at a higher temperature to form the final product [20,21]. In a few cases there is a requirement for two furnace systems in series to enhance the quality of crystallinity and morphology of the final product. The first furnace promotes the formation of microporous particles, and the second furnace further increases the density of the microporous particle and increases crystallinity. In spray pyrolysis, the final composition of materials is determined by the starting precursor solution composition. Furthermore, with a starting solution, the particle size can be easily controlled by adjusting the overall concentration of liquid precursors at a fixed sprayed droplet size. Although the spray pyrolysis method is a relatively simple and low-cost method to produce materials with high production rate in a single continuous process, it also has a few limitations, e.g., in the spray pyrolysis process, hollow and highly porous materials are generated. Moreover, to produce nanoparticles, it is required to have a lower overall concentration of starting precursor solution or introduce a smaller droplet size into the reactors; as a result, the production rate decreases.

3.2.6 Pechini and citrate gel methods

In the Pechini method, polybasic chelates are formed between α-hydroxycarboxylic acids containing at least one hydroxy group, for example, citric acid, with metallic ions [22]. The chelate undergoes polyesterification on heating with a polyfunctional alcohol, for example, ethylene glycol. On further heating, it produces a viscous resin and then a rigid, transparent, glassy gel. Finally, a fine oxide powder is generated.

This method is based on the esterification reaction between organic acid and alcohol. Higher hydrocarboxylic acid and polyfunctional alcohol yield the highly viscous polyester complex. The advantages of the Pechini method are the ability to prepare complex compositions, good homogeneity through mixing at the molecular leveling solution, and control of stoichiometry. The disadvantage of the Pechini sol-gel method is the slow reaction that takes place during synthesis. It is a time-consuming way of obtaining the final phosphor powder product at the nanoscale.

3.2.7 Hydrothermal synthesis

The hydrothermal technique is widely used in industrial processes for the dissolution of bauxite prior to the precipitation of gibbsite in the Bayer process and for the preparation of aluminosilicate zeolites. Many researchers have shown an interest in this method for the synthesis of materials but it is not as popular as the sol-gel process. This is surprising because hydrothermal synthesis offers a low-temperature, direct route to submicrometer size powders with a narrow distribution avoiding the calcination step required in sol-gel processing.

3.3 Characterization techniques

It is well known that to understand synthesized materials one needs to analyze the materials through different characterization techniques. These techniques provide critical information about the materials such as their structural, physical, and chemical properties. It also helps to understand the applicability of synthesized materials.

In this section we discuss a few characterization techniques most commonly employed for the analysis of energy materials. The discussion is categorized under different sections. In the first we explore structural and surface evaluation to understand the structural, physical, and chemical properties of energy materials, whereas in another section we discuss electrical, optical characterization and lifetime assessment generally used to understand the applicability of materials in energy devices. Moreover, throughout the discussion mathematical expression will be kept to a minimum so that new researchers and can adopt and grasp the techniques simply.

3.3.1 Structural evaluation

In materials science, structural evaluation is the same as that for the quality control process used in manufacturing in the field of engineering. It will provide information on the exact structure of materials, their crystal structures, position of functional groups, and other structural parameters that have a significant effect on the characteristics and properties of the materials. In modern techniques there are many tools available for the analysis of these properties. However, in the following section we discuss a few important and basic characterizations employed in energy materials required to understand the structure of synthesized materials.

3.3.1.1 Fourier transform infrared spectroscopy

In conventional infrared (IR) spectrophotometry, the IR beam is directed through the chamber and measured against a reference beam at each wavelength of the spectrum. During the process the entire spectral region should be scanned slowly to produce a good-quality spectrum. IR spectroscopy has been dramatically improved by the development of the Fourier transform method. It is very similar to nuclear magnetic resonance, which has been revolutionized by this method (Fig. 3.1).

As shown in the illustration, the basic principle of a Fourier transform infrared (FTIR) spectrophotometer is a Michelson interferometer built around the chamber. Radiation from an IR source is directed through the sample cell to a beam splitter. Half of the radiation is reflected from a fixed mirror while the other half is reflected from a mirror that is moved continuously over a distance of about 2.5 μm. When the two beams are recombined at the detector, an interference pattern is produced. A single scan of the entire distance takes about 2 s and is stored in the computer. So that several scans may be added, they must coincide exactly. Noticeably, this would be impossible considering the thermal fluctuations and vibrations in the laboratory. To solve this problem, a helium—neon laser is simultaneously directed through the Michelson interferometer and the interference pattern of the laser is used as a frequency reference.

The performance of an FTIR is dramatically superior to that of conventional instruments. Generally, only a small amount of sample will produce an excellent spectrum in a fraction of the time.

3.3.1.2 Principles of IR spectroscopy

The different regions of the electromagnetic spectrum will be used in this section to learn about the structure and reactions of organic molecules. For each spectroscopic method, it is helpful to understand how much energy corresponds to each wavelength and how this

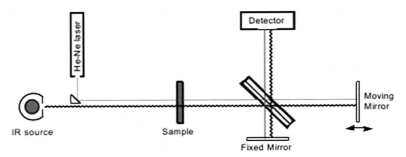

FIGURE 3.1 Block diagram of a Michelson interferometer used in a Fourier transform infrared spectrophotometer.

relates to the physical process after absorption of radiation. Organic molecules can absorb IR radiation between 4000 cm^{-1} and 400 cm^{-1}, which corresponds to absorption of energy between 11 and 1 kcal/mol. This amount of energy initiates transitions between vibrational states of bonds contained within the molecule.

3.3.1.3 X-ray diffraction

X-rays are electromagnetic radiation that typical have photon energies in the range of 100−100 keV. For diffraction applications, only short-wavelength X-rays in the range of a few angstroms to 0.1 Å, which means in the range of 1−120 keV energy, are used. Because the wavelength of X-rays is comparable to the size of atoms, they are ideally suited for probing the structural arrangement of atoms and molecules in a wide range of materials. The energetic X-rays can penetrate deep into the materials and provide information about the bulk structure.

X-rays primarily interact with electrons in atoms with which they collide, and some photons from the incident beam are deflected away from the original. If the wavelength of these scattered X-rays does not change, the process is called elastic scattering where only momentum transfer takes place. These diffracted X-rays are measured to extract information about the material, since they carry information about the electron distribution in materials. Diffracted waves from different atoms can interfere with each other and the resultant intensity distribution is strongly modulated by this interaction. If the atoms are arranged in a periodic fashion as in crystals, the diffracted waves will consist of sharp interference maxima with the same symmetry as in the distribution of atoms.

Measuring the diffraction pattern therefore allows us to deduce the distribution of atoms in a material. When certain geometric requirements are met, X-rays scattered from a crystalline solid can constructively interfere, producing a diffracted beam. In 1912, W.L. Bragg gave the relation, which elucidates the condition for constructive interference. The Bragg equation is defined as:

$$n\lambda = 2d\sin\theta \qquad (3.1)$$

where n denotes the order of diffraction, λ represents the wavelength, d is the interplanar spacing, and θ signifies the scattering angle. The distance between similar atomic planes in a crystal is known as d spacing and is measured in angstroms. The angle of diffraction is called the θ angle and is measured in degrees. For practical reasons, the diffractometer measures an angle twice that of the θ angle [23−25]. A schematic diagram of X-ray diffraction is shown in Fig. 3.2.

The diffraction pattern so obtained acts as a fingerprint of that crystalline substance. Therefore the crystalline phase of a material can also be identified by examining the diffraction pattern. The widths of the diffraction lines are closely related to (1) size, (2) size distribution, (3) defects, and (4) strain in crystals.

3.3.2 Surface evaluation

The outermost part of materials is always an important face of materials, as it provides information about atomic and electronic properties of materials. There are several techniques

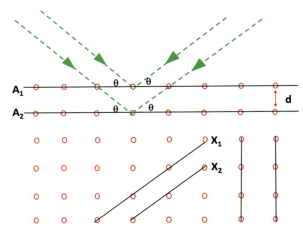

FIGURE 3.2 Schematic of Bragg's diffraction.

that have been developed to understand the surface of prepared materials. However, as discussed earlier, we will focus on only basic characterization that provides vital information.

3.3.2.1 Scanning electron microscopy

Scanning electron microscopy (SEM) is an important tool to investigate the surface and morphology of materials. By using it, we can estimate the diameter, length, thickness, density, shape, and orientation of materials. Fig. 3.3 shows a schematic diagram of the SEM apparatus. It uses electrons emitted from tungsten or lanthanum hexaboride (LaB_6) thermionic emitters for visualization of the surface of the sample. The filament is heated resistively by a current to achieve a temperature between 2000 and 2700 K. This results in an emission of thermionic electrons from the tip over an area of about 100×150 μm. The electron gun generates electrons and accelerates them to energy in the range 0.1–30 keV toward the sample. A series of lenses focuses the electron beam on the sample where it interacts with the sample to a depth of approximately 1 μm. When the electron beam impinges on the specimen, many types of signals are generated and any of these can be displayed as an image.

The two signals most often used to generate SEM images are secondary electrons (SEs) and backscattered electrons (BSEs). Most of the electrons are scattered at large angles (from 0 to 180 degrees) when they interact with the positively charged nucleus. These elastically scattered electrons, usually called BSEs, are used for SEM imaging. Some electrons are scattered inelastically due to the loss in kinetic energy upon their interaction with orbital shell electrons. Incident electrons may knock loosely bound conduction electrons out of the sample. These SEs along with BSEs are widely used for SEM topographical imaging. Both SE and BSE signals are collected when a positive voltage is applied to the collector screen in front of the detector. When a negative voltage is applied to the collector screen, only a BSE signal is captured because the low-energy SEs are repelled. These captured SEs and BSEs from the sample are sensed by detectors and these detectors transfer the detected electrons into an electronic signal, which is sent to a computer to display the image.

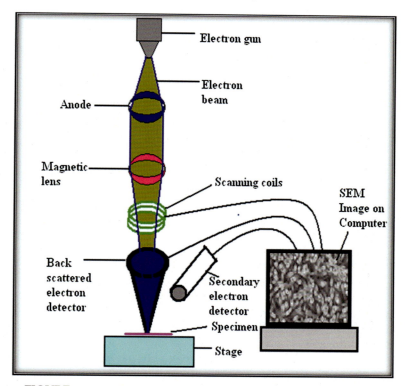

FIGURE 3.3 Schematic diagram of scanning electron microscopy (SEM).

3.3.2.2 Tunneling electron microscopy

In the electron diffraction process, if both the diffracted waves and the transmitted beam interfere with the plane of the image, a large image is formed known as a tunneling electron microscopy (TEM) photograph. TEM is used to obtain a direct image of the surface and particles of materials. It involves high-resolution imaging of the crystal lattice. TEM is commonly used to provide information on the shape and size of crystalline samples [26]. Generally, in this technique a beam of electrons is transmitted through an ultrathin sample and diffracted (scattered) by its atomic potential. The diffracted (scattered) beam contains information about electron distribution in the sample that is used to form an image. Thus TEM gives the original image of the sample that includes high-resolution imaging of the crystal lattice. In modern transmission electron microscopes, a range of corresponding capabilities interprets the method of analytical electron microscopy. There exist detectors that analyze inelastically scattered electrons, i.e., electron energy-loss spectroscopy, excited electromagnetic waves, i.e., energy dispersion spectroscopy, and Z-contrast. All these functions of TEM provide information on chemical compositions and local environments at the atomic level [27]. TEM is used to study transformations in between the amorphous state and the increased structural organized state of various materials, and vice versa. Amorphous samples do not show any different features in TEM because of lack of diffraction [28].

3.3.2.3 Atomic force microscopy

Atomic force microscopy (AFM) operates on the interaction between a fine tip that scans a surface and the forces from the sample. AFM is the most common method used in the field of materials science for imaging, measuring forces, and manipulating molecules on surfaces of materials. It can measure multiple types of forces, including mechanical contact force, van der Waals forces, capillary forces, chemical bonding, electrostatic forces, and magnetic forces depending of the mode selected for measurement during the experimental analysis. The most important advantage of AFM over electron microscopes (TEM and SEM) is that AFM does not require a vacuum; because of this it can be used at ambient conditions and in liquid samples as well [27]. Sample preparation for AFM is substantially simpler than that required by SEM or TEM and AFM can produce detailed three-dimensional maps rather than two-dimensional images without expensive and time-consuming sample preparation. The main feature of AFM is that it can work over nonconducting samples directly without any conductive coating. In general, AFM works in three different modes: contact mode, noncontact mode, and tapping mode (Fig. 3.4).

3.3.2.4 X-ray photoelectron spectroscopy

X-ray photoelectron spectroscopy (XPS), which is also called electron spectroscopy for chemical analysis, is the most widely used analytical technique to monitor the surface chemistry of solid materials due to its simplicity, flexibility, and sound theoretical basis. A typical XPS instrument includes an ultrahigh vacuum system, X-ray source, electron energy analyzer, and data acquisition system (Fig. 3.5).

XPS is typically performed by first taking a survey scan covering a range of 1000 eV. Then, smaller energy ranges indicative of specific features are subsequently scanned in higher resolution. Survey scans are often in low resolution and are used to identify as well as quantify in terms of atomic percentage of major elements present on the surface of materials. The *x*-axis is generally the "binding energy" and the *y*-axis is typically "intensity" or "number of counts."

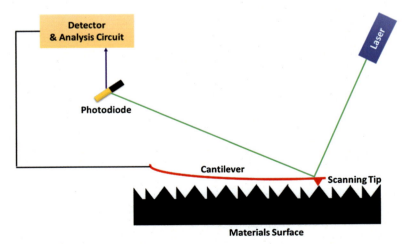

FIGURE 3.4 Schematic representation of the atomic force microscopy technique.

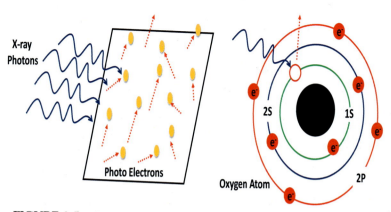

FIGURE 3.5 Illustration of the X-ray photoelectron spectroscopy process.

XPS is a much more surface-sensitive technique than attenuated total reflection FTIR spectroscopy [29,30].

3.3.3 Electrical characterization

Energy materials and devices continue to occupy the position of leading technology due to their importance in making any integrated electronic systems used in a wide range of applications from daily utilities to instrumentation used for medical diagnostics and environmental monitoring. Hence, it has become important to understand the electrical properties of the materials employed in the technology. In energy devices, it is preferred to have negligible standby power dissipation, good input—output isolation, surface potential control, and reliable operation. This is possible only if we have fast, nondestructive, accurate, and easy-to-use electrical characterization techniques that can give information about parameters such as carrier doping density, type and mobility of carriers, interface quality, oxide trap density, bulk defect density, contact and other parasitic resistances, etc. In this section we will discuss these techniques in summarized form.

3.3.3.1 Current—voltage characteristics

Current—voltage measurements of conventional semiconductor devices are possibly the simplest and most routine measurements performed. These can provide valuable information about the quality of materials used. Generally, they can be performed with different types of gates such as a simple diode, metal oxide semiconductor field effect transistor (MOSFET), and floating gates. For a simple gate like a p-n junction diode, the source—substrate or drain—substrate junctions can provide useful information on the quality of the junction. This means it will provide information about whether defects are present or not. This can be achieved because they give rise to generation—recombination currents or large parasitic resistances for contacts at the source, drain, or substrate terminals. This is easily seen from the current—voltage relation given by the sum of the diffusion (I_D) and recombination (I_R) currents:

$$I = I_D + I_R = I_{D0} \left[e^{\left(\frac{e\, V_D}{nk_b\, T} \right)} - 1 \right] + I_{R0}\, e^{\left(\frac{e\, V_D}{2k_b\, T} \right)} \tag{3.2}$$

where I_{D0} and I_{R0} are considered as the zero-bias diffusion and recombination currents, respectively, n is an ideality factor (generally 1), and V_D is the voltage across the intrinsic diode, which can be derived by using:

$$V_D = V_{\text{applied}} - I_D R_{\text{parasitic}} \tag{3.3}$$

To solve Eq. (3.1) a plot is drawn in between $\ln(I_D)$ versus V_D that separates out the diffusion and recombination current components. Moreover, by using Eq. (3.2) it is possible to calculate parasitic resistance in series with the intrinsic diode.

However, for MOSFETs, simple current−voltage measurements of the drain current versus gate voltage and drain current versus drain voltage are routinely examined to study their electrical characteristics. These can also be used to obtain useful information on the quality of the semiconductor, contacts, oxide, and semiconductor−oxide interface. On the other hand, the floating gate technique is another simple $I−V$ measurement in which the evolution of the drain current I_{DS} is monitored after the gate bias has been removed.

3.3.3.2 Hall effect

For a semiconductor material, resistivity is not the fundamental parameter; it is also related to or dependent on carrier density and carrier mobility available in the materials. For determining these parameters, the strength of the Hall effect is used. It directly determines the sheet carrier density by measuring the voltage generated transversely to the current flow direction in a semiconductor sample when a magnetic field is applied perpendicularly. Hence, together with a resistivity measurement technique, as well as the four-point probe or the van der Pauw technique, Hall measurements are used to determine the mobility of a semiconductor sample.

Generally, in the experimental field, the development of a characterization technique is related to its cost and ease of use. Since these practical characteristics are satisfied, the Hall effect measurement technique has become a very popular method for characterizing materials when specially shaped samples are required.

The basic principle of this Hall phenomenon is the deviation of some carriers from the current line because of the Lorentz force induced by the presence of a transverse magnetic field. The Lorentz force is given by the vector relation:

$$F_L = q(v \times B) = -qv_x B_z \tag{3.4}$$

where v_x is the carrier velocity in the x-direction. Assuming a homogeneous p-type semiconductor:

$$v_x = \frac{I}{qtW} \tag{3.5}$$

Significantly, an excess surface electrical charge develops on one side of the sample, and this generates an electric field in the y-direction E_y. In this case, the Hall voltage V_H is the magnetic Lorentz force F_L is balanced by the electric force F_E and hence we get:

$$F = F_L + F_E = -qv_x B_z + qE_y = 0 \text{ where } E_y = \frac{BI}{qtWp} \tag{3.6}$$

Hence, the Hall voltage is given by:

$$V_H = V_y = WE_y = \frac{BI}{qtp} \tag{3.7}$$

So, if the magnetic field B and the current I are known, then the measurement of the Hall voltage gives the hole concentration p_s by using:

$$p_s = p_t = \frac{BI}{qV_H} \tag{3.8}$$

If the conducting layer thickness t is known, then the bulk hole concentration can be determined and expressed as a function of the Hall coefficient R_H using Eq. (3.8) defined as:

$$R_H = \frac{tV_H}{BI} \text{ and } p = \frac{1}{qR_H} \tag{3.9}$$

Similarly, we can also carry out the same approach for n-type semiconductor materials, which becomes:

$$R_H = -\frac{tV_H}{BI} \text{ and } n = -\frac{1}{qR_H} \tag{3.10}$$

Now, if the bulk resistivity ρ is known/measured, then the carrier drift mobility can be obtained from:

$$\mu = \frac{|R_H|}{\rho} \tag{3.11}$$

If the sample thickness t and the mobility are known, then there are two main sample geometries commonly used in Hall effect measurements for the determination of carrier density. The first is by using the van der Pauw structure and the other is by using the Hall bar structure in which the Hall voltage is measured. However, for any kind of geometry used for Hall measurements, the crucial issue related to the offset voltage induced by the nonsymmetric positions of the contact is the biggest hurdle. This issue can be controlled by two different sets of measurement: in one set a magnetic field is in the same direction and in the other set a magnetic field is in the opposite direction (Fig. 3.6).

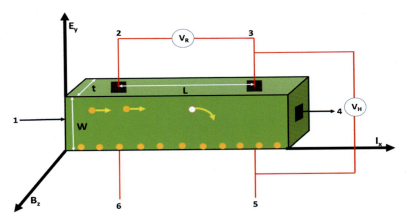

FIGURE 3.6 Representation of the Hall effect in a p-type bar-shaped semiconductor.

3.3.3.3 Quantum efficiency

Quantum efficiency (QE) is simply the ratio of the incident photon to the converted electron of a photosensitive device. In some special cases, it may be referred as the tunnel magneto resistance effect of a magnetic tunnel junction. However, the QE of a solar cell is a measure of the amount of current that the cell will produce when irradiated by photons of a particular wavelength. If the efficiency is incorporated over the complete solar spectrum, then it can be possible to evaluate the amount of current that the cell will produce when exposed to sunlight. The ratio between this energy-production value and the highest possible energy-production value for the cell gives the cell's overall energy conversion efficiency value. It is an important fact that during multiple exciton generation, more that 100% QE may be achieved. This will happen because the incident photons have more than twice the bandgap energy and can create two or more electron—hole pairs per incident photon.

There are two main types of quantum efficiencies considered while measuring in terms of solar cells or energy devices: external and internal quantum efficiencies. External quantum efficiency (EQE) is the ratio of the number of charge carriers collected by the solar cell/energy device to the number of incident photons. However, internal quantum efficiency (IQE) is the ratio of the number of charge carriers collected by the solar cell to the number of photons of incident photon absorbed by the cell. It is noticed that IQE is always greater than EQE. A lower value of IQE indicates low generation of photons by the active layer. To measure the IQE, one first measures the EQE of the solar device; then its transmission and reflection are measured and combined to produce IQE. Spectral response (SR) measurement needs to be done first before measuring the EQE value. SR is the ratio between the current generated by the cell and the power incident [31]. The relationship between EQE and SR is given by following equation:

$$\text{EQE}(\lambda) = \frac{hc}{q\lambda} \times \text{SR} \tag{3.12}$$

where h is Planck's constant, c is the speed of light, q is the electronic charge, and λ is the wavelength.

A typical setup for a QE measurement instrument consists of a tunable light source, detection system, bias light source, and accessories for proper beam manipulation and delivery. The main task of the system is to provide a nearly monochromatic light source with a typical band pass of 1–10 nm, tunable over the whole wavelength range at which the solar cell is active.

3.3.3.4 Electron impedance spectroscopy

This is a multifrequency AC electrochemical measurement technique that measures the electrical impedance or resistance of either a metal or solution interface of a material over a wide range of frequencies (from 1 mHz to 10 kHz). Electron impedance spectroscopy (EIS) determines polarization resistance in the low-frequency region, solution resistance in the high-frequency region, and capacitance of the double layer. However, polarization resistance is mostly used for determining the corrosion rate. Moreover, EIS offers more than resistance readings; it can estimate state-of-charge and capacity. Research laboratories have been using EIS for many years to evaluate battery characteristics. EIS is able to read each component of the Randles model individually; however, analyzing the values at different frequencies and correlating the data is an enormous task. Fuzzy logic and advanced digital signal processor technology have simplified this task. It is also said that EIS is very surface sensitive, which makes it more useful over other techniques because it can examine the changes in polymer layers due to swelling and surface changes due to protein adsorption or penetration of corrosion protection layers. As we know, resistance is the ratio of voltage or potential to current for a DC system, whereas impedance is the ratio of voltage or potential to current for an AC system. The nature of waves makes it necessary to define impedance with two parameters: total impedance Z and phase shift Φ.

These two parameters give rise to the two most popular plots for impedance spectra: the Bode plot and the Nyquist plot. In the Bode plot, there are actually two plots present in a single plot. The abscissa is a logarithmic scale of the frequency and one ordinate is the logarithm of the impedance Z, while the second ordinate is the phase shift Φ; however, in negative imaginary impedance, $-Z''$ is plotted versus the real part of impedance Z' (Fig. 3.7).

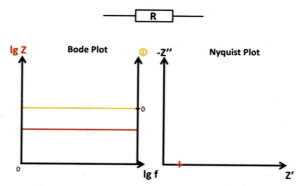

FIGURE 3.7   Electron impedance spectroscopy of a resistor in a schematic Bode and Nyquist plot.

3.3.3.5 *Capacitance measurements*

Capacitance—voltage ($C-V$) measurement is carried out to determine important physical and defect structure information of insulator and semiconductor materials. In general, the method is applied to a metal—oxide semiconductor or metal—semiconductor and energy device materials. In experimental analysis, material parameters such as high-frequency (HF), low-frequency (LF), or quasi-static $C-V$ measurements were examined to determine insulator thickness, doping concentration and profile, density of interface states, oxide charge density, and work function or barrier height. The $C-V$ measurement is made using three different methods. In the first method, i.e., the maximum—minimum HF capacitance method, HF capacitance under strong accumulation (C_O) and strong inversion (C_I) is used to determine the average doping density. In the second method, i.e., doping profile by HF and HF—LF capacitance methods, the doping profile in the depletion layer can be obtained by assuming series connection in between the depletion capacitance per unit area C_D and the oxide capacitance per unit area C_{OX}. In the last method, i.e., density of interface states, during the interface, traps change their charge state depending on whether they are filled or empty. It is observed that stretching of the $C-V$ curves occurs whenever interface trap occupancy varies with slow gate bias. Quantitative treatment of this *stretch-out* can be obtained from Gauss's law [32].

3.3.4 Optical properties

Any process that involves the interaction between electromagnetic radiations is known as optical properties such as absorption, diffraction, polarization, reflection, refraction, and scattering effects in materials. The electromagnetic spectrum is an important path for providing physical processes characteristic of various regions of interest involving the optical properties of energy materials. It is well known that most of the optical properties of energy materials are integrally related to the particular nature of their electronic band structures. Their electronic band structures are in turn related to the type of crystallographic structure, the particular atoms, and their bonding.

3.3.4.1 *Fluorescence spectroscopy*

Fluorescence spectroscopy is most commonly known as fluorimetry or spectrofluorometry. It is simply a type of electromagnetic spectroscopy used to analyze fluorescence emitted from materials. In the process of analysis, it involves a beam of light, generally ultraviolet light from a source that excites the electrons in molecules of certain compounds and causes them to emit light. Most commonly, visible light is emitted after excitation of materials, but not always. As we know, molecules of any material have various states denoted as energy levels, and fluorescence spectroscopy is one of the techniques concerned with electronic and vibrational levels. These states lie within the ground state, i.e., a low-energy state, and excited state, i.e., a higher-energy state. In the typical process of fluorescence, the sample or material is first irradiated by the source light and excited by absorbing a photon from its ground electronic state to one of the various vibrational states in the excited electronic state. These excited molecules lose their energy on collision with the other molecules and return to the lowest vibrational state from the excited electronic state.

In the experimental instrument of fluorescence emission measurement, the detection wavelength varies with respect to the constant excitation wavelength. On the other hand, during a fluorescence excitation measurement, the detection wavelength is fixed and the excitation wavelength is varied across a region of interest. The instruments used in the analysis of fluorescence are of two types: a filter fluorometer that uses filters to isolate the incident light and fluorescent light and a spectrofluorometer that uses a diffraction grating monochromator to isolate the incident light and fluorescent light. In both types the light from an excitation source passes through a filter or monochromator, and is incident on the sample. A proportion of the incident light is absorbed by the material and gives rise to the phenomenon of fluorescence. The fluorescent light from the material is emitted in all directions, passes through a second filter or monochromator, and reaches a detector, which is usually placed at 90 degrees to the incident light beam to minimize the risk of transmitted or reflected incident light reaching the detector [33].

3.3.4.2 Polarized light microscopy

As the name suggests, polarized light microscopy uses polarized light for the analysis of materials. However, it is similar to other microscopy techniques. Simple techniques include illumination of the material using polarized light. However, in this process, the directly transmitted light is blocked with a polarizer oriented at 90 degrees to the illumination. On the other hand, complex microscopy techniques that take advantage of polarized light include differential interference contrast microscopy and interference reflection microscopy. In a typical instrument, a polarizing plate is used to convert natural light into polarized light. Polarized light microscopy is generally used on materials having two different refractive indices, in which the polarized light interacts strongly with the material and generates a contrast with the background.

It is known that whenever polarized light passes through the material having two different refractive indices (i.e., birefringent materials) the phase difference between the fast and slow components varies with the thickness and wavelength of light used. Thus by using this technique we can calculate the refractive index using the thickness of material and vice versa. Moreover, the technique is capable of providing evidence of absorption color and optical path boundaries between materials having different refractive indices. This is similar to brightfield illumination, but in this technique, it is possible to distinguish between isotropic and anisotropic substances.

3.3.4.3 Spatial and time-resolved photoluminescence

Photoluminescence (PL) is a significant and contactless optical method employed to measure purity and crystalline quality, and identify certain impurities in materials for energy devices. Moreover, this a nondestructive method of analysis [23,34]. Using PL technique, identification of impurity in material becomes easier than its density [24]. It is well known that for PL measurement, there is a greater need for radiative recombination; however, this phenomenon is more difficult to achieve in an indirect energy gap material like silicon than it is in direct energy gap semiconductors like GaAs, InP, etc. A simple PL measurement will not provide information about when that process will take place on the molecule in the excited state, by which time the excited state is deactivated. To overcome these issues and to explain the excited state process exactly, time-resolved PL based on the PL technique is applied. This technique involves the following process to understand the excited state:

- **Radiative relaxation**: this process refers to the returning of molecules to the ground state with the emission of a photon.
- **Internal conversion or nonradiative relaxation**: in this process a molecule uses the intermediate vibrational sublevels to return to the lower electronic state.
- **Intersystem crossing**: this refers to the movement of molecules into a triplet state.
- **Conformational change: refers to change in the shape of a macromolecule, often induced by environmental factors**.

In the experimental setup of time-resolved PL measurement, the material is excited by a pulsed laser and the PL signals are detected by a photomultiplier tube and then analyzed by a good-quality photon-counting multichannel scaler. In this process, once the laser pulse is incident on the material, a large number of electron—hole pairs are generated, which will then decay with time due to recombination. Carrier lifetime and PL decay are dependent on the purity of the materials, this means it takes longer time to decay for the pure material and decay time decrease with increase in the impurity in materials. However, decay is affected by an effective lifetime, determined by bulk, surface, and interface recombination [25].

3.3.5 Lifetime assessment

The lifetime assessment of carriers—electrons and holes—is mostly employed for bipolar energy devices such as solar cells and imagers. The process provides information about the defect density and impurity level in the materials. There are many modern techniques to access the defect density of an energy device, but in the lifetime measurement technique there is no lower limit to determine the defect density [35]. The technique can detect defect densities as low as 10^9-10^{11} cm^{-3} in a simple, contactless way at room temperature. Lifetimes fall into two primary categories: recombination lifetimes and generation lifetimes. The concept of recombination lifetime holds when additional carriers decay as a result of recombination, while generation lifetime comes into play when there is a lack of carriers, as in the space-charge region of a reverse-biased device where the device tries to attain equilibrium [36].

3.3.5.1 Minority carrier lifetime measurement

In lifetime assessment, minority carrier lifetime is an important process because it has significant material parameters. The process is tremendously sensitive to minimum amounts of impurities or intrinsic defects in the energy materials. Thus it becomes a perfect parameter for direct characterization of material quality and process control. However, this is essential and important for the better performance of energy devices. As is known, minority carrier lifetime is defined as the average time taken for recombination by an excess of minority carrier. It is strongly dependent on the magnitude and type of recombination processes in the semiconductor. As per the literature, there are three main types of recombination, known as Shockley—Read—Hall (SRH) or multiphonon recombination, which takes place via defects, Auger recombination, which takes place via a three-particle process, and intrinsic or radiative recombination, which takes place via band to band, as illustrated in Fig. 3.8.

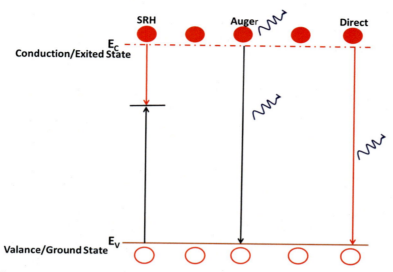

FIGURE 3.8 Energy scheme of the different types of recombination. *SRH*, Shockley–Read–Hall.

Minority carrier lifetime in the bulk depends accordingly on the number of defects present and on their recombination properties. For example, in silicon, SRH is the leading recombination mechanism, thus in silicon the lifetime can be as high as 1 ms. However, in a direct semiconductor like GaAs, radiative recombination is dominant and hence lifetime is only in the range of ns to μs. Besides the defect properties, minority carrier lifetime is dependent on the injection level (excess carrier concentration) and doping concentration [37,38].

3.4 Failure parameters

Many failure parameters occur during the processing of energy devices by using energy materials that can be divided into two stages during the process of development or after application or use. Failures that occur during development may be due to defects during production, unsuitable design or execution of the component, or failures during screening. It is well known that most device designs are revised before they are applied as electronic components. Usually, failures during this phase are caused by extensive pressures like high temperature called thermal overstress, high voltage/current called electrical overstress, humidity, vibration, and mechanical or thermal shock. However, after development, failures may be due to corrosion, electrical leakage, insulation breakdown, migration of metallic ions in the direction of current flow, cracking of the encapsulating material due to deterioration of the material, and cracks in the bond wires due to repeated stresses [39].

3.5 Failure analysis/damage evaluation

It is well known that, due to the evolution in technological research, today's electronic systems are becoming more complex and compact, and concepts of quality and reliability are most important. However, system component failures are still common. Failure in the system causes disruption in the service and costly downtime for repair, which affects the economy of operation.

Thus failure analysis (FA) is the tool to analyze the evolution of device reliability and can give valuable insight into the causes of failure, providing input for product improvement. There are several methods used for the FA of electronic components such as external visual examination, electrical measurement and testing, decapsulation, optical or electron microscopy, IR or X-ray examination, and destructive analysis [40].

3.6 Conclusions

As mentioned earlier, the materials and methods employed to store natural and artificial energy are termed energy materials and energy devices, respectively. In this chapter, we discussed synthesis methods employed for the preparation of these materials at the laboratory and larger scales. However, in the detailed discussion, focus was given to the bottom-up approach because of its suitability to produce required particle sizes and distribution. In a further section, structural, chemical, and physical characterizations were discussed in a simple way without the use of too much mathematical or theoretical expression, so that new researchers could understand them better. Moreover, the chapter explained atomic characterizations such as fluorescence spectroscopy, lifetime assessment to understand about electronic transitions in materials, and purity level of the materials. Lastly, the chapter gave a brief glimpse at the failure of materials during technological application and methods employed for failure assessment, which is the most important part of research and development.

References

[1] D.H. Manh, N.C. Thuan, P.T. Phong, I.V. Hong, N.X. Phuc, Magnetic properties of $La_{0.7}Ca_{0.3}MnO_3$ nanoparticles prepared by reactive milling, J. Alloys Compd. 479 (2009) 828−831.

[2] E.V. Sampathkumaran, K. Mukherjee, K.K. Iyer, N. Mohapatra, S.D. Das, Magnetism of fine particles of kondo lattices, obtained by high-energy ball-milling, J. Phys. Condens. Matter 23 (2011) 094209.

[3] Z. Chen, Y. Cao, J. Qian, X. Ai, H. Yang, Facile synthesis and stable lithium storage performances of Sn-sandwiched nanoparticles as a high capacity anode material for rechargeable Li batteries, J. Mater. Chem. 20 (2010) 7266−7271.

[4] S. Hallmann, M.J. Fink, B.S. Mitchell, Advances in Nanomaterials and Nanostructures The American Ceramic Society: Imprint, Ceramic Transactions Series, Part-I, vol. 229, 2011, pp. 129−142. Ch. 13.

[5] K. Zaghib, P. Charest, M. Dontigny, J.F. Labrecque, A. Mauger, C.A. Julien, New synthetic route of $LiFePO_4$ nanoparticles from molten ingot, ECS Trans. 33 (2011) 23−31.

[6] T. Laaksonen, P. Liu, A. Rahikkala, L. Peltonen, E.I. Kauppinen, J. Hirvonen, K. Järvinen, J. Raula, Intact nanoparticulate indomethacin in fast-dissolving carrier particles by combined wet milling and aerosol flow reactor methods, Pharmaceut. Res. 28 (2011) 2403−2411.

[7] S. Basa, T. Muniyappan, P. Karatgi, R. Prabhu, R. Pillai, Production and in vitro characterization of solid dosage form incorporating drug nanoparticles, Drug Dev. Ind. Pharm. 34 (2008) 1209−1218.

[8] V. Amendola, M. Meneghetti, G. Granozzi, S. Agnoli, S. Polizzi, P. Riello, A. Boscaini, C. Anselmi, G. Fracasso, M. Colombatti, C. Innocenti, D. Gatteschi, C. Sangregorio, Top-down synthesis of multifunctional iron oxide nanoparticles for macrophage labelling and manipulation, J. Mater. Chem. 21 (2011) 3803−3813.

[9] J.J. Kingsley, K.C. Patil, A novel combustion process for the synthesis of fine particle α-alumina and related oxide materials, Mater. Lett. 6 (1988) 427−432.

References

[10] S. Rahman, K. Nadeem, M. Anis-Ur-Rehman, M. Mumtaz, S. Naeem, I. Letofsky-Papst, Structural and magnetic properties of Zn Mg ferrite nanoparticles prepared using the co-precipitation method, Ceram. Int. 39 (2013) 5235–5239.

[11] S. Amiri, H. Shokrollahi, The role of cobalt ferrite magnetic nanoparticles in medical science, Mater. Sci. Eng. C 33 (2013) 1–8.

[12] A.H. Lu, E.L. Salabas, F. Schüth, Magnetic nanoparticles: synthesis, protection, functionalization, and application, Angew. Chem. Int. Ed. 46 (2007) 1222–1244.

[13] W.M. Yen, M.J. Weber, Inorganic Phosphors (Compositions, Preparations and Optical Properties), CRC Press imprint, Boca Raton London New York Washington, D.C, 2004 (Ch. 1).

[14] Y. Gogotsi, Nanomaterials Handbook, CRC Press imprint, Taylor and Francis Group Publicating, Boca Raton London New York Washington, D.C, 2006 (Ch.1 and Ch. 2).

[15] K.C. Patil, S.T. Aruna, S. Ekambaram, Combustion synthesis, Curr. Opin. Solid State Mater. Sci. 2 (1997) 158–165.

[16] S. Ekambaram, K.C. Patil, Synthesis and properties of rare earth doped lamp phosphors, Bull. Mater. Sci. 18 (1995) 921–930.

[17] N.S. Bajaj, C.B. Palan, S.K. Omanwar, Low temperature synthesis and improvement in optical properties of MgO: Tb^{3+}, Int. J. Mater. Sci. Eng. 3 (2015) 167–174.

[18] C.B. Palan, N.S. Bajaj, A. Soni, M.S. Kulkarni, S.K. Omanwar, Combustion synthesis and preliminary luminescence studies of $LiBaPO_4 : Tb^{3+}$ phosphor, Bull. Mater. Sci. 38 (2015) 1527–1531.

[19] R.S. Palaspagar, A.B. Gawande, R.P. Sonekar, S.K. Omanwar, Fluorescence properties of Tb^{3+} and Sm^{3+} activated novel $LiAl_7B_4O_{17}$ host via solution combustion synthesis, Mater. Res. Bull. 72 (2015) 215–219.

[20] K. Okuyama, I.W. Lenggoro, Preparation of nanoparticles via spray route, Chem. Eng. Sci. 58 (2003) 537–547.

[21] K. Okuyama, I.W. Lenggoro, M. Abdulla, F. Iskandar, Preparation of functional nanostructured particles by spray drying, Adv. Powder Technol. 17 (2006) 587–611.

[22] M.P. Pechini, Method of Preparing Lead and Alkaline Earth Titanates and Niobates and Coating Method Using the Same to Form a Capacitor, U S Patent, 3330697, 1967.

[23] H.B. Bebb, E.W. Williams, Photoluminescence I: theory, in: R.K. Willardson, A.C. Beer (Eds.), Semiconductor and Semimetals, vol. 8, Academic Press, New York, 1972, pp. 181–320.

[24] Y. Hayamizu, R. Hoshi, Y. Kitagawara, T. Takenaka, Novel evaluation methods of silicon epitaxial layer lifetimes by photoluminescence technique and surface charge analysis, J. Electrochem. Soc. Jpn. 63 (1995) 505–519.

[25] J.E. Park, D.K. Schroder, S.E. Tan, B.D. Choi, M. Fletcher, A. Buczkowski, F. Kirscht, Silicon epitaxial layer lifetime characterization, J. Electrochem. Soc. 148 (2001) G411–G419.

[26] G. Wang, Q. Peng, Y. Li, Lanthanide-doped nanocrystals: synthesis, optical-magnetic properties, and applications, Acc. Chem. Res. 44 (2011) 322–332.

[27] R. Andre, M.N. Tahir, T. Link, F.D. Jochum, U. Kolb, P. Theato, R. Berger, M. Wiens, H.C. Schroeder, W.E.G. Muller, Chemical mimicry: hierarchical 1D $TiO_2@ZrO_2$ core-shell structures reminiscent of sponge spicules by the synergistic effect of silicate in-r and silintaphin-1, Langmuir 27 (2011) 5464–5471.

[28] E. Karavas, M. Georgarakis, A. Docoslis, D. Bikiaris, Combining SEM, TEM, and micro-Raman techniques to differentiate between the amorphous molecular level dispersions and nano dispersions of a poorly water-soluble drug within a polymer matrix, Int. J. Pharm. 340 (2007) 76–83.

[29] F. Barrere, C.A. van Blitterswijk, K. de Groot, P. Layrolle, Nucleation of biomimetic Ca-P coatings on Ti_6Al_4V from a SBF X 5 solution: influence of magnesium, Biomaterials 23 (2002) 2211–2220.

[30] M.H. Wong, F.T. Cheng, H.C. Man, Characteristics, apatite-forming ability and corrosion resistance of Ni-Ti surface modified by AC anodization, Appl. Surf. Sci. 253 (2007) 7527–7534.

[31] W. Ananda, in: External Quantum Efficiency Measurement of Solar Cell, 15th International Conference on Quality in Research (QiR): International Symposium on Electrical and Computer Engineering, July 2017, Conference Proceeding IEEE, 2017, pp. 450–456 (978-602-50431-1-6).

[32] E.H. Nicollian, J.R. Brews, MOS (Metal Oxide Semiconductor) Physics and Technology, Wiley: Imprint, published by Wiley Classical Library, New York, 1982.

[33] https://orgchemboulder.com/Labs/Handbook/UV-Vis.pdf.

[34] D.K. Schroder, Semiconductor Material and Device Characterization, third ed., Wiley: Imprint, Wiley Interscience, New York, 2006.

[35] D.K. Schroder, Carrier lifetimes in silicon, IEEE Trans. Electron. Dev. 44 (1997) 160–170.

I. Fundamentals and overarching topics

[36] D.K. Schroder, The concept of generation and recombination lifetimes in semiconductors, IEEE Trans. Electron. Dev. 29 (1982) 1336–1338.

[37] S. Rein, Lifetime Spectroscopy - A Method of Defect Characterization in Silicon for Photovoltaic Applications, vol. 85, Springer: Imprint, Berlin Heidelberg, 2005.

[38] D.K. Schroder, Semiconductor Material and Device Characterization, 2nd ed., John Wiley & Sons: imprint, New York, 1998.

[39] E.A. Doyle, How parts fail: a fundamental explanation for any part failure exists, and it's up to the failure analyst to find the cause with modern sleuthing, IEEE Spectrum 10 (1981) 36–43.

[40] V. Lakshminarayanan, Minimizing Failures in Electronic Systems by Design, vol. 45 (16), EDN-Boston then Denver then Highlands Ranch Co, 2000, pp. 87–104.

CHAPTER 4

An extensive study of the adhesion and antifogging of the transparent polydimethylsiloxane/Sylgard coating system

A. Syafiq[1], B. Vengadaesvaran[1], N.A. Rahim[1], A.K. Pandey[2], A.R. Bushroa[3], K. Ramesh[4], S. Ramesh[5]

[1]Higher Institution Centre of Excellence (HICoE), UM Power Energy Dedicated Advanced Center (UMPEDAC), Level 4, Wisma R&D, University of Malaya, Jalan Pantai Baharu, Kuala Lumpur, Malaysia; [2]Research Centre for Nano-Materials and Energy Technology (RCNMET), School of Science and Technology, Sunway University, Petaling Jaya, Selangor, Malaysia; [3]AMMP Centre Level 8, Engineering Tower, Faculty of Engineering, University of Malaya, Kuala Lumpur, Malaysia; [4]Center for Ionics University of Malaya, Department of Physics, University of Malaya, Kuala Lumpur, Malaysia; [5]Center for Ionics University of Malaya, Department of Physics, Faculty of Science, University of Malaya, Kuala Lumpur, Malaysia

4.1 Introduction

In general, the condensation of fog vapor on a solid surface can occur in dropwise or filmwise condensation mode. In dropwise mode, the small mist droplets appear on the low-energy surface, while in filmwise condensation, the mist droplets spread as a thin film layer of water and exhibit very low contact angle on the high-energy surface. The shape of mist droplets on the solid surface can be changed by adjusting the surface topography and suitable chemical property of the coating [1]. According to Snell's law, a superhydrophilic coating displays better antifog performance as compared to hydrophobic or superhydrophobic surfaces since no light scattering occurs if the water contact angle (WCA) is below the critical value of 48.8 degrees [2]. The thin film of water allows the direct transmission of incident light through the transparent substrate. However, it should be noted that frost will be formed

Energy Materials
https://doi.org/10.1016/B978-0-12-823710-6.00001-7

Copyright © 2021 Elsevier Ltd. All rights reserved.

under harsh fog conditions and freeze the operation or whole system [3]. As the mist drop spreads (not rebounds) on the superhydrophilic coating, it moves slowly in the direction of the external gravity forces and has sufficient time to freeze. Unlike the hydrophilic coating, the fog droplet will slide off or roll off the hydrophobic coating and detach from the surface during the process of freezing. Thus fog droplets can be removed within a short time before freezing and the formation of ice will be delayed and avoided [4,5]. Due to this problem, the hydrophobic coating seems more reliable than the superhydrophilic coating under prolonged fog exposure.

In this study, the antifog coating system is synthesized by using Sylgard 184 elastomer and polydimethylsiloxane (PDMS) resin only. PDMS consists of high surface energy entities (partially ionic siloxane backbones) and low surface energy entities (methyl pendant groups) that dominate the surface behavior of fog droplets. An Si—C bond (in the PDMS chain) has no charge separation inside its covalent bond due to equivalent electrons between alkyl groups (CH_3) and silicon. Zero charge separation prohibits the penetration of water molecules and ameliorates the antifog property of PDMS. When PDMS is made up of longer alkyl chains, the chains exhibit stronger repulsive forces to the droplets [6—9]. In comparison to other silanes, PDMS possesses excellent water repellency, good chemical stability, high transparency [10], low toxicity, and is a more durable and inexpensive material [11].

This research study also gives serious attention to the development of a durable coating system in an outdoor environment. Most research studies have recommended the incorporation of nanoparticles in the polymer matrix because the nanoparticles can create capillary pressure, P_C, to deteriorate water hammer pressure, P_{EWH}, from rainfall impact. This P_{EWH} impact would detach the adhered nanoparticles easily due to weaker binder. Thereafter, a nanoparticle coating system would fail to retain its hydrophobic property and possess low lifespan after prolonged outdoor exposure [12]. The straightforward way to synthesize a strong binder is by increasing the number of crosslinkings of PDMS where the identical terminating methyl groups of PDMS can act as a binder [13]. Theoretically, the interaction between Sylgard 184 elastomer and PDMS would increase the number of methyl groups in the PDMS chain, improving the adhesive force of the coating system. The adhesion of coated substrate increases as the Sylgard composition increases due to larger crosslinking in the PDMS matrix. This statement is clearly in agreement with the scratch results and outdoor test observations where the PDMS/Sylgard-coated surface experiences reduced droplet impact when the weight percentage (wt%) of Sylgard resin has been raised above 10 wt%. For a practical application, a novel PDMS/Sylgard coating system can be promoted as a suitable binder to prepare an antifog coating with great mechanical durability against the scratch test and prolonged rainfall impact.

4.2 Methodology

4.2.1 Preparation of PDMS/Sylgard coating

Hydroxyl-terminated PDMS with a viscosity of 25 cSt (centistokes, a unit used to measure viscosity) was used as a surface modifier, purchased from Sigma Aldrich, Malaysia. Silicone elastomer, Sylgard 184, with a viscosity (base) at 5100 cP and viscosity (mixer) at 3500 cP, was purchased from Dow Corning, USA, and was employed as a binder. Ethanol absolute (analytical grade), C_2H_6O, with a viscosity (20°C) at 1.2 mPas and density of 0.79 g/cm^3, was purchased from Friendemann Schmidt, USA. The ethanol was employed as a solvent. The technical data of the solvent and resin used in the work are presented in Table 4.1.

PDMS resin is a nonpolar compound that consists of hydrocarbon chains [14]. Nonpolar covalent bond, Si—C, groups have an equal number of shared electrons between the alkyl groups (CH$_3$) and silicon. Due to the equal number of shared electrons, no charge separation occurs in the nonpolar covalent bond. Subsequently, the nonpolar bond restricts the substitution of water molecules into prepared coatings and contributes to their hydrophobic property. PDMS Sylgard 184 (Dow Corning Corporation) is a silicone elastomer that consists of a prepolymer base (Sylgard 184A) and a crosslinker curing agent (Sylgard 184B) that has been blended at a proposed 10:1 weight ratio. The crosslinking between each silicone elastomer is important to serve as components with useful physical properties. Crosslinking indicates that the silicone elastomer is a thermoset and cannot be used in normal thermoplastic processing equipment [15]. The preparation of PDMS/Sylgard coating systems is illustrated in Fig. 4.1.

TABLE 4.1 The specifications of used chemicals.

Chemicals	Specification	Crosslinking	Supplier(s)
Polydimethylsiloxane, hydroxyl terminated	CAS: 70131-57-8/481939 Viscosity: 25 cSt Density: 0.95 g/mL at 25°C Boiling point: 182°C Quantity: 500 mL	$(C_2H_6OSi)_n$	Sigma Aldrich, Malaysia
Sylgard 184 (base)	LOT: 0009155256 Cure time (25°C): 48 h Viscosity (base): 5100 cP Viscosity (mixer): 3500 cP Tensile strength: 980 PSI Refractive index @ 632.8 nm: 1.4225	$(CH_3)_3SiO[Si(CH_3)_2O]$ $nSi(CH_3)_3$	Dow Corning, USA
Sylgard 184 (curing agent)	LOT: 0009155256 Cure time (25°C): 48 h Viscosity (base): 5100 cP Viscosity (mixer): 3500 cP Tensile strength: 980 PSI Refractive index @ 632.8 nm: 1.4225	$(CH_3)_3SiO[Si(CH_3)_2O]$ $nSi(CH_3)_3$	Dow Corning, USA
Absolute ethanol	CAS: 64-17-5 Density: 0.79 g/cm^3 Solubility in water (20 degrees): miscible Viscosity (20°C): 1.2 mPas Boiling point: 78.3°C	C_2H_6O	Friendemann Schmidt, USA

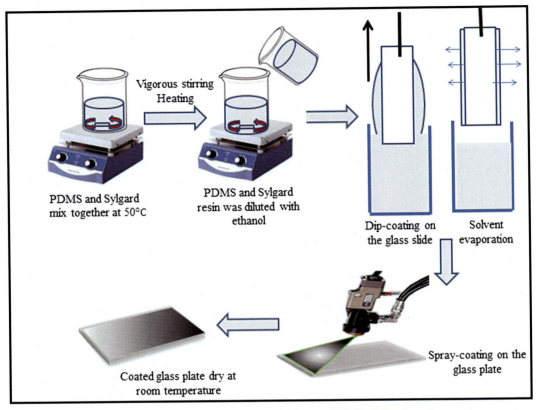

FIGURE 4.1 The preparation of polydimethylsiloxane (PDMS)/Sylgard coating systems.

The Sylgard resin was mixed with PDMS at different weight percentages of 5 wt%, 10 wt%, 15 wt%, and 20 wt%. The resin systems were labeled as S1, S2, S3, and S4. The mixture was then subjected to vigorous stirring by an automagnetic stirrer at 50°C for 1 h. Afterward, the PDMS/Sylgard was diluted with 240 mL of ethanol via constant stirring for approximately 20 min to blend the prepared resin inside the aqueous ethanol. All four formulated resins were fabricated on glass slides via dip coating. The prepared resin was fabricated on the glass plates via the spray-coating technique. The coated glass slides and glass plates were dried at ambient temperature. Details of the formulation of PDMS/Sylgard coating are presented in Table 4.2.

4.2.2 Characterization of the coating system

Field emission scanning electron microscopy (FESEM) FEI Quanta 450 FEG was employed to study nanoparticle dispersion within the polymeric matrix with an SDD EDS detector at 5 kV as the accelerating voltage. The images of static contact angle and dynamic contact angle were captured by an optical contact angle 15 EC instrument. A 5 μL water droplet was

4.2 Methodology

TABLE 4.2 The formulation of polydimethylsiloxane (PDMS)/Sylgard coating.

Coating system	Compatibility ratio	Weight percentage of Sylgard (wt%)
S1	PDMS resin: 20 g Sylgard 184: 10 g Ethanol: 240 mL	5
S2	PDMS resin: 20 g Sylgard 184: 20 g Ethanol: 240 mL	10
S3	PDMS resin: 20 g Sylgard 184: 30 g Ethanol: 240 mL	15
S4	PDMS resin: 20 g Sylgard 184: 40 g Ethanol: 240 mL	20

injected by an automated dispensing system at a velocity of 2 µL/s. Each measurement was recorded five times and the average of static contact angle value was used as the actual contact angle. An ATR-RX1 spectrometer (PerkinElmer, USA) was used to record the Fourier transform infrared (FTIR) spectra in transmittance mode. Perkin Elmer scan spectra accumulation in the wave number range of 400–4000 cm^{-1} at a resolution of 3.0 cm^{-1} was used to obtain all recorded spectra. The phase analysis of prepared hydrophobic coatings was captured by using X-ray diffraction (XRD, Philips PW1840, the Netherlands) at 30 kV and 35 mA. A collimated angle of 2θ was adjusted in the range from 5 to 90 degrees. An ultraviolet-visible LAMBDA 750 spectrometer (Perkin Elmer, USA) was used to characterize the transparency of coating in the spectral region of 400–1000 nm and the thickness of prepared coatings was measured by using an ellipsometer (M-2000, J.A. Wollam Co., USA). The adhesion strength of the coating systems was measured by a material nanoscratch tester (Wrexham, UK) before being subjected to an outdoor test, and the damaged profiles along the scratch tracks were determined by using a light optical microscope (Olympus BX61, Japan). The scratch hardness value of the coating system was calculated by using the formulation ASTM G171-03 and the width of the scratch was measured on the scratch track. A fogging test was carried out by placing the prepared coatings above a hot boiling bath that was heated to 130°C for a fogging time of 5 min. Afterward, the coated glass was placed under a piece of A4 paper to observe the adhered fog droplets on the surface. Photos of the adhered fog were taken by a mobile camera. The outdoor exposure test was conducted on the roof of the Wisma R&D building over a period of 4 months (from July to November). During the outdoor exposure test, heavy rainfall frequently occurred from August to November. The impact of rain droplets on the prepared coatings was investigated and analyzed. Any interior damage of the surface was inspected by using FESEM. The reduction in hydrophobicity of prepared coating systems was measured by a 15 EC optical contact angle machine.

4.3 Results and discussions

4.3.1 Water contact angle measurement

Organic materials with lower surface energy, including silane [16–18], fluorocarbon polymers [19,20], high-density polyethylene [21], and polysiloxane [22,23], are commonly used as hydrophobic agents. Siloxanes are known for their excellent hydrophobic composition due to their good thermal stability, elasticity, strong bond energy, and low surface tension [6,7,24–33]. The hydrophobicity of prepared coatings is well understood by analyzing the contact angle measurement. The contact angle data of four-blended systems is presented in Fig. 4.2.

From Fig. 4.2, it can be observed that the WCA of bare glass was 59.58 ± 1.33 degrees, denoting a hydrophilic surface. By incorporating the PDMS/Sylgard resin, the hydrophilic property of bare glass was altered to a hydrophobic surface as the WCA increased to 96.34 ± 0.42 degrees. It was observed that the WCA of PDMS/Sylgard coating was higher when the Sylgard composition increased above 5 wt% up to 10 wt%. The main factor that increases the contact angle to a higher value is purely due to hydrophobic chemical composition and low surface energy. However, the contact angle starts to decrease as the Sylgard composition increases above 15 wt%. The relationship between the WCA and the wt% of the Sylgard composition in the hybrid coating system is depicted in Fig. 4.3.

From Fig. 4.3, it can be seen that an elastomer with weak intermolecular forces and the intercalation of PDMS into the Sylgrad affect the preexisting surface energy of the PDMS, thus increasing the hydrophobicity of the coating [34]. The S1 resin obtains lower density of methyl groups due to the weak interaction of PDMS and Sylgard resin. In general, the methyl groups strongly interact with each other via covalent bonds and share an equal

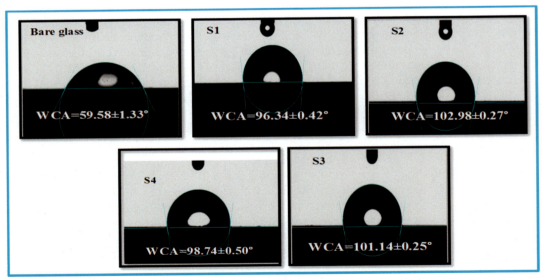

FIGURE 4.2 The water contact angle (WCA) of S1 coating, S2 coating, S3 coating, and S4 coating.

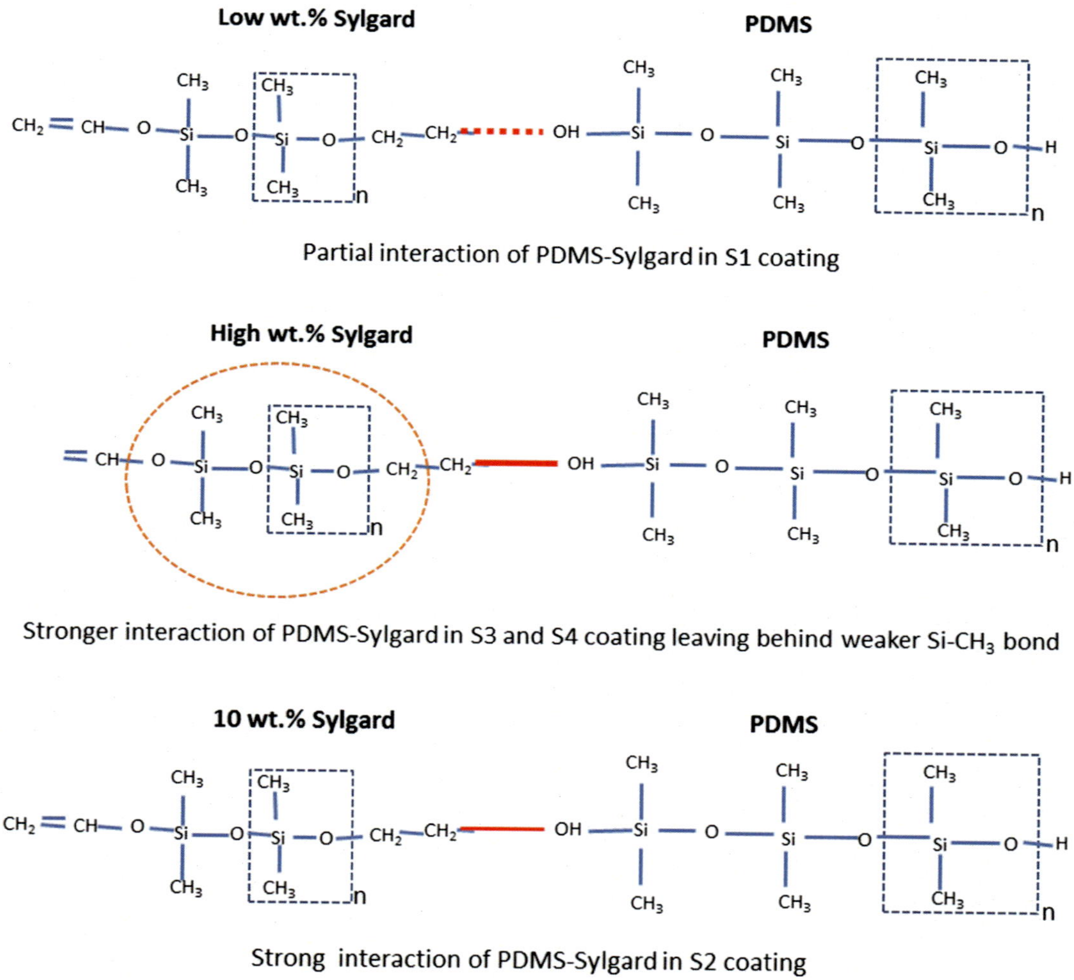

Low wt.% Sylgard — PDMS

Partial interaction of PDMS-Sylgard in S1 coating

High wt.% Sylgard — PDMS

Stronger interaction of PDMS-Sylgard in S3 and S4 coating leaving behind weaker Si-CH$_3$ bond

10 wt.% Sylgard — PDMS

Strong interaction of PDMS-Sylgard in S2 coating

FIGURE 4.3 The interaction of polydimethylsiloxane (PDMS)/Sylgard in S1, S2, S3, and S4 coatings.

number of electrons, which prevents the substitution of water molecules into prepared coatings [35]. By increasing the wt% of Sylgard up to 10 wt%, the hydrogen bond between PDMS and Sylgard increases. As a result, the higher density of methyl groups and longer alkyl chain polymers give greater restriction to water molecules on the S2-coated substrate [6–9]. The increase in the density of methyl groups with increasing Sylgard composition can be observed in FTIR spectra where the increment in density of methyl groups clearly depicts that the crosslinking between PDMS and Sylgard has taken place.

When the Sylgrad composition increases above 15 wt%, the PDMS/Sylgard resin is not soluble and tends to ball-up in ethanol solvent due to strong attractive forces between polymer chains [36]. The saturated PDMS/Sylgard chain forms a weaker Si—C bond and leaves behind an unreacted long alkyl chain, CH$_3$. Unreacted long alkyl chains may lead to a

4. An extensive study of the adhesion and antifogging of the transparent polydimethylsiloxane/Sylgard coating system

TABLE 4.3 Details of silicone composition and water contact angle.

Coating system	Sylgard composition (wt%)	Water contact angle (degrees)
Bare glass	—	59.58 ± 1.33
S1	5.0	96.34 ± 0.42
S2	10.0	102.98 ± 0.27
S3	15.0	101.14 ± 0.25
S4	20.0	98.74 ± 0.50

decrease in hydrophobicity on the S3- and S4-coated substrates [37]. Details of WCA are presented in Table 4.3. From this experiment, it can be concluded that the ideal weight percentage of Sylgrad that can be introduced into PDMS resin is 10 wt%.

4.3.2 FESEM analysis

The surface morphology of coating systems has been examined by using FESEM at the magnification of 1 μm. This is because the polymer coating systems exhibit blurry images at higher magnification due to electrical charging of the sample. FESEM images of a four-blended system above the glass substrate are presented in Fig. 4.4A−D. The energy dispersive X-ray analysis (EDX) spectrum was employed to analyze the wt% of the main elements such as Si, O, and C inside PDMS/Sylgard coating systems. The EDX spectrum of PDMS/Sylgard coating systems is presented in Fig. 4.4E−H.

From Fig. 4.4, the single-phase coatings show a smooth surface without any irregularities, indicating that the mixture of PDMS/Sylgard resin was coated uniformly on the glass substrate. The hydrophobicity of single-phase coatings was achieved by modifying with low surface energy materials, Sylgard, and PDMS resin. Both materials possess multiple bonds of nonpolar carbon atoms because carbon and hydrogen have similar electronegativity values [38]. Most of the organic compounds possess covalent bonds where a pair of electrons is shared by two atoms in the space between them. Each atom formally provides one electron to the bond and these electrons are also simultaneously attracted to the positive charges of both nuclei [39]. Hydrophobic forces are responsible for noncovalent intermolecular interactions in aqueous solution. The nonpolar groups experience no attractive forces when dissolving in one another, but the free energy of nonpolar groups is decreased due to increased entropy of the surrounding water molecules. There is no hydrophobic effect between water and alkanes; instead, there is just not enough hydrophilicity to break the hydrogen bonds of water [40].

Fig. 4.4E−F shows that the wt% of the C element increases with the increasing composition of the Sylgard resin up to 10 wt%, suggesting that long alkyl chains, CH_3, have covered the area of the coated surface. However, the wt% of the C element decreases on S3 and S4 coatings probably because saturated siloxane, Si−O−Si, chains left behind an unreacted long alkyl chain, Si−CH_3, due to strong attractive forces between polymer chains [36]. The formation of saturated Si−O−Si chains can be observed with the increase in wt% of Si and O elements. On the S1 coating system, the highest wt% of Si and O is incidental to the surface of uncrosslinked PDMS only.

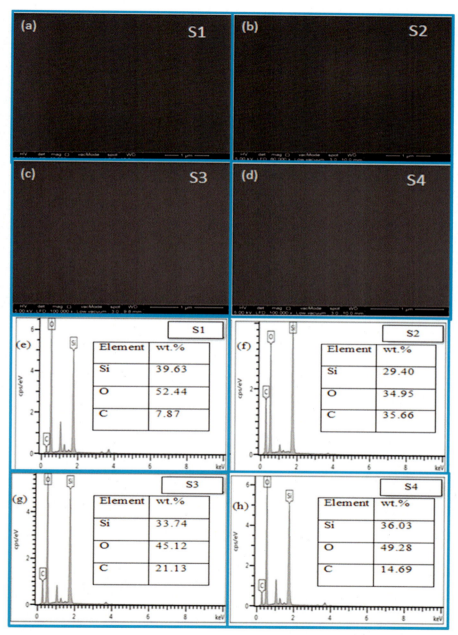

FIGURE 4.4 The surface morphology of (A) S1 coating, (B) S2 coating, (C) S3 coating, and (D) S4 coating. The energy dispersive X-ray analysis spectrum of (E) S1 coating, (F) S2 coating, (G) S3 coating, and (H) S4 coating.

4.3.3 FTIR analysis

Fig. 4.5 shows the FTIR spectra of pristine PDMS and prepared hybrid PDMS/Sylgard resin. The pristine PDMS resin exhibits peaks at 2963 and 2906 cm^{-1} due to asymmetrical vibration of CH$_3$ stretching and CH$_2$ stretching. FTIR absorption around 1258 cm^{-1} corresponds to the vibration of Si–CH$_3$ [41–43]. FTIR spectra around 1019 cm^{-1} and 786 cm^{-1} corresponds to Si–O–Si stretching and the Si–C band. The absorption FTIR peaks around 1019, 889, and 690 cm^{-1} are incidental to asymmetrical Si–O–Si stretching, Si–C stretching, and the deformation of CH out of plane [41,43–45]. The absence of OH groups verifies the hydroxyl-terminated PDMS resin.

The addition of 5 wt% of Sylgard in S1 resin reduces the stretching vibration of methyl groups (–CH$_3$, CH$_2$, and CH) around 2800–3000 cm^{-1} [31,33,46,47], the deformation of Si–CH$_3$ around 1260 cm^{-1} [48–50], and the stretching vibration of the Si–C band around 800–865 cm^{-1} [10,31,42]. The changes in these band intensities verify the crosslinking between PDMS and Sylgard resin through the methyl groups as observed in Fig. 4.6.

Crosslinking causes the splitting of Si–O–Si peaks into separated bands around 1000–12,000 cm^{-1} [31,32,51], which indicates the formation of a hybrid coating structure. Fig. 4.6 shows that the functional silanols, Si–OH, of PDMS are linked to siloxane bridges by methyl groups [52]. The Si–OH bond is linked to methyl groups, CH$_2$, via the hydrogen bond. During the crosslinking process, the double bond of CC opens up and forms a CH$_2$–OH bond to PDMS, thus increasing the chain length of linear PDMS polymers.

The crosslinking of C=C peaks can also be observed around 1645–1651 cm^{-1} [53]. When the wt% of Sylgard is greater, the peak intensity of C=C bands reduces to lower values due to stronger PDMS/Sylgard crosslinking. This statement is further supported by increments in peak intensity of the CH band as observed in FTIR spectra. The increment in the methyl group of PDMS/Sylgard resin can be observed in the wave number around 2970 cm^{-1} and

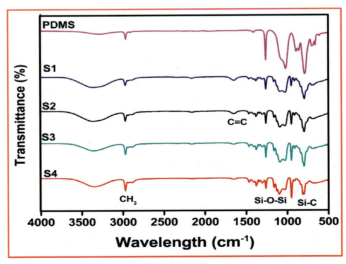

FIGURE 4.5 The infrared spectra of polydimethylsiloxane (PDMS) and PDMS/Sylgard coating systems.

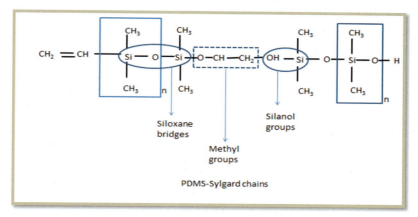

FIGURE 4.6 The crosslinking between polydimethylsiloxane (PDMS) and Sylgard chains.

950 cm^{-1}. On the other hand, the FTIR spectra show the Si—O—Si peaks almost form single peaks inside the S4 resin, indicating the saturated Si—O—Si network and leaving behind an unreacted long alkyl chain, which is confirmed by the presence of the shorter peak of the Si—CH$_3$ bond around 1260 cm^{-1} and Si—C bond around 790—816 cm^{-1}. An unreacted long alkyl chain may lead to lower hydrophobic property [37]. Details of the peaks are presented in Table 4.4.

4.3.4 The relationship between adhesion and crosslinking in resin

Before testing, all the coatings possessed elastic-to-plastic transition, L_{e-p} [54]. An optical microscope was utilized to observe the surface damage after the scratch impact. Failure

TABLE 4.4 The bands observed on all the polydimethylsiloxane/Sylgard coatings.

Bands	S1	S2	S3	S4
CH$_3$ stretching	2970.51	2970.16	2970.19	2970.55
Si—OH stretching	3348.93	3352.08	3326.65	3345.15
C=C bond	1651.02	1651.80	1648.55	1647.60
Deformation of CH$_3$	1409.26	1409.21	1410.79	1410.77
Deformation of Si—CH$_3$	1260.11	1259.82	1260.65	1258.89
Si—O—Si stretching	1030.15	1028.05	1026.45	1024.73
	1095.78	1089.88	1088.00	1080.43
Si—CH$_3$ bending	950.63	950.92	951.23	949.84
Si—C	799.24	798.50	816.16	793.43

points are defined by the abrupt changes on the scratch tracks, namely (1) L_{e-p}: elastic-to-plastic transition and (2) L_{c1}: cohesive failure due to parallel cracking [55]. The scratch hardness is calculated based on the ASTM G171-03 specification as shown in Eq. (4.1) [56]:

$$Scratch \ hardness, \ Hs_p = \frac{8P}{\Pi w^2} \tag{4.1}$$

where p is applied load, w is width of scratch, and H_{sp} is the scratch hardness number. The scratch hardness of all the PDMS/Sylgard coating systems is measured. The S3 coating exhibits the strongest scratch hardness at 9.81 GPa among the PDMS/Sylgard coating systems. This strong scratch hardness can be attributed to the saturated Si—O—Si bonding, which is strongly adhered to the glass substrate, consequently reducing the penetration of the indenter load. The strong coating undergoes slow propagation of cracks and delays the initial fracture behavior [57]. However, the S4 coating experiences a lower value of hardness than the S3 coating due to its crystalline and brittle surface. Details of scratch hardness are detailed in Table 4.5.

The results reveal that the adhesion of coatings increases when the wt% of Sylgard increases above 5. This is because a higher degree of crosslinking in the elastomer restricts the ability of chains to uncoil and raises the elasticity and stiffness of the polymer resin [36,58]. Besides that, the higher number of crosslinkings in PDMS would create a strong binder since the identical terminating methyl groups of PDMS can act as binder [13]. Wang et al. [59] stated that a highly covalent Si—O bond can prevent the effect of tensile stress on the glass substrates. At constant tensile stress, the cracks will undergo slow propagation and stop at a shorter distance. When the tensile stress is increased to a higher value, it will break the covalent bond and cause unstable crack propagation [59]. In this case, the glass substrate will undergo destructive fracture.

4.3.5 XRD analysis

X-ray diffraction is an important tool to determine the structural phase of coatings by analyzing their crystalline or amorphous region. Fig. 4.7 illustrates the XRD spectrogram of PDMS/Sylgard coating systems, namely S1, S2, S3, and S4 coatings. All the PDMS/Sylgard-blended coating systems have shown an amorphous nature around $2\theta = 11-15$

TABLE 4.5 The scratch hardness of polydimethylsiloxane/Sylgard coating systems.

Coating systems	Scratch width (nm)	Maximum load (mN)	Hardness (GPa)
S1	14.67	4000	4.73
S2	12.02	4000	7.05
S3	11.39	5000	9.81
S4	10.67	4000	8.94

4.3 Results and discussions

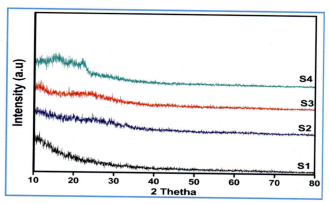

FIGURE 4.7 The diffraction peaks of (A) S1 coating, (B) S2 coating, (C) S3 coating, and (D) S4 coating.

degrees and around $2\theta = 22-25$ degrees. The diffraction angles of PDMS/Sylgard are very similar to the previous report [30,60]. In general, the transparent polymer is completely amorphous because the transparent polymer possesses very small crystallites and a low degree of crystallinity [36].

Fig. 4.7 clearly shows that the diffraction peaks show sharper peak intensity when the wt% of Sylgard resin increases in the PDMS matrix. In the S4 coating system, two of the strongest intensity peaks are located at $2\theta = 15.54$ and 22.06 degrees. The formation of the sharper peak in the S4 coating can be attributed to the perfect arrangement of the PDMS chain and the larger PDMS crystallites in a poor PDMS solvent, ethanol. Yang et al. [61] stated that ethanol molecules were dispersed as individual molecules in the PDMS matrix because ethanol solvent has a weak interaction with the PDMS resin and is unable to react with the intersegment bonds in the PDMS crystallites. The dispersed ethanol molecule increases the free volume of the PDMS system and organizes the PDMS chain into more perfect and larger PDMS crystallites [61]. Details of diffraction peaks of PDMS/Sylgard coating systems are presented in Table 4.6.

TABLE 4.6 Details of diffraction peaks of polydimethylsiloxane/Sylgard coating systems.

Coating system	Diffraction angle 2Θ	Phase ID	Coating phase
S1	11.46	—	Amorphous
S2	23.05	—	Amorphous
S3	11.85	—	Amorphous
	24.68	—	Amorphous
S4	15.54	—	Amorphous
	22.06	—	Amorphous

4.3.6 Rainfall impact

There are several droplet impact and rainfall tests that can examine surface durability, including the artificial rainfall impact sound pressure test (BS EN ISO 140-18), the rain test chamber (IEC 60,529-2001; GB 420-93), NTS drop testing (IEC 60,068), and the standard test method for drop test (ASTM D5276); however, rain is difficult to simulate because of the shape, size, and velocity of the droplets. First, the size of rain droplets is not homogeneous during rainfall where the maximum diameter of rain droplets is determined by the intensity of the rainfall. Second, the shape of rain droplets changes constantly because smaller droplets merge to form larger single droplets on the coated surface. Third, the terminal velocity of rain droplets is influenced by humidity, drop mass, temperature, and wind [62]. Fig. 4.8 presents the defects on the surface of PDMS/Sylgard-coated substrates that are caused by rainfall impact and dust particles after prolonged outdoor exposure.

Fig. 4.8A showed that the S1-coated glass was impacted directly large impact of rain droplets; later experienced large diameter of rain droplets due to liquid splashing. Liquid splashing occurs when the rain droplets exhibit large surface deformation and are later spread at maximum diameter during high-speed impact [63]. These splashes reduce drop infiltration and cause bombardment of the coated surface. When a droplet impinges on the surface, it spontaneously flattens and converts its kinetic energy into surface energy. Once all the kinetic energy has been transformed into the surface energy of the flattened droplet, the energy is dissipated and reverted to kinetic energy of the droplet. As a result, the droplet retracts and bounces in the vertical direction. This phenomenon is called the *reverse surface-to-kinetic energy conversion process* [64].

From Fig. 4.8B–D, it is observed that the PDMS/Sylgard-coated surface experiences reduced droplet impact when the wt% of Sylgard resin has been increased above 10 wt%, suggesting a reduced pinning regime during *bouncing-to-fragmentation transition* of the droplet [29]. This observation clearly agrees with the result of the mechanical property where the PDMS/Sylgard-coated surface becomes stiffer and more elastic at higher wt% of the Sylgard resin. The coated surface can reduce its contact time with the droplet and subsequently eliminates droplet splashing. This is because elastic substrate creates an upward force that causes detachment of the droplet before undergoing a complete *surface-to-kinetic energy conversion process*. Low stiffness substrate does not reduce the contact time because the coating system is unable to accelerate upward, and consequently the substrate experiences full droplet recoil before lift-off. The low stiff substrate can delay the formation of splash on its surface only [64].

Generally, the hydrophobic property of coated glass will experience degradation in an outdoor environment, which is caused by the presence of polar groups on the coated surface [65] and adhered moisture from the outdoor environment [66]. This is because the adhered moisture will cause the adsorption of new hydrophilic groups (e.g., *hydroxyls*, *hydroperoxides*, and *carbonyl*) on the coated surface [66]. The WCA of each coated glass substrate after outdoor exposure is presented in Fig. 4.9.

From Fig. 4.9, it is observed that the WCAs of all the PDMS/Sylgard-coated glass substrates were above 90 degrees, indicating that the PDMS/Sylgard-coated glass substrates can retain their hydrophobic property during prolonged outdoor exposure. This could be attributed to the great durability of PDMS coating under weathering and to its strong silicone backbone, which is not degraded by UV radiation and moisture effects [67]. The degradation

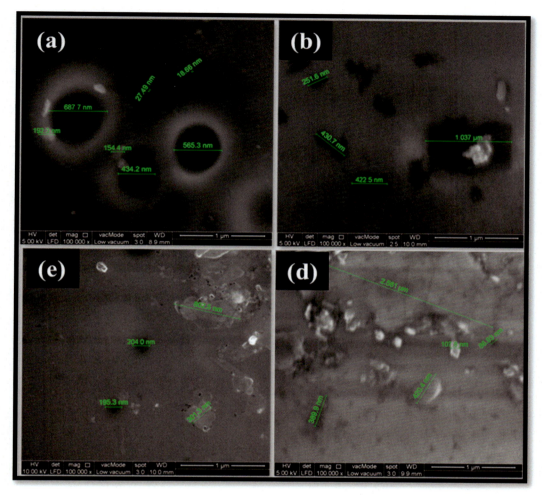

FIGURE 4.8 The surface defects on the (A) S1, (B) S2, (C) S3, and (D) S4 coatings after prolonged outdoor exposure.

rate in the WCA of PDMS/Sylgard coating systems was in the range of 1.46%–4.44% and it decreases with higher Sylgard composition. Details of the WCA of PDMS/Sylgard-coated glass substrates are presented in Table 4.7.

4.3.7 Antifog analysis

An antifog test was carried out by placing all the PDMS/Sylgard-coated glass substrates above the hot-water vapor of a boiling bath for 5 min. After 5 min of water steam exposure, the coated glass was placed under A4 paper to observe the adhered fog droplets on the coated surfaces. In this antifog test, fog exposure was conducted based on the previous report

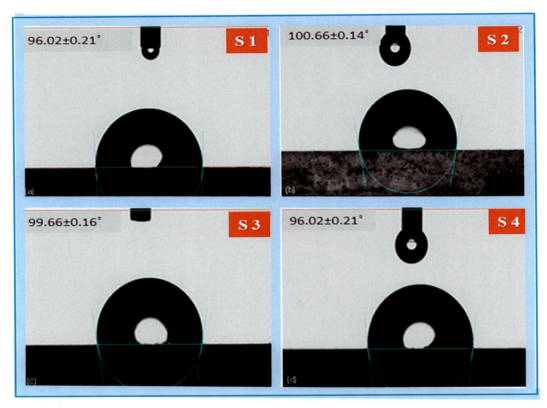

FIGURE 4.9 The water contact angle of polydimethylsiloxane/Sylgard-coated glass substrates after prolonged outdoor exposure.

[68]. According to the standard antifog surface, large fog droplets should not adhere to the hydrophobic glass, which appears to have a transparent surface. Fig. 4.10 depicts the antifog performance of bare glass and PDMS/Sylgard-coated glass.

TABLE 4.7 The average water contact angle of polydimethylsiloxane/Sylgard-coated glass substrates after prolonged outdoor exposure.

Coating system	Average water contact angle (Θ degrees) (before outdoor exposure)	Average water contact angle (Θ degrees) (after outdoor exposure)	Degradation rate (%)
S1	96.34 ± 0.42	92.06 ± 0.21	4.44
S2	102.98 ± 0.27	100.66 ± 0.14	2.25
S3	101.14 ± 0.25	99.66 ± 0.16	1.46
S4	98.74 ± 0.50	96.52 ± 0.23	2.22

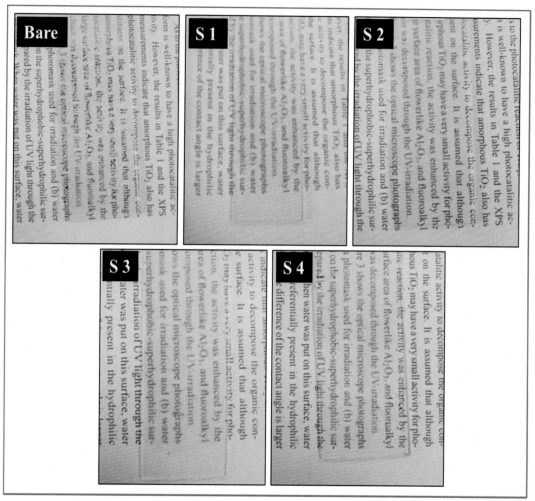

FIGURE 4.10 The antifog performance of bare glass, S1 coating, S2 coating, S3 coating, and S4 coating.

Fig. 4.10 shows that the bare glass substrate possesses a blurry image due to the adhesion of large fog droplets on its surface. This is because the large fog droplets immediately condensed on the surface of bare glass and scattered the incident light. All the PDMS/Sylgard-coated glass substrates have displayed clear images and were adhered by tiny mist droplets only, indicating the antifog performance of PDMS/Sylgard-coated glass substrates. The evaporation time of fog droplets for each coating system is presented in Table 4.8. From the table, the bare glass substrate has recorded a longer evaporation time compared to PDMS/Sylgard-coated glass substrate. In comparison, the S2 coating possesses the fastest evaporation time among the other PDMS/Sylgard-coated glass substrates where the mist droplets completely evaporated within 10 min.

TABLE 4.8 The evaporation time of droplets on the poly-dimethylsiloxane/Sylgard-coated substrates.

Coating system	Evaporation time of fog droplets (min)
Bare glass	25
S1	15
S2	10
S3	11
S4	13

From the obtained result, it can be concluded that the evaporation time of fog droplets is influenced by different methyl groups, Si—CH$_3$ bond, and Si—O—Si linkage on the PDMS/Sylgard coating. In general, PDMS has low interfacial tension against water due to the polarity of the siloxane backbone. When exposed to water, the silicone molecules orient to maximize the polar siloxane bond at the surface. When exposed to air, the silicone molecules orient to maximize the packing density of methyl groups at the surface. The interfacial tension value differs slightly with the length of the PDMS chain and can be reduced by the presence of silanol groups (Si—OH) on PDMS [69]. The S2 and S3 coatings possess higher amounts of nonpolar methyl groups, Si—CH$_3$, which give strong repulsion to polar mineral substrates and water droplets. On the other hand, the higher amount of polar siloxane Si—O—Si linkage on the S4 coating has a strong attraction to mineral substrates and water droplets [70]. The mist droplets were strongly adhered to the S4 surface and evaporated over a longer period.

4.4 Conclusion

Formulated PDMS/Sylgard coatings were synthesized by using simple fabrication methods and can be applied to a large glass panel throughout the spraying technique. Research work has focused on the rainfall impact and antifog property of coating systems after prolonged fog exposure. The adhesion test was examined via scratch hardness tests before placing in an outdoor environment. The current study has demonstrated the improvement of coating systems by simply altering the Sylgard composition in the PDMS matrix. The coated substrate with 15 wt% of Sylgard displays no failure points at 4000 mN (4 N), indicating that the coated surface becomes stiffer with larger Sylgard composition in the PDMS matrix due to higher methyl group composition in the PDMS resin. After 4 months of outdoor exposure, the coating systems showed a slight degradation rate in WCA in the range of 1.46%—4.44% and the degradation rate decreased with higher Sylgard composition. The coated glass panel also exhibited excellent antifog property where the tiny fog droplets only appeared after prolonged fog exposure.

Acknowledgment

The authors would like to express their utmost gratitude for the technical and financial assistance of the UM Power Energy Dedicated Advanced Centre (UMPEDAC) and the Higher Institution Centre of Excellence (HICoE) Program Research Grant, UMPEDAC - 2017 (MOHE HICOE - UMPEDAC).

References

[1] I.R. Durán, G. Laroche, Water drop-surface interactions as the basis for the design of anti-fogging surfaces: theory, practice, and applications trends, Adv. Colloid Interface Sci. 263 (2018) 68−94.

[2] I.R. Duran, G. Laroche, Current trends, challenges, and perspectives of anti-fogging technology: surface and material design, fabrication strategies, and beyond, Prog. Mater. Sci. 99 (2019) 106−186.

[3] Z. Han, X. Feng, Z. Jiao, Z. Wang, J. Zhang, J. Zhao, S. Niu, L. Ren, Bio-inspired antifogging PDMS coupled micro-pillared superhydrophobic arrays and SiO 2 coatings, RSC Adv. 8 (2018) 26497−26505.

[4] C. Antonini, M. Innocenti, T. Horn, M. Marengo, A. Amirfazli, Understanding the effect of superhydrophobic coatings on energy reduction in anti-icing systems, Cold Reg. Sci. Technol. 67 (2011) 58−67.

[5] P. Tourkine, M. Le Merrer, D. Quéré, Delayed freezing on water repellent materials, Langmuir 25 (2009) 7214−7216.

[6] L. Huo, P. Du, H. Zhou, K. Zhang, P. Liu, Fabrication and tribological properties of self-assembled monolayer of N-alkyltrimethoxysilane on silicon: effect of sam alkyl chain length, Appl. Surf. Sci. 396 (2017) 865−869.

[7] J. Israelachvili, R. Pashley, The hydrophobic interaction is long range, decaying exponentially with distance, Nature 300 (1982) 341−342.

[8] T. Textor, B. Mahltig, A sol−gel based surface treatment for preparation of water repellent antistatic textiles, Appl. Surf. Sci. 256 (2010) 1668−1674.

[9] H. Wang, J. Ding, Y. Xue, X. Wang, T. Lin, Superhydrophobic fabrics from hybrid silica sol-gel coatings: structural effect of precursors on wettability and washing durability, J. Mater. Res. 25 (2010) 1336−1343.

[10] S. Gao, X. Dong, J. Huang, S. Li, Y. Li, Z. Chen, Y. Lai, Rational construction of highly transparent superhydrophobic coatings based on a non-particle, fluorine-free and water-rich system for versatile oil-water separation, Chem. Eng. J. 333 (2018) 621−629.

[11] X. Liu, Y. Xu, K. Ben, Z. Chen, Y. Wang, Z. Guan, Transparent, durable and thermally stable PDMS-derived superhydrophobic surfaces, Appl. Surf. Sci. 339 (2015) 94−101.

[12] C. Li, Y. Sun, M. Cheng, S. Sun, S. Hu, Fabrication and characterization of a TiO_2/polysiloxane resin composite coating with full-thickness super-hydrophobicity, Chem. Eng. J. 333 (2018) 361−369.

[13] H. Liu, J. Huang, Z. Chen, G. Chen, K.-Q. Zhang, S.S. Al-Deyab, Y. Lai, Robust translucent superhydrophobic PDMS/PMMA film by facile one-step spray for self-cleaning and efficient emulsion separation, Chem. Eng. J. 330 (2017) 26−35.

[14] M.E. Callow, R.L. Fletcher, The influence of low surface energy materials on bioadhesion—a review, Int. Biodeterior. Biodegrad. 34 (1994) 333−348.

[15] A. Padsalgikar, Plastics in Medical Devices for Cardiovascular Applications, William Andrew, 2017.

[16] C. Wang, H. Yang, F. Chen, L. Peng, H.-fang Gao, L.-ping Zhao, Influences of $Vtms/SiO_2$ ratios on the contact angle and morphology of modified super-hydrophobic Silicon dioxide material by vinyl trimethoxy silane, Results Phys. 10 (2018) 891−902.

[17] S. Haghanifar, P. Lu, Md I. Kayes, S. Tan, K.-J. Kim, T. Gao, P. Ohodnicki, P.W. Leu, Self-cleaning, high transmission, near unity haze Ots/silica nanostructured glass, J. Mater. Chem. C 6 (34) (2018) 9191−9199.

[18] V.S. Smitha, P.M.A. Azeez, K.G. Warrier, B.N. Nair, U.N.S. Hareesh, Transparent and hydrophobic MTMS/GPTMS hybrid aerogel monoliths and coatings by sol-gel method: a viable remedy for oil-spill cleanup, Chemistry 3 (2018) 2989−2997.

[19] Q. An, Z. Lyu, W. Shangguan, B. Qiao, P. Qin, The synthesis and morphology of a perfluoroalkyl oligosiloxane@ SiO_2 resin and its performance in anti-fingerprint coating, Coatings 8 (2018) 100.

[20] F. Lei, J. Yang, B. Wu, L. Chen, H. Sun, H. Zhang, D. Sun, Facile design and fabrication of highly transparent and hydrophobic coatings on glass with anti-scratch property via surface dewetting, Prog. Org. Coating 120 (2018) 28−35.

[21] Y. Cheng, B. Wu, X. Ma, S. Lu, W. Xu, S. Szunerits, R. Boukherroub, Facile preparation of high density polyethylene superhydrophobic/superoleophilic coatings on glass, copper and polyurethane sponge for self-cleaning, corrosion resistance and efficient oil/water separation, J. Colloid Interface Sci. 525 (2018) 76−85.

[22] D. Wang, Z. Zhang, Y. Li, C. Xu, Highly transparent and durable superhydrophobic hybrid nanoporous coatings fabricated from polysiloxane, ACS Appl. Mater. Interfaces 6 (2014) 10014−10021.

[23] C.P. Xing, T. He, Y.J. Peng, R.H. Li, Synthesis and characterization of poly (methyl methacrylate)/polysiloxane composites and their coating properties, J. Appl. Polym. Sci. 135 (2018) 46358.

I. Fundamentals and overarching topics

4. An extensive study of the adhesion and antifogging of the transparent polydimethylsiloxane/Sylgard coating system

[24] J. Xie, J. Hu, X. Lin, L. Fang, F. Wu, X. Liao, H. Luo, L. Shi, Robust and anti-corrosive PDMS/SiO$_2$ superhydrophobic coatings fabricated on magnesium alloys with different-Sized SiO$_2$ nanoparticles, Appl. Surf. Sci. 457 (2018) 870−880.

[25] C. Cao, M. Ge, J. Huang, S. Li, S. Deng, S. Zhang, Z. Chen, K. Zhang, S.S. Al-Deyab, Y. Lai, Robust fluorine-free superhydrophobic PDMS−Ormosil@ fabrics for highly effective self-cleaning and efficient oil−water separation, J. Mater. Chem. 4 (31) (2016) 12179−12187.

[26] T. Zhu, S. Li, J. Huang, M. Mihailiasa, Y. Lai, Rational design of multi-layered superhydrophobic coating on cotton fabrics for UV shielding, self-cleaning and oil-water separation, Mater. Des. 134 (2017) 342−351.

[27] H.-S. Ahn, P.D. Cuong, S. Park, Y.-W. Kim, J.-C. Lim, Effect of molecular structure of self-assembled monolayers on their tribological behaviors in nano-and microscales, Wear 255 (7−12) (2003) 819−825.

[28] A. Mata, A.J. Fleischman, S. Roy, Characterization of Polydimethylsiloxane (PDMS) properties for biomedical micro/nanosystems, Biomed. Microdevices 7 (4) (2005) 281−293.

[29] S. Atherton, D. Polak, C.A.E. Hamlett, N.J. Shirtcliffe, G. McHale, S. Ahn, S.H. Doerr, R. Bryant, M.I. Newton, Drop impact behaviour on alternately hydrophobic and hydrophilic layered bead packs, Chem. Eng. Res. Des. 110 (2016) 200−208.

[30] K. Bendo Demétrio, M.J. Giotti Cioato, A. Moreschi, G.A. Oliveira, W. Lorenzi, F. Hehn de Oliveira, A. Vieira de Macedo Neto, P.R. Stefani Sanches, R.G. Xavier, L.A. Loureiro dos Santos, Polydimethylsiloxane/nano calcium phosphate composite tracheal stents: mechanical and physiological properties, J. Biomed. Mater. Res. B Appl. Biomater. 107 (3) (2019) 545−553.

[31] C. Hoppe, F. Mitschker, P. Awakowicz, D. Kirchheim, R. Dahlmann, T. de los Arcos, G. Grundmeier, Adhesion of plasma-deposited Silicon oxide barrier layers on PDMS containing polypropylene, Surf. Coating. Technol. 335 (2018) 25−31.

[32] G.L. Jadav, V.K. Aswal, H. Bhatt, J.C. Chaudhari, P.S. Singh, Influence of film thickness on the structure and properties of PDMS membrane, J. Membr. Sci. 415 (2012) 624−634.

[33] C. Kapridaki, P. Maravelaki-Kalaitzaki, TiO$_2$−SiO$_2$−PDMS nano-composite hydrophobic coating with self-cleaning properties for marble protection, Prog. Org. Coating 76 (2013) 400−410.

[34] S. Ammar, K. Ramesh, I.A.W. Ma, Z. Farah, B. Vengadaesvaran, S. Ramesh, A.K. Arof, Studies on SiO$_2$-hybrid polymeric nanocomposite coatings with superior corrosion protection and hydrophobicity, Surf. Coating. Technol. 324 (2017) 536−545.

[35] A. Syafiq, A.K. Pandey, V. Balakrishnan, S. Shahabuddin, N. Abd Rahim, Organic-inorganic composite nanocoatings with superhydrophobicity and thermal stability, in: Pigment & Resin Technology, 2018.

[36] C.S. Brazel, S.L. Rosen, Fundamental Principles of Polymeric Materials, John Wiley & Sons, 2012.

[37] X. Wang, Y. Gao, Effects of length and unsaturation of the alkyl chain on the hydrophobic binding of curcumin with tween micelles, Food Chem. 246 (2018) 242−248.

[38] J.D. Roberts, M.C. Caserio, Basic Principles of Organic Chemistry, WA Benjamin, Inc., 1977.

[39] G.A. Webb, Annual Reports on Nmr Spectroscopy, vol. 30, Academic Press, 1995.

[40] R.B. Silverman, M.W. Holladay, The Organic Chemistry of Drug Design and Drug Action, Academic press, 2014.

[41] C. Bai, X. Zhang, J. Dai, Synthesis and characterization of PDMS modified UV-curable waterborne polyurethane dispersions for soft tact layers, Prog. Org. Coating 60 (2007) 63−68.

[42] D. Maji, S.K. Lahiri, S. Das, Study of hydrophilicity and stability of chemically modified PDMS surface using Piranha and Koh solution, Surf. Interface Anal. 44 (2012) 62−69.

[43] F. Xiangli, Y. Chen, W. Jin, N. Xu, Polydimethylsiloxane (PDMS)/Ceramic composite membrane with high flux for pervaporation of ethanol− water mixtures, Ind. Eng. Chem. Res. 46 (2007) 2224−2230.

[44] N. Stafie, D.F. Stamatialis, M. Wessling, Effect of PDMS cross-linking degree on the permeation performance of Pan/PDMS composite nanofiltration membranes, Separ. Purif. Technol. 45 (2005) 220−231.

[45] S. Tanaka, Y. Chao, S. Araki, Y. Miyake, Pervaporation characteristics of pore-filling PDMS/PMHS membranes for recovery of ethylacetate from aqueous solution, J. Membr. Sci. 348 (2010) 383−388.

[46] I.O. Arukalam, M. Meng, H. Xiao, Y. Ma, E.E. Oguzie, Y. Li, Effect of perfluorodecyltrichlorosilane on the surface properties and anti-corrosion behavior of poly (dimethylsiloxane)-ZnO coatings, Appl. Surf. Sci. 433 (2018) 1113−1127.

[47] H.T. Kim, O.C. Jeong, PDMS surface modification using atmospheric pressure plasma, Microelectron. Eng. 88 (2011) 2281−2285.

References

[48] K.H.A. Bogart, S.K. Ramirez, L.A. Gonzales, G.R. Bogart, E.R. Fisher, Deposition of SiO 2 films from novel alkoxysilane/O 2 plasmas, J. Vac. Sci. Technol.: Vacuum, Surfaces, and Films 16 (1998) 3175–3184.

[49] K. Ren, D.A. Kagi, Study of water repellent effect of earth substrates impregnated with water-based silicones, J. Chem. Technol. Biotechnol. 63 (1995) 237–246.

[50] L. Téllez, J. Rubio, F. Rubio, E. Morales, J.L. Oteo, Ft-Ir study of the hydrolysis and polymerization of tetraethyl orthosilicate and polydimethyl siloxane in the presence of tetrabutyl orthotitanate, Spectrosc. Lett. 37 (2004) 11–31.

[51] K. Shahidi, D. Rodrigue, Gas transport and mechanical properties of PDMS-Tfs/Ldpe nanocomposite membranes, J. Polym. Res. 25 (2018) 179.

[52] V.V. Ganbavle, U.K.H. Bangi, S.S. Latthe, S.A. Mahadik, A.V. Rao, Self-cleaning silica coatings on glass by single step sol–gel route, Surf. Coating. Technol. 205 (2011) 5338–5344.

[53] V.V. Balakrishnan, Investigation on the Properties of Silicone Resins Blending with Acrylic Polyol Resins, Institut Pengajian Siswazah, Universiti Malaya, 2003.

[54] R. Alireza, Mechanical and Biological Evaluations of Smart Antibacterial Nanostructured Ti-6al-7nb Implant/ Alireza Rafieerad, University of Malaya, 2017.

[55] B. Vengadaesvaran, N. Arun, R. Chanthiriga, A.R. Bushroa, S. Ramis Rau, K. Ramesh, R. Vikneswaran, G.H.E. Alshabeeb, S. Ramesh, A.K. Arof, Scratch resistance enhancement of 3-glycidyloxypropyltrimethoxysilane coating incorporated with silver nanoparticles, Surf. Eng. 30 (2014) 177–182.

[56] M. Sarraf, B.A. Razak, A. Dabbagh, B. Nasiri-Tabrizi, N.H.A. Kasim, W.J. Basirun, Optimizing pvd conditions for electrochemical anodization growth of well-adherent Ta 2 O 5 nanotubes on Ti–6al–4v alloy, RSC Adv. 6 (2016) 78999–79015.

[57] D. Zhao, W. Wang, Z. Hou, Tensile initial damage and final failure behaviors of glass plain-weave fabric composites in on-and off-axis directions, Fibers Polym. 20 (2019) 147–157.

[58] F. Carrillo, S. Gupta, M. Balooch, S.J. Marshall, G.W. Marshall, L. Pruitt, C.M. Puttlitz, Nanoindentation of polydimethylsiloxane elastomers: effect of crosslinking, work of adhesion, and fluid environment on elastic modulus, J. Mater. Res. 20 (2005) 2820–2830.

[59] D. Wang, Q. Wang, Z. Wang, H. Jiang, Z. Zhang, P. Liu, C. Xu, L. Gao, Study on the long-term behaviour of glass fibre in the tensile stress field, Ceram. Int. 45 (9) (2019) 11578–11583.

[60] J.H. Kim, J.-Y. Hwang, H.R. Hwang, H.S. Kim, J.H. Lee, J.-W. Seo, U.S. Shin, S.-H. Lee, Simple and cost-effective method of highly conductive and elastic carbon nanotube/polydimethylsiloxane composite for wearable electronics, Sci. Rep. 8 (2018) 1375.

[61] H. Yang, Q.T. Nguyen, Y.D. Ding, Y.C. Long, Z. Ping, Investigation of poly (dimethyl siloxane)(PDMS)–solvent interactions by Dsc, J. Membr. Sci. 164 (2000) 37–43.

[62] L.A. Bartolome Marques, J.J.E. Teuwen, Prospective challenges in the experimentation of the rain erosion on the leading edge of wind turbine blades, Wind Energy 22 (1) (2019) 140–151.

[63] M. Song, J. Ju, S. Luo, Y. Han, Z. Dong, Y. Wang, Z. Gu, L. Zhang, H. Ruiran, L. Jiang, Controlling liquid splash on superhydrophobic surfaces by a vesicle surfactant, Sci. Adv. 3 (2017) e1602188.

[64] P.B. Weisensee, J. Tian, N. Miljkovic, W.P. King, Water droplet impact on elastic superhydrophobic surfaces, Sci. Rep. 6 (2016) 30328.

[65] C.-H. Xue, Z.-D. Zhang, J. Zhang, S.-T. Jia, Lasting and self-healing superhydrophobic surfaces by coating of polystyrene/SiO 2 nanoparticles and Polydimethylsiloxane, J. Mater. Chem. 2 (2014) 15001–15007.

[66] A. Davis, Y.H. Yeong, A. Steele, E. Loth, I.S. Bayer, Nanocomposite coating superhydrophobicity recovery after prolonged high-impact simulated rain, RSC Adv. 4 (2014) 47222–47226.

[67] D. Kronlund, M. Lindén, J.-H. Smått, A polydimethylsiloxane coating to minimize weathering effects on granite, Construct. Build. Mater. 124 (2016) 1051–1058.

[68] Q. Shang, Y. Zhou, Fabrication of transparent superhydrophobic porous silica coating for self-cleaning and anti-fogging, Ceram. Int. 42 (2016) 8706–8712.

[69] Y. Liu, Silicone Dispersions, first ed., CRC Press, Boca Raton, 2016, p. 387.

[70] A. Lork, I. König-Lumer, H. Mayer, Silicone resin networks: the structure determines the effect, Eur. Coating J. (2003) 132–137.

I. Fundamentals and overarching topics

SECTION II

Photovoltaic Materials and devices

CHAPTER 5

Crystalline silicon solar cells

R.M. Pujahari

Assistant Professor in Physics, Department of Applied Sciences, ABES Institute of Technology (ABESIT), Ghaziabad, Uttar Pradesh, India

5.1 Introduction

When a solar cell is struck by sunlight, the incident light is directly converted into energy without any mechanical activity or polluting by-product. Earlier solar cells have been in use for over two decades, initially for providing electrical power for spacecraft, but nowadays the use of solar cells has been extended to every walk of life. The reason behind the widespread use of solar cells is the dramatic revolution in manufacturing technology. This would allow solar cells to be produced at prices where they could make significant contributions to world energy demands.

Solar cells work by transforming sunlight directly into electricity using the electronic properties of a specific material called semiconductors.

This should also be economically and environmentally sustainable as well as socially appropriate. The current patterns of energy consumption are neither stable nor sustainable. The rising trend in fossil fuel consumption along with increased greenhouse gas emissions is threatening our secure energy supply.

Hence, the development of safe, stable, sustainable, and accessible energy sources in this century should be our first priority.

5.1.1 Physics of solar cells

A solar cell is a device that converts sunlight directly into electricity. Once solar cells are exposed to sunlight, photons with $h\nu$ of energy (where h is Planck's constant and is the incident radiation frequency) are absorbed in the cell and are greater than the bandgap (E_g) of the semiconducting material.

Throughout this process, a fraction of the photon energy E_g is used in the formation of electron–hole pairs and the excess energy (i.e., $h\nu - E_g$) is typically dissipated as the thermal energy supplied to the crystal. For the solar cell, light radiation of wavelengths (λ) greater than a certain value is also not appropriate [1].

Energy Materials
https://doi.org/10.1016/B978-0-12-823710-6.00004-2

Copyright © 2021 Elsevier Ltd. All rights reserved.

Such a wavelength is known as wavelength cut-off (λ_g), which is given as 1.24 (μm)/E_g (eV). Strictly speaking, energy photons ($h\nu \geq E_g + E_p$) can be absorbed in an indirect bandgap semiconductor, where E_p is the energy of the phonon.

THis is the real photovoltage of the cell, called the open-circuit voltage (V_{oc}). A current flows through it when the cell is connected to a finite load resistance R_L, and there is a decrease in voltage (V_L) through the load. If the load resistance connected across the diode terminals is zero, the current is optimal and is known as the short-circuit current (I_{SC}).

5.1.1.1 Space charge region

A detailed fabrication mechanism for a p-n junction is crucial to understanding solar cell function. A body of a semiconductor is the base unit of several products of a semiconductor in which two separate dopants are directly linked. It is called a p-n junction if a p-doped region fuses within the same lattice in an n-doped zone [2].

In a simple example, it is presumed that both dopants are of the same magnitude in silicon and combine with the same magnitude and fuse suddenly together.

For example, the side to the left $x < 0$ will be doped with boron atoms at a concentration of $N_A = 10^{16}$ atoms per cm^{-3}, making it p-conductive. On the other hand, the right-hand side $x > 0$ could be doped with phosphorous atoms at $N_D = 10^{16}$ atoms per cm^{-3}, which makes it n-conductive.

This flow of charge carriers continues until an equilibrium is established, i.e., until a field current of equal magnitude compensates for the diffusion flow.

5.1.1.2 The biased p-n junction

When electric voltage U_A is applied, the p-doped side is connected to the positive terminal and the n-doped side is the negative terminal the unloaded state of the diffusion current is being reduced from total current, which leads to further free charge carrier concentration. The np value is now greater than the value n_i^2.

This explains why the recombination rate in the space charge region increases simultaneously.

5.1.1.3 The illuminated infinite p-n junction

For an explanation it is considered an infinitely extended crystal with a p-n junction in which there are two doped regions.

Now, the entire crystal is illuminated in such a way that a homogeneous collection of load carrier pairs is made up of the crystal. In practice, this can be achieved if the silicon crystal at the edge of the band is exposed to infrared light. If the coefficient of absorption light is much higher, it means that absorption in the crystal is small and therefore almost the same [3].

The I−V (current−voltage) characteristic of an infinite solar cell: In the case of an illuminated infinite solar cell, the $I-V$ characteristic curve lies within the fourth quadrant (as per normal electrotechnical parameters).

In both cases, the voltages are similar but the current in the illuminated solar cell is negative, which means the current flows in the solar cell in a forward-biased diode direction.

The solar cell parameters, which are vital, are as follows:

Short-circuit current: In the case of a short-circuit condition of a solar cell, current is drawn, i.e., the cell does not have voltage. This current is symbolized as I_{SC} (from the current of the short-circuit configuration). $I_{SC} = -I_L$ means that the magnitude of the short-circuit current is equal to the absolute light-current value.

Open-circuit voltage: The voltage V_{OC} when the circuit is open can be assessed when the solar cell has no current. So:

$$V_{OC} = U_T \ln (I_L / I_0 + 1) \tag{5.1}$$

When current density is low, the value 1 can be neglected with respect to I_L/I_0 and $I_L \approx I_{SC}$, and the modified value of V_{OC} is:

$$V_{OC} = U_T \ln (I_{SC} / I_0) \tag{5.2}$$

The logarithm of the ratio of the short circuit to the dark current is proportional to open-circuit voltage.

Fill factor (FF): In the case of solar cells, optimum output power requires a sufficient load resistor R_S, which corresponds to the ratio V_M/I_M. At the optimum point of operation, V_M and I_M are the voltage and current, and maximum attainable power at the source is P_m. The peak performance ratio ($V_M I_M$) to the variable ($V_{OC} I_{SC}$) is called the solar cell's *FF*:

$$FF = V_M I_M / V_{OC} I_{SC} \tag{5.3}$$

The FF is so called because it helps to find out how much area the rectangle $V_M I_M$ fills under the $I-V$ characteristic relative to the rectangle $V_{OC} I_{SC}$ when graphically depicted [4].

Efficiency: The efficiency of a solar cell is the ratio of the electrical output produced by the photovoltaic cell to the luminous power that falls on it, as follows:

$$\eta = \frac{V_M I_M}{P_{light}} = \frac{FF \, V_{OC} I_{SC}}{P_{light}} \tag{5.4}$$

5.1.1.4 Real solar cells

The situation with modern solar cells is different. Such a cell's cross-section is shown in Fig. 5.1.

The material used to fabricate a solar cell, which is the base, is always p-doped. The n-doped region is called the emitter side.

Photocurrents in a real solar cell: Light is believed to enter on the emitter side for the measurement of photocurrents. Emitter current density, region of space charge current density, and base current density can be summarized as overall current density:

$$I_L(\lambda) = I_E(\lambda) + I_{SCR}(\lambda) + I_B(\lambda) \tag{5.5}$$

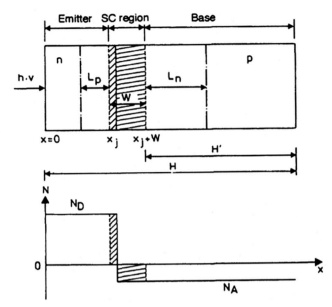

FIGURE 5.1 A typical real silicon solar cell cross-section.

and the total current density can be expressed as follows:

$$I_{L_{total}} = \int_{\lambda_{min}}^{\lambda_{max}} I_L(\lambda)d\lambda \qquad (5.6)$$

Ohmic resistance in real solar cells: There are two types of ohmic resistance: shunt resistance and series resistance.

Shunt resistance (R_P): Shunt resistance is generated along the edges of the solar cells from the leaking currents. Low parallel resistance can be caused by point defects in the p-n junction. These kinds of defects may be a hindrance for a p-n junction and occur at certain points during n-diffusion of the emitters. The base material is in electrical contact with the network of fingers resulting in a short circuit [5].

Series resistance (R_S): Series resistance has various components as follows:

- A metal—semiconductor contact resistor;
- Ohmic contact resistance in metals; and
- Semiconductor material with ohmic resistance.

Two-diode model: For solar cells, this model is very relevant. Both the *FF* and the efficiency depend on this.

The recombination rate formula is:

$$R = \frac{pn - n_i^2}{\tau_{p0}(n + n_1) + \tau_{n0}(p + p_1)} \qquad (5.7)$$

It can be assumed that:

- At the intrinsic Fermi level, the impurity level at in the mid-band region and in the of space charge region is distributed absolutely uniformly.
- Parameters such as lifetime and mobility are of equal magnitude on both sides of the dopant level at the p-n junction.

5.2 Efficient solar cells

At this point we can talk about reducing the overall cost of a photovoltaic array by replacing solar cells with high performance.

The construction costs of a solar array are composed of crystalline solar cells that can be divided into four major categories:

- Wafers in silicon;
- Process engineering;
- Manufacturing modules; and
- Preparation of land and electrical connections, etc.

On a percentage basis, cost breakdown is illustrated in Fig. 5.2 and is applicable to large solar arrays [6].

5.2.1 The role of high efficiency in solar cells

To obtain high efficiency in solar cells, the total amount of loss is to be reduced. Calculations related to efficiency calculation are purely on the basis of multicrystalline silicon (mC-Si) solar cells. Fig. 5.3 provides an idea about how to classify the various forms of solar cell losses. Those losses can be divided into two kinds.

It is clear that optical losses are responsible for both solar radiation reduction by shadowing light and reflection and inadequate absorption of long wavelength radiation [7]. Therefore electrical losses have a negative effect on both the voltage and current of the solar cell. The final types of losses depend on semiconductor physics and technology. Therefore minimizing these losses will be the prime target for achieving high efficiency in solar cells.

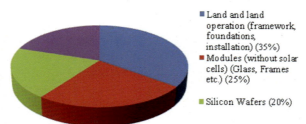

FIGURE 5.2   Analysis of cost of a solar array.

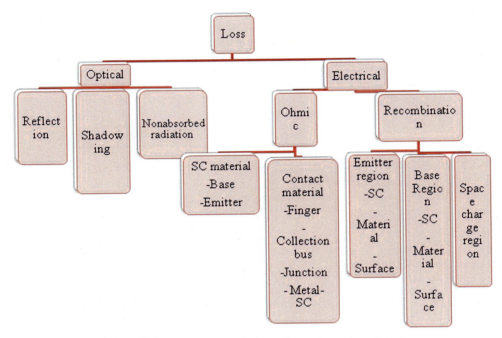

FIGURE 5.3 Flow chart of types of losses in a solar cell (SC).

5.2.2 Electrical losses

Recombination losses: In real solar cells, the photocurrents and saturation currents are utilized to analyze the recombination losses. The following factors also influence the recombination losses:

- The charge carrier's diffusion length;
- Concentration of dopants and dopant profiles; and
- The recombination velocity of the surface.

The back surface field: The back surface field, however, is obstructed by various restrictions for the reduction of reverse-side recombination. The solar cell back is fully shielded, and the recombination velocity of the surfaces is as high as required (high recombination velocities are required for ohmic metal–semiconductor contact).

One way to minimize S_n is by coating a significant portion of the back surface with a thermally induced SiO_2 film, thus passivating it electrically. This layer includes a number of holes of fixed diameter and distance for cell bonding, having a total size around 1%–4% of the total area [8].

Good ohmic contact under the contact point requires high-doping p^+, which also acts as a local back surface field (BSF). An efficient recombination of the charge carrier's surface at the Si–SiO_2 barrier and the metal area's minimal percentage contribution are required. SiO_2 layers are used for passivation to the emitter.

Photocurrent and saturation current from the emitter: Compared to the surface concentration of the dopant in high-efficiency cells of the order of 10^{19}cm^{-3}, the emitter is very small, around a few tenths of a micrometer. The length of diffusion of the holes as minority carriers, which are produced here, is much greater than the emitter thickness.

At the p-n junction, this unified approach is strengthened by the requisite electrical field, which is expected to be generated by the doping concentration gradient. In general, this sort of emitter is transparent in nature.

Schottky contact: If a metal contact has a pure, high-grade semiconductor surface, a similar potential barrier is produced that satisfies the thermal equilibrium requirement between different electron contributions [9].

It is evident that for metal−n-semiconductor contact:

$$q\phi_{Bn} = q[\phi_m - X_s] \tag{5.8}$$

and for p-type metal semiconductor contact:

$$q\phi_{Bp} = E_g - q[\phi_m - X_s] \tag{5.9}$$

where:

$q\phi_{Bn}$, $q\phi_{Bp}$ are the designated metal−semiconductor contact barrier heights
$q\phi_m$ is the metal work function
qX_S is the work function of the semiconductor

According to Eqs. (5.8) and (5.9), the height of the barrier will differ with the metal's work function.

Ohmic metal−semiconductor contact: Electrons will tunnel into the space charge region if a metal is deposited directly on a highly doped semiconductor.

A barrier with low height and a low work function $[\phi_m]$ metal is needed to materialize this. Therefore according to the theory, the following is found to be the contact resistance: Ω cm^2 for an n-semiconductor. The next possibility for electrons to be transferred when doping low is provided by the fact that the thermionic effect causes charge transport.

Using the tunneling effect:

$$\rho_c = \frac{k}{qTA^*} \exp\left[\frac{4\pi\sqrt{\varepsilon_{si}m}}{h} \frac{\phi_{Bn}}{\sqrt{N_D}}\right] \tag{5.10}$$

where A^* is Richardson's contact in which $A^* = A(m^*/m_n)$.

For the thermionic effect:

$$\rho_c = \frac{k}{qTA^*} \exp\left[\frac{q\phi_{Bn}}{kT}\right] \tag{5.11}$$

Due to less doping, contact resistance decreases, while in the other case, it depends only on the height of the barrier. According to Eq. (5.10), the left-hand side shows tunnel behavior and the right-hand side of Eq. (5.11) shows thermionic behavior [10].

Contact resistance [R₄] of a grid finger: The grid finger current path is shown in Fig. 5.4 The current density distribution is directly proportional to the degree of the common contact resistance ρ_C.

The present direction moves along the edge of the grid finger in the case of small values of ρ_C. Otherwise, high transition resistance would cause the current path to expand. This particular action as seen in Fig. 5.5 can be described using a network of resistance.

Then, longitudinal resistances and contact resistances dR_C and dR_E are measured with respect to layer dx as follows:

$$dR_E = \frac{\overline{\rho}dx}{lx_j} = \frac{R_\square dx}{l} \qquad (5.12)$$

where $R_\square = \overline{\rho}/x_j$ is designated as sheet resistance (in $\Omega\square$) in ohms per square. The contact resistance is then found to be:

$$dR_C = \frac{\rho_C}{ldx} \qquad (5.13)$$

where ρ_C (in Ω cm²) is the specific contact resistance.

The distribution of voltage within the grid finger (at right angles) is measured as:

$$U_{(x)} = U_0 \exp\left[-\left(\frac{x}{\sqrt{\rho_C/R_\square}}\right)\right] \qquad (5.14)$$

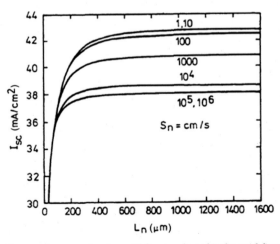

FIGURE 5.4 Present direction as shown under the grid finger where *l* is the grid finger length, *L* is the grid finger width, and *I* is the electrical current.

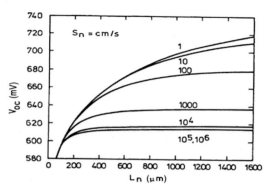

FIGURE 5.5 Distributed resistance network for the contact resistance metal–semiconductor.

If the transport length L_T is designated as $(\rho_C/R_\square)^{0.5}$, the voltage lowers to $1/e$ after distance L_T and the current underneath the grid finger decreases accordingly [11]. Therefore the transport length is an important variable for the current direction.

Ohmic resistance losses in semiconductors:

Base resistance: The resistance of the base material is R_2 of the solar cell, which in simple form is found to be:

$$R_2 = \rho_{Si} D.A \tag{5.15}$$

For *Si*, wafer A is the cell area having a resistivity of $\rho = 1$ Ωcm and thickness D of 200 μm, generating $R_2 = 0.020$ Ω, which is negligible in most cases.

Resistance in the emitter: The current distribution in the grid fingers and the emitter is illustrated in Fig. 5.6. Starting from the base, the current flows vertically first, and is then redirected horizontally through the grid fingers inside the emitter.

5.2.3 Optical losses

The different parameters that are responsible for optical losses in a solar cell are as follows:

- Depending on the wavelength, the Si surface reflects around 35%–50% of incident light.

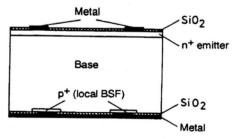

FIGURE 5.6 Direction of current in the emitter with left view, side view, and right view. *BSF,* Back surface field.

- Grid structure shadow loss starts at 3%–12% of light, which is based on the design.

Various technical steps are being introduced to avoid and reduce these losses.

5.2.3.1 Antireflection process

There are two ways in which an antireflection effect will materialize: by using an antireflection coating and by using a textured cell surface [12].

Antireflection using a thin coating layer: By the theory of "optical quarter wavelength," a reflection can be reduced from the solar cell surface. In this case, a penetrating light beam between the antireflection medium and the silicon is reflected in the barrier layer.

On entry to an optically denser medium, there is a phase shift of $\pi/2$, which the electromagnetic wave suffers due to continuity conditions and energy conservation. The antireflective layer thickness is set in such a way that the optical direction, i.e., the refractive index product and the thickness of the layer, is equal to a quarter of the wavelength. Due to the destructive interference of the light with a specific wavelength falling vertically is therefore absolutely disappeared.

5.2.3.2 Textured surfaces

Silicon has a pyramidal-structured surface morphology. Anisotropic etching will materialize these structures on <100> surfaces with a peak angle of 70.5 degrees. At an angle 35.25 degrees, a vertical beam hits the silicon [13,14].

The best performance with the so-called inverted pyramidal structure has been achieved in practice.

5.2.3.3 Nonabsorbed light losses

Nonabsorbed sunlight (AM1.5) is about 10% in a 200 μm thick silicon solar cell. To increase absorption, a reflecting metal at the back mirrors the light, which increases its path of absorption twice. To mitigate the loss, a better strategy is multiple reflections. The light limits itself to the crystal.

Thus Lambert's reflection is used to match the solar cells back surface with a so-called random surface structure. Because of the high refractive index of Si, radiation can escape only via a narrow leakage cone.

This leakage cone's angle Φ_C is calibrated as:

$$\sin^2\Phi_C = \frac{1}{n^2} \tag{5.16}$$

Using the value of $n_{Si} = 3.5$, it is found that $\Phi_C = 17$ degrees.

5.2.3.4 Losses due to shadowing contact fingers

Losses due to shadowing are often proportional to the finger count. The emitter losses are directly proportional to the third power of the distance between the fingers.

The width of the contact finger ranges between 30 and 100 μm at a given surface resistance of 100 Ω □. There is a total loss of 10–15 fingers.

5.2.4 High-efficiency solar cell structure

If all the factors discussed so far are incorporated and a solar cell structure is made, an efficient solar cell can be fabricated. The solar cell parameters of a high-efficiency solar cell can be listed as follows:

- The base thickness is maintained at about 200 μm, mostly due to production methods and handling techniques.
- The light-facing front surface is textured preferably with inverted pyramids.
- Except for point contacts, the cell's back surface is covered with an SiO_2 layer, which is thermally generated for passivation (about 1% −2% of the total surface area) [15].
- With a good antireflection effect on the emitter side (light entry), the average thickness of SiO_2 passivation at 100 nm is achieved.
- Contact with the back surface shows reflective behavior and contains accumulated aluminum vapor.
- For a cell of dimension 2×2 cm^2 (tapering away from the busbar), the grid fingers have a thickness of approximately 8 μm and a width of approximately 15 μm.
- The busbar is about 150 μm in width; the connection contact is situated at the middle of the busbar.
- The efficiency of the solar cells can be increased in the laboratory under AM1.5 from 23% to 24%, very similar to the theoretical efficiency attained by working on these steps.

5.2.5 Method of manufacture for high-performance solar cells

To fulfill the requirements to fabricate these types of solar cells, highly complex solar cell technologies are necessary. However, in these process sequence steps, high temperature, photolithography, chemical processes, and the compatibility of all these steps are also crucial [12].

For the manufacturing of solar cells, boron-doped p-type monocrystalline silicon wafers of area 5 or 6 inch pseudo-square, <100> orientation, and resistivity 0.5−5.0 Ω cm, are used. The saw damage removal process is done for pseudo-square mC-Si wafers by using newly developed NaOH−NaOCl solution at ≈80°C. The damage removal process is immediately followed by a normal texturization process using conventional hot alcoholic NaOH solution at a temperature of >82°C. Phosphorous diffusion will be carried out on the textured wafers in a diffusion furnace at >850°C using phosphorous oxychloride. The phosphosilicate glass coating is eliminated after diffusion in dilute hydrofluoric solution and then the diffused wafers undergo low-temperature oxidation for the growth of a thin passivating silicon dioxide layer. The oxidized wafers are then edge etched in an acid-etching process for the separation of front- and back-diffused junctions. The titanium oxide is coated by high-pressure chemical vapor deposition as an antireflection coating and the back and front surfaces are painted with silver-aluminum/aluminum and silver pastes, respectively. Finally, the printed wafers are fired in a belt furnace after the baking of individual pastes to achieve sufficient ohmic contact on both ends [10].

5.3 Silicon solar cells and fabrication processes

5.3.1 Technology to manufacture silicon

In the course of discussion, various technologies to manufacture solar cells have been discussed.

5.3.1.1 Base material

One of the materials most easily and commonly available on Earth is silicon. It occurs primarily in quartz and sand as silicon dioxide (SiO_2). The pulverized carbon and quartz with graphite are placed in a container (crucible). An arc allows them to fuse at around 1800°C. A reduction cycle then happens as per the formula:

$$SiO_2 + 2C \rightarrow Si + 2CO \tag{5.17}$$

The accumulated liquid silicon at the base of the crucible (14–15°C melting point) can then be drained off. The purity is roughly 98%. This is called metallurgical-grade silicon, and a large volume is used in the iron and aluminum industries.

5.3.1.2 Refractioning processes

In a fluidized-bed reactor, metallic metallurgical silicon is subjected to hydrochloric gas in the first step. The chemical reaction (exothermic reaction) yields trichlorosilane and hydrogen:

$$Si + 3HCl \rightarrow SiHCl_3 + H_2 \tag{5.18}$$

Because trichlorosilane is a liquid under 300°C, it can be easily isolated from hydrogen. The chlorides and product impurities also have to be isolated from the trichlorosilane. In the next step, the trichlorosilane is rendered in fractional distillation columns to be free of impurities [9].

5.3.1.3 The fabrication of material using polycrystalline silicon

High-purity silicon fabrication is materialized using equipment for the industrial precipitation of polysilicon. A combination of hydrogen (high purity) and trichlorosilane is placed in the reactor vessel. The silicon's hot surface is reduced and silicon is deposited on the surface of the rod by the reduction process from trichlorosilane. The procedure is carried out in line with the following formulation:

$$4SiHCl_3 + 2H_2 \rightarrow 3Si + SiCl_4 + 8HCl \tag{5.19}$$

Thus in an uninterrupted process, high purity polycrystalline silicon is manufactured as a rod having a diameter approximately 30 cm and a length 2 m.

5.3.1.4 Crystal pulling method

This consists of two processes: the crucible pulling process, also known as the Czochralski process, and the float zone pulling method.

The Czochralski process: In the Czochralski method, multicrystalline material is placed in a quartz crucible in the form of fragments obtained from polysilicon. This is stored in a crucible made of graphite, and by induction heating is melted under inert gas [16].

On seed crystal, this process is materialized using vertical pulling and by using the silicon rotary motion it emerges in monocrystalline shape.

The addition of highly doped silicon fragments allows the desired doping to be modified simultaneously, depending on the degree and form of conductivity.

The float zone pulling process: In the float zone pulling process, inert gas is induced in the pulling device is inside a container. A single crystal seed is liquefied at the bottom end by induction heating on the polycrystalline chain. Upon melting, the vertical movement of the induction coil while spinning propels the region of liquid silicon upward [17].

5.3.1.5 Fabrication of silicon wafers

A modern method has been developed that is simply a multiwire saw in which a multikilometer-long wire is passed through a rod made of crystals in multiple coils in a suspension, which is abrasive while being wrapped from one coil to another in turn.

In this case, thinner wafers can be processed and the losses due to sawing in comparison with the inner diameter saw process are decreased by around 30%.

5.3.2 Silicon solar cell technology

The technology used to produce crystalline silicon solar cells consists of four categories:

- P-n junction fabrication technologies;
- SiO_2 layering technologies;
- Electrical contact production technologies;
- Antireflection technologies.

5.3.2.1 P-n junction fabrication processes

The two main methods for manufacturing p-n junctions are ion implantation techniques and diffusion.

Diffusion technologies and diffusion theory: In the Si solid, the diffusion of solid substances is governed by Fick's second law, whose one-dimensional mathematical expression is as follows:

$$\frac{\partial N(x,t)}{\partial t} = D\frac{\partial^2 N(x,t)}{\partial x^2} \tag{5.20}$$

where:

$N(x,t)$ is the diffusing substances' concentration at point x and time t

D is simply the coefficient of diffusion, which is related to specific material and is temperature dependent

Diffusion technologies and diffusion theory: Ficks second law is applicable to the diffusion of solid substances in an Si solid, which recognizes quartz trays in single-dimensional form.

120
5. Crystalline silicon solar cells

Emitter diffusion processes: Because the starting wafers for solar cells are almost always doped with boron and are thus p-conductive ($10^{15}-10^{17}$ cm^{-3} concentrations), phosphorus is used to produce an n-doped emitter [18].

When the oxygen is added, the dopant gases react with the silicon surface at high temperatures of about 800°C. On the upper surface, silicon dioxide is formed as per the following relation:

$$Si + O_2 \rightarrow SiO_2$$

and phosphine is changed to phosphorous pentoxide:

$$2PH + 3O_2 \rightarrow P_2O_5 + H_2O \tag{5.21}$$

Liquid phosphorous glass is formed when P_2O_5 is mixed with the silicon dioxide that develops on the surface of the silicon that becomes the source of diffusion.

The BSF diffusion processes: As already described, high-efficiency solar cells need a so-called BSF. The needed p$^+$ doping is accomplished through boron diffusion. For this function, BBr$_3$ can act as the source of boron, which can be treated quite similarly to POCL$_3$.

5.3.2.2 Oxidation technologies

As described, the method of thermal oxidation (i.e., without the addition of water vapor) follows the following formula:

$$Si + O_2 \rightarrow SiO_2 \tag{5.22}$$

Oxygen diffuses over the forming SiO_2 sheet.

Wet oxidation (with water vapor) occurs by the formula:

$$2Si + O_2 + 2H_2O \rightarrow 2SiO_2 + 2H_2 \tag{5.23}$$

In this process, subsequently, chlorine can be added to the cycle in the form of trichloroethane ($C_2H_3C_{13}$):

$$C_2H_3Cl_3 + 2O_2 \rightarrow 3HCl + 2CO_2 \tag{5.24}$$

and:

$$4HCl + O_2 \rightarrow 2H_2O + 2Cl_2 \tag{5.25}$$

5.3.2.3 Auxiliary technologies

Auxiliary technologies such as etching, washing, and photolithography are important for mC-Si solar cell manufacturing.

Etching and cleaning methods: The various methods of etching and cleaning are necessary to produce semiconductor devices. As far as possible, the silicon wafers surface must be free from contaminants. Those contaminants can be of any type: atomic or molecular or ionic in nature [19].

Among the various processes of surface cleaning is RCA cleaning. This cycle is based on the use of hydrogen peroxide (H_2O_2) in the first place as an addendum to ammonium hydroxide (NH_4OH), which is a weak solution, and in the second place hydrochloric acid (HCl).

A low-cost chemical texture for an industrial mC-Si solar cell that includes NaOH—NaOCl provides a simple and low-cost alternative to production.

Photolithography: To initiate silicon dioxide in various masking processes, photolithography is implemented. As the photosensitive resistor is light sensitive, a thin photoresist film is spun on a silicon wafer in a yellow field [20].

5.3.2.4 Solar cell metallization

Many contact technologies are used for semiconductors.

Finger grid structuring: There are three methods that are generally used for the structure of a finger row.

Screens with opening as per needed pattern are used in the screen printing techniques. The glass is kept in contact with the cells, and is coated with a metal in the form of a paste. The template layout is passed onto the cells. A finger width as small as possible is ideal to reduce shadowing losses.

5.3.2.5 Antireflection technologies

Applying the antireflection coating (ARC): Updated technology for vacuum evaporation and thick film methods is designated to apply ARCs. Titanium dioxide (TiO_2) is vigorously used for a single layer of ARC. During the evaporation cycle, the refractive index can be changed within a limited range by selecting the evaporation rate and adding small quantities of oxygen [21].

5.4 Large area crystalline silicon solar cells

A fundamental prerequisite for the creation and advancement of solar cells is to calculate individual parameters and assess their complex relations.

5.4.1 Current—voltage characteristics

As per international agreement, solar cell efficiency should be calibrated within the range AM1.5.

In the laboratory, a sun simulator approximates the radiation. An ultrahigh-pressure xenon lamp acts as a light source that emits a white spectrum because the individual xenon spectral lines experience a high expanding pressure level. Spectral filters are used to minimize the high intensities that still exist in certain spectral ranges to the degree that they reach a reasonable approximation of the AM1.5 spectrum [22].

5.4.1.1 *Illuminated* **I–V** *curve measurement*

The *I–V* feature is calibrated using an electrical charge regulator to track the current from the solar cell gradually from nil up to the short-circuit current. The measured varying parameters are:

- Open-circuit voltage V_{OC};
- Short-circuit current I_{SC};
- Maximum current I_m and maximum voltage V_m;
- Fill factor *FF*; and
- Efficiency η.

5.4.1.2 *Dark current characteristic measurement*

More accurately, series resistance R_S can be measured by taking measurements under both illuminated conditions and dark conditions. A higher voltage (V_S) than the open-circuit voltage (V_{OC}) is required for dark current calculation to achieve a current that has the same value as the short-circuit current, since the additional voltage drop at the series resistance must be overcome [23].

Efficiency and its dependence on temperature: Assuming radiation is constant, and thus short-circuit voltage, the FF in this case is assumed to be constant. The comparative increase in output in the open-circuit voltage is then proportional to the relative increase in performance.

5.4.2 Spectral response of solar cells

5.4.2.1 *Front-illuminated solar cell spectral response*

For spectral reaction, the solar cell is irradiated with light from various spectra.

Spectral solar cell response relies on the accumulated charge carriers (the light source is a xenon high-pressure lamp).

The short-circuit current configuration must be the same as the intensity of the radiated light in between the intensity range from about 1/100 to 1/1000 sun. It cannot be a cell with a local back surface area, because recombination velocity depends on the number of load carriers produced and the stringent linearity of the short-circuit current is no longer preserved [24].

5.4.2.2 *Back surface illuminated solar cell spectral response*

Here, the test cell is not on the emitter side, but on the opposite side to it, as usual to the illuminated solar cell. The illuminated cell on both sides is called a bifacial cell.

Moreover, adding a BSF under the metal surface is also beneficial to further minimize recombination.

5.4.3 The Plasma chemical vapour deposition (PCVD) measurement technique

The process explains following advantages before illumination, measurement of short-circuit current and open-circuit voltage related to a solar cell. This principle has certain advantages over the static process as follows:

5.4 Large area crystalline silicon solar cells

- No absolute light intensity measurement is required;
- There must be no surface reflection; and
- The exact coefficient of absorption has to be determined.

The cell to be investigated is to be excited using laser pulses. The system is operated by an Nd–YAg laser, which produces light with a wavelength of 1064 nm [25].

The following parameters are calculated from the various states and distribution of the generated charge carriers:

- Effective surface recombination velocity of the emitter;
- Emitter saturation current (with good passivation);
- Diffusion length at the base; and
- Recombination velocity at the back surface.

5.4.4 The PCD method

Ohmic contacts must be affixed in the photocurrent decay (PCD) measuring technique to allow measurements.

According to the volume originating from the surface, a total separation of recombination values could not be materialized, but in the case of the PCD technique, this is possible. As in this case, ohmic contacts are not necessary.

As displayed in Fig. 5.7, the silicon wafer under experimentation is mounted in a resonant circuit of increased frequency, which operates at a frequency of 13.56 MHz at a distance of about 1 mm. In this way, the current related to decay is measured [26]. The data are stored and analyzed by a PC using a storage oscillograph. This measurement method helps to calculate two parameters: (1) the recombination velocity at the surface and (2) the saturation current of the emitter.

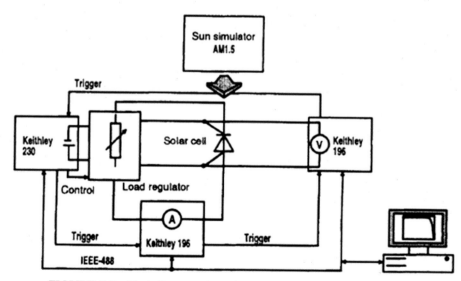

FIGURE 5.7 Block diagram of a photocurrent decay measuring apparatus.

5.4.4.1 Saturation current at the emitter

The superiority of the saturation current from the base decreases with the increasing diffusion time of the charge carriers and a lower recombination of the surface at the back makes saturation current more important.

Therefore the saturation current from the emitter area must be minimized for very high efficiency.

5.4.4.2 Determining recombination velocity at the surface

A sample wafer under investigation has SiO_2 on both sides. Because the emitter is not present, can be determined by sum of the reciprocal of high injection carrier lifetime ($1/\tau_{hli}$) and ratio of surface recombination velocity and thickness of the base (s/w).

The SiO_2 film is separated, and the Si wafer is placed in a special container, which is filled with hydrofluoric acid (in a Teflon case with a clear plastic sheet window) [27].

5.4.5 Atomic force microscopy

The atomic force microscope is a very high-resolution type of microscopy scanning probe that shows resolution in the range of fractions of a nanometer; it is more than 1000 times better than the optical diffraction limit. The atomic force microscope is one of the most important instruments for imaging, measuring, and manipulating nanoscale matter.

5.4.6 Scanning electron microscope

This uses electrons rather than light to form an image. After their introduction in the early 1950s, scanning electron microscopes have created new research areas within the fields of medical and physical sciences. It has enabled researchers to examine a far greater variety of specimens.

5.4.7 The KLA-tencor development series of stylus profilers

This offers a complete solution that meets the needs of the science and engineering community. The Surface Profiler Product Development Series was designed to meet diverse requirements by providing fully featured products, incorporating emerging technologies, and enhancing performance when necessary [28].

The key features of this instrument include:

- 30 mm scan length;
- Z sensor range up to 1.2 mm;
- 140 mm manual sample positioning stage;
- Force control of 0.03—10 mg; and
- Windows XP, Vista, and 7, etc. compatibility.

5.4.8 Zeta instruments

The following features are associated with Zeta Instruments, which are useful for multicrystalline solar cell research [29].

5.4.8.1 Images

The first phase in solving problems and making intelligent choices about the test surfaces is a well-defined true color and 3D image.

5.4.8.2 Dimensions

Precise measurements are made of items that can only be seen under a microscope. Measurements are made from a single scan using Zeta lateral lengths, step heights, and wall angle.

5.4.8.3 Metrology

Based on the millions of data points, measuring two-dimensional (2D) and three-dimensional (3D) roughness parameters provides precise topography knowledge compared to data from single line traces or much smaller fields of view in other devices. SoZeta instruments can do it [30].

5.4.8.4 Film thickness

The spectrometer method allows to determine the film thickness of single or multi layers. Now the topography of the surface and the films which make up the surface can be calculated at the same time.

5.4.8.5 Texture characterization

Zeta Instrument's optical profiling capability allows for the fast analysis of textured surfaces. 3D visualization offers a qualitative view of Earth, while feature size histograms, linear, and ISAL roughness measurements provide quantitative details on base- or acid-etched surface textures.

5.4.8.6 Finger contact profiling

Because of the large difference in reflectivity between the antireflection-coated textures and the metal contact, it is difficult to imagine solar finger metal contacts. The textured area can be 0.5% or less reflective, while metal reflectivity can surpass 90% [31].

5.4.8.7 A calculation of film thickness

Zeta Triple Play systems include a photospectrometer capable of measuring thicknesses of thin films deposited on textured surfaces, which are difficult to measure using conventional ellipsometric techniques. Ellipsometers perform very well on thin films deposited on smooth surfaces but they have trouble obtaining accurate readings on films deposited on rough, textured surfaces.

5.5 2D and 3D screen-printed metallization effect on industrial crystalline silicon solar cells

5.5.1 Surface analysis of NaOH—NaOCl-textured crystalline silicon solar cells

Surface regulation of mC-Si solar cells by texturing is one of the critical issues of mass production-level, high-performance, wide area, low-cost industrial cells. For crystalline silicon solar cells, the direction-dependent anisotropic alkaline texturization solution is standard.

126   5. Crystalline silicon solar cells

First, an isotropic etching process that contains an HF—HNO$_3$-deionized (DI) water-etching step followed by an HF—HNO$_3$-etching step provides a good choice for texture.

Second, it helps to develop shallow front junctions by growing a layer which is dead for cell blue response enhancement. Nonetheless, this process utilizes expensive equipment due to the involvement of control parameters and expensive chemicals and so it may only be suitable for very large-capacity photovoltaic plants [32,33].

A newly developed low-cost chemical texturing method for industrial mC-Si solar cells involving NaOH—NaOCl (sodium hypochlorite) offers a simple and low-cost alternative to processing. This optimized etching solution has no effect on the boundaries of mC-Si grains, and therefore has excellent isotropic etching features to boost parameters such as V_{oc} and FF.

5.5.1.1 Experimental procedure

The starting material for the experiment was Deutsche Solar's 125 × 125 mm square boron-doped p-type mC-Si wafers of base resistivity 0.5—3.0 qm from SOLSIX. MC-Si texturing was done using the solution NaOH—NaOCl. Thoroughly rinsed wafers in DI water were taken and dried after the final etching phase.

A representative wafer as sample was then analyzed using atomic force microscopy (AFM).

The wafers were then doped with phosphorus (P) as the source of the dopant using diffusion, which was appropriate with phosphorous oxychloride (POCl$_3$) [34]. After removal of thermal oxide and etching of the base, a silicon nitride (Si$_3$N$_4$) ARC layer was formed by the process of plasma-enhanced deposition of chemical vapors. Final solar cells are then completed by screen printed metallization and proper firing of contacts. The process flow displaying the complete cycle is shown in Fig. 5.8.

5.5.1.2 Impact on efficiency of solar cells

The key goal and intent of texturization is to increase the silicon solar cells performance. This can be achieved dramatically by reducing the losses in reflection and by generating a damage-free Si surface before diffusion. A new method consisting of an NaOH and NaOCl solution at 80—82°C can be used for texturization.

The chemical reaction is given as:

$$NaOCl \leftrightarrow Na^+ + OCl^-$$

NaOCl ionizes the hypochlorous (HOCl) and hypochlorite (OCl$^-$) ions in water. Due to the scarcity of OH$^-$, the silicon etching rate is very small. The presence of the higher concentration of NaOH in the Sodium Hydroxide and Sodium Hypo-Chloride (SHSHC) solution now has a better chance of polishing rather than grinding the silicon surface [35].

This dramatically decreases the probability of grain boundary isolation, and by reducing V_{oc} affects cell performance.

The AFM experiments on these wafers were performed over an area of 3 × 3 μm. Fig. 5.9 depicts our textured wafer as a three-dimensional AFM micrograph. On the wafer surface, excellent smoothness is observed. Here, the slow and guided nature of polishing leads to polished low-height peaks as seen from the AFM image.

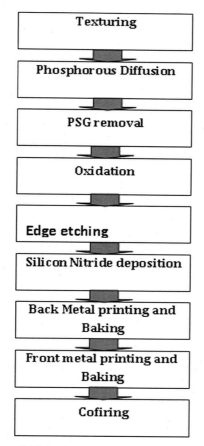

FIGURE 5.8 Solar cell fabrication process steps. *PSG*, Phosphorous silicate glass.

From two-dimensional analysis and roughness mapping (shown in Fig. 5.10), it is observed that the surface is more normal with bumps of large base during alkali polishing and a uniform flatter top surface. The area of each individual white mark (indicating a height surface greater than 75 nm) is also larger in Fig. 5.10. This indicates a greater flatness right at the peaks of certain bumps in the polished alkali-based samples [36].

AFM analysis of intragrain roughness reflects the surface roughness of the mC-Si polished wafer surface at the submicron level. This is expressed in Figs. 5.11 and 5.12.

This figure shows, as per standard height deviation, that the polished surface within the scanned area gives only 6.967 nm of the quantitative value of the roughness content of the polished surface (as expressed by the "R_q" parameter).

From the section analysis it has been observed that numbers of peaks of larger heights (Fig. 5.11) along each scanned path is much lesser through the scanned path. The detailed surface mapping statistics are given in Fig. 5.13. This indicates that in the NaOH−NaOCl polishing solution, values between the reference markers (RMS) are 4.572 nm as per standard deviation

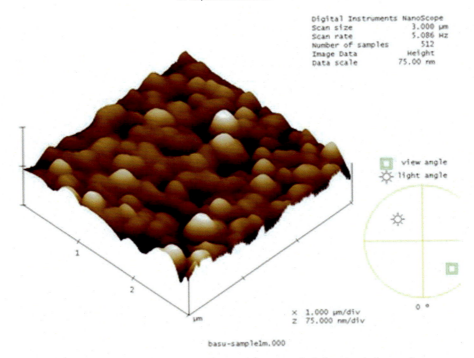

FIGURE 5.9 Three-dimensional surface roughness analysis by atomic force microscopy.

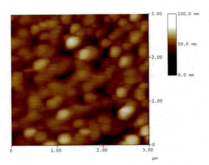

FIGURE 5.10 Two-dimensional analysis and roughness mapping.

of the height. For the NaOH–NaOCl polished sample, the difference between the highest and lowest points (R_{max}) on the sectional profile relative to the central line over the length of the roughness curve (L) is 16.281 nm and the peak height radius (R) is 530.75 nm [37].

All these qualities, and a few more parameters of roughness, are shown in Fig. 5.13.

Using our standard fabrication processing sequence, solar cells are fabricated as shown in Fig. 5.8. The illuminated current–voltage (LIV) characteristic of one of our best cells is measured and shown in Fig. 5.14.

5.5 2D and 3D screen-printed metallization effect on industrial crystalline silicon solar cells

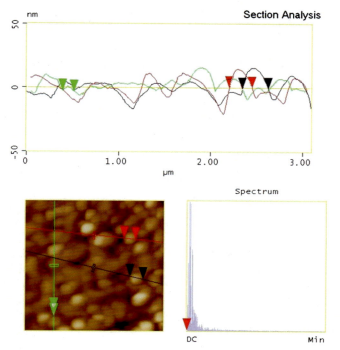

FIGURE 5.11 Image statistics after two-dimensional roughness analysis by atomic force microscopy

FIGURE 5.12 Section analysis and mapping by atomic force microscopy.

This figure shows higher V_{oc} and FF values. It is the product of optimized metallization of front and back metal contacts as seen in our AFM analysis in compliance with a smoother surface area. The serial resistance (R_s) value for this cell is 7.17 mOhm. An output of >15% is achieved with a moderate I_{SC} value (with an efficient Si_3N_4 ARC) [38].

This leakage current has much to do with surface uniformity and surface consistency. The grid fingers have a fair chance of losing their continuity during metallization. This results in nonuniform nucleation of charge carriers along the boundaries of the grain. At lower solar

L	246.09 nm
RMS	4.572 nm
lc	DC
Ra(lc)	3.513 nm
Rmax	16.281 nm
Rz	16.281 nm
Rz Cnt	2
Radius	530.75 nm
Sigma	1.037 nm

Surface distance	248.68 nm
Horiz distance(L)	246.09 nm
Vert distance	0.496 nm
Angle	0.115 °
Surface distance	123.37 nm
Horiz distance	123.05 nm
Vert distance	0.615 nm
Angle	0.286 °
Surface distance	284.41 nm
Horiz distance	281.25 nm
Vert distance	0.177 nm
Angle	0.036 °
Spectral period	DC
Spectral freq	0 /μm
Spectral RMS amp	0.015 nm

FIGURE 5.13 Statistical data of section analysis and mapping by atomic force microscopy.

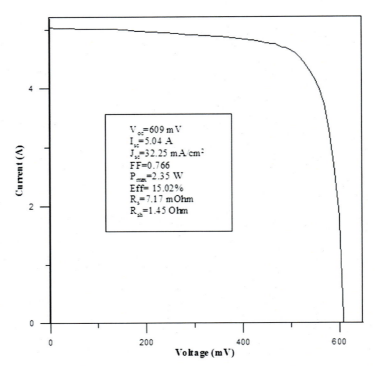

FIGURE 5.14 Illuminated $I-V$ characteristics of our solar cell.

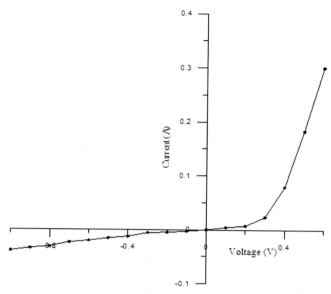

FIGURE 5.15 Dark $I-V$ characteristics of our solar cell.

intensity, this reasoning is dominated by high leakage current. The rate of leakage current will increase in both directions. This is quite relevant in the characteristic curve of our experimental dark current–voltage (DIV) as shown in Fig. 5.15. The surface is finer and the dark leakage current is less. This is very evident from Fig. 5.15: the leakage current is the bare minimum for the NaOH− NaOCl textured cells.

5.5.1.3 Conclusion

A nanometric surface roughness study by AFM for all the NaOH–NaOCl wet texturization or polishing processes for industrial mC-Si solar cells has been performed. The AFM study clearly indicates the superiority of the NaOH–NaOCl texturization process over the surface in terms of surface smoothness and also ascertains more efficient industrial solar cells coupled with a proper ARC.

5.5.2 Impact of 2D and 3D screen-printed metallization on industrial crystalline silicon solar cells

Different metallization schemes such as photolithography after vacuum evaporation, electroplating, and buried contact are used in commercial large area crystalline silicon photovoltaic plants; these are more expensive and time consuming for large-scale production processes. The screen-printing technique offers a cost-effective alternative to those complicated schemes that have already made widespread progress in the solar cell industry.

MC-Si solar cells are produced in a traditional industrial way in the present research. Both 2D and 3D stainless steel meshes are used during cell front screen printing. The research describes the effect of these screen-printing processes on the efficiency of the solar cell's electrical parameters by observing the characteristics of LIV and DIV to assess the 3D screens supremacy. An optical microscope and spectrophotometer also test meshing and the surface reflectivity of the cells.

5.5.2.1 Experimental process

Metallization: Two batches of 40 wafers each are taken after aluminum and silver—aluminum printing for front emitter metallization. The first 40 wafers are then screen printed with 2D 325 and the other 40 wafers with 3D stainless steel (SS) 325 displays on the front surfaces. One cell from each batch is taken as the representative of the 2D and 3D screen-printing processes after printing and cofiring [8].

Characterization: Tiny pieces of dimension 2×2 cm are cut from the representative cells of the 2D and 3D screen printing by using an Nd—YAG laser for spectrophotometer surface reflectance analysis. Wafer parts (as samples) are cleaned ultrasonically in isopropyl alcohol before surface analysis, followed by rinsing in DI water and drying.

Such samples are taken from the same areas of both cells, including only the grid lines. The cells also calculate the DIV and LIV characteristics. Their LIV characteristics with an AM1.5 global spectrum are measured under 1 sun intensity.

5.5.2.2 The impact on efficiency of solar cells due to metallization

An optical microscope detects the knitting of the SS wires, and the magnified images in the 2D-325 and 3D-325 displays are shown respectively in Fig. 5.16.

For the 2D screen, just as in Fig. 5.16, only two wires are used and the maximum possible knitting height is the total diameter of the two wires, although Fig. 5.16 indicates that the 3D

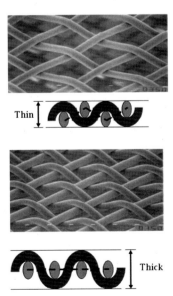

FIGURE 5.16 The knitting of stainless steel wires in 2D and 3D mesh.

5.5 2D and 3D screen-printed metallization effect on industrial crystalline silicon solar cells

TABLE 5.1 Comparison table of 3D and 2D stainless steel (SS) mesh parameters.

Mesh properties	Mesh type					
	2D			3D		
Mesh count (nos./inch)	325	400	500	325	400	500
SS wire diameter (μm)	28	23	19	28	23	19
Maximum metal height possible (μm)	45–56	35–46	25–36	77	66	53
Open area (%)	41.2	40.7	39.2	41.2	40.7	39.2
Maximum volume of the paste flow (cm^3/m^2)	23.07	19.54	14.11	31.72	26.86	20.78

mesh consists of three wires, hence the maximum possible knitting height is the thread, the metal strip region, and the Ag grid finger radius, respectively.

While R is set for a total of the three wires' diameters, ρ and l are fixed. Table 5.1 outlines a comparison of different screen parameters for the 2D and 3D displays for the 325, 400, and 500 mesh counts. The diameters of the SS wires are the same for both the 2D and 3D displays in all three applications, and the wires are small for higher mesh counts [11].

This results in the height of the Ag metal created by 3D mesh being greater than that of the usual 2D mesh. The metal front resistance is provided by:

$$R = \rho \frac{l}{A} = \rho \frac{l}{\pi r^2} \tag{5.26}$$

Here, R, l, A, and r are the resistance of the Ag metal front, the length of the metal specific form of Ag paste, area of paste and radius of cross-section, related to screen opening design and the value of is more compared between the 2D and 3D screen printed cell.

This simply reduces the value of R. For our front printing, the 3D-325 mesh is chosen and a silver grid finger height of about $\approx 12-15\ \mu m$ is achieved for Ag at the top, while for the 2D-325 mesh, this height is only $\approx 8-10\ \mu m$. Hence, this rise in metal height has an effect on the resistance of the cell series through the reduction of Ag front metal resistance [39].

The comparison graph for reflectivity is shown in Fig. 5.17. In the wavelength range of 300–1200 nm, there is a significant marginal decrease in average reflectivity from 14.60% to 14.47%, which is caused by the rise in finger height in the 3D printing.

This slight enhancement of the reflectivity is also caused by the minimization of the marginal shadowing loss on the 3D panel, and this fact contributes to the increase in the cell's short-circuit current as shown in the LIV features in Figs. 5.18 and 5.19.

The uniformity of metal coverage in the 3DSP cell is calculated by comparing the parameters of cell LIV, as shown in Table 5.2.

The substantial increase in FF of the 3DSP cell as opposed to the 2DSP cell clearly indicates a big impact on excellent emitter metallization. This upgrade of I_{sc} and FF results in cell

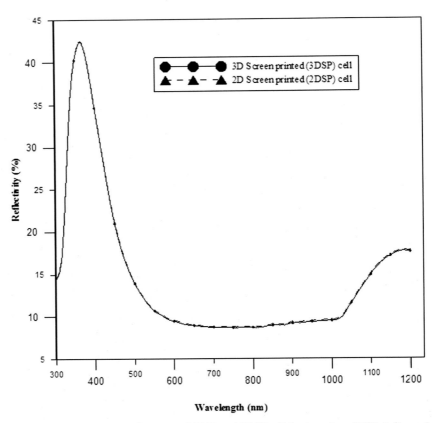

FIGURE 5.17 Variation of surface reflectance of 2DSP and 3DSP cell front surface. 2DSP, 2 dimensional stainless steel; 3DSP, 3 dimensional stainless steel.

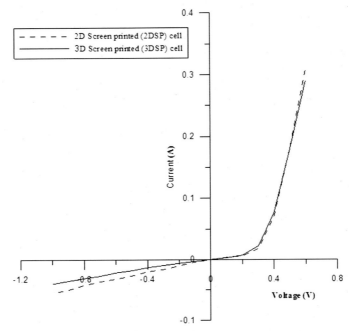

FIGURE 5.18 The illuminated current–voltage characteristic curves of the 2DSP and 3DSP cells.

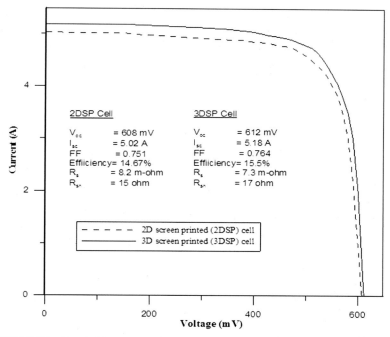

FIGURE 5.19 The dark current–voltage characteristic curves of the 2DSP and 3DSP cells.

efficiency improvement from 14.67% (2DSP) to 15.5% (3DSP) without any significant changes in cell manufacturing techniques. All the experimentally determined screen-printed parameters from Fig. 5.20A and B are listed in Table 5.3. From Table 5.3, we have for the 2DSP cell $l = 116$ mm, $a = 12.05$ μm, and $L = 196$ μm. Using $\rho_{met} = 4.8 \times 10^{-10}$ Ω m [11], the value of R_5 is calculated as 7.99 mΩ. For the 3DSP cell, with the experimental value from Table 5.3, the value of R_5 is calculated as 7.01 mΩ with $l = 118$ mm, $a = 15.03$ μm, $L = 179$ μm, and ρ_{met} remaining the same as in the case of the 2DSP cell. The theoretical value of R_s is calculated by adding R_3 and R_5, and the values for R_s are 8.1 and 7.1 mΩ for the 2DSP and 3DSP cells, respectively. The values of R_s as obtained in Table 5.3 are 8.1 and 7.1 mΩ, which are very close to the experimental values. This indicates that the experimental data for the height and width of fingers determine that the series resistance values of the 3DSP cells are 10%–13% lower compared to the 2DSP cells [40].

TABLE 5.2 Electrical parameters of solar cells fabricated using 2D and 3D screens.

Screen used	V_{oc} (V)	I_{sc} (A)	Fill factor	R_s (mΩ)	R_{sh} (Ω)	Efficiency (η) (%)
2D	0.608	5.02	0.751	8.2	15	14.67
3D	0.612	5.18	0.764	7.3	17	15.50

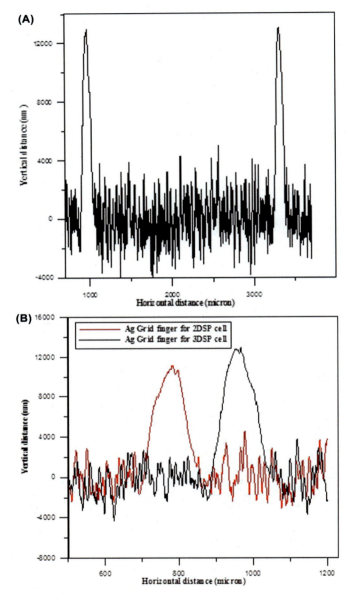

FIGURE 5.20 (A) Telestep profiles of two consecutive Ag grid fingers for a 3DSP cell in a single scan. (B) Telestep profiles of Ag grid fingers of 2DSP as well as 3DSP cells.

TABLE 5.3 2DSP and 3DSP screen-printed parameters as measured by stylus.

Screen used	d (mm)	A (μm)	L (μm)	R_3 (mΩ)	R_5 (mΩ)	R_s (mΩ)
2D	2.35	12.45	196	0.1674	7.99	8.16
3D	2.37	15.03	179	0.1659	7.01	7.18

The difference in L values for 2DSP and 3DSP cells as given in Table 5.3 is 17 μm.

The substantial increase in FF of the 3DSP cell as opposed to the 2DSP cell clearly indicates a big impact on excellent emitter metallization. This upgrade of I_{sc} and FF results in a cell efficiency increase from 14.7% (2DSP) to 16% (3DSP) without any significant difference in cell manufacturing techniques, reflecting its superiority [41,42].

5.5.2.3 Conclusion

The increase in metal heights of the emitter surface layer during 3D screen printing was very high. Lower front silver height reduced both the lack of shadowing and resistance of the metal. This resulted in a small reduction in reflectivity in the wavelength range from 300 to 1200 nm by 0.13%.

Low series resistance value (7.1 mΩ) of the 3DSP cell enhanced cell FF by up to 0.764. In the case of the 3DSP cell there was a reduction of shadowing loss of 0.7%, which was also responsible for the increase in cell efficiency. The cell output rose from 14.7% to 15.5% in a cumulative effect without any other improvement in the daily production line.

References

[1] R.A. Arndt, J.F. Allision, J.G. Haynos, A. Meulenberg, in: Proceedings of 11th IEEE International Specialist Conference, New York, 1975, p. 40.

[2] S.K. Ghandhi, Semiconductor Power Devices, Wiley, New York, 1977.

[3] D.H. Macdonald, A. Cuevas, M.J. Kerr, C. Samundselt, D. Ruby, S. Winderbaum, A. Leo, Sol. Energy 76 (2004) 277–283.

[4] S. Narayanan, Sol. Energy Mater. Sol. Cells 74 (2002) 107–115.

[5] Y. Nishimoto, T. Ishihara, K. Namba, J. Electrochem. Soc. 146 (1999) 457–461.

[6] J. Lindmayer, J.F. Allison, COSMAT Tech Rev 3 (1973) 1–22.

[7] G. Willeke, H. Nussbaumer, H. Bender, E. Bucher, Sol. Energy Mater. Sol. Cells 26 (1992) 345–356.

[8] U. Gangopadhyay, S.K. Dhungel, K. Kim, U. Manna, P.K. Basu, H.J. Kim, B. Karungaran, K. S Leo, J.S. Yow, J. Yi, Semicond. Sci. Technol. 20 (2005) 938–946.

[9] P.K. Basu, H. Dhasmana, N. Udayakumar, D.K. Thakur, Renew. Energy 34 (2009) 2571–2576.

[10] P.K. Basu, H. Dhasmana, D. Varandani, B.R. Mehta, D.K. Thakur, Sol. Energy Mater. Sol. Cells 93 (Oct, 2009) 1743–1748.

[11] E. Vazsonyi, Z. Vertesy, A. Toth, J. Szlufcik, J. Micromech. Microeng. 13 (2003) 165–169.

[12] U. Gangopadhyay, S.K. Dhungel, P.K. Basu, S.K. Dutta, H. Saha, J. Yi, Sol. Energy Mater. Sol. Cells 91 (2007) 285–289.

[13] A. Ebony, Y.H. Cho, M. Hilalii, A. Rohatgi, D. Ruby, Sol. Energy Mater. Sol. Cells 74 (2002) 51.

[14] M. Bohm, E. Urbanski, A.E. Delahoy, Z. Kiss, Sol. Cells 20 (1987) 155.

[15] P.K. Basu, B.C. Chakravarty, S.N. Singh, P. Dutta, R. Kesavan, Sol. Energy Mater. Sol. Cells 43 (1996) 15.

[16] J. Lindmayer, J.F. Allison, COSMAT Tech. Rev. 3 (1973) 1.

[17] J. Dietc, D. Helmreich, E. Sirtal, Crystals: Growth, Properties and Applications, 5, Springer-Verlag, 1981, p. 57.

138   5. Crystalline silicon solar cells

[18] W. Zulehner, D. Huber, Crystals: Growth, Properties and Applications, 8, Springer-Verlag, 1982, p. 92.

[19] J. Czochralski, Z. Phys, Chemie 92 (1977) 219.

[20] W. Zulehner, D. Huber, Crystals: Growth, Properties and Applications, 8, Springer-Verlag, 1982, p. 4.

[21] W. Zulehner, D. Huber, Crystals: Growth, Properties and Applications, 8, Springer-Verlag, 1982, p. 6.

[22] J. Dietl, D. Helmreich, E. Sirtal, Crystals: Growth, Properties and Applications, 5, Springer-Verlag, 1981, p. 73.

[23] J. Dietl, D. Helmreich, E. Sirtal, Crystals: Growth, Properties and Applications, 5, Springer-Verlag, 1981, p. 67.

[24] P. Schatzel, T. Zollner, R. Schindler, A. Eyer, in: Proc. 23rd IEEE PV Spec. Conf., Louisville, Kentucky, 1982, p. 78.

[25] F.V. Wald, Crystals: Growth, Properties and Applications, 5, Springer-Verlag, 1981, p. 157.

[26] R.S. Rouen, P.H. Robinson, J. Electrochem. Soc. 119 (1972) 747.

[27] S.C. Tsai, Proc. IEEE 57 (1969) 1499.

[28] R.B. Fair, S.C. Tsai, J. Electrochem. Soc. 127 (1977) 1107.

[29] K.O. Jeppson, D.J. Anderson, J. Electrochem. Soc. 136 (1986) 397.

[30] S.M. Hu, et al., J. Appl. Phys. 54 (1983) 6912.

[31] J. Knobloch, et al., in: Proc. 9th EC PV Solar Energy Conf., Freiburg, Germany, 1989, p. 777.

[32] A.W. Blakers, et al., in: Proc. 9th EC PV Solar Energy Conf., Freiburg, Germany, 1989, p. 328.

[33] H.F. Wolf, Silicon Semiconductor Data, Pergamon Press, 1976.

[34] E.M. Convell, Proc. IRE 46 (1958) 1281.

[35] R.G. Cosway, et al., J. Electrochemical Soc. 132 (1985) 151.

[36] W. Kern, D.A. Puotinen, RCA Review, Pergamon Press, 1970, p. 187.

[37] A.L. Bogenschutz, Atzpraxis fur Halbleiter, Hanser Verlag, 1967.

[38] A.M.Jaffry, Kluwer Academic Publishers, Dordrecht, 1997, pp.137.

[39] A. Heuberger, Micromechanik, Springer − Verlag, 1989.

[40] J.B. Price, Semiconductor Silicon, 1983, p. 339. Princeton, NJ.

[41] www.teiko-sino.com.

[42] P.K. Basu, R.M. Pujahari, H. Kaur, Devi Singh, D. Varandani, B.R. Mehta, Sol. Energy 84 (2010) 1658−1665.

CHAPTER 6

Introduction to solar energy and its conversion into electrical energy by using dye-sensitized solar cells

N.H. Rased, B. Vengadaesvaran, S.R.S. Raihan, N.A. Rahim

Higher Institution Centre of Excellence (HICoE), UM Power Energy Dedicated Advanced Center (UMPEDAC), Level 4, Wisma R&D, University of Malaya, Jalan Pantai Baharu, Kuala Lumpur, Malaysia

6.1 Introduction

Most countries in this world utilize renewable energy sources such as solar, wind, geothermal, hydro, and many more because they are more cost effective per kWh than fossil fuels. Solar energy is an unlimited renewable energy source that meets the demand for global renewable energy. In addition, a lot of researchers and scientists are keen to utilize solar energy in life applications. Professor M. Grätzel in 1991 [1] invented the low-cost dye-sensitized solar cell (DSSC), which belongs to the third generation of solar cells. DSSCs are attractive due to their simple device fabrication processes in ambient conditions and have reasonable cell efficiency for commercialization.

6.2 Solar cells

The most effective way to utilize solar energy is through photovoltaic device application, which uses photoenergy from the sun and converts it into electrical energy. Photovoltaic technology can be divided into four categories depending on its generation. First-generation solar cells are wafer based, and consist of crystalline silicon (Si), Ga—As, and a III—V single junction. There are single and multicrystalline silicon-based solar cells, which give high-power conversion efficiency. However, Si sources and fabrication processes are costly. Solar

Energy Materials
https://doi.org/10.1016/B978-0-12-823710-6.00006-6

Copyright © 2021 Elsevier Ltd. All rights reserved.

applications across all continents need innovation to increase the power conversion efficiency of solar cells and reduce the fabrication cost.

Second-generation solar cells, also known as thin-film solar cells, consist of p-n heterojunctions such as cadmium telluride, copper indium gallium diselenide, hydrogenated amorphous silicon, and copper zinc tin sulfide. This generation aims to enhance power conversion efficiency with low costs and target large-scale manufacturing processes. The main disadvantage of this generation is the complex fabrication process, which limits research into second-generation solar cells.

Next, organic photovoltaics are the third generation of solar cells, classified as DSSCs, quantum dot photovoltaics, and perovskite, which already fulfill the low-cost requirement. However, the main disadvantages of this generation of solar cells are its instability and low efficiency performance. Related to these limitations, the fundamentals applied to both materials and synthesis procedures need attention. Furthermore, emerging photovoltaics such as hybrid inorganic crystals with polymer matrices are gaining momentum.

From a Science Direct data source search, the total number of articles listed was around 8964, 2211, and 5975 from 2009 up to March 2020 by inserting the terms DSSC, photoanode DSSC, and counter electrode DSSC, respectively. However, the number of articles increased by the end of 2020 which 9524, 4264 and 6303 by inserted term DSSC, photoanode DSSC and counter electrode DSSC respectively, even most of researcher around the world faced a Coronavirus Pandemic COVID 19 started spread from the end February 2020. A different number of articles were listed if the term "photoanode DSSC" was used. The number of articles for each year are presented in Fig. 6.1. As shown in Fig. 6.1, there was increased research in DSSC applications in both theoretical or experimental research. By considering the efforts from researchers due to the large number of articles published until the end of 2020, this chapter proposes to provide the latest review information on fundamental materials and fabrication procedures that can be applied to fabricate electrodes for DSSC applications.

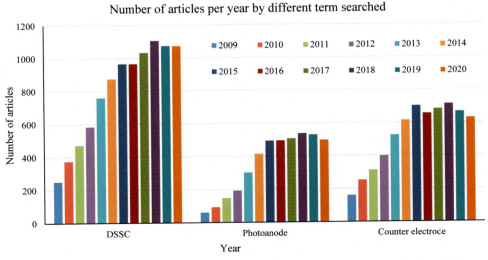

FIGURE 6.1 Number of articles per year sorted by term search. *DSSC*, Dye-sensitized solar cell. *Source: From Science Direct website.*

6.3 Dye-sensitized solar cells

The DSSC was first invented by Michael Grätzel and Brian O'Regan in 1991 [1]. It consists of four main components: photoanode, counter electrode (CE), dye sensitizer, and electrolyte, as shown in Fig. 6.2. The first DSSC device (with a power conversion efficiency of 7.12%) consisted of bare fluorine-doped tin oxide (FTO) as a CE, 15 nm nanoparticulate TiO_2 on FTO as a photoanode, trimeric ruthenium complex as a dye, and an electrolyte [1]. Another impressive power conversion efficiency of 12% was recorded in 2011 [2] and 13% in 2014 [3]. In 2015, a DSSC device with the highest power conversion efficiency of 14.3% was reported [4].

The CE acts as a catalyst for the redox couple reaction and accepts an electron from the external circuit. Two requirements for CE materials are high catalytic activity and electronic conductivity. There are various types of catalysts that have been used as a CE such as Pt, composites of inorganic materials, and recently carbon-based materials. The excellent flexibility and optoelectronic and electrocatalytic properties of graphene make it a potential material for photoanodes and CEs.

Common materials used to form photoanodes are semiconductor-based materials such as TiO_2, ZnO, and SnO_2. Nowadays, carbon-based materials have caught the attention of other researchers due to their low cost and easy-to-alter nanostructure. Different nanostructures exist due their synthesis procedures and composite materials such as zero-, first-, second-, and third-dimension structures noted as 0D, 1D, 2D, and 3D, respectively. Fundamentally, a high-efficiency DSSC can be obtained from the large surface-specific area of photoanodes due to better carrier transportation, high-density dye loading, and scattering of light to enhance the absorption of the sensitizer.

Besides dye, electrolyte also plays an important role in building efficient DSSCs. There are three types of electrolyte, depending on their state: liquid, quasi-solid-state, and solid electrolyte. Liquid electrolytes suffer from leakage, photodegradation of the dye, volatilization of the electrolyte, and corrosion oof the CE, which can decrease the power conversion efficiency of the DSSC. To counter this liquid electrolyte problem, quasi-solid and solid electrolytes have been researched [5]. Quasi-solid-state electrolytes have the ability to diffuse within nanoporous structures, simple processing techniques, long-term stability, and are cost-effective compared to solid electrolytes [6].

6.4 Working principle of DSSCs

In this section, the DSSC working principle will be explained. Generally, the conversion of light energy into electrical energy takes place when electrons from photoexcited dye

FIGURE 6.2 Illustration of four main components of a dye-sensitized solar cell.

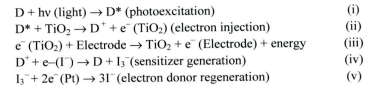

$$D + h\nu \text{ (light)} \rightarrow D^* \text{ (photoexcitation)} \quad \text{(i)}$$
$$D^* + TiO_2 \rightarrow D^+ + e^- \text{ (TiO}_2\text{) (electron injection)} \quad \text{(ii)}$$
$$e^- \text{ (TiO}_2\text{)} + \text{Electrode} \rightarrow TiO_2 + e^- \text{ (Electrode)} + \text{energy} \quad \text{(iii)}$$
$$D^+ + e-(I^-) \rightarrow D + I_3^- \text{ (sensitizer generation)} \quad \text{(iv)}$$
$$I_3^- + 2e^- \text{ (Pt)} \rightarrow 3I^- \text{ (electron donor regeneration)} \quad \text{(v)}$$

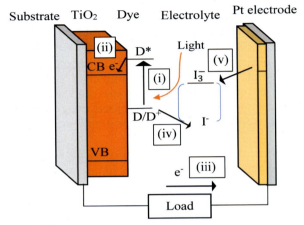

FIGURE 6.3 Illustration of working principle of a dye-sensitized solar cell. *CB*, Conduction band; *VB*, valence band.

molecules are injected into the conduction band (CB) of a semiconductor layer on the DSSC substrate. In these examples, titanium dioxide (TiO_2) materials were selected. Next, the electrons enter the external circuit and return to the deexcited ionized dye molecules through the redox mediator. The photoelectrochemical cycle of the DSSC starts upon illumination of light through the photoelectrode, and dye molecules (*D*) absorb light or photons (*hν*) forming electron—hole pairs (*D**) in the dye molecule, as shown in Eq. (6.1). Next, the electrons are released from the excited dye molecules and injected into the CB of TiO_2 resulting in a hole in the dye molecule, as mentioned in Eq. (6.2). Following Eq. (6.3), the flow of current produced as a result of the injected electrons in the CB is transported across the TiO_2 by diffusion toward the transparent conducting oxide substrate. Next, electrons from the redox reaction replace the holes in dye molecules, as in Eq. (6.4). The dye is subsequently regenerated from its oxidized state by electron transfer from iodide (I^-) ions in the electrolyte containing the (I^-/I_3^-) redox couple. Next, Eq. (6.5) shows the chemical equation for triiodide (I_3^-) ions formed in this regeneration process diffusing to the Pt-coated electrode, where they are reduced rapidly to I^-. The circuit is completed via electron migration through the external load, as shown in Fig. 6.3. To have a unit of charge collection efficiency at the photoelectrode back contact, the photoinjected electrons should escape from any recombination process.

6.5 Anatomy of DSSCs

This third generation of solar cell imply leaf anatomy in approaching to global knowledge either rural or urban area [7]. The working principle of the DSSC device is based on the photosynthesis concept and it is important to get the highest percentage of solar energy from incident photons to increase DSSC energy conversion. Interest in this type of solar cell is due to the highest energy efficiency obtained under indirect weak illumination. Hence, leaf part of plant were selected by Yun et al. [7] as anatomy features for DSSC.

Leaf anatomy was selected due to its principles of photosynthesis, including electron transfer and energy transfer [7—9]. The performance of DSSCs is also due to the dye used as a sensitizer. Various components of a plant such as the flower petals, leaves, and bark have been tested as pigment sensitizers together with other parameters to build up DSSCs. DSSC parallels photosynthesis in the use of a dye as the light harvester to produce excited electrons: TiO_2 replaces carbon dioxide as the electron acceptor, iodide/triiodide replaces water and oxygen as the electron donor and oxidation product, and a multilayer structure (similar to the thylakoid membrane) enhances both light absorption and electron collection efficiency.

Currently, leaf anatomy structure orientation was suggested by Yun et al. [7]. The light-trapping layer was created on the solar cell surface and microscale patterned the photoanode, which changed the light distribution within the cell as well as within the photoanode. The higher the use of incident photon, the higher the energy conversion efficiency will be. Yun et al. [7] mimicked the leaf array structure by inclining 3D arrays of DSSC submodules to maximize electricity production. Leaf anatomy structure, especially epidermis, and palisade structure, as well as leaf array structure, was applied to DSSCs by taking into consideration light distribution with oblique and vertical incident light.

6.6 Materials used as electrodes of DSSCs

6.6.1 Photoanode

One of the focuses today in DSSCs is to select photoanodes that are low cost and can reduce recombination reaction, increase dye pickup, scatter light, and transport charge well during DSSC operation. Photoanodes should transfer electrons quickly from dye to external circuit and there should be rapid electron injection from the dye. To increase dye pickup capacity, a high surface area is required. In addition, to enhance the diffusion of dye and electrolyte, the pore size should be carefully controlled and synthesized. In the same way, photoanodes must have high resistance to photocorrosion, a good electron acceptor, be able to absorb/scatter sunlight to increase the function of dye and the interface contact with the dye molecules, and the conductive layer on the substrate should be optimum [10]. Some photoanode materials reported by researchers and classified for this chapter are shown in Fig. 6.4.

6.6.1.1 Titanium dioxide-based materials

The most common semiconductor-based material used as a photoanode is TiO_2 due to its biocompatibility, low cost, abundance, nontoxicity, stable n-type semiconductor material, low intrinsic conductivity, and small size of the nanocrystalline particles. For instance, TiO_2 exists in three crystalline polymorphs: anatase, rutile, and brookite. The most common and stable polymorph of TiO_2 is rutile [11]. Both rutile and anatase have a tetragonal crystal structure, whereas brookite has an orthorhombic structure. The brookite phase is hard to obtain due to its metastable crystal structure (orthorhombic structure). Each anatase, rutile, and brookite TiO_2 phase has a bandgap of 3.2, 3.0, and 3.4 eV, respectively. Fundamentally,

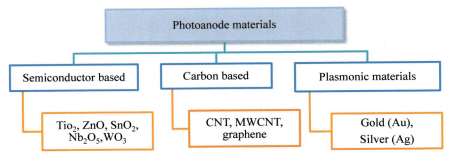

FIGURE 6.4 Summary of photoanode materials class. *CNT*, Carbon nanotube; *MWCNT*, multiwalled carbon nanotube.

anatase TiO_2 is more chemically stable compared to the rutile phase due to anatase's wider bandgap. Among others, TiO_2 (anatase) with its wide bandgap has advantages in forming a photoanode due to its easy availability, nontoxicity, good stability, and compatible optical and electronic properties.

Also, TiO_2 material gives good electrostatic shielding of the injected electrons from the oxidized dye molecule attached to the TiO_2 surface, thus reducing their recombination before reduction of the dye by the redox electrolyte. A direct bandgap of 3.66–3.88 eV and an indirect bandgap of 3.44–3.68 eV were obtained [12]. However, the capabilities of the TiO_2 anatase phase to absorb the solar spectrum in the range of ultraviolet or near-ultraviolet radiation can only capture about 4% of solar light. Thus dye sensitization of wide bandgap semiconductor surfaces by anchored dyes has provided a successful solution to extending the absorption range of the cells to the long wavelength region. Semiconductor materials act as sensitizers to facilitate light-reduced redox processes because of their conductive electronic structure, referred to as valence band (VB) and CB. In the quantum physics theory, a photon energy that exceeds or matches the energy gap of a semiconductor can excite an electron to the CB leaving a hole with a positive charge at the VB. These charges can either be transferred to the external circuit to provide electrical current or work as a catalyst to a certain chemical reaction [13].

DSSC performance can be enhanced by controlling the shapes and crystal facets of TiO_2. For rutile TiO_2, the (011) surface has a higher reactivity in catalytic reactions than other surfaces. Moreover, the (001) facet for anatase is more reactive than the (101) facet due to its being more thermodynamically stable. Equally important, the morphologies of TiO_2 also play an important role in enhancing DSSC performance. Highly crystalline TiO_2 nanoparticles (NPs) are beneficial because they provide a high surface area for better dye loading. However, electron mobility is poorly affected by the grain boundaries between the TiO_2 nanocrystals.

TiO_2 NPs are very small, have high transparency, and most of the incident photons are transmitted from the device without any work done [10]. TiO_2 NPs (10–20 nm), which have a specific surface area of 50–100 m^2/g, have been used in most commercial DSSCs; however, these TiO_2 NP layers show low light-scattering ability due to smaller particle size compared to the wavelength of incident light. Problems related to solar energy conversion efficiency to electrical energy of DSSC technology have been reported by Ahmad et al. [10].

The disadvantage of TiO_2 NP-based networks is random transportation of photocarriers resulting in the scattering of photocarriers caused by recombination at the grain boundaries. Next, low electron diffusion coefficients from surface states, defects, and grain boundaries act as electron trapping sites, which results in recombination of photocarriers and hence decreases the electron collection efficiency at the back contact. Therefore 1D structures such as nanowires (NWs), nanotubes, and nanorods enhance electron mobility and light harvesting via the light-scattering effect as well as reduce the recombination process in DSSCs. A comparative study of the electrodynamics between nanotubes and NP films showed that the electron transportation times were the same, whereas recombination was significantly slower by ~ 10 times in the nanotube films, leading to a significantly improved photocarrier collection efficiency for the nanotube photoanodes [11].

Moreover, Zubaidah et al. [14] interpreted the reduction of size in the TiO_2-reduced graphene oxide (rGO) layer due to the electron attraction between Ti^{4+} and the rGO surface. The DSSC efficiency of the TiO_2-rGO nanocomposite was 4.18%, which was higher than the pure TiO_2 sample (2.21%) because of the high photocatalytic activity during irradiation. Also, the author highlighted the lower charge recombination and faster charge transfer at the interface of the TiO_2-rGO DSSC. The effective path due to the 2D graphene bridge contributes to effective photogenerated electron separation and transport across the photoanode [14].

Besides carbon material, incorporation with metal materials also will alter the TiO_2/multi-walled carbon nanotube (MWCNT) properties. This is shown in chromium, Cr(III), doping in TiO_2/MWCNT nanocomposites, which results in the enhancement of physicochemical properties in relation to photovoltaic studies [15]. Average lifetime of the electrons in the excited state increases and interfacial charge-transfer resistance decreases after the doping of Cr(III) ions into the TiO_2 host lattice. The ternary composite-based DSSC shows the highest power conversion efficiency of 7.69%, which is higher than the undoped TiO_2/MWCNT-based DSSC (6.18%) [15].

Kanimozhi et al. [16] synthesized pure and cobalt-doped zinc oxide (Co-ZnO) 1D mesoporous nanofibrous photoanodes of DSSCs. The 5 wt% Co-doped ZnO nanofibers (NFs) gave a photocurrent density of 5.36 mA/cm^2, a fill factor of 0.731, and a power conversion efficiency of 2.97%, which had higher efficiency than using pure zinc oxide (1.63%) under AM1.5 illumination. The cobalt ions doped inside ZnO NFs obstructed the recombination of the excited electrons with the dye or the electrolyte [16]. Conductive ionic liquid (BVImI + LiI) was dropped onto TiO_2 paste substrate at a low temperature to modify the surface and CB level of the photoanode. The probable outcome was to increase the electron injection rate into TiO_2. As a result, current density value and power conversion efficiency of DSSCs were 12.46 mA/cm^2 and 4.84%, respectively, which was higher than the normal photoanode (11.56 mA/cm^2 and 4.35%) [17].

2D TiO_2 nanostructures. On other modified materials, the 0.5% TiO_2 on rGO sheets with a silver NP (rGO-Ag) photoanode showed 9.15% DSSC power conversion efficiency under solar irradiation 1.5AM [18]. The morphology of graphite predominantly appeared as uneven-sized flakes, whereas GO showed irregular flake-like structures. One fundamental point, after reduction by hydrazine hydrate, rGO agglomeration formed because oxygenic groups on the GO surface were removed leading to a decrease in the distance between layers of the GO. Moreover, irregularly shaped NPs with an average size of 40 nm of TiO_2 were obtained. The Ag particles became smaller on the surface of rGO due to electrostatic properties between the positively charged Ag^+ and negatively charged rGO [18]. This showed that the doping materials can change the nanostructure of TiO_2.

3D nanostructures. 3D mesoporous nano/microspheres with larger surface area possess better light-scattering properties [19]. Additionally, 3D nanostructures have a high surface area, high optical absorption through the scattering of light, and good electron transport properties [20]. 3D nanostructures can be illustrated by branched nanorods, bridge nanotubes, dendritic hollow structures, and hedgehog nanostructures. This 3D nanostructure can be obtained by modifying with other materials as well as by the synthesis method. The semiconductor electrode is usually a layer of nanocrystalline TiO_2, a thin film deposited on the conducting glass film with a thickness of ca. 5—30 mm, which plays an important role in both exciton dissertation and the electron transfer process. The porosity and morphology of the TiO_2 layer are dominant factors that determine the amount of dye molecules absorbed on its surface, which can provide an enormous area of reaction sites for the monolayer dye molecules to harvest incident light [21].

Mesoporous and bimodal TiO_2 thin films with different thicknesses will affect the DSSC's performance [23]. The mesoporous TiO_2 film was characterized by the adsorption of n-pentane. The 1-μm-thick Pluronic P123-templated mesoporous film was grown via layer-by-layer (three-layer) deposition, and showed a DSSC conversion efficiency of about 50% compared to traditional film DSSC of the same thickness made from randomly oriented anatase nanocrystals [22]. Another interested researcher produced mesoporous films containing small mesopores of 8—10 nm, which resulted from the use of Pluronic P123 surfactant (soft), which gave a high surface area. Unique bimodal porous structures comprising 8—10 nm mesopores and 60—70 nm macropores resulted from the use of P123 and 130 nm polystyrene beads. This author suggested the enhancement of electrolyte pore infiltration and light harvesting. Next, dual-templated TiO_2 films have lower charge-transfer resistance than soft TiO_2 electrodes as well as higher DSSC efficiency. The addition of ionic liquids to polymeric ionic liquids results in low viscosity, increased ionic conductivity, increased $I_3^-I_3$ diffusion rate, and enhanced photovoltaic performance in both soft- and dual-templated photoanodes [23].

Moreover, Shaban [24] reported the preparation and performance analysis of DSSCs using different common substrates such as Ti foil as flexible photoanode. Surface morphology of the treated Ti foil with hydrogen peroxide (H_2O_2) showed the formation of a TiO_2 nanostructure, which resulted in significant attachment of the TiO_2 layer to the Ti foil. The fabricated flexible photoanode DSSC had a power conversion efficiency of 1.00% under 1.5A.M solar radiation when back illuminated. In comparison, a DSSC with solid glass photoanode gave a power conversion efficiency of 0.53% (back illuminated) and 2.22% (front illuminated) [24].

TiO_2 has a problem with its light absorption capability because it can absorb only the ultraviolet region of the spectrum and has low electron mobility [10]. To improve visible light absorption ability and electron mobility, the bandgap of materials should be taken into consideration. Ion doping such as with magnesium (Mg) [25] and lanthanum (La) [25] is an alternative to adjust the position of both the CB and VB for photocatalysis applications. Certain materials have recently been introduced by a number of researchers [19] to reduce recombination resistance, improve light absorption capacity, enhance electron mobility, and prolong electron lifetime in photoanodes. Doping with an appropriate cation/anion alters the bandgap and can modulate its electrical and optical properties. The presence of Mg^{2+}

ions at Ti^{4+} sites in the TiO_2 lattice and the presence of La ions at the crystallite surfaces and crystallite interstitials increased generated oxygen vacancies and reduced particle sizes, respectively, compared to pure TiO_2 nanocrystalline particles [25].

Different nanostructures and materials of the photoanode can be modified by introducing other materials such as metal NPs, metal atoms, carbon materials, and metal oxides. These modified photoanodes are expected have high surface area, good electron transportation characteristics, and scattering effects.

6.6.1.2 Metal oxide/semiconductor materials

Various other metal oxides such as ZnO, SnO_2, Nb_2O_5, $SrTiO_3$, Zn_2SnO_4, and WO_3 have been investigated. ZnO showed promising results comparable to TiO_2 but although ZnO possesses better electron mobility than TiO_2, its efficiency is less compared to DSSCs manufactured using TiO_2. This reduced efficiency is due to low dye pickup and reduced stability of ZnO in an acidic environment. Metal oxide semiconducting materials act as electron acceptors and electronic conduction paths to facilitate photoexcited electrons because of their conductive electronic structure, referred to as VB and CB [21]. Moreover, incorporation with carbon materials such as carbon nanotubes (CNTs) and graphene in semiconductor photoanodes will provide transport for photogenerated electrons to enhance DSSC performance [19].

The first semiconductor photoanode DSSC was fabricated by using ZnO as reported by Tsubomura [26]. Next, O'Regan and Grätzel [1] reported higher power conversion efficiency DSSCs by using of mesoporous TiO_2 NPs. Later, TiO_2 became the most popular metal oxide for DSSCs, mainly due to its better photostability and performance, which is better than that of other metal oxides such as ZnO and SnO_2. The CB of SnO_2 with a high band-gap energy of 3.8 eV is more positive by about 500 mV than TiO_2. Therefore the rate of charge injection by SnO_2 is higher than ZnO and TiO_2.

A device that used a ZnO nanofibrous photoanode gave a power conversion efficiency of about 1.7% and a current density of 8.45 mA/cm^2, which is higher than the reported value for a fibrous photoelectrode since 2012 [27]. Based on fundamental theory, ZnO has dissimilar parity because the VB is composed of filled 3d orbitals. The CB consists of hybridized s-p orbitals and leads to a decrease in the probability of recombination. Research into the electron transport properties of ZnO and TiO_2 photoanodes for DSSCs revealed the same optimal thickness of 5 nm and showed equal efficiency of 4.4% [11]. DSSC efficiency can be increased using ZnO due to its high mobility of electrons. Otherwise, TiO_2 can increase DSSC efficiency by factors such as higher dye loading, low recombination rate, and fast electron injection. However, increasing the thickness of the layer will decrease the power conversion efficiency due to decreased surface area and increased optical loss.

TiO_2 has a higher adsorption rate of dye due to faster electron injection from the dye (a few picoseconds) compared to ZnO and SnO_2 (10—100 ps). Injection dynamics depend on the dye-binding models and density of states available for injection, which vary from site to site. The effective mass of an electron in TiO_2 (5—10 me) is higher than in ZnO ($\sim$0.3 me) and SnO_2 ($\sim$0.3 me). Therefore the available density of states in TiO_2 is almost two orders of magnitude higher than ZnO and SnO_2. Hence, faster electron injection takes place in TiO_2 than in both ZnO and SnO_2. The bulk conductivities of ZnO, TiO_2, and SnO_2 are in the order of $TiO_2 < ZnO < SnO_2$ [11].

6.6.1.3 Plasmonic-based materials

Plasmonic materials have attracted attention because their surfaces can localize incident light and extend the optical path length, which enhances DSSC performance. Examples of plasmonic materials are silica-coated Au nanocubes, Au and Ag core/shells, Zn, and Cu. The idea is that Au NPs will improve electron transfer in conjunction with their plasmonic and scattering effects.

For example, $SiO_2@Ag@TiO_2$ nanostructures were prepared, in which the coating of SiO_2 prevented the corrosion of Ag NPs by I^-/I_3 scattering and the surface plasmon effect. This led to an improvement in electrolyte and enabled enhanced light absorption. However, the effect of plasmonic materials incorporated with photoanode DSSCs is still controversial and systematic investigations into their precise role are required for the future [19].

Moreover, Ag NWs with a diameter of 60×10^{90} nm and a length of 1×10^2 mm were synthesized via the polyol reduction method. TiO_2 NFs with diameter 80–120 nm were prepared by electrospinning. The DSSC with a trilayer photoanode made with a composite of TiO_2 P25, Ag NWs, and TiO_2 NFs sandwiched between two TiO_2 P25 layers exhibited a power conversion efficiency of 9.74%, and open-circuit voltage (V_{oc}), short-circuit current density (J_{sc}), and fill factor of 727.4 mV, 19.8 mA/cm^2, and 67.6%, respectively, under an irradiance of 100 mW/cm^2. The efficiency of the reference DSSC with a TiO_2 P25/P25/P25 trilayer photoanode of the same thickness was found to be 6.69%. Efficiency enhancement of the DSSC with the Ag NW- and TiO_2 NF-incorporated composite trilayer photoanode compared to the reference DSSC was 45.6%. This evidence is due to the enhancement in short-circuit photocurrent density (J_{sc}) by the localized surface plasmon resonance effect of Ag NWs and the presence of TiO_2 NFs, which increased the light harvesting [28].

6.6.1.4 Carbon-based materials

The highest graphene sheet content of 83.86% is found in CH_3COOH pretreatment followed by a 1500°C heat treatment of monocotyledonous biochar materials, and its conductivity was measured at 84.69 S/cm [29]. Scanning electron microscopy images of graphene sheet content (>80%) carbon materials (GSCCM) samples such as surface erosion and the formation of graphite spherical grains are observed for biochar materials after acid pretreatment and heat treatment, respectively. The temperature necessary to generate many graphite spherical grains was found to begin at 800°C. The size of graphite spherical grains will increase when the temperate increases to 1500°C [29].

The disintegration of cellulose during the acid pretreatment process and the formation of graphitic crystals and oxide layers (2θ degrees = 29 degrees) were found in the GSCCM samples [29]. On the other hand, new ball-milled biochar (BM-biochar) was produced through the ball milling of pristine biochar derived from different biomass materials. BM-biochar had smaller particle size (140–250 nm compared to 0.5–1 mm for unmilled biochar), greater stability, and more oxygen-containing functional groups (2.2–4.4 mmol/g compared to 0.8–2.9 for unmilled biochar) than pristine biochar. The best electrocatalytic activity for the reduction of $Fe(CN)_6^{3-}$ due to a special structure facilitates electron transfer and reduces interface resistance [30].

6.6.2 Counter electrode

A CE is one of the important components for enhancing the energy conversion efficiency of a DSSC. Commonly, the CE catalyzes the redox reduction of the electrolyte. An important parameter that must be considered in CE materials selection is the catalytic rate. The reduction rate at the CE decreases without the presence of a catalyst at the surface of the glass, metal, or plastic substrate [31]. In fact, the role of a CE of a DSSC is to collect electrons from the external circuit and catalyze the reduction of redox electrolyte or transporting holes in the electrolyte. The primary requirements for a CE are high conductivity for charge transport, good electrocatalytic activity for reducing the redox couple, and excellent stability. Pt has high electrocatalytic activity for the reduction of redox couples in liquid electrolytes compared to Au and Ag, which are effective for hole transfer in solid-state electrolytes. However, Pt, Au, and Ag are expensive and have a corrosion effect in liquid electrolyte. Thus several alternative materials have been researched to replace Pt for CEs, as shown in Fig. 6.5. The use of a suitable catalyst helps in the reduction of the charge-transfer resistance at the CE, and thus helps to push the reaction in a forward direction [4].

6.6.2.1 Platinum-based materials

Pt is commonly used as a catalyst of CEs for redox mediators in electrolyte solution. However, Pt is a nonsustainable and expensive material for long-term use. At the same time, research into doping materials into Pt to form a composite Pt CE has also been performed. The expectation is to merge other materials doped into Pt to improve power conversion efficiency.

Bolhan et al. [32] researched the difference between the effect of Pt, rGO, and composite Pt-rGO using the stacked layer method. Pt and Pt-rGO thin films exhibit rougher surfaces compared to rGO thin film and Pt, with 10 μL rGO recording the highest current density at -3.075 mA/cm^2, indicating high catalytic activity at the CE [32] comparable to Pt CE material. Furthermore, Pattarith and Areerob [18] used doping material in Bolhan et al.'s [32] base materials and formed a silver-reduced graphene oxide (Ag-rGO-Pt) composite. The Ag-rGO-Pt CE recorded the highest power conversion efficiency of 9.15% under (1.5AM) solar irradiation. Interfacial charge transfer increased due to high electrocatalytic activity for the reduction of triiodide to iodide redox after being doped with Ag NPs [18].

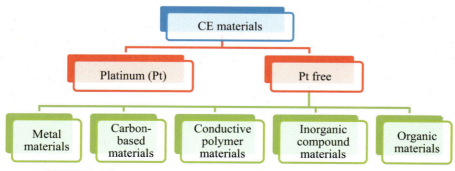

FIGURE 6.5 The summary flowchart of counter electrode (CE) materials.

6.6.2.2 Metal nanoparticle materials

Rather than use Pt material as the CE in a DSSC application, other metal materials have also been tested. A DSSC using an Au nanostructure as a CE for a DSSC gave 4.9% compared to Pt-based CEs at 4.8%, as recorded by Gullace et al. [33].

6.6.2.3 Carbon-based materials

As an alternative to Pt-based CEs, carbon-based materials have been used by DSSC researchers. The optical and electrochemical properties of carbon-based CE DSSCs can be affected by the type, concentration, and composition of the precursor used. The optimized low cost and high photoelectric conversion efficiency of carbon CE exhibits high transparency and enough catalytic activity for reduction [31]. Previous research has shown that carbon-based nanomaterials such as graphene and CNTs are promising as highly efficient CE materials.

Due to low cost, good electrocatalytic activity, high electrical conductivity, high thermal stability, and corrosion resistance, carbon materials such as carbon black, porous carbon, CNTs, and graphene have been used as CEs to enhance DSSC performance. Thus combining two carbon materials, for example, porous carbon/CNTs and CNT/graphene nanoribbons, can further amplify the electrocatalytic activity of CEs.

Carbon black

Carbon black is a potential material for DSSC CEs due to its crystallinity, electrocatalytic activity toward the I^-/I_3^- redox species, high surface area-to-volume ratio, and excellent conductivity. Previously, research has been conducted on the effects of layer thickness on the resistance and catalysis of CE materials. Murakami and Grätzel [34] found that when the thickness of carbon film increased (14.47 μm), the fill factor of DSSC increased, and charge transfer resistance of the CE decreased, giving 9.1% DSSC efficiency. This relationship is due to the high catalytic activity resulting from thick carbon black CEs. Moreover, carbon black shows around $2.96 \, \Omega/cm^2$ charge-transfer resistance with a high surface area [34]. This fundamental relation results in the potential of carbon black as a low-cost and corrosion-free CE for DSSC applications.

To fabricate a low-cost CE, a low-cost catalyst was applied to a low-cost substrate such as stainless steel or nickel. Metal substrate not prefer liquid state material such as Pt tri-iodide solution to be coated on the metal substrate to form counter electrode due to the risk of corrosion [34,34a] on the electrode's surface. Because of the corrosion problem of Pt counter electrode, carbon materials promise to fabricate anticorrosion substrate. Besides, carbon counter electrode gave large fill factor, low sheet resistance and high DSSC efficiency. For example, coating a carbon catalyst on stainless-steel substrate SUS-316 and SUS-304 gave the DSSC 9.15% and 8.86% efficiency, respectively [34]. Next, Murakami and Grätzel [34] showed that carbon black with low crystallinity will be more active than highly orientated carbon materials such as graphite and CNTs [34].

Research by Xu et al. [35] showed that glassy carbon (GC) with controlled crystallinity was used as catalytic material for a CE in a DSSC due to the fundamental understanding of catalytic activity for triiodide reduction. Low crystallinity of the GC provides high catalytic activity for the triiodide reduction and increased the power conversion efficiency of the DSSC. Active sites in the GC for catalysis were located at the edges of graphene stacks.

6.6 Materials used as electrodes of DSSCs

The increased graphene stacks and active sites in the GC were obtained because catalytic activity was enhanced [35].

Mesoporous carbon

The disadvantages of the catalytic activity of carbon compared with Pt can be solved by increasing the active surface area of the catalytic layer using a porous structure. Interestingly, among all carbon materials, mesoporous carbon materials have caught most researchers' attention due to narrow pore diameter and large internal surface area. The common method used to synthesize a mesoporous carbon structure is by the templated synthesis method (nanocasting), which gives interconnected pores and is homogeneous.

Porous materials are solids containing pores and voids, and have traditionally been categorized into mesoporous, and microporous materials. Microporous materials contain micropores, which have dimensions in the range from 2 to 50 nm, and microporous materials have pores greater than 50 nm. These definitions were formalized in the International Union of Pure and Applied Chemistry guidelines in 1985. Following the definition of nanoscale, which refers to a size range of approximately 1—100 nm, this terminology refers to materials with pores less than 100 nm in dimension.

Mesoporous carbon using ethylene glycol as a pore-forming agent was prepared by the carbonization of phenol resin and paste coated on a CE of a DSSC at low temperature. Thus a DSSC using mesoporous carbon reached a maximum of 5.65% efficiency for 0.6 g with 0.1 mL of surfactant, which was 46.5% higher than a DSSC using activated carbon as a catalyst and 95.4% of the value obtained using a DSSC with a Pt-CE under the same conditions. The enhanced DSSC performance improvement of the surfactant-modified mesoporous carbon CE was due to a uniform mesoporous carbon film. By adding the surfactant, the large average mesopore size was suitable for mass transport, and provided a high-surface-area GC structure that had a high content of active sites. An average pore size of 34.6 nm and apparent porosity of 44.9% were achieved [36].

On the other hand, there is another problem caused by oxygen deficiency in the TiO_2 structure that can create electron—hole pairs, and oxidizing holes can react with dye or be scavenged by iodide ions. These deficiencies can shorten the lifetime of a DSSC. Thus Wang et al. [37] suggested a CE for a DSSC by using nitrogen-doped mesoporous carbon. The DSSC achieved a power conversion efficiency of 7.02% with the liquid redox couple, whereas the power conversion efficiency of Pt CE was 7.26% under the same test conditions [37]. The nitrogen-doped mesoporous carbon samples had large pore volume, high surface area, and a bimodal mesopore structure at ∼3.5 and ∼16 nm (maximum) based on nitrogen adsorption analysis. As a result, electrochemical impedance spectroscopy tests revealed a low charge-transfer resistance of $1.45\ cm^2$ for nitrogen-doped mesoporous carbon electrode in the iodide/triiodide redox electrolyte. To summarize, excellent electrocatalytic activity for the iodide/triiodide redox reaction of nitrogen-doped mesoporous carbon electrode can be related to a combination of nitrogen doping, bimodal mesopore structure, and high surface area [37].

Much more interestingly, in 2018, 20 types of biowaste-derived carbon materials (BCMs) synthesized by facile one-step pyrolysis were developed as CEs of DSSCs. The biowastes selected for the BCM preparation were divided into three groups. The first group was seven leaves consisting of pine needles, camphor, palm, maple, poplar, Chinese fir, and red afterwood (*Loropetalum chinense* var. *rubrum*). The second group was 11 woods from weeping

II. Photovoltaic materials and devices

willow, phoenix, camphor, Chinese fir, maple, peach, poplar, cypress, tea-oil camellia, orange, and chinaberry. The third group consisted of two papers: filter paper and facial tissue. Fibrous BCM fabricated from the papers showed the highest efficiencies around 4.70% due to the superior catalytic activity and faster electron transportation for triiodide (I^-_3) reduction enhanced by the unique structural and morphological characteristics of the paper-derived BCM. The unique characteristics of these BCMs were hierarchical fibrous carbon skeletons at nanometer and micron scales, abundant exposed microcrystal edges and defects, rough surfaces at nanoscale with rich burrs and convex microstructures, and oxygen-containing surface functional groups. Furthermore, the efficiency of the DSSC employing the paper-derived BCM was equal to about 88% of the devices using the mesoporous carbon synthesized from the phenolic resin mixtures based on polymerization-induced phase separation DSSC efficiency from BCM prepared from woods and leaves in the ranges of 1.23%–1.91% and 1.07%–1.85%, respectively, which were higher than DSSCs from graphite at 0.77% [38]. In the same year, DSSCs consisting of carbon CEs from sucrose gave a power conversion efficiency of only 0.041% compared to carbon CEs from table sugar that exhibited 3.239%, which was almost equivalent to the 4.024% efficiency of Pt CE DSSCs [39]. Accordingly, carbon-based materials also have potential as CEs for DSSC applications.

Graphene

Other than carbon polymorphs, graphene-based materials have also captured much attention in Pt-free DSSCs. A single, atomic, thick layer of the mineral graphite is called graphene. Graphene is a compound that has a 2D sheet of atomically thick sp2 carbon atoms arranged in a hexagonal "honeycomb" lattice structure. Graphene has been demonstrated to be a promising CE material for DSSCs due to its excellent conductivity, high electrocatalytic activity, and remarkable transparency over the entire solar spectrum [31].

A submicrometer-sized, colloidal graphite (CG), paste-conducting electrode was doctor bladed to replace transparent conducting oxide (TCO) electrodes and as a catalytic material to replace Pt in DSSCs. The 9-μm-thick CG film has low resistivity. Cyclic voltammetry and electrochemical impedance spectroscopy studies clearly showed a decrease in the charge-transfer resistance with an increase in the thickness of the graphite layer from 3 to 9 μm. Under 1 sun illumination, DSSCs with CG as a catalyst on FTO TCO showed an energy conversion efficiency greater than 6.0% compared to Pt DSSCs. However, DSSCs with CG CE on TCO-free bare glass showed an energy conversion efficiency greater than 5.0%, which illustrated that the graphite layer could be used as a conducting layer or as a catalytic layer. The advantages of CG are that it can act dually as a substrate as well as an electrocatalyst, therefore successfully replacing both TCO and Pt [40]. Veerappan et al. [40] proved CG as an efficient CE for triiodide reduction in DSSCs.

There has been little research done on graphene-based composite materials such as graphene-polymer composite, graphene-metal composite, and graphene-carbon composite. Graphene-polymer composite acts as an effective CE. This combination is ideal due to the high catalytic activity of graphene and polymers as a conducting support material.

For a graphene-metal composite, the graphene material is incorporated with metal material such as nickel (Ni) and cobalt (Co) to form CEs for DSSCs. The main advantages of this type of composite CE are the increase in electron-transfer rate at the CE and its ability to act both as an active site for the electrocatalytic process and as a spacer between the graphene sheets, which speeds up the diffusion of the electrolyte; however, the graphene provides a fast diffusion pathway for the electrolyte by creating a brilliant electrode/electrolyte, which improves the electron-transfer rate at the electrolyte/CE interface. In addition, Au NPs with a radius ranging from 4 to 18 nm (mostly 11 nm) were stably and uniformly hybridized on the surface of rGO by dry plasma reduction. Increasing the number of Au NP-rGO layers decreased the transmittance and sheet resistance of the Au NP-rGO nanohybrid. A developed electrode based on the Au NP-rGO nanohybrid showed high electrochemical catalytic activity and high conductivity in comparison with the Au NP and GO electrodes [41].

Carbon nanotubes

CNT electrodes consist of a mesoporous and crystalline network of CNTs and can be used as free-standing, unipolar, and dual CE/current collectors in highly efficient and stable planar DSSCs. The power energy conversion efficiency of CNT fiber DSSCs reached 8.8% compared to DSSCs with Pt CEs (8.7%). The author successfully identified different processes involved in the use of CNT fiber CEs, namely bulk ion diffusion, ion diffusion inside the mesoporous electrode, and charge transfer at the CNT surface. These results provide clear directions for a further understanding of the fundamental catalytic properties of CNT fibers for improving their photoelectrochemical features [42].

Sebastián et al. [43] noted that the synthesis temperature, crystallinity, and surface area of materials affect DSSC efficiency. The carbon NFs, prepared at increasing temperatures in the range of 550–750°C, result in increasing diameters, increasing graphitization, and decreasing surface area. The low-temperature carbon NF CEs result in 24 nm average diameter, 183 m^2/g surface area, and 0.53 cm^3/g porosity. Fundamentally, electrical conductivity increases with the graphitization degree. However, a high compactness degree of thin nanofilaments results in low series resistance for the system. Because of this, low preparation temperature for carbon NF needs to be applied to enhance DSSC performance [43]. The research proved that electrochemical activity is affected by surface area and surface graphitization.

Other reasons for using a lightweight carbon electrode are temperature resistivity, ease of fabrication, and gentleness to skin . The graphene dip-coated carbon CE (Gr@CCE) possesses remarkable electrocatalytic activity toward the I_3^-/I_3 I_3^-/I_3 redox couple with low 0.79 Ω charge-transfer resistance. The sustainable design of carbon DSSCs has attained $\sim 6 \pm 0.5\%$ efficiency with a high photocurrent density of 18.835 mA/cm^2. The superior performance of DSSCs is due to low internal resistance, improved charge mobility, and better interfacial electrode contact [44].

Ternary palladium alloys PdNiCo and PdNiCo-rGO synthesized by Meyer et al. [45] showed that PdNiCo-rGO CEs could be a potential replacement for Pt CEs with reduced current density, peak-to-peak potential difference, charge-transfer resistance, and DSSC power conversion efficiency of 21 mA/cm^2, 0.12 mV, 0.726 Ω, and 4.36%, respectively [46]. To address complicated fabrication process issues, a highly efficient and low-cost 3D porous carbon composite constructed of dense and conductive graphite film as the bottom layer (2B pencil carbon (PC) layer), and porous carbon NP film as the top layer (candle carbon (smoke) (CC) layer) [47]. The results (short-circuit current density, open-circuit voltage) of DSSCs with PC, CC, CC/PC, and Pt CEs under solar simulator illumination (100 mW/cm^2) are 11.45 mA/cm^2, 0.72 V, 11.88 mA/cm^2, 0.73 V, 12.00 mA/cm^2, 0.75 V, and 13.46 mA/cm^2, 0.74 V, respectively. The fill factors of DSSCs with PC, CC, CC/PC, and Pt CEs are

56.09%, 59.80%, 65.28%, and 62.69%, respectively. Furthermore, the photoelectric conversion efficiency of PC, CC, CC/PC, and Pt DSSCs are 4.61%, 5.20%, 5.90%, and 6.26%, respectively. Those efficiency results showed that the CC/PC CEs deliver better photovoltaic performance. Particularly, the fill factor of DSSCs with CC/PC (65.28%) is 4.10% higher than that of DSSCs with commercial Pt (62.69%), and the photoelectric conversion efficiency of CC/PC-based DSSCs is as large as 5.90%, which reaches 94.2% of the Pt-based DSSCs (6.26%) [47]. The excellent performance of DSSCs with CC/PC CEs is attributed to the unique 3D porous structure, which can not only facilitate the transfer of electrons and ions, but also provides abundant catalytic sites. These synergistic effects greatly improved the DSSC conversion performance of CC/PC-based CEs.

Other complex research has been conducted on highly dispersive CNTs to bridge ordered mesoporous carbon. A new micronanostructured composite noted as N-doped ordered mesoporous carbon (NOMC)-Ni@NCNTs of Ni-encapsulated and N-doped carbon nanotubes (Ni@NCNTs) pinned on NOMC has been constructed by a two-step synthesis strategy. A 3D conductive scaffold was obtained by composite Ni@NCNTs. The pristine NOMC gave a conductivity of 20.4 S/cm, which was lower than NOMC-Ni@NCNT composite conductivity at 254.1 S/cm [47]. The NOMC-Ni@NCNT CE has excellent catalytic activity toward I^-_3 reduction, a low charge transfer resistance of 2.21 Ω, and a power conversion efficiency DSSC of 8.39%. Moreover, the NOMC-Ni@NCNT CE-based DSSC also results in electrochemical stability in corrosive/electrolyte with a remnant efficiency of 7.82% after 72 h of illumination. The Ni@NCNT conductive substrate helps to accelerate electron transportation among NOMC micron particles, and the amorphous NOMC with short-range mesopores helps to accelerate electrolyte diffusion [47].

6.6.2.4 Conductive polymer-based materials

Conducting polymers such as polyaniline (PANI), poly(3,4-ethylenedioxythiophene) (PEDOT) [48], and polypyrrole (PPy) have been adopted as CEs for DSSCs [31] due to their high conductivity, good stability, low cost, and high catalytic activity.

PANI is a well-known conductive polymer due to its easy synthesis, high conductivity, good environmental stability, transparency, and interesting redox properties. The larger the surface area, the greater the electrocatalytic activity. The lower the charge transfer resistance, the higher the electrocatalytic activity for the I^-_3/I redox reaction compared to the Pt electrode [31]. This was proved by Miranda et al. [49]. Hybrid hydrogels were prepared based on poly(acrylic acid-co-acrylamide) hydrogels and self-assembled nanostructured PANI. Three different nanostructured gel-PANI hybrid hydrogels were fabricated by various PANI concentrations to form CEs and gave conductivity values between 0.003 and 0.02 S/cm under AM1.5; the DSSC gave a power conversion efficiency above 2.0% [49]. From another angle, the addition of other doping materials into the polymer matrix can obviously alter the polymer properties as a base for CE DSSCs. The doping materials depend on targeted properties such as metal, anion, cation, nonmetal, or other materials. For example, composites made up of PANI and tungsten trioxide (WO_3) substrates as CEs gave a better DSSC efficiency (6.78%) than Pt-based CEs of DSSCs due to the high conductivity and electrocatalytic behavior of the PANI-WO_3 composite [50].

PANI-CNT and PPy-CNT CEs were prepared by putting horizontally pretreated FTO glass substrates in aniline-CNT and pyrrole-CNT solutions. Jin et al. [51] presented the molecular design of a PANI (PPy)-graphene complex for conducting gel electrolytes and CEs, aiming to reduce their charge-transfer resistance and enhance catalytic activity. DSSCs with PANI-CNT- and PPy-CNT-based CEs gave power conversion efficiencies of 8.23% and 7.69%, respectively, which result from good encapsulation of the 3D gel matrix and relatively good long-term stability over 15 days [51].

Fundamentally, adding graphene material in the polymer matrix reduces the charge-transfer resistance at the CE and improves the photovoltaic performance of the DSSC. Therefore Mehmood et al. [52] prepared PANI/graphene composites as CEs for DSSC applications. The DSSC constructed by PANI-9% graphene composite CE provided an efficiency of 7.45% and gave 20.1 Ω interface resistance between electrolyte and CE, which was slightly lower than Pt-based CE with a power efficiency of 7.63% and internal resistance of 19.2 Ω under similar conditions [52].

Moreover, another doping material called emeraldine salt (ES), a salt-based material, has been used as a dopant material in PANI to form a PANI-ES composite CE. Amalina et al. [53] varied the polymerization temperature from 34.85 to 74.85°C with respect to the standard low polymerization temperature at -0.15°C. The electrical conductivity profile changed from metal-like at low temperature to a semiconductor-like profile at high temperature. The morphology showed a globular shape at a temperature of -0.15°C, which was different to that at high temperature, and the material tends to form a nanorod structure. DSSC devices with 1.91% (highest) efficiency are obtained for PANI-ES polymerized at -0.15°C due to its high conductivity, and 1.15% (lowest) efficiency is obtained by PANI-ES at 54.85°C (due to its low conductivity because of the formation of a phenazine structure) [53].

In addition to the foregoing, PEDOT is also commonly used as a CE material. Rather than being used purely as a CE material for DSSCs, PEDOT also can be doped with other components such as poly(styrenesulfonate) (PSS) and polyoxometalate to increase the solubility and electrical conductivity in DSSCs. Much more interestingly, the composite among the conductive polymer itself also gives impact to the energy efficiency of DSSCs. For example, a PEDOT:PSS and PPy composite was deposited as the CE of a DSSC. As a result, the PEDOT:PSS-PPy film had low surface resistance, high conductivity, and good catalytic performance for the I^-/I_3^- electrolyte. The power conversion efficiency of the DSSC based on the PEDOT:PSS-PPy CE reached 7.60% under a simulated solar light illumination of 100 mW/cm^2 compared to a Pt electrode DSSC [54]. Moving on to other research, PEDOT:PSS film was employed as a CE in DSSCs. The morphology of the CE showed well-distributed spherical clusters with an average size of 214.04 nm [55].

Because PPy is the cheapest among the other three common conductive polymers, in 2014, a MWCNT-PPy composite film was synthesized and served as a CE in a DSSC to speed up the reduction of triiodide to iodide due to the unique rough surface structural properties consisting of numerous MWCNTs coated on PPy NPs, which guaranteed fast mass transport for the electrolyte. Furthermore, the MWCNT-PPy CE showed excellent electrocatalytic activity, electrochemical stability, and lower charge transfer resistance in comparison with a sputtered Pt CE. The DSSC assembled with the novel MWCNT-PPy CE exhibited 7.42% conversion efficiency under an illumination of 100 mW/cm^2, compared to the Pt electrode DSSC (6.85%) [56]. Therefore the MWCNT/PPy composite film can be considered as a promising alternative CE material for DSSCs, due to its high electrocatalytic performance and excellent electrochemical stability.

Pt-free composites

A composite CE, consisting of two or more components that combine the properties of each material into one, has been widely investigated. In addition, other composite-based DSSCs, such as PEDOT:PSS-TiN-CoS-TiS$_2$, have also shown comparable performance to devices fabricated using conventional Pt CEs [19]. However, zinc oxide (ZnO) normally used as a photoanode material had been use as a CE by Chew et al. [57]. Five types of CEs using two different materials were fabricated. Three types were ZnO based, whereas the other two were MWCNT CEs. The efficiencies achieved using ZnO and MWCNT-based CEs ranged from 0.46% to 7.07% [57] and the highest was an MWCNT CE DSSC. The results demonstrate that both materials are good candidate CEs for DSSC applications. A new conceptual nature-inspired fractal-based [58] CE has overcome the limitations of conventional planar designs by significantly increasing the number of active reaction sites, which enhances catalytic activity and improves DSSC performance. The fabrication of these innovative fractal designs is realized through cost-effective manufacturing techniques, including additive manufacturing and selective electrochemical codeposition processes. The results of the study suggest that the fractal-based CE performs better than conventional designs. Additionally, the fractal designs can overcome electrolyte leakage, fabrication costs, and DSSC scalability problems [58].

6.6.2.5 Inorganic compound-based materials

Transition-metal sulfides such as MoS$_2$ [17], MoIn$_2$S$_4$ [59], CoS [60], and ZnS [61] have potential as CE materials due to their low costs for large-scale DSSC use. However, the stability of inorganic compounds for DSSC applications still needs to be researched further [19].

Molybdenum disulfide (MoS$_2$) CEs show high reflectivity, which facilitates the absorbance of more photons and leads to the creation of more active edge sites exposed to I/I$_3$ redox couples in the electrolyte. This technique yields an efficiency of approximately 7.5%, which excels that of Pt-based (7.28%) electrodes [62]. On the other hand, Gurulakshmi et al. [17] fabricated a semitransparent MoS$_2$ flexible CE with a power conversion efficiency of 4.21%, whereas that of a Pt-based flexible DSSC is 6.08% [17]. According to Subalakshmi et al. [61], the binary metal sulfides CoS and NiS show clear hierarchical flowers and hexagonal microcages, respectively. The morphology of CoS is like the hierarchical flower taken from the flake part [61]. Interestingly, the intertwined flakes with open channels formed as flowers can provide more electrolyte diffusion as well as enhance the electrolyte/CE interfacial contact. Furthermore, a rapid charge transport pathway and increased electron transport rate at the electrolyte/CE interface are offered by the intertwined flakes. Another interesting morphology of NiS like hexagonal microcage consist of nanocolumns [61].

By comparison but using different composite materials, the binary ZnS shows clear spherical microball morphology composed of NPs with an average particle size of 10 nm, as reported by Subalakshmi et al. [61]. A nanoreticular structure with high electroconductivity and electrocatalysis has been synthesized from cobalt diselenide (CoSe$_2$) materials to form CEs for DSSCs [63]. The formation of CoSe$_2$ at -0.62 V is achieved through an inductive effect reaction with selenide (Se$_2^-$). Besides, CoSe$_2$ CEs have excellent electrocatalytic activity toward I$^-$/I$_3^-$ reduction reaction due to the electronic effect between Co and Se and electrochemically active sites. The CoSe$_2$ CE of DSSCs shows a power conversion efficiency of 3.96% under AM1.5G illumination, which is 1.6 times higher than that of commercial Pt DSSCs [63].

Ternary $Zn_{0.76}Co_{0.24}S$ materials show clear microsphere morphology covered with small NPs. As a result, more active sites are provided for adsorption of I_3^- species [61]. Furthermore, other ternary metal sulfide materials can be used, e.g., $CoNi_2S_4$. $CoNi_2S_4$ material shows flake morphology covered with a wispy lichen-like morphology. The nanocrystals (7—10 nm) provide a more active area for electrolyte ion diffusion. This results in more efficient I_3^- reduction reaction at the electrolyte/CE interface. $CoNi_2S_4$ shows excellent electrocatalytic activity and electrical conductivity compared to other CEs for the reduction of triiodide to iodide. The power conversion efficiency of $CoNi_2S_4$ CE DSSCs is $\sim 4.03\%$ compared to standard Pt-based DSSCs at around 4.59% [61]. Cheng et al. [59] researched ternary metal sulfide materials used in $MoIn_2S_4$ composite material. The $MoIn_2S_4$ material showed a petal-like structure with a charge transfer resistance of 19.90 Ω/cm. The power conversion efficiency for DSSCs obtained from $MoIn_2S_4$ CEs was 6.51% compared to Pt CE DSSCs (6.39%) under 100 mW/cm^2 illumination [59]. This concluded that ternary transition-metal sulfide material can serve as Pt-free CEs due to its high electrocatalytic activity and high DSSC performance.

Transition-metal carbides

Transition-metal carbides offer good sheet resistance and charge transfer resistance value [64]. An example of a common transition-metal carbide is SiC. SiC material (layered sheets and NPs) has greater potential due to its excellent catalytic activity. Research into Cr-doped SiC exhibited better activity to split the triiodide, but it was less capable of further splitting the iodine into monoiodides when compared with pure SiC and Pt-doped SiC slabs [64]. The outcome of this model shows the potential of transition-metal material doped into SiC slabs for use as CEs of DSSCs.

However, nanocomposites of iron carbide (Fe_3C) encaged in nitrogen-doped carbon were prepared by using simple carbothermal reduction of iron(II) oxalate (FeC_2O_4) nanowires in the presence of cyanamide (NH_2CN) at 600°C. The Fe_3C nanocomposite DSSCs had an efficiency of 7.36%, which is higher than Pt-based CEs (7.15%) under the same conditions. The good electrochemical performance was attributed to the synergistic effect of the combination of nitrogen-doped carbon, Fe_3C, and the 1D configuration, which endowed the nanocomposites with more interfacial active sites and improved electron-transfer efficiency for the reduction of I_3^-/I_3[65].

Transition-metal oxide

Metal oxide is also one of the best potential materials to replace Pt CEs due to its good thermal properties, cost effectiveness, and high catalytic activity. Examples of metal oxides discussed in this section are ZnO [66], WO_3 [67], and a-Fe_2O_3 [68]. For example, Wang [66] researched the comparison between ZnO and ZnO-PEDOT:PSS composite as a CE of DSSCs. The semiconductor nature of ZnO provides very low conductivity of charge transfer between electrolyte/oxide interfaces but the addition of PEDOT:PSS enhances its conductivity with a DSSC power conversion efficiency of 8.17% compared to ZnO CEs (0.777%) [66]. The next example of a metal oxide is tungsten oxide (WO_3). Basically, the vacancy of oxygen serves as an electron donor to enhance electron transfer, provide active sides for triiodine adsorption, and accelerate its reduction rate to generate iodide [69]. Previously, Cheng et al. (2013) reported simple hydrogen treatment to introduce the oxygen vacancies in catalytically passive commercial WO_3 (C-WO_3) applied as an active CE for DSSCs. As a result, the electrocatalytic activity and conductivity of DSSC were enhanced with a power conversion efficiency of 5.43% [67]. In addition, Zatirostami [50] reported the comparison between Pt, WO_3, and WO_3-PANI materials for CE DSSCs. The WO_3 layer formed on the ITO substrate showed a sponge-like morphology. The charge transport resistance value for Pt, WO_3, and WO_3-PANI nanocomposite was estimated to be 322, 820, and 201 Ω/cm^2, respectively [50]. The author claimed that composite WO_3 with a conducting polymer as the WO_3-PANI CE gave better results as mentioned in [50]. On the basis of theoretical predictions using first-principle quantum chemical calculations, Hou et al. [68] successfully suggested rust (a-Fe_2O_3) nanocrystals mainly bounded by (012) and (104) surfaces as a new CE catalyst due to their high electrocatalytic activity toward triiodide reduction compared to Pt. The author verified the fundamental theory of synthesized pure-phase Fe_2O_3 by the hydrothermal method and gave a DSSC conversion efficiency of 6.96%. This result is comparable to a Pt CE DSSC (7.32%) due to the excellent catalytic activity of a-Fe_2O_3.

Organic materials. Also attracting attention are nonmetal materials such as boron nitride (BN) as CEs for DSSCs. By applying a conductive binder of sulfonated poly(thiophene-3-[2-(2methoxyethoxy)ethoxy]-2,5-diyl) (s-PT), the BN-s-PT composite film DSSC was successfully formed and gave a power conversion efficiency of 9.21% compared to expensive Pt (8.11%) at 1 sun. The result is due to large active surface area and high intrinsic heterogeneous rate constant; the latter formed fast electron-transfer matrices [70].

6.7 Synthesis

The synthesis procedure for photoanodes of DSSCs should be simple, low cost, and safe, which contributes to the stability and high power conversion efficiency of DSSCs. This section is divided into the synthesis of TiO_2 and fabrication of photoanodes procedures. This tells us that the synthesis procedure is also one of the crucial parameters that we must tackle to build efficient DSSCs. This section illustrates a few synthesis procedures depending on materials and type of DSSC. The procedures are divided into two main categories: physical methods and chemical methods.

6.7.1 Synthesis of TiO_2 paste procedure

Commonly, a P25-based TiO_2 slurry is made by dissolving P25 NP materials in a *tert*-butanol and distilled water mixed solvent, which is stirred for 2 h and 30 min and then ultrasonicated to achieve a homogeneous slurry. The TiO_2 paste for flexible DSSCs is prepared by mixing P25 slurry with the commercially available T-L TiO_2 slurry (w:w 8:1); it is then stirred for 30 min and then ultrasonicated for 2 h, followed by ball milling for 48 h [17]. The TiO_2 P25 anatase nanopowder is preheated for 30 min in a furnace at 400°C to eliminate organic impurities and any moisture. The heated TiO_2 powder is mixed with 0.2 mL of distilled water and acetic acid, respectively. The mixture is ground in a ceramic mortar [24]. Another simple procedure reported to prepare TiO_2 paste from P25 TiO_2 powder is by using a mixture of anatase and rutile phases of the oxide [71]. Further details of the procedure to synthesize TiO_2 paste using Degussa P25 TiO_2 powder (0.25 g), which is ground for 30 min with 1.0 mL of 0.1-M HNO_3, Triton X-100 (a drop), and polyethylene glycol (0.05 g), can be found in Kumari et al. [28].

6.7.2 Deposition method and procedure

This section explains the deposition method, classified as physical and chemical methods. Details of the latest research are also discussed as examples.

6.7.3 Physical method

The physical method refers to physical action to prepare and deposit materials on the substrate surface. This section discusses a few deposition techniques and procedures applied by a number of researchers. Basically, the physical method depends on the precursor phase used, which is divided into the liquid phase and gas phase, as shown in Fig. 6.6.

6.7.3.1 Liquid-phase precursor

This method uses a precursor in a liquid phase, which is deposited by printing (doctor blade, screen, inkjet), coating (spin, dip, screen), electrospray deposition (ESD), or electrohydrodynamically depending on the expectation properties of the photoanode and CE. For deposited TiO_2 paste, the spin-coating and doctor-blade techniques were applied for most of the research depending on the parameters and results that were expected. Both have their own advantages and disadvantages, for example, a doctor blade is cheap, easy to handle, and has a rapid fabrication process. However, a doctor blade has some disadvantages such as low evaporation rate and low particle concentration. Also, different substrate surfaces give different effects to TiO_2 paste even when using the same method.

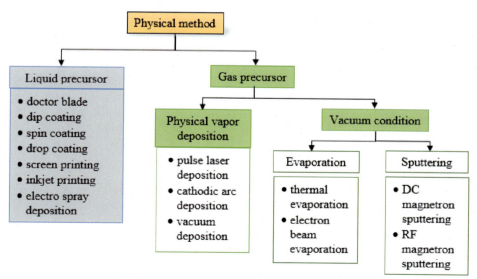

FIGURE 6.6 Flowchart of classification of the physical method. *DC*, Direct current; *RF*, radio frequency.

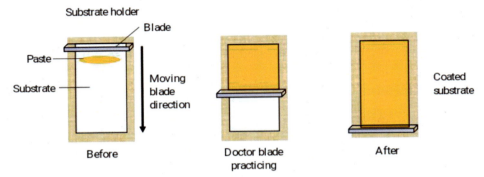

FIGURE 6.7 Illustration of the doctor-blade technique.

Doctor blade

Up to 2020, the low-cost, flexible, and easy-to-operate doctor-blade method has been one of the most popular application methods to form both electrodes of DSSCs. This method is flexible on any type of surface such as FTO glass substrate, ITO glass substrate, flexible film, Ti foil, and many more. The tape is used to control the thickness of the paste and to make sure the substrate is static as the blade is moved on the substrate. The blade placed on the substrate needs space to spread the paste over the substrate uniformly. A multilayer structure can be formed by continuously repeating the spreading action in the same direction and increasing the thickness of the paste. The basic idea of this procedure is shown in Fig. 6.7. The parameters to control during this technique are thickness of the coated layer, uniform spread of the coated layer, and time taken for the paste to dry or evaporate (thermally) normally by heat treatment.

Low-temperature conditions must be applied to cure the paste due to the melting temperature of the substrate, as mentioned by Gurulakshmi et al. [17]. The time taken to immerse the dye depends on the thickness and type of paste or substrate used. Next, Shaban [24] practiced almost the same procedure with Subalakshmi et al. [61] but a different substrate was used, this time Ti foil. This implies that the method is flexible on any solid substrate. A different strategy was practiced by Kumari et al. [28], which concentrated on the layer-by-layer strategy by using a doctor blade. A summary of the latest research using the doctor-blade procedure in the synthesis of electrodes for DSSCs is listed in Tables 6.1 and 6.2.

Dip coating

Two types of dip coating are commonly used: batch-type dip coating and continuous dip coating. Both types of dip-coating procedure involve five main steps: immersion, startup, deposition, drainage, and evaporation [73]. Continuous-type dip coating can only be applied to flexible substrates such as conducting plastic film or conducting metal sheets. If important parameters (speed, duration, angle of immersion) are not controlled well, then uneven, smooth, and thick layers will form. The procedure of dip coating can easily be illustrated as a substrate immersed in solution vertically for a period of time with controlled depth

6.7 Synthesis

TABLE 6.1 Doctor-blade application for a photoanode dye-sensitized solar cell.

Materials	Substrate	Dye	Dye immersion time (min)	Temperature (°C)	Duration (min)	References
TiO$_2$ nanoparticle	FTO glass	N719	1440	500.00	30	2011 [40]
TiO$_2$ paste	FTO glass	nd	nd	325.00	5	2019 [42]
			nd	450.00	15	
			nd	500.00	15	
TiO$_2$ nanopowder	FTO glass	N719	720	350.00	30	2019 [61]
TiO$_2$ nanoparticle	Glass	N719	nd	450.00	45	2019 [28]
TiO$_2$	FTO glass	N3	1440	449.85	120	2020 [72]
TiO$_2$ paste	Flexible PETP	N719	1080	120.00	10	2020 [17]
TiO$_2$ nanoparticle	FTO glass	N719	—	500.00	30	2011 [40]

FTO, Fluorine-doped tin oxide; *nd*, not determined; *PETP*, poly(ethylene terephthalate).

TABLE 6.2 Doctor-blade application for counter electrode of a dye-sensitized solar cell.

Materials	Substrate	Heated temperature (°C)	Duration (min)	Efficiency (%)	References
Pt-rGO	FTO glass	500	30	nd	2019 [32]
ZnO	FTO glass	450	30	0.94	2018 [57]
Graphite	Bare glass	nd	nd	5.9	2011 [40]

FTO, Fluorine-doped tin oxide; *nd*, not determined.

and speed of immersion, as shown in Fig. 6.8. Next, the substrate is pulled from the solution and dried to evaporate the solvent. As a result, a smooth thin layer is obtained on the substrate surface. Normally, the solution applied is a nanoparticle solution. By fundamental theory, the layer thickness of the materials can be predicted and controlled using the Landau—Levich equation [74], as shown in Eq. (6.1):

$$h = \frac{0.94 \, (n \cdot v)^{\frac{2}{3}}}{\gamma_{LV}^{\frac{1}{6}} (\rho \cdot g)^{\frac{1}{2}}} \tag{6.1}$$

where h is the thickness of the coated layer, n is the viscosity, v is the dragging speed, γ_{LV} is the liquid—vapor surface tension, ρ is the density, and g is the gravity constant. Film thickness can be controlled by substrate retraction and the relation between solution gravity and viscosity [75]. Thus for a thick coated layer, a high pull-up speed is required with fast drying using highly volatile solvents. Table 6.3 shows evidence of the dip-coating application.

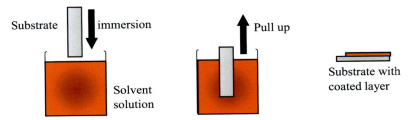

FIGURE 6.8 Schematic diagram for the dip-coating method.

TABLE 6.3 Dip-coating application.

Materials	Solvent	Substrate	Efficiency	Remarks	References
Mesoporous TiO$_2$	nd	F-doped SnO$_2$	nd	nd	2005 [22]
Graphene carbon counter electrode	Graphene ink	Carbon fiber wet-laid web counter electrode sheet	5.92	Flexible DSSC	2020 [44]
ZnO@TiO$_2$ core–shell	C$_{16}$H$_{36}$O$_4$Ti	ZnO-FTO glass	4.85	nd	2015 [76]

DSSC, Dye-sensitized solar cell; *FTO*, fluorine-doped tin oxide; *nd*, not determined.

Spin coating

Spin coating utilizes centrifugal forces to deposit the required material on the substrate. In general, a cleaned substrate is kept on the chuck of the spin-coating machine. It generates a vacuum below the substrate to fix it in position (labeled as vacuum chuck) and rotates the substrate at the setting speed. The solution or suspension material is dropped from above into the rotating substrate. The height between the dropped material and the surface of the substrate is important. Due to high centrifugal forces, the solution tends to spread out on the substrate and coat it completely [10]. A schematic representation of the spin-coating procedure is shown in Fig. 6.9. Other than involving the photoanode deposition method, this technique also involves the CE deposition procedure, as mentioned in Table 6.4, to control the thickness and uniformity of the CE coated layer. The procedures involved in spin coating are deposition, spin-up, spin-off, and evaporation. Another significant parameter was reported by Bandara et al. [77], which is that the number of TiO$_2$ layers deposited on the substrate will increase the uniformity and porosity structure of TiO$_2$.

Drop coating

The drop-coating technique is also known as the drop-cast method. The procedure operates by depositing consecutive drops of solution onto any type of surface and then waiting for the solvent to evaporate. Table 6.5 shows the drop-coating applications mostly used to fabricate CEs for DSSCs.

Screen printing

Basically, the screen-printing procedure utilizes a mask between substrate and paste to be applied with open patterns, which are carefully drafted on the mask. The openings in the

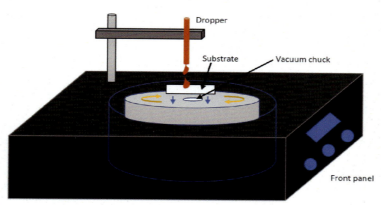

FIGURE 6.9 Schematic diagram of spin-coating equipment.

TABLE 6.4 Latest research into spin-coating applications.

Materials	Substrate	Speed of rotation (rpm)	Duration (s)	Efficiency (%)	Remarks	References
TiO$_2$	FTO glass	1000	1800	4.40	• Up to 6 layers • Reliable and consistent	2019 [77]
MWCNT	FTO	1000	1800	4.25	—	2018 [57]
Precursor solution	BL-coated FTO	1930, 2500, 6900	2, 2, 1800	5.80	• Equivalent to one set • One set, one layer • Runs for three and five layers	2019 [23]
PEDOT:PSS:DMSO	FTO glass	1000	3600	nd	• Ratio of 3:1	2016 [78]
PEDOT:PSS:DMSO + MWCNT	FTO glass	1000	3600	nd	• Highest conductivity • Highest transparency	

BL, Blocking layer; *FTO*, fluorine-doped tin oxide; *MWCNT*, multiwalled carbon nanotube; *nd*, not determined; *PEDOT:PSS:DMSO*, poly(3,4-ethylenedioxythiophene):poly(styrenesulfonate):dimethyl sulfoxide.

TABLE 6.5 Drop-coating applications.

Material	Solvent or precursor	Substrate	Temperature (°C)	Duration (s)	Efficiency (%)	References
Pt	H$_2$PtCl$_6$	TCO-coated glass	400	900	nd	2015 [71]
Pt	Chloroplatinic acid hexahydrate in 2-propanol	FTO glass	150	50	8.70	2019 [42]
Pt	H$_2$PtCl$_6$	FTO glass	450	1800	6.39	2020 [59]
Boron nitride	Boron nanoparticle and 2% sulfonated polythiophene ink	Flexible carbon cloth	50	7200	9.21	2020 [70]

FTO, Fluorine-doped tin oxide; *nd*, not determined; *TCO*, transparent conducting oxide.

164   6. Introduction to solar energy and its conversion into electrical energy by using dye-sensitized solar cells

mask or mesh allow the paste to attach to the substrate using squeegee pressure and to produce the desired pattern of the photoanode for DSSCs [10]. For example, the screen-printing procedure was applied to fabricate a TiO_2 photoanode [71] on three different types of substrates: flexible Willow Glass, FTO-coated glass, and ITO-coated glass. The TiO_2 paste was screen printed on the substrates for several cycles depending on the required layer thickness. The TiO_2 electrodes were then heated in an oven to 325–500°C (with different durations for different temperature steps). As a result, FTO-glass substrate gave the highest DSSC efficiency of 7.42% compared to ITO glass substrate, which was 3.09% efficient [71]. This implies that TiO_2 paste is highly recommended to be deposited on FTO glass substrate by using this technique. Tables 6.6 and 6.7 list research reported for screen-printing applications.

TABLE 6.6 Screen printing for photoanode dye-sensitized solar cells.

Materials	Substrate	Temperature (°C)	Duration (min)	Remarks	Efficiency (%)	References
TiO_2	Flexible Willow Glass	325–500	nd	Three layers	nd	2015 [71]
	FTO glass				7.42	
	ITO glass				3.09	
TiO_2 nanoparticle	FTO glass	80, 80, 80, 80, 470	5, 5, 5, 5, 30	• Five layers • 16 μm	3.89	2018 [79]
Transparent titania + polyester fiber	TCO glass	125, 325, 375, 450, 500	6, 5, 5, 15, 15	• Five layers • 8.5 μm	–	2020 [33]

FTO, Fluorine-doped tin oxide; *ITO*, indium tin oxide; *nd*, not determined; *TCO*, transparent conducting oxide.

TABLE 6.7 Screen printing for counter electrodes for dye-sensitized solar cells.

Materials	Substrate	Annealed temperature (°C)	Annealed duration (min)	Efficiency (%)	Remarks	References
Ag-rGO/Pt	FTO glass	450	30	nd	na	2020 [18]
Graphite	ITO	120	15	1.10	na	2019 [53]
PANI-ES	ITO	120	15	1.91	na	2019 [53]
Colloidal graphite	FTO glass	300	30	6.20	9 μm layer	2011 [40]
Pt	FTO glass	450	30	6.80	Commercial	2011 [40]

FTO, Fluorine-doped tin oxide; *ITO*, indium tin oxide; *na*, not applicable; *nd*, not determined; *PANI-ES*, polyaniline–emeraldine salt.

II. Photovoltaic materials and devices

Inkjet printing

In this procedure, the number of materials used can be controlled and saved. In addition, a combination of different materials and layer structures can also be altered. A typical inkjet printing (IJP) system consists of an ink cartridge to eject droplets for drop-on-demand formation of a printed pattern on a substrate. Note that the droplets can be generated from a small nozzle and positioned with separating spaces in planar x-y coordinates. Consequently, one layer of the droplet pattern is formed with an area using this simple deposition scheme. To increase the thickness of printed patterns, multiple layers are applied by repeating the same formation of droplet deposition. In this way, the thickness of a thin film of the layered structure can be tuned along the z-axis. Therefore a microstructure produced by the IJP method is fabricated in 3D x-y-z-coordinates [80]. For example, deposition of a thin film of TiO_2 as a photoanode of a DSSC on a glass substrate has been reported by Chen and Chen [80]. The experimental results showed that the short-circuit photocurrent can be tuned and optimized by changing the photoelectrode thickness and exposure area to the sun [80]. This implies that IJP is applicable to synthesize both electrodes of a DSSC.

Electrospray deposition

The idea behind ESD comes from the electrostatic painting technique, which involves the atomization of paint particles and directs the flow toward the grounded part to be painted. The ESD technique coats the charged solution particles uniformly on the substrate. This technique requires fewer NPs and wastage is approximately 5%–8%. As shown in Fig. 6.10, ESD requires a nozzle to atomize the droplets and connect to the power supply to charge the atomized droplets. Additionally, ESD needs a pump to pressurize the solution. These charged droplets are directed

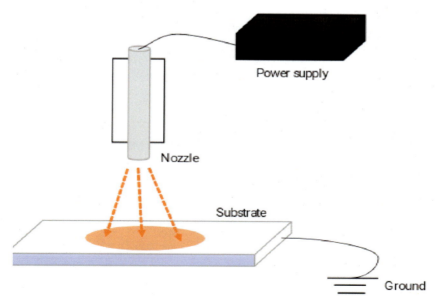

FIGURE 6.10 Schematic diagram of electrospray deposition.

to the faced substrate surface below it, as shown schematically in Fig. 6.10. Low [75] mentioned that ESD is commonly involved in synthesizing thin films less than 10 μm thick.

6.7.3.2 Gas-phase deposition method

As shown in Fig. 6.6, there is another physical deposition method that use gas phase precursor to deposit NPs on a substrate. These physical deposition method that use gas precurser can be divided into physical vapor deposition (PVD) method or method under vacuum condition such as evaporation or sputtering methods. Example of physical vapor deposition are vacuum deposition, pulse laser deposition and arc depositions techniques. On the other hand, examples methods under vacuum condition that use gas precursor are evaporation (thermal and electron beam evaporation) and sputtering (DC and RF magnetron sputtering) techniques.

Physical vapor deposition

Vacuum deposition Basically, PVD is the deposition of thin layers by sputtering and evaporating in a vacuum chamber. The film layers are obtained due to the condensation of atoms on a relatively low-temperature substrate [10]. For example, to synthesize graphene-TiO_2 film, a closed environment should be considered to prevent the escape of TiO_2 particles from the surface. The particles move in direct motion to the substrate when heat is supplied to the boat containing TiO_2, as shown in Fig. 6.11. As a result, TiO_2 is physically coated onto the graphene surface forming graphene-TiO_2 composite thin films on the substrate within a short deposition duration under closed chamber conditions [75].

Pulse laser deposition To grow high-quality multicomponent oxide thin films, the pulse laser deposition (PLD) procedure is employed by using short pulses of laser (1–30 ns) to evaporate the target. A schematic diagram of PLD is shown in Fig. 6.12. The PLD procedure begins by irradiating a small area of the order of a few square millimeters with short and high-energy pulses of laser beam. The target absorbs the laser energy and become heated and defected. This causes evaporation of the target composition. Then, the vaporized composition is directed toward the substrate and film growth takes place on the substrate surface. The composition of the vapors

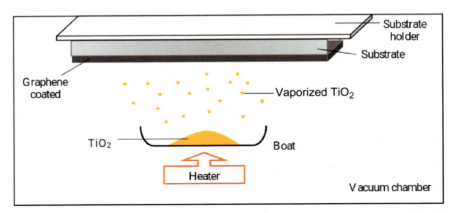

FIGURE 6.11 Schematic diagram for the physical vapor deposition method to deposit TiO_2 thin film.

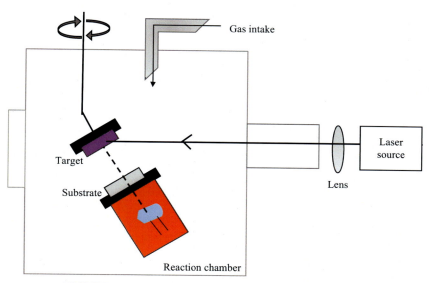

FIGURE 6.12 Schematic diagram for pulse laser deposition.

must be same as the composition of the target. The exact composition of the vapors greatly facilitates the growth of the desired phase on the substrate. Because no filaments and charged particles are present, film growth occurs in the presence of reactive gases, which encourages material growth through gas-phase reactions. An example procedure of the PLD of Au NPs on FTO glass begins with the settingup of the KrF excimer laser ($\lambda = 248$ nm, $\tau = 24$ ns). Next, a pure Au target is fixed on a rotating holder to avoid excessive damage of the target surface, which is then ablated at a laser fluence of 2.0 J/cm^2. Substrate on a substrate holder is located 35 mm from the target. Ablation is performed in a controlled Ar atmosphere at 100 Pa. Samples are deposited using 30,000 laser shots. The process takes place in a vacuum chamber with a residual pressure of more than 1.0×10^{-4} Pa [33].

Arc deposition Arc deposition, also called cathodic arc deposition, uses an electrical arc, which produces blast ions from the cathode [73]. Because the output of the vacuum arc is highly ionized, the process can control both the trajectory of the coating material and the energy with which ions impinge on the substrate. The material to be deposited is located within the hollow cathode and forms the anode. The anode material evaporates under low voltages (10–15 V) and a high-current ($\sim$100 A) electron beam, which along with trapped electrons ionize the vapors. Because the voltage is very low, the risk of a cathode being sputtered is minimized [10]. The advantages of this technique include its compatibility with industry, good film adhesion, excellent stoichiometric control, low temperature, multilayer compact coating, uniform film, and low voltage. However, the disadvantage is its inability to coat complex geometries [73].

Evaporation

Thermal evaporation Thermal evaporation uses heat energy to activate the deposition process. The source material is melted or sublimated through electrical heating, which converts

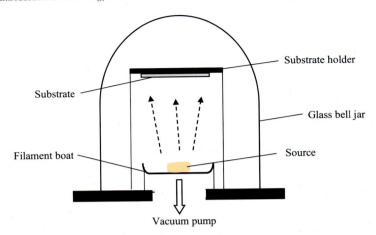

FIGURE 6.13 Schematic diagram of the thermal evaporation setup.

the materials into the gaseous phase [73]. A schematic diagram is shown in Fig. 6.13. As an example, ZnO nanorod thin film was grown on FTO glass substrate by using the thermal evaporation procedure. Briefly, the seeded substrate with its conductive surface facing down was placed on an alumina boat, including zinc acetate dehydrate powder, and annealed at 350°C in air for 9 h to form ZnO nanorod thin films [76]. The disadvantage of this procedure is poor adhesion properties of the deposited film/layer.

A procedure to synthesize TiO_2 paste was reported by Zalas and Jelak [72]. First, 3 g of TiO_2 nanopowder (3D nano) was mixed with 0.5 mL of acetic acid and 20 mL of 96% ethanol. Next, a solution containing 1.5 g of ethyl cellulose and 10 mL of α-terpineol in 13.5 g of 96% ethanol was prepared and then added to the former suspension. The mixture was sonicated for 1 h and then magnetically stirred overnight. Ethanol was slowly removed in a rotary evaporator and the obtained paste was ready to use. Different materials that can be used with this procedure are tungsten trioxide and PANI nanocomposite on an ITO substrate [81].

Electron beam evaporation Electron beam evaporation is one of the techniques that uses high-speed electrons to bombard the target source, as shown in Fig. 6.14. Kinetic energy of the electron beam (noted as E-beam in Fig. 6.14) is produced from the electron gun using electric and magnetic fields to shoot the target and vaporize the surrounding vacuum area. Once the substrate is heated by the radiation heating element, the surface atoms will have enough energy to leave the substrate. Meanwhile, the substrate will be coated once the thermal energy is less than 1 eV and the working distance is in the range of 0.3–1 m.

Sputtering Sputtering is a process where the accelerated ions are applied to expel the original particles on a target substrate via ion bombardment. This can be considered as the momentum transfer process of ions accelerated from the source to collide with substrate particles [10]. In addition, the electrical potential will cause the ions to accelerate and the ions will be reflected or absorbed by the substrate provided the kinetic energy is less than 5 eV. The substrate and lattice positions will be scratched if the kinetic energy is higher than that of the

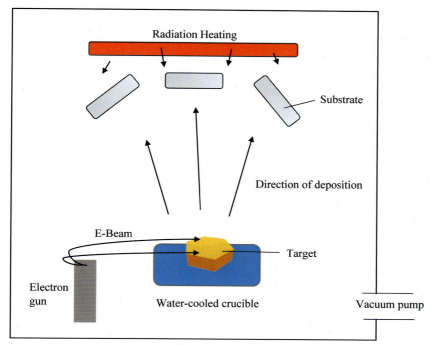

FIGURE 6.14 Schematic diagram for electron beam evaporation.

surface atom binding energy. There are two types of sputtering process discussed in this section: direct current (DC) magnetron sputtering and radio frequency (RF) magnetron sputtering.

DC sputtering is a controllable and low-cost sputtering technique. Furthermore, the DC sputtering method can increase the rate of deposition of the material with minimum damage to the substrate. Accordingly, this method compromises the ionization of the mixture of argon and nitrogen gas, while the positively charged sputtering gas effortlessly accelerates toward the conductive target materials causing the ejected target atoms to deposit on the substrate more easily. However, DC magnetron sputtering can only be applied for the conductive material and not for the dielectric target due to the termination of the discharge of insulator materials during the deposition process. In other words, the positively charged ions will be produced and accumulated on the surface of the dielectric or insulator films. Moreover, the synthesis rate of magnetron sputtering depends on the power of the discharge current and gas pressure in the working chamber, which allows the growth of semiconductor material membranes in the argon atmosphere [70]. As an example, DC magnetron sputtering was used to prepare ZnO nanofibrous material. First, the glass substrates were covered with tin oxide films and were cleaned in a mixed solution of Cr at room temperature. Afterward, substrate cleaning was followed by rinsing with abundant distilled water. The cleaned substrate was mounted on a working electrode for DC magnetron sputtering in a classical setup. A disc of 99.99% pure zinc served as the metal target. Argon working gas pressure was regulated in such a way to maintain a constant vacuum pressure of 5×10^{-3} Torr. The DC used was 0.12–0.15 A, the temperature was 210°C, and the deposition time was 22 min. Lastly, the nanofibrous ZnO layers grown on the substrate were introduced into a reactor and annealed at 480°C for 45 min in an oxygen atmosphere with a gas flow of 100 mL/min [27]. In addition, Pt material was also deposited by DC sputtering onto FTO glass and gave an efficiency of 8.11%, which was higher than carbon cloth, which gave an efficiency of 7.97% [70]. The result highlighted that the different substrates gave different DSSC performance when using the same deposition procedure.

RF magnetron sputtering uses alternating current (AC) as a power source and provides a direct pathway to the deposition of insulators. Generally, the frequency used in this technique is an alternating voltage at a specific frequency of 13.56 MHz within the frequency range of 1 kHz–103 MHz. When using a positive electric field, the positively charged ions are accelerated to the surface of the target and directly sputtered on the FTO/ITO substrate. The positively charged ions on the surface of the target are eliminated/neutralized by the electron bombardment force during the negative field moment. However, the high deposition rate of RF magnetron sputtering for FTO/ITO substrate deposition led to a mobility difference between the electrons and ions within the plasma region. Therefore high heating temperature is required to accelerate the sputtering process. RF magnetron sputtering deposition is generally only limited to smaller substrate quantities and sizes due to the cost of RF power supplies.

6.7.4 Chemical method

The chemical method can be classified into two main groups depending on the state of the precursors involved, which are the liquid-phase precursor and the gas-phase precursor. The classification procedure in this chapter can be understood by referring to Fig. 6.15.

6.7.4.1 Liquid-phase precursor

Sol-gel

The sol-gel technique, also known as the molecular-level mixing method, produces finer-sized oxide particles by the preparation of sol. The sol transforms into gel, which is then dried and crushed to form nanosized oxide particles. It is then directly coated on the substrate. Next, the drying step results in a fine coating. This method is very similar to doped NPs and composites. For example, a nanocrystalline TiO_2 film was synthesized by a sol-gel

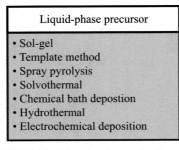 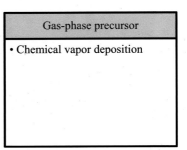

FIGURE 6.15 The chemical method group depends on the phase of the precursor used.

procedure and subsequently impregnated with TiCl$_4$ [22]. Furthermore, TiO$_2$-rGO nanocomposite film was also successfully formed through the sol-gel method by using titanium(IV) isopropoxide and rGO [14]. Moreover, a single sol-gel procedure was applied to form Cr@TiO$_2$/MWCNT by doping Cr(III) into TiO$_2$/MWCNTs [15]. In a rare case, a ZnO compact layer synthesized by a sol-gel method was introduced as a photoelectrode at the interface between the FTO substrate and a mesoporous ZnO layer [82].

Template method

The template method uses a textured, removable structure or platform on which the desired material can be deposited. After the material is successfully deposited, the template can be removed by heating or by acid leaching [10]. As an example, the Pluronic P123 templated mesoporous TiO$_2$ film was grown via layer-by-layer deposition by Markéta Zukalová [22].

Spray pyrolysis

The spray pyrolysis procedure starts by heating the substrate and then an NP solution is sprayed onto it. Solution droplets evaporate on contact with the substrate and a thin, adherent film results. No sophisticated or expensive equipment is needed. Parameter depositions that must be considered are temperature gradient, heating time, and concentration of solution.

Chemical bath deposition

Chemical bath deposition uses precursors of materials for deposits. It claims to be a simple chemical deposition method because it only needs substrate and solution in holding fixtures with stirring and heating. The procedure starts with the substrate immersed in precursor solution while the solution is hydrolyzed, heated, and stirred. As a result, particle nucleation and growth occur at the surface of the substrate. The advantages of this deposition method are stability, good uniformity, and good reproducibility [10]. The parameters that must be controlled are temperature and deposition time to control shape, morphology, size, and film thickness. For example, cobalt sulfide (CoS) has been deposited on FTO substrate using a microwave-assisted chemical bath with very low thermal treatment in the range of 100−120°C. This deposition method improves the electrocatalytic properties and exhibits a DSSC power conversion efficiency of 8.39% compared to 7.88% for commercial Pt [60]. In detail, first, a solution of 0.2 mol/L Co(NO$_3$)$_2$.$_6$H$_2$O and 0.2 mol/L CH$_3$CSNH$_2$ was prepared in 40 mL ethanol with a magnetic stirrer for 20 min at room temperature. Thioacetamide was chosen as the sulfur source because it reacts faster with the metal salts. Then, the solution and FTO were placed within a polytetrafluoroethylene reaction vessel and subjected to microwave radiation at a heating rate of 4°C/min until it reached100°C, and remained so for 30 min. Lastly, the product was cooled to room temperature [60].

Solvothermal

The solvothermal procedure is very similar to the hydrothermal procedure. However, the solvothermal procedure usually uses a nonaqueous precursor and allows size, shape distribution, and crystallinity of the nanostructure layer to be controlled. Yang et al. [83] successfully synthesized two phases of nickel sulfide (a-NiS and b-NiS) with two different solvents of alcohol and water, respectively. Solvothermal advantages are that the sphere-like shape for a-NiS and a cross-like shape composed of nanorods for b-NiS are uniform and well distributed, as well as their size. DSSCs with an a-NiS CE has better power conversion efficiency at 5.2% compared to b-NiS CE at 4.2% [83].

Different material but the same solvothermal technique was applied by Cheng et al. [59] to grow an $MoIn_2S_4$ layer on FTO substrates. The procedure involved 0.0696 g Na_2MoO_4, 0.169 g $InCl_3$, and 0.1394 g thioacetamide added to 30 mL of deionized water under ultrasonication for 2 h until all the reactants were dissolved and transferred into a 100-mL Teflon-lined stainless-steel autoclave. The conductive side of the FTO substrates were faced upward and placed into the solution. Then, the autoclave was placed in an oven and heated at 200°C for 15 h, while the FTO substrates covered with $MoIn_2S_4$ film were removed until the temperature dropped to room temperature and were dried in an oven at 60°C overnight [59].

Hydrothermal

Basically, the hydrothermal process uses water as solvent. Hou et al. [68] synthesized Fe_2O_3 using the hydrothermal procedure with iron(V) nitrate nonahydrate $Fe(NO_3)_3$ $9H_2O$ as precursor and sodium hydroxide aqueous solution as reaction-controlling agent. Next, a ternary metal sulfide ($CoNi_2S_4$ and $Zn_{0.76}Co_{0.24}S$)-based CE was prepared by a simple hydrothermal method, as reported by Subalakshmi et al. [61]. The synthesis procedure of $CoNi_2S_4$ film needed 1 M of thiourea, 0.07 M of cobalt chloride hexahydrate, 0.15 M of nickel chloride hexahydrate, and 0.05 mL of ethylenediamine dissolved in 30 mL of distilled water by magnetic stirring for 30 min. Then, the solution was transferred into a Teflon-lined stainless-steel autoclave. The FTO substrate was immersed in the reaction solution followed by the hydrothermal process at 180°C for 24 h and resulted in $CoNi_2S_4$-coated FTO. A similar procedure was carried out to prepare $Zn_{0.76}Co_{0.24}S$ film using 1.2 M of thiourea, 0.5 M of zinc chloride, 0.1 M of cobalt chloride hexahydrate, and 0.05 mL of ethylenediamine. To compare DSSC performance, binary metal sulfide CEs such as CoS, NiS, and ZnS were also prepared by the same experimental procedure using the respective metal precursors (chloride hexahydrate for CoS, nickel chloride hexahydrate for NiS, and zinc chloride for ZnS), ethylenediamine, and thiourea [61].

Hydrothermal fabrication of ternary palladium alloys PdNiCo and PdNiCo-rGO was reported by Meyer et al. [45]. The as-synthesized alloys consisted of agglomerated spherical particles. The PdNiCo-rGO CE could be a potential replacement for the Pt CE with a reduction in current density, peak-to-peak potential difference, charge-transfer resistance, and power conversion efficiency of $21\,mA/cm^2$, 0.12 mV, 0.726 Ω, and 4.36%, respectively [45]. Moreover, a petal-like $MoIn_2S_4$ thin film with excellent performance was also synthesized on FTO substrate by the hydrothermal method [59]. In addition, Arbab et al. [44] synthesized P25-S (TiO_2 nanospheres) and P25-R (TiO_2 nanorods) by the hydrothermal procedure [44].

Electrochemical deposition

Electrochemical deposition is also known as electrodeposition, electroless deposition, or electrophoretic deposition. For electrodeposition, an electric current is supplied to the electrodes, which are dipped in a suitable electrolyte/solution containing ions of deposited species. As an example, Gurulakshmi et al. [17] fabricated flexible CEs using the electrodeposition method from Pt- and MoS_2-based CEs. Both gave power conversion efficiencies of 6.08% and 4.21%, respectively [17]. With the same electrodeposition principle, Guo et al. [63] also synthesized a nanoreticular-structured cobalt diselenide ($CoSe_2$) CE with high electroconductivity and electrocatalytic results [63]. As details procedure example [17], a molybdenum disulfide CE was obtained using the electrodeposition procedure with 5 mM $(NH_4)_2MoS_4$ and 0.1 M KCl on a conductive poly(ethylene terephthalate) (PETP) substrate [17]. First, a constant potential of -1 V for 300 s was applied to PETP-coated FTO substrate, which acted as a working electrode, Ag/Ag^+ electrode as reference electrode, Pt wire as CE to obtain flexible, semitransparent MoS_2 film on PETP substrate. Hexachloroplatinic acid (H_2PtCl_6) was dispersed in distilled water and ultrasonicated for 10 min for dispersion and add 10 μL of H_2SO_4. The obtained light yellowish color platinic acid solution was dispersed over the flexible conductive PETP substrate and further reduced with $NaBH_4$ [17]. A PEDOT:PSS and PPy composite was deposited on an FTO substrate as CE of a DSSC by a facile electrochemical method [54]. Next, the MWCNT/PPy composite film was synthesized and fabricated on rigid FTO substrate by using a facile electrochemical polymerization route, and served as a CE in a DSSC [56].

However, electroless deposition is an autocatalytic chemical technique used to deposit ionic species present in the electrolyte on the workpiece without using electrical current. The process relies on the presence of reducing agent, which reacts with the ions for deposition [10]. Also, the electrophoretic deposition is focused on the movement of charged particles in a stable suspension. In the electrochemical deposition procedure involving the application of electric charge, the charged particles move toward an oppositely charged workpiece and are deposited there; they also lose their surface charge and are neutralized on the surface of the electrode [10].

6.7.4.2 Gas precursor

A gas precursor (gas phase) can be easily understood as activating gas atoms that use thermal energy or plasma sources to deposit any metal or compound over a substrate's surface. The most commonly used are metal—organic chemical vapor deposition and molecular beam epitaxy. Various chemical reactions such as reduction, pyrolysis, disproportionation, hydrolysis, oxidation, carburization, and nitridation can be used in the chemical vapor deposition (CVD) process to deposit a coating/thin-film layer.

Chemical vapor deposition

CVD reactions are controlled by thermodynamics, mass transport, chemistry of reaction, kinetics of reaction, temperature, pressure, and chemical activity of species present in the reaction chamber. Raj et al. [62] claimed that CVD is an easy fabrication method and proposed a new interesting technique by vertically inclining the molybdenum disulfide (MoS_2) on the FTO substrate. As a result, reflectivity of CEs increased. This technique resulted in an efficiency of about 7.5%, which was better than the Pt-based (7.28%) electrode [62]. Also, Monreal-Bernal et al. [42] applied the same type of technique to the synthesis of CNT fibers onto FTO glass in the presence of a hydrogen atmosphere at 1250°C. Sofyan et al. [39] successfully pyrolyzed carbon from the precursors of table sugar and sucrose by using the dehydrated precursors with sulfate acid and by using a pyrolysis process for CEs for DSSCs. The precursors of table sugar without the addition of a metal catalyst and sucrose with a catalyst produced carbon with particle sizes of 600—900 nm. On the other hand, new BM-biochar was produced using the ball milling of pristine biochar derived from different biomass materials at three pyrolysis temperatures (300, 450, and 600°C). BM-biochar had smaller particle size (140–250 nm compared to 0.5–1 mm for unmilled biochar), greater stability, and more oxygen-containing functional groups (2.2–4.4 mmol/g compared to 0.8–2.9 for unmilled biochar) than the pristine biochar [30].

6.8 Summary

Until the advent of new research, DSSCs needed crucial improvements to enhance their performance, especially photoanodes and CEs. Today's focus is on the best and cheapest materials for the synthesis of electrodes for DSSCs. A lot of research has been done to overcome the problems associated with photoanodes such as interfacial contact and reactions, light scattering, and absorption ability. In this chapter, improvements reported by other researchers such as new semiconductors as photoanodes, carbon-based alternative materials, and doping with other materials to form composites have been made. Also, this chapter explained a few common synthesis methods for photoanodes, which can enhance DSSC performance. In addition to CEs, more research is needed to replace electrocatalyst Pt materials with other cheaper materials such as carbon-based material, polymer-based material, or inorganic materials. In addition, alternatives for composite materials are also arousing interest. Besides optimum materials, design and best synthesis procedures are also being researched by the scientific community.

Acknowledgment

The authors would like to acknowledge the financial support of the Higher Institution Centre of Excellence (HICoE) Program Research and UM Power Dedicated Advanced Centre (UMPEDAC), HICOE-UMPEDAC-2018 Grant.

References

[1] B. O'Regan, M. Grätzel, A low-cost, high-efficiency solar cell based on dye-sensitized colloidal TiO_2 films, Nature 353 (1991) 737.

[2] A. Yella, H.W. Lee, H.N. Tsao, C.Y. Yi, A.K. Chandiran, Porphyrin-sensitized solar cells with cobalt (II/III)-based redox electrolyte exceed 12 percent efficiency (vol 334, pg 629, 2011), Science 334 (6060) (2011) 1203.

[3] S. Mathew, A. Yella, P. Gao, R. Humphry-Baker, B.F.E. Curchod, N. Ashari-Astani, I. Tavernelli, U. Rothlisberger, M.K. Nazeeruddin, M. Grätzel, Dye-sensitized solar cells with 13% efficiency achieved through the molecular engineering of porphyrin sensitizers, Nat. Chem. 6 (3) (2014) 242–247.

[4] K. Kakiage, Y. Aoyama, T. Yano, K. Oya, J.-i. Fujisawa, M. Hanaya, Highly-efficient dye-sensitized solar cells with collaborative sensitization by silyl-anchor and carboxy-anchor dyes, Chem. Commun. 51 (88) (2015) 15894–15897.

[5] Q.H. Li, Q.W. Tang, H.Y. Chen, H.T. Xu, Y.C. Qin, B.L. He, Z.C. Liu, S.Y. Jin, L. Chu, Quasi-solid-state dye-sensitized solar cells from hydrophobic poly(hydroxyethyl methacrylate/glycerin)/polyaniline gel electrolyte, Mater. Chem. Phys. 144 (3) (2014) 287–292.

[6] F.J. Li, F.Y. Cheng, J.F. Shi, F.S. Cai, M. Liang, J. Chen, Novel quasi-solid electrolyte for dye-sensitized solar cells, J. Power Sources 165 (2) (2007) 911–915.

[7] M.J. Yun, Y.H. Sim, S.I. Cha, D.Y. Lee, Leaf anatomy and 3-D structure mimic to solar cells with light trapping and 3-D arrayed submodule for enhanced electricity production, Sci. Rep. 9 (1) (2019) 10273.

References

[8] M.R. Narayan, Dye sensitized solar cells based on natural photosensitizers, Renew. Sustain. Energy Rev. 16 (2012) 208−215.

[9] Y. Koyama, T. Miki, X. Wang, H. Nagae, Dye-sensitized solar cells based on the principles and materials of photosynthesis: mechanisms of suppression and enhancement of photocurrent and conversion efficiency, Int. J. Mol. Sci. 10 (2009) 4575−4622.

[10] M. Ahmad, D.A. Pandey, N. Rahim, Advancements in the development of TiO 2 photoanodes and its fabrication methods for dye sensitized solar cell (DSSC) applications. A review, Renew. Sustain. Energy Rev. 77 (2017) 89−108.

[11] J.S. Shaikh, N.S. Shaikh, S.S. Mali, J.V. Patil, K.K. Pawar, P. Kanjanaboos, C.K. Hong, J.H. Kim, P.S. Patil, Nanoarchitectures in dye-sensitized solar cells: metal oxides, oxide perovskites and carbon-based materials, Nanoscale 10 (11) (2018) 4987−5034.

[12] M. Narayan, A. Raturi, Deposition and characterisation of titanium dioxide films formed by electrophoretic deposition, Int. J. Mater. Eng. Innovat. 3 (2012) 17−31.

[13] M.R. Hoffmann, S. M., W. Choi, D.W. Bahnemann, Environmental applications of semiconductor photocatalysis, Chem. Rev. 95 (1) (1995) 69−96.

[14] S. Zubaidah, C.w. Lai, J.C. Juan, S.B. Abd Hamid, Reduced graphene oxide−titania nanocomposite film for improving dye-sensitized solar cell (DSSCs) performance, Curr. Nanosci. 13 (2017) 494−500.

[15] A.G. Dhodamani, K.V. More, S.M. Patil, A.R. Shelke, S.K. Shinde, D.Y. Kim, S.D. Delekar, Synergistics of Cr(III) doping in TiO_2/MWCNTs nanocomposites: their enhanced physicochemical properties in relation to photovoltaic studies, Sol. Energy 201 (2020) 398−408.

[16] G. Kanimozhi, S. Vinoth, H. Kumar, E.S. Srinadhu, N. Satyanarayana, A novel electrospun cobalt-doped zinc oxide nanofibers as photoanode for dye-sensitized solar cell, Mater. Res. Express 6 (2) (2019).

[17] M. Gurulakshmi, A. Meenakshamma, G. Siddeswaramma, K. Susmitha, Y.P. Venkata Subbaiah, T. Narayana, M. Raghavender, Electrodeposited MoS_2 counter electrode for flexible dye sensitized solar cell module with ionic liquid assisted photoelectrode, Sol. Energy 199 (2020) 447−452.

[18] K. Pattarith, Y. Areerob, Fabrication of Ag nanoparticles adhered on RGO based on both electrodes in dye-sensitized solar cells (DSSCs), Renew. Wind, Water, Sol. 7 (1) (2020) 1.

[19] M.D. Ye, X.R. Wen, M.Y. Wang, J. Iocozzia, N. Zhang, C.J. Lin, Z.Q. Lin, Recent advances in dye-sensitized solar cells: from photoanodes, sensitizers and electrolytes to counter electrodes, Mater. Today 18 (3) (2015) 155−162.

[20] P. Varshney, M. Deepa, S.A. Agnihotry, K.C. Ho, Photo-polymerized films of lithium ion conducting solid polymer electrolyte for electrochromic windows (ECWs), Sol. Energy Mater. Sol. C 79 (4) (2003) 449−458.

[21] J.W. Gong, J. Liang, K. Sumathy, Review on dye-sensitized solar cells (DSSCs): fundamental concepts and novel materials, Renew. Sustain. Energy Rev. 16 (8) (2012) 5848−5860.

[22] A.t.Z. Marketa Zukalova´, L. Kavan, M.K. Nazeeruddin, L. Paul, M. Gratzel, Organized mesoporous TiO_2 films exhibiting greatly enhanced performance in dye-sensitized solar cells, Nano Lett. 5 (2005) 1789−1792.

[23] A.K. Bharwal, L. Manceriu, F. Alloin, C. Iojoiu, J. Dewalque, T. Toupance, C. Henrist, Bimodal titanium oxide photoelectrodes with tuned porosity for improved light harvesting and polysiloxane-based polymer electrolyte infiltration, Sol. Energy 178 (2019) 98−107.

[24] S. Shaban, S.S. Pandey, M.Q. Lokman, F. Ahmad, M.N. Hamidon, N.F.M. Sharif, Back-illuminated dye-sensitized solar cell flexible photoanode on titanium foil, ASM Sci. J. 12 (4) (2019).

[25] R.T. Ako, P. Ekanayake, A.L. Tan, D.J. Young, Z. Zhang, V. Chellappan, S.S. Gomathy, R.L.N. Chandrakanthi, Enhanced efficiency of dye-sensitized solar cells based on Mg and La co-doped TiO_2 photoanodes, Electrochem. Acta 178 (2015) 240−248.

[26] H. Tsubomura, M. Matsumura, Y. Nomura, T. Amamiya, Dye sensitised zinc oxide: aqueous electrolyte: platinum photocell, Nature 261 (1976) 402.

[27] O. Lupan, V.M. Guérin, L. Ghimpu, I.M. Tiginyanu, T. Pauporté, Nanofibrous-like ZnO layers deposited by magnetron sputtering and their integration in dye sensitized solar cells, Chem. Phys. Lett. 550 (2012) 125−129.

[28] M.G.C.M. Kumari, C.S. Perera, B.S. Dassanayake, M.A.K.L. Dissanayake, G.K.R. Senadeera, Highly efficient plasmonic dye-sensitized solar cells with silver nanowires and TiO_2 nanofibres incorporated multi-layered photoanode, Electrochim. Acta 298 (2019) 330−338.

[29] Y.J. Jang, Y.H. Jang, D.H. Kim, Nanostructured carbon-TiO2 shells onto silica beads as a promising candidate for the alternative photoanode in dye-sensitized solar cells, Sci. Adv. Mater. 7 (2015) 956−963.

[30] H. Lyu, Z. Yu, B. Gao, F. He, J. Huang, J. Tang, B. Shen, Ball-milled biochar for alternative carbon electrode, Environ. Sci. Pollut. Res. 26 (2019).

[31] J. Theerthagiri, A.R. Senthil, J. Madhavan, T. Maiyalagan, Recent progress in non-platinum counter electrode materials for dye-sensitized solar cells, ChemElectroChem 2 (7) (2015) 928−945.

[32] A. Bolhan, N.A. Ludin, N.S.M. Nor, M.A. Ibrahim, S. Sepeai, M.A.M. Teridi, K. Sopian, A. Zaharim, Catalytic performance of Pt/rGO using stacked layer technique for DSSC counter electrode, J. Kejuruter 31 (1) (2019) 115−122.

[33] S. Gullace, F. Nastasi, F. Puntoriero, S. Trusso, G. Calogero, A platinum-free nanostructured gold counter electrode for DSSCs prepared by pulsed laser ablation, Appl. Surf. Sci. 506 (2020) 144690.

[34] T.N. Murakami, M. Grätzel, Counter electrodes for DSC: application of functional materials as catalysts, Inorg. Chim. Acta. 361 (2008) 572−580.

[34a] E. Olsen, G. Hagen, L.S. Eric, Dissolution of platinum in methoxy propionitrile containing LiI/I_2, Sol. Energy Mater. Sol. Cells 63 (2000) 267−273.

[35] S. Xu, Y. Luo, W. Zhong, Investigation of catalytic activity of glassy carbon with controlled crystallinity for counter electrode in dye-sensitized solar cells, Sol. Energy 85 (11) (2011) 2826−2832.

[36] S.-j. Xu, Y.-f. Luo, W. Zhong, Z.-h. Xiao, An efficient dye-sensitized solar cell using surfactant-modified mesoporous carbon film as a counter electrode, N. Carbon Mater. 28 (4) (2013) 254−260.

[37] G. Wang, S. Kuang, D. Wang, S. Zhuo, Nitrogen-doped mesoporous carbon as low-cost counter electrode for high-efficiency dye-sensitized solar cells, Electrochim. Acta 113 (2013) 346−353.

[38] S. Xu, C. Liu, J. Wiezorek, 20 renewable biowastes derived carbon materials as green counter electrodes for dye-sensitized solar cells, Mater. Chem. Phys. 204 (2018) 294−304.

[39] N. Sofyan, Muhammad, A. Ridhova, A.H. Yuwono, A. Udhiarto, Characteristics of carbon pyrolyzed from table sugar and sucrose for Pt-less dssc counter electrode, Int. J. Technol. 9 (2) (2018) 372−379.

[40] G. Veerappan, K. Bojan, S.-W. Rhee, Sub-micrometer-sized graphite as a conducting and catalytic counter electrode for dye-sensitized solar cells, ACS Appl. Mater. Inter. 3 (3) (2011) 857−862.

[41] V.-D. Dao, S.-H. Jung, J.-S. Kim, Q.C. Tran, S.-A. Chong, L.L. Larina, H.-S. Choi, AuNP/graphene nanohybrid prepared by dry plasma reduction as a low-cost counter electrode material for dye-sensitized solar cells, Electrochim. Acta 156 (2015) 138−146.

[42] A. Monreal-Bernal, J.J. Vilatela, R.D. Costa, CNT fibres as dual counter-electrode/current-collector in highly efficient and stable dye-sensitized solar cells, Carbon 141 (2019) 488−496.

[43] D. Sebastián, V. Baglio, M. Girolamo, R. Moliner, M.J. Lázaro, A.S. Aricò, Carbon nanofiber-based counter electrodes for low cost dye-sensitized solar cells, J. Power Sources 250 (2014) 242−249.

[44] A.A. Arbab, M. Ali, A.A. Memon, K.C. Sun, B.J. Choi, S.H. Jeong, An all carbon dye sensitized solar cell: a sustainable and low-cost design for metal free wearable solar cell devices, J. Colloid Interface Sci. 569 (2020) 386−401.

[45] E. Meyer, J. Mbese, D. Mutukwa, N. Zingwe, Structural, morphological and electrochemical characterization of hydrothermally fabricated PdNiCo and PdNiCo-rGO alloys for use as counter electrode catalysts in DSSC, Materials 12 (19) (2019).

[46] S. Bashir, Y.Y. Teo, S. Ramesh, K. Ramesh, M.W. Mushtaq, Rheological behavior of biodegradable N-succinyl chitosan-g-poly (acrylic acid) hydrogels and their applications as drug carrier and in vitro theophylline release, Int. J. Biol. Macromol. 117 (2018) 454−466.

[47] Z. Chen, L. Fang, Y.F. Chen, Fabrication and photovoltaic performance of counter electrode of 3D porous carbon composite, Acta Phys. Sin. 68 (1) (2019).

[48] S. Nagarajan, P. Sudhagar, V. Raman, W. Cho, K.S. Dhathathreyan, Y.S. Kang, A PEDOT-reinforced exfoliated graphite composite as a Pt- and TCO-free flexible counter electrode for polymer electrolyte dye-sensitized solar cells, J. Mater. Chem. 1 (4) (2013) 1048−1054.

[49] D.O. Miranda, M.F. Dorneles, R.L. Orefice, A facile and low-cost route for producing a flexible hydrogel-PANI electrolyte/counter electrode applicable in dye-sensitized solar cells (DSSC), SN Appl. Sci. 1 (12) (2019).

[50] A. Zatirostami, A new electrochemically prepared composite counter electrode for dye-sensitized solar cells, Thin Solid Films (2020) 137926.

[51] X. Jin, L. You, Z.P. Chen, Q.H. Li, High-efficiency platinum-free quasi-solid-state dye-sensitized solar cells from polyaniline (polypyrrole)-carbon nanotube complex tailored conducting gel electrolytes and counter electrodes, Electrochim. Acta 260 (2018) 905−911.

References   **177**

[52] U. Mehmood, H. Asghar, F. Babar, M. Younas, Effect of graphene contents in polyaniline/graphene composites counter electrode material on the photovoltaic performance of dye-sensitized solar cells (DSSCSs), Sol. Energy 196 (2020) 132–136.

[53] A.N. Amalina, V. Suendo, M. Reza, P. Milana, R.R. Sunarya, D.R. Adhika, V.V. Tanuwijaya, Preparation of polyaniline emeraldine salt for conducting-polymer-activated counter electrode in dye sensitized solar cell (DSSC) using rapid-mixing polymerization at various temperature, Bull. Chem. React. Eng. 14 (3) (2019) 521–528.

[54] G. Yue, J. Wu, Y. Xiao, J. Lin, M. Huang, Z. Lan, Application of poly(3,4-ethylenedioxythiophene):polystyrenesulfonate/polypyrrole counter electrode for dye-sensitized solar cells, J. Phys. Chem. C 116 (34) (2012) 18057–18063.

[55] S.S. Shenouda, I.S. Yahia, H.S. Hafez, F. Yakuphanoglu, Facile and low-cost synthesis of PEDOT:PSS/FTO polymeric counter electrode for DSSC photosensor with negative capacitance phenomenon, Mater. Res. Express 6 (6) (2019).

[56] G. Yue, L. Wang, X.a. Zhang, J. Wu, Q. Jiang, W. Zhang, M. Huang, J. Lin, Fabrication of high performance multi-walled carbon nanotubes/polypyrrole counter electrode for dye-sensitized solar cells, Energy 67 (2014) 460–467.

[57] J.W. Chew, M.H. Khanmirzaei, A. Numan, F.S. Omar, K. Ramesh, S. Ramesh, Performance studies of ZnO and multi walled carbon nanotubes-based counter electrodes with gel polymer electrolyte for dye-sensitized solar cell, Mater. Sci. Semicond. Process. 83 (2018) 144–149.

[58] S. James, R. Contractor, Study on nature-inspired fractal design-based flexible counter electrodes for dye-sensitized solar cells fabricated using additive manufacturing, Sci. Rep. 8 (2018).

[59] R. Cheng, X. Gao, G. Yue, L. Fan, Y. Gao, F. Tan, Synthesis of a novel $MoIn_2S_4$ alloy film as efficient electrocatalyst for dye-sensitized solar cell, Sol. Energy 201 (2020) 116–121.

[60] L.T. Gularte, C.D. Fernandes, M.L. Moreira, C.W. Raubach, P.L. Jardim, S.S. Cava, In situ microwave-assisted deposition of CoS counter electrode for dye-sensitized solar cells, Sol. Energy 198 (2020) 658–664.

[61] K. Subalakshmi, K.A. Kumar, O.P. Paul, S. Saraswathy, A. Pandurangan, J. Senthilselvan, Platinum-free metal sulfide counter electrodes for DSSC applications: structural, electrochemical and power conversion efficiency analyses, Sol. Energy 193 (2019) 507–518.

[62] I. Raj, X. Xiuwen, W. Yang, F. Yang, L. Hou, Y. Li, Highly active and reflective MoS_2 counter electrode for enhancement of photovoltaic efficiency of dye sensitized solar cells, Electrochim. Acta 212 (2016) 614–620.

[63] X. Guo, R. Xu, D. Li, Y. Yang, Q. Tian, One-step electrodeposited $CoSe_2$ nano-reticular with high electroconductivity and electrocatalytic as a counter electrode for dye sensitized solar cell, J. Alloys Compd. (2020) 154712.

[64] A. Majid, I. Ullah, K.T. Kubra, S.U.D. Khan, S. Haider, First principles study of transition metals doped SiC for application as counter electrode in DSSC, Surf. Sci. 687 (2019) 41–47.

[65] H. Xu, C. Zhang, Z. Wang, S. Pang, X. Zhou, Z. Zhang, G. Cui, Nitrogen-doped carbon and iron carbide nanocomposites as cost-effective counter electrodes of dye-sensitized solar cells, J. Mater. Chem. 2 (13) (2014) 4676–4681.

[66] H. Wang, W. Wei, Y.H. Hu, Efficient ZnO-based counter electrodes for dye-sensitized solar cells, J. Mater. Chem. A 1 (2013) 6622–6628.

[67] L. Cheng, Y. Hou, B. Zhang, S. Yang, J.W. Guo, L. Wu, H.G. Yang, Hydrogen-treated commercial WO_3 as an efficient electrocatalyst for triiodide reduction in dye sensitized solar cells, Chem. Commun. 49 (2013) 5945–5947.

[68] Y. Hou, D. Wang, X.H. Yang, W.Q. Fang, B. Zhang, H.F. Wang, G.Z. Lu, P. Hu, H.J. Zhao, H.G. Yang, Rational screening low-cost counter electrodes for dye-sensitized solar cells, Nat. Commun. 4 (1) (2013) 1583.

[69] S. Deb, Electron spin resonance of defects in single crystal and thin films of tungsten trioxide, Phys. Rev. B 16 (1977) 1020.

[70] H.-T. Chen, Y.-J. Huang, C.-T. Li, C.-P. Lee, J.T.s. Lin, K.-C. Ho, Boron nitride/sulfonated polythiophene composite electro-catalyst as the TCO and Pt-free counter electrode for dye-sensitized solar cells: 21% at dim light, ACS Sustain. Chem. Eng. 8 (13) (2020) 5251–5259.

[71] S. Sheehan, P.K. Surolia, O. Byrne, S. Garner, P. Cimo, X. Li, D.P. Dowling, K.R. Thampi, Flexible glass substrate based dye sensitized solar cells, Sol. Energy Mater. Sol. C 132 (2015) 237–244.

6. Introduction to solar energy and its conversion into electrical energy by using dye-sensitized solar cells

[72] M. Zalas, K. Jelak, Optimization of platinum precursor concentration for new, fast and simple fabrication method of counter electrode for DSSC application, Optik (2020) 164314.

[73] S. Ahmadi, N. Asim, M.A. Alghoul, F.Y. Hammadi, K. Saeedfar, N.A. Ludin, S.H. Zaidi, K. Sopian, The role of physical techniques on the preparation of photoanodes for dye sensitized solar cells, Int. J. Photoenergy 2014 (2014) 19, https://doi.org/10.1155/2014/198734.

[74] L.E. Scriven, Physics and applications of DIP coating and spin coating, MRS Proc. 121 (1988) 717.

[75] F.W. Low, C.W. Lai, Recent developments of graphene-TiO 2 composite nanomaterials as efficient photoelectrodes in dye-sensitized solar cells: a review, Renew. Sustain. Energy Rev. 82 (2018) 103−125.

[76] X. Ji, W. Liu, Y. Leng, A. Wang, Facile synthesis of ZnO@TiO$_2$ core-shell nanorod thin films for dye-sensitized solar cells, J. Nanomater. 2015 (2015).

[77] T.M.W.J. Bandara, L.A. DeSilva, J.L. Ratnasekera, K.H. Hettiarachchi, A.P. Wijerathna, M. Thakurdesai, J. Preston, I. Albinsson, B.E. Mellander, High efficiency dye-sensitized solar cell based on a novel gel polymer electrolyte containing RbI and tetrahexylammonium iodide (Hex(4)NI) salts and multi-layered photoelectrodes of TiO$_2$ nanoparticles, Renew. Sustain. Energy Rev. 103 (2019) 282−290.

[78] A. Ramachandran, I. Jinchu, C.O. Sreekala, Studies on polymer based counter electrodes for DSSC application, in: 2016 International Conference on Electrical, Electronics, and Optimization Techniques (ICEEOT), 2016, pp. 4628−4630.

[79] W.O.S. Arsyad, H. Bahar, B. Prijamboedi, R. Hidayat, Revealing the limiting factors that are responsible for the working performance of quasi-solid state DSSCs using an ionic liquid and organosiloxane-based polymer gel electrolyte, Ionics 24 (3) (2018) 901−914.

[80] C. Chen, C. Chen, Inkjet printing and characterization of titanium dioxide and platinum electrodes for dye-sensitized solar cells (DSSCs), in: 2019 IEEE 46th Photovoltaic Specialists Conference (PVSC), 16−21 June 2019, 2019, pp. 0439−0442.

[81] A. Zatirostami, Increasing the efficiency of TiO$_2$-based DSSC by means of a double layer RF-sputtered thin film blocking layer, Optik 207 (2020) 164419.

[82] J. Gong, K. Sumathy, Q. Qiao, Z. Zhou, Review on dye-sensitized solar cells (DSSCs): Advanced techniques and research trends, Renew. Sustain. Energy Rev. 68 (2017) 234−246.

[83] X. Yang, L. Zhou, A. Feng, H. Tang, H. Zhang, Z. Ding, Y. Ma, M. Wu, S. Jin, G. Li, Synthesis of nickel sulfides of different phases for counter electrodes in dye-sensitized solar cells by a solvothermal method with different solvents, J. Mater. Res. 29 (8) (2014) 935−941.

CHAPTER 7

Dye-sensitized solar cells

Prashant K. Baviskar[1], Babasaheb R. Sankapal[2]

[1]Department of Physics, SN Arts, DJ Malpani Commerce and BN Sarda Science College (Autonomous), Sangamner, Maharashtra, India; [2]Nano Materials and Device Laboratory, Department of Physics, Visvesvaraya National Institute of Technology, Nagpur, Maharashtra, India

7.1 Introduction

An increasing global population requires the need to find alternative energy supplies to nonrenewable energy sources that we have been using for decades. Although in many developed countries solar energy is being used widely, it cannot reach the expectations needed to solve the energy demands of today and the future. Photovoltaics is a field of research and technology related to energy by converting sunlight directly into electricity. If module manufacturing cost can be significantly reduced, then it is possible to generate electricity using groundbreaking solar cell technology that can be sufficient to meet the costs associated with grid parity. There are few ways to reduce manufacturing costs as:

Moving away from:

1. clean environment deposition processes
2. high processing temperatures
3. long processing time, and
4. rigid, heavy, and expensive substrates.

And moving toward:

1. chemical deposition process,
2. low-temperature processing,
3. short processing times,
4. low-cost active materials, and
5. flexible, light, & inexpensive substrates.

Energy Materials
https://doi.org/10.1016/B978-0-12-823710-6.00020-0

Copyright © 2021 Elsevier Ltd. All rights reserved.

180   7. Dye-sensitized solar cells

"Higher efficiencies have been reported for a variety of solar cells manufactured by expensive techniques with clean room technologies and are mostly imported from abroad. Hence, there is a prevailing need to develop low-cost solar cells. Although efficiency may be less, solar cells can be used generally and, most importantly, they should be affordable to common peoples at lower cost. If solar cells are colorful, they can be used for many applications. These colorful solar cells can be used to replace every color that is contact with sunlight. This can serve two purposes: attractiveness and conversion of light to electricity. The ability to produce low-cost and colorful solar cells could be a game-changer, revolutionizing the field of solar energy generation and protecting human life as well as the environment against the global warming".

Presently, modern molecular photovoltaic (PV) materials have been developed for the production of low-cost solar cells. In this concern, dye-sensitized solar cells (DSSCs) have been attracting a lot of interest due to their moderate photoconversion efficiency and, importantly, low-cost production. In the last decade, the highest certified efficiency value reported by Grätzel et al. was more than 7% with TiO_2 and ruthenium metal dye [1]. The same group with Yella et al. achieved 12.3% efficiency for porphyrin-sensitized TiO_2 with cobalt (II/III)-based redox electrolyte [2], followed by 13% efficiency reported by the molecular engineering of porphyrin sensitizers [3]. Recently, a groundbreaking cell gave a power conversion efficiency of 13.6% for DSSCs by minimizing energy losses with a single dye [4]. DSSCs have numerous advantages: mainly low-cost production [5], they work under ambient light and show a remarkable efficiency of 28.9% [6], they are flexible [7], and see-through and multicolor options are also available [8]. Different dyes have been utilized for DSSC applications in combination with various wide bandgap semiconductor metal oxides (MOs) as photoanodes.

7.2 Construction and working of DSSCs

DSSCs provide an economically credible and technically simpler alternative to p-n junction based PV devices. In traditional silicon solar cells, the semiconductor performs the dual task of charge generation after light absorption and charge transport. These two functions are distinct in DSSCs where charge generation is achieved in the sensitizer dye after light absorption and charge transport is accomplished through a semiconductor metal oxide (MO). The sensitizer (dye) is anchored over a high surface area, wide bandgap semiconductor oxide and charge separation takes place at the interface via photo-induced electron injection from the dye into the conduction band of the semiconductor oxide. Carriers are transported in the conduction band of the semiconductor to the charge collector [9]. The use of wider absorption band sensitizers in conjunction with oxide films of nanocrystalline morphology permits the harvesting of a large fraction of sunlight. The maximum conversion of incident photons to electric current is achieved by the spectral range of dye extending from the ultraviolet to the near-infrared regions.

7.2.1 Device construction

Device construction consist of a compact/porous layer of n-type wide band-gap semiconductor MO (TiO_2/ZnO) coated over transparent conducting oxide (TCO)-coated glass substrate. Usually, fluorine or indium-doped tin oxide (FTO/ITO)-coated glass substrates are commonly used as TCO layers. This n-type semiconductor film consists of nanoparticles

II. Photovoltaic materials and devices

(diameter ~20–50 nm) leading to a high surface area of the monolayer for dye adsorption. The dye-sensitized photoanode and platinum-coated TCO counter electrodes are sandwiched together using spacers over dye-coated semiconductor MO film. The liquid electrolyte (I^-/I^{-3}) is incorporated between the sensitizer and the counter electrode by using capillary action. An exposed area of known dimensions is created by using a black-painted mask. Schematic depiction of the construction of a device-grade DSSC is represented in Fig. 7.1A.

7.2.2 Device workings

A schematic energy-level band diagram exhibiting the band edge positions of MO, ground (highest occupied molecular orbital (HOMO) and excited (lowest unoccupied molecular orbital (LUMO) levels of the dye, work functions of electrolyte, and platinum counter electrode is presented in Fig. 7.1B. In the case of DSSCs, liquid electrolyte works as a hole conductor. The currently accepted mechanism of photosensitization of a wide bandgap semiconductor MO by an adsorbed dye is illustrated in the Fig. 7.1B. After light illumination, the dye promotes electrons from the ground state (HOMO) to an excited state (LUMO). The MO film does not absorb much of the sunlight because of its wide bandgap (>3.2 eV). If the oxidation potential of the excited state of the dye is more negative than the semiconductor conduction band edge, a positively charged dye molecule (hole) is created by the excited dye molecules resulting in the injection of electrons into the MO film; this phenomenon produces the charge separation required for the PV effect. The oxidized dye is reduced to the ground state by electron donor I^- from the electrolyte and converted to I_3^-. Open-circuit voltage (V_{oc}) is thus expressed as the difference between I^-/I_3^- redox potential and the Fermi energy level close to the n-type semiconductor conduction band. The electrons return to the cell and are captured by a redox species at the second electrode (generally, a metal like platinum) to

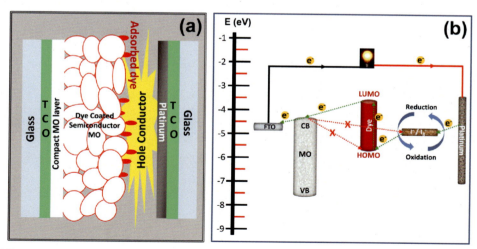

FIGURE 7.1 Schematic (A) construction and (B) working using the energy band diagram of a dye-sensitized solar cell. *CB*, Conduction band; *FTO*, fluorine-doped tin oxide; *HOMO*, highest occupied molecular orbital; *LUMO*, lowest unoccupied molecular orbital; *MO*, metal oxide; *TCO*, transparent conducting oxide; *VB*, valence band.

complete the circuit, which finally reduces the electrolyte to I^-, where it donates electrons to the positively charged dye anchored on the MO surface to counterbalance it, reverting the dye molecules to their original state. In this manner, the cycle is completed and electricity is generated [10–12].

The working principle of DSSCs is considerably different from that of the first two generations of solar cells. DSSC technology is also referred as artificial photosynthesis due to the way in which it mimics natural absorption of light energy. Fig. 7.2 schematically illustrates the fundamental processes occurring in DSSCs along with the typical time constants of the forward and recombination reactions, which include the following steps as described earlier [10–14].

Step 1: Initially, the dye is in its ground state (*D*). After illumination of light, dye absorbs the photon that determines the promotion of an electron in the excited state (*D**), Eq. 7.1.

Step 2: Injection of electrons from excited dye (*D**) into the conduction band of the metal oxide leads to the generation of oxidized dye (*D*$^+$), Eq. 7.2.

The injected electron is transported through the semiconductor MO driven by a chemical diffusion gradient [15] and is collected at the TCO electrode and then moved to the external circuit.

Step 3: The regeneration of oxidized dye (*D*$^+$) takes place by accepting electrons from the iodide (I^-) in the redox electrolyte, Eq. 7.3.

Step 4: Finally, electrons reach the counter electrode through an external circuit and react with the redox electrolyte turning it into its reduced form, i.e., the iodide (I^-) is regenerated by reduction of the triiodide (I_3^-), Eq. 7.4.

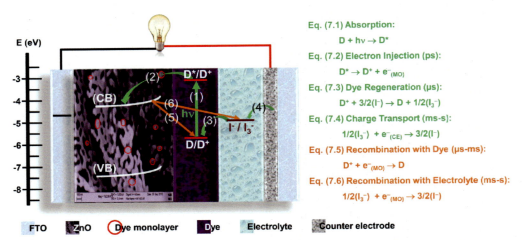

FIGURE 7.2 Kinetic electron transfer processes in dye-sensitized solar cells indicating typical time constants of the forward reactions and recombination reactions. *CB*, Conduction band; *FTO*, fluorine-doped tin oxide; *VB*, valence band.

In addition to these four steps, there are some unwanted processes that can also takes place during the operation of DSSCs which lead to the reduction of device performance.

Step 5: The injected electrons recombine with oxidized dye, Eq. 7.5.
Step 6: The injected electrons recombine with the oxidized state of the redox electrolyte, Eq. 7.6.

Table 7.1 shows the differences between DSSCs and traditional p-n junction solar cells.

7.3 Materials requirement for DSSCs

7.3.1 Substrate (TCO)

Commercially available TCO substrates such as FTO (F:SnO$_2$) or ITO(In:SnO$_2$)-coated glass substrates can be used. For flexible solar cells, ITO-coated poly(ethylene terephthalate) can be used.

7.3.2 Metal oxide semiconductor

A wide bandgap n-type MO semiconductor having a nanoporous structure with interconnected nanoparticles leading to high surface area for dye adsorption with appropriate

TABLE 7.1 How dye-sensitized solar cells are differ from traditional p-n junction solar cells.

Dye-sensitized solar cells	p-n junction solar cells
Two-step photovoltaic process, i.e. charge generation and charge transfer take place at two different places	One-step photovoltaic process, i.e. charge generation and charge transfer take place at the interface
Light absorption occurs in dye not in the semiconductor	Light absorption occurs in the semiconductor
Only majority carriers are present	Majority and minority carriers are present
Unipolar charge transport	Dipolar charge transport
Charge carrier separation due to chemical rather than electrical potential gradient	Charge carrier separation due to built in electric field
Recombination can be reduced	Recombination cannot be reduced
Large surface area	Small surface area
Less sensitive to angle of incidence of radiation and ambient temperature changes	More sensitive to angle of incidence of radiation and ambient temperature changes
Colorful	Black or blue in color
Low-cost device fabrication	High-cost device fabrication
Moderate efficiency	High efficiency

thickness and energy band alignment can be used. Prior to the nanoporous layer, a compact layer is sometimes necessary as a blocking layer to prevent a short circuit, i.e., direct contact between counter electrode and conducting TCO through the hole conductor. The material requirements of MOs are as follows: (1) the semiconductor should have a wide bandgap so that the complete visible part of the solar spectrum to pass through, (2) the energy level of the conduction band should be suitably set to accept electrons from the excited dye molecules (sensitizer), (3) the material needs to have high charge carrier mobility when formed as a porous film, and (4) films fabricated from the material need to deliver the surface area multiplicity necessary for efficient light harvesting from a monolayer of a surface-adsorbed sensitizer. Usually, TiO_2, ZnO, and Nb_2O_5 can be used as a wide bandgap n-type semiconductor.

7.3.3 Photosensitizer (dye)

Dyes employed in highly efficient DSSCs have to meet several requirements: organic dyes, which are sensitive to light (photosensitive) and can absorb a wider range of the sun's spectrum, the energy level of the ground/excited state, aromatic dyes (containing one or more benzene rings), constant rate of charge injection/recombination, and stability. Ru metal-based dyes (N3, N719, Z907, etc.) were frequently used as photosensitizer and were commonly used to anchor onto TiO_2 with the −NCS group. Also, the color of organic dyes is an attractive point for commercialization. A series of dyes with a carboxylic group (Eosin Y, Mercurochrome, Coumarin 343, Rose Bengal, Rhodamine B, D149, D102, etc.) were also used in DSSCs, mostly used with ZnO. The use of Ru metal-free organic dye as a sensitizer is a cost effective towards device fabrication.

7.3.4 Hole conductor

Three aspects are necessary for any hole conductor in DSSCs: (1) it must be able to regenerate the oxidized dye after the sensitizer has injected the electrons in the MO semiconductor, (2) it must be able to be deposited within the porous network of MO nanocrystalline material (in the case of solid-state DSSCs), (3) it must be transparent, or, if it absorbs light, it must be as efficient as the sensitizer in electron injection. The redox couple of iodide/triiodide is usually used as a liquid electrolyte, which serves as a hole conductor. Recently, several metal complexes designed for redox couples have been reported. Co-complexes showed smaller absorption coefficients with relatively high conversion efficiencies [16]. Cu complexes showed lower (more positive) redox potential with relatively high V_{oc} [17], whereas 2,20,7,70-tetrakis(N,N-di-p-methoxyphenylamine)-9,90-spirobifluorene (spiro-OMeTAD), poly(3,4-ethylenedioxythiophene) (PEDOT), poly(3-hexylthiophene-2,5-diyl) (P3HT), etc. as organic hole conductors along with p-type inorganic semiconductors such as CuSCN, CuI, NiO, etc. were used.

7.3.5 Counter electrode

Optimal properties of the counter electrode used in DSSCs are: low cost, high conductivity, catalytic activity, reflectivity, porosity, and surface area with optimum thickness. It should have high mechanical, chemical, and electrochemical stability, chemical corrosion resistance, an appropriate energy level that matches the potential of the redox electrolyte, and also good adhesion with TCO, etc. Platinum-coated TCO or graphite is generally used as a counter electrode. Recently, MOs & metal chalcogenides are being used to replace expensive platinum.

7.3.6 Sealing material

A sealing material is required to prevent leakage of the liquid electrolyte and evaporation of the solvent, which also separates the photoanode and counter electrode. Chemical and photochemical stability of the sealing material against the electrolyte component, iodine, and solvent is required. Surlyn (DuPont) is a well-known example; a copolymer of ethylene and methacrylic acid is used as a coating and packaging material for this purpose. Nowadays, glass frit is also used as a sealing material, which is thermally, mechanically, and chemically stable and can be coated by using screen printing technique [18].

7.4 Solar cell performance

7.4.1 Air mass 1.5 global

A standard solar cell is tested under air mass 1.5 global (AM1.5G) conditions. This is the energy flux of the sun after solar irradiation has passed through an atmosphere that is 1.5 times thicker than the atmosphere at the equator ($100\,\mathrm{mW/cm^2}$ or $1000\,\mathrm{W/m^2}$) when the solar zenith angle is 48.2 degrees.

7.4.2 Current—voltage characteristics

The PV performance of DSSCs is estimated from the current—voltage ($I{-}V$) curve. The output PV parameters of DSSCs are short-circuit current (I_{sc}), open-circuit potential (V_{oc}), fill factor (FF), and power conversion efficiency (η). These parameters are measured under the standard condition of $100\,\mathrm{mW/cm^2}$ incident solar radiation and the spectral power distribution of AM1.5G at a temperature of 25°C.

The $J{-}V$ characteristic of a DSSC is shown in Fig. 7.3. The photocurrent per unit active area (A) of the photoelectrode when the applied potential across the DSSCs is zero is called short-circuit current density (J_{sc}) and is $I_{sc}/A = J_{sc}$. The value of J_{sc} is effectively dependent on the efficiency of the charge injection from the excited dye to the conduction band of MO illuminated under standard conditions. It is assumed that the V_{oc} of a photoelectrochemical cell is nearly constant as the standard redox potential of electrolyte is constant. However, by tuning the flat band (V_{FB}) of MO by introducing MO composites or dopants into MOs, one can control the V_{oc} of a photoelectrochemical cell [19].

7.4.3 Output parameters of DSSCs

The key factors of a DSSC in terms of its components have been discussed. Now, think of a DSSC from a different perspective—how the output parameters (photocurrent, photovoltage, FF) depend on the cell components.

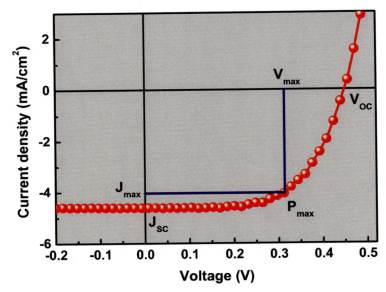

FIGURE 7.3 $J-V$ curve under 1 sun illumination and output parameters of dye-sensitized solar cells.

7.4.3.1 Photocurrent

The photocurrent (I_{sc}) physically expresses the total current drawn from the system, and will be higher if the dye can absorb over an extensive range of the solar spectrum.

How to enhance the photocurrent

- Minimize recombination by controlling the rate of back electron transfer from the semiconductor to the electrolyte (in particular, to the triiodide), which enhances greatly the rate of electron injection from dye to semiconductor.
- Prevent excessive electron build-up and, as a consequence, potential recombination by removing injected electrons as quickly as possible using back contact.
- Reduce recombination by selecting the right polarity of the Helmholtz field at the semiconductor/electrolyte interface [20].
- In the case of the tri-iodide, negatively charge the hole carrier, which can significantly reduce recombination by electron repulsion (increased scattering cross-section).

The energy level of the absorption edge of the dye should be as low as possible to allow absorption of maximum photons. However, under illumination, the LUMO level of the sensitizer must be above the conduction band of the semiconductor MO and the HOMO level must be below the actual redox level at the surface of the semiconductor. Also, the driving forces for charge separation, which are the difference between the HOMO and redox levels as well as that between the LUMO and conduction band level, should never be zero, which means that the absorption edge of the dye is always higher than the maximum achievable photovoltage, so that the dye absorption edge cannot be too small.

7.4.3.2 Photovoltage

Photovoltage (V_{oc}) physically reflects the potential difference between the Nernst potential of the redox mediator in the electrolyte and the Fermi level of the semiconductor MO. One can increase the value of V_{oc} by shifting the conduction band of the semiconductor MO more negatively by surface engineering or instead by shifting the mediator potential more positively with different redox couples.

How to improve the photovoltage

Under dark condition the Fermi level of the semiconductor should be away as possible from the MO conduction band leads to the electron injection. In principle, a higher voltage can be achieved by injecting electrons into a p-type material. However, there are some practical difficulties, such as when electron mobility in a hole conductor is very poor due to the presence of negative scattering centers [21]. The requirements from the light absorber for maximum photocurrent and photovoltage are contradictory in any PV cell. For maximum photovoltage, the energy level of the dye absorption edge should be as high as possible. The redox electrolyte should be chosen so that, under operating conditions, the redox potential of the electrolyte is slightly higher in energy than the HOMO level of the dye, while the LUMO level of the dye should be slightly higher than the semiconductor conduction band edge. This would be the ideal operating condition for achieving maximum photovoltage. Electrolyte diffusion should be optimized. This is challenging because of the porous semiconductor structure, which will tend to increase the concentration of the oxidized redox species at the photoelectrode, resulting in a shift of its potential (photovoltage reduction). This effect would also increase recombination, thereby affecting the overall cell performance. In general, photovoltage and fill factor (FF) increase with decrease in recombination.

7.4.3.3 Fill factor

Graphically, the FF represents the measure of "square" of the $I-V$ curve of a solar cell under illumination. Actually, the quality of the fabricated solar cell is stated by the FF and it increases with shunt resistance and decreases with series resistance and overpotentials [22].

How to optimize the fill factor

- To reduce recombinations: Various factors concerned with recombination were already discussed in the foregoing sections. Specifically, as the Fermi level in the semiconductor moves toward the conduction band, the charge distribution at the surface of the semiconductor particles will change, which affects the potential that leads to a increase in efficacy and a decrease in recombination.
- The counter electrode should be a good electrocatalyst for the oxidation reaction, which prevents the increasing loss of photovoltage because of overpotential at the counter electrode as the photocurrent increases.

FF is the most significant factor of a solar cell. It is the ratio of maximum power output ($P_{max} = J_{max} \cdot V_{max}$) and the product of J_{sc} and V_{oc}:

$$FF = \frac{J_{max} \cdot V_{max}}{J_{SC} \cdot V_{OC}}$$

188

7. Dye-sensitized solar cells

The FF of the solar cell is mainly affected by series resistance (R_s) of the cell. The value of R_s is related to the slope of the tangent line to the $I-V$ curve at zero V_{oc}. Series resistance is a combination of the resistance at the FTO glass, the charge-transfer resistance at the counter electrode and electrolyte interface, and Warburg impedance relative to the Nernst diffusion of the I_3^- species in the electrolyte (Z_n), which is the impedance related to charge transfer and recombination at the MO/dye/electrolyte interfaces [24].

7.4.3.4 Efficiency

The overall power conversion efficiency (η) of the PV cell is calculated from the short-circuit photocurrent density (J_{sc}), open-circuit photovoltage (V_{oc}), FF of the cell, and the incident light intensity ($P_{in} = 100 \text{ mW/cm}^2$). Therefore, from Fig. 7.3, it can be concluded that the more square-like $J-V$ curve is essential for achieving the maximum value of FF. The shunt resistance increases with reduction of electron loss between the FTO/electrolyte interface leading to increases in FF. On the counterpart, an increase in back electron transfer and charge recombination results in poor FF. Finally, efficiency (η) can be measured as the ratio of total power output (P_{out}) to solar power input (P_{in}):

$$\eta = \frac{P_{out}}{P_{in}} = \frac{J_{SC} \cdot V_{OC} \cdot FF}{P_{in}}$$

7.4.3.5 Series and shunt resistances

A parasitic effect causes power loss in a solar cell by internal resistances, namely series resistance (R_s) and shunt resistance (R_{sh}).

Ideally, $R_s = 0$ and $R_{sh} = \infty$. The series and shunt resistance can be estimated from the $J-V$ curve by the inverse of the slope near the open-circuit voltage (i.e., at $J = 0$ for R_s) and near the short-circuit current (i.e., at $V = 0$ for R_{sh}) using the following equations:

$$\frac{1}{R_s} = \left[\frac{dJ}{dV}\right]_{J=0}$$

$$\frac{1}{R_{sh}} = \left[\frac{dJ}{dV}\right]_{V=0}$$

7.4.4 Mott–Schottky

Mott–Schottky analysis of DSSCs under dark conditions was carried out to find the shift in flat band potential of semiconductor MO. In a semiconductor/redox electrolyte interface, equilibrium is attained as the electrons transfer from the valence band of the electrode to the redox system provided that the Fermi level of the semiconductor is above the electrolyte redox potential. Mott–Schottky analysis, which determines the depletion capacitance at the junction, can be applied to find the built-in bias at the photoanode/electrolyte interface.

The analysis is conducted in a three-electrode system with platinum wire as counter electrode, semiconductor MO-coated FTO substrate as working electrode, and Ag/AgCl in saturated NaCl as reference electrode. The bias-dependent interfacial capacitance is related to the flat band potential and can be calculated using the equation [23]:

$$\frac{1}{C^2} = \frac{2}{AN_d e \varepsilon \varepsilon_0} \left(E - E_{FB} - \frac{kT}{e} \right)$$

where C is the interfacial capacitance, A is the area of the semiconductor, N_d is the donor density, e is the elemental charge, ε and ε_0 are the dielectric constant and the vacuum permittivity of the semiconductor, E and E_{FB} are the applied and flat band potentials, T is the temperature, and k is the Boltzmann constant.

One can determine the flat band potential (E_{FB}) by extrapolating the Mott–Schottky plot to the X-axis (i.e., $1/C^2 = 0$):

$$E_{FB} = E - \frac{kT}{e}$$

7.4.5 Electrochemical impedance spectroscopy

Electrochemical impedance spectroscopy (EIS) is a powerful technique for measuring the electrical properties of materials and their interfaces. The charge transport properties of the DSSC are measured using EIS.

Two different representations are usually studied for presenting and analyzing EIS data, namely (1) the Nyquist plot, which is the plot of imaginary impedance (Z'') as a function of real impedance (Z') in the complex plane, and (2) the Bode phase plot, which is the graph of phase (θ) as a function of frequency (f) or some time logarithmic frequency ($\log f$).

Typically, three semicircles are observed in the Nyquist plot with increasing frequency:

1. The low-frequency region reflects the diffusion in the redox electrolyte (Warburg diffusion).
2. The mid-frequency region shows the charge-transfer resistance at the electrolyte/dye/ MO interface and the charge transport in the MO electrode.
3. The high-frequency region represents electrochemical reactions at the counter electrode (charge transfer).

The Bode phase plot shows the signature peak, which represents interfacial electron recombination toward the low-frequency region. Electron lifetime is calculated from the Bode plot using the formula $\tau_e = 1/2\pi f_{peak}$.

7.4.6 Quantum efficiency

An additional measure of DSSC performance is the monochromatic incident photon-to-current conversion efficiency (IPCE) or external quantum efficiency. It is defined as the number of incoming photons at one wavelength that are converted to electrons. It is also defined as the ratio of the number of charge carriers collected in the external circuit to the number of incident photons at a particular excitation wavelength. This measurement contains partial information of the solar cell related to quantum efficiencies of photocurrent generation processes. The following are the steps of the generation processes: (1) the light-harvesting efficiency, which is related to the ability of the dye in absorbing photons, (2) electron injection, which measures the charge transfer from LUMO of the dye to the MO conduction band, and (3) charge collection efficiency, which is the number of electrons that effectively reach the anode, avoiding recombination [25].

Internal quantum efficiency shows how efficiently the numbers of absorbed photons are converted into current. Internal quantum efficiency is obtained by taking the ratio of the IPCE number to the light-harvesting efficiency. It should be remembered that quantum efficiency (QE) measurement is carried out under short-circuit conditions, where electron lifetime is higher. To have efficient DSSCs, it is significant that the high fraction of QE is most important and that it can be achieved with the mesoporous MO having high surface area (a prerequisite for higher QE value).

QE describes how efficiently the light of a specific wavelength is converted into current using the following equation:

$$QE = \frac{\dfrac{h \cdot c}{\lambda} \cdot \dfrac{J_{SC}(\mathrm{mA \cdot cm^{-2}})}{q}}{P_{\mathrm{in}}(\mathrm{mW \cdot cm^{-2}})} = \frac{1240 \cdot J_{SC}(\mathrm{mA \cdot cm^{-2}})}{\lambda(nm) \cdot P_{\mathrm{in}}(\mathrm{mW \cdot cm^{-2}})}$$

where h is Planck's constant, c is the speed of light, λ is the wavelength of the incident light, P_{in} is the intensity of the incident light, and q is the elementary charge.

7.5 Performance of DSSCs

The PV performance of DSSCs mainly depends on four components, namely (1) photoanode material, (2) hole conductor, (3) counter electrode, and (4) light harvester (dye).

7.5.1 Photoanode materials

The mesoporous TiO_2 nanoparticle network is generally used, which shows high surface area for dye adsorption. There have been many alternative electrode materials to the traditional TiO_2, including ZnO [26–29], SnO_2 [30–32], ZrO_2 [33], Nb_2O_5 [34–36], $SrTiO_3$ [37–39], CdO [40], Zn_2SnO_4 [41–43], α-Bi_2O_3 [44], CdS [45], CuO [46], etc. Various electrode morphologies, such as nanorods [47–50], nanotubes [51–54], nanobelts [55–57], nanowires [58–62], nanotips [63], nanocombs [64], nanobeads [65,66], nanosheets [67,68], nanoflakes [69–71], nanotrees [72], nonoforests [73,74], tetrapods [75–77], and hierarchical structures [78,79] were tested to enhance the electron transport. There have also been reports studying core-shell or layer-by-layer structures to reduce charge recombination by depositing the other wide bandgap semiconductor oxide materials along with TiO_2 [80], namely ZnO [81–84], ZrO_2 [85,86], SnO_2 [87–89], Al_2O_3 [90–92], Nb_2O_5 [93–96], and $SrTiO_3$ [97–100].

7.5 Performance of DSSCs

Additional efforts have been made to boost the performance of DSSCs by broadband light confinement using multilayer hierarchical TiO_2 structures having different-sized particles and achieving 11.43% efficiency under 1 sun [101]. Extra efforts have been put forth to enhance the efficiency of DSSCs by depositing a relatively dense and compact layer of MO over the FTO substrate (prior to the deposition of mesoporous MO film) to restrict electron recombination between the FTO and redox electrolyte [102–106].

7.5.2 Light harvester (dye)

Apart from photoanode material, electrolyte, and counter electrode, dye as a sensitizer plays a vital role in DSSCs for the enhancement of efficiency. Dye molecules are used to harvest the sunlight and generate electrons, which then flow through the device. This allows an independent electron injection into the semiconductor conduction band and conversion of visible and near-infrared photons to electricity.

Ru metal-based dyes were generally used as photosensitizers with semiconductor MOs such as N3, N719, and Z907 dyes [107–110]. In contrast, metal-free organic dyes were used as an alternative to well-known Ru metal-based dyes resulting in notable efficiencies [68,111]. Conversely, the metal-free organic dyes have additional benefits, namely they are eco-friendly, cost effective, have a high molar extinction coefficient [112], and possess effortless photophysical and electrochemical tuning properties [113]. Otaka et al. reported multicolored DSSCs and observed 2.7% maximum efficiency for metal-free purple-colored dye-sensitized TiO_2 [114]. A chlorophyll derivative (chlorin e_6 dye)-sensitized TiO_2-based DSSC achieved 6.7% efficiency after molecular engineering of the dye [115]. A more than 2% efficient device was constructed using Eosin Y-sensitized TiO_2 containing Br^-/Br_3^- electrolyte and ZnO containing I^-/I_3^- electrolyte [116,117]. Rose Bengal dye is an alternative to the conventional ruthenium complex, which acts as a photosensitizer, and recorded 0.7% efficiency with ZnO and a $KI-I_2$ redox couple for DSSC applications [118]. Another class of dye, i.e., xanthene dye, emerged as a better alternative to Ru metal dyes and recorded comparable performance under illumination at 200 mW/cm^2. The xanthene dyes show very good stability properties under these conditions [119]. Squaraine sensitizers are a good option to cover the near-infrared region and produce 8.9% power conversion efficiency [120,121]. A high-power conversion efficiency of 12.5% was reported using indenoperylene dye [122]. A number of organic metal-free dyes are available, namely Rhodamine B [65], Coumarin 343 [123], and azo [124] dyes for DSSC applications. There is one more class of organic dye, i.e., indoline dyes (D102, D131, and D149) [125]. A highly efficient DSSC reported a 7.07% efficiency with D149 dye-sensitized ZnO nanosheet [67].

The limited spectral coverage of dyes (400–700 nm), which is smaller than that of crystalline silicon (500–1100 nm), is one of the most important and challenging factors attributed to the lower efficiency of DSSCs [126]. The efficiency of DSSCs can be enhanced by harvesting photons from the main part of the visible spectrum, which appears to be difficult with single metal-free low-cost dyes. Individual organic dyes can only absorb a relatively narrow range of the solar spectrum, which limits the efficiency of DSSCs. Hence, it is necessary to think about the advanced colorful approach.

7.5.3 Hole conductor

Similar to the photoanode, numerous efforts have been attempted to enhance the properties of hole conductors. So far, plenty of different redox couples as liquid electrolytes and solid hole transport materials (HTMs) have been studied. Ideally, the main role of electrolyte is to provide efficient charge transport without any significant light absorption in a noncorrosive environment. Since the advent of DSSCs, iodide/triiodide (I^-/I_3^-) electrolyte has been preferred as the redox complex. The use of I^-/I_3^- has many advantages such as favorable charge-transfer mechanism, fast regeneration of the oxidized dye with the iodide, and exceptionally slow back electron recombination in the MO with the triiodide [127]. The highest certified efficiency of $11.9 \pm 0.4\%$ was reported for DSSCs with I^-/I_3^- redox electrolyte [128]. However, the two-electron redox process is a major disadvantage of this redox system, which results in a large overpotential [129]. This overpotential leads to lower V_{oc}, which is the measure of the difference in the redox potential and Fermi energy level of electrons in the semiconductor MO under illumination. Consequently, to improve the V_{oc}, alternative redox mediators having more positive (lower) potentials are being sought, which can reduce the mismatch with the sensitizer. Other encouraging factors that prompt the search for substitutes for the I^-/I_3^- redox system are corrosive properties and its visible light absorption, which result in a 13% reduction in photocurrent [10]. Alternative redox mediators are, for example, ferrocene/ferrocenium (Fe(0/I)) [130,131], Ni(III/IV) [132,133], and the cobalt complex (Co(II/III)) [134–137]. The copper redox couples are promising substitutes, such as the Cu(I/II) redox process that can involve much lower reorganization energies that enable dye regeneration to proceed with reduced driving force [138–140]; however, many of them experience poor performance due to fast recombination kinetics stemming from the single electron redox reactions [11]. It is observed that Br^-/Br_3^- offers high impedance compared to I^-/I_3^- couples and produces higher V_{oc} [141–144]. Very high photovoltages exceeding 1 V were recorded using tris(bipyridine) cobalt-based electrolytes [145]. This groundbreaking cell achieved a high 12.3% power conversion efficiency with cobalt(II/III)-based redox electrolytes [2].

The use of redox liquid electrolyte creates several technical problems such as: evaporation of solvent, corrosion, photoreaction, degradation, imperfect sealing, and hence the device stability; these can be overcome by replacing the liquid electrolyte with p-type solid HTM, which forms a complete solid-state dye-sensitized solar cell (SS-DSSC). HTM should ideally possess high hole mobility, have high transparency, should not degrade the sensitizer, and be able to interface with the photoanode. Xu et al. suggested using solid HTMs instead of liquid electrolytes in DSSCs, which can solve the solvent evaporation and sealing problems and also improve the stability of the device. They reported SS-DSSC using multilayer TiO_2-coated zinc oxide (ZnO) nanowire arrays sensitized with Z907 dye and spiro-OMeTAD as HTM and observed 5.65% efficiency [146]. A maximum efficiency of 10.2% was reported by Chung et al. for SS-DSSCs using N719 dye-sensitized TiO_2 with $CsSnI_3$ as a hole transport layer [147]. Recently, many research groups have developed different kinds of HTMs as an alternative to traditional liquid electrolytes, for example, organic materials, namely spiro-OMeTAD [148,149], which is one of the most extensively used HTMs for the development of SS-DSSCs, PEDOT [150–152], P3HT [153–158], pentacene [159], polypyrrole [160–162], polyaniline [163–165], polythiophene [166], poly(triphenyldiamine)s [167], and polycarbazoles [168,169]. In addition, some p-type inorganic semiconductors such as Cs_2SnI_6 [170−172], CuI [173−175], CuBr [176], CuSCN [177−182], and NiO [183−186] have also been examined, but they usually experience partial stability, low hole conductivity, incomplete filling, or reduced coverage of the material over mesopore semiconductor MOs and thus, resulting in comparatively poor efficiencies.

7.5.4 Counter electrode

The counter electrode in DSSCs is an essential constituent that collects electrons from the external circuit and catalyzes the redox reduction in an electrolyte. Hence, the counter electrode has a significant influence on the performance, cost, and long-term stability of PV devices.

The counter electrode mainly performs two major tasks: (1) it acts as a catalyst by reducing the redox species, which are the mediators for regeneration of dye after electron injection, and (2) it collects the positive carrier (hole) from HTMs in a DSSC. To improve the performance of DSSCs, the main focus is to enhance the values of I_{sc}, V_{oc}, and FF.

In general, platinum (Pt) is used as a counter electrode but several attempts have been made to find economical alternatives to replace Pt as it is becoming a scarce and expensive noble metal [187]. According to literature surveys, there are numerous promising Pt-free materials for counter electrodes in DSSCs, including graphite [188], graphene [189], reduced graphene oxide [190], carbon nanotubes [191,192], carbon nanofibers [193], and polymers [160,194−202]. Also, there are some metal chalcogenides and oxides such as nickel sulfide (NiS) [203], cobalt sulfide (CoS) [204], CuS [205], MoS_2 [206], $FeSe_2$ [207], α-Fe_2O_3 [208], α-MoO_3 [209], as well as transition-metal carbides [210,211], nitrides [212], selenides [213], etc. that are useful in DSSCs as counter electrode.

7.6 Advanced colorful approach

An excellent potential approach is to use a combination of two or more metal-free low-cost sensitizers that show absorption in the shorter and longer wavelength regions of the visible part of the solar spectrum. This idea of mixing two or more dyes having different absorption regions leads to new color dyes and will help toward improvement in device performance. The approach to obtain wide optical absorption expanding throughout the visible region is to use a blend of two or more dyes that conflict with each other in their spectral response. The idea is promising as the optical properties are found to be additive for two or more sensitizers; this opens up possibilities for testing several combinations of different dyes [214]. However, two diverse organic dyes showing different absorptions in the visible region of the solar spectrum were combined to develop a cocktail dye. This cocktail dye is responsible for the enhancement of absorption spectrum in the visible region. If the absorption spectrum of the dye increases, then light-harvesting ability also increases. This corresponds to improvement in the performance of DSSCs [9]. Several reports are also available on the sensitization of cocktail dye prepared by mixing ruthenium complex and ruthenium metal-free organic dyes with TiO_2 photoelectrodes [215]. It is known that

194   7. Dye-sensitized solar cells

organic sensitizers have higher molar absorption and excitation coefficient than transition-metal complexes, i.e., outstanding photophysical properties [216]. A ZnO tetrapod sensitized with Mercurochrome and a mixture of Mercurochrome and N719 dye shows enhancement in the J_{sc} value due to extra light harvested by the mixed dye [217]. Cao et al. recorded 13.1% certified efficiency for TiO_2 using novel cosensitization [218]. The highly efficient DSSC exhibited 14.3% for collaborative sensitization using silyl-anchor and carboxy-anchor dyes [219]. There are numerous reports available on the enrichment in performance of the DSSC by using a combination of two or more dyes as a cocktail/blend/mix dye sensitized over TiO_2, ZnO, Nb_2O_5, etc. Table 7.2 summarizes the literature reports available on the enhancement of PV parameters of DSSCs with a combination of two or more dyes having contrast absorption spectra [8, 220−247].

Recently, 4.6% efficiency was reported by Sirimanne's group for cocktail dye (a mixture of N719 and black dye) with TiO_2 film [220]. A report is also available for cocktail dye-sensitized ZnO film prepared by mixing expensive metal-free organic indoline dyes. Magne et al. achieved a maximum efficiency of 4.53% for nanoporous ZnO sensitized using cocktail dye with coadsorbent and observed the improvement in performance of PV parameters [221]. Mimicking a simple idea of mixing two individual colors, for example, yellow and red to get orange, has been explored for an advanced colorful approach to enhance the performance of devices. An efficiency boost using the sensitization of blended dye over ZnO films with "Basic idea, advance approach". It is observed that the efficiency for individual dyes ranging from 0.02% to 1.98%, whereas an improvement in efficiency up to 2.45% was recorded for blended dye prepared by mixing Eosin-Y and Coumarin 343 dyes [222]. Recently, the enhancement in efficiency was observed with the same concept of mixing Coumarin 343 and Mercurochrome dyes as sensitizer over an Nb_2O_5 photoanode [223].

Fig. 7.4 is a pictorial representation of an advanced colorful approach to make rainbow-colored solar cells and Fig. 7.5 demonstrates attractive applications of DSSCs to replace any color exposed to sunlight, which can have dual application: light to electricity and attractive color.

7.7 DSSC limitations, commercialization, and future prospects

A DSSC has been the center of attraction for many researchers since its discovery (1991) by Prof. Grätzel. However, there are two main challenges for the commercialization of DSSC technology that have to be overcome. This involves large-area device fabrication and long-term stability with liquid electrolyte. There are a few reports available on large-area fabrication of DSSCs [247,248], but the issue with long-term stability is still a major concern [249]. One can overcome the stability issue of DSSCs by fabricating the device using polymer gel electrolyte or solid HTMs [250−253]. The influence of a photoelectrode and sealant materials as well as module design, DSSC efficiency, and stability were investigated and highlighted by Nursam et al. [254].

G24 Innovations is the world's first commercial company to use DSSC solar technology [255]. The launch of world's first solid-state DSSC module (RICOH EH DSSC series) was announced by Ricoh Company Ltd [256].

II. Photovoltaic materials and devices

7.7 DSSC limitations, commercialization, and future prospects 195

TABLE 7.2 Summary of the literature reports for enhancement of photovoltaic performance of dye-sensitized solar cells with a combination of two or more dyes.

Photoanode	Dye	η (%)	References
TiO_2	Hibiscus	0.41	[224]
	Eosin Y	1.53	
	Mixed dye	2.02	
TiO_2	SQ	2.4	[225]
	Cy0	0.8	
	Cy1	2.3	
	Cy0 + Cy1 + SQ	3.1	
TiO_2	Dye A	1.3	[226]
	Dye B	2.9	
	Mixed dye	3.4	
TiO_2	Red	3.36	[227]
	Blue	1.56	
	Red + blue	4.00	
TiO_2	N719	3.8	[220]
	Black dye	3.0	
	Mixed dye	4.6	
TiO_2	RhCL	0.63	[228]
	N3	2.37	
	N3 + RhCL	4.74	
TiO_2	SQ2	2.81	[229]
	JH-1	5.1	
	SQ2 + JH-1	6.31	
TiO_2	SQ2	4.11	[230]
	JD1	5.44	
	JD1 + SQ2	6.36	
TiO_2	Zn-tri-PcNc1	2.38	[231]
	DH44	5.16	
	Zn-tri-PcNc1 + DH44	6.61	
TiO_2	N719	4.9	[232]

(Continued)

II. Photovoltaic materials and devices

196
7. Dye-sensitized solar cells

TABLE 7.2 Summary of the literature reports for enhancement of photovoltaic performance of dye-sensitized solar cells with a combination of two or more dyes.—cont'd

Photoanode	Dye	η (%)	References
TiO$_2$	N749	3.9	[233]
	Panchromatic dye	7.1	
	SQ1	4.23	
	JK2	7.0	
	SQ1 + JK2	7.43	
TiO$_2$	Y1,	3.44	[234]
	TP2A	4.26	
	HSQ4	5.78	
	Y1 + TP2A + HSQ4	7.48	
TiO$_2$	T4BTD-A	6.4	[235]
	HSQ4	5.8	
	T4BTD-A + HSQ4	7.7	
TiO$_2$	Z907	5.08	[236]
	SQ2	1.39	
	Z907 + SQ2	7.83	
TiO$_2$	GD3	4.3	[237]
	N719	4.1	
	NG-cocktail	8.1	
TiO$_2$	K-60	6.26	[238]
	Y1	—	
	K-60 + Y1	8.19	
TiO$_2$	SPSQ2	3.95	[239]
	N3	7.13	
	SPSQ2 + N3	8.20	
TiO$_2$	D-dye	7.1	[240]
	N719	8.26	
	N719 + D-dye	8.34	
TiO$_2$	N719	6.18	[241]
	CSGR2	—	
	N719 + CSGR2	8.4	

II. Photovoltaic materials and devices

TABLE 7.2 Summary of the literature reports for enhancement of photovoltaic performance of dye-sensitized solar cells with a combination of two or more dyes.—cont'd

Photoanode	Dye	η (%)	References
TiO_2	D35	5.5	[242]
	Dyenamo blue	7.3	
	D35 + dyenamo blue	8.7	
TiO_2	LD12	7.5	[243]
	CD5	5.7	
	LD12 + CD5	9.0	
TiO_2	D131	5.06	[244]
	Black dye	10.04	
	D131 + black dye	11.0	
TiO_2	C106	9.5	[245]
	D131	5.6	
	C106 + D131	11.1	
TiO_2	Black dye	10.7	[257]
	Y1	—	
	Black dye + Y1	11.4	
ZnO	Eosin Y	1.98	[222]
	Coumarin 343	1.16	
	Eosin Y + Coumarin 343	2.45	
ZnO	D131	1.77	[221]
	D149	3.89	
	D149 + D131	4.1	
ZnO	Rose Bengal	2.4	[8]
	Eosin Y	1.5	
	Rhodamine-B	0.6	
	Fast green	0.4	
	Acridine orange	0.06	
	Blended dye	7.9	
Pd-ZnO	Eosin Y	1.31	[246]
	D131	1.76	

(*Continued*)

TABLE 7.2 Summary of the literature reports for enhancement of photovoltaic performance of dye-sensitized solar cells with a combination of two or more dyes.—cont'd

Photoanode	Dye	η (%)	References
Nb$_2$O$_5$	D358	2.37	[223]
	D131 + Eosin Y	2.66	
	D131 + D358	3.03	
	C343	0.35	
	MC	0.64	
	C343 + MC	0.70	

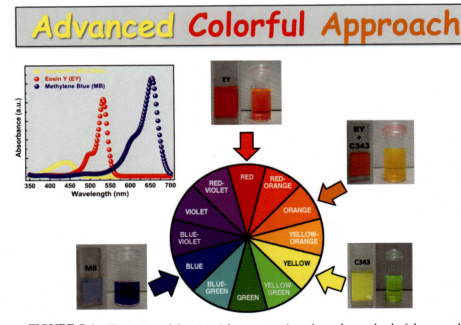

FIGURE 7.4 Illustration of the pictorial representation of an advanced colorful approach.

7.8 Summary and scope

In 1991, Grätzel and coworkers started working on the development of efficient, novel DSSCs by mimicking the natural photosynthesis process using TiO$_2$, Ru-complex dye as light sensitizer, polyiodide redox electrolyte, and platinum counter electrode. Researchers from all over the world have intensively investigated DSSC mechanisms, making efforts to understand the correlation between DSSC components, new materials for photoanodes, sensitizers

FIGURE 7.5 Demonstration of the attractive applications of dye-sensitized solar cells [222].

(dyes), hole conductors (redox liquid electrolyte or solid HTM), and counter electrodes for the fabrication of efficient DSSCs. Maximum power conversion efficiencies of ~13.1% under 1 sun and 32% under ambient light for laboratory cells have been attained so far with TiO_2 cosensitized-based DSSCs. Organic dye sensitizers, such as indoline, xanthene, coumarin, azo, and carbazole dyes, also show comparable performance to that of Ru metal-based dyes. In addition, the advanced colorful approach will help to further improve DSSC performance. The enhancement in performance is observed for MO semiconductor-based solar cells sensitized with a cocktail/blend of dyes prepared by mixing individual dyes resulting in the broadening of absorption spectra in the visible region. This basic idea of mixing two or more dyes not only helps to boost the efficiency of DSSCs, but also utilizes the making of rainbow-colored solar cells. It will be possible to fabricate commercial DSSCs for indoor applications in the near future.

DSSC research has concentrated almost entirely on glass (substrate)-based technology, which is now very near commercialization. Also, the fundamental research related to physicochemical phenomena in DSSCs has been completed with traditional glass substrate-based solar cells. A divergence from the conventional path would therefore be a study of DSSCs on conducting plastic substrate. This will be useful for the production of flexible and colorful solar cells. The liquid-state DSSC has some disadvantages like liquid evaporation and imperfect

200   7. Dye-sensitized solar cells

sealing. Hence, attempts can be made to develop SS-DSSCs by replacing liquid electrolyte with a solid hole conductor.

References

[1] B. O'Regan, M. Gratzel; A low-cost, high-efficiency solar cell based on dye-sensitized colloidal TiO_2 films, Nature 353(1991)737–740.

[2] A. Yella, H.W. Lee, H.N. Tsao, C. Yi, A.K. Chandiran, M.K. Nazeeruddin, E.W.G. Diau, C.Y. Yeh, S.M. Zakeeruddin, M. Gratzel, Porphyrin-sensitized solar cells with cobalt (II/III)–based redox electrolyte exceed 12 percent efficiency, Science 334 (2011) 629–634.

[3] S. Mathew, A. Yella, P. Gao, R. Humphry-Baker, B.F.E. Curchod, N. Ashari-Astani, I. Tavernelli, U. Rothlisberger, M.K. Nazeeruddin, M. Gratzel, Dye-sensitized solar cells with 13% efficiency achieved through the molecular engineering of porphyrin sensitizers, Nat. Chem. 6 (2014) 242–247.

[4] L. Zhang, X. Yang, W. Wang, G.G. Gurzadyan, J. Li, X. Li, J. An, Z. Yu, H. Wang, B. Cai, A. Hagfeldt, L. Sun, 13.6% efficient organic dye-sensitized solar cells by minimizing energy losses of the excited state, ACS Energy Lett. 4 (2019) 943–951.

[5] R.K. Kokal, S. Bhattacharya, L.S. Cardoso, P.B. Miranda, V.R. Soma, P. Chetti, D. Melepurath, S.S.K. Raavi, Low cost 'green' dye sensitized solar cells based on new Fuchsin dye with aqueous electrolyte and platinum-free counter electrodes, Sol. Energy 188 (2019) 913–923.

[6] M. Freitag, J. Teuscher, Y. Saygili, X. Zhang, F. Giordano, P. Liska, J. Hua, S.M. Zakeeruddin, J.E. Moser, M. Gratzel, A. Hagfeldt, Dye-sensitized solar cells for efficient power generation under ambient lighting, Nat. Photon. 11 (2017) 372–378.

[7] B. Li, F. Huang, J. Zhong, J. Xie, M. Wen, Y. Peng, Fabrication of flexible dye-sensitized solar cell modules using commercially available materials, Energy Technol. 4 (2016) 536–542.

[8] S. Rani, P.K. Shishodia, R.M. Mehra, Development of a dye with broadband absorbance in visible spectrum for an efficient dye-sensitized solar cell, J. Renew. Sustain. Energy 2 (2010) 043103.

[9] M. Gratzel, Dye-sensitized solar cells, J. Photochem. Photobiol. C Photochem. Rev. 4 (2003) 145–153.

[10] G. Boschloo, A. Hagfeldt, Characteristics of the iodide/triiodide redox mediator in dye-sensitized solar cells, Acc. Chem. Res. 42 (2009) 1819–1826.

[11] A. Hagfeldt, G. Boschloo, L. Sun, L. Kloo, H. Pettersson, Dye-sensitized solar cells, Chem. Rev. 110 (2010) 6595–6663.

[12] S. Mahalingam, H. Abdullah, Electron transport study of indium oxide as photoanode in DSSCs: a review, Renew. Sustain. Energy Rev. 63 (2016) 245–255.

[13] A. Listorti, B. O'Regan, J.R. Durrant, Electron transfer dynamics in dye-sensitized solar cells, Chem. Mater. 23 (2011) 3381–3399.

[14] U. Mehmood, A. Al-Ahmed, F.A. Al-Sulaiman, M.I. Malik, F. Shehzad, A.U.H. Khan, Effect of temperature on the photovoltaic performance and stability of solid state dye-sensitized solar cells: a review, Renew. Sustain. Energy Rev. 79 (2017) 946–959.

[15] J.N. Clifford, E. Martinez-Ferrero, A. Viterisi, E. Palomares, Sensitizer molecular structure-device efficiency relationship in dye sensitized solar cells, Chem. Soc. Rev. 40 (2011) 1635–1646.

[16] H. Nusbaumer, J.E. Moser, S.M. Zakeeruddin, M.K. Nazeeruddin, M. Gratzel, $Co^{II}(dbbip)_2^{2+}$ complex rivals tri-iodide/iodide redox mediator in dye-sensitized photovoltaic cells, J. Phys. Chem. B 105 (2001) 10461–10464.

[17] S. Hattori, Y. Wada, S. Yanagida, S. Fukuzumi, Blue copper model complexes with distorted tetragonal geometry acting as effective electron-transfer mediators in dye-sensitized solar cells, J. Am. Chem. Soc. 127 (2005) 9648–9654.

[18] R. Sastrawan, J. Beier, U. Belledin, S. Hemming, A. Hinsch, R. Kern, C. Vetter, F.M. Petrat, A. Prodi-Schwab, P. Lechner, W. Hoffmann, A glass frit-sealed dye solar cell module with integrated series connections, Sol. Energy Mater. Sol. Cells 93 (2009) 820–824.

[19] X. Lu, X. Mou, J. Wu, D. Zhang, L. Zhang, F. Huang, F. Xu, S. Huang, Improved-performance dye-sensitized solar cells using Nb-doped TiO_2 electrodes: efficient electron injection and transfer, Adv. Funct. Mater. 20 (2010) 509–515.

[20] D. Cahen, G. Hodes, M. Grätzel, J.F. Guillemoles, I. Riess, Nature of photovoltaic action in dye-sensitized solar cells, J. Phys. Chem. B 104 (2000) 2053–2059.

II. Photovoltaic materials and devices

References

[21] D. Cahen, M. Grätzel, J.F. Guillemoles, G. Hodes, Dye-sensitized Solar Cells: Principles of Operation, Electrochemistry of Nanomaterials, Wiley-VCH, Weinheim, 201–228.

[22] N. Kutlu, Investigation of electrical values of low-efficiency dye-sensitized solar cells (DSSCs), Energy 199 (2020) 117222.

[23] H.Y. Wang, J. Chen, S. Hy, L. Yu, Z. Xu, B. Liu, High-surface-area mesoporous TiO_2 microspheres via one-step nanoparticle self-assembly for enhanced lithium-ion storage, Nanoscale 6 (2014) 14926–14931.

[24] S. Thomas, T.G. Deepak, G.S. Anjusree, T.A. Arun, S.V. Nair, A.S. Nair, A review on counter electrode materials in dye-sensitized solar cells, J. Mater. Chem. 2 (2014) 4474–4490.

[25] D. Gentilini, D. D'Ercole, A. Gagliardi, A. Brunetti, A. Reale, T. Brown, A. di Carlo, Analysis and simulation of incident photon to current efficiency in dye sensitized solar cells, Superlattice. Microst. 47 (2010) 192–196.

[26] J.A. Anta, E. Guillen, R. Tena-Zaera, ZnO-based dye-sensitized solar cells, J. Phys. Chem. C 116 (2012) 11413–11425.

[27] R. Vittal, K.C. Ho, Zinc oxide based dye-sensitized solar cells: a review, Renew. Sustain. Energy Rev. 70 (2017) 920–935.

[28] J. Cheng, J. Ma, Y. Ma, C. Zhou, Y. Qiang, X. Zhou, J. Yang, H. Shiab, Y. Xie, Highly efficient ZnO-based dye-sensitized solar cells with low-cost Co-Ni/carbon aerogel composites as counter electrodes, New J. Chem. 42 (2018) 16329–16334.

[29] Y. Xie, X. Zhou, H. Mi, J. Ma, J. Yang, J. Cheng, High efficiency ZnO-based dye-sensitized solar cells with a 1H,1H,2H,2H-perfluorodecyltriethoxysilane chain barrier for cutting on interfacial recombination, Appl. Surf. Sci. 434 (2018) 1144–1152.

[30] A. Birkel, Y.G. Lee, D. Koll, X.V. Meerbeek, S. Frank, M.J. Choi, Y.S. Kang, K. Char, W. Tremel, Highly efficient and stable dye-sensitized solar cells based on SnO_2 nanocrystals prepared by microwave-assisted synthesis, Energy Environ. Sci. 5 (2012) 5392–5400.

[31] S.A. Arote, V.A. Tabhane, H.M. Pathan, Enhanced photovoltaic performance of dye sensitized solar cell using SnO_2 nanoflowers, Opt. Mater. 75 (2018) 601–606.

[32] R. Vasanthapriya, N. Neelakandeswari, N. Rajasekaran, K. Uthayarani, M. Chitra, S. Sathiesh kumar, Synthesis and characterisation of SnO_2 nanostructures for dye-sensitized solar cells, Mater. Lett. 220 (2018) 218–221.

[33] S. Bhand, D. Chadar, K. Pawar, M. Naushad, H. Pathan, S. Salunke-Gawali, Benzo[α]phenothiazine sensitized ZrO_2 based dye sensitized solar cell, J. Mater. Sci. Mater. Electron. 29 (2018) 1034–1041.

[34] Y.C. Chen, Y.C. Chang, C.M. Chen, The study of blocking effect of Nb_2O_5 in dye-sensitized solar cells under low power lighting, J. Electrochem. Soc. 165 (2018) F409.

[35] X. Liu, R. Yuan, Y. Liu, S. Zhu, J. Lin, X. Chen, Niobium pentoxide nanotube powder for efficient dye-sensitized solar cells, New J. Chem. 40 (2016) 6276–6280.

[36] R.A. Rani, A.S. Zoolfakar, M. Rusop, Photovoltaic performance of dye-sensitized solar cells based nanoporous-network Nb_2O_5, AIP Conf. Proc. 2151 (2019) 020030.

[37] P. Jayabal, V. Sasirekha, J. Mayandi, K. Jeganathan, V. Ramakrishnan, A facile hydrothermal synthesis of $SrTiO_3$ for dye sensitized solar cell application, J. Alloys Compd. 586 (2014) 456–462.

[38] Y. Okamoto, Y. Suzuki, Perovskite-type $SrTiO_3$, $CaTiO_3$ and $BaTiO_3$ porous film electrodes for dye sensitized solar cells, J. Ceram. Soc. Jpn. 122 (2014) 728–731.

[39] S. Gholamrezaei, M. Salavati-Niasari, An efficient dye sensitized solar cells based on $SrTiO_3$ nanoparticles prepared from a new amine-modified sol-gel route, J. Mol. Liq. 243 (2017) 227–235.

[40] R.S. Mane, H.M. Pathan, C.D. Lokhande, S.H. Han, An effective use of nanocrystalline CdO thin films in dye-sensitized solar cells, Sol. Energy 80 (2006) 185–190.

[41] S.H. Choi, D. Hwang, D.Y. Kim, Y. Kervella, P. Maldivi, S.Y. Jang, R. Demadrille, I.D. Kim, Amorphous zinc stannate (Zn_2SnO_4) nanofibers networks as photoelectrodes for organic dye-sensitized solar cells, Adv. Funct. Mater. 23 (2013) 3146–3155.

[42] P.P. Das, A. Roy, S. Das, P.S. Devi, Enhanced stability of Zn_2SnO_4 with N719, N3 and eosin Y dye molecules for DSSC application, Phys. Chem. Chem. Phys. 18 (2016) 1429–1438.

[43] P.P. Das, A. Roy, S. Agarkar, P.S. Devi, Hydrothermally synthesized fluorescent Zn_2SnO_4 nanoparticles for dye sensitized solar cells, Dyes Pigments 154 (2018) 303–313.

[44] M.J.J. Fatima, C.V. Niveditha, S. Sindhu, α-Bi_2O_3 photoanode in DSSC and study of the electrode-electrolyte interface, RSC Adv. 5 (2015) 78299–78305.

[45] B. Sankapal, A. Tirpude, S. Majumder, P. Baviskar, 1-D electron path of 3-D architecture consisting of dye loaded CdS nanowires: dye sensitized solar cell, J. Alloys Compd. 651 (2015) 399–404.

7. Dye-sensitized solar cells

[46] T. Jiang, M. Bujoli-Doeuff, Y. Farre, Y. Pellegrin, E. Gautron, M. Boujtita, L. Cario, S. Jobic, F. Odobel, CuO nanomaterials for p-type dye-sensitized solar cells, RSC Adv. 6 (2016) 112765–112770.

[47] K. Fan, W. Zhang, T. Peng, J. Chen, F. Yang, Application of TiO$_2$ fusiform nanorods for dye-sensitized solar cells with significantly improved efficiency, J. Phys. Chem. C 115 (2011) 17213–17219.

[48] X. Fang, Y. Li, S. Zhang, L. Bai, N. Yuan, J. Ding, The dye adsorption optimization of ZnO nanorod-based dye-sensitized solar cells, Sol. Energy 105 (2014) 14–19.

[49] S. Pace, A. Resmini, I.G. Tredici, A. Soffientini, X. Li, S. Dunn, J. Briscoe, U. Anselmi-Tamburini, Optimization of 3D ZnO brush-like nanorods for dye-sensitized solar cells, RSC Adv. 8 (2018) 9775–9782.

[50] X. Hu, H. Wang, ZnO/Nb$_2$O$_5$ core/shell nanorod array photoanode for dye-sensitized solar cells, Front. Opto-electron. 11 (2018) 285–290.

[51] A.B.F. Martinson, J.W. Elam, J.T. Hupp, M.J. Pellin, ZnO nanotube based dye-sensitized solar cells, Nano Lett. 7 (2007) 2183–2187.

[52] M. Ye, X. Xin, C. Lin, Z. Lin, High efficiency dye-sensitized solar cells based on hierarchically structured nanotubes, Nano Lett. 11 (2011) 3214–3220.

[53] P. Roy, S.P. Albu, P. Schmuki, TiO$_2$ nanotubes in dye-sensitized solar cells: higher efficiencies by well-defined tube tops, Electrochem. Commun. 12 (2010) 949–951.

[54] R. Ranjusha, P. Lekha, K.R.V. Subramanian, V.N. Shantikumar, A. Balakrishnan, Photoanode activity of ZnO nanotube based dye-sensitized solar cells, J. Mater. Sci. Technol. 27 (2011) 961–966.

[55] X.D. Wang, Y. Ding, C.J. Summers, Z.L. Wang, Large-scale synthesis of six-nanometer-wide ZnO nanobelts, J. Phys. Chem. B 108 (2004) 8773–8777.

[56] J. Fan, Z. Li, W. Zhou, Y. Miao, Y. Zhang, J. Hu, G. Shao, Dye-sensitized solar cells based on TiO$_2$ nanoparticles/nanobelts double-layered film with improved photovoltaic performance, Appl. Surf. Sci. 319 (2014) 75–82.

[57] Y. Wang, Z. Li, Y. Cao, F. Li, W. Zhao, X. Liu, J. Yang, Fabrication of novel Ag-TiO$_2$ nanobelts as a photoanode for enhanced photovoltage performance in dye sensitized solar cells, J. Alloys Compd. 677 (2016) 294–301.

[58] J.B. Baxter, E.S. Aydil, Nanowire-based dye-sensitized solar cells, Appl. Phys. Lett. 86 (2005) 053114.

[59] J.B. Baxter, E.S. Aydil, Dye-sensitized solar cells based on semiconductor morphologies with ZnO nanowires, Sol. Energy Mater. Sol. Cell. 90 (2006) 607–622.

[60] C.H. Ku, J.J. Wu, Chemical bath deposition of ZnO nanowire-nanoparticle composite electrodes for use in dye-sensitized solar cells, Nanotechnology 18 (2007) 505706.

[61] M. McCune, W. Zhang, Y. Deng, High efficiency dye-sensitized solar cells based on three-dimensional multi-layered ZnO nanowire arrays with "caterpillar-like" structure, Nano Lett. 12 (2012) 3656–3662.

[62] M.C. Kao, H.Z. Chen, S.L. Young, C.C. Lin, C.Y. Kung, Structure and photovoltaic properties of ZnO nanowire for dye-sensitized solar cells, Nanoscale Res. Lett. 7 (2012) 260.

[63] Z.L. Wang, Zinc oxide nanostructures: growth, properties and applications, J. Phys. Condens. Matter 16 (2004) R829.

[64] A. Umar, Growth of comb-like ZnO nanostructures for dye-sensitized solar cells application, Nanoscale Res. Lett. 4 (2009) 1004–1008.

[65] P.K. Baviskar, J.B. Zhang, V. Gupta, S. Chand, B.R. Sankapal, Nanobeads of zinc oxide with rhodamine B dye as a sensitizer for dye sensitized solar cell application, J. Alloys Compd. 510 (2012) 33–37.

[66] D. Hwang, J.S. Jin, H. Lee, H.J. Kim, H. Chung, D.Y. Kim, S.Y. Jang, D. Kim, Hierarchically structured Zn$_2$SnO$_4$ nanobeads for high-efficiency dye-sensitized solar cells, Sci. Rep. 4 (2015) 7353.

[67] C.Y. Lin, Y.H. Lai, H.W. Chen, J.G. Chen, C.W. Kung, R. Vittal, K.C. Ho, Highly efficient dye-sensitized solar cell with a ZnO nanosheet-based photoanode, Energy Environ. Sci. 4 (2011) 3448–3455.

[68] P. Baviskar, R. Gore, A. Ennaoui, B. Sankapal, Cactus architecture of ZnO nanoparticles network through simple wet chemistry: efficient dye sensitized solar cells, Mater. Lett. 116 (2014) 91–93.

[69] J. Mou, W. Zhang, J. Fan, H. Deng, W. Chen, Facile synthesis of ZnO nanobullets/nanoflakes and their applications to dye-sensitized solar cells, J. Alloys Compd. 509 (2011) 961–965.

[70] C. Zhao, J. Zhang, Y. Hu, N. Robertson, P.A. Hu, D. Child, D. Gibson, Y.Q. Fu, In-situ microfluidic controlled, low temperature hydrothermal growth of nanoflakes for dye-sensitized solar cells, Sci. Rep. 5 (2015) 17750.

[71] S.Y. Han, M.S. Akhtar, I. Jung, O.B. Yang, ZnO nanoflakes nanomaterials via hydrothermal process for dye sensitized solar cells, Mater. Lett. 230 (2018) 92–95.

[72] S.Y. Kuo, J.F. Yang, F.I. Lai, Improved dye-sensitized solar cell with a ZnO nanotree photoanode by hydrothermal method, Nanoscale Res. Lett. 9 (2014) 203.

References

[73] S.H. Ko, D. Lee, H.W. Kang, K.H. Nam, J.Y. Yeo, S.J. Hong, C.P. Grigoropoulos, H.J. Sung, Nanoforest of hydrothermally grown hierarchical ZnO nanowires for a high efficiency dye-sensitized solar cell, Nano Lett. 11 (2011) 666−671.

[74] R. Ghosh, M.K. Brennaman, T. Uher, M.R. Ok, E.T. Samulski, L.E. McNeil, T.J. Meyer, R. Lopez, Nanoforest Nb_2O_5 photoanodes for dye-sensitized solar cells by pulsed laser deposition, ACS Appl. Mater. Interfaces 3 (2011) 3929−3935.

[75] Y.K. Mishra, R. Adelung, ZnO tetrapod materials for functional applications, Mater. Today 21 (2018) 631−651.

[76] W.H. Chiu, C.H. Lee, H.M. Cheng, H.F. Lin, S.C. Liao, J.M. Wu, W.F. Hsieh, Efficient electron transport in tetrapod-like ZnO metal-free dye-sensitized solar cells, Energy Environ. Sci. 2 (2009) 694−698.

[77] Y.F. Hsu, Y.Y. Xi, C.T. Yip, A.B. Djurisic, W.K. Chan, Dye-sensitized solar cells using ZnO tetrapods, J. Appl. Phys. 103 (2008) 083114.

[78] J.Y. Liao, J.W. He, H. Xu, D.B. Kuang, C.Y. Su, Effect of TiO_2 morphology on photovoltaic performance of dye-sensitized solar cells: nanoparticles, nanofibers, hierarchical spheres and ellipsoid spheres, J. Mater. Chem. 22 (2012) 7910−7918.

[79] A. Omar, H. Abdullah, Electron transport analysis in zinc oxide-based dye-sensitized solar cells: a review, Renew. Sustain. Energy Rev. 31 (2014) 149−157.

[80] S. Xuhui, C. Xinglan, T. Wanquan, W. Dong, L. Kefei, Performance comparison of dye-sensitized solar cells by using different metal oxide- coated TiO_2 as the photoanode, AIP Adv. 4 (2014) 031304.

[81] M. Hamadanian, V. Jabbari, Investigation on the energy conversion of dye-sensitized solar cells based on TiO_2 core/shell using metal oxide as a barrier layer, Appl. Sol. Energy 47 (2011) 281−288.

[82] V.S. Manikandan, A.K. Palai, S. Mohanty, S.K. Nayak, Eosin-Y sensitized core-shell TiO_2-ZnO nano-structured photoanodes for dye-sensitized solar cell applications, J. Photochem. Photobiol. B Biol. 183 (2018) 397−404.

[83] M.E. Yeoh, K.Y. Chan, Efficiency enhancement in dye-sensitized solar cells with ZnO and TiO_2 blocking layers, J. Electron. Mater. 48 (2019) 4342−4350.

[84] K.A. Bhatti, M.I. Khan, M. Saleem, F. Alvi, R. Raza, S. Rehman, Analysis of multilayer based TiO_2 and ZnO photoanodes for dye-sensitized solar cells, Mater. Res. Express 6 (2019) 075902.

[85] K.S. Pawar, P.K. Baviskar, A.B. Nadaf, S. Salunke-Gawali, H.M. Pathan, Layer-by-layer deposition of TiO_2-ZrO_2 electrode sensitized with Pandan leaves: natural dye-sensitized solar cell, Mater. Renew. Sustain. Energy 8 (2019) 12.

[86] M.A. Waghmare, N.I. Beedri, P.K. Baviskar, H.M. Pathan, A.U. Ubale, Effect of ZrO_2 barrier layers on the photovoltaic parameters of rose bengal dye-sensitized TiO_2 solar cell, J. Mater. Sci. Mater. Electron. 30 (2019) 6015−6022.

[87] J. Gong, H. Qiao, S. Sigdel, H. Elbohy, N. Adhikari, Z. Zhou, K. Sumathy, Q. Wei, Q. Qiao, Characteristics of SnO_2 nanofiber/TiO_2 nanoparticle composite for dye-sensitized solar cells, AIP Adv. 5 (2015) 067134.

[88] Z. Li, Y. Zhou, R. Sun, Y. Xiong, H. Xie, Z. Zou, Nanostructured SnO_2 photoanode-based dye-sensitized solar cells, Chin. Sci. Bull. 59 (2014) 2122−2134.

[89] A. Pang, X. Sun, H. Ruan, Y. Li, S. Dai, M. Wei, Highly efficient dye-sensitized solar cells composed of TiO_2@SnO_2 core-shell microspheres, Nano Energy 5 (2014) 82−90.

[90] H. Choi, S. Kim, S.O. Kang, J. Ko, M.S. Kang, J.N. Clifford, A. Forneli, E. Palomares, M.K. Nazeeruddin, M. Gratzel, Stepwise cosensitization of nanocrystalline TiO_2 films utilizing Al_2O_3 layers in dye-sensitized solar cells, Angew. Chem. 47 (2008) 8259−8263.

[91] V. Ganapathy, B. Karunagaran, S.W. Rhee, Improved performance of dye sensitized solar cells with TiO_2/alumina core-shell formation using atomic layer deposition, J. Power Sources 195 (2010) 5138−5143.

[92] J.Y. Kim, S.H. Kang, H.S. Kim, Y.E. Sung, Preparation of highly ordered mesoporous Al_2O_3/TiO_2 and its application in dye-sensitized solar cells, Langmuir 26 (2010) 2864−2870.

[93] K. Eguchi, H. Koga, K. Sekizawa, K. Sasaki, Nb_2O_5-Based composite electrodes for dye-sensitized solar cells, J. Ceram. Soc. Jpn. 108 (2000) 1067−1071.

[94] E. Barea, X. Xu, V. Gonzalez-Pedro, T. Ripolles-Sanchis, F. Fabregat-Santiago, J. Bisquert, Origin of efficiency enhancement in Nb_2O_5 coated titanium dioxide nanorod based dye sensitized solar cells, Energy Environ. Sci. 4 (2011) 3414−3419.

[95] R. Elangovan, N.G. Joby, P. Venkatachalam, Performance of dye-sensitized solar cells based on various sensitizers applied on TiO_2-Nb_2O_5 core/shell photoanode structure, J. Solid State Electrochem. 18 (2014) 1601−1609.

II. Photovoltaic materials and devices

204   7. Dye-sensitized solar cells

[96] N.I. Beedri, P.K. Baviskar, V.P. Bhalekar, C.V. Jagtap, Innamudin, A.M. Asiri, S.R. Jadkar, H.M. Pathan, N3-Sensitized TiO_2/Nb_2O_5: a novel bilayer structure for dye-sensitized solar-cell application, Phys. Status Solidi A 215 (2018) 1800236.

[97] Y. Diamant, S.G. Chen, O. Melamed, A. Zaban, Core−Shell nanoporous electrode for dye sensitized solar cells: the effect of the $SrTiO_3$ shell on the electronic properties of the TiO_2 core, J. Phys. Chem. B 107 (2003) 1977−1981.

[98] I. Hod, M. Shalom, Z. Tachan, S. Ruhle, A. Zaban, $SrTiO_3$ recombination-inhibiting barrier layer for type II dye-sensitized solar cells, J. Phys. Chem. C 114 (2010) 10015−10018.

[99] S. Wu, X. Gao, M. Qin, J.M. Liu, S. Hu, $SrTiO_3$ modified TiO_2 electrodes and improved dye-sensitized TiO_2 solar cells, Appl. Phys. Lett. 99 (2011) 042106.

[100] E. Guoa, L. Yin, Tailored $SrTiO_3/TiO_2$ heterostructures for dye-sensitized solar cells with enhanced photoelectric conversion performance, J. Mater. Chem. 3 (2015) 13390−13401.

[101] Y.J. Chang, E.H. Kong, Y.C. Park, H.M. Jang, Broadband light confinement using a hierarchically structured TiO_2 multi-layer for dye-sensitized solar cells, J. Mater. Chem. 1 (2013) 9707−9713.

[102] G.S. Selopal, N. Memarian, R. Milan, I. Concina, G. Sberveglieri, A. Vomiero, Effect of blocking layer to boost photoconversion efficiency in ZnO dye-sensitized solar cells, ACS Appl. Mater. Interfaces 6 (2014) 11236−11244.

[103] L. Li, C. Xu, Y. Zhao, S. Chen, K.J. Ziegler, Improving performance via blocking layers in dye-sensitized solar cells based on nanowire photoanodes, ACS Appl. Mater. Interfaces 7 (2015) 12824−12831.

[104] Y. Feng, J. Chen, X. Huang, W. Liu, Y. Zhu, W. Qin, X. Mo, A ZnO/TiO_2 composite nanorods photoanode with improved performance for dye-sensitized solar cells, Crystal 51 (2016) 548−553.

[105] T.S. Bramhankar, S.S. Pawar, J.S. Shaikh, V.C. Gunge, N.I. Beedri, P.K. Baviskar, H.M. Pathan, P.S. Patil, R.C. Kambale, R.S. Pawar, Effect of Nickel−Zinc Co-doped TiO_2 blocking layer on performance of DSSCs, J. Alloys Compd. 817 (2020) 152810.

[106] Y.H. Sim, M.J. Yun, S.I. Cha, D.Y. Lee, Preparation 8.5%-efficient submodule using 5%-efficient DSSCs via three-dimensional angle array and light-trapping layer, NPG Asia Mater. 12 (2020) 14.

[107] P.K. Baviskar, W. Tan, J. Zhang, B.R. Sankapal, Wet chemical synthesis of ZnO thin films and sensitization to light with N3 dye for solar cell application, J. Phys. Appl. Phys. 42 (2009) 125108.

[108] L. Francis, A. Nair, R. Jose, S. Ramakrishna, V. Thavasi, E. Marsano, Fabrication and characterization of dye-sensitized solar cells from rutile nanofibers and nanorods, Energy 36 (2011) 627−632.

[109] K. Portillo-Cortez, A. Martinez, A. Dutt, G. Santana, N719 derivatives for application in a dye-sensitized solar cell (DSSC): a theoretical study, J. Phys. Chem. 123 (2019) 10930−10939.

[110] Y. Huang, W.C. Chen, X.X. Zhang, R. Ghadari, X.Q. Fang, T. Yu, F.T. Kong, Ruthenium complexes as sensitizers with phenyl-based bipyridine anchoring ligands for efficient dye-sensitized solar cells, J. Mater. Chem. C 6 (2018) 9445−9452.

[111] R.S. Mane, H.M. Nguyen, T. Ganesh, N. Kim, S.B. Ambade, S.H. Han, A novel HMP-2 dye of high extinction coefficient designed for enhancing the performance of ZnO platelets, Electrochem. Commun. 11 (2009) 752−755.

[112] M. Subbaiah, R. Sekar, E. Palani, A. Sambandam, One-pot synthesis of metal free organic dyes containing different acceptor moieties for fabrication of dye- sensitized solar cells, Tetrahedron Lett. 54 (2013) 3132−3136.

[113] H. Otaka, M. Kira, K. Yano, S. Ito, H. Mitekura, T. Kawata, F. Matsui, Multicolored dye-sensitized solar cells, J. Photochem. Photobiol. Chem. 164 (2004) 67−73.

[114] X.F. Wang, H. Tamiaki, O. Kitao, T. Ikeuchi, S.I. Sasaki, Molecular engineering on a chlorophyll derivative, chlorin e_6, for significantly improved power conversion efficiency in dye-sensitized solar cells, J. Power Sources 242 (2013) 860−864.

[115] Z.S. Wang, K. Sayama, H. Sugihara, Efficient Eosin Y dye-sensitized solar cell containing Br^-/Br_3^- electrolyte, J. Phys. Chem. B 109 (2005) 22449−22455.

[116] T. Yoshida, M. Iwaya, H. Ando, T. Oekermann, K. Nonomura, D. Schlettwein, D. Wohrle, H. Minoura, Improved photoelectrochemical performance of electrodeposited ZnO/EosinY hybrid thin films by dye re-adsorption, Chem. Commun. 40 (2004) 400−401.

[117] P. Baviskar, A. Ennaoui, B. Sankapal, Influence of processing parameters on chemically grown ZnO films with low cost Eosin-Y dye towards efficient dye sensitized solar cell, Sol. Energy 105 (2014) 445−454.

[118] B. Pradhan, S.K. Batabyal, A.J. Pal, Vertically aligned ZnO nanowire arrays in Rose Bengal-based dye-sensitized solar cells, Sol. Energy Mater. Sol. Cell. 91 (2007) 769−773.

II. Photovoltaic materials and devices

References

[119] E. Guillen, F. Casanueva, J.A. Anta, A.V. Poot, G. Oskam, R. Alcantara, C.F. Lorenzo, J.M. Calleja, Photovoltaic performance of nanostructured zinc oxide sensitized with xanthene dyes, J. Photochem. Photobiol. Chem. 200 (2008) 364–370.

[120] F.M. Jradi, X. Kang, D. O'Neil, G. Pajares, Y.A. Getmanenko, P. Szymanski, T.C. Parker, M.A. El-Sayed, S.R. Marder, Near-infrared asymmetrical squaraine sensitizers for highly efficient dye sensitized solar cells: the effect of π-bridges and anchoring groups on solar cell performance, Chem. Mater. 27 (2015) 24802487.

[121] G.M. Shivashimpi, S.S. Pandey, R. Watanabe, N. Fujikawa, Y. Ogomi, Y. Yamaguchi, S. Hayase, Effect of nature of anchoring groups on photosensitization behavior in unsymmetrical squaraine dyes, J. Photochem. Photobiol. Chem. 273 (2014) 1–7.

[122] Z. Yao, M. Zhang, H. Wu, L. Yang, R. Li, P. Wang, A donor/acceptor indenoperylene dye for highly efficient organic dye-sensitized solar cells, J. Am. Chem. Soc. 137 (2015) 3799–3802.

[123] M. Giannouli, M. Fakis, Interfacial electron transfer dynamics and photovoltaic performance of TiO_2 and ZnO solar cells sensitized with Coumarin 343, J. Photochem. Photobiol. Chem. 226 (2011) 42–50.

[124] K. Nakajima, K. Ohta, H. Katayanagi, K. Mitsuke, Photoexcitation and electron injection processes in azo dyes adsorbed on nanocrystalline TiO_2 films, Chem. Phys. Lett. 510 (2011) 228–233.

[125] T. Yoshida, J. Zhang, D. Komatsu, S. Sawatani, H. Minoura, T. Pauporte, D. Lincot, T. Oekermann, D. Schlettwein, H. Tada, D. Wohrle, K. Funabiki, M. Matsui, H. Miura, H. Yanagi, Electrodeposition of inorganic/organic hybrid thin films, Adv. Funct. Mater. 19 (2009) 17–43.

[126] V. Kumar, R. Gupta, A. Bansal, Role of chenodeoxycholic acid as co-additive in improving the efficiency of DSSCs, Sol. Energy 196 (2020) 589–596.

[127] S. Zhang, X. Yang, Y. Numata, L. Han, Highly efficient dye-sensitized solar cells: progress and future challenges, Energy Environ. Sci. 6 (2013) 1443–1464.

[128] M.A. Green, E.D. Dunlop, J. Hohl-Ebinger, M. Yoshita, N. Kopidakis, A.W.Y. Ho- Baillie, Solar cell efficiency tables (version 55), Prog. Photovoltaics Res. Appl. 28 (2020) 3–15.

[129] K.K.S. Lau, M. Soroush, Overview of Dye-Sensitized Solar Cells, Elsevier Academic Press, 2019, pp. 1–49.

[130] T. Daeneke, T.H. Kwon, A.B. Holmes, N.W. Duffy, U. Bach, L. Spiccia, High-efficiency dye-sensitized solar cells with ferrocene-based electrolytes, Nat. Chem. 3 (2011) 211–215.

[131] T. Daeneke, A.J. Mozer, T.H. Kwon, N.W. Duffy, A.B. Holmes, U. Bach, L. Spiccia, Dye regeneration and charge recombination in dye-sensitized solar cells with ferrocene derivatives as redox mediators, Energy Environ. Sci. 5 (2012) 7090–7099.

[132] T.C. Li, A.M. Spokoyny, C. She, O.K. Farha, C.A. Mirkin, T.J. Marks, J.T. Hupp, Ni(III)/(IV) bis(dicarbollide) as a fast, noncorrosive redox shuttle for dye-sensitized solar cells, J. Am. Chem. Soc. 132 (2010) 4580–4582.

[133] B. Pashaei, H. Shahroosvand, P. Abbasi, Transition metal complex redox shuttles for dye-sensitized solar cells, RSC Adv. 5 (2015) 94814–94848.

[134] N.Y. Nia, P. Farahani, H. Sabzyan, M. Zendehdel, M. Oftadeh, A combined computational and experimental study of the $[Co(bpy)_3]^{2+/3+}$ complexes as one-electron outer-sphere redox couples in dye-sensitized solar cell electrolyte media, Phys. Chem. Chem. Phys. 16 (2014) 11481–11491.

[135] Y. Hao, W. Yang, L. Zhang, R. Jiang, E. Mijangos, Y. Saygili, L. Hammarstrom, A. Hagfeldt, G. Boschloo, A small electron donor in cobalt complex electrolyte significantly improves efficiency in dye-sensitized solar cells, Nat. Commun. 7 (2016) 13934.

[136] A. Yella, S. Mathew, S. Aghazada, P. Comte, M. Gratzel, M.K. Nazeeruddin, Dye-sensitized solar cells using cobalt electrolytes: the influence of porosity and pore size to achieve high-efficiency, J. Mater. Chem. C 5 (2017) 2833–2843.

[137] K.Y. Chen, P.A. Schauer, B.O. Patrick, C.P. Berlinguette, Correlating cobalt redox couples to photovoltage in the dye-sensitized solar cell, Dalton Trans. 47 (2018) 11942–11952.

[138] J. Li, X. Yang, Z. Yu, G.G. Gurzadyan, M. Cheng, F. Zhang, J. Cong, W. Wang, H. Wang, X. Li, L. Kloo, M. Wang, L. Sun, Efficient dye-sensitized solar cells with [copper(6,6'-dimethyl-2,2'-bipyridine)$_2$]$^{2+/1+}$ redox shuttle, RSC Adv. 7 (2017) 4611–4615.

[139] C. Dragonetti, M. Magni, A. Colombo, F. Melchiorre, P. Biagini, D. Roberto, Coupling of a copper dye with a copper electrolyte: a fascinating springboard for sustainable dye-sensitized solar cells, ACS Appl. Energy Mater. 1 (2018) 751–756.

[140] A. Colombo, C. Dragonetti, F. Fagnani, D. Roberto, F. Melchiorre, P. Biagini, Improving the efficiency of copper-dye-sensitized solar cells by manipulating the electrolyte solution, Dalton Trans. 48 (2019) 9818–9823.

206
7. Dye-sensitized solar cells

[141] P. Suri, R.M. Mehra, Effect of electrolytes on the photovoltaic performance of a hybrid dye sensitized ZnO solar cell, Sol. Energy Mater. Sol. Cell. 91 (2007) 518−524.

[142] C. Teng, X. Yang, S. Li, M. Cheng, A. Hagfeldt, L.Z. Wu, L. Sun, Tuning the HOMO energy levels of organic dyes for dye-sensitized solar cells based on Br^-/Br_3^- electrolytes, Chem. A Eur. J. 16 (2010) 13127−13138.

[143] K. Kakiage, T. Tokutome, S. Iwamoto, T. Kyomen, M. Hanaya, Fabrication of a dye-sensitized solar cell containing a Mg-doped TiO_2 electrode and a Br_3^-/Br^- redox mediator with a high open-circuit photovoltage of 1.21 V, Chem. Commun. 49 (2013) 179−180.

[144] K. Kakiage, H. Osada, Y. Aoyama, T. Yano, K. Oya, S. Iwamoto, J.I. Fujisawa, M. Hanaya, Achievement of over 1.4 V photovoltage in a dye-sensitized solar cell by the application of a silyl-anchor coumarin dye, Sci. Rep. 6 (2016) 35888.

[145] Y. Hao, W. Yang, M. Karlsson, J. Cong, S. Wang, X. Li, B. Xu, J. Hua, L. Kloo, G. Boschloo, Efficient dye-sensitized solar cells with voltages exceeding 1 V through exploring tris(4-alkoxyphenyl)amine mediators in combination with the tris(bipyridine) cobalt redox system, ACS Energy Lett. 3 (2018) 1929−1937.

[146] C. Xu, J. Wu, U.V. Desai, D. Gao, High-efficiency solid-state dye-sensitized solar cells based on TiO_2-coated ZnO nanowire arrays, Nano Lett. 12 (2012) 2420−2424.

[147] I. Chung, B. Lee, J. He, R.P.H. Chang, M.G. Kanatzidis, All solid- state dye-sensitized solar cells with high efficiency, Nature 485 (2012) 486−489.

[148] N.O.V. Plank, I. Howard, A. Rao, M.W.B. Wilson, C. Ducati, R.S. Mane, J.S. Bendall, R.R.M. Louca, N.C. Greenham, H. Miura, R.H. Friend, H.J. Snaith, M.E. Welland, Efficient ZnO nanowire solid-state dye-sensitized solar cells using organic dyes and core-shell nanostructures, J. Phys. Chem. C 113 (2009) 18515−18522.

[149] C.P. Lee, C.T. Li, K.C. Ho, Use of organic materials in dye-sensitized solar cells, Mater. Today 20 (2017) 267−283.

[150] X. Liu, Y. Cheng, L. Wang, L. Cai, B. Liu, Light controlled assembling of iodine-free dye-sensitized solar cells with poly(3,4- ethylenedioxythiophene) as a hole conductor reaching 7.1% efficiency, Phys. Chem. Chem. Phys. 14 (2012) 7098−7103.

[151] Y.C. Li, S.R. Jia, Z.Y. Liu, X.Q. Liu, Y. Wang, Y. Cao, X.Q. Hu, C.L. Peng, Z. Li, Fabrication of PEDOT films via a facile method and their application in Pt-free dye-sensitized solar cells, J. Mater. Chem. 5 (2017) 7862−7868.

[152] W.N.A.W. Khalit, M.N. Mustafa, Y. Sulaiman, Synergistic effect of poly(3,4-ethylenedioxythiophene), reduced graphene oxide and aluminium oxide) as counter electrode in dye-sensitized solar cell, Result Phys. 13 (2019) 102355.

[153] T.H. Lee, D.Z. Sun, X. Zhang, H.J. Sue, X. Cheng, Solid-state dye-sensitized solar cell based on semiconducting nanomaterials, J. Vac. Sci. Technol. B 27 (2009) 3073−3077.

[154] T.H. Lee, H.J. Sue, X. Cheng, Solid-state dye-sensitized solar cells based on ZnO nanoparticle and nanorod array hybrid photoanodes, Nanoscale Res. Lett. 6 (2011) 517.

[155] W. Zhang, R. Zhu, F. Li, Q. Wang, B. Liu, High-performance solid-state organic dye sensitized solar cells with P3HT as hole transporter, J. Phys. Chem. C 115 (2011) 7038−7043.

[156] H. Wang, G. Liu, X. Li, P. Xiang, Z. Ku, Y. Rong, M. Xu, L. Liu, M. Hu, Y. Yang, H. Han, Highly efficient poly(3-hexylthiophene) based monolithic dye-sensitized solar cells with carbon counter electrode, Energy Environ. Sci. 4 (2011) 2025−2029.

[157] S. Ahmad, T. Bessho, F. Kessler, E. Baranoff, J. Frey, C. Yi, M. Gratzel, M.K. Nazeeruddin, A new generation of platinum and iodine free efficient dye-sensitized solar cells, Phys. Chem. Chem. Phys. 14 (2012) 10631−10639.

[158] S.K. Das, D. Yamashita, Y. Ogomi, S.S. Pandey, K. Yoshino, S. Hayase, Single-step fabrication of all-solid dye-sensitized solar cells using solution-processable precursor, Phys. Status Solidi (A) 210 (2013) 1846−1850.

[159] G.K.R. Senadeera, P.V.V. Jayaweera, V.P.S. Perera, K. Tennakone, Solid-state dye-sensitized photocell based on pentacene as a hole collector, Sol. Energy Mater. Sol. Cell. 73 (2002) 103−108.

[160] J. Wu, Q. Li, L. Fan, Z. Lan, P. Li, J. Lin, S. Hao, High-performance polypyrrole nanoparticles counter electrode for dye-sensitized solar cells, J. Power Sources 181 (2008) 172−176.

[161] S. Lu, S. Wang, R. Han, T. Feng, L. Guo, X. Zhang, D. Liu, T. He, The working mechanism and performance of polypyrrole as a counter electrode for dye-sensitized solar cells, J. Mater. Chem. 2 (2014) 12805−12811.

[162] P. Gemeiner, J. Kulicek, M. Mikula, M. Hatala, L. Svorc, L. Hlavata, M. Micusik, M. Omastova, Polypyrrole-coated multi-walled carbon nanotubes for the simple preparation of counter electrodes in dye-sensitized solar cells, Synth. Met. 210 (2015) 323−331.

[163] Q. Qin, J. Tao, Y. Yang, Preparation and characterization of polyaniline film on stainless steel by electrochemical polymerization as a counter electrode of DSSC, Synth. Met. 60 (2010) 1167−1172.

References

[164] B. He, Q. Tang, T. Lianga, Q. Lib, Efficient dye-sensitized solar cells from polyaniline-single wall carbon nanotube complex counter electrodes, J. Mater. Chem. 2 (2014) 3119–3126.

[165] M.U. Shahid, N.M. Mohamed, A.S. Muhsan, R. Bashiri, A.E. Shamsudin, S.N.A. Zaine, Few-layer graphene supported polyaniline (PANI) film as a transparent counter electrode for dye-sensitized solar cells, Diam. Relat. Mater. 94 (2019) 242–251.

[166] D. Gebeyehu, C.J. Brabec, N.S. Sariciftci, D. Vangeneugden, R. Kiebooms, D. Vanderzande, F. Kienberger, H. Schindler, Hybrid solar cells based on dye-sensitized nanoporous TiO_2 electrodes and conjugated polymers as hole transport materials, Synth. Met. 125 (2001) 279–287.

[167] C. Jager, R. Bilke, M. Heim, D. Haarer, H. Karickal, M. Thelakkat, Hybrid solar cells with novel hole transporting poly (triphenyldiamine)s, Synth. Met. 121 (2001) 1543–1544.

[168] J. Wagner, J. Pielichowski, A. Hinsch, K. Pielichowski, D. Bogdał, M. Pajda, S.S. Kurek, A. Burczyk, New carbazole-based polymers for dye solar cells with hole-conducting polymer, Synth. Met. 146 (2004) 159–165.

[169] J. Deng, L. Guo, Q. Xiu, L. Zhang, G. Wen, C. Zhong, Two polymeric metal complexes based on polycarbazole containing complexes of 8-hydroxyquinoline with Zn(II) and Ni(II) in the backbone: synthesis, characterization and photovoltaic applications, Mater. Chem. Phys. 133 (2012) 452–458.

[170] B. Lee, C.C. Stoumpos, N. Zhou, F. Hao, C. Malliakas, C.Y. Yeh, T.J. Marks, M.G. Kanatzidis, R.P.H. Chang, Airstable molecular semiconducting iodosalts for solar cell applications: Cs_2SnI_6 as a hole conductor, J. Am. Chem. Soc. 136 (2014) 15379–15385.

[171] L. Peedikakkandy, J. Naduvath, S. Mallick, P. Bhargava, Lead free, air stable perovskite derivative Cs_2SnI_6 as HTM in DSSCs employing TiO_2 nanotubes as photoanode, Mater. Res. Bull. 108 (2018) 113–119.

[172] B. Lee, Y. Ezhumalai, W. Lee, M.C. Chen, C.Y. Yeh, T.J. Marks, R.P.H. Chang, Cs_2SnI_6-encapsulated multidye-sensitized all-solid-state solar cells, ACS Appl. Mater. Interfaces 11 (2019) 21424–21434.

[173] L. Yang, Z. Zhang, S. Fang, X. Gao, M. Obat, Influence of the preparation conditions of TiO_2 electrodes on the performance of solid-state dye-sensitized solar cells with CuI as a hole collector, Sol. Energy 81 (2007) 717–722.

[174] H. Sakamoto, S. Igarashi, K. Niume, M. Nagai, Highly efficient all solid state dye-sensitized solar cells by the specific interaction of CuI with NCS groups, Org. Electron. 12 (2011) 1247–1252.

[175] N. Kato, S. Moribe, M. Shiozawa, R. Suzuki, K. Higuchi, A. Suzuki, M. Sreenivasu, K. Tsuchimoto, K. Tatematsu, K. Mizumoto, S. Doi, T. Toyoda, Improved conversion efficiency of 10% for solid-state dye-sensitized solar cells utilizing P-type semiconducting CuI and multi-dye consisting of novel porphyrin dimer and organic dyes, J. Mater. Chem. 6 (2018) 22508–22512.

[176] K. Tennakone, G.K.R. Senadeera, D.B.R.A. De Silva, I.R.M. Kottegoda, Highly stable dye-sensitized solid-state solar cell with the semiconductor $4CuBr \cdot 3S(C_4H_9)_2$ as the hole collector, Appl. Phys. Lett. 77 (2000) 2367–2369.

[177] E.V.A. Premalal, G.R.R.A. Kumara, R.M.G. Rajapakse, M. Shimomura, K. Murakami, A. Konno, Tuning chemistry of CuSCN to enhance the performance of TiO_2/N719/CuSCN all-solid-state dye-sensitized solar cell, Chem. Commun. 46 (2010) 3360–3362.

[178] Y. Selk, M. Minnermann, T. Oekermann, M. Wark, J. Caro, Solid-state dye-sensitized ZnO solar cells prepared by low temperature methods, J. Appl. Electrochem. 41 (2011) 445–452.

[179] E.V.A. Premalal, N. Dematage, G.R.R.A. Kumara, R.M.G. Rajapakse, M. Shimomura, K. Murakami, A. Konno, Preparation of structurally modified, conductivity enhanced-p-CuSCN and its application in dye-sensitized solid-state solar cells, J. Power Sources 203 (2012) 288–296.

[180] U.V. Desai, C. Xu, J. Wu, D. Gao, Solid-state dye-sensitized solar cells based on ordered ZnO nanowire arrays, Nanotechnology 23 (2012) 205401.

[181] K.J. Chen, A.D. Laurent, F. Boucher, F. Odobel, D. Jacquemin, Determining the most promising anchors for CuSCN: ab initio insights towards p-type DSSCs, J. Mater. Chem. 4 (2016) 2217–2227.

[182] P.K. Baviskar, Low-cost solid-state dye-sensitized solar cell based on ZnO with CuSCN as a hole transport material using simple solution chemistry, J. Solid State Electrochem. 21 (2017) 2699–2705.

[183] J. Bandara, H. Weerasinghe, Solid-state dye-sensitized solar cell with p-type NiO as a hole collector, Sol. Energy Mater. Sol. Cell. 85 (2005) 385–390.

[184] Y.M. Lee, C.H. Lai, Preparation and characterization of solid n-TiO_2/p-NiO hetrojunction electrodes for all-solid-state dye sensitized solar cells, Solid State Electron. 53 (2009) 1116–1125.

[185] H. Wang, W. Wei, Y.H. Hu, NiO as an efficient counter electrode catalyst for dye-sensitized solar cells, Top. Catal. 57 (2014) 607–611.

II. Photovoltaic materials and devices

[186] S. Kakroo, K. Surana, B. Bhattacharya, Electrodeposited MnO_2-NiO composites as a Pt free counter electrode for dye-sensitized solar cells, J. Electron. Mater. 49 (2020) 2197–2202.

[187] A. Roy, P.S. Devi, S. Karazhanov, D. Mamedov, T.K. Mallick, S. Sundaram, A review on applications of Cu_2ZnSnS_4 as alternative counter electrodes in dye-sensitized solar cells, AIP Adv. 8 (2018) 070701.

[188] Y.S. Wei, Q.Q. Jin, T.Z. Ren, Expanded graphite/pencil-lead as counter electrode for dye-sensitized solar cells, Solid State Electron. 63 (2011) 76–82.

[189] L. Kavan, J.H. Yum, M.K. Nazeeruddin, M. Gratzel, Graphene nanoplatelet cathode for Co(III)/(II) mediated dye-sensitized solar cells, ACS Nano 5 (2011) 9171–9178.

[190] H. Zheng, C.Y. Neo, X. Mei, J. Qiu, J. Ouyang, Reduced graphene oxide films fabricated by gel coating and their application as platinum-free counter electrodes of highly efficient iodide/triiodide dye-sensitized solar cells, J. Mater. Chem. 22 (2012) 14465–14474.

[191] W.J. Lee, E. Ramasamy, D.Y. Lee, J.S. Song, Efficient dye-sensitized solar cells with catalytic multiwall carbon nanotube counter electrodes, ACS Appl. Mater. Interfaces 1 (2009) 1145–1149.

[192] F. Yu, Y. Shi, W. Yao, S. Han, J. Ma, A new breakthrough for graphene/carbon nanotubes as counter electrodes of dye-sensitized solar cells with up to a 10.69% power conversion efficiency, J. Power Sources 412 (2019) 366–373.

[193] G. Wang, S. Kuang, W. Zhang, Helical carbon nanofiber as a low-cost counter electrode for dye-sensitized solar cells, Mater. Lett. 174 (2016) 14–16.

[194] S. Farooq, A.A. Tahir, U. Krewer, A.H.A. Shah, S. Bilal, Efficient photocatalysis through conductive polymer coated FTO counter electrode in platinum free dye sensitized solar cells, Electrochim. Acta 320 (2019) 134544.

[195] K. Saranya, M. Rameez, A. Subramania, Developments in conducting polymer based counter electrodes for dye-sensitized solar cells - an overview, Eur. Polym. J. 66 (2015) 207–227.

[196] W. Wei, H. Wang, Y.H. Hu, A review on PEDOT-based counter electrodes for dye-sensitized solar cells, Int. J. Energy Res. 38 (2014) 1099–1111.

[197] S.S. Shenouda, I.S. Yahia, H.S. Hafez, F. Yakuphanoglu, Facile and low-cost synthesis of PEDOT:PSS/FTO polymeric counter electrode for DSSC photosensor with negative capacitance phenomenon, Mater. Res. Express 6 (2019) 065004.

[198] J. Kwon, V. Ganapathy, Y.H. Kim, K.D. Song, H.G. Park, Y. Jun, P.J. Yoo, J.H. Park, Nanopatterned conductive polymer films as a Pt, TCO-free counter electrode for low-cost dye-sensitized solar cells, Nanoscale 5 (2013) 7838–7843.

[199] Q. Li, J. Wu, Q. Tang, Z. Lan, P. Li, J. Lin, L. Fan, Application of microporous polyaniline counter electrode for dye-sensitized solar cells, Electrochem. Commun. 10 (2008) 1299–1302.

[200] H. Wang, Q. Feng, F. Gong, Y. Li, G. Zhou, Z. Wang, In situ growth of oriented polyaniline nanowires array for efficient cathode of Co(III)/Co(II) mediated dye-sensitized solar cell, J. Mater. Chem. 1 (2013) 97–104.

[201] W. Hou, Y. Xiao, G. Han, D. Fu, R. Wu, Serrated, flexible and ultrathin polyaniline nanoribbons: an efficient counter electrode for the dye-sensitized solar cell, J. Power Sources 322 (2016) 155–162.

[202] S. Jeon, C. Kim, J. Ko, S. Im, Spherical polypyrrolena noparticles as a highly efficient counter electrode for dye-sensitized solar cells, J. Mater. Chem. 21 (2011) 8146–8151.

[203] H.C. Sun, D. Qin, S.Q. Huang, X.Z. Guo, D.M. Li, Y.H. Luoand, Q.B. Meng, Dye-sensitized solar cells with NiS counter electrodes electrodeposited by a potential reversal technique, Energy Environ. Sci. 4 (2011) 2630–2637.

[204] L.T. Gularte, C.D. Fernandes, M.L. Moreira, C.W. Raubach, P.L.G. Jardim, S.S. Cava, In situ microwave-assisted deposition of CoS counter electrode for dyesensitized solar cells, Sol. Energy 198 (2020) 658–664.

[205] S.A. Patil, N. Mengal, A.A. Memon, S.H. Jeong, H.S. Kim, CuS thin film grown using the one pot, solution-process method for dye-sensitized solar cell applications, J. Alloys Compd. 708 (2017) 568–574.

[206] D. Vikraman, S.A. Patil, S. Hussain, N. Mengal, H.S. Kim, S.H. Jeon, J. Jung, H.S. Kim, H.J. Park, Facile and cost-effective methodology to fabricate MoS_2 counter electrode for efficient dye-sensitized solar cells, Dyes Pigments 151 (2018) 7–14.

[207] W. Wang, X. Pan, W. Liu, B. Zhang, H. Chen, X. Fang, J. Yao, S. Dai, $FeSe_2$ films with controllable morphologies as efficient counter electrodes for dye-sensitized solar cells, Chem. Commun. 50 (2014) 2618–2620.

[208] Y. Hou, D. Wang, X.H. Yang, W.Q. Fang, B. Zhang, H.F. Wang, G.Z. Lu, P. Hu, H.J. Zhao, H.G. Yang, Rational screening low-cost counter electrodes for dye-sensitized solar cells, Nat. Commun. 4 (2013) 1583–1590.

[209] P.S. Tamboli, C.V. Jagtap, V.S. Kadam, R.V. Ingle, R.S. Vhatkar, S.S. Mahajan, H.M. Pathan, Spray pyrolytic deposition of α-MoO_3 film and its use in dye-sensitized solar cell, Appl. Phys. A 12 (2018) 4339.

References

209

[210] C. Gao, Q. Han, M. Wu, Review on transition metal compounds based counter electrode for dye-sensitized solar cells, J. Energy Chem. 27 (2018) 703—712.

[211] J. Jin, Z. Wei, X. Qiao, H. Fan, L. Cui, Substrate-mediated growth of vanadium carbide with controllable structure as high performance electrocatalysts for dye-sensitized solar cells, RSC Adv. 7 (2017) 26710—26716.

[212] G.R. Li, J. Song, G.L. Pan, X.P. Gao, Highly Pt-like electrocatalytic activity of transition metal nitrides for dye-sensitized solar cells, Energy Environ. Sci. 4 (2011) 1680—1683.

[213] J. Guo, S. Liang, Y. Shi, C. Hao, X. Wang, T. Ma, Transition metal selenides as efficient counter electrode materials for dye-sensitized solar cells, Phys. Chem. Chem. Phys. 17 (2015) 28985—28992.

[214] A.K. Jana, B.B. Bhowmik, Enhancement in power output of solar cells consisting of mixed dyes, J. Photochem. Photobiol. Chem. 122 (1999) 53—56.

[215] S.K. Balasingam, M. Lee, M.G. Kang, Y. Jun, Improvement of dye-sensitized solar cells towards the broader light harvesting of the solar spectrum, Chem. Commun. 49 (2013) 1471—1487.

[216] B.E. Hardin, E.T. Hoke, P.B. Armstrong, J.H. Yum, P. Comte, T. Torres, J.M.J. Frechet, M.K. Nazeeruddin, M. Gratzel, M.D. McGehee, Increased light harvesting in dye-sensitized solar cells with energy relay dyes, Nat. Photon. 3 (2009) 406—411.

[217] W. Chen, Y. Qiu, S. Yang, Branched ZnO nanostructures as building blocks of photoelectrodes for efficient solar energy conversion, Phys. Chem. Chem. Phys. 14 (2012) 10872—10881.

[218] Y. Cao, Y. Liu, S.M. Zakeeruddin, A. Hagfeldt, M. Gratzel, Direct contact of selective charge extraction layers enables high-efficiency molecular photovoltaics, Joule 2 (2018) 1108—1117.

[219] K. Kakiage, Y. Aoyama, T. Yano, K. Oya, J. Fujisawa, M. Hanaya, Highly-Efficient dye-Sensitized solar cells with collaborative sensitization by silyl-anchor and carboxy-anchor dyes, Chem. Commun. 51 (2015) 15894—15897.

[220] C.S.K. Ranasinghe, W.M.N.M.B. Wanninnayake, G.R.A. Kumara, R.M.G. Rajapakshe, P.M. Sirimanne, An enhancement of efficiency of a solid-state dye- sensitized solar cell due to cocktail effect of N719 and black dye, Optik 125 (2014) 813—815.

[221] C. Magne, M. Urienb, T. Pauporte, Enhancement of photovoltaic performances in dye-sensitized solar cells by co-sensitization with metal-free organic dyes, RSC Adv. 3 (2013) 6315—6318.

[222] P.K. Baviskar, D.P. Dubal, S. Majumder, A. Ennaoui, B.R. Sankapal, "Basic idea, advance approach": efficiency boost by sensitization of blended dye on chemically deposited ZnO films, J. Photochem. Photobiol. Chem. 318 (2016) 135—141.

[223] N.I. Beedri, P.K. Baviskar, M. Mahadik, S.R. Jadkar, J.S. Jang, H.M. Pathan, Efficiency enhancement for cocktail dye sensitized Nb_2O_5 photoanode towards dye sensitized solar cell, Eng. Sci. 8 (2019) 76—82.

[224] G. Richhariya, A. Kumar, Fabrication and characterization of mixed dye: natural and synthetic organic dye, Opt. Mater. 79 (2018) 296—301.

[225] K. Sayama, S. Tsukagoshi, T. Mori, K. Hara, Y. Ohga, A. Shinpou, Y. Abe, S. Suga, H. Arakawa, Efficient sensitization of nanocrystalline TiO_2 films with cyanine and merocyanine organic dyes, Sol. Energy Mater. Sol. Cells 80 (2003) 47—71.

[226] M. Guo, P. Diao, Y.J. Ren, F. Meng, H. Tian, S.M. Cai, Photoelectrochemical studies of nanocrystalline TiO_2 co-sensitized by novel cyanine dyes, Sol. Energy Mater. Sol. Cell. 88 (2005) 23—35.

[227] C.H. Lee, S.A. Kim, M.R. Jung, K.-S. Ahn, Y.S. Han, J.H. Kim, Aggregation control of organic sensitizers for panchromatic dye co-sensitized solar cells, Jpn. J. Appl. Phys. 53 (2014) 08NC04.

[228] V. Saxena, P. Veerender, A.K. Chauhan, P. Jha, D.K. Aswal, S.K. Gupta, Efficiency enhancement in dye sensitized solar cells through cosensitization of TiO_2 nanocrystalline electrodes, Appl. Phys. Lett. 100 (2012) 133303.

[229] H. Lee, J. Kim, D.Y. Kim, Y. Seo, Co-sensitization of metal free organic dyes in flexible dye sensitized solar cells, Org. Electron. 52 (2018) 103—109.

[230] L.Y. Lin, M.H. Yeh, C.P. Lee, J. Chang, A. Baheti, R. Vittal, K.R.J. Thomas, K.C. Ho, Insights into the co-sensitizer adsorption kinetics for complementary organic dye-sensitized solar cells, J. Power Sources 247 (2014) 906—914.

[231] L. Yu, K. Fan, T. Duan, X. Chen, R. Li, T. Peng, Efficient panchromatic light harvesting with Co-sensitization of zinc phthalocyanine and bithiophene-based organic dye for dye-sensitized solar cells, ACS Sustain. Chem. Eng. 2 (2014) 718—725.

[232] J. Lim, M. Lee, S.K. Balasingam, J. Kim, D. Kim, Y. Jun, Fabrication of panchromatic dye-sensitized solar cells using pre-dye coated TiO_2 nanoparticles by a simple dip coating technique, RSC Adv. 3 (2013) 4801—4805.

II. Photovoltaic materials and devices

210   7. Dye-sensitized solar cells

[233] J.H. Yum, S.R. Jang, P. Walter, T. Geiger, F. Nuesch, S. Kim, J. Ko, M. Gratzel, M.K. Nazeeruddin, Efficient co-sensitization of nanocrystalline TiO_2 films by organic sensitizers, Chem. Commun. (2007) 4680–4682.

[234] A. Islam, T.H. Chowdhury, C. Qin, L. Han, J.J. Lee, I.M. Bedja, M. Akhtaruzzaman, K. Sopian, A. Mirloupe, N. Leclerc, Panchromatic absorption of dye sensitized solar cells by co-sensitization of triple organic dyes, Sustain. Energy Fuels 2 (2018) 209–214.

[235] A. Islam, M. Akhtaruzzaman, T.H. Chowdhury, C. Qin, L. Han, I.M. Bedja, R. Stalder, K.S. Schanze, J.R. Reynolds, Enhanced photovoltaic performances of dye-sensitized solar cells by Co-sensitization of benzothiadiazole and squaraine-based dyes, ACS Appl. Mater. Interfaces 8 (2016) 4616–4623.

[236] M. Younas, M.A. Gondal, U. Mehmood, K. Harrabi, Z.H. Yamani, F.A. Al-Sulaiman, Performance enhancement of dye-sensitized solar cells via cosensitization of ruthenizer Z907 and organic sensitizer SQ2, Int. J. Energy Res. 42 (2018) 3957–3965.

[237] F. Huang, D. Chen, L. Cao, R.A. Caruso, Y.B. Cheng, Flexible dye-sensitized solar cells containing multiple dyes in discrete layers, Energy Environ. Sci. 4 (2011) 2803–2806.

[238] G. Koyyada, S. Shome, M. Chandrasekharam, G.D. Sharma, S.P. Singh, High performance dye-sensitized solar cell from cocktail solution of ruthenium dye and metal free organic dye, RSC Adv. 6 (2016) 41151–41155.

[239] G.H. Rao, A. Venkateswararao, L. Giribabu, L. Han, I. Bedja, R.K. Gupta, A. Islam, S.P. Singh, Near-infrared squaraine co-sensitizer for high-efficiency dye-sensitized solar cells, Phys. Chem. Chem. Phys. 18 (2016) 4279–14285.

[240] M.C. Sil, L.S. Chen, C.W. Lai, C.C. Chang, C.M. Chen, Enhancement of solar efficiency of dye-sensitized solar cell by molecular engineering of organic dye incorporating N-alkyl attached 1, 8-naphthalimide derivative, J. Mater. Chem. C (2020), https://doi.org/10.1039/D0TC01388A. Advance Article.

[241] G. Koyyada, R.K. Chitumalla, S. Thogiti, J.H. Kim, J. Jang, M. Chandrasekharam, J.H. Jung, A new series of EDOT based Co-sensitizers for enhanced efficiency of cocktail DSSC: a comparative study of two different anchoring groups, Molecules 24 (2019) 3554.

[242] Y. Hao, Y. Saygili, J. Cong, A. Eriksson, W. Yang, J. Zhang, E. Polanski, K. Nonomura, S.M. Zakeeruddin, M. Gratzel, A. Hagfeldt, G. Boschloo, Novel blue organic dye for dye-sensitized solar cells achieving high efficiency in cobalt-based electrolytes and by Co-sensitization, ACS Appl. Mater. Interfaces 8 (2016) 32797–32804.

[243] C.M. Lan, H.P. Wu, T.Y. Pan, C.W. Chang, W.S. Chao, C.T. Chen, C.L. Wang, C.Y. Lin, E.W.G. Diau, Enhanced photovoltaic performance with co-sensitization of porphyrin and an organic dye in dye-sensitized solar cells, Energy Environ. Sci. 5 (2012) 6460–6464.

[244] R.Y. Ogura, S. Nakane, M. Morooka, M. Orihashi, Y. Suzuki, K. Noda, High-performance dye-sensitized solar cell with a multiple dye system, Appl. Phys. Lett. 94 (2009) 073308.

[245] L.H. Nguyen, H.K. Mulmudi, D. Sabba, S.A. Kulkarni, S.K. Batabyal, K. Nonomura, M. Gratzel, S.G. Mhaisalkar, A selective co-sensitization approach to increase photon conversion efficiency and electron lifetime in dye-sensitized solar cells, Phys. Chem. Chem. Phys. 14 (2012) 16182–16186.

[246] L. Zhang, A. Konno, Development of flexible dye-sensitized solar cell based on pre-dyed zinc oxide nanoparticle, Int. J. Electrochem. Sci. 13 (2018) 344–352.

[247] S. Casaluci, M. Gemmi, V. Pellegrini, A.D. Carlo, F. Bonaccorso, Graphene-based large area dye-sensitized solar cell modules, Nanoscale 8 (2016) 5368–5378.

[248] A. Hegazy, N. Kinadjian, B. Sadeghimakki, S. Sivoththaman, N.K. Allam, E. Prouzet, TiO_2 nanoparticles optimized for photoanodes tested in large area Dye-sensitized solar cells (DSSC), Sol. Energy Mater. Sol. Cell. 153 (2016) 108–116.

[249] P.M. Sommeling, M. Spath, H.J.P. Smit, N.J. Bakker, J.M. Kroon, Long-term stability testing of dye-sensitized solar cells, J. Photochem. Photobiol. Chem. 164 (2004) 137–144.

[250] G.G. Sonai, A. Tiihonen, K. Miettunen, P.D. Lund, A.F. Nogueira, Long-term stability of dye-sensitized solar cells assembled with cobalt polymer gel electrolyte, J. Phys. Chem. C 121 (2017) 17577–17585.

[251] A. Maalinia, H.A. Moghaddam, E. Nouri, M.R. Mohammadi, Long-term stability of dye-sensitized solar cells using a facile gel polymer electrolyte, New J. Chem. 42 (2018) 13256–13262.

[252] A. Azmar, R.H.Y. Subban, T. Winie, Improved long-term stability of dye-sensitized solar cell employing PMA/PVAc based gel polymer electrolyte, Opt. Mater. 96 (2019) 109349.

[253] Y.J. Jang, S. Thogiti, K.Y. Lee, J.H. Kim, Long-term stable solid-state dye-sensitized solar cells assembled with solid-state polymerized hole-transporting material, Crystals 9 (2019) 452.

II. Photovoltaic materials and devices

References

[254] N.M. Nursam, J. Hidayat, L. Muliani, P.N. Anggraeni, L. Retnaningsih, N. Idayanti, From cell to module: fabrication and long-term stability of dye-sensitized solar cells, IOP Conf. Ser. Mater. Sci. Eng. 214 (2017) 012007.

[255] https://www.businesswire.com/news/home/20091012005499/en/G24-Innovations-Ships-World%E2%80%99s-Commercial-Application-DSSC.

[256] https://solarpv.expert/2020/02/06/ricoh-to-launch-the-worlds-first-solid-state-dye-sensitized-solar-cell-modules/.

[257] L. Han, A. Islam, H. Chen, C. Malapaka, B. Chiranjeevi, S. Zhang, X. Yang, M. Yanagida, High-efficiency dye-sensitized solar cell with a novel co-adsorbent, Energy Environ. Sci. 5 (2012) 6057−6060.

CHAPTER

8

Potential of nanooxidic materials and structures of photoanodes for DSSCs

Markus Diantoro[1,2], Siti Wihdatul Himmah[1], Thathit Suprayogi[1], Ulwiyatus Sa'adah[1], Arif Hidayat[1,2], Nandang Mufti[1,2], Nasikhudin[1,2]

[1]Department of Physics, Faculty of Mathematics and Natural Science, Universitas Negeri Malang, Malang, East Java, Indonesia; [2]Center of Advanced Materials for Renewable Energy (CAMRY), Universitas Negeri Malang, Malang, East Java, Indonesia

8.1 Introduction

Total world energy consumption is expected to increase from 19 to 30 TW in a 35-year span from 2015 to 2050 [1—3]. This relentless energy consumption is due to global population increase and fast growth in the industrial sector. Conventional fossil fuel or nonrenewable energy sources will remain the largest energy source at more than 75% even in 2050 [1,4,5]. In spite of this, the availability of such nonrenewable crude oil sources in nature is restricted. By the next couple of decades, increasing population and uncontrolled industrial growth will cause fossil-based oil and natural gas reserves to be exhausted. Besides, fossil fuel is also the main contributor to many ecological issues such as climate change, air contamination, acid rain, and temperature changes [6—8]. Therefore unlimited alternative energy resources are needed. The most promising alternative energy source is solar energy because of its abundant availability in nature and low negative environmental impact [9,10]. Earth receives about 4,000,000 EJ of solar energy every year [11]. For reference, $1\ EJ = 10^{18}\ J$. Solar radiation takes about 8 min to arrive on Earth from the sun's surface. When solar radiation reaches Earth, its energy is not evenly dispensed across the globe. Solar radiation hits Earth near the equator more than at the poles. Hence, there is an energy surplus at the equator and an energy deficit at the poles.

Energy Materials
https://doi.org/10.1016/B978-0-12-823710-6.00013-3

Several factors determine the intensity of solar radiation entering Earth's atmosphere. These include latitude, geographical variations, and climate. Countries in the tropics and subtropics receive a higher amount of solar radiation and thus have significant potential to harness solar energy [12]. The average energy received by Earth's atmosphere from the sun is about 342 W/m^2. Of this, 30% of solar energy will be dissipated and reflected, while 70% (239 W/m^2) of solar energy will be harvested and captured. Solar radiation throughout the world varies from 60 to 250 W/m^2. Theoretically, solar energy radiation has great potential to meet energy needs throughout the world if the technology for harvesting energy from the sun is well accessible [13].

In general, solar power technology is classified into two categories, namely solar thermal technology and solar cell technology. In the former, solar energy is converted to heat, which is widely used on an industrial scale to warm air or water. This technology has several advantages, namely reducing dependence on fossil fuels and greenhouse gas emissions [14]. However, the application of this technology to various industrial processes is still ineffective. Another solar energy technology most widely developed at this time is solar cell technology. In solar cell technology, energy from the sun is transformed into electrical energy. Depending on the material used, solar cell technology is grouped into three generations: (1) Silicon crystal technology is the first-generation technology that uses simple or multicrystalline silicon. Silicon is the most popular commercial solar cell material used because it has several favorable characteristics such as an ideal bandgap ($E_g = 1.1$ eV), abundant availability in nature, and is generally stable. The highest efficiencies that can be produced by silicon-based solar cells are 18.7% for simple crystalline silicon [15] and 22.3% for multicrystalline silicon [16]. (2) Thin-film solar cells constitute the second-generation technology. These include (a) cadmium telluride (CdTe); (b) copper indium selenide and copper indium gallium selenide (CIGS), and (c) amorphous silicon (a-Si) and microamorphous silicon. Thin-film CIGS polycrystalline solar cells have attracted much attention because they can work well in weak light conditions, have lower production cost than that of silicon solar cells, and are suitable for flexible and tandem solar cells. The efficiency produced by CIGS polycrystalline thin-film-based devices is 11.72% [17]. First-generation solar cells can demonstrate high energy conversion efficiency. However, both first- and second-generation solar cells require a long fabrication process and cause various environmental problems. Therefore in the next generation, there are many comprehensive studies on environmentally friendly solar cells. (3) Third-generation solar cells include organic-based technology, namely dye-sensitized solar cells (DSSCs) [8]. A lot of research has been carried out on DSSC efficiency improvements over the last 5 years. The factor that most influences the efficiency produced by the DSSC device is the optimization of the photoanode material. Generally, DSSC photoanodes are made from nanostructured metal oxides. The synthesized materials can differ in morphology with various shapes and sizes. 0D-nanostructured metal oxides, for example, ZnO, SnO_2, and TiO_2, have been widely investigated as photoanode materials. This is because they exhibit good photocorrosion stability and electronic properties [18,19]. Nanostructured materials have a high specific surface area so that the process of electron transport becomes faster. This is related to the shorter average distance for electron movement or the quantum confinement effect. This effect can be observed when the electron wavelength is larger than the size of the particle. From an optical perspective, nanostructures form photonic energy bands that are proven capable of manipulating light through photonic localization or optical confinement. The peculiar properties of nanomaterials have aroused much interest in and investigations into DSSC applications. However, the disadvantage of 0D-structured material is that it has a high electron transport resistance so that it can hamper the performance of DSSC devices [20]. Unlike 0D structures, 1D materials such as fibers [21] and rods [22] show a much faster process of electron transport. Metal oxides with different structures (0D and 1D) certainly offer different material characteristics so that they will affect the efficiency of energy conversion exhibited by DSSC devices.

8.2 DSSCs

DSSCs, commonly referred to as "Grätzel cells," are photoelectrochemical-based solar cells that utilize dye molecules to produce electrical energy directly from sunlight [23]. In 1991, O'Regan and Grätzel introduced DSSCs. A DSSC has many advantages. Its light-to-electricity conversion efficiency is relatively high and is comparable to that of conventional solar cell efficiency. It is easy to fabricate and can still work well in diffused light conditions such as at dawn or even in cloudy weather [24—26]. At present, DSSCs have experienced very rapid development with an increase in conversion efficiency from 7.1% in 1991 to 14.3% in 2015 [27—29]. Fig. 8.1 shows an illustrative design of the components of a DSSC [30].

In general, a DSSC consists of four main components.

8.2.1 Photoanode

The photoanode is the most important DSSC component. It enables dye molecules to adsorb to the porous photoanode. In other words, it acts as a matrix for adherence of dye molecules. In preparing the photoanode, a thin film of metal oxide is coated on a conductive indium tin oxide (ITO) or fluorine tin oxide (FTO) glass substrate. The metal oxide materials that are coated on the ITO or FTO substrates include TiO_2 [31], SnO_2 [32], ZnO [33], Nb_2O_5 [34], and $SrTiO_3$ [35]. These metal oxides have a wide bandgap. In addition to metal oxides, conductive polymers such as polyaniline are also widely used as photoanode layers in DSSCs because their electrical and optical properties can be adjusted [36—42]. However, the semiconductors and conducting polymers absorb a small portion of ultraviolet light. Therefore they

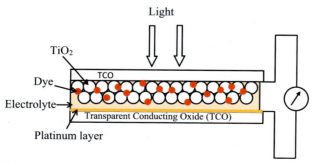

FIGURE 8.1 Illustrative design of the constituent components of a dye-sensitized solar cell.

are immersed in a dye solution so that the dye molecules bind covalently to the photoanode surface. The porous metal oxide structure with the higher surface area will allow a larger number of dye molecules to adhere to the metal oxide surface [23]. Hence, to achieve optimal dye molecular adsorption and fast electron transport processes, the photoanode or rather the metal oxide must have several characteristics, including (1) a large surface area for maximum dye loading, so that the absorption of sunlight occurs effectively; (2) a relatively high layer transparency to reduce photon loss; (3) good mobility for fast electron transport; (4) fewer grain boundary trapping sites to minimize electron recombination with the redox mediator in the electrolytes; and (5) hydroxyl or defect groups for attachment with dye molecules [30,43,44].

8.2.2 Dyes

The dye material is important for the absorption and transformation of photons into electrical energy [45]. For good DSSC performance, the dye molecules must be able to absorb photons, convert photons into electron—hole pairs, and transport charge effectively. To perform these processes efficiently, the dye molecules must have several requirements, including (1) high molar extinction coefficient especially in the visible wavelength region, which may extend into the near-infrared; (2) the position of lowest unoccupied molecular orbital (LUMO) should be a little higher than the conduction bands of metal oxides, whereas the position of highest occupied molecular orbital (HOMO) should be far from the conduction bands of metal oxides and lower than the redox potential of electrolyte to ensure efficient charge injection and regeneration of dye molecules by redox mediators from electrolyte solutions; (3) good solubility and photostability [43,46]; and (4) anchoring groups such as carbonyl (CO), hydroxyl (−OH), and carboxyl (−COOH) groups for bonding with metal oxides. Currently, the dyes used in DSSCs contain ruthenium (Ru). These include N3 [47], N719 [48], black dye [49], and Z-907 [50]. This metal-based dye has several advantages, namely high photon—electron conversion efficiency, high ability to absorb sunlight, and good transfer of ligand charges. However, this type of dye harms the environment so there is a lot of development related to organometallic-based dyes. This type of dye is environmentally friendly compared to Ru-based dyes. Organometallic dyes can either be made synthetically or extracted from plants such as chlorophyll from the leaves of *Pterocarpus indicus* [51]. A mixture of fruit and vegetable extracts has also been widely used as a coloring agent in DSSCs such as spinach and purple cabbage extracts [52], spinach and dragon fruit [42], purple cabbage and beets [53], pomegranate and berries [24], etc. The structures of N3, N719, black dye, and Z-907 can be found in Ref. [54].

8.2.3 Electrolyte

In DSSCs, electrolyte functions as a carrier and medium for charge transport to enable the electron cycle and regeneration of oxidized dye molecules [27,43]. Based on their physical conditions, electrolytes are classified into three categories, namely solid [54], quasi-solid [55], and liquid [56] electrolytes. An iodide/triiodide redox couple or mediator is most popularly used in electrolytes. The excellent performance of liquid electrolytes based on redox pairs is due to their fast photoanode penetration, very fast dye molecule regeneration, and low recombination loss. However, liquid electrolytes are easily subjected to evaporation of volatile iodide ions, which can cause a decrease in charge carrier concentration that will lead to cell degradation. In addition, toxic leakage from solvents can also cause environmental hazards [30,46].

8.2.4 Counter electrode

The counter electrode enables the return of electrons to the oxidized dye. An important criterion that the counter or opposing electrode material must have is low load transfer resistance. Likewise, the material must have good chemical and electrochemical stabilities. Platinum film-based counter electrodes have been shown to demonstrate excellent performance. The platinum film has several advantages, including high electrical conductivity and transmittance, high corrosion resistance, and low load transfer resistance [57]. On absorption of light by the dye molecules, electrons from the ground state (D) of the HOMO are promoted to the excited state (D^*) in the LUMO (Eq. 8.1). The excited electron(s) will be driven into the oxide semiconductor on oxidation of the dye molecules. In this situation, electrons and holes will be separated. The electrons traverse the oxide semiconductor and exit to the external circuit. The holes remain in the oxidized dye molecules (Eq. 8.2). The excited electrons travel to the counter electrode via the electrically connected external circuit (Eq. 8.3). The electrons that reach the counter electrode react with the triiodide ions that have diffused toward the cathode or counter electrode. The triiodide ion is then reduced to an iodide ion when it accepts electrons that have reached the counter electrode. The iodide ion then diffuses to the photoanode. The oxidized dye molecule is regenerated to a neutral state when it receives electrons from the iodide ion [58] (Eq. 8.4) and is oxidized into a triiodide ion, which again diffuses to the counter electrode where it will again be reduced as shown in Eq. (8.5). Thus the iodide and triiodide ions have undergone a reduction—oxidation reaction and hence the I^-/I_3^- couple is referred to as the redox couple or mediator. Fig. 8.2 shows the working principle of a DSSC device [59].

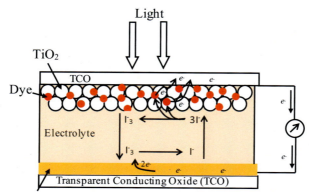

FIGURE 8.2 Working principle of a dye-sensitized solar cell.

$$S + h\upsilon \rightarrow S^* \ (\text{Absorption}) \tag{8.1}$$

$$S^* + TiO_2 \rightarrow e^- (TiO_2) + S^+ \ (\text{Injection process}) \tag{8.2}$$

$$e^- (TiO_2) + C.E \rightarrow TiO_2 + e^-_{(C.E)} + \text{ electrical energy (Energy generation)} \tag{8.3}$$

$$S^+ + \frac{3}{2} I^- \rightarrow S + \frac{1}{2} I_3^- \ (\text{Regeneration of dye}) \tag{8.4}$$

$$\frac{1}{2} I_3^- + e^-_{(C.E)} \rightarrow \frac{3}{2} I^- + C.E \ (\text{Regeneration reaction}) \tag{8.5}$$

8.3 Nanooxidic photoanode materials for DSSCs (TiO₂, ZnO, and SnO₂)

Titanium (IV) oxide (TiO_2) is n-type. It is used as an electron acceptor in solar cell technology. TiO_2 has a wide bandgap and high surface area [4,60]. However, these metal oxides show high levels of electron recombination, short diffusion hole lengths, and are unable to absorb visible light (absorption threshold is equivalent to 380 nm) [61−65]. In bulk form, TiO_2 is a polymorph material consisting of three phases: anatase, rutile, and brookite. The anatase and rutile phases have tetragonal structures. The brookite phase has an orthorhombic structure. Some details about the crystal structures of TiO_2 are listed in Table 8.1.

Based on Table 8.2, the crystallite size of TiO_2 nanoparticle films is 15.46 nm. Jamalullail and colleagues [66] conducted a similar study related to TiO_2-based nanoparticle photoanodes deposited on the ITO glass conductive layer in the form of a paste and obtained a grain size of 18.76 nm. Sasipriya and colleagues also conducted similar studies, but the conductive

TABLE 8.1 Crystal structures of TiO_2.

Parameter	Anatase	Rutile	Brookite
Crystal system	Tetragonal	Tetragonal	Orthorhombic
Space group	*I41/a m d*	*P42/m n m*	*Pbca*
a (Å)	3.872	4.590	9.180
b (Å)	3.872	4.590	5.430
c (Å)	9.616	2.960	5.164
Atomic position of Ti	0.500; 0.750; 0.375	0.000; 0.000; 0.000	0.128; 0.099; 0.862
Atomic position of O	0.500; 0.750; 0.162	0.327; 0.327; 0.000	O1: 0.009; 0.148; 0.183 O2: 0.231; 0.111; 0.536

TABLE 8.2 Rietveld refinement of TiO$_2$ nanoparticle film.

Parameter	Model COD 2310710	Refinement
Crystal system	Tetragonal	
Space group	*I41/a m d*	
$a = b$ (Å)	3.872	3.785
c (Å)	9.616	9.505
Atomic position of Ti	0.500; 0.750; 0.375	0.500; 0.750; 0.375
Atomic position of O	0.500; 0.750; 0.162	0.500; 0.750; 0.164
Crystallite size (nm)		15.46

substrate used was FTO glass and obtained a much larger grain size of around 21 nm [67]. A TiO$_2$ film-based DSSC with smaller grain size will show a much higher energy conversion efficiency when compared to a TiO$_2$ film-based DSSC with larger grain size. Higher energy conversion efficiency can be associated with weak backscattering light and increased adsorption of dye molecules [68]. Smaller grain size indicates a high surface area, thereby increasing surface roughness, electron transfer velocity, and current density [69].

The stability of the anatase phase is higher than the rutile phase when the TiO$_2$ nanoparticle size is less than 14 nm. The difference in energy between the anatase and rutile phases is only about 2–10 kJ/mol. In addition, the electron transport process in the anatase phase is faster. This is because of its much higher packing density. Thus anatase TiO$_2$ is widely used as a photoanode material in DSSCs. The anatase TiO$_2$ metal oxide has been successfully coated via the doctor-blade method on ITO glass. The DSSC has been assembled following the arrangement shown in Fig. 8.3.

The doctor-blade method is widely used in preparing the photoanode in DSSCs because it is more economical and can minimize nanoparticle semiconductor material loss by about 5% compared to spin coating [70]. Fig. 8.4 is a diffraction pattern of TiO$_2$ nanoparticle films. TiO$_2$ peaks indicate an anatase phase in the Bragg's planes of (011), (013), (004), (112), (200), (015), (211), (213), (024), (116), (220), (017), (215), and (301).

Based on refinement analysis using the Rietica software, several parameters are obtained as shown in Table 8.2.

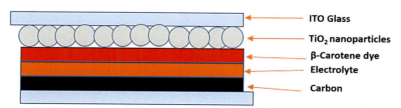

FIGURE 8.3 The assembled design of a dye-sensitized solar cell.

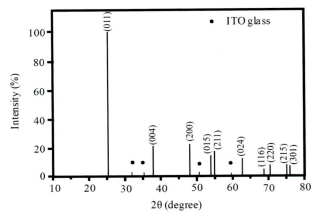

FIGURE 8.4 Diffraction pattern of TiO$_2$ nanoparticle film.

FIGURE 8.5 Surface micrograph of TiO$_2$ nanoparticle film.

Fig. 8.5 shows a surface morphology of TiO$_2$ nanoparticle film, which indicates that the TiO$_2$ layer has a round, porous shape with agglomeration, as shown in previous studies [71].

The porous surface morphology facilitates better diffusion of dye and electrolyte molecules. Agglomeration occurs due to the very small grain size of TiO$_2$ [24]. TiO$_2$ nanoparticle-based film has exhibited a high surface area. However, dye molecules have limited light-collecting efficiency. This is because the light-scattering ability of TiO$_2$ is weak. Trapping sites present at the grain boundaries prevent rapid electron transfer and this can result in electrons reducing the triiodide ions in the electrolyte into iodide ions [72]. Therefore TiO$_2$ nanoparticles need to be composited with other semiconducting oxides, for example, tin (IV) oxide (SnO$_2$) and zinc oxide (ZnO).

TiO$_2$ has relatively low electron mobility (10^{-1}–10^{-3} cm^2/Vs). The mobility of ZnO nanoparticles is even higher, being more than 100 cm^2/Vs. This advantage provides electron transport facilities that are more efficient and can reduce recombination rates. ZnO metal oxides

TABLE 8.3 Crystal structure of ZnO.

Parameter	Rocksalt	Zincblende	Wurtzite
Crystal system	Cubic	Cubic	Hexagonal
Space group	$Fm3m$	$F4(bar)3m$	$P63mc$
a (Å)	4.270	4.629	3.249
b (Å)	4.270	4.629	3.249
c (Å)	4.270	4.629	5.203
Atomic position of Zn	0.500; 0.500; 0.500	0.000; 0.000; 0.000	0.018; 0.018; 0.005
Atomic position of O	0.000; 0.000; 0.000	0.250; 0.250; 0.250	0.018; 0.018; 0.005

have the same electron affinity with TiO_2, but the energy gap of ZnO is slightly larger than TiO_2, that is, 3.3 and 3.2 eV, respectively [73]. Besides, the electronic diffusion rate of ZnO is much faster than TiO_2 [74,75]. ZnO has three crystal structures, namely two cubic structures (rocksalt and zincblende phases) and hexagonal structures (wurtzite phase). At room temperature, the wurtzite phase is the most stable crystal structure compared to the other phases. Details about the ZnO crystal structure are listed in Table 8.3.

Efforts to improve DSSC performance were also focused on substituting TiO_2 with ZnO in the photoanode through increased electron collection, faster electron transport, and decreased electron recombination. However, ZnO is unstable in acid-based dyes and the efficiency of electron injection in ruthenium-based dyes is low. This lowered the efficiency of the DSSC [75]. Therefore photoanodes with two or more oxides such as TiO_2 with ZnO have received a lot of attention due to the favorable properties of these oxides. ZnO has high electron transport rate capability and TiO_2 has high electron injection efficiency in ruthenium-based dyes, so it is expected that the combination of these two materials can enhance DSSC efficiency. TiO_2/ZnO (TZ) nanocomposite has been coated successfully on FTO glass substrate through screen printing with the arrangement shown in Fig. 8.6.

Screen printing exhibits superior photoanode film thickness compared to other coating methods. This is because during the deposition process the metal oxide layer is mechanically pressed using a squeegee so that the metal oxide enters the mesh hole and attaches to the surface of the conductive substrate [70]. The independent variable was the TiO_2 mass (0.02, 0.04, 0.06, and 0.08 g) added in the nanocomposite with a constant mass of ZnO of 0.12 g. Fig. 8.7 shows diffraction patterns of TZ nanocomposite films.

The TiO_2 peaks are less visible since the mass of TiO_2 nanoparticles is very small. The peaks due to FTO glass are more dominant. Peaks with Miller indices of (110), (101), (200), (211), (220), (310), and (301) are characteristic of FTO glass diffraction films. Tables 8.4 and 8.5 show the refinement analysis of TiO_2 and ZnO parameters using Rietica software, respectively.

Addition of TiO_2 nanoparticles into the nanocomposite caused a decrease in ZnO grain size to 26.1 nm and an increase in TiO_2 grain size to 14.8 nm in the 0.08 TZ sample. Films with small grain size show higher surface area. The high surface area enables more dye

222　　　8. Potential of nanooxidic materials and structures of photoanodes for DSSCs

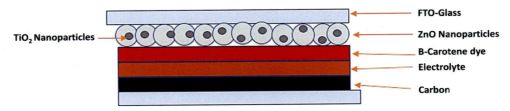

FIGURE 8.6　Illustration of a dye-sensitized solar cell design based on TiO$_2$/ZnO nanocomposite films.

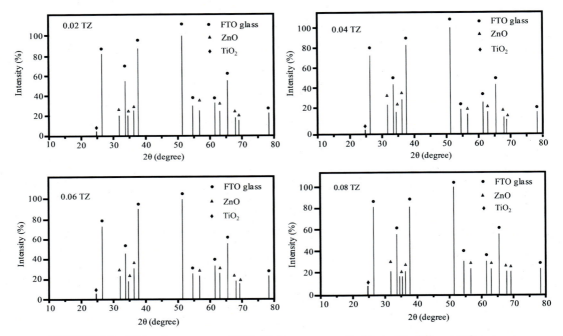

FIGURE 8.7　Diffraction patterns of TiO$_2$/ZnO nanocomposite films with different TiO$_2$ masses.

molecules to be adsorbed. If most of the dye molecules are successfully adsorbed by the TZ nanocomposite film, the number of photons absorbed will also increase and the DSSC performance will be improved [76].

Fig. 8.8 shows the surface micrographs of 0.06 and 0.08 TZ nanocomposite films. Qualitatively, TZ nanocomposite films have a round shape, are porous, and undergo agglomeration, as shown in previous studies [78]. Likewise, the 0.08 TZ film (Fig. 8.8B) also appears to have a smaller particle size compared to the 0.06 TZ film (Fig. 8.8A). Surface morphology of the film was quantitatively analyzed using ImageJ software. The particle diameters for the 0.06 and 0.08 TZ nanocomposite films were 91.35 and 75.78 nm, respectively. The small particle diameter causes an increase in surface porosity so that the dye molecules adsorbed by the TZ nanocomposite film become more numerous.

II. Photovoltaic materials and devices

8.3 Nanooxidic photoanode materials for DSSCs (TiO_2, ZnO, and SnO_2)

TABLE 8.4 Rietveld refinement of TiO_2.

Parameter	Model COD 2310710	Refinement results			
		0.02 TZ	0.04 TZ	0.06 TZ	0.08 TZ
Crystal system	Tetragonal				
Space group	*I41/a m d*				
$a = b$ (Å)	3.872	3.7744	3.752	3.812	3.964
c (Å)	9.616	9.4307	9.607	9.613	9.261
Atomic position of Ti	0.500; 0.750; 0.375	0.500; 0.750; 0.375	0.500; 0.750; 0.375	0.500; 0.750; 0.375	0.500; 0.750; 0.375
Atomic position of O	0.500; 0.750; 0.162	0.500; 0.750; 0.162	0.500; 0.750; 0.116	0.500; 0.750; 0.234	0.500; 0.750; 0.157
Crystallite size (nm)		6.6	4.9	7.8	14.8

TZ, TiO_2/ZnO.

TABLE 8.5 Rietveld refinement of ZnO.

Parameter	Model COD 9004178	Refinement result			
		0.02 TZ	0.04 TZ	0.06 TZ	0.08 TZ
Crystal system	Hexagonal				
Space group	*P63mc*				
$a = b$ (Å)	3.249	3.2490	3.257	3.247	3.246
c (Å)	5.203	5.1978	5.217	5.199	5.195
Atomic position of Zn	0.333; 0.667; 0.000	0.333; 0.667; 0.000	0.333; 0.667; 0.000	0.333; 0.667; 0.000	0.333; 0.667; 0.000
Atomic position of O	0.333; 0.667; 0.380	0.333; 0.667; 0.382	0.333; 0.667; 0.382	0.333; 0.667; 0.393	0.333; 0.667; 0.394
Crystallite size (nm)		36.8	27.7	33.3	26.1

TZ, TiO_2/ZnO.

Fig. 8.9 depicts the absorbance spectra of 0.06 and 0.08 TZ nanocomposite films. From the figure, it can be observed that addition of TiO_2 mass increased absorbance especially at wavelengths greater than 350 nm. The films showed a wide absorbance from 235 to 260 nm with peaks at 292 and 324 nm for the 0.06 and 0.08 TZ films, respectively. The absorbance spectrum also shows a peak shift toward a higher wavelength (visible light) with more TiO_2 mass. Shifting the absorbance peak toward the wavelength of visible light may be due to the finer surface morphology of the TZ nanocomposite film, fewer grain boundaries, and better homogeneity of nanoparticles, and has also been reported by other similar studies [77,78]. Increased absorbance is an advantage for DSSC performance.

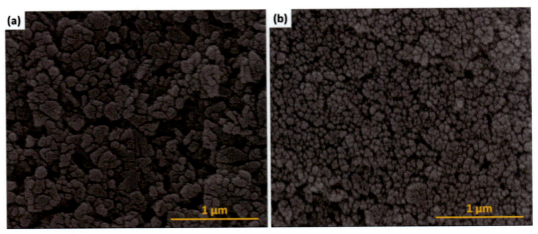

FIGURE 8.8 Surface micrographs of the nanocomposite films: (A) 0.06 TZ and (B) 0.08 TZ.

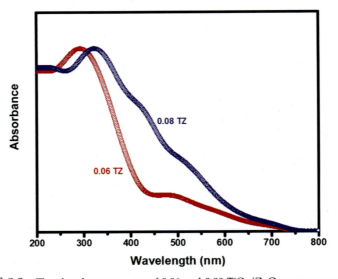

FIGURE 8.9 The absorbance spectra of 0.06 and 0.08 TiO$_2$/ZnO nanocomposite films.

Fig. 8.10 is an energy bandgap analysis using the Tauc plot equation given by Eq. (8.6).

$$\alpha(v)hv = B(hv - E_g)^m \tag{8.6}$$

The absorption coefficient is represented by $\alpha(v)$. $\alpha(v)$ is defined by Beer–Lambert's law. The photon energy is hv. m is the type of transition ($m = 1/2, 3/2, 2,$ and 3) for the transition of direct allowed, direct forbidden, and indirect forbidden types of semiconductor. E_g represents the optical gap. B is a constant [79]. Thus TiO$_2$ nanoparticle addition decreased the

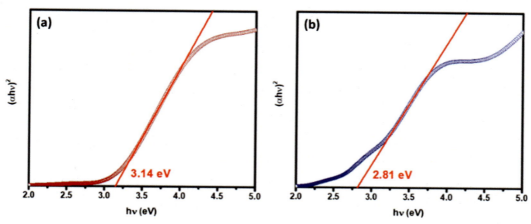

FIGURE 8.10 Bandgap energy of 0.06 and 0.08 TiO$_2$/ZnO nanocomposite films.

bandgap of the TZ nanocomposite film from 3.14 eV (0.06 TZ) to 2.81 eV (0.08 TZ), which may be due to the higher crystallinity of the 0.08 TZ film compared to the 0.06 TZ film. The smaller energy gap for the higher TiO$_2$-containing material has reduced the electron excitation energy from the valence to the conduction band [80].

Besides ZnO, SnO$_2$ has also been widely used as a photoanode material on DSSCs. SnO$_2$ is n-type with a wide energy gap of 3.6 eV at 300 K and can produce fewer holes to minimize degradation of the dye molecules and facilitate long-term DSSC stability [81–83]. The electron mobility of SnO$_2$, which is between 100 and 200 cm^2/Vs, is also greater than that of TiO$_2$, which is only between 10^{-1} and 10^{-3} cm^2/Vs. In addition, SnO$_2$ has faster electron diffusion ability than TiO$_2$ so that the charge recombination resistance becomes higher [84,85]. SnO$_2$ has a rutile crystalline structure. More detailed information about the crystal structure of SnO$_2$ is listed in Table 8.6.

TiO$_2$/SnO$_2$ (TS) nanocomposites have also been successfully screen printed on FTO glass in the order shown in Fig. 8.11. The independent variable used is the TS wt% ratio (91:9%; 82:18%; 73:27%, and 64:36%). Fig. 8.12 shows the diffraction patterns of TS nanocomposite

TABLE 8.6 Crystal structure of SnO$_2$.

Parameter	Rutile
Crystal system	Tetragonal
Space group	P4$_2$/m n m
a (Å)	4.738
b (Å)	4.738
c (Å)	3.186
Atomic position of Sn	0.000; 0.000; 0.000
Atomic position of O	0.306; 0.306; 0.000

8. Potential of nanooxidic materials and structures of photoanodes for DSSCs

FIGURE 8.11 Design illustration of dye-sensitized solar cell-based TiO$_2$/SnO$_2$ nanocomposite films.

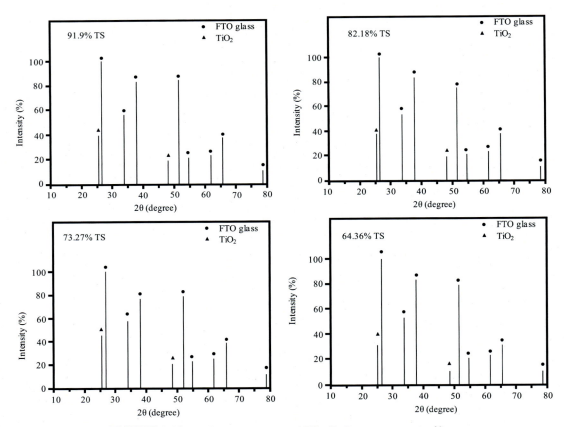

FIGURE 8.12 Diffraction patterns of TiO$_2$/SnO$_2$ nanocomposite films.

II. Photovoltaic materials and devices

8.3 Nanooxidic photoanode materials for DSSCs (TiO_2, ZnO, and SnO_2)

TABLE 8.7 Rietveld refinement of TiO_2.

Parameter	Model COD 2310710	Refinement			
		91:9% TS	82:18% TS	73:27% TS	64:36% TS
Crystal system	Tetragonal				
Space group	I41/a m d				
$a = b$ (Å)	3.80	3.859	3.810	3.868	3.866
c (Å)	9.47	9.664	10.23	9.575	9.612
Volume (Å^3)		143.6	143.3	148.5	143.9

TS, TiO_2/SnO_2.

TABLE 8.8 Rietveld refinement of SnO_2.

Parameter	Model COD 9007533	Refinement			
		91:9% TS	82:18% TS	73:27% TS	64:36% TS
Crystal system	Tetragonal				
Space group	P42/m n m				
$a = b$ (Å)	4.73	4.767	4.756	4.762	4.754
c (Å)	3.18	3.196	3.195	3.203	3.198
Volume (Å^3)		72.3	72.6	72.3	72.6

TS, TiO_2/SnO_2.

films showing peaks due to SnO_2 nanoparticles and FTO glass. Tables 8.7 and 8.8 show the refinement analysis of TiO_2 and SnO_2 parameters using Rietica software, respectively.

Fig. 8.13 shows the surface micrographs of 91:9% and 82:18% TS nanocomposite films showing that the TS nanocomposite layer has a round shape, is porous, and experiences agglomeration, as also shown in previous studies [86]. On the same magnification scale (100k×), the particle size of the 91:9% film TS is smaller when compared to the 82:18% TS film. Therefore in 91:9% TS film, more agglomeration was formed. Qualitatively, the porosity of the 91:9% TS film is also greater than that of the 82:18% TS film so that the 91:9% TS film can absorb the dye molecules optimally.

Fig. 8.14 represents the absorbance spectra of 91:9% and 82:18% TS nanocomposite films. The spectra show that both samples have two dominant absorption peaks, namely in between 230 and 600 nm. The absorbance peaks for the 91:9% and 82:18% TS nanocomposite films were between 282 and 312 nm, respectively, in the ultraviolet region. The absorbance peak in the range of 420 −600 nm is characteristic of β-carotene pigment. The use of this pigment is intended so that the electron excitation energy derived from TiO_2 and SnO_2 nanoparticles becomes smaller, which will enable the pigment to collect more visible light. The absorbance spectrum of the 82:18% TS film also showed a slight red shift [87].

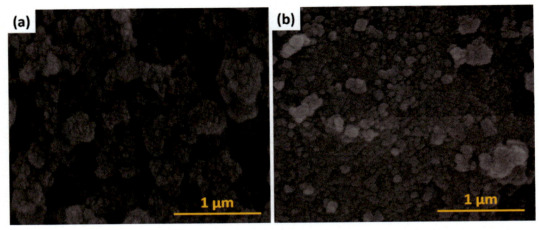

FIGURE 8.13 Surface micrographs of the nanocomposite film: (A) 91:9% and (B) 82:18% TiO$_2$/SnO$_2$.

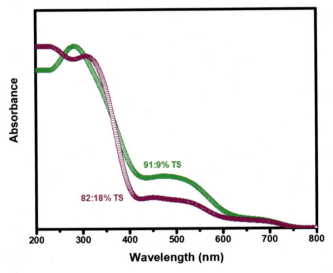

FIGURE 8.14 Absorbance spectra of 91:9% and 82:18% TiO$_2$/SnO$_2$ nanocomposite films.

Fig. 8.15 is an energy bandgap analysis of the 91:9% and 82:18% TS nanocomposite films using the Tauc plot equation as written in Eq. (8.6). Based on the analysis, the energy gaps of the 91:9% and 82:18% TS nanocomposite films are 3.14 and 3.18 eV, respectively. This energy gap increase may be attributed to the crystallinity of the 82:18% TS film, which is lower than the 91:9% TS film, so that the electron energy required for excitation is larger [80].

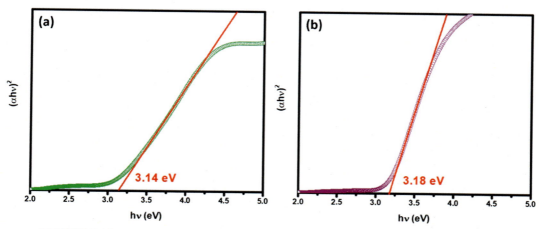

FIGURE 8.15 Bandgap energy of nanocomposite films: (A) 91:9% and (B) 82:18% TiO_2/SnO_2.

8.4 1D construction for photoanode material (fibers and rods)

The porous morphology of photoanode material with 0D structure can increase the efficiency of energy conversion generated by DSSCs. In addition, the 0D-structured material exhibits a high surface area for better adsorption of dye molecules and light. However, the surface of 0D material tends to form a defect that can behave as the center of the electrons trap, thus it will certainly disrupt the electron transport process because the trapping and detrapping phenomena cannot be avoided. Simply stated, the use of 0D-structured material as a photoanode in DSSC showed high electron flow resistivity and high probability of electron recombination [88–90]. Considerable efforts have been prepared to design a better electron transport pathway with an increase in electron collection; one of these is developing the morphology of different photoanode materials. 1D-structured materials such as fibers and rods have been shown to facilitate superior electron transport processes by providing direct electron transport pathways [73,91]. In addition, 1D-structured material shows electron transport hundreds of times faster, longer electron lifetime, and superior light scattering compared to 0D-structured material [92–94]. The most widely used methods for modifying 1D-structured material are electrospinning and hydrothermal.

The electrospinning method has been used successfully to modify the structure of fibers or fiber mats. The technique uses electrostatic forces to yield fine fibers from a polymer solution [95,96]. Generally, conductive polymer-based materials such as polyacrylonitrile (PAN) are used as precursors in the synthesis of fibers or fiber mats. PAN is a conductive polymer that possesses high thermal stability, low density, good electrical conductivity, elasticity strength, and high elasticity modulus [97,98]. The structure of PAN-based fibers that are composited with TiO_2 was successfully deposited on FTO glass via electrospinning, see Fig. 8.16. The independent variable used was the addition of TiO_2 nanoparticle mass (10% and 25%) in TiO_2/PAN (abbreviated as TP). Fig. 8.17 presents the diffraction patterns of TP-based fibers that show the presence of two dominant peaks in the angular 2θ range of 10–30 degrees. The peaks at 16.87 and 25.22 degrees are related to the characteristics of

230   8. Potential of nanooxidic materials and structures of photoanodes for DSSCs

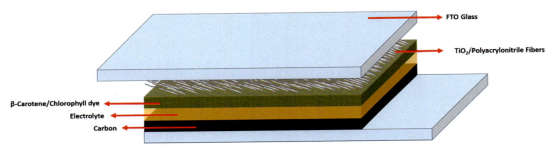

FIGURE 8.16 Illustration of a dye-sensitized solar cell sandwich based on TiO$_2$/PAN fiber film.

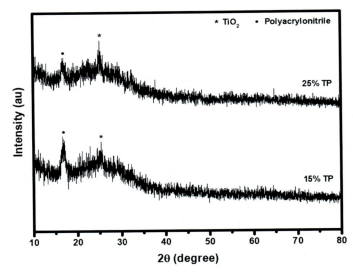

FIGURE 8.17 Diffraction patterns of TiO$_2$/PAN composite fibers.

PAN and TiO$_2$ metal oxide, respectively. Based on phase analysis using Rietica software, both samples exhibit a phase match with the anatase TiO$_2$ diffraction pattern. However, only TiO$_2$ peaks in the hkl plane (011) appear, while diffraction peaks in other hkl planes show amorphous patterns. The peak intensity of PAN diffraction at 15% TP is higher than that at 25% TP. In contrast, the TiO$_2$ diffraction peak intensity at 15% TP is lower compared to 25% TP. This case relates to the addition of TiO$_2$ mass, where more TiO$_2$ mass was introduced to 25% TP. The fibers exhibit large surface area-to-volume ratios indicating the ability to adsorb a larger number of dye molecules [95]. Some information on the crystal structure of 25% TP is listed in Table 8.9.

Fig. 8.18 shows a surface micrograph of 25% TP film, which shows clearly that electrospinning TP composites have succeeded in forming nonwoven fiber structures.

Agglomerations at some regions on the fibers have been identified as TiO$_2$ nanoparticles. It is known that nanoparticles show a stronger inclination to agglomerate compared to submicron particles. This is attributed to the higher surface area-to-volume ratio of the

TABLE 8.9 Parameters of TiO_2.

Parameters	Model COD 2310710	Refinement result values
Crystal system	Tetragonal	
Space group	*I41/a m d*	
$a = b$ (Å)	3.872	3.930
c (Å)	9.616	9.834
Atomic position of Ti	0.500; 0.750; 0.375	0.500; 0.750; 0.375
Atomic position of O	0.500; 0.750; 0.162	0.500; 0.750; 0.132
Crystal size (nm)		5.280

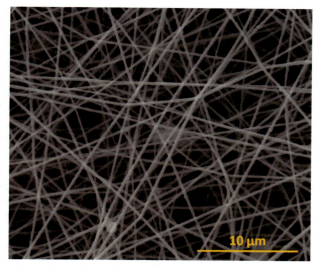

FIGURE 8.18 Surface micrograph of 25% TiO_2/PAN film.

nanoparticles. The agglomerates are the result of van der Waals attraction between nanoparticles and the spontaneous TiO_2 nanoparticle agglomeration on dissolution in aqueous media [99–101]. Using ImageJ software, the average fiber diameter of the 25% TP film is 229 nm. However, the average fiber diameter of the 25% TP film is smaller compared to similar studies conducted by Chen and coworkers who reported an average fiber diameter of around 273 nm [102]. Introducing TiO_2 nanoparticles to the composite reduces the viscosity of the solution and increases its electrical conductivity so that the average diameter of fibers decreases [95,103]. The structure of fibers shows high porosity characteristics with very small pore sizes [95,104]. The smaller average fiber diameter for the 25% TP film induced a larger surface area per unit mass and film porosity enabling more dye molecules to be adsorbed. Therefore it is anticipated that the 25% TP film will exhibit higher energy conversion efficiency.

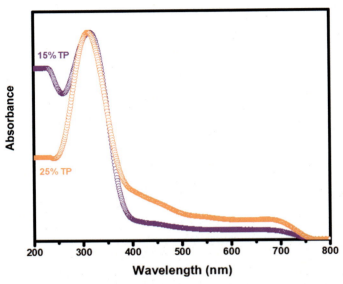

FIGURE 8.19 The absorption spectra of 15% and 25% TiO$_2$/PAN films.

Fig. 8.19 represents the absorbance spectra of 15% and 25% TP films. Both samples have a dominant absorbance between 240 and 400 nm. The 15% and 25% TP films show peaks at 313 and 310 nm, respectively, in the near-ultraviolet region. Both films have almost the same absorbance peak, except for a slight shift. The addition of TiO$_2$ nanoparticles into the composite resulted in a blue shift. This shift is due to the size of grains of 25% TP film [105]. The shift in the absorbance peak is also related to the absorbance limit of each material. In the nanoparticle spectrum, there is a hypochromic shift toward the blue. This may be due to the presence of nitrogen atoms in the PAN polymer. In the polymer spectrum, there is a bathochromic shift toward the visible that may be due to the presence of TiO$_2$ nanoparticles [106]. For these samples, the shift to lower wavelengths of the absorbance peak resulted in the decrease in energy gap of the TP composite-based photoanode film [107].

Fig. 8.20 depicts the energy gap for 15% and 25% TP films. The figure shows that the energy gap can be reduced from 3.41 to 3.18 eV for the 15% and 25% TP films, respectively, on addition of TiO$_2$ nanoparticles to the polymer matrix. In a similar study by Tanski and coworkers, a much higher energy gap of 3.83 eV was obtained for TP composite fibers [106].

However, 1D structures such as fibers possess a much lower surface area than 0D structures, so that the efficiency generated by DSSC devices will be lower [89,94]. Other than the morphology of TiO$_2$, ZnO morphology was also widely studied by researchers. ZnO's different morphology critically influenced several characteristics such as microstructure, energy gap, and surface defects, which are important requirements to maximize the operation of the photoanode material in DSSCs. Therefore modification of hybrid systems by mixing 0D and 1D structural materials has been widely proposed such as ZnO nanoparticles/ZnO nanorods.

ZnO nanoparticles showing morphological shapes of globular aggregates with high surface area and porous structures have been widely applied as photoanode materials on DSSCs.

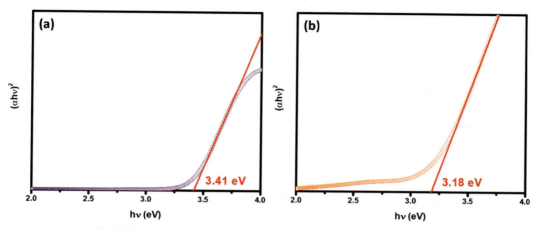

FIGURE 8.20 Energy gap of (A) 15% and (B) 25% TiO$_2$/PAN films.

The ZnO nanorods have been shown to have high electron collection characteristics due to the availability of direct electron conduction pathways and increased electron transport [73]. ZnO hybrid film nanoparticles/nanorods have successfully demonstrated good performance as a photoanode on DSSCs. Initially, ZnO nanoparticles were deposited on the FTO glass conductive layer using a screen-printing method, then scaled with temperature variations of 150 and 300°C. Furthermore, the growth of ZnO nanorods is based on the hydrothermal method, where temperature, pressure, and solution concentration affect the type and morphology of the resulting nanostructures. The process of growing ZnO nanorods was done by immersing the ZnO nanoparticle film into a mixture of precursors of zinc nitrate hexahydrate for 4 h at 100°C. The arrangement of the ZnO nanoparticle/nanorod-based DSSC device is shown in Fig. 8.21.

Fig. 8.22 displays the diffraction patterns of ZnO nanorods at 150 and 300°C annealing temperatures.

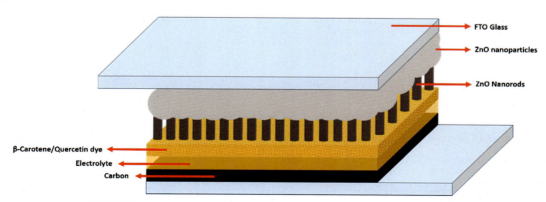

FIGURE 8.21 Illustration of a ZnO nanorod-based dye-sensitized solar cell.

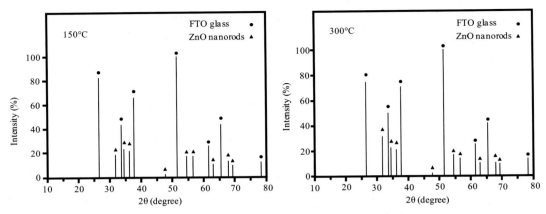

FIGURE 8.22 Diffraction patterns of ZnO nanorod films.

The intensity of ZnO nanorods is relatively lower than the intensity of conductive FTO glass substrate at the two temperatures. The diffraction peaks of ZnO nanorods were observed at the hkl planes of (010), (002), (011), (012), (110), (111), (013), (020), (112), (021), (004), and (022), which means that the ZnO nanorod film has not been perfectly oriented. ZnO nanorods were identified in the wurtzite phase. The highest peak diffraction intensity was observed at the plane hkl of (011) with an angle of 2θ around 36.3 degrees. ZnO nanorod film annealed at 300°C shows that the hkl plane of (002) shifted to a larger 2θ angle. The shift may be due to internal pressure, film defects, and low densification in the ZnO nanorods [108,109]. In the nanoscale, an increase in pressure results in a force directed toward the interior of the crystal so that the lattice distance from ZnO nanorods decreases. Table 8.10 shows in detail the parameters after refinement analysis using Rietica software for ZnO nanorod films annealed at 150 and 300°C.

TABLE 8.10 Several parameters of refinement analysis.

Parameter	Model COD 9004178	Refinement result 150°C	Refinement result 300°C
Crystal system	Hexagonal		
Space group	P63mc		
$a = b$ (Å)	3.249	3.244	3.250
c (Å)	5.203	5.199	5.206
Atomic position of Zn	0.333; 0.667; 0.000	0.333; 0.667; 0.000	0.333; 0.667; 0.000
Atomic position of O	0.333; 0.667; 0.382	0.333; 0.667; 0.396	0.333; 0.667; 0.410
Crystal size (nm)		44.13	35.49

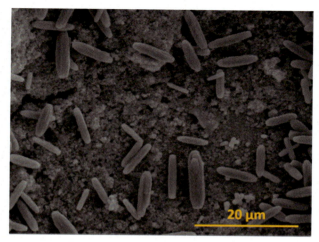

FIGURE 8.23 Micrograph of the surface of ZnO nanoparticle/nanorod film annealed at 150°C.

In general, a material's grain size increases with an increase in annealing temperature [110,111]. However, in this study, the grain size of ZnO nanoparticles/nanorods annealed at 150°C is greater than that of ZnO nanoparticles/nanorods annealed at 300°C. This is due to the annealing time of 1 h, which is insufficient to increase the grain size although the annealing temperature has increased.

Fig. 8.23 visualizes a surface micrograph of ZnO nanoparticle/nanorod film annealed at 150°C. Rods are shown to have formed, although they are not perfectly oriented yet. These unoriented rods may be caused by ZnO layers of nanoparticles that are too thick so that they do not have enough room to grow. In addition, orientation of the rods that are not in the same direction is also related to the hkl plane, where the highest intensity in the diffraction pattern is in the hkl plane (011), causing the rods to grow unidirectionally. If the highest intensity is in the plane hkl of (002), it will show rods that grow in the same direction and that are perfectly oriented [112].

Fig. 8.24 is the absorbance spectra of ZnO nanoparticle/nanorod film that has been annealed at 150 and 300°C, respectively. The spectra show two dominant absorbance peaks between 225 and 410 nm. In addition, an increase in annealing temperature resulted in the peaks being red shifted [113–115]. Higher annealing temperatures also led to an increase in film absorbance [116], as seen in Fig. 8.24, in the wavelength range from 200 to about 350 nm.

Fig. 8.25 depicts the $(\alpha h\nu)^2$ versus $h\nu$ plots to determine the energy gap of ZnO nanoparticle/nanorod film annealed at 150 and 300 °C. The energy gap of ZnO nanoparticle/nanorod film decreased with increased annealing temperature. The energy gap of ZnO nanoparticle/nanorod film annealed at 150°C was 3.37 eV compared to the energy gap of ZnO nanoparticle/nanorod film annealed at 300°C, which was 3.22 eV [116,117]. The conductivity of ZnO nanoparticle/nanorod film annealed at 300°C was higher than that annealed at 150°C [118]. Increasing the annealing temperature of the ZnO nanoparticle/nanorod film leads to an increase in oxygen vacuum providing more free electrons resulting in conductivity enhancement.

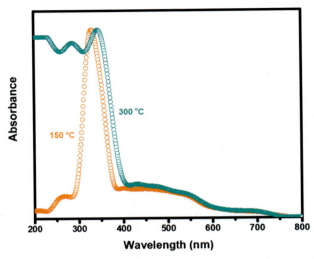

FIGURE 8.24 Absorbance spectra of ZnO nanoparticle/nanorod film.

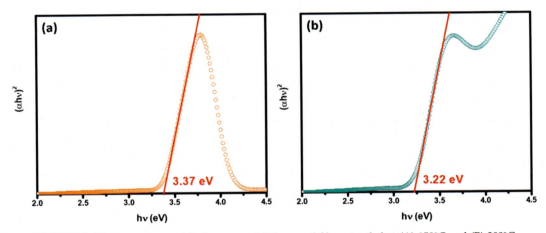

FIGURE 8.25 Energy gap of ZnO nanoparticle/nanorod film annealed at (A) 150°C and (B) 300°C.

8.5 Performance of DSSCs

From $J-V$ characteristics, DSSC efficiency can be calculated. $I-V$ characteristics were determined using the Keithley 2602A electrometer at 10.54 mW/cm^2 LED lighting intensity. The $J-V$ characteristics exhibited by a DSSC are similar to that of diode characteristics showing the presence of electron donors and acceptors (p-n junction). In obtaining the $J-V$ curve, several important parameters affect DSSC conversion efficiency. These include the open-circuit voltage (V_{oc}) and short-circuit current (I_{sc}) [119]. V_{oc} is the potential that is obtained when the DSSC device does not generate current. V_{oc} is also related to the energy difference between the Fermi level of the oxide semiconductor and the Nernst potential of the redox couple in the electrolyte. Mathematically, the relationship between V_{oc} and efficiency, η, of a DSSC is given by Eq. (8.7).

$$\eta = \frac{I_{sc} \times V_{oc} \times FF}{P_{input}} \tag{8.7}$$

Here FF is the fill factor, V_{oc} is the open-circuit voltage, I_{sc} is the short-circuit current, and P_{input} is the electric power flowing into the DSSC.

From Eq. (8.7), V_{oc} is directly proportional to η. Thus, theoretically, η increases with an increase in V_{oc}. I_{sc} is related to the efficiency of electron transfer, scattering of light, and light absorption by the dye molecules. I_{sc} also shows a relationship that is directly proportional to η. Besides V_{oc} and I_{sc}, FF is also another important parameter for efficient operation of the DSSC.

Fig. 8.26 shows the DSSC J–V characteristics with a TiO_2/ZnO nanocomposite photoanode. The TiO_2 composition is varied. Table 8.11 shows the parameters obtained from the J–V characteristics. V_{oc} showed an increase with increasing mass of TiO_2 nanoparticles in the TZ nanocomposites up to a TZ ratio of 0.06:0.12. The other samples showed the fraction of TiO_2 associated with the samples code, while the ZnO fraction was kept constant at 0.12. However, for the DSSC with 0.08 TZ film, all three parameters decreased. The energy conversion efficiency of the DSSC also showed a similar behavior. For the highest mass of TiO_2 nanoparticles (0.08 TZ) added, DSSC efficiency and V_{oc} decreased. The reduction in efficiency and V_{oc} at 0.08 TZ indicated that the efficiencies and V_{oc} passed the optimum point, where the optimum efficiency of the DSSC was at 0.06 TZ. Similar research was carried out by Ahmad and his colleagues, who synthesized the nanocomposite materials using the wet

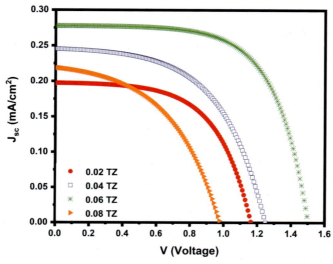

FIGURE 8.26 J–V curve of a TiO_2/ZnO nanocomposite-based dye-sensitized solar cell.

TABLE 8.11 Several parameters of TiO$_2$/ZnO nanocomposite-based dye-sensitized solar cells.

Composition of photoanode	J_{sc} (mA/cm^2)	V_{oc} (mV)	FF	η (%)
0.02 TZ	0.242	1020	0.679	1.591
0.04 TZ	0.236	1222	0.769	2.111
0.06 TZ	0.303	1470	0.723	3.061
0.08 TZ	0.239	1088	0.529	1.307

FF, Fill factor; TZ, TiO$_2$/ZnO.

incipient wetness impregnation technique. The efficiency produced by a DSSC with a nanocomposite TZ photoanode was 8.10% using illumination from an IV-5 solar simulator [120]. The increase in J_{sc} of TZ nanocomposites was due to the faster electron transport and shorter electron transfer distances. V_{oc} and FF also increased with ZnO impregnation. This was due to a higher electron recombination resistance. DSSC efficiency with a TZ photoanode was higher than the efficiency of a DSSC with a photoanode comprising only TiO$_2$ or pure ZnO.

Fig. 8.27 depicts the DSSC J–V characteristics with a TS photoanode with different TS wt% concentrations. Table 8.12 shows the parameters from the J–V characteristics. A DSSC with a TS nanocomposite photoanode shows increased J_{sc} and efficiency with increased SnO$_2$ percentage. So does V_{oc}. The highest addition of TiO$_2$ shows a lower V_{oc}.

Conversely, the highest addition of SnO$_2$ mass produces high V_{oc}. Xie and colleagues also succeeded in researching the synthesis of a TS scattering layer material using the hydrothermal technique. The efficiency generated by the TS-based DSSC as photoanode material was 6.99%

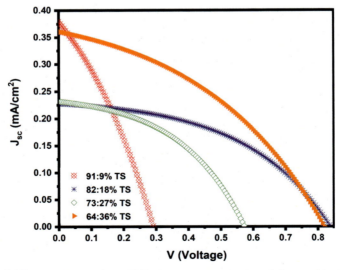

FIGURE 8.27 J–V curves of a TiO$_2$/SnO$_2$ nanocomposite-based dye-sensitized solar cell.

TABLE 8.12 Several parameters of TiO_2/SnO_2 nanocomposite-based dye-sensitized solar cells.

Composition of photoanode	J_{sc} (mA/cm^2)	V_{oc} (mV)	FF	η (%)
91:9% TS	0.187	406	0.658	0.292
82:18% TS	0.208	638	0.619	0.476
73:27% TS	0.210	867	0.599	0.548
64:36% TS	0.388	806	0.452	0.819

FF, Fill factor; *TS*, TiO_2/SnO_2.

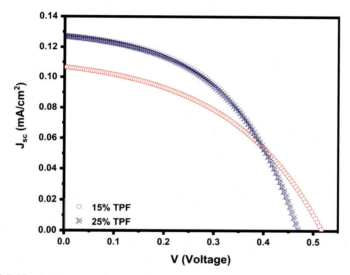

FIGURE 8.28 $J-V$ curves of a TiO_2/polyacrylonitrile fiber-based dye-sensitized solar cell.

under AM1.5 illumination [121]. The efficiency produced by a DSSC using TiO_2-based/SnO_2 photoanode material was higher than the efficiency produced by pure TS alone. Increased efficiency was caused by the ability of light scattering from the TS material.

Fig. 8.28 shows the $J-V$ characteristics of a DSSC with a TiO_2/polyacrylonitrile fiber (TPF) film photoanode. The mass of TiO_2 was varied in the photoanode. Table 8.13 shows the parameters obtained from the $J-V$ characteristics. A 1D structure of TPF film optimized the efficiency produced by the DSSC. The structure of nanofibers can reduce the agglomeration of TiO_2 nanoparticles and increase the porosity of the film so that the dye molecules can be adsorbed more optimally.

On the other hand, the reduction in the energy gap of the TPF film leads to an increase in the energy conversion efficiency produced by the DSSC. Bandgap energy is the energy needed by electrons to be excited to the conduction band [104]. Therefore reduction in energy gap will lead to an increase in photon number absorbed by the TPF photoanode and an

TABLE 8.13 Several parameters of TiO$_2$/polyacrylonitrile fiber-based dye-sensitized solar cells.

Composition of photoanode	J_{sc} (mA/cm^2)	V_{oc} (V)	FF	η (%)
15% TPF	0.425	518	1.730	0.905
25% TPF	0.508	470	1.899	1.074

FF, Fill factor; TPF, polyacrylonitrile fiber.

increase in electron number excited from the ground state. Jo and his colleagues successfully studied TiO$_2$ nanofibers synthesized using electrospinning. The efficiency of the DSSC with TPF-based photoanode was 3.8% under AM1.5 illumination. However, the DSSC using a TPF photoanode with TiO$_2$ rutile crystal structure exhibited a lower V_{oc} compared to the DSSC using the TPF photoanode. The same is true for the current density. It is widely known that more dye molecules can adsorb on to the surface of materials with higher specific surface area. Thus fibers result in greater J_{sc} and efficiency. J_{sc} increases with the presence of light-scattering layers in the photoanode that result in increased light harvesting. TPF with a TiO$_2$ rutile phase acts as a barrier and prevents electron recombination with the triiodide ions of the redox couple. At the same time, electron injection into the conduction band of the TPF is enhanced [122].

Fig. 8.29 shows the $J-V$ characteristics exhibited by two DSSCs using a nanoparticle/nanorod ZnO photoanode annealed at 150 and 300°C, respectively. Table 8.14 lists the variables obtained from the $J-V$ curves of the DSSCs using the photoanode. The increase in annealing temperature led to the J_{sc} and V_{oc} of the DSSC devices decreasing. This condition may be caused by a surface defect in the film. The increased annealing temperature on ZnO

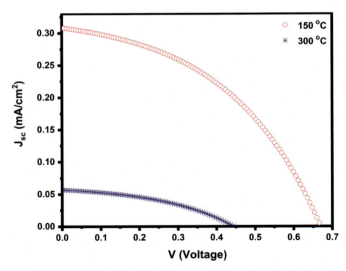

FIGURE 8.29 $J-V$ curves of ZnO nanoparticle/nanorod-based dye-sensitized solar cells.

TABLE 8.14 Several parameters of ZnO nanoparticle/nanorod-based dye-sensitized solar cells.

Composition of photoanode	J_{sc} (mA/cm^2)	V_{oc} (mV)	FF	η (%)
150°C ZnO nanoparticles/nanorods	0.364	654	0.385	0.873
300°C ZnO nanoparticles/nanorods	0.069	357	0.787	0.186

FF, Fill factor.

nanoparticle/nanorod photoanode film showed a decrease in η. This result is consistent with the work of Chou and Hsu in the range of 250−350°C annealing temperature [123]. However, at higher annealing temperature, efficiency could increase. Pandey and colleagues also successfully researched the synthesis of ZnO nanoparticles using aqueous chemical methods. The efficiency produced by the DSSC using the photoanode annealed at 300°C was 1.83% under illumination of 1 sun or AM1.5 [124].

8.6 Summary

Nanooxidic materials such as TiO_2, ZnO, and SnO_2 or their composites and 1D structures such as fibers and nanorods have been used as photoanodes in DSSCs. In general, TiO_2 nanoparticles composited with other metal oxides showed a decrease in grain size compared to TiO_2 pure nanoparticles, thus creating a higher surface area. The surface morphology of the nanocomposite film was round shaped, porous, and agglomerated. The porosity and higher surface area of the nanocomposite film increased the adsorption of dye molecules. If most of the dye molecules were successfully adsorbed, the nanocomposite film showed a wider absorbance spectrum over the range of ultraviolet light and visible wavelengths. The decrease in the energy gap of the nanocomposite film resulted in the energy conversion efficiency of DSSCs increasing. Solar cell parameters such as J_{sc}, V_{oc}, and FF also affected the energy conversion efficiency of DSSCs, where an increase in J_{sc}, V_{oc}, and FF led to an increase in DSSC efficiency.

Acknowledgments

This work was funded by the Hibah PNBP-KBK Universitas Negeri Malang awarded in 2020.

References

[1] K.K.S. Lau, M. Soroush, Overview of Dye-Sensitized Solar Cells, Elsevier Inc., 2019, https://doi.org/10.1016/B978-0-12-814541-8.00001-X.

[2] R. Paul, Chapter 16. Automated Energy Storage Using Carbon Nanostructured Materials, Elsevier Inc., 2019, https://doi.org/10.1016/B978-0-12-814083-3.00016-0.

[3] A. Carella, F. Borbone, R. Centore, Research progress on photosensitizers for DSSC, Front. Chem. 6 (2018) 1−24, https://doi.org/10.3389/fchem.2018.00481.

[4] S. Shen, J. Chen, M. Wang, X. Sheng, X. Chen, X. Feng, S.S. Mao, Titanium dioxide nanostructures for photoelectrochemical applications, Prog. Mater. Sci. 98 (2018) 299−385, https://doi.org/10.1016/j.pmatsci.2018.07.006.

[5] N. Ali, A. Hussain, R. Ahmed, M.K. Wang, C. Zhao, B.U. Haq, Y.Q. Fu, Advances in nanostructured thin film materials for solar cell applications, Renew. Sustain. Energy Rev. 59 (2016) 726–737, https://doi.org/10.1016/j.rser.2015.12.268.

[6] A.K. Pandey, M.S. Ahmad, N.A. Rahim, V.V. Tyagi, R. Saidur, Natural Sensitizers and Their Applications in Dye-Sensitized Solar Cell, 2019.

[7] N. Abas, A.R. Kalair, N. Khan, A. Haider, Z. Saleem, M.S. Saleem, Natural and synthetic refrigerants, global warming: a review, Renew. Sustain. Energy Rev. 90 (2018) 557–569, https://doi.org/10.1016/j.rser.2018.03.099.

[8] P.G.V. Sampaio, M.O.A. González, Photovoltaic solar energy: conceptual framework, Renew. Sustain. Energy Rev. 74 (2017) 590–601, https://doi.org/10.1016/j.rser.2017.02.081.

[9] M. Dasari, R.P. Balaraman, P. Kohli, Photovoltaics and Nanotechnology as Alternative Energy, 2018, https://doi.org/10.1007/978-3-319-76090-2.

[10] M. Yu, W.D. McCulloch, Z. Huang, B.B. Trang, J. Lu, K. Amine, Y. Wu, Solar-powered electrochemical energy storage: an alternative to solar fuels, J. Mater. Chem. 4 (2016) 2766–2782, https://doi.org/10.1039/c5ta06950e.

[11] S. Khan, A. Ahmad, F. Ahmad, M. Shafaati Shemami, M. Saad Alam, S. Khateeb, A comprehensive review on solar powered electric vehicle charging system, Smart Sci. 6 (2018) 54–79, https://doi.org/10.1080/23080477.2017.1419054.

[12] J. Khan, M.H. Arsalan, Solar power technologies for sustainable electricity generation - a review, Renew. Sustain. Energy Rev. 55 (2016) 414–425, https://doi.org/10.1016/j.rser.2015.10.135.

[13] E. Kabir, P. Kumar, S. Kumar, A.A. Adelodun, K.H. Kim, Solar energy: potential and future prospects, Renew. Sustain. Energy Rev. 82 (2018) 894–900, https://doi.org/10.1016/j.rser.2017.09.094.

[14] L. Kumar, M. Hasanuzzaman, N.A. Rahim, Global advancement of solar thermal energy technologies for industrial process heat and its future prospects: a review, Energy Convers. Manag. 195 (2019) 885–908, https://doi.org/10.1016/j.enconman.2019.05.081.

[15] J.K. Lee, J.S. Lee, Y.S. Ahn, G.H. Kang, H.E. Song, M.G. Kang, Y.H. Kim, C.H. Cho, Simple pretreatment processes for successful reclamation and remanufacturing of crystalline silicon solar cells, Prog. Photovoltaics Res. Appl. 26 (2018) 179–187, https://doi.org/10.1002/pip.2963.

[16] F. Schindler, A. Fell, R. Müller, J. Benick, A. Richter, F. Feldmann, P. Krenckel, S. Riepe, M.C. Schubert, S.W. Glunz, Towards the efficiency limits of multicrystalline silicon solar cells, Sol. Energy Mater. Sol. Cell. 185 (2018) 198–204, https://doi.org/10.1016/j.solmat.2018.05.006.

[17] H. Li, F. Qu, H. Luo, X. Niu, J. Chen, Y. Zhang, H. Yao, X. Jia, H. Gu, W. Wang, Engineering CIGS grains qualities to achieve high efficiency in ultrathin $Cu(In_xGa_{1-x})Se_2$ solar cells with a single-gradient band gap profile, Result Phys. 12 (2019) 704–711, https://doi.org/10.1016/j.rinp.2018.12.043.

[18] S. Hoang, P.X. Gao, Nanowire array structures for photocatalytic energy conversion and utilization: a review of design concepts, assembly and integration, and function enabling, Adv. Energy Mater. 6 (2016), https://doi.org/10.1002/aenm.201600683.

[19] T.S. Bramhankar, S.S. Pawar, J.S. Shaikh, V.C. Gunge, N.I. Beedri, P.K. Baviskar, H.M. Pathan, P.S. Patil, R.C. Kambale, R.S. Pawar, Effect of nickel–zinc co-doped TiO_2 blocking layer on performance of DSSCs, J. Alloys Compd. (2019) 152810, https://doi.org/10.1016/j.jallcom.2019.152810.

[20] C. Cavallo, F. Di Pascasio, A. Latini, M. Bonomo, D. Dini, Nanostructured semiconductor materials for dye-sensitized solar cells, J. Nanomater. 2017 (2017), https://doi.org/10.1155/2017/5323164.

[21] M.S. Mahmoud, M.S. Akhtar, I.M.A. Mohamed, R. Hamdan, Y.A. Dakka, N.A.M. Barakat, Demonstrated photons to electron activity of S-doped TiO_2 nanofibers as photoanode in the DSSC, Mater. Lett. 225 (2018) 77–81, https://doi.org/10.1016/j.matlet.2018.04.108.

[22] I. Iwantono, S.K. Md Saad, F. Anggelina, A. Awitdrus, M.A. Ramli, A.A. Umar, Enhanced charge transfer activity in Au nanoparticles decorated ZnO nanorods photoanode, Phys. E Low-dimens. Syst. Nanostruct. 111 (2019) 44–50, https://doi.org/10.1016/j.physe.2019.03.001.

[23] K. Sharma, V. Sharma, S.S. Sharma, Dye-sensitized solar cells: fundamentals and current status, Nanoscale Res. Lett. 13 (2018), https://doi.org/10.1186/s11671-018-2760-6.

[24] W. Ghann, H. Kang, T. Sheikh, S. Yadav, T. Chavez-Gil, F. Nesbitt, J. Uddin, Fabrication, optimization and characterization of natural dye sensitized solar cell, Sci. Rep. 7 (2017) 1–12, https://doi.org/10.1038/srep41470.

[25] K. Fan, J. Yu, W. Ho, Improving photoanodes to obtain highly efficient dye-sensitized solar cells: a brief review, Mater. Horizons 4 (2017) 319–344, https://doi.org/10.1039/c6mh00511j.

[26] J. Gong, K. Sumathy, Q. Qiao, Z. Zhou, Review on dye-sensitized solar cells (DSSCs): advanced techniques and research trends, Renew. Sustain. Energy Rev. 68 (2017) 234–246, https://doi.org/10.1016/j.rser.2016.09.097.

References

[27] H. Iftikhar, G.G. Sonai, S.G. Hashmi, A.F. Nogueira, P.D. Lund, Progress on electrolytes development in dye-sensitized solar cells. https://doi.org/10.3390/ma12121998, 2019.

[28] K. Zeng, Y. Lu, W. Tang, S. Zhao, Q. Liu, W. Zhu, H. Tian, Y. Xie, Efficient solar cells sensitized by a promising new type of porphyrin: dye-aggregation suppressed by double strapping, Chem. Sci. 10 (2019) 2186–2192, https://doi.org/10.1039/C8SC04969F.

[29] M. Diantoro, A. Hidayat, Z.A.I. Supardi, Electron diffusion model based on I-V data fitting as the calculation method for DSSC internal parameters electron diffusion model based on I-V data fitting as the calculation method for DSSC internal parameters, IOP Conf. Ser.: Mater. Sci. Eng. (2019), https://doi.org/10.1088/1757-899X/515/1/012016.

[30] D. Sengupta, P. Das, B. Mondal, K. Mukherjee, Effects of doping, morphology and film-thickness of photoanode materials for dye sensitized solar cell application - a review, Renew. Sustain. Energy Rev. 60 (2016) 356–376, https://doi.org/10.1016/j.rser.2016.01.104.

[31] A. Taleb, F. Mesguich, A. Hérissan, C. Colbeau-Justin, X. Yanpeng, P. Dubot, Optimized TiO_2 nanoparticle packing for DSSC photovoltaic applications, Sol. Energy Mater. Sol. Cell. 148 (2016) 52–59, https://doi.org/10.1016/j.solmat.2015.09.010.

[32] Y. Rui, H. Xiong, B. Su, H. Wang, Q. Zhang, J. Xu, P. Müller-Buschbaum, Liquid-liquid interface assisted synthesis of SnO_2 nanorods with tunable length for enhanced performance in dye-sensitized solar cells, Electrochim. Acta 227 (2017) 49–60, https://doi.org/10.1016/j.electacta.2017.01.004.

[33] M.H. Jung, High efficiency dye-sensitized solar cells based on the ZnO nanoparticle aggregation sphere, Mater. Chem. Phys. 202 (2017) 234–244, https://doi.org/10.1016/j.matchemphys.2017.09.034.

[34] R. Panetta, A. Latini, I. Pettiti, C. Cavallo, Synthesis and characterization of Nb_2O_5 mesostructures with tunable morphology and their application in dye-sensitized solar cells, Mater. Chem. Phys. 202 (2017) 289–301, https://doi.org/10.1016/j.matchemphys.2017.09.030.

[35] S. Gholamrezaei, M. Salavati-Niasari, An efficient dye sensitized solar cells based on $SrTiO_3$ nanoparticles prepared from a new amine-modified sol-gel route, J. Mol. Liq. 243 (2017) 227–235, https://doi.org/10.1016/j.molliq.2017.08.031.

[36] M. Diantoro, D. Purwaningtyas, N. Muthoharoh, A. Hidayat, A. Taufiq, A. Fuad, The influence of iron- and copper- doped of PANi thin film on their structure and dielectric properties, AIP Conf. Proc. 1454 (2011) 268–271, https://doi.org/10.1063/1.4730737.

[37] M. Diantoro, T. Suprayogi, A. Taufiq, A. Fuad, ScienceDirect the effect of PANI fraction on photo anode based on TiO_2-PANI/ITO DSSC with β -carotene as dye sensitizer on its structure , absorbance, and efficiency, Mater. Today 17 (2019) 1197–1209, https://doi.org/10.1016/j.matpr.2019.05.345.

[38] M. Diantoro, I.N. Fitriana, F. Parasmayanti, Nasikhudin, A. Taufiq, Sunaryono, N. Mufti, H. Nur, Crystallinity and electrical conductivity of PANI-Ag/Ni film: the role of ultrasonic and silver doped, IOP Conf. Ser. Mater. Sci. Eng. 202 (2017), https://doi.org/10.1088/1757-899X/202/1/012005.

[39] M. Diantoro, F. Rohmiani, A.A. Mustikasari, Sunaryono, fabrication of PANI/Ag/AgCl/ITO-PET flexible film and its crystallinity and electrical properties, IOP Conf. Ser. Mater. Sci. Eng. 367 (2018), https://doi.org/10.1088/1757-899X/367/1/012019.

[40] M. Diantoro, Kholid, A.A. Mustikasari, Yudiyanto, the influence of SnO_2 nanoparticles on electrical conductivity, and transmittance of PANI-SnO_2 films, IOP Conf. Ser. Mater. Sci. Eng. 367 (2018), https://doi.org/10.1088/1757-899X/367/1/012034.

[41] M. Diantoro, M.Z. Masrul, A. Taufiq, Effect of TiO_2 nanoparticles on conductivity and thermal stability of PANI-TiO_2/glass composite film, J. Phys. Conf. 1011 (2018), https://doi.org/10.1088/1742-6596/1011/1/012065.

[42] S.W. Himmah, U. Sa'adah, A.D. Iswatin, M. Diantoro, A. Hidayat, Z.A. Imam Supardi, The effect of spin coating rotation on the optoelectronic properties of PANI/TiO_2/FTO-Glass photoanode, IOP Conf. Mater. Sci. Eng. 515 (2019) 012084, https://doi.org/10.1088/1757-899x/515/1/012084.

[43] M.E. Yeoh, K.Y. Chan, Recent advances in photo-anode for dye-sensitized solar cells: a review, Int. J. Energy Res. 41 (2017) 2446–2467, https://doi.org/10.1002/er.3764.

[44] C.C. Raj, R. Prasanth, A critical review of recent developments in nanomaterials for photoelectrodes in dye sensitized solar cells, J. Power Sources 317 (2016) 120–132, https://doi.org/10.1016/j.jpowsour.2016.03.016.

[45] B. Cerda, R. Sivakumar, M. Paulraj, Natural dyes as sensitizers to increase the efficiency in sensitized solar cells, J. Phys. Conf. 720 (2016), https://doi.org/10.1088/1742-6596/720/1/012030.

8. Potential of nanooxidic materials and structures of photoanodes for DSSCs

[46] M. Ye, X. Wen, M. Wang, J. Iocozzia, N. Zhang, C. Lin, Z. Lin, Recent advances in dye-sensitized solar cells: from photoanodes, sensitizers and electrolytes to counter electrodes, Mater. Today 18 (2015) 155–162, https://doi.org/10.1016/j.mattod.2014.09.001.

[47] S. Bhattacharya, A. Pal, A. Jana, J. Datta, Synthesis and characterization of CdS nanoparticles decorated TiO_2 matrix for an efficient N3 based dye sensitized solar cell (DSSC), J. Mater. Sci. Mater. Electron. 27 (2016) 12438–12445, https://doi.org/10.1007/s10854-016-5298-3.

[48] C.Y. Cai, S.K. Tseng, M. Kuo, K.Y. Andrew Lin, H. Yang, R.H. Lee, Photovoltaic performance of a N719 dye based dye-sensitized solar cell with transparent macroporous anti-ultraviolet photonic crystal coatings, RSC Adv. 5 (2015) 102803–102810, https://doi.org/10.1039/c5ra21194h.

[49] E. Nosheen, S.M. Shah, H. Hussain, G. Murtaza, Photo-sensitization of ZnS nanoparticles with renowned ruthenium dyes N3, N719 and Z907 for application in solid state dye sensitized solar cells: a comparative study, J. Photochem. Photobiol. B Biol. 162 (2016) 583–591, https://doi.org/10.1016/j.jphotobiol.2016.07.033.

[50] M. Younas, M.A. Gondal, U. Mehmood, K. Harrabi, Z.H. Yamani, F.A. Al-Sulaiman, Performance enhancement of dye-sensitized solar cells via cosensitization of ruthenizer Z907 and organic sensitizer SQ2, Int. J. Energy Res. 42 (2018) 3957–3965, https://doi.org/10.1002/er.4154.

[51] M. Diantoro, D. Maftuha, T. Suprayogi, M. Reynaldi Iqbal, Solehudin, N. Mufti, A. Taufiq, A. Hidayat, R. Suryana, R. Hidayat, Performance of Pterocarpus indicus willd leaf extract as natural dye TiO_2-dye/ITO DSSC, Mater. Today. Proc. 17 (2019) 1268–1276, https://doi.org/10.1016/j.matpr.2019.06.015.

[52] S. Roa, S. Radhakrishnan, P. Manidurai, Synthesis of solar cells sensitized using natural photosynthetic pigments & study for the cell performance under different synthesis parameters, J. Phys. Conf. 720 (2016), https://doi.org/10.1088/1742-6596/720/1/012035.

[53] D. Sinha, D. De, A. Ayaz, Performance and stability analysis of curcumin dye as a photo sensitizer used in nanostructured ZnO based DSSC, Spectrochim. Acta Mol. Biomol. Spectrosc. 193 (2018) 467–474, https://doi.org/10.1016/j.saa.2017.12.058.

[54] Y. Cao, Y. Saygili, A. Ummadisingu, J. Teuscher, J. Luo, N. Pellet, F. Giordano, S.M. Zakeeruddin, J.E. Moser, M. Freitag, A. Hagfeldt, M. Grätzel, 11% efficiency solid-state dye-sensitized solar cells with copper(II/I) hole transport materials, Nat. Commun. 8 (2017) 1–8, https://doi.org/10.1038/ncomms15390.

[55] M. Suzuka, N. Hayashi, T. Sekiguchi, K. Sumioka, M. Takata, N. Hayo, H. Ikeda, K. Oyaizu, H. Nishide, A quasi-solid state DSSC with 10.1% efficiency through molecular design of the charge-separation and -transport, Sci. Rep. 6 (2016) 1–7, https://doi.org/10.1038/srep28022.

[56] P. Gu, D. Yang, X. Zhu, H. Sun, P. Wangyang, J. Li, H. Tian, Influence of electrolyte proportion on the performance of dye-sensitized solar cells, AIP Adv. 7 (2017), https://doi.org/10.1063/1.5000564.

[57] J. Wu, Z. Lan, J. Lin, M. Huang, Y. Huang, L. Fan, G. Luo, Y. Lin, Y. Xie, Y. Wei, Counter electrodes in dye-sensitized solar cells, Chem. Soc. Rev. 46 (2017) 5975–6023, https://doi.org/10.1039/c6cs00752j.

[58] N.A. Ludin, A.M. Al-Alwani Mahmoud, A. Bakar Mohamad, A.A.H. Kadhum, K. Sopian, N.S. Abdul Karim, Review on the development of natural dye photosensitizer for dye-sensitized solar cells, Renew. Sustain. Energy Rev. 31 (2014) 386–396, https://doi.org/10.1016/j.rser.2013.12.001.

[59] F. Bella, C. Gerbaldi, C. Barolo, M. Grätzel, Aqueous dye-sensitized solar cells, Chem. Soc. Rev. 44 (2015) 3431–3473, https://doi.org/10.1039/c4cs00456f.

[60] A. Kumar, Different methods used for the synthesis of TiO_2 based nanomaterials: a review, Am. J. Nano Res. Appl. 6 (2018) 1, https://doi.org/10.11648/j.nano.20180601.11.

[61] A. Eshaghi, A.A. Aghaei, Effect of TiO_2-graphene nanocomposite photoanode on dye-sensitized solar cell performance, Bull. Mater. Sci. 38 (2015) 1177–1182, https://doi.org/10.1007/s12034-015-0998-5.

[62] C.K. Lim, Y. Wang, L. Zhang, Facile formation of a hierarchical TiO_2-SnO_2 nanocomposite architecture for efficient dye-sensitized solar cells, RSC Adv. 6 (2016) 25114–25122, https://doi.org/10.1039/c5ra25772g.

[63] P.C. Ricci, C.M. Carbonaro, R. Corpino, D. Chiriu, L. Stagi, Surface effects and phase stability in metal oxides nanoparticles under visible irradiation, AIP Conf. Proc. 1624 (2014) 104–110, https://doi.org/10.1063/1.4900464.

[64] J.R. De Lile, S.G. Kang, Y.A. Son, S.G. Lee, Investigating polaron formation in anatase and brookite TiO 2 by density functional theory with hybrid-functional and DFT + U methods, ACS Omega 4 (2019) 8056–8064, https://doi.org/10.1021/acsomega.9b00443.

[65] T. Suprayogi, M.Z. Masrul, M. Diantoro, A. Taufiq, The effect of annealing temperature of ZnO compact layer and TiO 2 mesoporous on photo-supercapacitor performance, IOP Conf. Ser. Mater. Sci. Eng. 515 (2019), https://doi.org/10.1088/1757-899X/515/1/012006.

II. Photovoltaic materials and devices

References

[66] J. Nurnaeimah, S.M. Ili, N.N. Mohd, M. Norsuria, The effect of temperature on anatase TiO_2 photoanode for dye sensitized solar cell, Solid State Phenom. 273 (2018) 146−153, https://doi.org/10.4028/www.scientific.net/SSP.273.146.

[67] S. Kathirvel, H.S. Chen, C. Su, H.H. Wang, C.Y. Li, W.R. Li, Preparation of smooth surface TiO_2 photoanode for high energy conversion efficiency in dye-sensitized solar cells, J. Nanomater. 2013 (2013), https://doi.org/10.1155/2013/367510.

[68] M.J. Jeng, Y.L. Wung, L.B. Chang, L. Chow, Particle size effects of TiO_2 layers on the solar efficiency of dye-sensitized solar cells, Int. J. Photoenergy 2013 (2013), https://doi.org/10.1155/2013/563897.

[69] F.H. Ali, D.B. Alwan, Effect of particle size of TiO_2 and additive materials to improve dye sensitized solar cells efficiency, J. Phys. Conf. 1003 (2018), https://doi.org/10.1088/1742-6596/1003/1/012077.

[70] S. Ahmadi, N. Asim, M.A. Alghoul, F.Y. Hammadi, K. Saeedfar, N.A. Ludin, S.H. Zaidi, K. Sopian, The role of physical techniques on the preparation of photoanodes for dye sensitized solar cells, Int. J. Photoenergy 2014 (2014), https://doi.org/10.1155/2014/198734.

[71] S.N. Sadikin, M.Y.A. Rahman, A.A. Umar, T.H.T. Aziz, Improvement of dye-sensitized solar cell performance by utilizing graphene-coated TiO_2 films photoanode, Superlattice. Microst. 128 (2019) 92−98, https://doi.org/10.1016/j.spmi.2019.01.014.

[72] I.S. Mohamad, S.S. Ismail, M.N. Norizan, S.A.Z. Murad, M.M.A. Abdullah, ZnO photoanode effect on the efficiency performance of organic based dye sensitized solar cell, IOP Conf. Ser. Mater. Sci. Eng. 209 (2017), https://doi.org/10.1088/1757-899X/209/1/012028.

[73] R.A. Wahyuono, C. Schmidt, A. Dellith, J. Dellith, M. Schulz, M. Seyring, M. Rettenmayr, J. Plentz, B. Dietzek, ZnO nanoflowers-based photoanodes: aqueous chemical synthesis, microstructure and optical properties, Open Chem. 14 (2016) 158−169, https://doi.org/10.1515/chem-2016-0016.

[74] R. Vittal, K.C. Ho, Zinc oxide based dye-sensitized solar cells: a review, Renew. Sustain. Energy Rev. 70 (2017) 920−935, https://doi.org/10.1016/j.rser.2016.11.273.

[75] L.T. Yan, F.L. Wu, L. Peng, L.J. Zhang, P.J. Li, S.Y. Dou, T.X. Li, Photoanode of dye-sensitized solar cells based on a ZnO/TiO_2 composite film, Int. J. Photoenergy 2012 (2012) 1−5, https://doi.org/10.1155/2012/613969.

[76] Q. Zhang, D. Myers, J. Lan, S.A. Jenekhe, G. Cao, Applications of light scattering in dye-sensitized solar cells, Phys. Chem. Chem. Phys. 14 (2012) 14982−14998, https://doi.org/10.1039/c2cp43089d.

[77] L. Zhao, M. Xia, Y. Liu, B. Zheng, Q. Jiang, J. Lian, Structure and photocatalysis of TiO_2/ZnO double-layer film prepared by pulsed laser deposition, Mater. Trans. 53 (2012) 463−468, https://doi.org/10.2320/matertrans.M2011345.

[78] N.A.M. Asib, A.N. Afaah, A. Aadila, M. Rusop, Z. Khusaimi, Studies of surface morphology and optical properties of ZnO nanostructures grown on different molarities of TiO_2 seed layer, AIP Conf. Proc. 1733 (2016), https://doi.org/10.1063/1.4948871.

[79] Y. Qiu, S. Lu, S. Wang, X. Zhang, S. He, T. He, High-performance polyaniline counter electrode electropolymerized in presence of sodium dodecyl sulfate for dye-sensitized solar cells, J. Power Sources 253 (2014) 300−304, https://doi.org/10.1016/j.jpowsour.2013.12.061.

[80] P.K. Jain, M. Salim, D. Kaur, Structural and optical properties of pulsed laser deposited ZnO/TiO_2 and TiO_2/ZnO thin films, Optik 126 (2015) 3260−3262, https://doi.org/10.1016/j.ijleo.2015.07.127.

[81] U. Sa'Adah, A. Hidayat, S. Solehudin, N. Mufti, M. Diantoro, Band gap shift and electrical conductivity of (Ag-$xSnO_2$)NPs-β-carotene thin film, J. Phys. Conf. 1093 (2018), https://doi.org/10.1088/1742-6596/1093/1/012032.

[82] M. Diantoro, U. Sa, A. Hidayat, Effect of SnO_2 Nanoparticles on Band Gap Energy of x (SnO 2) -y (Ag) -β-Carotene/FTO Thin Film, 2018, pp. 0−10.

[83] M. Diantoro, L.A. Sari, T. Istirohah, A.D. Kusumawati, Nasikhudin, Sunaryono, Control of dielectric constant and anti-bacterial activity of PVA-PEG/x-SnO_2 nanofiber, IOP Conf. Ser. Mater. Sci. Eng. (2018), https://doi.org/10.1088/1757-899X/367/1/012012.

[84] Z. Li, Y. Zhou, R. Sun, Y. Xiong, H. Xie, Z. Zou, Nanostructured SnO_2 photoanode-based dye-sensitized solar cells, Chin. Sci. Bull. 59 (2014) 2122−2134, https://doi.org/10.1007/s11434-013-0079-3.

[85] M. Liu, J. Yang, S. Feng, H. Zhu, J. Zhang, G. Li, J. Peng, Hierarchical double-layered SnO_2 film as a photoanode for dye-sensitized solar cells, New J. Chem. 37 (2013) 1002−1008, https://doi.org/10.1039/c3nj40962g.

[86] Musyaro'Ah, I. Huda, W. Indayani, B. Gunawan, G. Yudhoyono, Endarko, Fabrication and characterization dye sensitized solar cell (DSSC) based on TiO_2/SnO_2 composite, AIP Conf. Proc. 1788 (2017), https://doi.org/10.1063/1.4968315.

246   8. Potential of nanooxidic materials and structures of photoanodes for DSSCs

[87] X. Mao, R. Zhou, S. Zhang, L. Ding, L. Wan, S. Qin, Z. Chen, J. Xu, S. Miao, High efficiency dye-sensitized solar cells constructed with composites of TiO_2 and the hot-bubbling synthesized ultra-small SnO_2 nanocrystals, Sci. Rep. 6 (2016) 1−10, https://doi.org/10.1038/srep19390.

[88] S. Pace, A. Resmini, I.G. Tredici, A. Soffientini, X. Li, S. Dunn, J. Briscoe, U. Anselmi-Tamburini, Optimization of 3D ZnO brush-like nanorods for dye-sensitized solar cells, RSC Adv. 8 (2018) 9775−9782, https://doi.org/10.1039/c7ra13128c.

[89] T.H. Lee, H.J. Sue, X. Cheng, Solid-state dye-sensitized solar cells based on ZnO nanoparticle and nanorod array hybrid photoanodes, Nanoscale Res. Lett. 6 (2011) 1−8, https://doi.org/10.1186/1556-276X-6-517.

[90] W. Peng, L. Han, Z. Wang, Hierarchically structured ZnO nanorods as an efficient photoanode for dye-sensitized solar cells, Chem. Eur J. 20 (2014) 8483−8487, https://doi.org/10.1002/chem.201402250.

[91] M.H. Lai, M.W. Lee, G.J. Wang, M.F. Tai, Photovoltaic performance of new-structure ZnO-nanorod dye-sensitized solar cells, Int. J. Electrochem. Sci. 6 (2011) 2122−2130.

[92] K.K. Wong, A. Ng, X.Y. Chen, Y.H. Ng, Y.H. Leung, K.H. Ho, A.B. Djurišić, A.M.C. Ng, W.K. Chan, L. Yu, D.L. Phillips, Effect of ZnO nanoparticle properties on dye-sensitized solar cell performance, ACS Appl. Mater. Interfaces 4 (2012) 1254−1261, https://doi.org/10.1021/am201424d.

[93] B.G. Zhai, L. Yang, Y.M. Huang, Improving the efficiency of dye-sensitized solar cells by growing longer ZnO nanorods on TiO_2 photoanodes, J. Nanomater. 2017 (2017), https://doi.org/10.1155/2017/1821837.

[94] Z. Sun, J.H. Kim, Y. Zhao, D. Attard, S.X. Dou, Morphology-controllable 1D-3D nanostructured TiO_2 bilayer photoanodes for dye-sensitized solar cells, Chem. Commun. 49 (2013) 966−968, https://doi.org/10.1039/c2cc37212f.

[95] Nasikhudin, E.P. Ismaya, M. Diantoro, A. Kusumaatmaja, K. Triyana, Preparation of PVA/TiO_2 composites nanofibers by using electrospinning method for photocatalytic degradation, IOP Conf. Ser. Mater. Sci. Eng. 202 (2017), https://doi.org/10.1088/1757-899X/202/1/012011.

[96] T. Istirohah, S. Wihdatul, M. Diantoro, ScienceDirect fabrication of aligned PAN/TiO 2 fiber using electric electrospinning (EES), Mater. Today. Proc. 13 (2019) 211−216, https://doi.org/10.1016/j.matpr.2019.03.216.

[97] X. Huang, Fabrication and properties of carbon fibers, Materials 2 (2009) 2369−2403, https://doi.org/10.3390/ma2042369.

[98] J. Yao, C.W.M. Bastiaansen, T. Peijs, High strength and high modulus electrospun nanofibers, Fibers 2 (2014) 158−187, https://doi.org/10.3390/fib2020158.

[99] G. Li, L. Lv, H. Fan, J. Ma, Y. Li, Y. Wan, X.S. Zhao, Effect of the agglomeration of TiO_2 nanoparticles on their photocatalytic performance in the aqueous phase, J. Colloid Interface Sci. 348 (2010) 342−347, https://doi.org/10.1016/j.jcis.2010.04.045.

[100] S.H. Othman, S. Abdul Rashid, T.I. Mohd Ghazi, N. Abdullah, Dispersion and stabilization of photocatalytic TiO 2 nanoparticles in aqueous suspension for coatings applications, J. Nanomater. 2012 (2012), https://doi.org/10.1155/2012/718214.

[101] M.A. Ashraf, W. Peng, Y. Zare, K.Y. Rhee, Effects of size and aggregation/agglomeration of nanoparticles on the interfacial/interphase properties and tensile strength of polymer nanocomposites, Nanoscale Res. Lett. 13 (2018), https://doi.org/10.1186/s11671-018-2624-0.

[102] K.N. Chen, F.N.I. Sari, J.M. Ting, Multifunctional TiO_2/polyacrylonitrile nanofibers for high efficiency PM2.5 capture, UV filter, and anti-bacteria activity, Appl. Surf. Sci. 493 (2019) 157−164, https://doi.org/10.1016/j.apsusc.2019.07.020.

[103] L. Shahreen, G.G. Chase, Effects of electrospinning solution properties on formation of beads in TiO_2 fibers with PdO particles, J. Eng. Fibers Fabr. 10 (2015) 136−145, https://doi.org/10.1177/155892501501000308.

[104] U. Sa'adah, S.W. Himmah, T. Suprayogi, M. Diantoro, S. Sujito, N. Nasikhudin, The effect of time deposition of PAN/TiO_2 electrospun on photocurrent performance of dye-sensitized solar cell, Mater. Today Proc. 13 (2019) 175−180, https://doi.org/10.1016/j.matpr.2019.03.210.

[105] L. Zhao, J. Yu, Controlled synthesis of highly dispersed TiO_2 nanoparticles using SBA-15 as hard template, J. Colloid Interface Sci. 304 (2006) 84−91, https://doi.org/10.1016/j.jcis.2006.08.042.

[106] T. Tański, W. Matysiak, Ł. Krzemiński, Analysis of optical properties of TiO_2 nanoparticles and PAN/TiO_2 composite nanofibers, Mater. Manuf. Process. 32 (2017) 1218−1224, https://doi.org/10.1080/10426914.2016.1257129.

[107] P. Sanjay, E. Chinnasamy, K. Deepa, J. Madhavan, S. Senthil, Synthesis, structural, morphological and optical characterization of TiO_2 and Nd^{3+} doped TiO_2 nanoparticles by sol gel method: a comparative study for

II. Photovoltaic materials and devices

References **247**

photovoltaic application, IOP Conf. Ser. Mater. Sci. Eng. 360 (2018), https://doi.org/10.1088/1757-899X/360/1/012011.

[108] F.K. Konan, B. Hartiti, H.J.T. Nkuissi, A. Boko, Optical-structural characteristics of i-ZnO thin films deposited by chemical route, J. Mater. Environ. Sci. 2508 (2019) 1003−1010.

[109] K. Li, Z. Wei, X. Zhu, W. Zhao, X. Zhang, J. Jiang, Microstructure and optical properties of ZnO nanorods prepared by anodic arc plasma method, J. Appl. Biomater. Funct. Mater. 16 (2018) 105−111, https://doi.org/10.1177/2280800017751492.

[110] Y.X. Zhuang, X.L. Zhang, X.Y. Gu, Effect of annealing on microstructure and mechanical properties of Al0.5CoCrFeMoxNi high-entropy alloys, Entropy 20 (2018), https://doi.org/10.3390/e20110812.

[111] N.M. Ahmed, F.A. Sabah, H.I. Abdulgafour, A. Alsadig, A. Sulieman, M. Alkhoaryef, The effect of post annealing temperature on grain size of indium-tin-oxide for optical and electrical properties improvement, Result Phys. 13 (2019) 102159, https://doi.org/10.1016/j.rinp.2019.102159.

[112] N. Chamangard, H. Asgharzadeh, Growth of ZnO nanostructures on polyurethane foam using the successive ionic layer adsorption and reaction (SILAR) method for photocatalytic applications, CrystEngComm 18 (2016) 9103−9112, https://doi.org/10.1039/C6CE01777K.

[113] T.M. Hammad, J.K. Salem, R.G. Harrison, The influence of annealing temperature on the structure, morphologies and optical properties of ZnO nanoparticles, Superlattice. Microst. 47 (2010) 335−340, https://doi.org/10.1016/j.spmi.2009.11.007.

[114] J. Wu, Y. Zhao, C.Z. Zhao, L. Yang, Q. Lu, Q. Zhang, J. Smith, Y. Zhao, Effects of rapid thermal annealing on the structural, electrical, and optical properties of Zr-doped ZnO thin films grown by atomic layer deposition, Materials 9 (2016), https://doi.org/10.3390/ma9080695.

[115] J. Ungula, B.F. Dejene, H.C. Swart, Effect of annealing on the structural, morphological and optical properties of Ga-doped ZnO nanoparticles by reflux precipitation method, Result Phys. 7 (2017) 2022−2027, https://doi.org/10.1016/j.rinp.2017.06.019.

[116] Nandani, A. Supriyanto, A.H. Ramelan, F. Nurosyid, Effect of annealing temperature on optical properties of TiO$_2$ 18 NR-T type thin film, J. Phys. Conf. 1011 (2018), https://doi.org/10.1088/1742-6596/1011/1/012016.

[117] S. Sanjeev, D. Kekuda, Effect of annealing temperature on the structural and optical properties of zinc oxide (ZnO) thin films prepared by spin coating process, IOP Conf. Ser. Mater. Sci. Eng. 73 (2015), https://doi.org/10.1088/1757-899X/73/1/012149.

[118] G. Zhang, K. Lu, X. Zhang, W. Yuan, M. Shi, H. Ning, R. Tao, X. Liu, R. Yao, J. Peng, Effects of annealing temperature on optical band gap of sol-gel tungsten trioxide films, Micromachines 9 (2018), https://doi.org/10.3390/mi9080377.

[119] M. Diantoro, T. Suprayogi, A. Hidayat, A. Taufiq, A. Fuad, R. Suryana, Shockley's equation fit analyses for solar cell parameters from I-V curves, Int. J. Photoenergy 2018 (2018), https://doi.org/10.1155/2018/9214820.

[120] W. Ahmad, U. Mehmood, A. Al-Ahmed, F.A. Al-Sulaiman, M.Z. Aslam, M.S. Kamal, R.A. Shawabkeh, Synthesis of zinc oxide/titanium dioxide (ZnO/TiO$_2$) nanocomposites by wet incipient wetness impregnation method and preparation of ZnO/TiO$_2$ paste using poly(vinylpyrrolidone) for efficient dye-sensitized solar cells, Electrochim. Acta 222 (2016) 473−480, https://doi.org/10.1016/j.electacta.2016.10.200.

[121] F. Xie, Y. Li, J. Dou, J. Wu, M. Wei, Facile synthesis of SnO 2 coated urchin-like TiO 2 hollow microspheres as efficient scattering layer for dye-sensitized solar cells, J. Power Sources 336 (2016), https://doi.org/10.1016/j.jpowsour.2016.10.061.

[122] M.S. Jo, J.S. Cho, X.L. Wang, E.M. Jin, S.M. Jeong, D.W. Kang, Improving of the photovoltaic characteristics of dye-sensitized solar cells using a photoelectrode with electrospun porous TiO$_2$ nanofibers, Nanomaterials 9 (2019), https://doi.org/10.3390/nano9010095.

[123] H.-T. Chou, H.-C. Hsu, The effect of annealing temperatures to prepare ZnO seeds layer on ZnO nanorods array/TiO 2 nanoparticles photoanode, Solid State Electron. 116 (2016) 15−21, https://doi.org/10.1016/j.sse.2015.11.004.

[124] P. Pandey, M. Ramzan, P. Fozia, Effects of annealing temperature optimization on the efficiency of ZnO nanoparticles photoanode based dye sensitized solar cells, J. Mater. Sci. Mater. Electron. 28 (2017) 1537−1545, https://doi.org/10.1007/s10854-016-5693-9.

CHAPTER 9

Perovskite solar cells

Amol Nande[1], Swati Raut[2], S.J. Dhoble[2]

[1]Guru Nanak College of Science, Ballarpur, Maharashtra, India; [2]Department of Physics, RTM Nagpur University, Nagpur, Maharashtra, India

9.1 Introduction

In the 19th and 20th centuries, the human race increased fossil fuel energy consumption to satisfy its demand for electrical energy. Unfortunately, fossil fuels are limited but global demand for energy for day-to-day use continues to grow. Moreover, carbon emissions have also grown on average by 2% [1]. Furthermore, fossil fuels have many disadvantages regarding extraction, transportation, and consumption. There are high risks involved in the transportation and accidental spills of fossil fuels, which can cause very serious problems. Therefore humankind has to find alternative sources of easily available and clean energy. The available options for clean and safe energy are solar energy, wind energy, biomass, and geothermal energy. Of these energy sources, solar energy is easily available and can be directly converted to electrical energy [2].

Solar energy offers an easy solution to produce electricity in a clean and safe way. The materials or devices that convert solar energy to electrical energy are termed photovoltaic or solar cells. Daryl Chapin, Gerald Pearson, and Calvin Fuller from Bell's Laboratory invented a silicon-based solar cell that ultimately revolutionized the photovoltaics industry [3,4]. Due to advancements in fundamental physics and device fabrication along with silicon wafer-based solar technology, other thin film technologies such as GaAs, $Cu(In,Ga)Se_2$, CdS, $CuZnSnSe_2$, CdTe, organic molecule and polymer solar cells, quantum dot solids, dye-sensitized solar cells, and perovskite solar cells (PVSCs) have been established and researchers are intensively investigating materials and systems for efficient solar cells [5—8]. The major photovoltaics industries still dominate silicon photovoltaics, but it has high manufacturing and installation costs [9—11], although it is expected that the cost of silicon photovoltaics will decrease. But for now, PVSCs are a potentially proven candidate for photovoltaic technology. The materials used are cheap and easily processable organic—inorganic hybrid perovskite semiconductors used as absorber materials [11—14]. This is the third generation of solar cells and the first work on this subject was published in 2009 by Kojima et al. [15]. Researchers have seen variations in power conversion efficiencies from 3.78% to 28% and they are expecting to see

Energy Materials
https://doi.org/10.1016/B978-0-12-823710-6.00002-9

Copyright © 2021 Elsevier Ltd. All rights reserved.

further improvements in efficiencies in the coming years [13,16,17]. This chapter discusses the photovoltaic and solar cell properties of perovskite materials. First, the chapter explains perovskite materials, where we focus on their common structure and the literature devoted to inorganic and organic perovskite materials. This will lead to synthesis techniques, especially for perovskite solar cells, which consist of topics such as device anatomy, requirements of each layer, working principles, and characterization techniques. Later, the chapter discusses the fabrication approach and device evaluation of PVSCs. The chapter concludes with key challenges and the future outlook of PVSCs.

9.2 Perovskite materials

Perovskite was first used for calcium titanium oxide (CaTO$_3$) and is named after Lev Perovski. Later, perovskite was used to explain any material having the same structure as CaTO$_3$. To express perovskite material, the more generalized formula ABX_3 is deployed, where A (such as metal ions and organic cations) and B (such as metal ions) are cations and X (such as oxides and halides) is the anion. Also, their valences are in a ratio of 1:2:1 with a cubic structure of conventional perovskite. Cations A are larger than cations B, as shown in Fig. 9.1 [18]. The figure shows that perovskite materials have a cubic crystal structure where the smallest cation B is surrounded by an octahedral anion. Twelvefold octahedral coordination with 12 X anion neighbors is covered by cation A. It is well known that when discussing the stability of perovskite structures, including oxides and halides, Goldschmidt's tolerance factor is widely accepted. This is given as $t = (r_A + r_X)/[\sqrt{2}(r_B + r_X)]$, where, r_A, r_B,

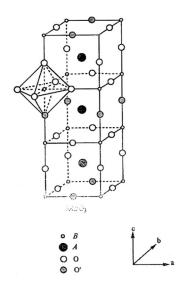

FIGURE 9.1 Ideal perovskite crystal structure in which B cations are linked at the corners of an octahedron to form a cubic lattice [18]. *Used with permission granted from Copyright © 2001, American Chemical Society.*

and r_X are the ionic radii of A, B, and X, respectively, and for all perovskite structures t values should be in the range of 0.75—1 so as to maintain stability [18,19]. However, for a few materials, the t range is not the ideal condition for the formation of perovskite structures, because for some materials even if t is in the most favorable range, no stable perovskite structure is formed [19,20]. A detailed study of the stability of perovskite materials is discussed by Li and coworkers [19,20]. According to their work, the modified range of t values is 0.813—1.107, excluding $CsBeF_3$. Further formation on perovskite crystal structures in AX—BX_2 complex halide systems the tolerance factor is not the necessary factor (as for solar cells application perovskite materials with halides as anions are more useful [7,12,21]). The octahedral factor—if anion X and cation B form a sixfold coordination octahedral structure, then the ratios of r_B/r_X are in the range of 0.377 and 0.895 for perovskite halides (except for $TlMnI_3$ and $CsBeI_3$) [19,22]. Therefore tolerance and octahedral factors are used to span the two-dimensional structure and are considered as the criteria of formability of halide perovskite materials. These two factors can be summarized in a two-dimensional plot, which is an effective model to predict perovskite materials, as shown in Fig. 9.2. From the figure it is clear that the materials with perovskite structure are in the rectangular dotted box represented by open circles, which obeys both conditions (a list of the compounds is present in the cited article) [19].

The ability of perovskite materials to create cation- and anion-deficient structures makes them special and these materials show extraordinary physical properties such as magnetic, electrical, superconducting, optical, ferroelectric, photovoltaic, and photocatalytic properties

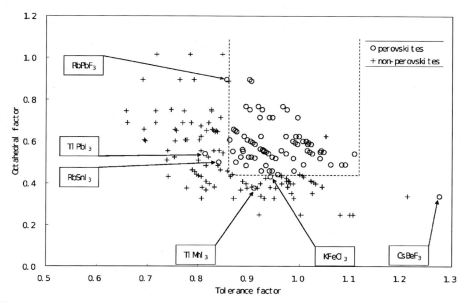

FIGURE 9.2 Tolerance versus octahedral map for perovskite compounds [19]. *Used with permission of the International Union of Crystallography.*

[11,12,18]. Perovskite materials have either oxygen anions (oxides) or halogens. Perovskite materials with halides are of great interest due to their exceptional optoelectronic properties. Therefore it is observed that the materials have halides (Cl, Br, I, and F) at the X position in ABX_3 and their optoelectronic properties make them better candidates for solar cell application compared to oxides. Also, the published work shows that highly efficient perovskite materials for solar cells contain lead (Pb) or tin (Sn) at position B. These compounds have low bandgaps that enhance photon absorption in visible and near-infrared regions. One can tune the perovskite bandgap by varying the composition of perovskite elements. Also, by changing any elements in the ABX_3 perovskite stoichiometry, the bandgap can be tuned throughout the visible spectrum. The energy bandgap depends on the size of cation A, and it is observed that by increasing the size of A, the bandgap size decreases and vice versa [23–25]. Also, one can tune the bandgap by keeping the same composition of materials and by changing the quantum confinement in perovskite nanoparticles. There are organometallic perovskite materials for which the experimental bandgap value of ~1.6 eV is an ideal bandgap value for single junction solar cells.

Another interesting parameter used for photovoltaics is absorption coefficient, which determines the intensity of light as it passes through a given material. It is observed that perovskite materials, especially organometallic perovskite materials, have higher absorption coefficient compared to other photovoltaic materials [26–28]. Along with absorbance coefficient, carrier diffusion length also plays an important role in the performance of the solar cell device; the higher the carrier diffusion length, the better the performance of the device. Literature dictates that perovskite materials exhibit longer carrier diffusion lengths compared to traditional semiconductors or solar materials; for example, the single crystal of MAPbI$_3$ showed ~175 µm [29]. Thus this large diffusion length allows a planar heterojunction with a selective charge interface for perovskite materials. This gives perovskite materials both p-type and n-type interfaces, which form p-i-n or n-i-p structures. The performance of these structures is similar to mesoporous structures. Thus perovskite materials are both excellent light absorbers and electron and hole transporters, which make them outstanding candidates for solar cells.

9.3 Synthesis techniques

The structure and morphology of perovskite materials or layers play a vital role in their inherent properties and hence their solar cell performance. The homogeneity and uniformity of the photovoltaic grains are also responsible for the performance of solar cells. The crystalline nature, grain size, and quality of perovskite thin film depend highly on depositing parameter and substrate. Here, we explain the most commonly used techniques to deposit dense and uniform perovskite layers. The perovskite absorber can be deposited by solution-based methods (such as dip coating, spin coating, inkjet printing, doctor blading, etc.), vapor-based methods (such as thermal evaporation, chemical vapor deposition, sputtering, pulsed laser deposition, sequential evaporation, etc.), and hybrid vapor–solution-based deposition methods. Some of these are described next.

9.3.1 Solution-based methods

From earlier published works, it is observed that one-step spin coating was used to deposit perovskite thin films on mesoporous substrate using a blended precursor solution of AX and BX_2. However, spin coating is a slow process and hence it is difficult to deposit a uniform layer of perovskite materials [30,31]. Due to the slow process, the nucleation rate automatically decreases, which often causes pinhole structures in film or uneven films [32,33]. Slow nucleation in the simple spin-coating technique or nucleation and growth of perovskite films can be induced by adding antisolvents such as toluene or chlorobenzene during spinning of the precursor solution. Therefore this technique is called the antisolvent dripping method [34]. It is observed that due to fast nucleation, more uniform films are produced, which increases the power conversion efficiency (PCE) values of PVSCs [35,36].

The nucleation and growth of perovskite thin films also increases when hot precursor solution is deposited on the hot substrate. This method is termed the hot casting method [37]. Perovskite thin films with larger grain size can be produced using the hot casting method. This suggests that a bigger grain size would have fewer bulk defects, which improve the charge carrier mobility and PCE value of PVSCs [37]. This is one of the important and most used methods because it provides control over two-dimensional crystallographic orientation, which facilitates efficient charge transport and good stability against light sources [38]. Furthermore, it is possible to deposit two- and three-dimensional hybrid structures using this method, which opens a new scope for the PVSC field. This provides high and efficient charge transport, and hence PCE value and stability rise [39,40].

Another approach to deposit the perovskite layer is the sequential two-step deposition method. In PVSCs, two-step deposition follows spin coating of AX and BX_2 followed by thermal annealing to form a uniform layer by the interfusion of perovskite grains [37,41]. In this method, perovskite morphology can be controlled by precursor materials, spin-coating parameter, and annealing conditions. This process also provides better PCE values than simple spin-coating methods [42].

9.3.2 Vacuum-based deposition

In the solution processing method, it is virtually impossible to obtain completely homogeneous, dense, and compact thin film. This difficulty can be overcome by depositing perovskite film using vacuum-based methods. This process also forms a dense and uniform thin film over a large area. It also provides control over substrate temperature, thickness of films, and deposits even on rough surfaces. This kind of method is necessary to develop large-scale PVSCs with high PCE values [43]. In this method, AX and BX_2 are coevaporated to a synthesized pure ABX_3 thin layer [44]. Coevaporation can be achieved by the thermal evaporation method or pulse laser method. In all deposition methods, the perovskite-deposited layer would be uniform, free from solution contamination, and independent of smoothness of substrates.

9.3.3 Hybrid solution and vapor method

In this method, BX_2 is deposited on the electrode or substrate via sputtering or thermal evaporation. Later, a solution of AX is added by spin coating to form the perovskite layer

254

9. Perovskite solar cells

[45]. This method is used to deposit perovskite thin film over a large surface. In this process, one could replace the spin-coating step with the inkjet printing technique or the dip-coating technique. This method allows comparatively fast formation of perovskite even at room temperature, and controls deposition rate and substrate temperature to deposit BX_2. By varying these two conditions, the formation of size, shape, and arrangement of grains can be controlled. This will provide a controlled porous nanostructured BX_2 thin film, which defines the formation and structure of the perovskite layer when AX is deposited onto it.

9.4 Perovskite solar cell

As discussed in earlier sections, there are many alternatives to traditional silicon-based solar cells but the materials should match power conversion efficiency, long-term stability, and cost effectiveness. PVSCs are promising materials for photovoltaic cells due to their PCE and low costs [16,21,46]. Due to their crystal structure, availability of cations and anions, high absorbance coefficient, large diffusion length, and bandgap comparable to visible and near-infrared photon region, these materials show significant optical properties, which are used in diode, superconductor, and transistor technologies [47–51].

As discussed earlier, $CH_3NH_3PbBr_3$ (MAPb) cells reported 2.2% efficiency and a sudden rise in efficiency was observed when bromine was replaced by iodine [15]. Afterward, iodine-based perovskite was used as a sensitizer in photoelectrochemical cells and an observed PCE value of ~3.8% was documented for this kind of system. However, when a hole transporting layer was used in the device, the PCE suddenly increased to ~9.7% [31]. Later, when the evolution of PVSCs started, researchers began modifying PVSCs to increase the efficiency and stability of the devices by modifying perovskite composition. The literature depicts various previous compositions of photovoltaic cells, which contained perovskite materials; a summary of selected perovskite materials used from 2018 with their device PCE values and hole transport layers is provided in Table 9.1.

9.4.1 Inorganic perovskite

Due to the poor thermal instability of organic cations, inorganic materials for PVSCs have been studied intensively due to their visible light absorbance with bandgap and thermal stability [69]. Inorganic materials like cesium lead halides, mixed cesium lead halide, cesium tin halide, cesium germanium halide, cesium tin–germanium halide, and silver–bismuth halide have also been studied for PVSCs. Inorganic perovskite materials such as $CsPbI_3$-based solar cells demonstrated a PCE value of ~17% and stability up to 500 h [70,71]. Mixed cesium lead halide perovskite, e.g., $CsPbI_{3-x}Br_x$ solar cells, has better thermal stability and bandgap. For example, $CsPbI_{3-x}Br_x$ perovskite carries thermal stability from $CsPbBr_3$ and bandgap from $CsPbI_3$. Mixed halide perovskite inorganic materials when used as a gradient annealing method in combination with the antisolvent method formed large grains and hence produced high-quality film. This material had an efficiency of ~16% and better stability [72,73] than earlier types.

Due to the poisonous nature of Pb, researchers began working on lead-free inorganic perovskite materials. Tin (Sn) or germanium (Ge) were used in place of Pb and iodine and

II. Photovoltaic materials and devices

9.4 Perovskite solar cell

TABLE 9.1 A summary of perovskite materials used since 2018 along with their efficiency and hole transport materials used for the device.

Perovskite	Hole transport material	Efficiency (%)	References
$CsPbI_3$ quantum dots	Spiro-MeOTAD	70	[52]
$CsPbI_{3-x}Br_x$	Metal oxide	Above 7	[21]
$CsPbI_{2+x}Br_{1-x}$		14.45	[53]
$(FA_{0.79}MA_{0.16}Cs_{0.05})_{0.97}Pb(I_{0.84}Br_{0.16})_{2.97}$	LiTFSI	22.70	[54]
$FAPbI_3$	Spiro-MeOTAD	20.64	[55]
Cesium−formamidinium−methylammonium triple cation PVSCs $[(FAPbI_3)_{1-x}(MAPbBr_3)_x]$	Spiro-MeOTAD	20.80	[46]
Formamidinium−lead triiodide perovskite films	Spiro-MeOTAD	20.64	[55]
$CH_3NH_3PbI_3$ films	m-PEDOT:PSS/spiro-MeOTAD	15.57/22.70	[56−58]
$(PEA)_2(MA)_{n-1}Pb_nI_{3n+1}$	m-PEDOT:PSS	11.01	[59]
$Cs_2AgBiBr_6$ film	Poly(3-hexylthiophene-2,5-diyl) (P3HT)	1.44	[17]
$CH_3NH_3PbI_3$ layer sandwich by carbon	Spiro − MeOTAD, PTTA, P3HT	17	[60]
$FA_{0.75}MA_{0.25}Sn_{1-x}Ge_xI_3$	PEDOT:PSS	20	[61]
$Cs_xFA_{1-x}PbI_3$	Spiro-MeOTAD	14.16	[62]
$HTAB_x(FAPbI_3)_{0.95}(MAPbBr_3)_{0.05}$	Poly(3-hexylthiophene)	22.70	[63]
$MAPbI_3/MAPbI_xBr_{3-x}$ perovskite stacking structure	Carbon-based solar cells	16.2	[64]
CsPbI2Br films	Carbon-based solar cells	10.95	[65]
$FA_{0.85}MA_{0.15}PbI_3$	Tris(pentafluorophenyl) boron	21.6	[66]
$CsPbBr_3$	No	9.92	[67]
$[CH(NH2)_2]_{0.9}Cs_{0.1}PbI_3$	Spiro-MeOTAD	20.60	[68]

Abbreviations: *CsPbI3*, Cesium lead iodide; *Spiro-OmeTAD*, 2,2′,7,7′-Tetrakis[N,N-di(4-methoxyphenyl)amino]-9,9′-spirobifluorene; *Spiro-MeOTAD*, 2,2′,7,7′-Tetrakis[N,N-di(4-methoxyphenyl)amino]-9,9′-spirobifluorene; *PEDOT:PSS*, poly(3,4-ethylenedioxythiophene) polystyrene sulfonate; *P3HT*, Poly(3-hexylthiophene-2,5-diyl); *PTTA*, purified terephthalic acid.

bromine were used as a halide in the perovskite combination. Sn and Ge have good optoelectronic properties similar to Pb. However, the perovskite structure of these used materials is less stable and hence efficiency is less as compared to previously studied perovskite materials. Further studies are required to understand the optical and electrical properties of Sn or Ge used for perovskite materials [73−76]. A mixture of Sn and Ge is used instead of monometallic counterparts due to its stable narrow bandgap absorbers [77]. Instead of

256   9. Perovskite solar cells

two-dimensional structures, three-dimensional structures (like Cs_2TiI_6, Rb_2TiI_6, In_2TiI_6, $AgBi_2I_7$, etc.) are efficient candidate PVSCs because of their higher electronic dimensionality and promising light absorbency.

Thus all inorganic perovskites are stable under ambient conditions and they do not undergo chemical degradation like organic perovskites. The efficiency of inorganic perovskites has reached 17% in recent years.

9.4.2 Organometal perovskite

Miyasaka and coworkers reported and provided new directions for solar cells [15]. They reported $MAPbI_3$ and $MAPbBr_3$ perovskite as sensitizers in dye-sensitized solar cells with electrolyte solution. This kind of structure is termed organometal perovskite. The efficiency of the first PVSCs was 3.81%. PVSC materials are easy to prepare and deposit but there was a problem with their stability. This was resolved when the hole transport layer changed to spiro-MeOTAD.

Later experiments started to increase the efficiency and stability of PVSCs. Therefore apart from changing other layers' components, researchers also started experimenting with components of $MAPbI_3$. In other cations, including ethylammonium (EA) and formamidinium (FA), the radius of the cation $r_{MA} < r_{FA} < r_{EA}$ was investigated for a replacement for methylammonium (MA). Replacing MA with a larger cation disorders the three-dimensional arrangement resulting in a two-dimensional orthorhombic crystalline morphology with larger bandgap [78]. It was observed that when MA (2.17 Å) was replaced with FA (2.53 Å), it modified the bandgap of the perovskite. $FAPbI_3$ has a small bandgap compared to $MAPbI_3$; therefore the absorption wavelength increases [79]. Replacing MA with FA increases the tolerance factor and hence enhances thermal stability.

Mixed A cation perovskites like $MAFA_{1-x}PbI_3$, by changing the composition proportion of MA to FA, eventually attain an efficiency of 14.9% [80]. Mixed cation perovskites like $FA_{0.85}Cs_{0.15}PbI_3$ PVSCs showed better performance and stability against $FAPbI_3$ [81]. Three-dimensional perovskites like $(PEA)_2(MA)_2Pb_3I_{10}$ improved stability [82]. Other experiments changed the B cation from Pb to Sn or Ge in $MAPbI_3$ perovskite. It is observed that by changing the cation to Sn, the bandgap reduced to 1.17 eV [83]. Later, changing the halide from iodine to another halide or mixed halide again affected the bandgap of the perovskite and hence increased the absorption wavelength of the perovskite layer [84]. Theoretical calculations showed that lead-free organometallic perovskites (like $MASnI_3$, $MASrI_3$, $MABiSI_2$, $MABi_{0.5}Tl_{0.5}I_3$, and $MACaI_3$) are prominent candidates for PVSCs [85].

The commonly studied perovskite materials for photovoltaic applications are $MAPbI_3$, $MAPbI_{3-x}Cl_x$, $MAPbBr_3$, $MAPbI_{3-3x}Br_{3x}$, $MASnI_3$, $HC(NH_2)_2PbI_3$, etc. Although the efficiency of organometallic PVSCs is matched with traditional solar cells, degradation of organic molecules affects their stability. Ambient conditions like temperature, humidity, and oxygen affect the degradation of organic molecules. The efficiency and stability of perovskite materials also depend on hole transport and electron transport materials. Details of these two layers are discussed further in this chapter.

II. Photovoltaic materials and devices

9.5 Fabrication approach and device anatomy

A PVSC device schematic is shown in Fig. 9.3, where the front contact or electrode on the transparent glass substrate is deposited. An electron transport material layer is deposited on top of the front contact to increase the mobility or transport of the photogenerated electrons in the perovskite material layer, which is directly deposited on the electron transport material layer. Mobility of the photogenerated holes is controlled by a hole transport material layer, which is deposited onto the perovskite material layer; finally a back electrode is deposited onto the hole transport layer.

There are several methods and approaches for synthesizing PVSCs, of which three approaches are well known: (1) mesoporous approach, (2) planer n-i-p approach, and (3) planer p-i-n approach.

9.5.1 Mesoporous perovskite solar cells

Dye-sensitized solar cells are generally prepared using mesoporous perovskite solar cells. As discussed earlier and shown in Fig. 9.4A, a transparent electron transport layer like TiO_2 or ZnO of 50–70 nm thickness and a mesoporous layer like Al_2O_3 of 150–200 nm thickness are deposited on the transparent glass substrate. When the perovskite material layer is deposited onto the mesoporous layer, the perovskite fills the mesopores. A spiro-MeOTAD layer is deposited on top of the perovskite layer, which acts like a hole transport layer. Finally, a metal electrode like gold, silver, or aluminum is deposited, which also acts like an opaque layer of the device [86].

It is observed that the mesoporous layer has significant impact on the performance of the solar cell device. This forms a matrix of perovskite materials, which enhances the generation of more electron and hole pairs. Furthermore, the comparatively higher thickness of the perovskite layer also allows light to absorb more effectively [87,88]. The mesopores do not form a uniform perovskite layer and hence do not control the growth of perovskite grains. This leads to the formation of highly disordered phases [89]. These disordered phases limit the open-circuit voltages and short-circuit density of the devices. The literature shows that by decreasing the thickness of the mesoporous layer, the crystallinity of the perovskite materials and hence the performance of the devices are increased [90,91]. The study also shows

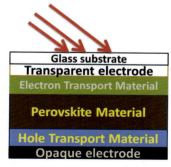

FIGURE 9.3 Simple representation of a perovskite solar cell device.

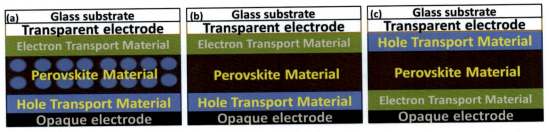

FIGURE 9.4 (A) Schematic for a mesoporous perovskite solar cell (PVSC) indicating different layers of the device. (B) Schematic for the n-i-p approach to the synthesis of the PVSC indicating different layers. (C) Schematic for the p-i-n approach to the synthesis of the PVSC indicating different layers.

that if the thickness of nanoporous materials is increased up to certain values (~300 nm), the number of pores filled with perovskite materials will also increase, which causes an increase in pore-filling fraction. Therefore an increase in charge carriers is observed and hence the transportation of charge carriers increases. This leads to increases in efficiencies at the electron transport layer interface and hence there is a rise in the PCE of PVSC devices [32,92]. The highest PCE recorded for MA chloride PVSCs, which are prepared using the mesoporous approach, is 24.02% [93].

9.5.2 N-i-p structure approach

It is observed that for PVSCs, without separating the interface between electron transport layer and perovskite material layer, the perovskite is capable of efficient ambipolar charge transport [90,91]. Therefore PVSCs do not need mesoscopic configuration and hence we can discard the mesoporous layers. This will significantly reduce the size of the device, and also the high sintering step to deposit the mesoporous layer is no longer required. If the same device arrangement is maintained as explained in the last section by excluding the mesoporous layer, an n-i-p structure is formed.

In this approach, a transparent substrate like glass or indium tin oxide (ITO) is coated with a transparent electrode and later an electron transport layer (ionic dopant layer) is deposited on top of it, as shown in Fig. 9.4B. A perovskite layer is deposited on the electron transport layer and this follows the deposition of the hole transport layer. Finally, a metal electrode is deposited. In this approach, the PCE value can be tuned by changing the composition of the perovskite layer and the interface between the perovskite layer and the hole–electron transport layer [94]. This is the most efficient approach to construct PVSC devices in which the hole transport layer with ionic dopants is deposited on top of the perovskite materials, which acts like a window layer. These hole transport layers have parasite absorption and they accelerate the degradation of PVSCs due to mobile ionic dopants [95].

9.5.3 Planar p-i-n approach

In this approach, the hole transport layer is deposited before the perovskite layer opposite the n-i-p structure, as shown in Fig. 9.4C. Therefore it is also called an inverted architecture.

In this case, a p-type conducting layer (such as the conducting polymer poly(3,4-ethylenedioxythiophene) polystyrene sulfonate [PEDOT:PSS]) is deposited on the transparent electrode (which is deposited on the transparent glass or ITO substrate). The peroxide layer is deposited atop the hole transport layer and on top of the perovskite layer, and the electron transport layer is deposited. Finally, the metal or opaque electrode is deposited on it. Recent work has shown that p-type semiconductors like PEDOT:PSS, poly(triaryl amine) (PTAA), NiO, Cu_2O, CuSCN, etc. are used as hole transfer materials to construct PVSCs with the p-i-n approach. The planar p-i-n approach is effective at harnessing high-energy photons and is compatible with an ionic dopant-free transport layer [96–99].

The p-i-n approach (inverted planar structure) is considered the best choice for the fabrication of perovskite:silicon tandems [100]. However, the inverted approach of PVSCs with an inorganic hole transport layer affects the PCE of the devices. The architecture decreases the PCE value for the inorganic hole transport layer; the highest certified value for the p-i-n approach with an inorganic hole transport layer is 19.2%, which is very small compared to the n-i-p approach [101]. Also, the stability of these kinds of cells is less than 20 h [102]. Therefore the efficiency of this approach is increased using organometallic perovskite as an organic hole transport layer [103]. As far as we know, a methylamine–dimethylamine lead iodate $(MA_{1-x}DMA_xPbI_0)$ perovskite structure used to form a p-i-n structure for PVSCs showed the highest PCE of 21.60% [100].

9.6 Requirement of each layer

The working PVSCs is explained as — when photons are incident on substrate it passes from the transparent electrode to absorbing layer like perovskite layer. Incident photons are absorbed by the absorbing layer; excitons are generated and propagate to the perovskite–electron transport layer interface. At the interface, excitons are separated into electrons and holes. The electrons are collected by the electron–hole layer and holes are collected by hole transport layers. The PCE of PVSCs depends on transparency of the electrode or photoanode, mobility, and collection of photogenerated charges. Also, a metal electrode in the topmost layer in PVSCs, which is directly in contact with the environment, affects the stability of devices. Therefore it is necessary to understand the requirements and properties of each layer.

9.6.1 Photoanode and electron transport layer

As shown in Figs. 9.3 and 9.4, the transparent anode layer is deposited on the ITO or glass transparent substrate. The electron transport layer is deposited on top of the anode layer; sometimes the electron transport layer acts like a transparent anode in PVSC devices. It plays an important role in the efficient working of PVSCs, as it provides sufficient charge collection and formic ohmic contact with the electrode.

The main function of the electron layer in PVSCs is to extract electrons from the absorbing layer, transport them to the cathode, and minimize recombination of holes–electrons [104,105]. This can be done by selecting electron transport materials that can reduce the

260

9. Perovskite solar cells

potential barrier (between electron transport layer and perovskite layer) for electrons and block the holes [105,106]. The common electron transport materials are TiO_2, SnO_2, ZnO, fullerene butyric acid methyl ester (PCBM), tin-doped zinc oxide, $SrTiO_3$, B_2S_3, $BaSnO_3$, and Indene—C60 bisadduct as well as doped metal oxides [107—110]. The basic requirements to qualify for the layer are high electron mobility, easy dissolution in organic solvent, inability to absorb ultraviolet light, and good air stability. Also, there must be good alignment between the perovskite layer and the electron transport layer [111].

The electron transport layer can be deposited using atomic layer deposition, electrochemical deposition, the solution-processed method, the hydrothermal method, radiofrequency sputtering, and electrodeposition [112—115]. These methods, like atomic layer depositions, facilitate large surface deposition of the electron transport layer; the large surface area of the electron transport layer (ZnO, TiO_2, and SnO_2) allows better contact with the perovskite layer. Better contact provides efficient extraction of electrons and hence increases the PCE value of PVSCs. To increase the extraction ability and improve electron-transporting properties, metal oxide electrodes are doped with n-type impurity [115]. Because of their high electron mobility, transparency to ultraviolet-visible light, and high electron affinity, carbon materials like fullerene and its derivatives are used as electron transport layers [116,117].

It is observed that this layer directly affects the stability and performance of the PVSC device. For better performance, the layer should be resistant to air and humidity, transparent to ultraviolet-visible light, and have high ability to extract electrons from the perovskite layer, high electron mobility, high tolerance to holes to avoid recombination of electron—hole pairs, and zero degradation over time. It can be concluded that the electron transport layer should have negligible degradation because degradation in the layer causes a decrease in carrier collection and damages ohmic contact with the electrode. Therefore efficiency of PVSCs decreases.

9.6.2 Absorption layer

In PVSCs, perovskite materials are used as an absorption layer. When energy photons are closer to their bandgap, the photons are absorbed by the perovskite layer. The electrons jump from the valence band to the conduction band; hence, electron—photon pairs are generated [87,118]. The efficiency of PVSCs depends on several factors, such as migration of electrons toward the electron transport layer and holes toward the hole transport layer. It also depends on structure, the components of perovskite materials, and stability of the layer. Structures and their stabilities are well explained in Section 2.

The stability of perovskite directly affects the PCE values of the device. Long-term stability of the perovskite layer depends on ambient conditions like temperature or thermal, moisture, and structural stability. Perovskite materials like $MAPbI_3$ are hygroscopic in nature and create a degradation process in which the methylamine group is lost and PbI_2 is formed [47,119]. This happens due to high polarity of a water molecule and it attributes to structural ability. Research work showed that mesostructured materials like TiO_2 help to stabilize ambient conditions [88]. Apart from humidity, temperature also plays a vital role in layer stability, as the perovskite materials' phase changes from tetragonal to cubic at 54—56°C [120]. A study showed that for efficient solar cells, PVSCs must be stable up to 85°C, but perovskite

II. Photovoltaic materials and devices

materials decompose to PbI_2 when they are heated at 85°C in a nitrogen environment [121]. Therefore for efficient performance of PVSCs, the absorbance layer should be tolerant to the moisture and temperature of the environment.

9.6.3 Hole transport layer

Hole transport layers mainly work as blockade layers for electrons between perovskite and transporting holes, and protect the perovskite layer and electrode with reduced carrier recombination to boost efficiency of PVSCs [106,122].

Initially, the redox couple iodine—triiodine in an organic liquid electrolyte can be found in PVSCs, but the redox couple is highly corrosive, volatile, and photoreactive. These properties decrease the stability and performance of the device [15,123]. A lot of research is happening to find appropriate materials for the hole transport layer. The hole transport layer affects the PCE values of the device. Solid-state organics and p-type conducting materials or polymers are used as hole transport layers to improve PCE values of the devices [13,19,52,55]. Spiro-MeOTED and PCBM/piezoforce microscopy are other strong candidates for the hole transport layer as the efficiency values are high for these materials. However, the market price of these materials is quite high, so PVSCs synthesized using these materials are also high. Inorganic materials or composites like CuI or C are hole conductors and replace hole-conducting layers.

Initially in PVSCs, organic electrolyte containing lithium halide and a similar halogen formed the hole-transporting medium [15]. It is observed that this solution base can be replaced by spiro-MeOTAD as a solid-state hole transport material [30]. It is also observed that the PCE value increases to 9.70% as well as the stability of the device [15]. At this point, PVSCs started following standard solar cell structures (as shown in Figs. 9.3 and 9.4). Afterward, researchers started performing experiments by changing hole transport materials, the composition of perovskite, and electron transport layer to increase the efficiency and stability of PVSCs. Other materials for the hole transport layer are PEDOT:PSS, MoO_3, PTAA, $CuGaO_2$, CuSCN, and NiO_x [26,33,57,122]. Metal oxides are considered to be better than organic layers due to their nontoxicity and ambient condition stability [57,124]. Cheap polymers like PEDOT:PSS with high conductivity are used as replacements for regular hole transport material. These materials have good processability, high thermal stability, good mechanical flexibility, sufficient optical transparency, and proper energy levels. Polymer hole transport layers also make flexible PVSCs in inverted devices [125,126].

The hole transport layer affects the stability and efficiency of PVSCs. It is necessary to have a highly transparent layer so that it does not affect the perovskite layer and allows the transportation of holes. It should also stop electron—hole recombination and protect the perovskite layer from external factors. A high-performance hole transport layer is necessary to increase the performance and stability of PVSC devices.

9.6.4 Metal electrode

Metal electrodes directly affect the stability and operation of PVSCs; the electrode is the uppermost layer and is closer to the environment [106,127]. Initially, the electrode perovskite

262   9. Perovskite solar cells

absorber layer was deposited in the PVSC device; therefore the electrode could protect or decrease the transfer of moisture from air to the perovskite layer. Metals, like silver, aluminum, and gold, are the most commonly used electrodes in PVSCs. Metal electrodes have stability issues as well as affect the PCE values. Silver and aluminum electrodes undergo corrosion from ion migration from perovskite layers. These generated ions from metal halides have stability issues with perovskite cells [128−130].

Other elements like chromium and carbon are used for metal electrodes. A chromium electrode has been used in MAPbBr$_3$ PVSC devices. According to studies [131,132], chromium increases the interfacial resistance and reduces the amount of interfacial recombination. Sometimes a chromium layer is deposited between the metal electrode and perovskite layer, which prevents diffusion of the metal electrode into the perovskite layer, and leads to stable PCE. As the layer is chemically inert toward iodine, it does react with PVSCs [133,134]. Researchers have also started working on hole transport layer-free PVSC devices by using carbon electrodes [60,65]. A thick carbon electrode protects the perovskite layer from humidity and increases the stability of the device. PVSCs with porous carbon back electrodes usually exhibit exceptional long-term stability under light and ambient conditions [135].

Thus to fabricate stable PVSCs, diffusion barrier layers should be inserted to separate the metal electrodes and perovskite layer. Although electrodes do not directly affect the efficiency of PCE values of PVSCs, they affect their stability.

9.7 Working mechanism of PVSCs or device operation

In PVSCs, the perovskite layer acts as an absorber layer, which absorbs light and gives rise to free electrons and holes. These photogenerated free charge carriers diffuse and drift under the influence of an electric field. The holes move toward the hole transport layer while electrons move toward the electron transport layer. The efficiency of extraction depends on the transporting ability of holes and electrons toward their respective layers. When holes and electrons are collected by respective electrodes, the electrons dissipate energy and produce power before returning to the device at the opposite electrode and recombine with the hole. If electrons are not extracted by the electron transport layer from the perovskite layer, then it will eventually recombine with the hole. This is nothing but radiative transition, which emits a photon of energy equal to the bandgap of the perovskite material.

Fig. 9.5 depicts the general working principle of PVSC devices. When light falls on PVSCs, perovskite acts as an absorber layer and absorbs light leading to the creation of free carriers (electrons−holes). Charge separation occurs through the injection of photogenerated electrons like TiO$_2$ into the electron transport layer, e.g., spiro-MeOTAD. When a large number of photons are incident on PVSCs, a number of electrons are generated in the perovskite absorber. Therefore electrons are captured by the electron transport layer from which electrons move externally to the metal electrodes. However, photogenerated holes are extracted from the perovskite layer and these electrons flow to the metal electrode (such as gold, silver, or aluminum). The hole−electron recombines externally to produce electrical potential due to the absorption of light [25,28]. At zero cell voltage, the cell current reaches its maximum limiting value and is called a short-circuit current. This depends on the production of charge

II. Photovoltaic materials and devices

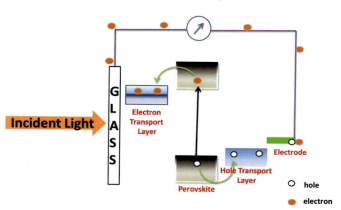

FIGURE 9.5 Schematic diagram showing the working principle of perovskite solar cells.

carrier in the perovskite layers and the extraction of charge carriers by the respective charge carrier layers. This directly affects the efficiency and stability of PVSCs. Thus it is concluded that the device configuration of PVSCs is a perovskite layer sandwiched between the electron and hole blocking layer. This configuration leads to stable and efficient photovoltaic devices.

9.8 Characterization technique

This section deals with specific characterizations and supporting examples from the literature used for PVSC devices. Also, a brief introduction to general characterization is discussed, which is used to analyze PVSC devices; detailed information on these general devices is provided in Chapter 3.

9.8.1 Specific characterization

9.8.1.1 Current density–voltage characteristics

This is the most important characterization of solar cells as it estimates solar cell efficiency. Therefore this technique is discussed in detail. In this measurement, PVSC electrodes are connected, a certain bias voltage ramp is applied, and simultaneously the current produced is measured. The measurements are taken in both dark and light conditions. In the dark, PVSCs behave like a p-n junction diode with the characteristic that in reverse bias no current should pass through it, while in forward bias a huge current (and sharp increase after the breakdown region) is observed. In the presence of light, the diode shows similar behavior to the normal diode after breakdown voltage, but before the finite negative current, reverse bias is observed before the breakdown region. This produces power in solar cells when $J-V$ measurements are performed in the presence of light.

From these measurements, important parameters like short-circuit current, open-circuit voltage, and maximum power are estimated. Short-circuit density is the current density produced by the cell at 0 V and it is the maximum current produced by the illuminating solar cell. Open-circuit current density is the voltage at which no net current flows through the device and it depends on the bandgap of the absorber, in this case the bandgap of the perovskite layer. The power of the solar cell is estimated by combining applied voltage and current produced in the PVSCs. Maximum power of the devices is the product of the current maximum power point and voltage maximum power point. One can also estimate the fill factor of the solar cells; it is the ratio of maximum power of the devices to the product of short-circuit current density and open-circuit voltage.

The $J-V$ characteristics of PVSC change with the illumination time of solar cells. The performance generally increases with the increase in illumination time as there is an increase in fill factor and open-circuit voltage. This effect is termed the light-soaking effect. This effect is due to the presence of mobile ions (AB^+ and I^-) inside the perovskite layer [136,137]. When the $J-V$ scan is performed from negative to positive voltage and vice versa, PVSCs show a difference in response time. This phenomenon is termed hysteresis [65,92]. Apart from scan speed, temperature, and prescanning conditions, hysteresis depends on the movement of ions in the perovskite layer and the PVSC architecture [92,138–140]. If hysteresis is only because of movement of ions in the perovskite layer, then hysteresis can be tuned by changing either the electron–hole transport layer or the device architecture [136,141,142].

The inherent hysteresis and soaking effect change the $J-V$ characteristics and therefore change the PCE values of PVSCs. Therefore it is obviously an essential tool for PVSCs [142].

9.8.2 Time-resolved photoluminescence

When light is incident on the perovskite layer, an incident photon is absorbed by the layer. Thus electrons jump from the valence band to the conduction band and electron–hole pairs are generated in it. The bandgap of perovskite materials is in the visible near-infrared region. Therefore these materials absorb incident photons and generate electron photon pairs, which combine either by radiative or nonradiative transition with liberation of energy in the form of photons. This gives rise to the photoluminescence effect in perovskite materials [143–145]. It is seen that the efficiency of PVSCs depends on the combination of photogenerated holes and electrons produced in perovskite layers [11,28]. Thus it is useful to study radiative and nonradiative transitions in the perovskite method from which the recombination time can be estimated.

Time-resolved spectroscopy characterizes the dynamics of radiative and nonradiative transition in both organic and inorganic materials. With this method, the intensity of luminescence material depends on the energy of incident photons. Therefore it is necessary to adjust the laser so that it can emit one photon at a time or per pulse. During this measurement, when a pulse leaves the laser, time measurements are simultaneously started by the instrument. The emitted photon excites the studied sample as soon as photons are detected by the detector. The procedure starts with the measurement of time-resolved decays at a fixed number of wavelengths across the emission spectrum, which is discovered from photoluminescence measurements. The intensity decays once the sample is excited by the source and is quicker at shorter wavelengths due to spectral relaxation but slower at longer wavelengths [146]. This technique is useful to study the recombination and injection dynamics of solar cells using time-resolved photoluminescence and photocurrent measurements [147]. This technique is also used for investigating the carrier recombination process, electrical performance, and intrinsic physical processes in solar cell devices [148–151].

9.8.3 General characterization

This section deals with general characterizations required to analyze devices. Structural, surface, and optical characterization like absorption and photoluminescence spectroscopy are discussed here.

9.8.3.1 X-ray diffraction spectrometer

This is a well-known technique for discovering the crystal structure of the deposited material. The X-rays are diffracted in a specific direction by crystal atoms. By measuring diffracted rays and angles, one can decide the crystal structure of given materials. Along with three-dimensional crystal structures, lattice parameters, chemical bonds, and disorders can be estimated. The principle of the X-ray diffraction (XRD) spectrometer is Bragg's law, $2d \sin \theta = n\lambda$, where d is the interplanar distance, θ is the angle of diffraction, n is the order of diffraction, and λ is the wavelength of the X-ray. One can estimate the crystal structure and orientation of the examined sample. One more measurement that can be carried out using XRD is low-angle XRD or X-ray reflectivity. This measurement is recorded between angles of 0 and 5 degrees, which allow us to estimate the thickness of each layer along with roughness of the interface.

9.8.3.2 Scanning electron microscopy

Scanning electron microscopy (SEM) is an important technique used to study the surface morphology of thin films and devices. It is even possible to estimate grain size using SEM. However, the top view only provides surface morphology of the top layer, as shown in Fig. 9.6A. It is possible to study surface morphology even up to 10 nm. The literature dictates

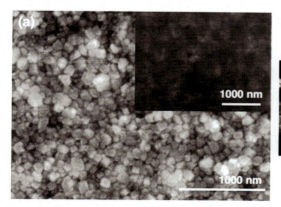
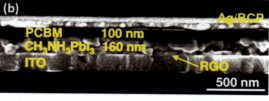

FIGURE 9.6 (A) An illustration of surface morphology using scanning electron microscopy for perovskite solar cell (PVSC) devices. (B) Illustration of cross-sectional analysis of a PVSC device. *ITO*, Indium tin oxide; *PCBM*, butyric acid methyl ester; *RGO*, reduced graphine oxide. *Copied with permission. Copyright © 2014 Elsevier.*

that in the interface between layers, the crystallinity of absorbing layers and uniformity of layers directly affect the stability and performance of PVSC devices [3,6,21,55,92].

It is important to study the interface between the layers as well as to discover the uniformity and exact thickness of each layer in the devices. The interface between layers and an estimate of the exact thickness of each layer cross-sectionally can be studied using cross-section SEM images [152,153]. Fig. 9.6B shows the cross-section of a PVSC device. This confirms the fabrication quality of the device and the uniformity and thickness of each layer [153].

9.8.3.3 Transmission electron microscopy

It is well known that the PCE values of PVSCs are dependent on the grain size of the perovskite layer. To provide the exact grain size of perovskite materials, transmission electron microscopy (TEM) is the best technique to study and discover the crystal structure of particular grains. By this method, the nature and functionality of PVSCs can be studied [154–156].

Fig. 9.7A shows a TEM image of a PVSC device used for studying particle size analysis. By studying their electron diffraction patterns, one can study the transformation occurring in nanocrystals or perovskite induced by an electron beam at low temperature [157]. Using TEM, one can image and study the domain patterns of perovskite nanocrystals [156,158]. Orientation and disorientation of adjacent grains are characterized and studied using TEM [156,159]. PVSC thickness of each layer, grain boundaries, orientation, and interface between each layer are studied and imaged (as shown in Fig. 9.7B) using this instrument.

9.8.3.4 Atomic force microscopy measurement

For very high-resolution surface analysis, atomic force microscopy (AFM) is used; this is a very sophisticated instrument with very high resolving power up to the order of a nanometer. Information is collected using an extremely small piezoelectric mechanical probe with precise movement on electronic command. AFM consists of a cantilever with a sharp tip at its end, which is used to scan the sample surface. The cantilever is made of silicon or silicon nitride and the tip has a curvature scaled in nanometers. The working principle is

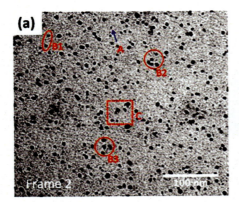
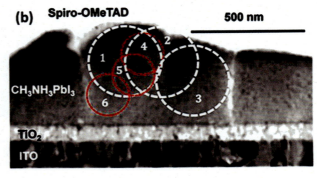

FIGURE 9.7 Standard examples showing particle size analysis and cross-sectional morphology using transmission electron microscopy [156]. *ITO*, Indium tin oxide.

to obtain images according to Hooke's law, which acts between the force exerted by the tip on the surface of the sample and the deflection produced in the cantilever. Using this instrument, one can estimate thickness, morphology, and roughness of the film.

Fig. 9.8A and E depicts illustrations for typical examples of AFM images of perovskite materials. In AFM, one can also scan electric potential variation using conducting cantilevers. One can map $I-V$ characteristics with imaging using conducting AFM studies. Studies show that conductivity is high at gain boundaries; an example is shown in Fig. 9.8A and E, where the boundaries shown in greenish color represent conducting areas. Moreover, Fig. 9.8A shows the overlapping of grains, which causes hysteresis in $I-V$ characteristics (Fig. 9.8B–D), but Fig. 9.8E shows that higher conductivity near the boundaries provides continuous pathways for free electrons (for details of $I-V$ characteristics studies refer to the cited paper) [160]. This clearly suggests that AFM is an efficient tool to study grain boundaries, size of grains, and mapping of $I-V$ characteristics studies.

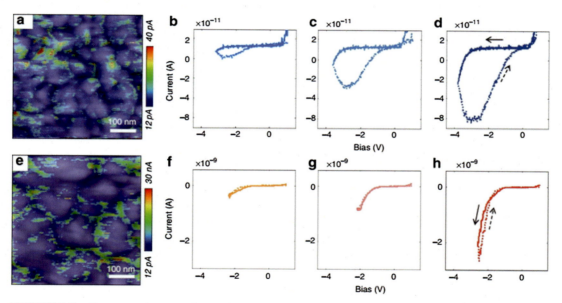

FIGURE 9.8 Contact mode atomic force microscopy study of the hysteresis of ion relationships for control films (A–D) and hybrid films (E–H) with $I-V$ characteristics [160]. *Copyright © 2015, Springer Nature.*

9.9 Device evaluation

PVSCs strongly depend on the composition of perovskite, hole transport materials, electron transport materials, and their interfaces. The evaluation of PVSC devices depends on the roughness of the interface, the cost of the devices, PCE values, and stability. When PVSCs were introduced to the world they were highly unstable and PCE values were very low. However, progress in PVSCs happened when the electrolyte-based hole transport layer was replaced by spiro-MeOTAD, a solid-state material. This increased the stability as well as

268   9. Perovskite solar cells

PCE values of PVSCs [15,30,31]. Instead of using a single halide in perovskite materials, a mixed halide $CH_3NH_3PbI_{3-x}Cl_x$ composition was used. It was observed that PCE values and stability increased [31,161]. Furthermore, modification of the nanoporous layer, which replaced conducting material (TiO_2) with nonconducting material (Al_2O_3), improved the open-circuit voltage of PVSCs, which increased the PCE to 10.90% [161]. Later, it was shown that the performance of PVSC devices increased for $CH_3NH_3PbI_{3-x}Br_x$ mixed halide perovskites. The result was that efficiency increased for lower concentrations of Br, and for higher concentrations perovskite films provided better stability against humidity [162]. Subsequently, researchers tried to change deposition methods: instead of single-step deposition they used two-step deposition for perovskite films, which improved their morphology [163]. The efficiency of PVSCs further increased when solvent deposition was replaced by thermal evaporation and $CH_3NH_3PbI_{3-x}Cl_x$ mixed perovskite. The observed efficiency for this was 15.40% [164]. Further increase in efficiency of PVSC devices was reported for poly(triarylamine) as a hole transport material and $CH_3NH_3PbI_{3-x}Br_x$ mixed halide perovskites [25,163]. There followed a further modification in deposition technique, and cations of larger radii were used, which had a symmetrical cubic phase and increased the t factor. This modification improved the devices. Researchers also tried different proportions of inorganic cations and halide anions in mixed perovskite, which allowed their properties to be tuned [66,165,166].

Around 2014, experiments began with electron transport layers. The reported electron transport material was TiO_2, which was replaced by fluorine-doped TiO_2 and a TiO_2 combination. Later, this combination was replaced by ITO as transparent material combined with a thin layer of zinc oxide [115]. Other metal oxides (like ZrO_2, NiO_2, SiO_2, and SnO_2) and nanocore shell nanoparticles (such as Al_2O_3/ZnO, TiO_2/MgO, and WO_3/TiO_2) were used as electron transport material and for charge retardation or to avoid recombination of photogenerated holes and electrons, which enhanced PCE values [127,167−169]. As discussed in an earlier chapter, materials with a lower highest occupied molecular orbital band but a higher lowest unoccupied molecular orbital band than the active perovskite layer can be used as electron extraction or electron transport layer [170]. Researchers also tried to make metal oxide-free PVSC devices. Organic materials like P3HT, PCBM composites to achieve better efficiency in p-i-n architecture of the PVSCs device [127,171−173]. It is observed that when organic material was used as an electron transport layer and a perovskite layer was deposited onto it, perovskite materials crystallized well to form compact, uniform, smooth, and pinhole-free film [174]. So, an organic electron transport layer provides flexible PVSC devices and uniform perovskite films, and increases stability and PCE values of PVSC devices. The flexible PVSC device even fabricates under ambient air and humidity providing a PCE value of ~21% and a high open-circuit voltage of 1.15 V [46].

In 2006, dye-sensitized solar cells based on an organic—metal hybrid semiconductor within a nanoporous TiO_2 layer achieved a PCE of 2.2% [25]. Miyasaka suggested the use of organic—metal perovskite ($CH_3NH_3PbI_3$) for solar cells [15]. As discussed earlier in this chapter, the crystal structure of perovskite is defined as ABX_3; Miyasaka's group chose A to be the organic molecule, B as Pb—metal, and X as Br—halide. The observed efficiency value was ~3.8% as shown in Fig. 9.9. In the last decade, researchers have extensively worked on PVSC; the progress in PCE values is shown in Fig. 9.9, which was plotted by taking values from published research papers and published solar efficiency tables every year. So far, the maximum recordable power efficiency achieved for PVSCs is 25.2% [175]. The overall performance of PVSCs is shown in Fig. 9.9. It is clear that perovskites are the fastest developing

II. Photovoltaic materials and devices

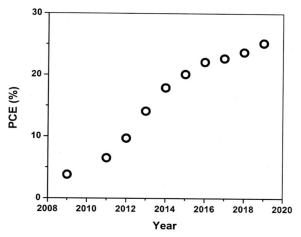

FIGURE 9.9 Year versus power conversion efficiency for perovskite solar cells. *PCE*, Power conversion efficiency. *Data taken from research papers and from a solar efficiency table published every year.*

solar cell technology. However, more work has to be performed to increase their stability and bring them to the commercial world.

The PCE values of PVSCs depend on many parameters: quality of perovskite film, electron transport layer, hole transport layer, metal electrode, and interface between layers. A high-quality polycrystalline perovskite layer and continuous defect-free junctions with electron transport layer and hole transport layer are essential to obtain stable and high-performance solar cells [176]. It is obvious that uniform grains with a minimal area of grain boundaries provide faster transportation of charge carriers which increases the performance of PVSCs [177]. The uniform grains can be achieved by using double cation perovskites like formamidinium/MA, which provide better thermal stability but have high hysteresis in $J-V$ characterizations. Table 9.2 summarizes notable PVSC devices with their PCE values, cutoff voltage, and current density.

It is also observed that although perovskite materials have excellent defect-tolerant properties, the ionic migration in perovskite makes their characterization more difficult. Ionic migration in perovskite layers also affects the interface between perovskite and the charge transport layers [176]. This is also one of the reasons for the presence of hysteresis in $J-V$ characteristics. Later, studies showed that hysteresis decreases if lithium, potassium, or sodium and their bis(-trifluoromethane) sulfonimide salt are doped to the TiO_2 electron transport layer in triple cation perovskite solar cells; also, PCE values increased [188,189]. Surface analysis showed that sulfur chemically bridges TiO_2 and perovskite at their interface by forming Ti—S—Pb bonding, thereby forming a structural continuity. This decreases grain boundary effects; hence, the hysteresis effect decreases and the PCE value increases to 21.10% in this case [190]. Another factor to improve PVSCs is the presence of hysteresis in $J-V$ characteristics.

Based on the discussions, it can be concluded that for superior performance and better stability of PVSCs (1) high optical absorption coefficient of perovskite thin film, (2) long carrier diffusion length and suppressed recombination, (3) defect tolerance, and (4) well-balanced charge transfer are needed.

270

9. Perovskite solar cells

TABLE 9.2 Notable perovskite solar cell devices with power conversion efficiency, cutoff voltages, and short-current density.

Fabricated devices	Power consumption efficiency (%)	V_{OC} (V)	J_{SC} (mA/cm^2)	References
FTO/bl-TiO$_2$/mp-TiO$_2$/MAPb(I$_{1-x}$Br$_x$)$_3$/Au	12.3	0.87−1.13	5−18	[162]
Glass/FTO/TiO$_2$/TiO$_2$/CH$_3$NH$_3$PbI$_3$/HTM/Au	13	0.992	17.1	[163]
FTO-glass/bl-TiO$_2$/mp-TiO$_2$/FAPbI$_3$/PTAA/Au	20.1	1.06	24.7	[178]
FTO/d-TiO$_2$/mp-TiO$_2$/NBH/P$_3$HT/Au	22.70	1.152	24.97	[63]
ITO/SnO$_2$/FAPbI$_3$/spiro-MeOTAD/Au	21.06	1.06	22.70	[179]
Perovskite/HIT tandem cell	23.58	1.6506	18.093	[180]
FTO/c-TiO$_2$/mesoTiO$_2$/MAPbI$_{3-x}$Br$_x$/spiro-MeOTAD/Ag	13.56	0.97	22.05	[181]
ITO/PEDOT:PSS/MAPbI$_3$/PC61BM/Al	14.6	1.01	21.0	[182]
FTO/c-TiO$_2$/meso-TiO$_2$/MAPbI$_3$/spiro-MeOTAD/Au	4.69	0.86	11.11	[183]
ITO/PEDOT:PSS/MAPbI$_{3-x}$Cl$_x$/PC61BM/Ag	10.83	0.899	17.10	[184]
Glass/Bi−TiO$_2$/Mp−TiO$_2$ + SnO$_2$/(FAPbI$_3$)$_{0.875}$(MAPbBr$_3$)$_{0.125}$(CsPbI$_3$)$_{0.1}$/spiro-MeOTAD/Au	21.6	1.18		[185]
FTO/compact-TiO$_2$(c-TiO$_2$)/mesoporous TiO$_2$ (mp-TiO$_2$)/Cs$_{0.05}$FA$_{0.80}$MA$_{0.15}$Pb(I$_{0.85}$Br$_{0.15}$)$_3$/spiro-MeOTAD/Au	22	1.19	23.2	[186]
FTO/SnO$_2$/CsPbBr$_3$/carbon	10.71	1.588	7.64	[187]
CH$_3$NH$_3$PbBr$_3$/TiO$_2$ and CH$_3$NH$_3$PbBr$_3$/TiO$_2$	3.13/3.81	0.96/0.61	5.57/11.0	[15]

Abbreviations: *CsPbI3*, Cesium lead iodide; *Spiro-OmeTAD*, 2,2′,7,7′-Tetrakis[N,N-di(4-methoxyphenyl)amino]-9,9′-spirobifluorene; *Spiro-MeOTAD*, 2,2′,7,7′-Tetrakis[N,N-di(4-methoxyphenyl)amino]-9,9′-spirobifluorene; *PEDOT:PSS*, poly(3,4-ethylenedioxythiophene) polystyrene sulfonate; *P3HT*, Poly(3-hexylthiophene-2,5-diyl); *PTTA*, purified terephthalic acid; *Peht*, Phosphatidylethanol; *FTO*, Fluorine doped tin oxide; *ITO*, Indium tin oxide; *Perovskite/HIT tandem cell*, Perovskite/ Heterojunction amorphous/Crystalline Silicon tandem cell.

9.10 Key challenges

Solar cells are considered to be the best alternative for fossil fuels and other energy sources. However, traditional solar cells have good PCE but installation and production charges are quite high. So, PVSCs are ideal alternatives. Although PVSCs are certified with high PCEs such as 25.20%, they still face problems such as toxicity, long-term stability, and cost effectiveness. As highly efficient PVSCs mostly contain Pb atoms, which are toxic and can cause serious problems like cancer and mental retardation, it has been a big challenge to make Pb-free PVSCs with high PCE values and better stability. At the same time, devices should have long-term stability in terms of temperature and efficiency because Si cell stability is more than 20 years. Moreover, perovskite absorbers and charge transporting layers are sensitive to moisture and temperature, which may affect the long-term stability of PVSCs. Furthermore, for commercialization, these solar cells should be cost effective for everyone.

II. Photovoltaic materials and devices

It is also difficult to deposit high-quality perovskite films on planar surfaces with minimum hysteresis in voltage—current measurements, and they also have high photovoltaic performance. Also, in PVSCs, metal back contacts are used, which are not transparent to visible and near-infrared light. In stacking or tandem applications, the metal contacts must be replaced by highly transparent and conductive contacts through which low-energy photons must pass to perovskite cells. In addition to these, it is observed that in PVSCs, light incident on the cell thorough the substrate. Therefore, glass substrate is used. However, the observed efficiency of the device is increased when light photons directly fall on PVSCs. Therefore it is necessary to consider the appropriate choice of substrate, which does not sacrifice the efficiency of the device.

Thus there are many challenges but ample scope for research work to improve PVSCs by increasing their stability up to 25 years. This can be achieved by appropriate choice of substrate and the use of visible near-infrared transparent electrodes, which will allow low-energy photons to the cell. Also, it is important to find a replacement for Pb to reduce toxicity of the solar cell.

The main challenges to develop a process to fabricate PVSCs with high efficiency and negligible hysteresis and improve ultraviolet to near-infrared transparent devices. The devices should have long-term operational stability under ambient temperature, at least up to ~100°C. The extraction efficiency of charge carrier layers needs to increase but there will be a minimum recombination of photogenerated holes—electrons. Fabrication and material costs should be reduced considerably, thereby reducing the cost of the solar cells.

9.11 Future outlook

Over the last decade, the development of PVSCs has provided huge scope and another promising candidate in the field of photovoltaics. Perovskite materials exhibit impressive performance in solar cells due to their tunable bandgap, which converts a significant portion of the solar cell into photocurrent. They also possess large charge-carrier mobilities, which make them capable of effectively transporting the resultant photocurrent to charge a selective layer of electrodes. By comparing these properties with other existing photovoltaic materials makes them a better option for solar cell devices. In spite of all these benefits, all highly efficient PVSCs are Pb-based hybrid perovskites. So, researchers are now working on lead-free and organic-free PVSCs. However, uniform and defect-free films similar to Pb perovskites are still a challenge. This seems to be the first step toward PVSC improvement in the near future. Another challenge is to establish methods to create a voidless high-quality interface at the junction between perovskite and charge-carrier transport layers. Interface improvement methods have been developed for Pb-based PVSCs but Pb-free PVSCs have to be developed.

Apart from these, PVSCs are easy to fabricate and optimize, and even mass production is possible. Also, they provide high efficiency and thermal stability. However, the practical durability of devices in operation is widely affected by photoinduced degradation, humidity, and oxygen penetration in materials. To compete with traditional solar cells, PVSCs should be durable at least for 20 years. However, these perovskites are a better option for relatively short-life applications, e.g., in the car and electronic industries. For these applications, the requirements are low cost of modules, Pb-free PVSCs, light weight, and stability up to 10 years.

Also, to replace traditional solar cells, PVSCs should be Pb free (to avoid pollution), highly efficient, thermally stable, and have long-term stability. These can be achieved by modifying perovskite films so that absorption increases, the quality of the carrier transporting layer is maintained to increase the extraction of charge carriers and avoid degradation of the perovskite layer, and the quality of electrodes is preserved so that external recombination can easily happen and be protected from the ambient environment.

9.12 Conclusion

It is interesting to note that perovskite materials were invented over a century ago. These materials and their thin films have witnessed technological applications in the fields of transistors, light-emitting diodes, spintronics, and superconductivity. Due to their outstanding optical properties and absorbance in the visible near-infrared region, it has been noticed that they could be used in solar cell devices. This possibility is enhanced due to high efficiency, low-cost starting materials and an easy fabrication approach. Both inorganic and organometallic perovskites have proved their acceptance as absorbance layers in solar cells and could replace traditional solar cell materials [14,127].

The wide range bandgap and high absorption coefficient of perovskite materials makes them an ideal absorber material for solar cells. Also, the literature shows that perovskites have high carrier mobilities and long carrier lifetime resulting in long carrier diffusion length. High diffusion length decreases the possibility of recombination of photogenerated carriers. These devices can be synthesized effectively by two approaches: n-i-p and p-i-n junction structures. Another approach (which is not discussed here) is the perovskite tandem approach, which is proven to have good efficiency and stability. However, some important issues like hysteresis in $J-V$ characteristics, long-term stability, and toxicity need to be addressed.

These PVSCs are easy to fabricate and have strong solar absorption and low nonradiative recombination rate and efficiency, making them a replacement for traditional solar cells. PVSCs also have high carrier mobility to increase the overall PCE values. However, better performing PVSCs contain Pb, which is highly toxic. Hence, this raises the toxicity issue during the fabrication, development, and disposal of devices. Also, most of the PVSCs undergo degradation when they are exposed to moisture. Although the efficiency of organometallic PVSCs is good, the materials are highly unstable in humid environments. It is possible to replace organometallic perovskites with inorganic perovskites. Still, a lot of research is required to obtain commercial PVSC devices. However, it is still considered a better candidate for traditional solar cells than other semiconductor solar cells.

References

[1] V. Smil, Energy Transitions: Global and National Perspectives, ABC-CLIO, 2016.
[2] S. Sareen, H. Haarstad, Legitimacy and Accountability in the Governance of Sustainable Energy Transitions, Elsevier, 2020.

References 273

[3] K. Chopra, P. Paulson, V. Dutta, Thin-film solar cells: an overview, Prog. Photovoltaics Res. Appl. 12 (2-3) (2004) 69−92.

[4] J. Perlin, Silicon Solar Cell Turns 50, National Renewable Energy Lab., Golden, CO.(US), 2004.

[5] F. Kessler, D. Hariskos, S. Spiering, E. Lotter, H. Huber, R. Wuerz, CIGS thin film photovoltaic—approaches and challenges, in: High-Efficient Low-Cost Photovoltaics, Springer, 2020, pp. 175−218.

[6] N. Shrivastava, H. Barbosa, K. Ali, S. Sharma, Materials for solar cell applications: an organic overview and polymer-based of TiO_2, ZnO, upconverting solar cells, in: Solar Cells: From Materials to Device Technology, vol. 55, 2020.

[7] R. Patidar, D. Burkitt, K. Hooper, D. Richards, T. Watson, Slot-die coating of perovskite solar cells: an overview, Mater. Today Commun. 22 (2020) 100808.

[8] P. Fonteyn, S. Lizin, W. Maes, The evolution of the most important research topics in organic and perovskite solar cell research from 2008 to 2017: a bibliometric literature review using bibliographic coupling analysis, Sol. Energy Mater. Sol. Cell. 207 (2020) 110325.

[9] S. Group, International Technology Roadmap for Photovoltaic (ITRPV)-Results, Itrpv, 2017.

[10] J. Jean, P.R. Brown, R.L. Jaffe, T. Buonassisi, V. Bulović, Pathways for solar photovoltaics, Energy Environ. Sci. 8 (4) (2015) 1200−1219.

[11] A. Rajagopal, K. Yao, A.K.Y. Jen, Toward perovskite solar cell commercialization: a perspective and research roadmap based on interfacial engineering, Adv. Mater. 30 (32) (2018) 1800455.

[12] M. Anaya, G. Lozano, M.E. Calvo, H. Míguez, ABX3 perovskites for tandem solar cells, Joule 1 (4) (2017) 769−793.

[13] L. Qiu, L.K. Ono, Y. Qi, Advances and challenges to the commercialization of organic—inorganic halide perovskite solar cell technology, Mater. Today Energy 7 (2018) 169−189.

[14] Z. Li, Y. Zhao, X. Wang, Y. Sun, Z. Zhao, Y. Li, H. Zhou, Q. Chen, Cost analysis of perovskite tandem photovoltaics, Joule 2 (8) (2018) 1559−1572.

[15] A. Kojima, K. Teshima, Y. Shirai, T. Miyasaka, Organometal halide perovskites as visible-light sensitizers for photovoltaic cells, J. Am. Chem. Soc. 131 (17) (2009) 6050−6051.

[16] M.A. Green, E.D. Dunlop, D.H. Levi, J. Hohl-Ebinger, M. Yoshita, A.W. Ho-Baillie, Solar cell efficiency tables (version 54), Prog. Photovoltaics Res. Appl. 27 (7) (2019) 565−575.

[17] C. Wu, Q. Zhang, Y. Liu, W. Luo, X. Guo, Z. Huang, H. Ting, W. Sun, X. Zhong, S. Wei, The dawn of lead-free perovskite solar cell: highly stable double perovskite Cs2AgBiBr6 film, Adv. Sci. 5 (3) (2018) 1700759.

[18] M. Pena, J. Fierro, Chemical structures and performance of perovskite oxides, Chem. Rev. 101 (7) (2001) 1981−2018.

[19] C. Li, X. Lu, W. Ding, L. Feng, Y. Gao, Z. Guo, Formability of ABX3 (X= F, Cl, Br, I) halide perovskites, Acta Crystallogr. Sect. B Struct. Sci. 64 (6) (2008) 702−707.

[20] C. Li, K.C.K. Soh, P. Wu, Formability of ABO3 perovskites, J. Alloys Compd. 372 (1−2) (2004) 40−48.

[21] J. Lin, M. Lai, L. Dou, C.S. Kley, H. Chen, F. Peng, J. Sun, D. Lu, S.A. Hawks, C. Xie, Thermochromic halide perovskite solar cells, Nat. Mater. 17 (3) (2018) 261−267.

[22] G.S. Rohrer, Structure and Bonding in Crystalline Materials, Cambridge University Press, 2001.

[23] P. Gao, M. Grätzel, M.K. Nazeeruddin, Organohalide lead perovskites for photovoltaic applications, Energy Environ. Sci. 7 (8) (2014) 2448−2463.

[24] A.R.b.M. Yusoff, M.K. Nazeeruddin, Organohalide lead perovskites for photovoltaic applications, J. Phys. Chem. Lett. 7 (5) (2016) 851−866.

[25] M.A. Green, A. Ho-Baillie, H.J. Snaith, The emergence of perovskite solar cells, Nat. Photon. 8 (7) (2014) 506.

[26] P. Qin, S. Tanaka, S. Ito, N. Tetreault, K. Manabe, H. Nishino, M.K. Nazeeruddin, M. Grätzel, Inorganic hole conductor-based lead halide perovskite solar cells with 12.4% conversion efficiency, Nat. Commun. 5 (1) (2014) 1−6.

[27] X. Qiu, B. Cao, S. Yuan, X. Chen, Z. Qiu, Y. Jiang, Q. Ye, H. Wang, H. Zeng, J. Liu, From unstable CsSnI3 to air-stable Cs2SnI6: a lead-free perovskite solar cell light absorber with bandgap of 1.48 eV and high absorption coefficient, Sol. Energy Mater. Sol. Cell. 159 (2017) 227−234.

[28] T.-B. Song, Q. Chen, H. Zhou, C. Jiang, H.-H. Wang, Y.M. Yang, Y. Liu, J. You, Y. Yang, Perovskite solar cells: film formation and properties, J. Mater. Chem. 3 (17) (2015) 9032−9050.

[29] Q. Dong, Y. Fang, Y. Shao, P. Mulligan, J. Qiu, L. Cao, J. Huang, Electron-hole diffusion lengths> 175 μm in solution-grown CH3NH3PbI3 single crystals, Science 347 (6225) (2015) 967−970.

[30] H.-S. Kim, C.-R. Lee, J.-H. Im, K.-B. Lee, T. Moehl, A. Marchioro, S.-J. Moon, R. Humphry-Baker, J.-H. Yum, J.E. Moser, Lead iodide perovskite sensitized all-solid-state submicron thin film mesoscopic solar cell with efficiency exceeding 9%, Sci. Rep. 2 (2012) 591.

II. Photovoltaic materials and devices

9. Perovskite solar cells

[31] M.M. Lee, J. Teuscher, T. Miyasaka, T.N. Murakami, H.J. Snaith, Efficient hybrid solar cells based on meso-superstructured organometal halide perovskites, Science 338 (6107) (2012) 643—647.

[32] G.E. Eperon, V.M. Burlakov, P. Docampo, A. Goriely, H.J. Snaith, Morphological control for high performance, solution-processed planar heterojunction perovskite solar cells, Adv. Funct. Mater. 24 (1) (2014) 151—157.

[33] J.H. Heo, S.H. Im, J.H. Noh, T.N. Mandal, C.-S. Lim, J.A. Chang, Y.H. Lee, H.-j. Kim, A. Sarkar, M.K. Nazeeruddin, Efficient inorganic—organic hybrid heterojunction solar cells containing perovskite compound and polymeric hole conductors, Nat. Photon. 7 (6) (2013) 486.

[34] M. Xiao, F. Huang, W. Huang, Y. Dkhissi, Y. Zhu, J. Etheridge, A. Gray-Weale, U. Bach, Y.B. Cheng, L. Spiccia, A fast deposition-crystallization procedure for highly efficient lead iodide perovskite thin-film solar cells, Angew. Chem. Int. Ed. 53 (37) (2014) 9898—9903.

[35] N.J. Jeon, J.H. Noh, W.S. Yang, Y.C. Kim, S. Ryu, J. Seo, S.I. Seok, Compositional engineering of perovskite materials for high-performance solar cells, Nature 517 (7535) (2015) 476—480.

[36] D. Bi, C. Yi, J. Luo, J.-D. Décoppet, F. Zhang, S.M. Zakeeruddin, X. Li, A. Hagfeldt, M. Grätzel, Polymer-templated nucleation and crystal growth of perovskite films for solar cells with efficiency greater than 21%, Nat. Energy 1 (10) (2016) 1—5.

[37] W. Nie, H. Tsai, R. Asadpour, J.-C. Blancon, A.J. Neukirch, G. Gupta, J.J. Crochet, M. Chhowalla, S. Tretiak, M.A. Alam, High-efficiency solution-processed perovskite solar cells with millimeter-scale grains, Science 347 (6221) (2015) 522—525.

[38] H. Tsai, W. Nie, J.-C. Blancon, C.C. Stoumpos, R. Asadpour, B. Harutyunyan, A.J. Neukirch, R. Verduzco, J.J. Crochet, S. Tretiak, High-efficiency two-dimensional Ruddlesden—Popper perovskite solar cells, Nature 536 (7616) (2016) 312—316.

[39] Z. Wang, Q. Lin, F.P. Chmiel, N. Sakai, L.M. Herz, H.J. Snaith, Efficient ambient-air-stable solar cells with 2D—3D heterostructured butylammonium-caesium-formamidinium lead halide perovskites, Nat. Energy 2 (9) (2017) 17135.

[40] G. Grancini, C. Roldán-Carmona, I. Zimmermann, E. Mosconi, X. Lee, D. Martineau, S. Narbey, F. Oswald, F. De Angelis, M. Graetzel, One-Year stable perovskite solar cells by 2D/3D interface engineering, Nat. Commun. 8 (1) (2017) 1—8.

[41] Z. Xiao, Q. Dong, C. Bi, Y. Shao, Y. Yuan, J. Huang, Solvent annealing of perovskite-induced crystal growth for photovoltaic-device efficiency enhancement, Adv. Mater. 26 (37) (2014) 6503—6509.

[42] Y. Wang, S. Li, P. Zhang, D. Liu, X. Gu, H. Sarvari, Z. Ye, J. Wu, Z. Wang, Z.D. Chen, Solvent annealing of PbI_2 for the high-quality crystallization of perovskite films for solar cells with efficiencies exceeding 18%, Nanoscale 8 (47) (2016) 19654—19661.

[43] C. Momblona, L. Gil-Escrig, E. Bandiello, E.M. Hutter, M. Sessolo, K. Lederer, J. Blochwitz-Nimoth, H.J. Bolink, Efficient vacuum deposited pin and nip perovskite solar cells employing doped charge transport layers, Energy Environ. Sci. 9 (11) (2016) 3456—3463.

[44] C.Y. Chen, H.Y. Lin, K.M. Chiang, W.L. Tsai, Y.C. Huang, C.S. Tsao, H.W. Lin, All-vacuum-deposited stoichiometrically balanced inorganic cesium lead halide perovskite solar cells with stabilized efficiency exceeding 11%, Adv. Mater. 29 (12) (2017) 1605290.

[45] F. Fu, L. Kranz, S. Yoon, J. Löckinger, T. Jäger, J. Perrenoud, T. Feurer, C. Gretener, S. Buecheler, A.N. Tiwari, Controlled growth of PbI_2 nanoplates for rapid preparation of $CH_3NH_3PbI_3$ in planar perovskite solar cells, Phys. Status Solidi 212 (12) (2015) 2708—2717.

[46] T. Singh, T. Miyasaka, Stabilizing the efficiency beyond 20% with a mixed cation perovskite solar cell fabricated in ambient air under controlled humidity, Adv. Energy Mater. 8 (3) (2018) 1700677.

[47] B.R. Sutherland, E.H. Sargent, Perovskite photonic sources, Nat. Photon. 10 (5) (2016) 295.

[48] J. Hou, X. Yin, Y. Fang, F. Huang, W. Jiang, Novel red-emitting perovskite-type phosphor CaLa1— xMgM' O6: xEu3+ (M'= Nb, Ta) for white LED application, Opt. Mater. 34 (8) (2012) 1394—1397.

[49] T. He, Q. Huang, A. Ramirez, Y. Wang, K. Regan, N. Rogado, M. Hayward, M. Haas, J. Slusky, K. Inumara, Superconductivity in the non-oxide perovskite MgCNi 3, Nature 411 (6833) (2001) 54—56.

[50] D.B. Mitzi, C.D. Dimitrakopoulos, L.L. Kosbar, Structurally tailored organic— inorganic perovskites: optical properties and solution-processed channel materials for thin-film transistors, Chem. Mater. 13 (10) (2001) 3728—3740.

[51] Y. Wu, J. Li, J. Xu, Y. Du, L. Huang, J. Ni, H. Cai, J. Zhang, Organic—inorganic hybrid CH 3 NH 3 PbI 3 perovskite materials as channels in thin-film field-effect transistors, RSC Adv. 6 (20) (2016) 16243—16249.

[52] J.A. Christians, A.R. Marshall, Q. Zhao, P. Ndione, E.M. Sanehira, J.M. Luther, In perovskite quantum dots. A new absorber for perovskite-perovskite tandem solar cells, in: 2018 IEEE 7th World Conference on Photovoltaic

References 275

Energy Conversion (WCPEC)(A Joint Conference of 45th IEEE PVSC, 28th PVSEC & 34th EU PVSEC), IEEE, 2018, pp. 81–84.

[53] H. Bian, D. Bai, Z. Jin, K. Wang, L. Liang, H. Wang, J. Zhang, Q. Wang, S.F. Liu, Graded bandgap CsPbI2+xBr1−x perovskite solar cells with a stabilized efficiency of 14.4%, Joule 2 (8) (2018) 1500–1510.

[54] J.A. Christians, P. Schulz, J.S. Tinkham, T.H. Schloemer, S.P. Harvey, B.J.T. de Villers, A. Sellinger, J.J. Berry, J.M. Luther, Tailored interfaces of unencapsulated perovskite solar cells for> 1,000 hour operational stability, Nat. Energy 3 (1) (2018) 68–74.

[55] J.-W. Lee, Z. Dai, T.-H. Han, C. Choi, S.-Y. Chang, S.-J. Lee, N. De Marco, H. Zhao, P. Sun, Y. Huang, 2D perovskite stabilized phase-pure formamidinium perovskite solar cells, Nat. Commun. 9 (1) (2018) 1–10.

[56] C. Zuo, D. Vak, D. Angmo, L. Ding, M. Gao, One-step roll-to-roll air processed high efficiency perovskite solar cells, Nano Energy 46 (2018) 185–192.

[57] W. Nie, H. Tsai, J.C. Blancon, F. Liu, C.C. Stoumpos, B. Traore, M. Kepenekian, O. Durand, C. Katan, S. Tretiak, Critical role of interface and crystallinity on the performance and photostability of perovskite solar cell on nickel oxide, Adv. Mater. 30 (5) (2018) 1703879.

[58] M. Kim, S.G. Motti, R. Sorrentino, A. Petrozza, Enhanced solar cell stability by hygroscopic polymer passivation of metal halide perovskite thin film, Energy Environ. Sci. 11 (9) (2018) 2609–2619.

[59] X. Zhang, G. Wu, W. Fu, M. Qin, W. Yang, J. Yan, Z. Zhang, X. Lu, H. Chen, Orientation regulation of phenylethylammonium cation based 2D perovskite solar cell with efficiency higher than 11%, Adv. Energy Mater. 8 (14) (2018) 1702498.

[60] N. Ahn, I. Jeon, J. Yoon, E.I. Kauppinen, Y. Matsuo, S. Maruyama, M. Choi, Carbon-sandwiched perovskite solar cell, J. Mater. Chem. 6 (4) (2018) 1382–1389.

[61] N. Ito, M.A. Kamarudin, D. Hirotani, Y. Zhang, Q. Shen, Y. Ogomi, S. Iikubo, T. Minemoto, K. Yoshino, S. Hayase, Mixed Sn–Ge perovskite for enhanced perovskite solar cell performance in air, J. Phys. Chem. Lett. 9 (7) (2018) 1682–1688.

[62] Y. Jiang, M.R. Leyden, L. Qiu, S. Wang, L.K. Ono, Z. Wu, E.J. Juarez-Perez, Y. Qi, Combination of hybrid CVD and cation exchange for upscaling Cs-substituted mixed cation perovskite solar cells with high efficiency and stability, Adv. Funct. Mater. 28 (1) (2018) 1703835.

[63] E.H. Jung, N.J. Jeon, E.Y. Park, C.S. Moon, T.J. Shin, T.-Y. Yang, J.H. Noh, J. Seo, Efficient, stable and scalable perovskite solar cells using poly (3-hexylthiophene), Nature 567 (7749) (2019) 511–515.

[64] J. Liu, Q. Zhou, N.K. Thein, L. Tian, D. Jia, E.M.J. Johansson, X. Zhang, In situ growth of perovskite stacking layers for high-efficiency carbon-based hole conductor free perovskite solar cells, J. Mater. Chem. 7 (22) (2019) 13777–13786.

[65] X. Zhang, Y. Zhou, Y. Li, J. Sun, X. Lu, X. Gao, J. Gao, L. Shui, S. Wu, J.-M. Liu, Efficient and carbon-based hole transport layer-free CsPbI2Br planar perovskite solar cells using PMMA modification, J. Mater. Chem. C 7 (13) (2019) 3852–3861.

[66] F. Qian, S. Yuan, Y. Cai, Y. Han, H. Zhao, J. Sun, Z. Liu, S. Liu, Novel surface passivation for stable FA0.85MA0.15PbI3 perovskite solar cells with 21.6% efficiency, Solar RRL 3 (7) (2019) 1900072.

[67] W. Zhang, X. Liu, B. He, Z. Gong, J. Zhu, Y. Ding, H. Chen, Q. Tang, Interface engineering of imidazolium ionic liquids toward efficient and stable CsPbBr3 perovskite solar cells, ACS Appl. Mater. Interfaces 12 (4) (2020) 4540–4548.

[68] R. Ishikawa, K. Ueno, H. Shirai, Improved efficiency of methylammonium-free perovskite thin film solar cells by fluorinated ammonium iodide treatment, Org. Electron. 78 (2020) 105596.

[69] B. Zhao, S.-F. Jin, S. Huang, N. Liu, J.-Y. Ma, D.-J. Xue, Q. Han, J. Ding, Q.-Q. Ge, Y. Feng, Thermodynamically stable orthorhombic γ-CsPbI3 thin films for high-performance photovoltaics, J. Am. Chem. Soc. 140 (37) (2018) 11716–11725.

[70] P. Wang, X. Zhang, Y. Zhou, Q. Jiang, Q. Ye, Z. Chu, X. Li, X. Yang, Z. Yin, J. You, Solvent-controlled growth of inorganic perovskite films in dry environment for efficient and stable solar cells, Nat. Commun. 9 (1) (2018) 1–7.

[71] Y. Wang, T. Zhang, M. Kan, Y. Zhao, Bifunctional stabilization of all-inorganic α-CsPbI3 perovskite for 17% efficiency photovoltaics, J. Am. Chem. Soc. 140 (39) (2018) 12345–12348.

[72] C. Liu, W. Li, C. Zhang, Y. Ma, J. Fan, Y. Mai, All-inorganic CsPbI2Br perovskite solar cells with high efficiency exceeding 13%, J. Am. Chem. Soc. 140 (11) (2018) 3825–3828.

[73] T. Miyasaka, A. Kulkarni, G.M. Kim, S. Öz, A.K. Jena, Perovskite solar cells: can we go organic-free, lead-free, and dopant-free? Adv. Energy Mater. (2019) 1902500.

II. Photovoltaic materials and devices

276
9. Perovskite solar cells

[74] W. Li, J. Li, J. Li, J. Fan, Y. Mai, L. Wang, Addictive-assisted construction of all-inorganic CsSnIBr 2 mesoscopic perovskite solar cells with superior thermal stability up to 473 K, J. Mater. Chem. 4 (43) (2016) 17104–17110.

[75] L.-J. Chen, C.-R. Lee, Y.-J. Chuang, Z.-H. Wu, C. Chen, Synthesis and optical properties of lead-free cesium tin halide perovskite quantum rods with high-performance solar cell application, J. Phys. Chem. Lett. 7 (24) (2016) 5028–5035.

[76] A.K. Jena, A. Kulkarni, T. Miyasaka, Halide perovskite photovoltaics: background, status, and future prospects, Chem. Rev. 119 (5) (2019) 3036–3103.

[77] M.-G. Ju, J. Dai, L. Ma, X.C. Zeng, Lead-free mixed tin and germanium perovskites for photovoltaic application, J. Am. Chem. Soc. 139 (23) (2017) 8038–8043.

[78] M.I.H. Ansari, A. Qurashi, M.K. Nazeeruddin, Frontiers, opportunities, and challenges in perovskite solar cells: a critical review, J. Photochem. Photobiol. C Photochem. Rev. 35 (2018) 1–24.

[79] C.C. Stoumpos, C.D. Malliakas, M.G. Kanatzidis, Semiconducting tin and lead iodide perovskites with organic cations: phase transitions, high mobilities, and near-infrared photoluminescent properties, Inorg. Chem. 52 (15) (2013) 9019–9038.

[80] N. Pellet, P. Gao, G. Gregori, T.Y. Yang, M.K. Nazeeruddin, J. Maier, M. Grätzel, Mixed-organic-cation Perovskite photovoltaics for enhanced solar-light harvesting, Angew. Chem. Int. Ed. 53 (12) (2014) 3151–3157.

[81] Z. Li, M. Yang, J.-S. Park, S.-H. Wei, J.J. Berry, K. Zhu, Stabilizing perovskite structures by tuning tolerance factor: formation of formamidinium and cesium lead iodide solid-state alloys, Chem. Mater. 28 (1) (2016) 284–292.

[82] L.N. Quan, M. Yuan, R. Comin, O. Voznyy, E.M. Beauregard, S. Hoogland, A. Buin, A.R. Kirmani, K. Zhao, A. Amassian, Ligand-stabilized reduced-dimensionality perovskites, J. Am. Chem. Soc. 138 (8) (2016) 2649–2655.

[83] F. Hao, C.C. Stoumpos, R.P. Chang, M.G. Kanatzidis, Anomalous band gap behavior in mixed Sn and Pb perovskites enables broadening of absorption spectrum in solar cells, J. Am. Chem. Soc. 136 (22) (2014) 8094–8099.

[84] E. Mosconi, A. Amat, M.K. Nazeeruddin, M. Grätzel, F. De Angelis, First-principles modeling of mixed halide organometal perovskites for photovoltaic applications, J. Phys. Chem. C 117 (27) (2013) 13902–13913.

[85] T.J. Jacobsson, M. Pazoki, A. Hagfeldt, T. Edvinsson, Goldschmidt's rules and strontium replacement in lead halogen perovskite solar cells: theory and preliminary experiments on $CH_3NH_3SrI_3$, J. Phys. Chem. C 119 (46) (2015) 25673–25683.

[86] J. Ali, Y. Li, P. Gao, T. Hao, J. Song, Q. Zhang, L. Zhu, J. Wang, W. Feng, H. Hu, Interfacial and structural modifications in perovskite solar cells, Nanoscale 12 (10) (2020) 5719–5745.

[87] A. Karavioti, E. Vitoratos, E. Stathatos, Improved performance and stability of hole-conductor-free mesoporous perovskite solar cell with new amino-acid iodide cations, J. Mater. Sci. Mater. Electron. (2020) 1–9.

[88] Y. Xiang, J. Zhuang, Z. Ma, H. Lu, H. Xia, W. Zhou, T. Zhang, H. Li, Mixed-phase mesoporous TiO 2 film for high efficiency perovskite solar cells, Chem. Res. Chin. Univ. 35 (1) (2019) 101–108.

[89] J.J. Choi, X. Yang, Z.M. Norman, S.J. Billinge, J.S. Owen, Structure of methylammonium lead iodide within mesoporous titanium dioxide: active material in high-performance perovskite solar cells, Nano Lett. 14 (1) (2014) 127–133.

[90] T. Leijtens, G.E. Eperon, S. Pathak, A. Abate, M.M. Lee, H.J. Snaith, Overcoming ultraviolet light instability of sensitized TiO 2 with meso-superstructured organometal tri-halide perovskite solar cells, Nat. Commun. 4 (1) (2013) 1–8.

[91] T. Leijtens, B. Lauber, G.E. Eperon, S.D. Stranks, H.J. Snaith, The importance of perovskite pore filling in organometal mixed halide sensitized TiO2-based solar cells, J. Phys. Chem. Lett. 5 (7) (2014) 1096–1102.

[92] H.J. Snaith, A. Abate, J.M. Ball, G.E. Eperon, T. Leijtens, N.K. Noel, S.D. Stranks, J.T.-W. Wang, K. Wojciechowski, W. Zhang, Anomalous hysteresis in perovskite solar cells, J. Phys. Chem. Lett. 5 (9) (2014) 1511–1515.

[93] M. Kim, G.-H. Kim, T.K. Lee, I.W. Choi, H.W. Choi, Y. Jo, Y.J. Yoon, J.W. Kim, J. Lee, D. Huh, H. Lee, S.K. Kwak, J.Y. Kim, D.S. Kim, Methylammonium chloride induces intermediate phase stabilization for efficient perovskite solar cells, Joule 3 (9) (2019) 2179–2192.

[94] H. Zhou, Q. Chen, G. Li, S. Luo, T.-b. Song, H.-S. Duan, Z. Hong, J. You, Y. Liu, Y. Yang, Interface engineering of highly efficient perovskite solar cells, Science 345 (6196) (2014) 542–546.

[95] N.J. Jeon, H. Na, E.H. Jung, T.-Y. Yang, Y.G. Lee, G. Kim, H.-W. Shin, S.I. Seok, J. Lee, J. Seo, A fluorene-terminated hole-transporting material for highly efficient and stable perovskite solar cells, Nat. Energy 3 (8) (2018) 682–689.

II. Photovoltaic materials and devices

References

[96] Y. Zheng, J. Kong, D. Huang, W. Shi, L. McMillon-Brown, H.E. Katz, J. Yu, A.D. Taylor, Spray coating of the PCBM electron transport layer significantly improves the efficiency of pin planar perovskite solar cells, Nanoscale 10 (24) (2018) 11342–11348.

[97] J. Sun, J. Lu, B. Li, L. Jiang, A.S. Chesman, A.D. Scully, T.R. Gengenbach, Y.-B. Cheng, J.J. Jasieniak, Inverted perovskite solar cells with high fill-factors featuring chemical bath deposited mesoporous NiO hole transporting layers, Nano Energy 49 (2018) 163–171.

[98] A.M. Elseman, M.S. Selim, L. Luo, C.Y. Xu, G. Wang, Y. Jiang, D.B. Liu, L.P. Liao, Z. Hao, Q.L. Song, Efficient and stable planar n-i-p perovskite solar cells with negligible hysteresis through solution-processed Cu_2O nanocubes as a low-cost hole-transport material, ChemSusChem 12 (16) (2019) 3808–3816.

[99] I. Zimmermann, P. Gratia, D. Martineau, G. Grancini, J.-N. Audinot, T. Wirtz, M.K. Nazeeruddin, Improved efficiency and reduced hysteresis in ultra-stable fully printable mesoscopic perovskite solar cells through incorporation of CuSCN into the perovskite layer, J. Mater. Chem. 7 (14) (2019) 8073–8077.

[100] H. Chen, Q. Wei, M.I. Saidaminov, F. Wang, A. Johnston, Y. Hou, Z. Peng, K. Xu, W. Zhou, Z. Liu, Efficient and stable inverted perovskite solar cells incorporating secondary amines, Adv. Mater. 31 (46) (2019) 1903559.

[101] N. NREL, Best Research-Cell Efficiencies, National Renewable Energy Laboratory, Golden, Colorado, 2019.

[102] Y. Wu, F. Xie, H. Chen, X. Yang, H. Su, M. Cai, Z. Zhou, T. Noda, L. Han, Thermally stable MAPbI3 perovskite solar cells with efficiency of 19.19% and area over 1 cm^2 achieved by additive engineering, Adv. Mater. 29 (28) (2017) 1701073.

[103] D. Luo, W. Yang, Z. Wang, A. Sadhanala, Q. Hu, R. Su, R. Shivanna, G.F. Trindade, J.F. Watts, Z. Xu, Enhanced photovoltage for inverted planar heterojunction perovskite solar cells, Science 360 (6396) (2018) 1442–1446.

[104] A. Bera, A.D. Sheikh, M.A. Haque, R. Bose, E. Alarousu, O.F. Mohammed, T. Wu, Fast crystallization and improved stability of perovskite solar cells with Zn_2SnO_4 electron transporting layer: interface matters, ACS Appl. Mater. Interfaces 7 (51) (2015) 28404–28411.

[105] F. Xia, Q. Wu, P. Zhou, Y. Li, X. Chen, Q. Liu, J. Zhu, S. Dai, Y. Lu, S. Yang, Efficiency enhancement of inverted structure perovskite solar cells via oleamide doping of PCBM electron transport layer, ACS Appl. Mater. Interfaces 7 (24) (2015) 13659–13665.

[106] R. Wang, M. Mujahid, Y. Duan, Z.K. Wang, J. Xue, Y. Yang, A review of perovskites solar cell stability, Adv. Funct. Mater. 29 (47) (2019) 1808843.

[107] Z. Zhu, Y. Bai, X. Liu, C.C. Chueh, S. Yang, A.K.Y. Jen, Enhanced efficiency and stability of inverted perovskite solar cells using highly crystalline SnO_2 nanocrystals as the robust electron-transporting layer, Adv. Mater. 28 (30) (2016) 6478–6484.

[108] X. Liu, C.-C. Chueh, Z. Zhu, S.B. Jo, Y. Sun, A.K.-Y. Jen, Highly crystalline Zn 2 SnO 4 nanoparticles as efficient electron-transporting layers toward stable inverted and flexible conventional perovskite solar cells, J. Mater. Chem. 4 (40) (2016) 15294–15301.

[109] S. Yang, H. Kou, J. Wang, H. Xue, H. Han, Tunability of the band energetics of nanostructured $SrTiO_3$ electrodes for dye-sensitized solar cells, J. Phys. Chem. C 114 (9) (2010) 4245–4249.

[110] D.-B. Li, L. Hu, Y. Xie, G. Niu, T. Liu, Y. Zhou, L. Gao, B. Yang, J. Tang, Low-temperature-processed amorphous Bi2S3 film as an inorganic electron transport layer for perovskite solar cells, ACS Photonics 3 (11) (2016) 2122–2128.

[111] L. Zhu, J. Ye, X. Zhang, H. Zheng, G. Liu, X. Pan, S. Dai, Performance enhancement of perovskite solar cells using a La-doped BaSnO 3 electron transport layer, J. Mater. Chem. 5 (7) (2017) 3675–3682.

[112] Q. Zhang, C.S. Dandeneau, X. Zhou, G. Cao, ZnO nanostructures for dye-sensitized solar cells, Adv. Mater. 21 (41) (2009) 4087–4108.

[113] F.J. Ramos, M.C. López-Santos, E. Guillén, M.K. Nazeeruddin, M. Grätzel, A.R. Gonzalez-Elipe, S. Ahmad, Perovskite solar cells based on nanocolumnar plasma-deposited ZnO thin films, ChemPhysChem 15 (6) (2014) 1148–1153.

[114] P. Zhang, J. Wu, T. Zhang, Y. Wang, D. Liu, H. Chen, L. Ji, C. Liu, W. Ahmad, Z.D. Chen, Perovskite solar cells with ZnO electron-transporting materials, Adv. Mater. 30 (3) (2018) 1703737.

[115] D. Liu, T.L. Kelly, Perovskite solar cells with a planar heterojunction structure prepared using room-temperature solution processing techniques, Nat. Photon. 8 (2) (2014) 133–138.

[116] C.-Z. Li, C.-C. Chueh, H.-L. Yip, J. Zou, W.-C. Chen, A.K.-Y. Jen, Evaluation of structure–property relationships of solution-processible fullerene acceptors and their n-channel field-effect transistor performance, J. Mater. Chem. 22 (30) (2012) 14976–14981.

[117] C.A. Reed, R.D. Bolskar, Discrete fulleride anions and fullerenium cations, Chem. Rev. 100 (3) (2000) 1075–1120.

9. Perovskite solar cells

[118] W.S. Yang, B.-W. Park, E.H. Jung, N.J. Jeon, Y.C. Kim, D.U. Lee, S.S. Shin, J. Seo, E.K. Kim, J.H. Noh, Iodide management in formamidinium-lead-halide—based perovskite layers for efficient solar cells, Science 356 (6345) (2017) 1376—1379.

[119] B. Philippe, B.-W. Park, R. Lindblad, J. Oscarsson, S. Ahmadi, E.M. Johansson, H.k. Rensmo, Chemical and electronic structure characterization of lead halide perovskites and stability behavior under different exposures. A photoelectron spectroscopy investigation, Chem. Mater. 27 (5) (2015) 1720—1731.

[120] Y. Rong, L. Liu, A. Mei, X. Li, H. Han, Beyond efficiency: the challenge of stability in mesoscopic perovskite solar cells, Adv. Energy Mater. 5 (20) (2015) 1501066.

[121] T. Supasai, N. Rujisamphan, K. Ullrich, A. Chemseddine, T. Dittrich, Formation of a passivating $CH_3NH_3PbI_3$/PbI_2 interface during moderate heating of $CH_3NH_3PbI_3$ layers, Appl. Phys. Lett. 103 (18) (2013) 183906.

[122] A. Krishna, A.C. Grimsdale, Hole transporting materials for mesoscopic perovskite solar cells—towards a rational design? J. Mater. Chem. 5 (32) (2017) 16446—16466.

[123] J.-H. Im, C.-R. Lee, J.-W. Lee, S.-W. Park, N.-G. Park, 6.5% efficient perovskite quantum-dot-sensitized solar cell, Nanoscale 3 (10) (2011) 4088—4093.

[124] S. Shao, J. Liu, J. Bergqvist, S. Shi, C. Veit, U. Würfel, Z. Xie, F. Zhang, In situ formation of MoO_3 in PEDOT: PSS matrix: a facile way to produce a smooth and less hygroscopic hole transport layer for highly stable polymer bulk heterojunction solar cells, Adv. Energy Mater. 3 (3) (2013) 349—355.

[125] F. Hou, Z. Su, F. Jin, X. Yan, L. Wang, H. Zhao, J. Zhu, B. Chu, W. Li, Efficient and stable planar heterojunction perovskite solar cells with an MoO 3/PEDOT: PSS hole transporting layer, Nanoscale 7 (21) (2015) 9427—9432.

[126] S. Sun, T. Salim, N. Mathews, M. Duchamp, C. Boothroyd, G. Xing, T.C. Sum, Y.M. Lam, The origin of high efficiency in low-temperature solution-processable bilayer organometal halide hybrid solar cells, Energy Environ. Sci. 7 (1) (2014) 399—407.

[127] K. Mahmood, S. Sarwar, M. Mehran, Current status of electron transport layers in perovskite solar cells: materials and properties, RSC Adv. 7 (28) (2017) 17044—17062.

[128] H. Back, G. Kim, J. Kim, J. Kong, T.K. Kim, H. Kang, H. Kim, J. Lee, S. Lee, K. Lee, Achieving long-term stable perovskite solar cells via ion neutralization, Energy Environ. Sci. 9 (4) (2016) 1258—1263.

[129] Y. Han, S. Meyer, Y. Dkhissi, K. Weber, J.M. Pringle, U. Bach, L. Spiccia, Y.-B. Cheng, Degradation observations of encapsulated planar CH 3 NH 3 PbI 3 perovskite solar cells at high temperatures and humidity, J. Mater. Chem. 3 (15) (2015) 8139—8147.

[130] T.Y. Yang, G. Gregori, N. Pellet, M. Grätzel, J. Maier, The significance of ion conduction in a hybrid organic—inorganic lead-iodide-based perovskite photosensitizer, Angew. Chem. Int. Ed. 54 (27) (2015) 7905—7910.

[131] J.T. Tisdale, E. Muckley, M. Ahmadi, T. Smith, C. Seal, E. Lukosi, I.N. Ivanov, B. Hu, Dynamic impact of electrode materials on interface of single-crystalline methylammonium lead bromide perovskite, Adv. Mater. Interface 5 (18) (2018) 1800476.

[132] H. Wei, Y. Fang, P. Mulligan, W. Chuirazzi, H.-H. Fang, C. Wang, B.R. Ecker, Y. Gao, M.A. Loi, L. Cao, Sensitive X-ray detectors made of methylammonium lead tribromide perovskite single crystals, Nat. Photon. 10 (5) (2016) 333.

[133] A. Guerrero, J. You, C. Aranda, Y.S. Kang, G. Garcia-Belmonte, H. Zhou, J. Bisquert, Y. Yang, Interfacial degradation of planar lead halide perovskite solar cells, ACS Nano 10 (1) (2016) 218—224.

[134] K. Domanski, J.-P. Correa-Baena, N. Mine, M.K. Nazeeruddin, A. Abate, M. Saliba, W. Tress, A. Hagfeldt, M. Grätzel, Not all that glitters is gold: metal-migration-induced degradation in perovskite solar cells, ACS Nano 10 (6) (2016) 6306—6314.

[135] A. Mei, X. Li, L. Liu, Z. Ku, T. Liu, Y. Rong, M. Xu, M. Hu, J. Chen, Y. Yang, A hole-conductor—free, fully printable mesoscopic perovskite solar cell with high stability, Science 345 (6194) (2014) 295—298.

[136] Y. Yuan, Q. Wang, Y. Shao, H. Lu, T. Li, A. Gruverman, J. Huang, Electric-field-driven reversible conversion between methylammonium lead triiodide perovskites and lead iodide at elevated temperatures, Adv. Energy Mater. 6 (2) (2016) 1501803.

[137] T. Zhang, S.H. Cheung, X. Meng, L. Zhu, Y. Bai, C.H.Y. Ho, S. Xiao, Q. Xue, S.K. So, S. Yang, Pinning down the anomalous light soaking effect toward high-performance and fast-response perovskite solar cells: the ion-migration-induced charge accumulation, J. Phys. Chem. Lett. 8 (20) (2017) 5069—5076.

[138] H.-S. Kim, N.-G. Park, Parameters affecting I—V hysteresis of $CH_3NH_3PbI_3$ perovskite solar cells: effects of perovskite crystal size and mesoporous TiO_2 layer, J. Phys. Chem. Lett. 5 (17) (2014) 2927—2934.

References

[139] L.K. Ono, S.R. Raga, S. Wang, Y. Kato, Y. Qi, Temperature-dependent hysteresis effects in perovskite-based solar cells, J. Mater. Chem. 3 (17) (2015) 9074–9080.

[140] E. Zimmermann, K.K. Wong, M. Müller, H. Hu, P. Ehrenreich, M. Kohlstädt, U. Würfel, S. Mastroianni, G. Mathiazhagan, A. Hinsch, Characterization of perovskite solar cells: towards a reliable measurement protocol, Apl. Mater. 4 (9) (2016) 091901.

[141] D.A. Egger, A.M. Rappe, L. Kronik, Hybrid organic–inorganic perovskites on the move, Acc. Chem. Res. 49 (3) (2016) 573–581.

[142] Y. Yuan, J. Huang, Ion migration in organometal trihalide perovskite and its impact on photovoltaic efficiency and stability, Acc. Chem. Res. 49 (2) (2016) 286–293.

[143] M. Abdi-Jalebi, Z. Andaji-Garmaroudi, S. Cacovich, C. Stavrakas, B. Philippe, J.M. Richter, M. Alsari, E.P. Booker, E.M. Hutter, A.J. Pearson, Maximizing and stabilizing luminescence from halide perovskites with potassium passivation, Nature 555 (7697) (2018) 497–501.

[144] M.I. Saidaminov, J. Almutlaq, S. Sarmah, I. Dursun, A.A. Zhumekenov, R. Begum, J. Pan, N. Cho, O.F. Mohammed, O.M. Bakr, Pure Cs_4PbBr_6: highly luminescent zero-dimensional perovskite solids, ACS Energy Lett. 1 (4) (2016) 840–845.

[145] Y. Wang, J. He, H. Chen, J. Chen, R. Zhu, P. Ma, A. Towers, Y. Lin, A.J. Gesquiere, S.T. Wu, Ultrastable, highly luminescent organic–inorganic perovskite–polymer composite films, Adv. Mater. 28 (48) (2016) 10710–10717.

[146] C. Albrecht, J.R. Lakowicz, Principles of fluorescence spectroscopy, Anal. Bioanal. Chem. 390 (5) (2008) 1223–1224.

[147] T. Handa, D.M. Tex, A. Shimazaki, A. Wakamiya, Y. Kanemitsu, Charge injection mechanism at heterointerfaces in $CH_3NH_3PbI_3$ perovskite solar cells revealed by simultaneous time-resolved photoluminescence and photocurrent measurements, J. Phys. Chem. Lett. 8 (5) (2017) 954–960.

[148] D.M. Tex, M. Imaizumi, Y. Kanemitsu, Analyzing the electrical performance of a solar cell with time-resolved photoluminescence: methodology for fast optical screening, Phys. Rev. Appl. 7 (1) (2017) 014019.

[149] D.M. Tex, T. Ihara, H. Akiyama, M. Imaizumi, Y. Kanemitsu, Time-resolved photoluminescence measurements for determining voltage-dependent charge-separation efficiencies of subcells in triple-junction solar cells, Appl. Phys. Lett. 106 (1) (2015) 013905.

[150] S.D. Stranks, V.M. Burlakov, T. Leijtens, J.M. Ball, A. Goriely, H.J. Snaith, Recombination kinetics in organic-inorganic perovskites: excitons, free charge, and subgap states, Phys. Rev. Appl. 2 (3) (2014) 034007.

[151] D.W. DeQuilettes, W. Zhang, V.M. Burlakov, D.J. Graham, T. Leijtens, A. Osherov, V. Bulović, H.J. Snaith, D.S. Ginger, S.D. Stranks, Photo-induced halide redistribution in organic–inorganic perovskite films, Nat. Commun. 7 (1) (2016) 1–9.

[152] Z. Xiao, C. Bi, Y. Shao, Q. Dong, Q. Wang, Y. Yuan, C. Wang, Y. Gao, J. Huang, Efficient, high yield perovskite photovoltaic devices grown by interdiffusion of solution-processed precursor stacking layers, Energy Environ. Sci. 7 (8) (2014) 2619–2623.

[153] J.-S. Yeo, R. Kang, S. Lee, Y.-J. Jeon, N. Myoung, C.-L. Lee, D.-Y. Kim, J.-M. Yun, Y.-H. Seo, S.-S. Kim, S.-I. Na, Highly efficient and stable planar perovskite solar cells with reduced graphene oxide nanosheets as electrode interlayer, Nano Energy 12 (2015) 96–104.

[154] S. Cacovich, L. Cinà, F. Matteocci, G. Divitini, P.A. Midgley, A. Di Carlo, C. Ducati, Gold and iodine diffusion in large area perovskite solar cells under illumination, Nanoscale 9 (14) (2017) 4700–4706.

[155] Z. Fan, H. Xiao, Y. Wang, Z. Zhao, Z. Lin, H.-C. Cheng, S.-J. Lee, G. Wang, Z. Feng, W.A. Goddard III, Layer-by-layer degradation of methylammonium lead tri-iodide perovskite microplates, Joule 1 (3) (2017) 548–562.

[156] Y. Zhou, H. Sternlicht, N.P. Padture, Transmission electron microscopy of halide perovskite materials and devices, Joule 3 (3) (2019) 641–661.

[157] Z. Dang, J. Shamsi, Q.A. Akkerman, M. Imran, G. Bertoni, R. Brescia, L. Manna, Low-temperature electron beam-induced transformations of cesium lead halide perovskite nanocrystals, ACS Omega 2 (9) (2017) 5660–5665.

[158] D. Zhang, Y. Zhu, L. Liu, X. Ying, C.-E. Hsiung, R. Sougrat, K. Li, Y. Han, Atomic-resolution transmission electron microscopy of electron beam–sensitive crystalline materials, Science 359 (6376) (2018) 675–679.

[159] D.M. Saylor, B. El Dasher, T. Sano, G.S. Rohrer, Distribution of grain boundaries in SrTiO3 as a function of five macroscopic parameters, J. Am. Ceram. Soc. 87 (4) (2004) 670–676.

[160] J. Xu, A. Buin, A.H. Ip, W. Li, O. Voznyy, R. Comin, M. Yuan, S. Jeon, Z. Ning, J.J. McDowell, Perovskite–fullerene hybrid materials suppress hysteresis in planar diodes, Nat. Commun. 6 (1) (2015) 1–8.

II. Photovoltaic materials and devices

9. Perovskite solar cells

[161] S.D. Stranks, G.E. Eperon, G. Grancini, C. Menelaou, M.J. Alcocer, T. Leijtens, L.M. Herz, A. Petrozza, H.J. Snaith, Electron-hole diffusion lengths exceeding 1 micrometer in an organometal trihalide perovskite absorber, Science 342 (6156) (2013) 341–344.

[162] J.H. Noh, S.H. Im, J.H. Heo, T.N. Mandal, S.I. Seok, Chemical management for colorful, efficient, and stable inorganic–organic hybrid nanostructured solar cells, Nano Lett. 13 (4) (2013) 1764–1769.

[163] J. Burschka, N. Pellet, S.-J. Moon, R. Humphry-Baker, P. Gao, M.K. Nazeeruddin, M. Grätzel, Sequential deposition as a route to high-performance perovskite-sensitized solar cells, Nature 499 (7458) (2013) 316–319.

[164] M. Liu, M.B. Johnston, H.J. Snaith, Efficient planar heterojunction perovskite solar cells by vapour deposition, Nature 501 (7467) (2013) 395–398.

[165] J.-H. Im, J. Chung, S.-J. Kim, N.-G. Park, Synthesis, structure, and photovoltaic property of a nanocrystalline 2H perovskite-type novel sensitizer (CH 3 CH 2 NH 3) PbI 3, Nanoscale Res. Lett. 7 (1) (2012) 353.

[166] S. Pang, H. Hu, J. Zhang, S. Lv, Y. Yu, F. Wei, T. Qin, H. Xu, Z. Liu, G. Cui, $NH_2CH\ NH_2PbI_3$: an alternative organolead iodide perovskite sensitizer for mesoscopic solar cells, Chem. Mater. 26 (3) (2014) 1485–1491.

[167] K. Mahmood, B.S. Swain, A.R. Kirmani, A. Amassian, Highly efficient perovskite solar cells based on a nanostructured WO 3–TiO 2 core–shell electron transporting material, J. Mater. Chem. 3 (17) (2015) 9051–9057.

[168] A.K. Chandiran, M. Abdi-Jalebi, A. Yella, M.I. Dar, C. Yi, S.A. Shivashankar, M.K. Nazeeruddin, M. Grätzel, Quantum-confined ZnO nanoshell photoanodes for mesoscopic solar cells, Nano Lett. 14 (3) (2014) 1190–1195.

[169] G.S. Han, H.S. Chung, B.J. Kim, D.H. Kim, J.W. Lee, B.S. Swain, K. Mahmood, J.S. Yoo, N.-G. Park, J.H. Lee, Retarding charge recombination in perovskite solar cells using ultrathin MgO-coated TiO 2 nanoparticulate films, J. Mater. Chem. 3 (17) (2015) 9160–9164.

[170] L. Wang, W. Fu, Z. Gu, C. Fan, X. Yang, H. Li, H. Chen, Low temperature solution processed planar heterojunction perovskite solar cells with a CdSe nanocrystal as an electron transport/extraction layer, J. Mater. Chem. C 2 (43) (2014) 9087–9090.

[171] Q. Wang, Y. Shao, Q. Dong, Z. Xiao, Y. Yuan, J. Huang, Large fill-factor bilayer iodine perovskite solar cells fabricated by a low-temperature solution-process, Energy Environ. Sci. 7 (7) (2014) 2359–2365.

[172] L.-C. Chen, J.-C. Chen, C.-C. Chen, C.-G. Wu, Fabrication and properties of high-efficiency perovskite/PCBM organic solar cells, Nanoscale Res. Lett. 10 (1) (2015) 312.

[173] C. Li, F. Wang, J. Xu, J. Yao, B. Zhang, C. Zhang, M. Xiao, S. Dai, Y. Li, Z.a. Tan, Efficient perovskite/fullerene planar heterojunction solar cells with enhanced charge extraction and suppressed charge recombination, Nanoscale 7 (21) (2015) 9771–9778.

[174] W. Ke, G. Fang, J. Wan, H. Tao, Q. Liu, L. Xiong, P. Qin, J. Wang, H. Lei, G. Yang, Efficient hole-blocking layer-free planar halide perovskite thin-film solar cells, Nat. Commun. 6 (1) (2015) 1–7.

[175] M.A. Green, E.D. Dunlop, J. Hohl-Ebinger, M. Yoshita, N. Kopidakis, A.W. Ho-Baillie, Solar cell efficiency tables (version 55), Prog. Photovoltaics Res. Appl. 28 (2019) (NREL/JA-5900-75827).

[176] T. Miyasaka, Lead halide perovskites in thin film photovoltaics: background and perspectives, Bull. Chem. Soc. Jpn. 91 (7) (2018) 1058–1068.

[177] H.D. Kim, H. Ohkita, Potential improvement in fill factor of lead-halide perovskite solar cells (solar RRL 7/2017), Solar RRL 1 (6) (2017) 1770121.

[178] W.S. Yang, J.H. Noh, N.J. Jeon, Y.C. Kim, S. Ryu, J. Seo, S.I. Seok, High-performance photovoltaic perovskite layers fabricated through intramolecular exchange, Science 348 (6240) (2015) 1234–1237.

[179] Q. Jiang, Z. Chu, P. Wang, X. Yang, H. Liu, Y. Wang, Z. Yin, J. Wu, X. Zhang, J. You, Planar-structure perovskite solar cells with efficiency beyond 21%, Adv. Mater. 29 (46) (2017) 1703852.

[180] K.A. Bush, A.F. Palmstrom, J.Y. Zhengshan, M. Boccard, R. Cheacharoen, J.P. Mailoa, D.P. McMeekin, R.L. Hoye, C.D. Bailie, T. Leijtens, 23.6%-efficient monolithic perovskite/silicon tandem solar cells with improved stability, Nat. Energy 2 (4) (2017) 1–7.

[181] W. Zhu, C. Bao, F. Li, T. Yu, H. Gao, Y. Yi, J. Yang, G. Fu, X. Zhou, Z. Zou, A halide exchange engineering for $CH_3NH_3PbI_3-$ xBrx perovskite solar cells with high performance and stability, Nano Energy 19 (2016) 17–26.

[182] J. You, L. Meng, T.-B. Song, T.-F. Guo, Y.M. Yang, W.-H. Chang, Z. Hong, H. Chen, H. Zhou, Q. Chen, Improved air stability of perovskite solar cells via solution-processed metal oxide transport layers, Nat. Nanotechnol. 11 (1) (2016) 75.

[183] G. Niu, W. Li, F. Meng, L. Wang, H. Dong, Y. Qiu, Study on the stability of CH 3 NH 3 PbI 3 films and the effect of post-modification by aluminum oxide in all-solid-state hybrid solar cells, J. Mater. Chem. 2 (3) (2014) 705–710.

II. Photovoltaic materials and devices

References

[184] S. Shao, Z. Chen, H.-H. Fang, G. Ten Brink, D. Bartesaghi, S. Adjokatse, L. Koster, B. Kooi, A. Facchetti, M. Loi, N-type polymers as electron extraction layers in hybrid perovskite solar cells with improved ambient stability, J. Mater. Chem. 4 (7) (2016) 2419–2426.

[185] H. Baig, H. Kanda, A.M. Asiri, M.K. Nazeeruddin, T. Mallick, Increasing efficiency of perovskite solar cells using low concentrating photovoltaic systems, Sustain. Energy Fuels 4 (2020) 528–537.

[186] Y. Li, J. Shi, J. Zheng, J. Bing, J. Yuan, Y. Cho, S. Tang, M. Zhang, Y. Yao, C.F.J. Lau, Acetic acid assisted crystallization strategy for high efficiency and long-term stable perovskite solar cell, Adv. Sci. 7 (5) (2020) 1903368.

[187] Y. Zhao, J. Duan, Y. Wang, X. Yang, Q. Tang, Precise stress control of inorganic perovskite films for carbon-based solar cells with an ultrahigh voltage of 1.622 V, Nano Energy 67 (2020) 104286.

[188] Z. Tang, T. Bessho, F. Awai, T. Kinoshita, M.M. Maitani, R. Jono, T.N. Murakami, H. Wang, T. Kubo, S. Uchida, Hysteresis-free perovskite solar cells made of potassium-doped organometal halide perovskite, Sci. Rep. 7 (1) (2017) 1–7.

[189] D.-Y. Son, S.-G. Kim, J.-Y. Seo, S.-H. Lee, H. Shin, D. Lee, N.-G. Park, Universal approach toward hysteresis-free perovskite solar cell via defect engineering, J. Am. Chem. Soc. 140 (4) (2018) 1358–1364.

[190] T. Singh, S. Öz, A. Sasinska, R. Frohnhoven, S. Mathur, T. Miyasaka, Sulfate-Assisted interfacial engineering for high yield and efficiency of triple cation perovskite solar cells with alkali-doped TiO_2 electron-transporting layers, Adv. Funct. Mater. 28 (14) (2018) 1706287.

SECTION III

Electrochemical energy conversion and storage

CHAPTER

10

Layered and spinel structures as lithium-intercalated compounds for cathode materials

Z.I. Radzi[1], B. Vengadaesvaran[1], S. Ramesh[2], N.A. Rahim[1]

[1]Higher Institution Centre of Excellence (HICoE), UM Power Energy Dedicated Advanced Center (UMPEDAC), Level 4, Wisma R&D, University of Malaya, Jalan Pantai Baharu, Kuala Lumpur, Malaysia; [2]Center for Ionics University of Malaya, Department of Physics, Faculty of Science, University of Malaya, Kuala Lumpur, Malaysia

10.1 Introduction

To possess excellent electrochemical performance in lithium−ion batteries, several characteristics of cathode materials can be classified as ideal, as shown in Fig. 10.1 [1].

It is almost impossible for cathode materials to have all the described ideal characteristics. Battery manufacturers need to review all aspects to meet the growing demands of lithium−ion batteries. To date, only three parent types of cathode material have been used. In the early 1980s, John B. Goodenough, who recently won the Nobel Prize for Chemistry pioneered the practical use of lithium cobalt oxide ($LiCoO_2$) as lithium-intercalated compounds in lithium−ion batteries with high-energy density [2]. In 1984, spinel $LiMn_2O_4$ was discovered by Thackeray's research group, and was also used for other cathode materials for lithium−ions batteries [3]. Later in 1997, $LiFePO_4$ was developed by John B. Goodenough as an alternative choice for $LiCoO_2$ cathode material [4]. After various findings, cathode materials can be classified into three basic materials: layered $LiMO_2$, spinel LiM_2O_4, and olivine $LiMPO_4$, where M = transition metal (TM) elements such as Co, Ni, and Mn. Fig. 10.2 [5] and Table 10.1 show the electrochemical properties of cathode materials. Due to the high demand for high-voltage cathode materials, a compressive summary of layered and spinel structures is discussed in this chapter.

Energy Materials
https://doi.org/10.1016/B978-0-12-823710-6.00005-4

10. Layered and spinel structures as lithium-intercalated compounds for cathode materials

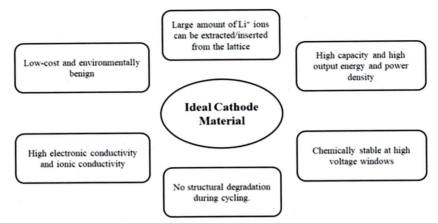

FIGURE 10.1 The characteristics of ideal cathode materials.

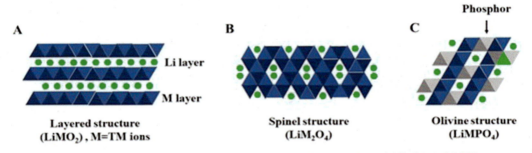

FIGURE 10.2 Models of (A) layered LiMO$_2$, (B) spinel LiM$_2$O$_4$, and (C) olivine LiMPO$_4$.

TABLE 10.1 Comparison of cathode materials for lithium—ion batteries.

Cathode material	LiCoO$_2$	LiNi$_x$Co$_y$Mn$_z$O$_2$	LiNi$_x$Co$_y$Al$_z$O$_2$	LiMn$_2$O$_4$	LiFePO$_4$
Structure		Layered		Spinel	Olivine
Specific capacity (mAh/g)	272	272	160	148	170
Operating voltage (V)	3.7	3.6	3.6	4	3.2
Stability	Good	Rather good	Poor	Good	Very good
Cycle life	Good	Medium	Good	Poor	Good
Application	Small	Small, medium, and large	Medium	Medium and large	Medium and large

III. Electrochemical energy conversion and storage

10.2 Layered structures

10.2.1 Lithium cobalt oxide

LiCoO$_2$ was discovered by John B. Goodenough due to the ability of Li$^+$ ions to be extracted/inserted through electrochemical reaction; additionally, the process was reversible. Thus the attraction of LiCoO$_2$ as a feasible cathode material was proposed for the first commercialization of Sony's lithium—ion batteries [2,6]. LiCoO$_2$ replaced lithium metal batteries because of its ideal safety property. LiCoO$_2$ can be synthesized into two structures depending on the treatment temperatures. LiCoO$_2$ prepared at high temperature (>800°C) has an O3-type layered structure, which is usually designated as HT-LiCoO$_2$, as shown in Fig. 10.3. Another structure is LT-LiCoO$_2$, which can be prepared at low temperatures (~400°C). LT-LiCoO$_2$ results in a spinel structure, in which the tetrahedral and octahedral sites are occupied with cation distribution of [Li$_2$]$_{16c}$[Co$_2$]$_{16d}$O$_4$ [7]. However, the poor cycling performance of LT-LiCoO$_2$ makes it unfavorable for cathode materials. If LT-LiCoO$_2$ is heated to 800°C, cations will reorder and adapt to the O3 phase.

O3-type LiCoO$_2$ is the most thermodynamically stable phase, and exhibits an O3 stacking order, in which LiO$_6$ octahedra share edges only with CoO$_6$ octahedra in the layer above. Extraction of Li$_x$CoO$_2$ results in several phase changes associated with the variation of lattice parameters [8]. The first order transition occurs when the insulating phase of Li$_x$CoO$_2$ transforms into the metallic Li$_x$CoO$_2$ phase in the region between $0.75 \leq x \leq 0.93$. This involves a large expansion of the c-lattice parameter of hexagonal structure due to an increase in interlayer distance between layers during Li removal, resulting from the repulsive forces between negatively charged oxygen ions [9—12]. The reversibility capacity of O3-LiCoO$_2$ is limited when $x = 0.5$, which gives a practical specific capacity of about 140 mAh/g [13]. This is attributed to the phase transformation from hexagonal O3- to O1-type monoclinic Li$_x$CoO$_2$, resulting from Li/vacancy ordering at $x = 0.5$ [2,14,15]. During deintercalation, the mixture of

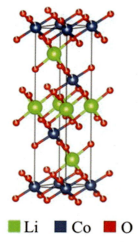

■ Li   ■ Co   ■ O

FIGURE 10.3 The crystal structure of LiCoO$_2$ in a hexagonal cell unit. *TM*, Transition metal.

O3- and O1-type structures of Li_xCoO_2 is formed, known as the H1−3 structure [10,11]. With further deintercalation ($x < 0.5$), this stable monoclinic O1 mixes with a metastable end member of the CoO_2 phase with an O1-type structure. Inside the structure of CoO_2, O−Co−O layers are held together by van der Waals force to form an ABAB stacking sequence with close-packed oxygen layers [16]. An abrupt lattice distortion occurs at the c-axis, indicating an activated stabilization mechanism present in Li_xCoO_2, which results in the stabilization of a delithiated CoO_2 structure. At $x = 0$, the O1-Li_xCoO_2 structure is often mixed with the metastable P3-type Li_xCoO_2.

O2-$LiCoO_2$ is a metastable phase that was first introduced by Delmas et al. The O2-type structure can only be prepared by ion exchange from P2-$Na_{0.7}CoO_2$ due to a thermodynamically unstable phase [17]. LiO_6 and CoO_6 octahedra share edges and faces in the O2 structure [17−19]. Extensive studies have been conducted to investigate the structural changes during the lithiation/delithiation process. In particular, Carlier et al. reported the phase transformation of O2-$LiCoO_2$ supported by ex situ X-ray diffraction (XRD) measurements and/or electron diffraction analysis [19−22]. This O2-$LiCoO_2$ has limited reversibility due to phase transformation until $x = 0.5$, similar to O3-type structures. However, hexagonal O2 reversibly transforms to the orthorhombic $T^{\#}2$ phase in the region between $0.52 \leq x \leq 0.72$ through several single-phase and two-phase reactions [20−22]. With the formation of O6-type Li_xCoO_2 during further Li removal between $0.33 \leq x \leq 0.42$, the hexagonal O2* phase reappears at the end of the process [20]. The overall process of phase transformations of O2- and O3-type structures of $LiCoO_2$ can be represented by the schematic diagram in Fig. 10.4. Mendiboure et al. suggested that O2-$LiCoO_2$ can evolve into O3-$LiCoO_2$ by heat treatment at 400°C [23]. In terms of electrochemical properties, there is no "winner" between O2- and O3-type layered structures. From differential scanning calorimetry (DSC) measurements done by Paulsen et al. [19], O2-$LiCoO_2$ charged to 4.2 V does not induce any serious problems, and this concludes that the O2 system is as safe as the O3 system at the same cutoff voltage. However, O2-$LiCoO_2$ is quite difficult to synthesize.

Kinetic Monte Carlo simulations done by Van der Ven and Ceder predicted that the lithium diffusion coefficient of $LiCoO_2$ is anticipated with a divacancy mechanism and remains valid at high Li concentration where the divacancy is low. Lithium diffusion coefficient is highly affected by varying Li concentrations due to the change of electronic properties inside Li_xCoO_2. The divacancy mechanism is illustrated in Fig. 10.5, which allows Li^+ ions to pass through an adjacent tetrahedral site with the lowest energy barrier [24,25].

$LiCoO_2$ suffers limited reversible capacity since only 0.5 Li^+ ions can be extracted/inserted without inducing structural transformation that impedes lithium diffusion [26]. Besides, it has been observed that the presence of hydrofluoric acid (HF) in electrolyte solution can cause serious dissolution of active material into the electrolyte. It originates from the decomposition of $LiPF_6$, triggered by the presence of water species in electrolyte. Thus low specific capacity and high capacity fading of $LiCoO_2$ are the primary concerns that need to be resolved for improvement by many approaches.

Modification of the surface of $LiCoO_2$ by covering with a coating layer can improve structural stability and prevent side reactions between the active material and electrolyte during charge/discharge as illustrated in Fig. 10.6A and B [27,28]. Aboulaich et al. developed a coating of fluoride metal, MF_3 (M = Al and Ce), on an $LiCoO_2$ surface with thicknesses of about 12 and 40 nm, respectively. MF_3-coated $LiCoO_2$ delivers a capacity of about

10.2 Layered structures

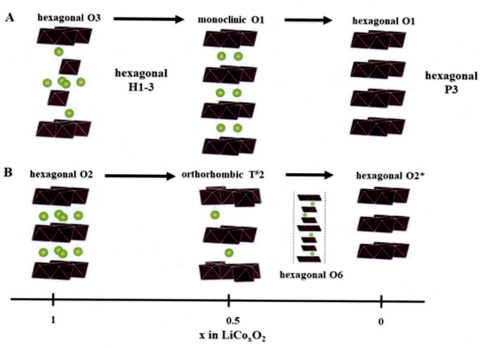

FIGURE 10.4 Illustration of the phase transformations of (A) O3- and (B) O2-type LiCoO$_2$ in the region between $0.0 \leq x \leq 1.0$.

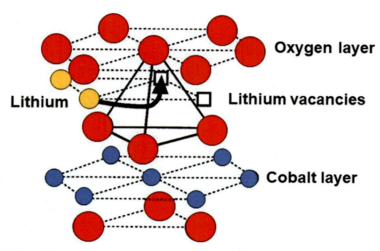

FIGURE 10.5 Schematic diagram of the divacancies mechanism in an LiCoO$_2$ structure.

III. Electrochemical energy conversion and storage

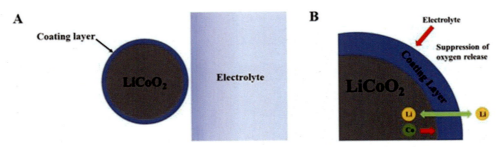

FIGURE 10.6 (A and B) Schematic illustrations of a coating layer to suppress the side reaction between LiCoO$_2$ and electrolyte.

160 mAh/g at 4.6 V cutoff potential, while the pristine LiCoO$_2$ can only achieve 105 mAh/g capacity and undergoes rapid capacity fading under the same conditions [29]. This proves that coating layers can not only retard the side reaction but also enhance the electrochemical properties of LiCoO$_2$ at higher voltage. However, one question that still remains is "Does the coating layer induce changes of lattice parameters of LiCoO$_2$ during cycling and resulting phase transitions?" Cho et al. proposed that the thin film of LiCo$_{1-x}$M$_x$O$_2$ (M = Zr, Al, Ti, and B) solid solution is beneficial to mitigate the lattice parameter changes and cycle-limiting phase transitions of LiCoO$_2$ upon the Li extraction/insertion process. In contrast, Chen et al. found that lattice parameters are almost independent of coated LiCoO$_2$ material upon electrochemical cycling. Unlike pristine LiCoO$_2$, the better electrochemical behavior manifested by ZrO$_2$-coated LiCoO$_2$ could be attributed to the reduction in interfacial area between cathode and electrolyte, preventing oxygen deficiencies [30]. A coating layer is shown to be an effective method to suppress oxygen release [31,32]. Chang et al. mentioned that 0.21 wt% ZnO-coated LiCoO$_2$ showed higher onset temperature of oxygen evolution compared with pristine LiCoO$_2$ confirmed by DSC and thermogravimetric analysis/differential thermal analysis. This indicates that the ZnO layer can inhibit the decomposition reaction of LiCoO$_2$ with the electrolyte and a reduced amount of heat will be generated during cycling [31]. Hence, it is concluded that the coating approach can improve the thermal safety of LiCoO$_2$ at higher voltage. In another study, the variety of coating procedures was also discussed to resolve the high cost and difficulties in integrating coating with cathode material issues. For example, Fey et al. established comparison studies among different procedures for the production of TiO$_2$-coated LiCoO$_2$. Both procedures have shown the suppression of the phase transitions (hexagonal/monoclinic phase transitions) upon cycling. However, the results showed that the mechanothermal coating process exhibited improvement of cyclability compared with the sol-gel coating process. This is because the mechanothermal process used an anatase form of TiO$_2$ rather than the rutile form used in the sol-gel process. Thus, the electrochemical properties of coated LiCoO$_2$ are controlled by the nature form of the coating material developed by mechano-thermal process. In addition, the mechanothermal coating process offered a facile, low-cost, and environmentally friendly process to prepare coated LiCoO$_2$ cathode materials [33].

Instead of integrating a coating layer, metal ion doping of LiCoO$_2$ has received wide attention to minimize Co dissolution due to structural degradation. This could be the reason for the significant variation of lattice parameter c in LiCoO$_2$ [34]. Various metal ions have been studied to understand their electrochemical behavior, such as Ni [35], Mn [36], Cr [37], Al [34], B [26], Rh [38], and Fe [39]. The small amount of metal doping enhances capacity retention after long cycles, prevents the complete extraction of Li$^+$ ions, and therefore inhibits structural damage [1].

In conclusion, layered LiCoO$_2$ has attracted special interest for high-power lithium–ion batteries because of its high theoretical capacity and high volumetric energy density. Therefore, at present, there is extensive effort to achieve 220 mAh/g by focusing on coating layer and metal ion doping methods for high-voltage applications (≥ 4.2 V).

10.2.2 Lithium nickel oxide

Among the layered cathode materials, lithium nickel oxide (LiNiO$_2$) is the one that has attracted particular attention. Nickel, a aITM, is found abundantly in almost all habitats and is cheaper compared with cobalt. LiNiO$_2$ is identified as a hexagonal crystal structure with a lattice parameter of $a = 2.87594$ Å and $c = 14.1977$ Å, as shown in Fig. 10.7 [40]. It has a potential platform of about 3.7 V and a theoretical capacity of 275 mAh/g. It has been reported that the maximum capacity that could be reached is 200 mAh/g, which corresponds to $x = 0.72$ in Li$_{1-x}$NiO$_2$ [41]. LiNiO$_2$ undergoes four transitions:

Rhombohedral (H1): $0.0 \leq x \leq 0.25$
Monoclinic (M): $0.25 \leq x \leq 0.55$
Rhombohedral (H2): $0.55 \leq x \leq 0.75$
H1 + H2: $0.75 \leq x \leq 1.0$

Thus the highest stability for Li insertion that can be achieved is up to $x = \sim 0.75$ due to small changes of interlayer distance between NiO$_2$ sheets in the range from ~ 4.73 to ~ 4.80 Å. It is noteworthy that further charging can induce a 0.3 Å decrease in interlayer

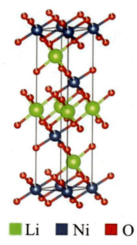

FIGURE 10.7 The crystal structure of LiNiO$_2$ in a hexagonal cell unit.

distance due to the appearance of a new H3 phase accompanied with smaller ionic radius of Ni^{IV+}, which gives poor cycling performance [41]. Since its discovery in 1990, $LiNiO_2$ was expected to replace highly toxic and expensive $LiCoO_2$ for lithium—ion battery applications.

However, two main drawbacks have been inherited by $LiNiO_2$: the high vapor pressure of $LiNiO_2$ at higher temperature and accelerated oxidation of electrolyte. $LiNiO_2$ can be easily decomposed into nonstoichiometric composition of lithium-deficient $Li_{1-x}NiO_2$ during high-temperature treatment [41—43]. Some Ni reduces to Ni^{II+} to compensate for the loss of Li during synthesis, and Ni^{II+} will occupy an Li^+ site because of small variations of ionic radius between Li^+ (0.9 Å) and Ni^{II+} (0.83 Å) [44]. This results in a serious partial exchange of Li/Ni, in which the presence of Ni^{II+} can significantly hinder Li^+ diffusion and deteriorate cycling performance [42]. $LiNiO_2$ shows high thermal instability upon cycling, which corresponds to accelerated electrolyte oxidation followed by the structural collapse of Li_xNiO_2 [45]. It was found that Li_xCoO_2 possesses more thermal stability characteristics than Li_xNiO_2 when at low Li concentration. In addition, it was demonstrated that extra Li^+ ions can be inserted into the chemical $LiNiO_2$ structure so that the mixture is expected to have the composition "Li_2NiO_2 and $LiNiO_2$" to prevent lithium deficiency [46]. A series of studies was conducted to investigate the effect of metal ion doping such as Mn [47], Co [48], and Al [49] to improve its thermal stability and reduce its partial exchange of Li/Ni. A schematic diagram of partial exchange of Li/Ni is represented in Fig. 10.8.

10.2.3 Lithium manganese oxide

Another well-known $LiMO_2$ family member is lithium manganese oxide ($LiMnO_2$). Its crystal structure belongs to two structures: orthorhombic and rhombohedral structures, as shown in Fig. 10.9. O-$LiMnO_2$ possesses a thermodynamic stability structure, in which LiO_6 and MnO_6 are arranged in zigzag layers. $LiMnO_2$ exhibits a small monoclinic distortion from the ideal rhombohedral structure due to Jahn—Teller distortion. Monoclinic $LiMnO_2$ is

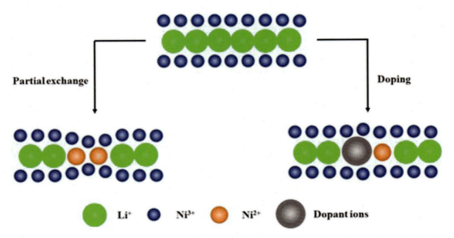

FIGURE 10.8 Schematic diagram of partial exchange of Li/Ni and the effect of metal ion doping in an $LiNiO_2$ structure.

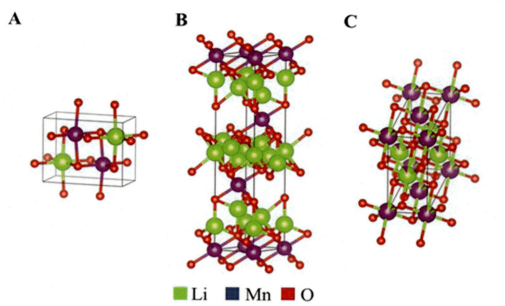

FIGURE 10.9 The crystal structure of LiMnO$_2$ in (A) orthorhombic, (B) rhombohedral, and (C) monoclinic cell units.

isostructural to LiCoO$_2$, where Li$^+$ ions occupy octahedral sites between MnO$_6$ layers. However, it is thermodynamically unstable and thereby it can be suppressed by ion exchange from α-NaFeO$_2$ [50,51].

Both polymorphs will transform into spinel structures upon cycling between 2 and 4.3 V. This results in poor cycling performance accompanied by voltage drop because of its low crystallinity in this phase. The spinel phase is induced at early electrochemical Li extraction and the process is usually completed within a few cycles. This phase transformation from layered to spinel-like phase is an unavoidable reaction and a common phenomenon in almost all layered structure cathode materials since it only requires simple cation diffusion before the spinel-like phase is initiated [52–54]. From the voltage profile, the voltage plateau at 3 V is longer compared with the 4 V voltage plateau and this indicates that LiMnO$_2$ is classified to 3 V class cathode materials [55,56].

To suppress the phase transformations, the most popular method is metal ion doping. For example, the doping of Al [57] and Cr [58] in Mn sites acts as pillar ions stabilizing the layered structure without bringing abrupt lattice distortion. However, some cases exhibit unavoidable layered-to-spinel-like phase reaction at the initial cycle but cyclability and capacity are greatly enhanced, which delivers ~220 mAh/g specific capacity at long cycles [59,60].

10.2.4 Lithium nickel manganese oxide

To possess high structural stability with high capacity and avoid the harmful and expensive Co element, the immediate prospect of mixing LiNiO$_2$ and LiMnO$_2$ to form a solid solution was first reported by Ohzuku et al. [61]. Mn stabilizes the structure since it is inert in structure and Ni is an electrochemically active ion that contributes to capacity. The role of Mn ions was confirmed when Kang et al. attempted to substitute completely Mn ions with Ti, resulting in a rock salt structure formation [62]. Ni and Mn ions have oxidation numbers of +2 and +4, respectively [63], and thus Jahn–Teller distortion and phase transformations can be avoided upon charge/discharge. Mn could mitigate the cation mixing between Li and Ni, which always occurred due to similar ion radius (Fig. 10.10).

The arrangement of Li, Ni, and Mn in the layer can be illustrated by the "flower" pattern shown in Fig. 10.11. It has been proposed that Li$^+$ ions in TM layer of lithium nickel manganese oxide (LiNi$_{0.5}$Mn$_{0.5}$O$_2$) are surrounded by a hexagonal ring of Mn^{IV+} ions, which resembles the structure of Li$_2$MnO$_3$ [64,65]. As a consequence, the honeycombs of Mn^{IV+} are surrounded by larger Ni^{II+} honeycombs and therefore no close contact neighbor between Li$^+$ and Ni^{II+} exists in the TM layer.

LiNi$_{0.5}$Mn$_{0.5}$O$_2$ exhibits good electrochemical properties due to the advantages of Mn and Ni ions. During the charging/discharging process, only Ni ions remain active to undo the redox reaction, which largely contributes to the capacity of LiNi$_{0.5}$Mn$_{0.5}$O$_2$. Recently, Li et al. claimed that LiNi$_{0.5}$Mn$_{0.5}$O$_2$ can deliver about 180 mAh/g of capacity and maintain high retention over 100 cycles at a voltage window of 2.5–4.5 V. For the rate capability test, 74.2% of capacity is reserved when cycled at 15°C and sustains its capacity for 1000 cycles [66]. XRD and neutron diffraction (ND) studies done by Lu et al. suggested 8%–10% displacement of Li$^+$ by Ni^{II+} between 3a and 3b sites [67–69], which is low compared to LiNiO$_2$. However, the existence of partial exchange is still the main drawback that impedes Li$^+$ ion movements. In another study, Kang et al. proposed a new synthesis strategy by employing an "ion-exchange" technique that was able to reduce Ni ions in Li layers to 4% [70].

LNMO inherits additional characteristics from LiNiO$_2$ compared with LiMnO$_2$. As an example, reversible-phase transition between H1 to H2 is observed for LiNi$_{0.5}$Mn$_{0.5}$O$_2$, similar to the LiNiO$_2$ system. Nevertheless, emergence of the H3 phase when charged above 4.3 V is suppressed, which often happens in the LiNiO$_2$ system [71]. In addition, LiNi$_{0.5}$Mn$_{0.5}$O$_2$ displays better thermal stability, lower thermal runaway, and less

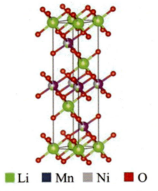

■ Li ■ Mn ■ Ni ■ O

FIGURE 10.10 The crystal structure of LiNi$_{0.5}$Mn$_{0.5}$O$_2$.

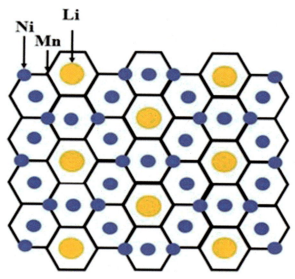

FIGURE 10.11 Transition metal layer in an LiNi$_{0.5}$Mn$_{0.5}$O$_2$ structure, exhibiting the "flower" pattern.

decomposition of electrolyte upon cycling compared with LiCoO$_2$ and LiNiO$_2$ [72]. Li$_x$-Ni$_{0.5}$Mn$_{0.5}$O$_2$ still exhibits high onset temperature of oxygen release at low and high Li concentrations, resulting high structural stability [73].

10.2.5 Lithium nickel cobalt manganese oxide

Many attempts have been made to suppress the partial exchange of Li/Ni, including the introduction of an LiCoO$_2$ structure in layered LiNiO$_2$–LiMnO$_2$ solid solution, as shown in Fig. 10.12. Despite the high cost and environmental issues of the Co element, Co ions have attracted much attention due high electronic conductivity. In 2014, Li et al. demonstrated that the conductivity of Li$_x$Ni$_y$Mn$_y$Co$_{1-2y}$O$_2$ increases with increasing Co content. An NCM-333 (Ni:Mn:Co = 3:3:3) sample shows the highest lithium diffusion coefficient measured by the galvanostatic intermittent titration technique compared with a 444 > 992 > 550 sample in sequence [74].

LiNi$_x$Co$_y$Mn$_z$O$_2$ was discovered by Yiu et al. [75] and is widely studied today. Like LiNi$_{0.5}$Mn$_{0.5}$O$_2$, NCM is also isostructural to LiCoO$_2$ [76]. Ni, Co, and Mn ions are distributed randomly in the octahedral site, as illustrated in Fig. 10.13. However, unavoidable reactions like partial exchange of Li/Ni can still be fully suppressed. If we look into XRD measurements, the intensity ratio of (003)/(104) is used as an indicator for degree of partial exchange. NCM exhibits a strong layered structure if the intensity ratio exceeds 1.2. Apart from that, the split (018)/(110) and (006)/(012) peaks should be well pronounced for the same purpose [77]. NCM exhibits higher capacity compared with LiCoO$_2$, which is 160 mAh/g below 4.3 V. The formal charge of Ni, Mn, and Co is +2, +3, and +4, respectively. The redox couple of Co^{III+}/Co^{IV+} occurs at higher potential compared with Ni^{II+}/Ni^{IV+}. Thus the extra capacity can be

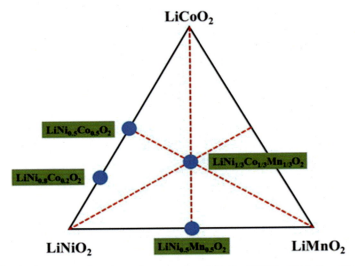

FIGURE 10.12 Triangular phase diagram of an LiCoO$_2$–LiNiO$_2$–LiMnO$_2$ solid solution.

attributed to the additional redox reaction by Co^{III+}/Co^{IV+} [78]. Tetravalent Mn remains inactive and acts as a structural stabilizer during the Li extraction/insertion process [76,79]. The absence of Mn^{III+} ions indicates that Jahn–Teller distortion can be avoided in NCM to suppress manganese dissolution. The introduction of Co ions in the TM layer can mitigate the partial exchange of Li/Ni up to 1%–6% [79–84].

Just like another family member of layered structures, NCM shows small variations of lattice volume upon cycling. However, zero strain of material can be observed in the range from $x = 0$ to $x = 0.67$ with Li$_{1-x}$Ni$_{1/3}$Mn$_{1/3}$Co$_{1/3}$O$_2$ notation. Thus it can offer competent rate

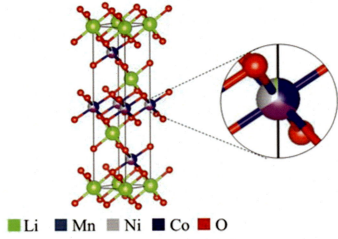

FIGURE 10.13 The crystal structure of LiNi$_{1-y-z}$Co$_y$Mn$_z$O$_2$.

III. Electrochemical energy conversion and storage

capability with fewer safety issues at low Li concentration compared with $LiCoO_2$ and $LiNiO_2$. A capacity of 200 mAh/g can be achieved if the NCM is charged to higher voltage, but poor cycle life is inevitable.

Besides tailoring the amount of Co, high content of Ni in NCM can improve the reversibility of capacity in $LiNi_{1-2x}Co_xMn_xO_2$. This is because high content of Ni induces full oxidation of Ni^{II+} to Ni^{IV+}, whereas Co^{III+} hardly oxidizes further without involving the oxidation of O^{2-} as Mn^{IV+} is still inactive [85–87]. To confirm this, initial discharge capacity of $LiNi_{1-2x}Co_xMn_xO_2$ is enhanced with an increase in Ni content due to higher activity of $Ni^{III+/IV+}$, $Co^{III+/IV+}$, and $Mn^{III+/IV+}$. The highest Ni amount with $x = 0.1$ can deliver 200 mAh/g, while at $x = 0.3$, the capacity is only 165 mAh/g [86].

Ni-rich layered structures encounter oxygen release from the lattice in the temperature range from 150 to 300°C at low Li concentration [88–90]. Noh et al. suggested that by lowering Ni content, thermal stability is improved because a reduced amount of heat is generated, which is confirmed by DSC measurement. Thus $LiNi_{1/3}Co_{1/3}Mn_{1/3}O_2$ or so-called "NCM-333" shows the best electrochemical properties and thermal stability among $LiNi_{1-2x}Co_xMn_xO_2$ solid solution [85,91]. Severe capacity fading is commonly found in Ni-rich compositions despite high discharge capacity. This originates from volumetric change of the electrode and phase instability, in which layered structures tend to transform into a rock salt structure [91]. Consequently, tailoring the number of TM ions still needs to be done to establish direct correlation between the optimum compositions with high capacity and thermal stability.

Until now, a coating approach has been the popular strategy to improve thermal stability and capacity retention rather than metal ion doping. The coating approach can directly protect the surface structure from side reaction with electrolyte and thus stabilize the solid-electrolyte interphase (SEI) layer. There are a variety of coating materials such as metal oxides [92], metal hydroxide phosphate [93], metal fluorides [94], and carbon [95].

10.2.6 Lithium nickel cobalt aluminum

Ni site $LiNiO_2$ can be partially replaced by Co and Al ions. Co ions helps to minimizing the partial exchange of Li/Ni. Al^{III+} ions play the same role as Mn^{IV+} ions, which are electrochemically inactive, and stabilize the structure when deeply delithiated. The positive effect of using Al^{III+} ions is that the occupation of Al^{III+} ions in the Ni site can limit Li^+ ion removal and thereby prevent the structure from overcharging [96]. Previous studies proposed that a low amount of Al and Co in $LiNiO_2$ is sufficient to offer new values. The new system of $LiNi_{0.7}Co_{0.3-x}Al_xO_2$ has improved thermal stability compared with $LiNiO_2$ [97,98]. Since then, it has received much attention and has been commercialized and used by Tesla. $LiNi_{0.8}Co_{0.15}Al_{0.05}O_2$ is the common composition of lithium nickel cobalt aluminum (NCA) that had been discussed and studied widely. NCA, with a theoretical capacity of 279 mAh/g, can achieve reversible capacity up to $\sim$199 mAh/g in practical use. NCA delivers higher capacity compared with $LiCoO_2$ and NCM-333, and high energy and power density. The crystal structure of NCA is represented in Fig. 10.14.

However, NCA inherits the problem of oxygen gas generation in the overcharge state. Overlithiated NCA can drive the migration of oxygen continuously from the bulk to the

298 10. Layered and spinel structures as lithium-intercalated compounds for cathode materials

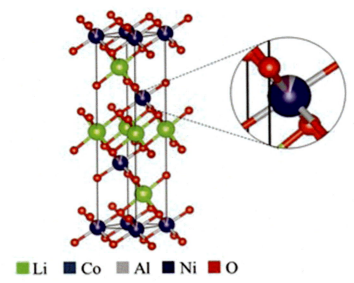

FIGURE 10.14 The crystal structure of LiNi$_{1-y-z}$Co$_y$Co$_z$O$_2$.

surface of particles and thus reduce oxygen levels in the bulk of particles. Low oxygen areas can simply transform into a rock salt core as domains covered by the spinel thin layer phase. The formation of rock salt in the core structure triggers structural instability [99], as illustrated in Fig. 10.15.

To suppress oxygen evolution, the structural stability of NCA can be suppressed by surface coating, just like NCM. For example, Lee et al. mentioned that an AlF$_3$ coating layer could protect the surface from being contacted directly with electrolyte. This prohibits oxygen evolution and deaccelerates the reactivity of Ni^{IV+}. This shows that the cycling performance of NCA is greatly enhanced by the AlF$_3$ coating layer with 84.7% capacity retention for 100 cycles at an elevated temperature of 55°C [100,101]. Other coating layers used for the same purpose are FeF$_3$ [102], AlPO$_4$ [103], Li$_2$O−ZrO$_2$ [104], SiO$_2$ [104], and Co$_3$O$_4$ [104]. NCA seems meet to meet all primary characteristics as promising cathode materials for next-

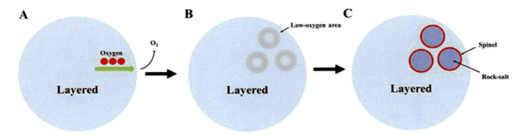

FIGURE 10.15 (A−C) Schematic diagram of the formation of rock salt domains in the NCA structure due to oxygen release.

III. Electrochemical energy conversion and storage

generation lithium—ion batteries. In 2005, NCA, provided by Panasonic, was commercialized by Tesla and was demonstrated to be satisfactory for energy storage in electric vehicles and hybrid electric vehicles.

10.2.7 Lithium-rich and manganese-rich oxide

Solid solutions between $Li[Li_{1/3}Mn_{2/3}]O_2$ and $LiMO_2$ (shown in Fig. 10.16) have attracted global attention as new cathode materials due to their extraordinarily high capacities when they are charged to over 4.6 V [105—107]. High capacity is achieved because electrochemically inactive Li_2MnO_3 is activated by extracting Li_2O from the lattice and transforming it into MnO_2, which is electrochemically active. Any inactive Li_2MnO_3 component in the composites plays the role of a reservoir for any additional lithium that can diffuse from TM layers into adjacent Li layers. The randomly distributed Li_2MnO_3 components will perform as solid electrolyte elements in the structure to facilitate Li^+ ion transport [105]. Li-rich and Mn-rich layered structure (also called LMR) belongs to the monoclinic Li_2MnO_3 structure and rhombohedral $LiMO_2$ structure [108]. The crystal structure of $xLi_2MnO_3 \cdot (1-x) LiMO_2$ is shown in Fig. 10.17.

Thackeray's group discovered LMR materials with two-component notation in 1991. It started when these researchers developed layered $Li_{2-x}MnO_{3-x/2}$ ($0 < x < 2$) by extracting Li_2O from Li_2MnO_3 using a chemical leach technique. This obtained the layered structure with the chemical formula $Li_{1.09}Mn_{0.91}O_2$ or $0.2Li_2MnO_3-0.8LiMnO_2$. It was found to be more stable compared with $LiMnO_2$ during the lithiation/delithiation process.

Despite high reversible capacity and high charge voltage, there are several problems that prevent its commercialization, e.g., low coulombic efficiency during the first cycle due to the

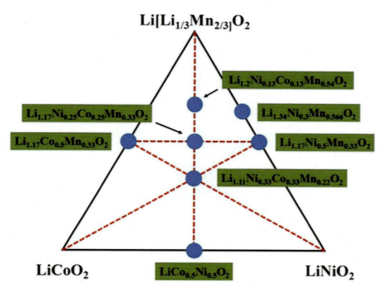

FIGURE 10.16 Triangular phase diagram of $LiCoO_2$—$LiNiO_2$—$Li[Li_{1/3}Mn_{2/3}]O_2$ solid solution.

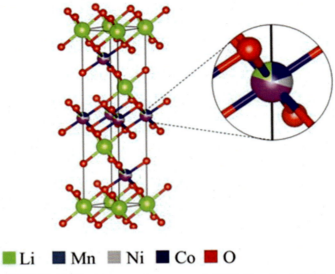

FIGURE 10.17 The crystal structure of xLi$_2$MnO$_3$ · (1−x) LiMO$_2$.

extraction of Li$_2$O from the Li$_2$MnO$_3$ component. The process is irreversible and causes high irreversible capacity associated with reduced energy density. Besides, this irreversible reaction is capable of increasing the polarization resistance and ruining the surface of the cathode [108]. Thus several solutions have been taken into consideration. Kang et al. used acid treatment to leach Li$_2$O from Li$_2$MnO$_3$. It eliminates the first irreversible capacity; however, it causes severe capacity fading and low rate capability. Wu and Manthiram attempted to modify the surface structures using various coating materials such as Al$_2$O$_3$, CeO$_2$, ZrO$_2$, SiO$_2$, ZnO, and AlPO$_4$. The results show that a small amount of Al$_2$O$_3$ is the most effective coating layer to decrease irreversible capacity loss (IRC) as lows as 41 mAh/g with 94% capacity retention for 30 cycles [109]. Wang et al. used the dual-coating approach with AlPO$_4$ and Al$_2$O$_3$, which delivers IRC values as low as 26 mAh/g with a discharge capacity of about 300 mAh/g [110]. LMR exhibits poor rate capability because of slow movement of Li$^+$ ions. The Li$_2$MnO$_3$ component in the LMR lattice controls most of the Li$^+$ ions' mobility but the LiMO$_2$ component still shows unsatisfactory rate capability [108]. Thus the direct way to facilitate Li transportation is by tailoring the morphology to shorten the Li diffusion path with a low-energy barrier. For example, Ma et al. synthesized nanofiber-decorated Li$_{1.2}$Mn$_{0.54}$Ni$_{0.13}$Co$_{0.13}$O$_2$ with improved initial coulombic efficiency and cyclic stability. It achieved 80.5% capacity retention after 100 cycles. The nanofiber-decorated sample showed larger differences in capacity by increasing the C-rate compared with the pristine sample. This could be attributed to faster electron and Li$^+$ ion transportation with the provided larger surface area of the nanofibers [111]. Another critical issue faced by LMR is rapid voltage degradation after extensive cycling. This is related to phase transformation from layered structure to spinel-like structure [112]. This can be well explained by Reed et al., who proposed that phase transformation requires a small effort of one-fourth of the TM layer from

3b sites to move into the empty 3a sites of the Li layer. The occupied 3a site becomes the new 16d position of the spinel without disturbing the framework of closed-paced oxygen arrays [113]. Extensive efforts have been devoted to mitigating voltage degradation, including coating materials and metal ion doping. However, it takes a great deal of effort to overcome completely because the voltage degradation mechanism does not have a clear consensus and is still being debated (Fig. 10.18).

10.3 Spinel structures

10.3.1 Lithium manganese oxide

LiMn$_2$O$_4$ exhibits spinel structure characteristics and was proposed as a cathode material for lithium—ion batteries in 1983 [3]. LiMn$_2$O$_4$ materials attract much attention compared with LiCoO$_2$ because of their cheapness, nontoxicity, and higher potential versus Li (3.0—4.5 V) with a theoretical capacity of 128 mAh/g [114—118]. LiMn$_2$O$_4$ has a cubic structure (space group *Fd-3m*) with a lattice parameter of $a = 8.242$ Å [3], as shown in Fig. 10.19. The octahedral site is occupied by Mn ions and one-quarter of them are located in Li layers that occupy the tetrahedral site. The tetrahedral site shares its face with the vacant octahedral site left by one-quarter of Mn ions in the TM layers. The special feature of the structure is the 3D connected diffusion path, which is designed to increase Li$^+$ transportation. In LiMn$_2$O$_4$, Mn ions are distributed randomly, and half of them exist as Mn^{III+} and half as Mn^{IV+}.

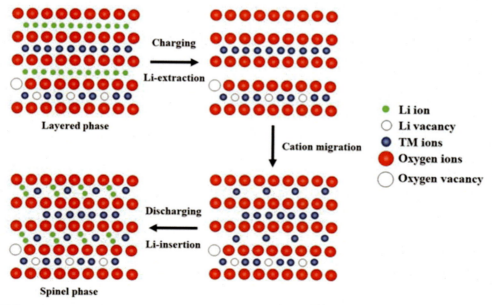

FIGURE 10.18 Illustration of layered-to-spinel-like phase formation in xLi$_2$MnO$_3$ · (1−x) LiMO$_2$. *TM*, Transition metal.

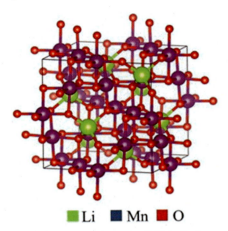

FIGURE 10.19 The crystal structure of LiMn$_2$O$_4$ in a cubic unit cell.

Experimentally, the achieved capacity is ∼140 mAh/g with an Mn^{III+}/Mn^{IV+} redox couple at the ∼4 V region [119,120]. In addition, LiMn$_2$O$_4$ can deliver high-energy density similar to LiCoO$_2$ and has drawn much attention as a 4 V class cathode material in lithium–ion batteries. However, LiMn$_2$O$_4$ still suffers from deteriorating cycling performance at elevated temperatures above 55°C [120–122]. Two main reasons have been proposed:

1. Dissolution of Mn ions into organic electrolyte generated due to the presence of HF through disproportional reaction. Mn^{II+} ions dissolve into electrolyte in the 4 V region. The mechanism reaction can be explained as follows [114,123,124]:

$$2\,\text{Mn}^{III+}\,(\text{solid}) \rightarrow \text{Mn}^{IV+}\,(\text{solid}) + \text{Mn}^{II+}\,(\text{solution})$$

2. The Jahn–Teller effect induces phase transformation from cubic to tetragonal structures in the 3 V region [114,123,125]. This occurs due to an imbalance ratio of Mn^{IV+}/Mn^{III+} since the average valence number of Mn ions should be equal to or greater than 3.5 [126,127]. A high content of Mn^{III+} ions triggers severe capacity fading and contributes to Mn dissolution.

To suppress capacity fading and cyclic stability, cations such as Co, Cr, Fe, Ni, Al, and Ru [128–133] have been substituted onto Mn ions in octahedral sites. The cations occupy 16d sites to stabilize the spinel structure from lattice deformation [134–136]. Another interesting approach was demonstrated by Reddy et al. by developing Li-doped LiMn$_2$O$_4$. It was suggested that the occupation of Li$^+$ ions in the 16c site and vacancy of Li$^+$ ions in the 8a site can decrease the Mn^{III+} content to suppress the Jahn–Teller effect. Li-doped LiMn$_2$O$_4$ displayed excellent capacity stability for 100 cycles at a current density of 360 mA/g [137]. Just like other cathode materials, adapting coating materials shows their effectiveness in protecting the lattice structure from collapsing. Mn^{II+} ions formed through disproportion reaction ions cannot penetrate into organic electrolyte. Various coating layers have been introduced to modify the surface of LiMn$_2$O$_4$, for example, Al$_2$O$_3$, B$_2$O$_3$, ZnO, CoO, MgO, ZrO$_2$, CeO$_2$, La$_2$O$_3$, and carbon [138−148].

10.3.2 Lithium nickel manganese oxide

Lithium nickel manganese oxide (LiNi$_{0.5}$Mn$_{1.5}$O$_4$) has emerged as an excellent cathode material for next-generation lithium−ion batteries. Since its reported discovery, it has been widely used for various applications especially for portable electronics devices [149,150]. LiNi$_{0.5}$Mn$_{1.5}$O$_4$ is designed to exhibit high operating voltage (4.75 vs. Li), excellent rate capability, and unique 3D channel structure. Due to its high operating voltage, the energy density can increase as high as ∼610 Wh/kg, which is higher than any commercially available cathode materials in today's market [151,152]. For thermal safety, LiNi$_{0.5}$Mn$_{1.5}$O$_4$ has an onset temperature up to 250°C, which is acceptable [153]. LiNi$_{0.5}$Mn$_{1.5}$O$_4$ shows the same spinel structure as LiMn$_2$O$_4$, in which Ni ions occupy a part of Mn ions. LiNi$_{0.5}$Mn$_{1.5}$O$_4$ crystallized into two different cubic structures: (1) disordered spinel LiNi$_{0.5}$Mn$_{1.5}$O$_{4-\delta}$ (space group Fd-3m) and (2) ordered spinel LiNi$_{0.5}$Mn$_{1.5}$O$_4$ (space group P4$_3$32). In the disordered structure, Mn ions exist with two valence numbers of Mn^{III+} and Mn^{IV+}. Li and O occupy 8a and 32e sites, respectively, whereas Ni and Mn occupy 16d sites with random distribution. In an ordered structure, the valence number of Ni ions is +2, so all Mn ions should have a valence number of +4. Ni and Mn ions occupy 4b and 12d sites, respectively, while Li ions occupy 8c sites. O ions occupy 8c and 24e sites [154−156]. The crystal structures of LiNi$_{0.5}$Mn$_{1.5}$O$_4$ in two different cubic structures are shown in Fig. 10.20.

Two redox reactions take place at 4.1 and 4.7 V, corresponding to Mn^{III+}/Mn^{IV+} and Ni^{II+}/Ni^{IV+}, respectively, during electrochemical reaction. This indicates the importance of the Ni^{II+}/Ni^{IV+} redox couple to elevate the operating voltage from 4.1 to 4.7 V. Such a high operating

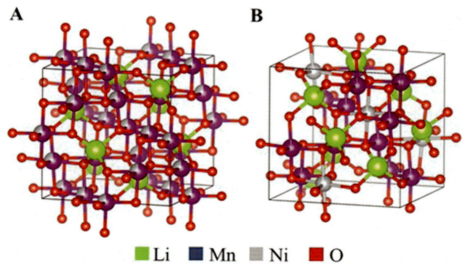

FIGURE 10.20 The crystal structure of (A) disordered LiNi$_{0.5}$Mn$_{1.5}$O$_{4-\delta}$ and (B) ordered LiNi$_{0.5}$Mn$_{1.5}$O$_4$.

304   10. Layered and spinel structures as lithium-intercalated compounds for cathode materials

voltage offers great promise for high-energy density. This means that $LiNi_{0.5}Mn_{1.5}O_4$ can be easily coupled with anode material that possesses better thermal safety and higher voltage such as $Li_4Ti_5O_{12}$ [157,158]. However, there are some detrimental issues that need to be overcome and thus limit its commercialization. The impurity $Li_xNi_{1-x}O$ is commonly found in the $LiNi_{0.5}Mn_{1.5}O_4$ phase [159—161]. The impurity phase results from oxygen deficiencies in $LiNi_{0.5}Mn_{1.5}O_4$ during high-temperature treatment and continues to deteriorate the electrochemical performance of $LiNi_{0.5}Mn_{1.5}O_4$ [149,162—164]. Liu et al. claimed that the annealing process can inhibit the formation of the $Li_xNi_{1-x}O$ phase. $LiNi_{0.5}Mn_{1.5}O_4$ prepared by the solid-state method with a postannealing step can reduce the $Li_xNi_{1-x}O$ phase from 8.5% to 1.6%, which was confirmed by Rietveld refinement. The impurity $Li_xNi_{1-x}O$ only reduces the specific capacity but no dramatic poor cycling performance was observed [165]. Apart from that, it is difficult to synthesize $LiNi_{0.5}Mn_{1.5}O_4$ with appropriate amounts of Mn^{III+} ions. If all Mn ions have +4 valence number, $LiNi_{0.5}Mn_{1.5}O_4$ with an ordered state tends to show poor electrochemical properties. In contrast, higher amounts of Mn^{III+} can trigger Mn dissolution into organic electrolyte by the Jahn—Teller effect [166,167]. So, it is important in controlling the amount of Mn^{III+}. Thus many researchers have attempted to perform careful control of the number of Mn^{III+} ions such as controlling the cooling rate after high-temperature treatment, partial substitution, and acidic treatment [166,168—170]. Another critical issue faced by high-voltage cathode material is oxidative electrolyte decomposition, forming an unstable SEI layer, which increases cell impedance. Mn dissolution rises in the presence of HF species formed by the small traces of water with the electrolyte [171,172]. Most previous studies proposed the coating layer technique because of its effectiveness to act as an HF scavenger and physical protector for $LiNi_{0.5}Mn_{1.5}O_4$ [172—174].

10.4 Summary

Lithium—ion batteries prompted progressive studies before their commercialization in the market. This includes consumer electronics, electric vehicles, industry and stationary energy, and others. Cathode materials have emerged as an important component to meet growing demands that require higher energy and power density and high operating voltage. $LiCoO_2$, as a pioneer in cathode materials, is still being investigated extensively, including the optimization of synthesis conditions aimed at nanoparticles with a variety of morphologies [175—179]. Besides, thermal stability is one the most demanding issues. As a consequence, various coating materials have been developed to increase thermal safety alongside higher capacity [30—33]. However, in terms of cost production, $LiCoO_2$ still has to contend with the issue of the expensive Co element. It is 30% more expensive compared with other cathode materials with Ni and Mn elements. This limits its application for electric vehicle and industrial applications. Due to this intolerable issue, Ni- and/or Mn-based cathode materials are preferred. Ni- and/or Mn-based cathode materials are only commercialized in small-sized lithium—ion batteries. Due to the limitation of energy density, Ni- and/or Mn-based cathode materials are unable to penetrate the electric vehicle market. Poor cycling performance at elevated temperatures has been improved, thanks to the additives in electrolyte [180,181]. At present, NCM and NCA are the most promising cathode materials since they are widely applicable for all devices. High spinel $LiNi_{0.5}Mn_{1.5}O_4$ cathode materials will also be used for

III. Electrochemical energy conversion and storage

commercial deployment in the large-scale energy storage field. The oxidative electrolyte decomposition has also been suppressed but still needs further attention.

In this chapter, we reviewed the layered and spinel structure cathode materials since their emergence in "rocking chair" lithium–ion batteries 50 years ago. The detrimental issues faced by these cathode materials and the strategies that were encountered were also discussed. Intriguingly, we can conclude that the electrochemical properties of cathode materials can be significantly enhanced by (1) metal-doping ions with cations and/or anions, (2) surface modifications, and (3) tailoring morphology. However, the cost and simplicity of synthesis should also be taken into account. Different applications require different demands on lithium–ion batteries. Thus new cathode materials will be pursued and hopefully will tick all the boxes in the future.

"It is definitely true that the fundamental enabling technology for electric cars is lithium–ion as a cell chemistry technology. In the absence of that I don't think it's possible to make an electric car that is competitive with a gasoline car." —*Elon Musk, CEO of Tesla Motors and SpaceX.*

References

[1] M.S. Whittingham, Lithium batteries and cathode materials, Chem. Rev. 104 (10) (2004) 4271–4302.

[2] K. Mizushima, P. Jones, P. Wiseman, J.B. Goodenough, Li_xCoO_2 (0< x<-1): a new cathode material for batteries of high energy density, Mater. Res. Bull. 15 (6) (1980) 783–789.

[3] M. Thackeray, P. Johnson, L. De Picciotto, P. Bruce, J. Goodenough, Electrochemical extraction of lithium from $LiMn2O4$, Mater. Res. Bull. 19 (2) (1984) 179–187.

[4] A.K. Padhi, K.S. Nanjundaswamy, J.B. Goodenough, Phospho-olivines as positive-electrode materials for rechargeable lithium batteries, J. Electrochem. Soc. 144 (4) (1997) 1188–1194.

[5] J. Gao, S.-Q. Shi, H. Li, Brief overview of electrochemical potential in lithium ion batteries, Chin. Phys. B 25 (1) (2015) 018210.

[6] K. Ozawa, Lithium-ion rechargeable batteries with $LiCoO_2$ and carbon electrodes: the $LiCoO_2/C$ system, Solid State Ionics 69 (3–4) (1994) 212–221.

[7] R. Gummow, D. Liles, M. Thackeray, Spinel versus layered structures for lithium cobalt oxide synthesised at 400 C, Mater. Res. Bull. 28 (3) (1993) 235–246.

[8] N. Imanishi, M. Fujiyoshi, Y. Takeda, O. Yamamoto, M. Tabuchi, Preparation and 7Li-NMR study of chemically delithiated $Li1-xCoO_2$ (0<x<0.5), Solid State Ionics 118 (1) (1999) 121–128.

[9] L. Pinsard-Gaudart, V.-C. Ciomaga, O. Dragos, R. Guillot, N. Dragoe, Growth and characterisation of $LixCoO_2$ single crystals, J. Cryst. Growth 334 (1) (2011) 165–169.

[10] J.N. Reimers, J. Dahn, Electrochemical and in situ X-ray diffraction studies of lithium intercalation in Li x CoO_2, J. Electrochem. Soc. 139 (8) (1992) 2091–2097.

[11] T. Ohzuku, A. Ueda, Solid-state redox reactions of $LiCoO_2$ (R3m) for 4 volt secondary lithium cells, J. Electrochem. Soc. 141 (11) (1994) 2972.

[12] S. Levasseur, M. Menetrier, E. Suard, C. Delmas, Evidence for structural defects in non-stoichiometric HT-$LiCoO_2$: electrochemical, electronic properties and 7Li NMR studies, Solid State Ionics 128 (1–4) (2000) 11–24.

[13] X. Lu, Y. Sun, Z. Jian, X. He, L. Gu, Y.-S. Hu, H. Li, Z. Wang, W. Chen, X. Duan, New insight into the atomic structure of electrochemically delithiated O3-Li (1−x) CoO2 (0≤ x≤ 0.5) nanoparticles, Nano Lett. 12 (12) (2012) 6192–6197.

[14] A. Van der Ven, M.K. Aydinol, G. Ceder, First-principles evidence for stage ordering in Li x CoO_2, J. Electrochem. Soc. 145 (6) (1998) 2149.

[15] Z. Chen, Z. Lu, J. Dahn, Staging phase transitions in Li x CoO_2, J. Electrochem. Soc. 149 (12) (2002) A1604–A1609.

[16] G. Amatucci, J. Tarascon, L. Klein, CoO_2, the end member of the li x CoO_2 solid solution, J. Electrochem. Soc. 143 (3) (1996) 1114.

[17] C. Delmas, J.-J. Braconnier, P. Hagenmuller, A new variety of $LiCoO_2$ with an unusual oxygen packing obtained by exchange reaction, Mater. Res. Bull. 17 (1) (1982) 117–123.

[18] C. Delmas, C. Fouassier, P. Hagenmuller, Structural classification and properties of the layered oxides, Phys. B+C 99 (1–4) (1980) 81–85.

[19] J. Paulsen, J. Mueller-Neuhaus, J. Dahn, Layered $LiCoO_2$ with a different oxygen stacking (O_2 structure) as a cathode material for rechargeable lithium batteries, J. Electrochem. Soc. 147 (2) (2000) 508.

[20] D. Carlier, I. Saadoune, M. Ménétrier, C. Delmas, Lithium electrochemical deintercalation from O 2 $LiCoO_2$ structure and physical properties, J. Electrochem. Soc. 149 (10) (2002) A1310–A1320.

[21] Y. Shao-Horn, F. Weill, L. Croguennec, D. Carlier, M. Menetrier, C. Delmas, Lithium and vacancy ordering in T# 2– Li x CoO_2 derived from O_2-type $LiCoO_2$, Chem. Mater. 15 (15) (2003) 2977–2983.

[22] D. Carlier, L. Croguennec, G. Ceder, M. Menetrier, Y. Shao-Horn, C. Delmas, Structural study of the T# 2-Li x CoO_2 (0.52< x≤ 0.72) phase, Inorg. Chem. 43 (3) (2004) 914–922.

[23] A. Mendiboure, C. Delmas, P. Hagenmuller, New layered structure obtained by electrochemical deintercalation of the metastable $LiCoO_2$ (O_2) variety, Mater. Res. Bull. 19 (10) (1984) 1383–1392.

[24] A. Van der Ven, G. Ceder, M. Asta, P. Tepesch, First-principles theory of ionic diffusion with nondilute carriers, Phys. Rev. B 64 (18) (2001) 184307.

[25] A. Van der Ven, G. Ceder, Lithium diffusion in layered Li x CoO_2, Electrochem. Solid State Lett. 3 (7) (2000) 301–304.

[26] R. Alcantara, P. Lavela, J. Tirado, R. Stoyanova, E. Zhecheva, Structure and electrochemical properties of boron-doped $LiCoO_2$, J. Solid State Chem. 134 (2) (1997) 265–273.

[27] J. Cho, Improvement of structural stability of LiCoO[sub 2] cathode during electrochemical cycling by sol-gel coating of SnO[sub 2], Electrochem. Solid State Lett. 3 (8) (1999) 362.

[28] J. Cho, Y.J. Kim, B. Park, Novel $LiCoO_2$ cathode material with Al_2O_3 coating for a Li ion cell, Chem. Mater. 12 (12) (2000) 3788–3791.

[29] A. Aboulaich, K. Ouzaouit, H. Faqir, A. Kaddami, I. Benzakour, I. Akalay, Improving thermal and electrochemical performances of $LiCoO_2$ cathode at high cut-off charge potentials by MF3 (M=Ce, Al) coating, Mater. Res. Bull. 73 (2016) 362–368.

[30] Z. Chen, J. Dahn, Effect of a ZrO_2 coating on the structure and electrochemistry of Li x CoO_2 when cycled to 4.5 V, Electrochem. Solid State Lett. 5 (10) (2002) A213–A216.

[31] W. Chang, J.-W. Choi, J.-C. Im, J.K. Lee, Effects of ZnO coating on electrochemical performance and thermal stability of $LiCoO_2$ as cathode material for lithium-ion batteries, J. Power Sources 195 (1) (2010) 320–326.

[32] J. Cho, T.-G. Kim, C. Kim, J.-G. Lee, Y.-W. Kim, B. Park, Comparison of Al_2O_3-and $AlPO_4$-coated $LiCoO_2$ cathode materials for a Li-ion cell, J. Power Sources 146 (1–2) (2005) 58–64.

[33] G.T.-K. Fey, C.-Z. Lu, T.P. Kumar, Y.-C. Chang, TiO_2 coating for long-cycling $LiCoO_2$: a comparison of coating procedures, Surf. Coating. Technol. 199 (1) (2005) 22–31.

[34] S.-T. Myung, N. Kumagai, S. Komaba, H.-T. Chung, Effects of Al doping on the microstructure of $LiCoO_2$ cathode materials, Solid State Ionics 139 (1) (2001) 47–56.

[35] C. Delmas, I. Saadoune, Electrochemical and physical properties of the $LixNi1–yCo_yO_2$ phases, Solid State Ionics 53–56 (1992) 370–375.

[36] C. Julien, M.A. Camacho-Lopez, T. Mohan, S. Chitra, P. Kalyani, S. Gopukumar, Combustion synthesis and characterization of substituted lithium cobalt oxides in lithium batteries, Solid State Ionics 135 (1) (2000) 241–248.

[37] C.D.W. Jones, E. Rossen, J.R. Dahn, Structure and electrochemistry of $LixCryCo1–yO_2$, Solid State Ionics 68 (1) (1994) 65–69.

[38] B.V. Chowdari, S. Madhavi, G. Subba Rao, S. Li, Synthesis and Cathodic Properties of $LiCo1-yRhyO_2$ (0≤ Y≤ 0.2) and $LiRhO_2$, 2001.

[39] M. Tabuchi, K. Ado, H. Kobayashi, H. Sakaebe, H. Kageyama, C. Masquelier, M. Yonemura, A. Hirano, R. Kanno, Preparation of $LiCoO_2$ and $LiCo1–xFe_xO_2$ using hydrothermal reactions, J. Mater. Chem. 9 (1) (1999) 199–204.

[40] H.-H. Ryu, G.-T. Park, C.S. Yoon, Y.-K. Sun, Suppressing detrimental phase transitions via tungsten doping of LiNiO 2 cathode for next-generation lithium-ion batteries, J. Mater. Chem. 7 (31) (2019) 18580–18588.

[41] T. Ohzuku, A. Ueda, M. Nagayama, Electrochemistry and structural chemistry of $LiNiO_2$ (R3m) for 4 volt secondary lithium cells, J. Electrochem. Soc. 140 (7) (1993) 1862.

[42] R. Kanno, H. Kubo, Y. Kawamoto, T. Kamiyama, F. Izumi, Y. Takeda, M. Takano, Phase relationship and lithium deintercalation in lithium nickel oxides, J. Solid State Chem. 110 (2) (1994) 216–225.

References

[43] W. Li, J. Reimers, J. Dahn, Crystal structure of Li x Ni 2− x O 2 and a lattice-gas model for the order-disorder transition, Phys. Rev. B 46 (6) (1992) 3236.

[44] A. Rougier, P. Gravereau, C. Delmas, Optimization of the composition of the Li1− z Ni1+ z O 2 electrode materials: structural, magnetic, and electrochemical studies, J. Electrochem. Soc. 143 (4) (1996) 1168−1175.

[45] T. Ohzuku, H. Komori, M. Nagayama, K. Sawai, T. Hirai, Electrochemical characteristics of $LiNiO_2$, Chem. Express 6 (1991) 161.

[46] J. Dahn, U. von Sacken, C. Michal, Structure and electrochemistry of $Li1\pm yNiO_2$ and a new Li_2NiO_2 phase with the Ni (OH) 2 structure, Solid State Ionics 44 (1−2) (1990) 87−97.

[47] M. Guilmard, L. Croguennec, C. Delmas, Effects of manganese substitution for nickel on the structural and electrochemical properties of $LiNiO_2$, J. Electrochem. Soc. 150 (10) (2003) A1287−A1293.

[48] E. Zhecheva, R. Stoyanova, Stabilization of the layered crystal structure of $LiNiO_2$ by Co-substitution, Solid State Ionics 66 (1−2) (1993) 143−149.

[49] M. Guilmard, A. Rougier, M. Grüne, L. Croguennec, C. Delmas, Effects of aluminum on the structural and electrochemical properties of $LiNiO_2$, J. Power Sources 115 (2) (2003) 305−314.

[50] P.G. Bruce, A.R. Armstrong, Synthesis of layered $LiMnO_2$ as an electrode for rechargeable lithium batteries, Nature 381 (1996) 499−500.

[51] Z.-F. Huang, F. Du, C.-Z. Wang, D.-P. Wang, G. Chen, Low-spin Mn 3+ ion in rhombohedral LiMnO 2 predicted by first-principles calculations, Phys. Rev. B 75 (5) (2007) 054411.

[52] Y.I. Jang, B. Huang, H. Wang, D.R. Sadoway, Y.M. Chiang, Electrochemical cycling-induced spinel formation in high-charge-capacity orthorhombic $LiMnO_2$, J. Electrochem. Soc. 146 (9) (1999) 3217.

[53] J. Cho, Stabilization of spinel-like phase transformation of o-$LiMnO_2$ during 55° C cycling by Sol− gel coating of CoO, Chem. Mater. 13 (12) (2001) 4537−4541.

[54] J.-M. Kim, H.-T. Chung, Electrochemical characteristics of orthorhombic $LiMnO_2$ with different degrees of stacking faults, J. Power Sources 115 (1) (2003) 125−130.

[55] R. Gummow, D. Liles, M.M. Thackeray, Lithium extraction from orthorhombic lithium manganese oxide and the phase transformation to spinel, Mater. Res. Bull. 28 (12) (1993) 1249−1256.

[56] X.-C. Tang, L.-P. He, Z.-Z. Chen, D.-Z. Jia, Application of the low-heating solid-state reaction method in preparation of multi-metal oxides composite—synthesis, structure and electrochemical properties of rhombohedral $LiMnO_2$ cathode material for lithium-ion batteries, J. Inorg. Mater. 18 (2) (2003) 313−319.

[57] Y.M. Chiang, D.R. Sadoway, Y.I. Jang, B. Huang, H. Wang, High capacity, temperature-stable lithium aluminum manganese oxide cathodes for rechargeable batteries, Electrochem. Solid State Lett. 2 (3) (1998) 107.

[58] I. Davidson, R. McMillan, J. Murray, Rechargeable cathodes based on $Li_2Cr_xMn_2−xO_4$, J. Power Sources 54 (2) (1995) 205−208.

[59] W.K. Pang, J.Y. Lee, Y.S. Wei, S.H. Wu, Preparation and characterization of Cr-doped $LiMnO_2$ cathode materials by Pechini's method for lithium ion batteries, Mater. Chem. Phys. 139 (1) (2013) 241−246.

[60] S.-H. Park, Y.-S. Lee, Y.-K. Sun, Synthesis and electrochemical properties of sulfur doped-$Li_xMnO_2−ySy$ materials for lithium secondary batteries, Electrochem. Commun. 5 (2) (2003) 124−128.

[61] T. Ohzuku, Y. Makimura, Layered lithium insertion material of LiNi1/2Mn1/2O2: a possible alternative to $LiCoO_2$ for advanced lithium-ion batteries, Chem. Lett. 30 (8) (2001) 744−745.

[62] K. Kang, D. Carlier, J. Reed, E.M. Arroyo, G. Ceder, L. Croguennec, C. Delmas, Synthesis and electrochemical properties of layered Li0. 9Ni0. 45Ti0. 55O2, Chem. Mater. 15 (23) (2003) 4503−4507.

[63] J. Reed, G. Ceder, Charge, potential, and phase stability of layered Li (Ni0. 5Mn0. 5) O 2, Electrochem. Solid State Lett. 5 (7) (2002) A145−A148.

[64] C.P. Grey, W.-S. Yoon, J. Reed, G. Ceder, Electrochemical activity of Li in the transition-metal sites of O3 Li [Li (1− 2x)/3Mn (2−x)/3Ni x] O 2, Electrochem. Solid State Lett. 7 (9) (2004) A290−A293.

[65] W.-S. Yoon, Y. Paik, X.-Q. Yang, M. Balasubramanian, J. McBreen, C.P. Grey, Investigation of the local structure of the LiNi0. 5Mn0. 5 O 2 cathode material during electrochemical cycling by X-ray absorption and NMR spectroscopy, Electrochem. Solid State Lett. 5 (11) (2002) A263−A266.

[66] J. Li, L. Wan, C. Cao, A high-rate and long cycling life cathode for rechargeable lithium-ion batteries: hollow LiNi0.5Mn0.5O2 nano/micro hierarchical microspheres, Electrochim. Acta 191 (2016) 974−979.

[67] Z. Lu, D. MacNeil, J. Dahn, Layered cathode materials Li [Ni x Li (1/3− 2x/3) Mn (2/3− x/3)] O 2 for lithium-ion batteries, Electrochem. Solid State Lett. 4 (11) (2001) A191−A194.

[68] Z. Lu, J.R. Dahn, Understanding the anomalous capacity of Li/Li [Ni x Li (1/3− 2x/3) Mn (2/3− x/3)] O 2 cells using in situ X-ray diffraction and electrochemical studies, J. Electrochem. Soc. 149 (7) (2002) A815−A822.

[69] Z. Lu, L. Beaulieu, R. Donaberger, C. Thomas, J. Dahn, Synthesis, structure, and electrochemical behavior of Li [Ni x Li1/3− 2x/3Mn2/3− x/3] O 2, J. Electrochem. Soc. 149 (6) (2002) A778−A791.

[70] K. Kang, Y.S. Meng, J. Bréger, C.P. Grey, G. Ceder, Electrodes with high power and high capacity for rechargeable lithium batteries, Science 311 (5763) (2006) 977−980.

[71] X.-Q. Yang, J. McBreen, W.-S. Yoon, C.P. Grey, Crystal structure changes of LiMn0. 5Ni0. 5O2 cathode materials during charge and discharge studied by synchrotron based in situ XRD, Electrochem. Commun. 4 (8) (2002) 649−654.

[72] Y. Makimura, T. Ohzuku, Lithium insertion material of LiNi1/2Mn1/2O2 for advanced lithium-ion batteries, J. Power Sources 119 (2003) 156−160.

[73] N. Yabuuchi, Y.-T. Kim, H.H. Li, Y. Shao-Horn, Thermal instability of cycled Li x Ni0. 5Mn0. 5O2 electrodes: an in situ synchrotron X-ray powder diffraction study, Chem. Mater. 20 (15) (2008) 4936−4951.

[74] Z. Li, C. Ban, N.A. Chernova, Z. Wu, S. Upreti, A. Dillon, M.S. Whittingham, Towards understanding the rate capability of layered transition metal oxides LiNiyMnyCo1−2yO2, J. Power Sources 268 (2014) 106−112.

[75] Z. Liu, A. Yu, J.Y. Lee, Synthesis and characterization of LiNi1− x− yCoxMnyO2 as the cathode materials of secondary lithium batteries, J. Power Sources 81 (1999) 416−419.

[76] Y. Koyama, I. Tanaka, H. Adachi, Y. Makimura, T. Ohzuku, Crystal and electronic structures of superstructural Li1− x [Co1/3Ni1/3Mn1/3] O2 (0≤ x≤ 1), J. Power Sources 119 (2003) 644−648.

[77] R. Gummow, M. Thackeray, W. David, S. Hull, Structure and electrochemistry of lithium cobalt oxide synthesised at 400 C, Mater. Res. Bull. 27 (3) (1992) 327−337.

[78] D. MacNeil, Z. Lu, J.R. Dahn, Structure and electrochemistry of Li [Ni x Co1− 2x Mn x] O 2 (0 x 1/2), J. Electrochem. Soc. 149 (10) (2002) A1332−A1336.

[79] J.-M. Kim, H.-T. Chung, The first cycle characteristics of Li [Ni1/3Co1/3Mn1/3] O2 charged up to 4.7 V, Electrochim. Acta 49 (6) (2004) 937−944.

[80] M.-H. Lee, Y.-J. Kang, S.-T. Myung, Y.-K. Sun, Synthetic optimization of Li [Ni1/3Co1/3Mn1/3] O2 via co-precipitation, Electrochim. Acta 50 (4) (2004) 939−948.

[81] P. Whitfield, I. Davidson, L. Cranswick, I. Swainson, P. Stephens, Investigation of possible superstructure and cation disorder in the lithium battery cathode material LiMn1/3Ni1/3Co1/3O2 using neutron and anomalous dispersion powder diffraction, Solid State Ionics 176 (5−6) (2005) 463−471.

[82] N. Yabuuchi, Y. Koyama, N. Nakayama, T. Ohzuku, Solid-state chemistry and electrochemistry of LiCo1/ 3Ni1/ 3Mn1/ 3O2 for advanced lithium-ion batteries II. Preparation and characterization, J. Electrochem. Soc. 152 (7) (2005) A1434−A1440.

[83] D.-C. Li, T. Muta, L.-Q. Zhang, M. Yoshio, H. Noguchi, Effect of synthesis method on the electrochemical performance of LiNi1/3Mn1/3Co1/3O2, J. Power Sources 132 (1−2) (2004) 150−155.

[84] J. Choi, A. Manthiram, Role of chemical and structural stabilities on the electrochemical properties of layered LiNi1/ 3Mn1/ 3Co1/ 3O2 cathodes, J. Electrochem. Soc. 152 (9) (2005) A1714−A1718.

[85] Y.-K. Sun, S.-T. Myung, H. Bang, B.-C. Park, S.-J. Park, N.-Y. Sung, Physical and electrochemical properties of Li [Ni0. 4Co x Mn0. 6− x] O2 (x= 0.1−0.4) electrode materials synthesized via coprecipitation, J. Electrochem. Soc. 154 (10) (2007) A937−A942.

[86] K.-S. Lee, S.-T. Myung, K. Amine, H. Yashiro, Y.-K. Sun, Structural and electrochemical properties of layered Li [Ni1− 2x Co x Mn x] O2 (x= 0.1−0.3) positive electrode materials for Li-ion batteries, J. Electrochem. Soc. 154 (10) (2007) A971−A977.

[87] J. Cho, T.-J. Kim, J. Kim, M. Noh, B. Park, Synthesis, thermal, and electrochemical properties of AlPO4-coated LiNi0. 8Co0. 1Mn0. 1 O 2 cathode materials for a Li-Ion cell, J. Electrochem. Soc. 151 (11) (2004) A1899−A1904.

[88] M. Guilmard, L. Croguennec, D. Denux, C. Delmas, Thermal stability of lithium nickel oxide derivatives. Part I: Li x Ni1. 02O2 and Li x Ni0. 89Al0. 16O2 (x= 0.50 and 0.30), Chem. Mater. 15 (23) (2003) 4476−4483.

[89] I. Belharouak, D. Vissers, K. Amine, Thermal stability of the Li(Ni[sub 0.8]Co[sub 0.15]Al[sub 0.05])O[sub 2] cathode in the presence of cell components, J. Electrochem. Soc. 153 (11) (2006) A2030.

[90] H.J. Bang, H. Joachin, H. Yang, K. Amine, J. Prakash, Contribution of the structural changes of LiNi0. 8Co0. 15Al0. 05O2 cathodes on the exothermic reactions in Li-ion cells, J. Electrochem. Soc. 153 (4) (2006) A731−A737.

[91] H.-J. Noh, S. Youn, C.S. Yoon, Y.-K. Sun, Comparison of the structural and electrochemical properties of layered Li[NixCoyMnz]O2 (x = 1/3, 0.5, 0.6, 0.7, 0.8 and 0.85) cathode material for lithium-ion batteries, J. Power Sources 233 (2013) 121−130.

References

309

[92] W. Liu, X. Li, D. Xiong, Y. Hao, J. Li, H. Kou, B. Yan, D. Li, S. Lu, A. Koo, Significantly improving cycling performance of cathodes in lithium ion batteries: the effect of Al_2O_3 and $LiAlO_2$ coatings on LiNi0. 6Co0. 2Mn0. 2O2, Nano Energy 44 (2018) 111−120.

[93] S.-W. Lee, M.-S. Kim, J.H. Jeong, D.-H. Kim, K.Y. Chung, K.C. Roh, K.-B. Kim, Li3PO4 surface coating on Ni-rich LiNi0. 6Co0. 2Mn0. 2O2 by a citric acid assisted sol-gel method: improved thermal stability and high-voltage performance, J. Power Sources 360 (2017) 206−214.

[94] X. Xiong, Z. Wang, X. Yin, H. Guo, X. Li, A modified LiF coating process to enhance the electrochemical performance characteristics of LiNi0. 8Co0. 1Mn0. 1O2 cathode materials, Mater. Lett. 110 (2013) 4−9.

[95] Q. Hou, G. Cao, P. Wang, D. Zhao, X. Cui, S. Li, C. Li, Carbon coating nanostructured-LiNi1/3Co1/3Mn1/3O2 cathode material synthesized by chemical vapor deposition method for high performance lithium-ion batteries, J. Alloys Compd. 747 (2018) 796−802.

[96] T. Ohzuku, T. Yanagawa, M. Kouguchi, A. Ueda, Innovative insertion material of LiAl1/4Ni3/4O2 (Rm) for lithium-ion (shuttlecock) batteries, J. Power Sources 68 (1) (1997) 131−134.

[97] S. Madhavi, G.S. Rao, B. Chowdari, S. Li, Effect of aluminium doping on cathodic behaviour of LiNi0. 7Co0. 3O2, J. Power Sources 93 (1−2) (2001) 156−162.

[98] M. Guilmard, C. Pouillerie, L. Croguennec, C. Delmas, Structural and electrochemical properties of LiNi0. 70Co0. 15Al0. 15O2, Solid State Ionics 160 (1−2) (2003) 39−50.

[99] H. Zhang, F. Omenya, M.S. Whittingham, C. Wang, G. Zhou, formation of an anti-core−shell structure in layered oxide cathodes for Li-ion batteries, ACS Energy Lett. 2 (11) (2017) 2598−2606.

[100] S.-U. Woo, C.S. Yoon, K. Amine, I. Belharouak, Y.-K. Sun, Significant improvement of electrochemical performance of AlF_3-coated Li [Ni0. 8Co0. 1Mn0. 1] O2 cathode materials, J. Electrochem. Soc. 154 (11) (2007) A1005−A1009.

[101] S.-H. Lee, C.S. Yoon, K. Amine, Y.-K. Sun, Improvement of long-term cycling performance of Li[Ni0.8Co0.15Al0.05]O2 by AlF_3 coating, J. Power Sources 234 (2013) 201−207.

[102] W. Liu, X. Tang, M. Qin, G. Li, J. Deng, X. Huang, FeF3-coated LiNi0.8Co0.15Al0.05O2 cathode materials with improved electrochemical properties, Mater. Lett. 185 (2016) 96−99.

[103] R. Qi, J.-L. Shi, X.-D. Zhang, X.-X. Zeng, Y.-X. Yin, J. Xu, L. Chen, W.-G. Fu, Y.-G. Guo, L.-J. Wan, Improving the stability of LiNi0.80Co0.15Al0.05O2 by AlPO4 nanocoating for lithium-ion batteries, Sci. China Chem. 60 (9) (2017) 1230−1235.

[104] P. Zhou, Z. Zhang, H. Meng, Y. Lu, J. Cao, F. Cheng, Z. Tao, J. Chen, SiO 2-coated LiNi 0.915 Co 0.075 Al 0.01 O 2 cathode material for rechargeable Li-ion batteries, Nanoscale 8 (46) (2016) 19263−19269.

[105] M.M. Thackeray, S.-H. Kang, C.S. Johnson, J.T. Vaughey, R. Benedek, S. Hackney, Li 2 MnO 3-stabilized LiMO 2 (M= Mn, Ni, Co) electrodes for lithium-ion batteries, J. Mater. Chem. 17 (30) (2007) 3112−3125.

[106] L. Zhang, K. Takada, N. Ohta, K. Fukuda, T. Sasaki, Synthesis of (1− 2x) LiNi1/2Mn1/2O2· xLi [Li1/3Mn2/3] O2· xLiCoO2 (0≤ x≤ 0.5) electrode materials and comparative study on cooling rate, J. Power Sources 146 (1−2) (2005) 598−601.

[107] S.J. Jin, K.S. Park, M.H. Cho, C.H. Song, A.M. Stephan, K.S. Nahm, Effect of composition change of metals in transition metal sites on electrochemical behavior of layered Li [Co1− 2x (Li1/3Mn2/3) x (Ni1/2Mn1/2) x] O2 solid solutions, Solid State Ionics 177 (1−2) (2006) 105−112.

[108] H. Yu, H. Zhou, High-energy cathode materials (Li_2MnO_3−$LiMO_2$) for lithium-ion batteries, J. Phys. Chem. Lett. 4 (8) (2013) 1268−1280.

[109] Y. Wu, A. Manthiram, Effect of surface modifications on the layered solid solution cathodes (1−z) Li [Li1/3Mn2/3] O2−(z) Li [Mn0. 5− yNi0. 5− yCo2y] O2, Solid State Ionics 180 (1) (2009) 50−56.

[110] Q. Wang, J. Liu, A.V. Murugan, A. Manthiram, High capacity double-layer surface modified Li [Li 0.2 Mn 0.54 Ni 0.13 Co 0.13] O 2 cathode with improved rate capability, J. Mater. Chem. 19 (28) (2009) 4965−4972.

[111] D. Ma, P. Zhang, Y. Li, X. Ren, Li1.2Mn0.54Ni0.13Co0.13O2-Encapsulated carbon nanofiber network cathodes with improved stability and rate capability for Li-ion batteries, Sci. Rep. 5 (1) (2015) 11257.

[112] H. Yu, H. Kim, Y. Wang, P. He, D. Asakura, Y. Nakamura, H. Zhou, High-energy 'composite'layered manganese-rich cathode materials via controlling Li 2 MnO 3 phase activation for lithium-ion batteries, Phys. Chem. Chem. Phys. 14 (18) (2012) 6584−6595.

[113] J. Reed, G. Ceder, A. Van Der Ven, Layered-to-spinel phase transition in Li x MnO2, Electrochem. Solid State Lett. 4 (6) (2001) A78.

[114] Y. Xia, Y. Zhou, M. Yoshio, Capacity fading on cycling of 4 V Li/LiMn2 O 4 Cells, J. Electrochem. Soc. 144 (8) (1997) 2593−2600.

[115] K. Oikawa, T. Kamiyama, F. Izumi, B.C. Chakoumakos, H. Ikuta, M. Wakihara, J. Li, Y. Matsui, Structural phase transition of the spinel-type oxide $LiMn_2O_4$, Solid State Ionics 109 (1−2) (1998) 35−41.

[116] Y.-S. Han, H.-G. Kim, Synthesis of $LiMn_2O_4$ by modified Pechini method and characterization as a cathode for rechargeable $Li/LiMn_2O_4$ cells, J. Power Sources 88 (2) (2000) 161−168.

[117] Y.-W. Tsai, R. Santhanam, B.-J. Hwang, S.-K. Hu, H.-S. Sheu, Structure stabilization of $LiMn_2O_4$ cathode material by bimetal dopants, J. Power Sources 119 (2003) 701−705.

[118] W.T. Jeong, J.H. Joo, K.S. Lee, Improvement of electrode performances of spinel $LiMn_2O_4$ prepared by mechanical alloying and subsequent firing, J. Power Sources 119 (2003) 690−694.

[119] D. Guyomard, J.-M. Tarascon, Li metal-free rechargeable LiMn2 O 4/carbon cells: their understanding and optimization, J. Electrochem. Soc. 139 (4) (1992) 937.

[120] M. Thackeray, W. David, P. Bruce, J.B. Goodenough, Lithium insertion into manganese spinels, Mater. Res. Bull. 18 (4) (1983) 461−472.

[121] J. Goodenough, M. Thackeray, W.F. David, P. Bruce, Lithium insertion/extraction reactions with manganese oxides, Rev. Chim. Miner. 21 (4) (1984) 435−455.

[122] W. David, M. Thackeray, P. Bruce, J.B. Goodenough, Lithium insertion into β MnO_2 and the rutile-spinel transformation, Mater. Res. Bull. 19 (1) (1984) 99−106.

[123] D. Aurbach, M. Levi, K. Gamulski, B. Markovsky, G. Salitra, E. Levi, U. Heider, L. Heider, R. Oesten, Capacity fading of $LixMn_2O_4$ spinel electrodes studied by XRD and electroanalytical techniques, J. Power Sources 81 (1999) 472−479.

[124] Y. Shin, A. Manthiram, Factors influencing the capacity fade of spinel lithium manganese oxides, J. Electrochem. Soc. 151 (2) (2004) A204−A208.

[125] B. Xu, D. Qian, Z. Wang, Y.S. Meng, Recent progress in cathode materials research for advanced lithium ion batteries, Mater. Sci. Eng. R Rep. 73 (5−6) (2012) 51−65.

[126] S. Martinez, I. Sobrados, D. Tonti, J.M. Amarilla, J. Sanz, Chemical vs. electrochemical extraction of lithium from the Li-excess Li 1.10 Mn 1.90 O 4 spinel followed by NMR and DRX techniques, Phys. Chem. Chem. Phys. 16 (7) (2014) 3282−3291.

[127] Q. Tong, Y. Yang, J. Shi, J. Yan, L. Zheng, Synthesis and storage performance of the doped LiMn2O4 spinel, J. Electrochem. Soc. 154 (7) (2007) A656−A667.

[128] K. Shaju, G.S. Rao, B. Chowdari, Li ion kinetic studies on spinel cathodes, Li (M 1/6 Mn 11/6) O 4 (M= Mn, Co, CoAl) by GITT and EIS, J. Mater. Chem. 13 (1) (2003) 106−113.

[129] L. Guohua, H. Ikuta, T. Uchida, M. Wakihara, The spinel phases LiM y Mn2− y O 4 (M= Co, Cr, Ni) as the cathode for rechargeable lithium batteries, J. Electrochem. Soc. 143 (1) (1996) 178.

[130] A. Sakunthala, M. Reddy, S. Selvasekarapandian, B. Chowdari, P.C. Selvin, Synthesis of compounds, Li (MMn11/6) O4 (M= Mn1/6, Co1/6,(Co1/12Cr1/12),(Co1/12Al1/12),(Cr1/12Al1/12)) by polymer precursor method and its electrochemical performance for lithium-ion batteries, Electrochim. Acta 55 (15) (2010) 4441−4450.

[131] N. Sharma, M. Reddy, G. Du, S. Adams, B. Chowdari, Z. Guo, V.K. Peterson, Time-dependent in-situ neutron diffraction investigation of a Li (Co0. 16Mn1. 84) O4 cathode, J. Phys. Chem. C 115 (43) (2011) 21473−21480.

[132] S.-T. Myung, S. Komaba, N. Kumagai, Enhanced structural stability and cyclability of Al-doped LiMn2 O 4 spinel synthesized by the emulsion drying method, J. Electrochem. Soc. 148 (5) (2001) A482−A489.

[133] M. Reddy, S.S. Manoharan, J. John, B. Singh, G.S. Rao, B. Chowdari, Synthesis, characterization, and electrochemical cycling behavior of the Ru-doped spinel, Li [Mn2− x Ru x] O4 (x= 0, 0.1, and 0.25), J. Electrochem. Soc. 156 (8) (2009) A652−A660.

[134] M.A. Kebede, M.J. Phasha, N. Kunjuzwa, L.J. Le Roux, D. Mkhonto, K.I. Ozoemena, M.K. Mathe, Structural and electrochemical properties of aluminium doped $LiMn_2O_4$ cathode materials for Li battery: experimental and ab initio calculations, Sustain. Energy Technol. Assess. 5 (2014) 44−49.

[135] B. Hwang, R. Santhanam, D. Liu, Y. Tsai, Effect of Al-substitution on the stability of $LiMn_2O_4$ spinel, synthesized by citric acid sol−gel method, J. Power Sources 102 (1−2) (2001) 326−331.

[136] S. Yang, J. Jia, L. Ding, M. Zhang, Studies of structure and cycleability of $LiMn_2O_4$ and LiNd0. 01Mn1. 99O4 as cathode for Li-ion batteries, Electrochim. Acta 48 (5) (2003) 569−573.

References

[137] M. Reddy, M.S. Raju, N. Sharma, P. Quan, S.H. Nowshad, H.-C. Emmanuel, V. Peterson, B. Chowdari, Preparation of Li1. 03Mn1. 97O4 and Li1. 06Mn1. 94O4 by the polymer precursor method and X-ray, neutron diffraction and electrochemical studies, J. Electrochem. Soc. 158 (11) (2011) A1231−A1236.

[138] S. Lim, J. Cho, PVP-Assisted ZrO_2 coating on $LiMn_2O_4$ spinel cathode nanoparticles prepared by MnO_2 nanowire templates, Electrochem. Commun. 10 (10) (2008) 1478−1481.

[139] J. Cho, VO_x-coated LiMn 2 O 4 nanorod clusters for lithium battery cathode materials, J. Mater. Chem. 18 (19) (2008) 2257−2261.

[140] D. Arumugam, G.P. Kalaignan, Synthesis and electrochemical characterization of nano-CeO_2-coated nanostructure $LiMn_2O_4$ cathode materials for rechargeable lithium batteries, Electrochim. Acta 55 (28) (2010) 8709−8716.

[141] D. Arumugam, G.P. Kalaignan, Synthesis and electrochemical characterizations of nano-La_2O_3-coated nanostructure $LiMn_2O_4$ cathode materials for rechargeable lithium batteries, Mater. Res. Bull. 45 (12) (2010) 1825−1831.

[142] X. Luan, D. Guan, Y. Wang, Enhancing high-rate and elevated-temperature performances of nano-sized and micron-sized $LiMn_2O_4$ in lithium-ion batteries with ultrathin surface coatings, J. Nanosci. Nanotechnol. 12 (9) (2012) 7113−7120.

[143] S.-X. Zhao, X.-F. Fan, Y.-F. Deng, C.-W. Nan, Structure and electrochemical performance of single-crystal Li1. 05Ni0. 1Mn1. 9O3. 98F0. 02 coated by Li−La−Ti−O solid electrolyte, Electrochim. Acta 65 (2012) 7−12.

[144] G. Amatucci, A. Blyr, C. Sigala, P. Alfonse, J. Tarascon, Surface treatments of Li1+ xMn2− xO4 spinels for improved elevated temperature performance, Solid State Ionics 104 (1−2) (1997) 13−25.

[145] J. Cho, T.-J. Kim, Y.J. Kim, B. Park, Complete blocking of Mn^{3+} ion dissolution from a $LiMn_2O_4$ spinel intercalation compound by Co_3O_4 coating, Chem. Commun. (12) (2001) 1074−1075.

[146] J.-M. Han, S.-T. Myung, Y.-K. Sun, Improved electrochemical cycling behavior of ZnO-coated Li1. 05Al0. 1Mn1. 85O3. 95F0. 05 spinel at 55 C, J. Electrochem. Soc. 153 (7) (2006) A1290−A1295.

[147] C. Li, H. Zhang, L. Fu, H. Liu, Y. Wu, E. Rahm, R. Holze, H. Wu, Cathode materials modified by surface coating for lithium ion batteries, Electrochim. Acta 51 (19) (2006) 3872−3883.

[148] J.W. Fergus, Recent developments in cathode materials for lithium ion batteries, J. Power Sources 195 (4) (2010) 939−954.

[149] Q. Zhong, A. Bonakdarpour, M. Zhang, Y. Gao, J. Dahn, Synthesis and electrochemistry of LiNi x Mn2− x O 4, J. Electrochem. Soc. 144 (1) (1997) 205.

[150] K. Amine, H. Tukamoto, H. Yasuda, Y. Fujita, Preparation and electrochemical investigation of LiMn2−xMexO4 (Me: Ni, Fe, and x= 0.5, 1) cathode materials for secondary lithium batteries, J. Power Sources 68 (2) (1997) 604−608.

[151] D. Liu, W. Zhu, J. Trottier, C. Gagnon, F. Barray, A. Guerfi, A. Mauger, H. Groult, C. Julien, J.B. Goodenough, Spinel materials for high-voltage cathodes in Li-ion batteries, RSC Adv. 4 (1) (2014) 154−167.

[152] C. Julien, A. Mauger, Review of 5-V electrodes for Li-ion batteries: status and trends, Ionics 19 (7) (2013) 951−988.

[153] S.-W. Oh, S.-H. Park, J.-H. Kim, Y.C. Bae, Y.-K. Sun, Improvement of electrochemical properties of LiNi0. 5Mn1. 5O4 spinel material by fluorine substitution, J. Power Sources 157 (1) (2006) 464−470.

[154] G. Blasse, Ferromagnetism and ferrimagnetism of oxygen spinels containing tetravalent manganese, J. Phys. Chem. Solid. 27 (2) (1966) 383−389.

[155] D. Gryffroy, R. Vandenberghe, Cation distribution, cluster structure and ionic ordering of the spinel series LiNi0. 5Mn1. 5− xTixO4 and LiNi0. 5− yMgyMn1. 5O4, J. Phys. Chem. Solid. 53 (6) (1992) 777−784.

[156] N. Amdouni, K. Zaghib, F. Gendron, A. Mauger, C. Julien, Magnetic properties of LiNi0. 5Mn1. 5O4 spinels prepared by wet chemical methods, J. Magn. Magn Mater. 309 (1) (2007) 100−105.

[157] W.K. Pang, N. Sharma, V.K. Peterson, J.-J. Shiu, S.-h. Wu, In-situ neutron diffraction study of the simultaneous structural evolution of a LiNi0. 5Mn1. 5O4 cathode and a $Li_4Ti_5O_{12}$ anode in a LiNi0. 5Mn1. 5O4 | | $Li_4Ti_5O_{12}$ full cell, J. Power Sources 246 (2014) 464−472.

[158] H. Xiang, X. Zhang, Q. Jin, C. Zhang, C. Chen, X. Ge, Effect of capacity matchup in the LiNi0. 5Mn1. 5O4/ $Li_4Ti_5O_{12}$ cells, J. Power Sources 183 (1) (2008) 355−360.

[159] L. Xiao, Y. Zhao, Y. Yang, X. Ai, H. Yang, Y. Cao, Electrochemical properties of nano-crystalline LiNi 0.5 Mn 1.5 O 4 synthesized by polymer-pyrolysis method, J. Solid State Electrochem. 12 (6) (2008) 687−691.

[160] H. Fang, Z. Wang, B. Zhang, X. Li, G. Li, High performance LiNi0. 5Mn1. 5O4 cathode materials synthesized by a combinational annealing method, Electrochem. Commun. 9 (5) (2007) 1077−1082.

III. Electrochemical energy conversion and storage

[161] H. Wu, C.V. Rao, B. Rambabu, Electrochemical performance of LiNi0. 5Mn1. 5O4 prepared by improved solid state method as cathode in hybrid supercapacitor, Mater. Chem. Phys. 116 (2−3) (2009) 532−535.

[162] K. Amine, H. Tukamoto, H. Yasuda, Y. Fujita, A new three-volt spinel Li1+ x Mn1. 5Ni0. 5 O 4 for secondary lithium batteries, J. Electrochem. Soc. 143 (5) (1996) 1607−1613.

[163] R. Alcantara, M. Jaraba, P. Lavela, J. Tirado, Optimizing preparation conditions for 5 V electrode performance, and structural changes in Li1− xNi0. 5Mn1. 5O4 spinel, Electrochim. Acta 47 (11) (2002) 1829−1835.

[164] A. Caballero, M. Cruz, L. Hernán, M. Melero, J. Morales, E.R. Castellón, Oxygen deficiency as the origin of the disparate behavior of LiM0. 5Mn1. 5 O 4 (M= Ni, Cu) nanospinels in lithium cells, J. Electrochem. Soc. 152 (3) (2005) A552−A559.

[165] G.Q. Liu, L. Wen, X. Wang, B.Y. Ma, Effect of the impurity LixNi1−xO on the electrochemical properties of 5V cathode material $LiNi_{0.5}Mn_{1.5}O_4$, J. Alloys Compd. 509 (38) (2011) 9377−9381.

[166] J. Zheng, J. Xiao, X. Yu, L. Kovarik, M. Gu, F. Omenya, X. Chen, X.-Q. Yang, J. Liu, G.L. Graff, Enhanced Li+ ion transport in LiNi 0.5 Mn 1.5 O 4 through control of site disorder, Phys. Chem. Chem. Phys. 14 (39) (2012) 13515−13521.

[167] J. Xiao, X. Chen, P.V. Sushko, M.L. Sushko, L. Kovarik, J. Feng, Z. Deng, J. Zheng, G.L. Graff, Z. Nie, High-performance LiNi0. 5Mn1. 5O4 spinel controlled by Mn^{3+} concentration and site disorder, Adv. Mater. 24 (16) (2012) 2109−2116.

[168] A. Ito, D. Li, Y. Lee, K. Kobayakawa, Y. Sato, Influence of Co substitution for Ni and Mn on the structural and electrochemical characteristics of LiNi0. 5Mn1. 5O4, J. Power Sources 185 (2) (2008) 1429−1433.

[169] C. Locati, U. Lafont, L. Simonin, F. Ooms, E. Kelder, Mg-doped LiNi0. 5Mn1. 5O4 spinel for cathode materials, J. Power Sources 174 (2) (2007) 847−851.

[170] J.S. Park, K.C. Roh, J.-W. Lee, K. Song, Y.-I. Kim, Y.-M. Kang, Structurally stabilized LiNi0. 5Mn1. 5O4 with enhanced electrochemical properties through nitric acid treatment, J. Power Sources 230 (2013) 138−142.

[171] N.P. Pieczonka, Z. Liu, P. Lu, K.L. Olson, J. Moote, B.R. Powell, J.-H. Kim, Understanding transition-metal dissolution behavior in LiNi0. 5Mn1. 5O4 high-voltage spinel for lithium ion batteries, J. Phys. Chem. C 117 (31) (2013) 15947−15957.

[172] J.-H. Kim, N.P. Pieczonka, Z. Li, Y. Wu, S. Harris, B.R. Powell, Understanding the capacity fading mechanism in LiNi0. 5Mn1. 5O4/graphite Li-ion batteries, Electrochim. Acta 90 (2013) 556−562.

[173] R. Alcantara, M. Jaraba, P. Lavela, J. Tirado, X-ray diffraction and electrochemical impedance spectroscopy study of zinc coated LiNi0. 5Mn1. 5O4 electrodes, J. Electroanal. Chem. 566 (1) (2004) 187−192.

[174] A. Eftekhari, Improving cyclability of 5 V cathodes by electrochemical surface modification, Chem. Lett. 33 (5) (2004) 616−617.

[175] T. Tsuji, T. Kakita, T. Hamagami, T. Kawamura, J. Yamaki, M. Tsuji, Preparation of nanoparticles of $LiCoO_2$ using laser ablation in liquids, Chem. Lett. 33 (9) (2004) 1136−1137.

[176] J.P. Park, J.Y. Park, C.H. Hwang, M.-h. Choi, J.E. Kim, K.M. Ok, I.-W. Shim, Synthesis of $LiCoO_2$ nanoparticles by a sonochemical method under the multibubble sonoluminescence conditions, Bull. Kor. Chem. Soc. 31 (2) (2010) 327−330.

[177] F. Jiao, K.M. Shaju, P.G. Bruce, Synthesis of nanowire and mesoporous low-temperature $LiCoO_2$ by a post-templating reaction, Angew. Chem. Int. Ed. 44 (40) (2005) 6550−6553.

[178] L. Xue, Q. Zhang, X. Zhu, L. Gu, J. Yue, Q. Xia, T. Xing, T. Chen, Y. Yao, H. Xia, 3D $LiCoO_2$ nanosheets assembled nanorod arrays via confined dissolution-recrystallization for advanced aqueous lithium-ion batteries, Nano Energy 56 (2019) 463−472.

[179] Y. Ou, J. Wen, H. Xu, S. Xie, J. Li, Ultrafine $LiCoO_2$ powders derived from electrospun nanofibers for Li-ion batteries, J. Phys. Chem. Solid. 74 (2) (2013) 322−327.

[180] N. Imachi, Y. Kodama, I. Yoshida, I. Nakane, S. Narukawa, Non-aqueous Electrolyte Cell Having a Positive Electrode with Ti-Attached $LiCoO_2$, Google Patents, 2002.

[181] C. Amemiya, J. Kurihara, M. Yonezawa, Improvement of cycle and storage performances of Li-ion rechargeable batteries with Mn-spinel cathode for HEV applications, NEC Res. Dev. 42 (2) (2001) 241−245.

CHAPTER 11

Prospects and challenges in the selection of polymer electrolytes in advanced lithium–air batteries

M.Z. Kufian[1], Z. Osman[1,2]

[1]Centre for Ionics University of Malaya (C.I.U.M), Faculty of Science, Universiti Malaya, Kuala Lumpur, Malaysia; [2]Department of Physics, Faculty of Science, Universiti Malaya, Kuala Lumpur, Malaysia

11.1 Introduction

Lithium–air batteries (LABs) comprise three major components, namely an Li anode, an electrolyte, and an air cathode. This chapter focuses on the mechanism of LABs and materials selection. Fig. 11.1A and B represents the number of publications reported from 1996 to 2018

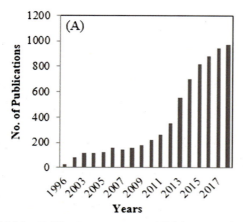 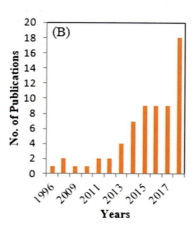

FIGURE 11.1 Publication progress on (A) lithium–air batteries (LABs) and (B) LABs with gel polymer electrolytes as ion transfer medium. Indexed in Scopus.

314   11. Prospects and challenges in the selection of polymer electrolytes

on LABs and also on LABs using gel polymer electrolytes (GPEs) as a medium for ion transport. It is clear that interest in LABs has increased over the span of more than 20 years. Research on GPEs for LABs application is still limited, although there have been many publications since 1996. The success of GPEs in lithium-ion batteries (LIBs) is widely known, although success has not reached the commercialization stage. GPEs are known to have good electrolyte—electrode stability [1,2], good ionic conductivity [2], and are leakproof.

11.2 Lithium—air batteries

LIBs with high specific energy are in increasing demand due to the advancement in the technologies of mobile electronics and electric vehicles (EVs). Unfortunately, the presently available LIB has relatively low energy density. This has forced researchers to find new advanced lithium battery technologies with higher energy density. The LAB is a potential candidate to replace LIB technology since these cells can theoretically exhibit maximum capacities of 3862 mAh/g [3]. The specific energy of LABs also depends on the electrolyte used. LABs using nonaqueous electrolyte can theoretically deliver a specific energy of ≈ 3500 Wh/kg [4]. This includes the oxygen mass during the discharge process and is much higher than the specific energy of LIBs, which is about 10% of the maximum theoretical specific energy of LABs [4—6]. In real energy storage situations, the specific energy of LABs may decrease due to the volume and mass of the LAB cell. According to Bruce et al. [4], LABs have specific energies between 500 and 900 Wh/kg.

The cathode of LABs is O_2 from the air. It is free and environmentally friendly. The fundamental battery chemistry of LABs is totally different from traditional LIB technology. Unlike LIBs, no transition-metal intercalation framework [7—10] is required in LABs. The process in LABs obeys a simple reversible exchange chemistry between Li^+ and O^{2-} [4].

11.3 Types of LABs

Depending on the nature of the electrolytes, LABs can be categorized as:

1. Nonaqueous LABs;
2. Aqueous LABs; and
3. Solid-state LABs.

11.3.1 Nonaqueous LABs

Nonaqueous LABs can contain either liquid or gel electrolyte. Abraham and Jiang [11] were the first to report on LABs in 1996. The medium for ion transport was a GPE. The authors used polyacrylonitrile (PAN) as the polymer host, which was dissolved in ethylene carbonate (EC) and propylene carbonate (PC). Configuration of the lithium—air cell was Li/PAN-lithium hexafluorophosphate ($LiPF_6$) in EC and a PC/air cathode and was operated at 1000 and 500 mA/m^2, respectively. The operating voltage was between 2.5 and 4.1 V. Hassoun et al. [12] also reported on LABs using GPE with polyethylene oxide (PEO) as polymer

III. Electrochemical energy conversion and storage

host. The ionic conductivity of the PEO-based GPE was of the order of 10^{-4} S/cm at room temperature (RT) [13]. The energy density of LABs using nonaqueous electrolyte is 3500 Wh/kg [4] following the reaction:

$$2Li + O_2 \rightarrow Li_2O_2 \tag{11.1}$$

Studies on nonaqueous LABs have received wide-scale interest. R&D has also escalated particularly for EVs. There is no technical basis to support the high energy densities and high specific areal capacity (area-normalized specific capacity) required for nonaqueous LABs to function efficiently [14]. Many challenges remain to be sorted out. These include electrolyte stability with $O_2^{\bullet-}$ nucleophilic attack [15], discharge reaction products that "disturb" lithium oxidation as in Eq. (11.1) [15,16], a high overpotential during operation [14], poor cyclability [17,18], and elimination of water and CO_2 from the air [19]. The working of a LAB with nonaqueous electrolyte is shown schematically in Fig. 11.2 [4].

$$\text{Anode: } Li \rightleftharpoons Li^+ + e^- \tag{11.2}$$
$$\text{Cathode: } O_2 + e^- \rightleftharpoons O_2^- \tag{11.3}$$
$$O_2^- + Li^+ \rightleftharpoons LiO_2 \tag{11.4}$$
$$LiO_2 + Li^+ + e^- \rightleftharpoons Li_2O_2 \tag{11.5}$$
$$\text{Overall reaction: } O_2 + 2Li \rightleftharpoons Li_2O_2 \tag{11.6}$$

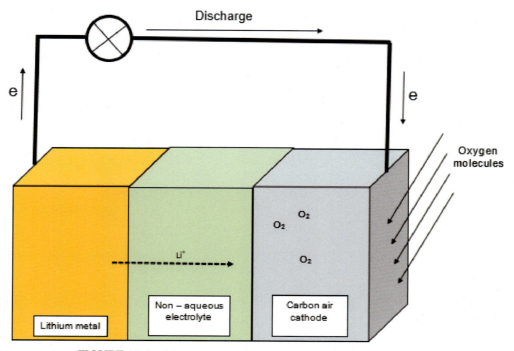

FIGURE 11.2 Schematic diagram of a nonaqueous lithium−air battery.

11.3.2 Aqueous LABs

As in nonaqueous LABs, aqueous LABs are also rechargeable. Visco et al. proposed this battery in 2004 [19]. The cathode of the battery is porous. The Li^+-ion conducting medium is an aqueous electrolyte and the anode is, of course, lithium.

As shown in Fig. 11.3, a solid protection film separates the Li metal from the aqueous electrolyte. This film is water stable and allows Li^+ to pass through. Two main issues in this type of LAB are energy density and stability of the water-stable protection film [20]. In aqueous LABs, H_2O also takes part in the reaction. Reaction (11.7) indicates this:

$$4Li + O_2 + 6H_2O \rightarrow 4(LiOH \cdot H_2O) \qquad (11.7)$$

The specific energy of the battery was computed following Eq. (11.7) [21]. The aqueous solvent can therefore reduce the battery's energy content. When the solubility limit of 2.5 g LiOH in 100 g of H_2O is reached at 25°C [21], LiOH will precipitate and fill the porous air electrode [20]. The phenomenon will prevent oxygen from outside reacting with the lithium ions. This will cause LAB death. To prevent LiOH from filling the pores in the air cathode, a lot of water is necessary. However, an attempt to fulfill this requirement will further reduce the specific energy.

The specific energy for aqueous LABs is 1910 Wh/kg [22]. This value is lower than the specific energy of LABs with nonaqueous electrolytes. The LiOH product dissolves in the

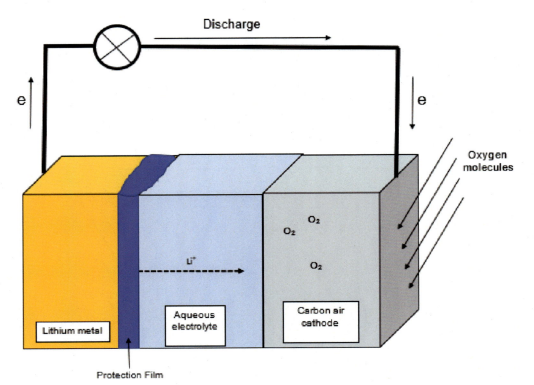

FIGURE 11.3 Schematic diagram of an aqueous lithium−air battery.

electrolyte. A water-impermeable and water-stable solid protection film such as lithium aluminum titanium phosphate (LATP) [23,24] separates the anode and the aqueous electrolyte. The separator must have good mechanical strength and high Li^+ conductivity. The specific area or volume capacity decreases when the mass or volume of the separator increases.

Aqueous LABs can consist of electrolytes that are alkaline or acidic. The electrode reactions are:

Anode:

$$Li \rightleftharpoons Li^+ + e^- \qquad (11.8)$$

Cathode:

$$(Alkaline)\ O_2 + 2H_2O + 4e^- \rightleftharpoons 4OH^- \qquad (11.9)$$
$$(Acid)\ O_2 + 4H^+ \rightleftharpoons 2H_2O \qquad (11.10)$$

11.3.3 Solid-state LABs

Kumar et al. introduced solid-state electrolyte in LABs in 2010 [25]. The battery consisted of an air electrode, an Li^+ solid electrolyte, and an anode of lithium metal, as shown in Fig. 11.4. LABs incorporated with solid-state electrolyte are nonflammable and do not suffer

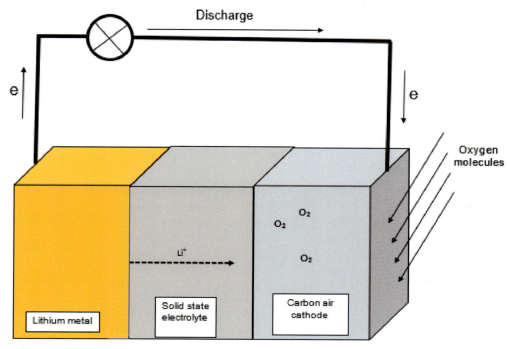

FIGURE 11.4 Schematic diagram of a solid-state electrolyte lithium–air battery.

318   11. Prospects and challenges in the selection of polymer electrolytes

from explosions as do nonaqueous electrolyte LABs. However, the conductivity of solid-state batteries with LATP-glass ceramic is low. The value is of the order 10^{-4} S/cm at RT [26] and interface contact of LATP-Li is not stable. To improve LATP-Li contact, a buffer layer will be added in between lithium and the conducting glass ceramic. PEO-lithium bis(trifluoromethanesulfonyl)imide (LiTFSI) polymer is an example of the buffer [27]. Although the buffer improved the LATP-Li contact, ionic conductivity decreased. This leds to an increase in resistance, which resulted in low capacity and efficiency [27].

Many reviews [16,17,20,22,28] on LABs have been published. These publications have reported the development and stages achieved. In this chapter, emphasis is focused on the research direction of polymer electrolytes in LABs.

11.4 Electrolytes

This section discusses the electrolyte component in the LAB system. Electrolytes are an essential component of any energy application. Electrolytes act as a medium in the transport of positive ions (in our case Li^+) between cathode and anode. In general, the electrolyte consists of lithium salts and organic solutions such as ether and ester groups (1,2-dimethoxyethane, tetrahydrofuran, and 1,3-dioxolane) [29—31] and carbonate groups (EC, PC, diethyl carbonate) [32—34]. From the literature, most electrolytes in electrochemical devices are liquids. However, liquid electrolytes pose problems of (1) leakage, which can cause electrode terminal corrosion, (2) low energy and low power density for the battery as these parameters are dependent on the total mass of the battery, and (3) limited temperature operation since the solvent chosen for dissolving the salt has its own boiling point.

To overcome the drawbacks of liquid electrolytes, polymer electrolytes (PEs) have been used. PEs exhibit excellent design flexibility, reasonable mechanical, chemical, electrochemical, and thermal stability, good electrolyte/electrode interfacial contact, and better safety features compared to liquid electrolytes. PEs have been prepared in solid form, quasi-solid state, or gel and can be added with fillers such as Al_2O_3, TiO_2, etc. to form composite polymer electrolytes. Studies on producing PEs with good performance that are suitable for device applications are still being conducted worldwide.

11.4.1 Solid polymer electrolyte

Solid polymer electrolytes (SPEs) are solvent-free electrolytes based on polymers. These polymers include poly(vinyl chloride) (PVC) [35], poly(ethylene oxide) (PEO) [36], chitosan [37], carboxymethyl cellulose (CMC) [38], poly(methyl methacrylate) (PMMA) [39], etc. These polymers have lone pair electrons. They are promising materials because ion conduction hosts can find application in LIBs. The electrolyte in these batteries currently used in cell phones and laptops is in the liquid state. In SPEs, ion conduction occurs in a phase created by the dissolution of salts with small lattice energy with polymers of high molecular weight.

Solution casting is the technique usually used to produce thin-film solid electrolytes [40—43]. SPEs function in a similar way to liquid electrolytes, but the RT ionic conductivity, σ, is low, being less than 10^{-3} S/cm [44—46]. However, SPEs have several advantages over liquid electrolytes. They are:

- Space saving since the electrolyte can be made small in size and miniaturize the devices;
- Weightless compared to liquid electrolytes;
- Easy to process;
- Flexible;
- Resistant to pressure; and
- Safe since they are not volatile.

It is to be noted that in their low conductivity state, SPEs are not suitable as electrolytes for device applications. Fig. 11.5 shows the ionic conductivity of solid, gel, and liquid electrolytes at RT compiled from the literature [47–57]. From the figure, it is obvious that GPEs exhibit higher conductivity compared to SPEs. Some GPEs exhibit RT conductivity that is comparable to the conductivity of liquid electrolytes.

Table 11.1 lists the conductivity of SPEs at RT. SPEs can achieve RT conductivity of 10^{-4} S/cm. This value of conductivity is insufficient for application in devices.

11.4.2 Gel polymer electrolytes

Due to the low conductivity for ion transport of SPE, i.e., less than 1 mS/cm, researchers have diverted their research to GPEs. A GPE is formed when an electrolyte in the liquid state

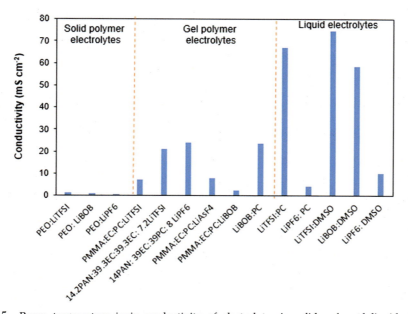

FIGURE 11.5 Room temperature ionic conductivity of electrolytes in solid, gel, and liquid states. *DMSO*, dimethyl sulfoxide; *EC*, ethylene carbonate; *LiBOB*, lithium bis(oxalato)borate; *LiPF$_6$*, lithium hexafluorophosphate; *LiTFSI*, lithium bis(trifluoromethanesulfonyl)imide; *PAN*, polyacrylonitrile; *PC*, propylene carbonate; *PEO*, poly(ethylene oxide); *PMMA*, poly(methyl methacrylate).

320   11. Prospects and challenges in the selection of polymer electrolytes

TABLE 11.1   Room temperature (298 K) ionic conductivity of solid polymer electrolytes containing lithium salts.

Solid polymer electrolyte	Conductivity, σ (S/cm)	References
PEO:PVDF:LiClO$_4$:SN	9.5×10^{-5}	[46]
PVA:gum Arabic:LiClO$_4$	1.6×10^{-4}	[58]
PVDF:PEO:LiClO$_4$	2.01×10^{-5}	[59]
CMC:LiBF$_4$	8.20×10^{-6}	[60]
60PMMA:40CH$_3$COOLi (in wt%)	8.21×10^{-5}	[61]
SPEEK-g-PEG:LiClO$_4$	$\sim 10^{-5}$	[62]
38PVA:62LiTFSI (in wt%)	1.0×10^{-4}	[63]
35.71PVDF−HFP:64.29LiCF$_3$SO$_3$ (in wt%)	1.64×10^{-4}	[64]
60PVA:40LiClO$_4$ (in wt%)	3.22×10^{-5}	[65]
74PAN:26LiCF$_3$SO$_3$ (in wt%)	3.04×10^{-4}	[66]
75PVA:25LiCF$_3$SO$_3$ (in mole %)	7.06×10^{-4}	[67]
35PVC:65LiCF$_3$SO$_3$ (in wt%)	2.40×10^{-5}	[68]
PVDF:LiCF$_3$SO$_3$	3.40×10^{-5}	[69]

CH$_3$COOLi, lithium acetate; *CMC*, sodium carboxymethyl cellulose; *LiBF$_4$*, lithium tetrafluoroborate; *LiCF$_3$SO$_3$*, lithium trifluoromethanesulfonate; *LiClO$_4$*, lithium perchlorate; *LiTFSI*, lithium bis(trifluoromethanesulfonyl)imide; *PMMA*, poly(methyl methacrylate); *PVA*, polyvinyl alcohol; *PVDF*, polyvinylidene fluoride; *PVDF−HFP*, poly(vinylidene fluoride-co-hexafluoropropylene); *SN*, succinonitrile.

is entrapped in a polymer network. Its physical state and ionic transport properties are between that of liquid and solid electrolytes [47]. By entrapping the liquid electrolyte in the polymer host, leakage is prevented [70]. In preparing the GPEs, plasticizers are used as solvents. Ion transport occurs via the encaged liquid phase. Fig. 11.6 shows the benefits of GPEs [71].

Previous studies have reported that conductivity is improved by gelling the SPE. Table 11.2 lists a comparison of RT conductivity between SPEs and GPEs. From the table, it can be observed that GPEs have a higher conductivity compared to SPEs. Yang and coresearchers [77] reported in 1996 that the conductivity of PAN:LiClO$_4$ SPE at RT was 6.51×10^{-7} S/cm. Song and coresearchers [78] improved the ionic transport of the PAN:LiClO$_4$ system by adding EC and PC to the polymer-salt system and achieved an ionic conductivity of 1.70 mS/cm at RT. Van Schalkwijk and Scrosati [73] added EC and dimethyl carbonate (DMC) to the PAN:LiClO$_4$ system and obtained an RT conductivity of 3.90 mS/cm, which is higher than the GPE reported by Song and coresearchers [78]. The relationship between conductivity and viscosity is written as [79] $\sigma = \frac{Zne^2}{6\pi r\eta}$, where ion valency is represented by Z, n is the concentration of free ions, e is the electron charge, r is the ionic radius, and η is

III. Electrochemical energy conversion and storage

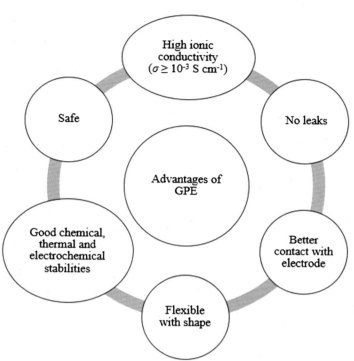

FIGURE 11.6 Advantages of gel polymer electrolyte (GPE).

TABLE 11.2 A comparison between solid polymer electrolyte and gel polymer electrolyte room temperature conductivity.

Solid polymer electrolyte		Gel polymer electrolyte	
Electrolyte	Conductivity, σ (S/cm)	Electrolyte	Conductivity, σ (S/cm)
PMMA:LiClO$_4$ [72]	7.00×10^{-8}	PMMA:EC:PC:LiClO$_4$ [73]	7.00×10^{-4}
PAN:LiCF$_3$SO$_3$ [66]	5.48×10^{-7}	PAN:EC:PC:LiCF$_3$SO$_3$ [74]	1.20×10^{-3}
PMMA:LiCF$_3$SO$_3$ [75]	9.88×10^{-4}	PMMA:EC:PC:LiCF$_3$SO$_3$ [76]	5.50×10^{-3}
PAN:LiClO$_4$ [77]	6.51×10^{-7}	PAN:EC:PC:LiClO$_4$ [78]	1.70×10^{-3}
		PAN:EC:DMC:LiClO$_4$ [73]	3.90×10^{-3}

DMC, dimethyl carbonate; *EC*, ethylene carbonate; *PAN*, polyacrylonitrile; *PC*, propylene carbonate; *PMMA*, poly(methyl methacrylate).

the electrolyte viscosity. The PAN:LiClO$_4$ system containing EC and PC exhibited a lower conductivity compared to the same polymer-salt system containing EC and DMC. This has been attributed to the viscosity of PC, which is 2.54 cP [80]. The viscosity of PC (2.54 cP) is higher than the viscosity of DMC (0.59 cP) [81].

Another example reported by Osman et al. [66] showed that the conductivity of 88 wt% PAN:12 wt% LiCF$_3$SO$_3$ SPE without a plasticizer system is 5.48×10^{-7} S/cm. Perera et al. [74] reported that the addition of 79.62 wt% of EC and PC (w/w = 1) to 8.59 wt% PAN:11.79 wt% LiCF$_3$SO$_3$ improved conductivity of the sample. With plasticizer, the conductivity can be enhanced to 1.20×10^{-3} S/cm.

11.5 Polymer host

To date, many polymers have been used as host in preparing SPEs and GPEs. One of the preferred polymers is PMMA. PMMA is an acrylic or acrylic glass, or Plexiglas, which is used as an alternative material to replace glass. PMMA is also a hydrophobic synthetic polymer. Due to its hydrophobicity, it can also be used in lighting, automotive and transportation, electronics, and medical and healthcare. The chemical formula of PMMA is $(C_5O_2H_8)_n$. The PMMA monomer has a molecular structure as shown in Fig. 11.7.

PMMA polymer was discovered by Appetecchi et al. [48,82]. These researchers compared the electrochemical stability of PMMA- and PAN-based GPEs added with lithium perchlorate, LiClO$_4$, as lithium-ion donor with a lithium electrode. They monitored the electrochemical stability of Li/PMMA GPE and Li/PAN systems by observing how impedance of an Li/GPE/Li changed with time. The temporal changes of lithium/PMMA and lithium/PAN interfacial resistance can be monitored through an electrochemical stability test. It has been clearly shown that the interfacial resistance of PAN-based cells increased with the continuous expansion of a passive film on the lithium anode with storage time. However, for PMMA-based cells, the interfacial resistance slowly approached a constant value with increased storage time indicating that the passivation layer on the surface of the lithium metal electrode had fully developed. From this argument, the interfacial stability between electrode and electrolyte is better manifested by PMMA GPE compared to PAN GPE. Apart from that, the stability of a polymer with a LAB discharge product, for example, Li$_2$O$_2$ also needs to be considered. Chibueze and coresearchers [83] reported the stability of several polymers with

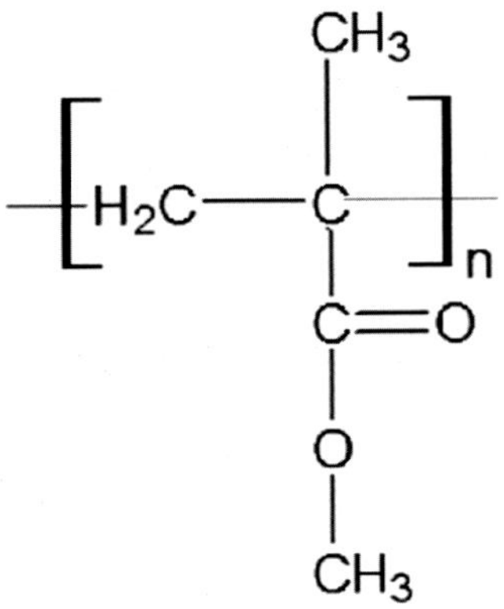

FIGURE 11.7   Molecular structure of poly(methyl methacrylate) monomer.

III. Electrochemical energy conversion and storage

TABLE 11.3 Chemical reactivity observation of polymers dissolved in dimethylformamide solution (except poly(vinylidene fluoride) [PDVF] dissolved in *N,N*-dimethylacetamide) exposed with Li_2O_2 [83].

Polymer	Observation
PAN	Polymer solution changed from clear and transparent to a yellow color
PMMA	No change observed
PVDF	Polymer solution changed from clear and transparent to slight orange and darker after 3 days
PVDF-HFP	Polymer solution changed from clear and transparent to slight orange and darker after 3 days

PAN, polyacrylonitrile; *PMMA*, poly(methyl methacrylate); *PVDF-HFP*, poly(vinylidene fluoride-*co*-hexafluoropropylene).

Li_2O_2 such as PAN, PMMA, poly(vinylidene fluoride) (PVDF), and poly(vinylidene fluoride-*co*-hexafluoropropylene) (PVDF-HFP). Except for PVDF, which dissolved in *N,N*-dimethylacetamide (DMAc), all other polymers dissolved in dimethylformamide (DMF). When Li_2O_2 was added to the polymer-solvent mixture, say PAN-DMF mixture, the PAN-DMF-Li_2O_2 gel polymer mixture turned yellowish from the clear and transparent PAN-DMF mixture within minutes. After 1 day, the solution turned an orange color and at the end of 2 days attained a permanent dark red hue. However, for PMMA-DMF-Li_2O_2 gel polymer mixture, the solution appeared stable without discoloration. Table 11.3 shows the chemical reactivity of PAN-, PMMA-, PVDF-, and PVDF-HFP-based solutions. From the table, only PMMA polymer does not demonstrate any chemical reactivity visually when added with Li_2O_2.

To confirm the chemical reactivity of polymer-solvent-Li_2O_2 gels, several measurements such as UV-Vis and Fourier transform infrared (FTIR) spectroscopy were conducted for every mixture [72].

11.5.1 PAN

UV-Vis measurement revealed that absorbance increased at the end of the chemical activity reaction of PAN-DMF-Li_2O_2 due to PAN degradation. FTIR spectroscopy indicated an intensity reduction at the $C\equiv N$ wavenumber from PAN as the reaction with Li_2O_2 was prolonged. This decrease revealed that Li_2O_2 attacked the $C\equiv N$ group, which led to the formation of a conjugated imine-like ($-C=N-C=N-$) species that accounts for the PAN-DMF-Li_2O_2 color changes. A new band that peaked at $2195\ cm^{-1}$, which can be assigned to a β-amino nitrile, was observed [84].

11.5.2 PVDF and PVDF-HFP

UV-Vis absorbance was also observed to increase with prolonged polymer-solvent mixture reaction with Li_2O_2. This was also attributed to PVDF and PVDF-HFP degradation. The polymer-solvent-Li_2O_2 mixture changed from a clear solution to dark after 3 days of storage time. FTIR spectra of the PVDF-DMAc-Li_2O_2 and the PVDF-HFP-DMF-Li_2O_2 observed a new vibration peak at around $1650\ cm^{-1}$ after 72 h storage time. This new peak can be assigned to conjugated alkene-like degradation products.

11.5.3 PMMA

FTIR spectra of the PMMA-DMF-Li$_2$O$_2$ polymer solution do not show any new vibration peak in comparison with PMMA-DMF polymer solution as control sample. This analysis supported stability of the PMMA-DMF-Li$_2$O$_2$ mixture.

11.6 Type of salt

For good electrolyte performance, the lithium salt that is to provide the cationic carriers must possess the following requirements:

1. Easy dissolution for good solubility in the solvent;
2. Wide and stable potential window;
3. Good stability with lithium metal anode; and
4. Good stability with Li$_2$O$_2$ discharge product.

A variable of crucial importance to consider for nonaqueous electrolytes is salt solubility. This is related to several factors such as (1) dielectric constant of the solvent, (2) polarity of the solvent, and (3) lattice energy of the salt [85]. There are many lithium-ion donors for lithium-ion or lithium–air batteries. In this chapter, emphasis is placed on two types of salt, namely lithium bis(oxalato)borate (LiBOB) and lithium bis(trifluoromethanesulfonyl)imide (LiTFSI). For the sake of comparison, the widely used lithium hexafluorophosphate (LiPF$_6$) is also included in the discussion. Table 11.4 lists useful information for the three types of salt. Anionic radius is one of the important parameters that need to be considered. The radius for LiPF$_6$ salt is 2.6 Å and that of LiBOB and LiTFSI salts are 3.2 and 3.3 Å, respectively. This implies that the radius of PF$_6^-$ is smaller than BOB$^-$ and TFSI$^-$ anions. According to Hassan et al. [86], ionic conductivity of GPEs is higher when salts of larger anion size are used. This may also be due to the lattice energy in between the Li$^+$-anion species. Lattice energy is the energy required to separate 1 mol of salt into its cations and anions. From Coulomb's law, the magnitude of the electrostatic attractive or repulsive force between the anion and cation is given as:

$$F = k\frac{q^+q^-}{r^2} \tag{11.11}$$

TABLE 11.4 Some information on LiBOB, LiTFSI, and LiPF$_6$ salts.

Properties	LiBOB	LiTFSI	LiPF$_6$
Molecule structure			
Molecular weight (g/mol)	193.79	287.09	151.91
Anion radius (Å)	3.2 [86]	3.3 [86]	2.6 [86]
Volume (Å^3)	143.7 [86]	147.6 [86]	72.6 [86]

LiBOB, lithium bis(oxalato)borate; *LiPF$_6$*, lithium hexafluorophosphate; *LiTFSI*, lithium bis(trifluoromethanesulfonyl)imide.

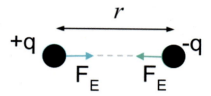

FIGURE 11.8 Coulomb's law.

where the Coulomb constant is designated as k. The cationic and anionic charges are represented by q^+ and q^-, respectively. The distance between the charges is indicated by r. The situation is depicted in Fig. 11.8. From this equation, we can predict that the lattice energy will decrease with the charge magnitude and the ionic size.

Since $TFSI^-$, BOB^-, and PF_6^- anions have similar charges (-1), prediction of lattice energy depends only on the ion size. In Table 11.4, the volume and radius of anions follow $TFSI^- > BOB^- > PF_6^-$. So, from this hypothesis, the lattice energy of an ionic compound containing $TFSI^-$ anions is less than BOB^-, followed by PF_6^-. From this argument, we can predict that salts with lower lattice energy will produce higher ionic conductivity since the salt will be easily dissociated into ions [86]. Table 11.5 lists the ionic conductivities of PEO:LiTFSI, PEO:LiBOB, and PEO:LiPF$_6$ being 1.43 S/cm × 10^{-4} S/cm, 8.45 S/cm × 10^{-5} S/cm, and

TABLE 11.5 Room temperature conductivity values of some solid polymer electrolytes, gel polymer electrolytes, and liquid electrolytes.

Electrolyte	Conductivity, σ (S/cm)	References
Solid polymer electrolytes		
PEO:LiTFSI	1.43×10^{-4}	
PEO:LiBOB	8.45×10^{-5}	[86]
PEO:LiPF$_6$	4.18×10^{-5}	
Gel polymer electrolytes		
PMMA:EC:PC:LiTFSI	0.70×10^{-3}	[73]
PMMA:EC:PC:LiBOB	0.25×10^{-3}	[51]
Liquid electrolytes		
LiBOB:PC	2.38×10^{-3}	[87]
LiTFSI:PC	6.71×10^{-3}	
LiPF$_6$:PC	0.41×10^{-3}	[88]
LiTFSI:DMSO	7.48×10^{-3}	[89]
LiBOB:DMSO	5.86×10^{-3}	[89]

DMSO, dimethyl sulfoxide; *EC*, ethylene carbonate; *LiBOB*, lithium bis(oxalato)borate; *LiPF$_6$*, lithium hexafluorophosphate; *LiTFSI*, lithium bis(trifluoromethanesulfonyl)imide; *PC*, propylene carbonate; *PEO*, poly(ethylene oxide); *PMMA*, poly(methyl methacrylate).

326   11. Prospects and challenges in the selection of polymer electrolytes

$4.18 \, S/cm \times 10^{-5} \, S/cm$, respectively. Similar observations were made by Van Schalkwijk and Scrosati [73] and Kufian et al. [51] on conductivity studies of PMMA:EC:PC gel electrolyte and PMMA-based gel electrolyte both of which contained LiTFSI salt. The conductivity of electrolytes containing LITFSI was higher than GPEs containing LiBOB. This is also true for liquid electrolytes. Electrolytes in the liquid state containing dimethyl sulfoxide (DMSO) exhibited a higher conductivity compared to liquid electrolytes containing PC since DMSO has a higher dielectric constant than PC [80]. Apart from that, LiTFSI also showed good stability against Li_2O_2 [90] compared to LiBOB and $LiPF_6$. However, LITFSI, which is usually used in LIBs, would corrode the aluminum current collector [91].

11.7 Anion trapping agent

High cationic conductivity and transference number of lithium ions are also crucial parameters that need consideration if the electrolyte is to be applied in battery applications. The transference number of lithium ion or t_{Li^+} is the proportion of the Li^+ current to the electric current contributed by Li^+ and the anion of the salt. An Li^+-ion transference number that is almost unity implies that the main ionic carrier is the Li^+. A high t_{Li^+} can reduce electrolyte concentration polarization while the battery is in operation. This can result in a higher power density. To improve these parameters, several strategies have been adopted. These include:

1. Introducing large and heavy anions [51,86];
2. Ceramic fillers [92–95];
3. Ionic liquids [95,96]; and
4. Addition of anion trapping agent [97–103].

In this chapter, focus is only on anion trapping agent as the strategy to improve the lithium-ion transference number. The increase or decrease in t_{Li^+} depends on the amount of free Li^+ in the electrolyte. The existence of a substantial number of contact ion pairs and ionic triplets in electrolyte will reduce free lithium cations and this will lower the lithium-ion transference number. Previous studies used, e.g., calix [4]arene [101], as the anion trapping agent in the low dielectric constant of polyethers and low lithium salts concentration, where ion pairs and aggregates were in lower quantities in electrolytes with low salt concentrations. However, most studies use high salt concentration to achieve high ionic conductivity. Thus there is the possibility of a large quantity of ion pairs and aggregates present. Addition of anion trapping agent or receptor in the electrolyte system will therefore prevent anion movement [94]. Hence, this will ensure that there will be more lithium ions in the electrolyte. There will be fewer contact ions and triplets as the Li^+ ions cannot interact with the anions to form these entities. To the best of our knowledge, reports on the use of anion trapping agent in GPEs and SPEs are scarce. Calix [4]arene [101] and calix [6]pyrrole [99] have been used as anion trapping agents in electrolyte systems. It is to be noted that the addition of this trapping agent would reduce ionic conductivity, since there are fewer contributing ions as the anions have been immobilized. The decrease in the conductivity is also attributed to increased viscosity of the sample [101]. Table 11.6 presents values of ionic conductivity of electrolytes that contain calix [4]arene [101] and calix [6]pyrrole [99].

III. Electrochemical energy conversion and storage

11.7 Anion trapping agent
327

TABLE 11.6 Room temperature conductivity (298 K) of composite gel polymer electrolytes with lithium-based salt.

Gel polymer electrolyte			Composite gel polymer electrolyte		
Sample	Conductivity, σ (S/cm)	References	Sample	Conductivity, σ (S/cm)	References
PEO:LiI:DME	0.33×10^{-3}	[100]	PEO:LiI:DME:calix [4] arene	0.19×10^{-3}	[100]
PEO-LiTFSI	1.45×10^{10}	[99]	PEO-LiTFSI-calix [6] pyrrole	7.29×10^{8}	[99]
PEO-LiBF$_4$	2.34×10^{-4}	[101]	PEO-LiBF$_4$-calix [6] pyrrole	9.65×10^{-5}	[101]

DME, dimethoxyethane; *LiBF$_4$*, lithium tetrafluoroborate; *LiI*, lithium iodide; *LiTFSI*, lithium bis(trifluoromethanesulfonyl)imide; *PEO*, poly(ethylene oxide).

TABLE 11.7 Values of t_{Li^+} for electrolytes with a poly(ethylene oxide) (PEO) matrix. The electrolytes were doped with lithium salts and contained calix-6-pyrrole anion trapping agent [99].

Type of the electrolyte	Molar fraction of calix-6-pyrrole	Lithium-ion transference number
(PEO)$_{20}$LiI	0	0.25
(PEO)$_{20}$LiI	0.125	0.56
(PEO)$_{20}$LiAsF$_6$	0	0.44
(PEO)$_{20}$LiAsF$_6$	0.5	0.84
(PEO)$_{20}$LiBF$_4$	0	0.32
(PEO)$_{20}$LiBF$_4$	0.125	0.78
(PEO)$_{20}$LiBF$_4$	0.25	0.81
(PEO)$_{20}$LiBF$_4$	0.5	0.85
(PEO)$_{100}$LiBF$_4$	0.25	0.95
(PEO)$_{100}$LiBF$_4$	1	0.92
(PEO)$_{20}$LiCF$_3$SO$_3$	0	0.45
(PEO)$_{20}$LiCF$_3$SO$_3$	0.125	0.68

Table 11.7 lists values of t_{Li^+} for electrolytes with PEO host that contain calix-6-pyrrole anion trapping agent as reported by Hekselman et al. [99]. The addition of 0.125 mol fraction of the receptor increased the t_{Li^+} for PEO$_{20}$LiI, PEO$_{20}$LiBF$_4$, and PEO$_{20}$LiCF$_3$SO$_3$. The increase in t_{Li^+} is particularly clear for the electrolytes with LiAsF$_6$ and LiBF$_4$ salts.

11.8 Solvents

The type of solvents for GPEs is also important. Organic solvents that have been used to dissolve salts include ethers, sulfoxides, amides, nitriles, and carbonates. LABs are operated in open environments exposed to air or oxygen; organic solvents may not be suitable in LAB systems. Organic solvents for LABs should possess [104]:

1. Small vapor pressure;
2. Wide electrochemical aperture;
3. Good chemical stability; and
4. High diffusivity of oxygen and solubility of Li salt.

The favored solvent in LAB is the ether group. This is because of the following properties [17]:

1. Low volatility;
2. Compatibility with lithium;
3. Decomposition at more than 4.5 V versus Li/Li$^+$; and
4. Economy.

Solvent stability against $O_2^{\bullet-}$ nucleophilic attack must be considered in selecting the appropriate solvent for LABs. Nucleophilic attack of $O_2^{\bullet-}$ is the attack by $O_2^{\bullet-}$ toward functional groups such as carbonyls (CO) and sulfoxides (SO) [105]. This attack will cause the degradation or decomposition of the solvent used in electrolytes [14,15]. In PC solvent, nucleophilic attack of $O_2^{\bullet-}$ occurs at the O-alkyl carbon [15].

Researchers have reported that when using ether-based electrolytes in LABs, the main discharge product is Li_2O_2 [90]. A computational study [15] has proven that ethers show good stability against nucleophilic attack by superoxides. Among the ethers, tetraethylene glycol dimethyl ether (TEGDME) has low volatility and is suitable as a solvent for LABs; see Fig. 11.9.

TEGDME is a polar solvent with a wide potential window and dissolves Li salt efficiently [101]. Its molecular structure is depicted in Fig. 11.9. Christy et al. [16] investigated LAB using an a-MnO$_2$/N-GNF-catalyzed cathode. They also used different organic solvents. Good rechargeability (~99%), cyclability, and stability were observed in cells employing TEGDME. There is still no electrolyte that exhibits good stability for application in LABs. Currently,

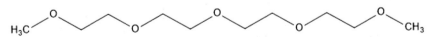

FIGURE 11.9 Molecular structure of tetraethylene glycol dimethyl ether polymer.

TEGDME is the closest to meeting stringent requirements. From the literature [102], LiTFSI-TEGDME-(PVDF-HFP) electrolyte revealed electrochemical stability as high as 4.6 V versus Li/Li^+. Without the polymer component, the electrolyte decomposition voltage was lower. The search for a suitable solvent for use in LABs is a challenge.

11.9 Summary

The parameters for a good electrolyte in LABs have been discussed. LABs with nonaqueous electrolytes still showed a higher specific energy of 3500 Wh/kg when compared to LABs with aqueous and solid-state electrolytes. Moreover, nonaqueous electrolytes have good chemical stability with Li_2O_2. Since liquid electrolytes are prone to leaks, only SPEs would be a suitable choice. However, the low conductivity of SPEs at the present state of development forced the use of GPEs. To achieve the performance of LABs with nonaqueous electrolytes, polymer, solvent, and salt have to be properly selected. Thus a combination of PMMA, TEGDME solvent, and LiTFSI salt is suggested. Addition of anion trapping agents or receptors can also help to enhance t_{Li^+} and the total performance of LABs.

Acknowledgments

We thank the support of the University of Malaya, PPP grant PG011-2013B, and Prof. A.K. Arof for sharing ideas and comments.

References

[1] M.H. Abdul Rahaman, M.U. Khandaker, Z.R. Khan, M.Z. Kufian, I.S. Noor, A.K. Arof, Effect of gamma irradiation on poly(vinylidene difluoride)-lithium bis(oxalato)borate electrolyte, Phys. Chem. Chem. Phys. 16 (2014) 11527–11537.

[2] A. Manuel Stephan, K.S. Nahm, Review on composite polymer electrolytes for lithium batteries, Polymer 16 (2006) 12.

[3] J.P. Zheng, R.Y. Liang, M. Hendrickson, E.J. Plichta, Theoretical energy density of Li–air batteries, J. Electrochem. Soc. 155 (2008) A432.

[4] P.G. Bruce, S.A. Freunberger, L.J. Hardwick, J.M. Tarascon, Li-O_2 and Li-S batteries with high energy storage, Nat. Mater. 11 (2011) 19–29.

[5] J.G.P. Kurzweil, Overview of Batteries for Future Automobiles, 2017.

[6] H.R.J.P. Tan, X.B. Zhu, L. An, C.Y. Jung, M.C. Wu, L. Shi, W. Shyy, T.S. Zhao, Advances and challenges in lithium-air batteries, Appl. Energy 204 (2017).

[7] A.K. Arof, M.Z. Kufian, N. Aziz, N.A.M. Nor, K.H. Arifin, Electrochemical properties of $LiMn_2O_4$ prepared with tartaric acid chelating agent, Ionics 23 (2017) 1663–1674.

[8] N.D. Phillip, A.S. Westover, C. Daniel, G.M. Veith, Structural degradation of high voltage lithium nickel manganese cobalt oxide (NMC) cathodes in solid-state batteries and implications for next generation energy storage, ACS Appl. Energy Mater. 3 (2020) 1768–1774.

[9] P.M. Zehetmaier, A. Cornelis, F. Zoller, B. Boller, A. Wisnet, M. Doblinger, D. Bohm, T. Bein, D. Fattakhova-Rohlfing, Nanosized lithium-rich cobalt oxide particles and their transformation to lithium cobalt oxide cathodes with optimized high-rate morphology, Chem. Mater. 31 (2019) 8685–8694.

[10] R. Methekar, S. Anwani, Manufacturing of lithium cobalt oxide from spent lithium-ion batteries: a cathode material, Adv. Intell. Syst. 757 (2019) 233–241.

[11] K.M. Abraham, Z. Jiang, A polymer electrolyte-based rechargeable lithium/oxygen battery, J. Electrochem. Soc. 143 (1996) 1–5.

[12] G.A. Elia, J. Hassoun, A polymer lithium-oxygen battery, Sci. Rep. 5 (2015) 12307.

[13] D.-J.J. Hassoun, Y.-K. Sun, B. Scrosati, A lithium ion battery using nanostructured Sn—C anode, LiFePO$_4$ cathode and polyethylene oxide-based electrolyte, Solid State Ionics 202 (2011) 4.

[14] N. Imanishi, O. Yamamoto, Rechargeable lithium—air batteries: characteristics and prospects, Mater. Today 17 (2014) 24—30.

[15] V.S. Bryantsev, V. Giordani, W. Walker, M. Blanco, S. Zecevic, K. Sasaki, J. Uddin, D. Addison, G.V. Chase, Predicting solvent stability in aprotic electrolyte Li—air batteries: nucleophilic substitution by the superoxide anion radical (O$_2^{\bullet-}$), J. Phys. Chem. A 115 (2011) 12399—12409.

[16] M. Christy, A. Arul, Z. Awan, K. Moon, M. Oh, A. Stephan, K.-S. Nahm, Role of solvents on the oxygen reduction and evolution of rechargeable Li-O$_2$ battery, J. Power Sources 342 (2017) 825—835.

[17] H. Guo, W. Luo, J. Chen, S. Chou, H. Liu, J. Wang, Review of electrolytes in nonaqueous lithium—oxygen batteries, Adv. Sustainable Syst. 2 (2018) 1700183.

[18] B.D. McCloskey, R. Scheffler, A. Speidel, D.S. Bethune, R.M. Shelby, A.C. Luntz, On the efficacy of electrocatalysis in nonaqueous Li—O$_2$ batteries, J. Am. Chem. Soc. 133 (2011) 18038—18041.

[19] J. Zhang, W. Xu, X. Li, W. Liu, Air dehydration membranes for nonaqueous lithium—air batteries, J. Electrochem. Soc. 157 (2010) A940.

[20] P. Tan, H.R. Jiang, X.B. Zhu, L. An, C.Y. Jung, M.C. Wu, L. Shi, W. Shyy, T.S. Zhao, Advances and challenges in lithium-air batteries, Appl. Energy 204 (2017) 780—806.

[21] D.R. Lide (Ed.) National Institute of Standards and Technology, CRC Handbook of Chemistry and Physics, eightyfourth edition, CRC Press LLC, Boca Raton, 2003, 2616 pp. $139.95. ISBN 0-8493-0484-9, J. Am. Chem. Soc. 126 (2004), 1586.

[22] N. Imanishi, A.C. Luntz, P. Bruce, The Lithium Air Battery: Fundamentals, Springer, 2014.

[23] C. Peng, Y. Kamiike, Y. Liang, K. Kuroda, M. Okido, Thin-film NASICON-type Li$_{1+x}$Al$_x$Ti$_{2-x}$(PO$_4$)$_3$ solid electrolyte directly fabricated on a graphite substrate with a hydrothermal method based on different Al sources, ACS Sustainable Chem. Eng. 7 (2019) 10751—10762.

[24] M. Kotobuki, M. Koishi, Preparation of Li$_{1.5}$Al$_{0.5}$Ti$_{1.5}$(PO$_4$)$_3$ solid electrolyte via a sol—gel route using various Al sources, Ceram. Int. 39 (2013) 4645—4649.

[25] B. Kumar, J. Kumar, R. Leese, J.P. Fellner, S.J. Rodrigues, K.M. Abraham, A solid-state, rechargeable, long cycle life lithium—air battery, J. Electrochem. Soc. 157 (2010) A50.

[26] Z. Tao, I. Nobuyuki, T. Yasuo, Y. Osamu, Aqueous lithium/air rechargeable batteries, Chem. Lett. 40 (2011) 668—673.

[27] T. Zhang, N. Imanishi, S. Hasegawa, A. Hirano, J. Xie, Y. Takeda, O. Yamamoto, N. Sammes, Li/Polymer Electrolyte/Water stable lithium-conducting glass ceramics composite for lithium—air secondary batteries with an aqueous electrolyte, J. Electrochem. Soc. 155 (2008) A965.

[28] M. Balaish, A. Kraytsberg, Y. Ein-Eli, A critical review on lithium—air battery electrolytes, Phys. Chem. Chem. Phys. 16 (2014) 2801—2822.

[29] Y. Matsuda, M. Morita, Organic electrolyte solutions for rechargeable lithium batteries, J. Power Sources 20 (1987) 273—278.

[30] Z. Lin, Q. Xia, W. Wang, W. Li, S. Chou, Recent research progresses in ether- and ester-based electrolytes for sodium-ion batteries, InfoMat 1 (2019) 376—389.

[31] X. Ren, L. Zou, S. Jiao, D. Mei, M.H. Engelhard, Q. Li, H. Lee, C. Niu, B.D. Adams, C. Wang, J. Liu, J.-G. Zhang, W. Xu, High-concentration ether electrolytes for stable high-voltage lithium metal batteries, ACS Energy Lett. 4 (2019) 896—902.

[32] J. Arai, Nonflammable methyl nonafluorobutyl ether for electrolyte used in lithium secondary batteries, J. Electrochemi. Soc. 150 (2003).

[33] A.M. Haregewoin, E.G. Leggesse, J.-C. Jiang, F.-M. Wang, B.-J. Hwang, S.D. Lina, Comparative study on the solid electrolyte interface formation by the reduction of alkyl carbonates in lithium ion battery, Electrochim. Acta 136 (2014) 11.

[34] R. Miao, J. Yang, Z. Xu, J. Wang, Y. Nuli, L. Sun, A new ether-based electrolyte for dendrite-free lithium-metal based rechargeable batteries, Sci. Rep. 6 (2016) 21771.

[35] J. Mindemark, M.J. Lacey, T. Bowden, D. Brandell, Beyond PEO—alternative host materials for Li$^+$-conducting solid polymer electrolytes, Prog. Polym. Sci. 81 (2018) 114—143.

[36] S. Ramesh, G.B. Teh, R.-F. Louh, Y.K. Hou, P.Y. Sin, L.J. Yi, Preparation and characterization of plasticized high molecular weight PVC-based polymer electrolytes, Sadhana 35 (2010) 87—95.

References   331

[37] Z. Osman, Z.A. Ibrahim, A.K. Arof, Conductivity enhancement due to ion dissociation in plasticized chitosan based polymer electrolytes, Carbohydr. Polym. 44 (2001) 167–173.

[38] A.S. Samsudin, M.A. Saadiah, Ionic conduction study of enhanced amorphous solid bio-polymer electrolytes based carboxymethyl cellulose doped NH_4Br, J. Non-Cryst. Solids 497 (2018) 19–29.

[39] S. Ramesh, L.Z. Chao, Investigation of dibutyl phthalate as plasticizer on poly(methyl methacrylate)–lithium tetraborate based polymer electrolytes, Ionics 17 (2011) 29–34.

[40] A.L. Tipton, M.C. Lonergan, M.A. Ratner, D.F. Shriver, T.T.Y. Wong, K. Han, Conductivity and dielectric constant of PPO and PPO-based solid electrolytes from Dc to 6 GHz, J. Phys. Chem. 98 (1994) 4148–4154.

[41] K.P. Singh, R.P. Singh, P.N. Gupta, Polymer based solid state electrochromic display device using PVA complex electrolytes, Solid State Ionics 78 (1995) 223–229.

[42] S. Rajendran, T. Uma, Experimental investigations on $PVC-LiAsF_6-DBP$ polymer electrolyte systems, J. Power Sources 87 (2000) 218–222.

[43] S. Ramesh, A.H. Yahaya, A.K. Arof, Dielectric behaviour of PVC-based polymer electrolytes, Solid State Ionics 152–153 (2002) 291–294.

[44] G.P. Kalaignan, M.-S. Kang, Y.S. kang, Effects of compositions on properties of $PEO-KI-I_2$ salts polymer electrolytes for DSSC, Solid State Ionics 177 (2006) 1091–1097.

[45] Z. Lin, X. Guo, Y. Yang, M. Tang, Q. Wei, H. Yu, Block copolymer electrolyte with adjustable functional units for solid polymer lithium metal battery, J. Energy Chem. 52 (2021) 67–74.

[46] H. Wang, C. Lin, X. Yan, A. Wu, S. Shen, G. Wei, J. Zhang, Mechanical property-reinforced PEO/PVDF/ $LiClO_4$/SN blend all solid polymer electrolyte for lithium ion batteries, J. Electroanal. Chem. 869 (2020) 114156.

[47] A.R. Karuppasamy K, S. Alwin, S. Balakumar, X. Sahaya Shajan, A review on PEO based solid polymer electrolytes (SPEs) complexed with LiX (X=Tf, BOB) for rechargeable lithium ion batteries, Mater. Sci. Forum (2014) 22.

[48] F. Croce, L. Settimi, B. Scrosati, D. Zane, Nanocomposite, PEO-LiBOB polymer electrolytes for low temperature, lithium rechargeable batteries, J. N. Mater. Electrochem. Syst. 9 (2006).

[49] W. Gorecki, M. Jeannin, E. Belorizky, C. Roux, M. Armand, Physical properties of solid polymer electrolyte PEO(LiTFSI) complexes, J. Phys. Condens. Matter 7 (1995) 6823–6832.

[50] I. Gunathilaka, L.R.A.K. Bandara, A.K. Arof, M. Careem, V. Seneviratne, Electrical and structural studies of a LiBOB-based gel polymer electrolyte, Ionics 23 (2016) 2669–2675.

[51] M.Z. Kufian, M.F. Aziz, M.A. Shukur, A.S. Rahim, N. Ariffin, N.E.A. Shuhaimi, S.R. Majid, R. Yahya, A.K. Arof, PMMA–LiBOB gel electrolyte for application in lithium ion batteries, Solid State Ionics 208 (2012) 36–42.

[52] M. Dahbi, F. Ghamouss, F. Tran-Van, D. Lemordant, M. Anouti, Comparative study of EC/DMC LiTFSI and $LiPF_6$ electrolytes for electrochemical storage, J. Power Sources 196 (2011) 9743–9750.

[53] Z. Wang, B. Huang, H. Huang, L. Chen, R. Xue, F. Wang, Infrared spectroscopic study of the interaction between lithium salt LiC_1O_4 and the plasticizer ethylene carbonate in the polyacrylonitrile-based electrolyte, Solid State Ionics 85 (1996) 143–148.

[54] P.A.R.D. Jayathilaka, M.A.K.L. Dissanayake, I. Albinsson, B.E. Mellander, Dielectric relaxation, ionic conductivity and thermal studies of the gel polymer electrolyte system PAN/EC/PC/LiTFSI, Solid State Ionics 156 (2003) 179–195.

[55] L.-Z. Fan, T. Xing, R. Awan, W. Qiu, Studies on lithium bis(oxalato)-borate/propylene carbonate-based electrolytes for Li-ion batteries, Ionics 17 (2011) 491–494.

[56] F. Chen, J. Hu, Z. Chen, Z. Yang, N. Gu, Determination and correlation of solubilities of lithium bis(oxalate) borate in six different solvents from (293.15 to 363.15) K, J. Chem. Eng. Data 59 (2014) 1614–1618.

[57] M. Morita, F. Tachihara, Y. Matsuda, Dimethyl sulfoxide-based electrolytes for rechargeable lithium batteries, Electrochim. Acta 32 (1987) 299–305.

[58] C.M. Cholant, L.U. Krüger, R.D.C. Balboni, M.P. Rodrigues, F.C. Tavares, L.L. Peres, W.H. Flores, A. Gündel, A. Pawlicka, C.O. Avellaneda, Synthesis and characterization of solid polymer electrolyte based on poly(vinyl alcohol)/gum Arabic/$LiClO_4$, Ionics 26 (2020) 2941–2948.

[59] R.J. Sengwa, P. Dhatarwal, Predominantly chain segmental relaxation dependent ionic conductivity of multiphase semicrystalline PVDF/PEO/$LiClO_4$ solid polymer electrolytes, Electrochim. Acta 338 (2020) 135890.

[60] S. Gupta, P.K. Varshney, Effect of plasticizer on the conductivity of carboxymethyl cellulose-based solid polymer electrolyte, Polym. Bull. 76 (2019) 6169–6178.

III. Electrochemical energy conversion and storage

332
11. Prospects and challenges in the selection of polymer electrolytes

[61] S. Kurapati, S.S. Gunturi, K.J. Nadella, H. Erothu, Novel solid polymer electrolyte based on PMMA:CH_3COOLi effect of salt concentration on optical and conductivity studies, Polym. Bull. 76 (2019) 5463−5481.

[62] M. Guo, M. Zhang, D. He, J. Hu, X. Wang, C. Gong, X. Xie, Z. Xue, Comb-like solid polymer electrolyte based on polyethylene glycol-grafted sulfonated polyether ether ketone, Electrochim. Acta 255 (2017) 396−404.

[63] G. Ek, F. Jeschull, T. Bowden, D. Brandell, Li-ion batteries using electrolytes based on mixtures of poly(vinyl alcohol) and lithium bis(triflouromethane) sulfonamide salt, Electrochim. Acta 246 (2017).

[64] V. Aravindan, P. Vickraman, P. Kumar, Polyvinylidene fluoride−hexafluoropropylene (PVdF−HFP)-based composite polymer electrolyte containing $LiPF_3(CF_3CF_2)_3$, J. Non-Cryst. Solids 354 (2008) 3451−3457.

[65] C.S. Lim, R. T subramaniam, S.R. Majid, The effect of antimony trioxide on poly (vinyl alcohol)-lithium perchlorate based polymer electrolytes, Ceram. Int. 39 (2013) 745−752.

[66] Z. Osman, K. Md Isa, L. Othman, N. Kamarulzaman, Studies of ionic conductivity and dielectric behavior in polyacrylonitrile based solid polymer electrolytes, Defect Diffusion Forum (2011) 312−315.

[67] J. Malathi, M. Kumaravadivel, G. Brahmanandhan, M. Hema, B. Rangasamy, S. Subramanian, Structural, thermal and electrical properties of $PVA-LiCF_3SO_3$ polymer electrolyte, J. Non-Cryst. Solids 356 (2010) 2277−2281.

[68] R. Subban, A. Ahmad, N. Kamarulzaman, A.M.M. Ali, Effects of plasticiser on the lithium ionic conductivity of polymer electrolyte $PVC-LiCF_3SO_3$, Ionics 11 (2005) 442−445.

[69] N.S. Mohamed, A.K. Arof, Conductivity studies of $LiCF_3SO_3$-doped and DMF-plasticized PVDF-based solid polymer electrolytes, Phys. Status Solidi {a} 201 (2004) 3096−3101.

[70] M.A.K.L. Dissanayake, C.A. Thotawatthage, G.K.R. Senadeera, T.M.W.J. Bandara, W.J.M.J.S.R. Jayasundera, B.E. Mellander, Efficiency enhancement by mixed cation effect in dye-sensitized solar cells with PAN based gel polymer electrolyte, J. Photochem. Photobiol., A Chem. 246 (2012) 29−35.

[71] A. Manuel Stephan, Review on gel polymer electrolytes for lithium batteries, Eur. Polym. J. 42 (2006) 21−42.

[72] N. Shukla, A.K. Thakur, Role of salt concentration on conductivity optimization and structural phase separation in a solid polymer electrolyte based on $PMMA-LiClO_4$, Ionics 15 (2009) 357−367.

[73] W. Van Schalkwijk, B. Scrosati, Advances in Lithium Ion Batteries Introduction, Advances in Lithium-Ion Batteries, Springer, Boston, MA, 2002, pp. 1−5.

[74] L.D. Kumudu Perera, W.S.K. Bandaranayake, P.W.S.K. Bandaranayake, K. West, Preparation of Pan: EC: PC: LiTf Polymer Electrolytes and Characterization of Li/Pan: EC: PC: LiTf/PPy: Dbs Cells, Solid State Ionics − Materials and Devices − 7th Asian Conference, World Scientific, 2000, p. 5.

[75] S. Ramesh, K. Wong, Conductivity, dielectric behaviour and thermal stability studies of lithium ion dissociation in poly (methyl methacrylate)-based gel polymer electrolytes, Ionics 15 (2009) 249−254.

[76] S. Rajendran, T. Uma, Characterization of plasticized $PMMA-LiBF_4$ based solid polymer electrolytes, Bull. Mater. Sci. 23 (2000) 27−29.

[77] C.R. Yang, J.T. Perng, Y.Y. Wang, C.C. Wan, Conductive behaviour of lithium ions in polyacrylonitrile, J. Power Sources 62 (1996) 89−93.

[78] J.Y. Song, Y.Y. Wang, C.C. Wan, Review of gel-type polymer electrolytes for lithium-ion batteries, J. Power Sources 77 (1999) 183−197.

[79] M.Z. Kufian, S.R. Majid, Performance of lithium-ion cells using 1 M $LiPF_6$ in EC/DEC (v/v=1/2) electrolyte with ethyl propionate additive, Ionics 16 (2010) 409−416.

[80] S.-I. Tobishima, K. Hayashi, Y. Nemoto, J.-I. Yamaki, Ethylene carbonate/propylene carbonate/2-methyl-tetrahydrofuran ternary mixed solvent electrolyte for rechargeable lithium/amorphous V_2O_5-P_2O_5 cells, Electrochim. Acta 42 (1997) 1709−1716.

[81] J.O. Besenhard, Handbook of Battery Materials, John Wiley & Sons, 2008.

[82] G.B. Appetecchi, F. Croce, B. Scrosati, Kinetics and stability of the lithium electrode in poly(methylmethacrylate)-based gel electrolytes, Electrochim. Acta 40 (1995) 991−997.

[83] C.V. Amanchukwu, J.R. Harding, Y. Shao-Horn, P.T. Hammond, Understanding the chemical stability of polymers for lithium−air batteries, Chem. Mater. 27 (2015) 550−561.

[84] M.M. Coleman, R.J. Petcavich, Fourier transform infrared studies on the thermal degradation of polyacrylonitrile, J. Polym. Sci. Polym. Phys. Ed. 16 (1978) 821−832.

[85] N. Xin, Y. Sun, M. He, C.J. Radke, J.M. Prausnitz, Solubilities of six lithium salts in five non-aqueous solvents and in a few of their binary mixtures, Fluid Phase Equil. 461 (2018) 1−7.

References

[86] N. Hasan, M. Pulst, M.H. Samiullah, J. Kressler, Comparison of Li^+-ion conductivity in linear and crosslinked poly(ethylene oxide), J. Polym. Sci. B Polym. Phys. 57 (2019) 21−28.

[87] W. Xu, J.-P. Belieres, C.A. Angell, Ionic conductivity and electrochemical stability of poly[oligo(ethylene glycol) oxalate]−lithium salt complexes, Chem. Mater. 13 (2001) 575−580.

[88] H.-Y. Song, M.-H. Jung, S.-K. Jeong, Improving electrochemical performance at graphite negative electrodes in concentrated electrolyte solutions by addition of 1,2-dichloroethane, Appl. Sci. 9 (2019) 4647.

[89] W. Xu, C.A. Angell, Weakly coordinating anions, and the exceptional conductivity of their nonaqueous solutions, Electrochem. Solid State Lett. 4 (2000) E1.

[90] J. Chen, C. Chen, T. Huang, A. Yu, LiTFSI concentration optimization in TEGDME solvent for lithium−oxygen batteries, ACS Omega 4 (2019) 20708−20714.

[91] T. Ma, G.-L. Xu, Y. Li, L. Wang, X. He, J. Zheng, J. Liu, M.H. Engelhard, P. Zapol, L.A. Curtiss, J. Jorne, K. Amine, Z. Chen, Revisiting the corrosion of the aluminum current collector in lithium-ion batteries, J. Phys. Chem. Lett. 8 (2017) 1072−1077.

[92] F. Croce, L. Persi, B. Scrosati, F. Serraino-Fiory, E. Plichta, M.A. Hendrickson, Role of the ceramic fillers in enhancing the transport properties of composite polymer electrolytes, Electrochim. Acta 46 (2001) 2457−2461.

[93] A. Manuel Stephan, K.S. Nahm, Review on composite polymer electrolytes for lithium batteries, Polymer 47 (2006) 5952−5964.

[94] C. Capiglia, J. Yang, N. Imanishi, A. Hirano, Y. Takeda, O. Yamamoto, Composite polymer electrolyte: the role of filler grain size, Solid State Ionics 154 (2002) 7−14.

[95] K. He, J.-W. Zha, D. Peng, H.S. Cheng, C. Liu, Z.-M. Dang, R. Li, Tailored highly cycling performance in solid polymer electrolyte with perovskite-type $Li_{0.33}La_{0.557}TiO_3$ nanofibers for all-solid-state lithium ion battery, Dalton Trans. 48 (2019).

[96] E. Karpierz, L. Niedzicki, T. Trzeciak, M. Zawadzki, M. Dranka, J. Zachara, G.Z. Żukowska, A. Bitner-Michalska, W. Wieczorek, Ternary mixtures of ionic liquids for better salt solubility, conductivity and cation transference number improvement, Sci Rep-Uk 6 (2016) 35587.

[97] B. Qiao, G.M. Leverick, W. Zhao, A.H. Flood, J.A. Johnson, Y. Shao-Horn, Supramolecular regulation of anions enhances conductivity and transference number of lithium in liquid electrolytes, J. Am. Chem. Soc. 140 (2018) 10932−10936.

[98] D.-S. Guo, Y. Liu, Supramolecular chemistry of p-Sulfonatocalix[n]arenes and its biological applications, Acc. Chem. Res. 47 (2014) 1925−1934.

[99] M.K. Hekselman, A. Plewa-Marczewska, G.Z. Żukowska, E. Sasim, W. Wieczorek, M. Siekierski, Effect of calix [6]pyrrole anion receptor addition on properties of PEO-based solid polymer electrolytes doped with LiTf and LiTfSI salts, Electrochim. Acta 55 (2010) 1298−1307.

[100] A. Blazejczyk, M. Szczupak, W. Wieczorek, P. Cmoch, G.B. Appetecchi, B. Scrosati, R. Kovarsky, D. Golodnitsky, E. Peled, Anion-binding calixarene receptors: synthesis, microstructure, and effect on properties of polyether electrolytes, Chem. Mater. 17 (2005) 1535−1547.

[101] M. Kalita, M. Bukat, M. Ciosek, M. Siekierski, S.H. Chung, T. Rodríguez, S.G. Greenbaum, R. Kovarsky, D. Golodnitsky, E. Peled, D. Zane, B. Scrosati, W. Wieczorek, Effect of calixpyrrole in PEO−$LiBF_4$ polymer electrolytes, Electrochim. Acta 50 (2005) 3942−3948.

[102] F. Zhou, D.R. MacFarlane, M. Forsyth, Boroxine ring compounds as dissociation enhancers in gel polyelectrolytes, Electrochim. Acta 48 (2003) 1749−1758.

[103] A. Sołgała, M. Kalita, G. Żukowska, Study of neutral species coordination by macrocyclic anion receptors using FTIR spectroscopy, Electrochim. Acta 53 (2007) 1541−1547.

[104] B. Li, Y. Liu, X. Zhang, P. He, H. Zhou, Hybrid polymer electrolyte for Li−O_2 batteries, Green Energy Environ. 4 (2019) 3−19.

[105] X. Yao, Q. Dong, Q. Cheng, D. Wang, Why do lithium−oxygen batteries fail: parasitic chemical reactions and their synergistic effect, Angew. Chem. Int. Ed. 55 (2016) 11344−11353.

CHAPTER
12

Li-S ion batteries: a substitute for Li-ion storage batteries

Kalpana R. Nagde[1], S.J. Dhoble[2]

[1]Department of Physics, Institute of Science, Nagpur, Maharashtra, India; [2]Department of Physics, RTM Nagpur University, Nagpur, Maharashtra, India

12.1 Introduction

After the discovery of electricity, there was a need to find effective methods to store energy for use on demand. A device that stores energy is known as an accumulator or battery. Different forms of energy are available in nature, which include radiation energy, electricity, gravitational potential energy, chemical energy, and kinetic energy. Energy storage also involves converting energy from different forms to more conveniently storable forms. There are some technologies that store energy for a short period, while others can store it for longer. Some examples of energy storage devices are the rechargeable battery, hydroelectric dam, fossil fuel storage, and mechanical, electrical, biological, electrochemical, thermal, and chemical storage devices. During the 20th century, electricity was generated by burning fossil fuel. Due to pollution from fossil fuels, researchers have concentrated their attention on renewable energy sources like solar and wind energy [1]. At the beginning of 21st century, portable devices were in demand all over the world. In this chapter, attention is given to electrochemical devices, which include batteries.

12.2 Energy storage materials

Materials for chemical and electrochemical energy storage are key for a diverse range of applications, including batteries, hydrogen storage, sunlight conversion into fuels, and thermal energy storage. The urgent need for energy storage materials for a sustainable and carbon-free society is the main stimulant for the new dawn in the development of functional materials for energy storage and conversion. Hydride-based all-solid-state batteries, which are considered to be safer, cheaper, and more abundant, while potentially having higher

Energy Materials
https://doi.org/10.1016/B978-0-12-823710-6.00008-X

Copyright © 2021 Elsevier Ltd. All rights reserved.

336 12. Li-S ion batteries: a substitute for Li-ion storage batteries

energy densities compared to Li-ion batteries (LIBs), are achievable. New reactive hydride composite systems for hydrogen storage applications with operating conditions near room temperature are available nowadays. There are different types of energy storage materials depending on their applications:

1. Active materials for energy storage that require a certain structural and chemical flexibility, for instance, as intercalation compounds for hydrogen storage or as cathode materials.
2. Novel catalysts that combine high (electro-) chemical stability and selectivity.
3. Solid-state ionic conductors for batteries and fuel cells.

Any electrochemical storage device consists of two electrodes known as cathode and anode which is separated by an ionically conducting but electronically insulating material known as electrolyte. The electrodes exhibit properties such as good electronic conductivity, good stability, and high catalytic activity [2]. The electrode on which the reduction reaction takes place is known as the cathode and where the oxidation reaction takes place is known as the anode. Different examples of electrochemical energy storage and conversion systems are batteries and fuel cells, which convert energy into electricity. Electrolytic capacitors and supercapacitors are used in batteries and are coupled with specific energy and specific power by the battery chemistry. On the other hand, in fuel cells, energy density by using a particular fuel is decoupled from power, which in turn is ordered by the kinetic and transport properties of cell components. Such decoupling processes play a vital role in many applications [3].

Most batteries and fuel cells consist of carbon-based materials, which are considered ideal candidates for a wide range of technical applications. Graphite, which is a form of carbon, is widely used in LIBs and serves as the active anode component. The intercalation of lithium into graphite consists of different phase changes, which are accompanied by complex surface reactions. These reactions lead to the formation of a protective layer on the graphite surface known as solid-electrolyte-interphase (SEI) layer. This layer performs a vital role in the operation and lifetime of an LIB and has been therefore intensively studied by different researchers. In the case of lithium—air and vanadium-redox-flow battery types; the interfaces of the carbon electrode play a leading role in electrochemical reactions. The structural diversity of carbon materials with their rich surface chemistry allows chemists to fine tune the material properties and to analyze the electrochemical surface processes of carbon electrodes. In the case of a battery, the three different components, namely anode, cathode, and electrolyte, are studied extensively. Researchers have demonstrated different types of materials, which can be used as suitable candidates depending on their applications. Different types of energy storage materials for anode, cathode, and electrolyte are explained in detail in the following sections.

12.2.1 Anode materials

Carbon is considered a suitable anode candidate material for all types of batteries due to its layer structure. Nowadays, more research is being conducted to find other more suitable candidate materials, which include elemental Li, Li alloys, or a nanostructured host material such as Si nanowire [4]. Li is the most electropositive element in the periodic table and has a high specific capacity of about 3860 mAh g, hence elemental Li is considered the best anode material. However, elemental Li has a number of disadvantages regarding safety, reliability, durability as well as chemical stability and reactivity issues with the battery electrolyte [4]. Chan et al. used Si nanowires and nanosize Sn anodes to improve the durability of a battery [5,6].

Other electropositive metals such as Na, K, Ca, and Mg are more abundant in nature. Therefore batteries based on Na, K, Ca, and Mg chemistries also provide cost-effective alternatives to existing Li-based batteries. More attention has been paid to Na, as it is the lightest element after Li. But it has been reported that Li is superior s compared to Na and K [7]. Silicon insertion has been demonstrated to enhance anode material performance. However, its capacity for fading and degradation after cycling has hindered its commercialization [8]. Morita and Takami used Si nanoparticles embedded in a silicon dioxide and carbon matrix for the development of a high-capacity anode [9]. Anode materials for lithium batteries have been reviewed by several investigators. Several elements of the group IV (C, Si, Ge, Sn) and group V (P, Sb) and their composites, including Ag and Mg, have been investigated [10]. Carbon-based nanoparticles with sizes ranging from 40 to 80 nm are prepared by electrochemical reduction and used as anode materials for LIBs [11]. Doi et al. prepared lithium titanate ($Li_4/3Ti_5/3O_4$) nanoparticles with an average diameter of 12 nm by electrospray pyrolysis to improve battery performance [12]. Lithium alloy anodes have also been investigated by Weppener et al. It has been reported that mixed ionic/electronic conductivity is the most important property, which includes both electrons and Li^+ ions. This is the most desirable property which can be used for reversible electrodes [13].

12.2.2 Cathode material

There are multiple cathode materials available and the task of scientists is to choose suitable materials that offer better performance for LIBs. Cobalt was used as the first primary active component of the cathode. Nowadays, cobalt has been partially substituted with nickel. These include nickel cobalt aluminum, core—shell gradient, spinel-based lithium ion, cobalt-based lithium ion, and nickel cobalt manganese. Several researchers have used the cathode along with anode in the form of oxide or phosphate based materials for lithium-ion battery [14]. Recently, it has been found that the doping of sulfur and potassium in lithium—manganese spinel enhanced Li-ion mobility in a battery [15]. Iron disulfide (FeS_2) has been widely used as cathode material for nonrechargeable LIBs. In a recent report, Xu et al. used a biomass-carbon-FeS_2 nanocomposite as a cathode material [16]. Rapulenyane et al. reported $Li_{0.2}Mn_{0.6}Ni_{0.2}O_2$, a lithium and manganese-rich cathode, for a battery [17].

The layered and spinel structure cathodes with general formulae $LiMO_2$ (M = Co, Ni, Mn) and LiM_2O_4 (M = Mn, Ni), respectively, have been found to be good conductors for Li^+ ions, and have been extensively studied. It also provide sufficient electronic conductivity due to the electronic transitions between the oxidation states of the multivalent transition metal cations present in the structure. It has also been reported that electronic conductivities of these materials can be improved by appropriate doping [18]. Now a days, Olivine structure ($LiMPO_4$) where M is a transition metal such as Fe has attracted attention due to lower cost among all the cathode materials [19].

12.2.3 Electrolyte materials

The organic carbonate-based solvent used in nonaqueous electrolyte leads to the formation of a thin SEI on the anode interface [20]. This layer blocks the transport of both Li^+ ions and electrons due to organic and inorganic components. There are different types of electrolyte material, which have been studied extensively. Of these materials, ceramic solid electrolytes are the materials that are applicable for high-temperature applications. Therefore there are several types of lithium-ion conducting inorganic ceramics that have been investigated for use in LIBs [21−23]. Sulfide compounds, which are crystalline, amorphous, and partially crystalline in nature, have also been used as lithium-ion conductors. $Li_2S-P_2S_5$ glass sulfide electrolytes have been studied by researchers in batteries, which consist of lithium−indium anodes [8−10]. Perovskite (La,Li)TiO_3 is another lithium-ion-conducting oxide reported in the literature [24]. Some oxides forming a garnet-related structure such as $Li_5La_3Ta_2O_{12}$ also have good lithium-ion conductivity [25,26]. The conductivities of some of the lithium-ion-conducting phosphate compounds have been studied by researchers [27−29]. The commonly used phosphate electrolyte is $Li_{1+x}Ti_{2-x}Al_x$ $(PO_4)_3$, to which silicon has been added to replace some of the phosphorus to improve the property of electrolyte in LIBs [30].

On the other hand, it has been studied that polymer electrolytes have advantages over ceramic electrolytes due to good processability, adaptability, dimensional stability, and safety. These electrolytes also have the ability to stop lithium dendrite formation [31]. Some solid polymer electrolytes such as poly(ethylene oxide) have been used for lithium-ion-conducting electrolytes. A polymer gel has been incorporated in an organic liquid electrolyte to form polymer gel electrolytes [32]. The conduction mechanism in polymer gels has been found to be similar to that in liquid electrolytes. Poly(methyl methacrylate) is another polymer used in gel electrolytes that possesses a low conductivity [33]. The polymer electrolyte polyacrylonitrile is a commonly used polymer for lithium-ion-conducting electrolytes [34].

12.3 Batteries

Batteries are well-known electrochemical devices that store electrical energy. Italian physicist Alessandro Volta discovered and fabricated the first electrochemical battery known as the voltaic pile in 1800 [35]. It consisted of a stack of copper and zinc plates separated by brine-soaked paper disks for the production of a steady current. At that time; Volta was not knowing the exact reason for producing the voltage. These early batteries had problems of voltage fluctuation and an inability to provide current for a long time. In 1836, British chemist John Frederic Daniell invented the Daniell cell [36]. It consisted of a copper pot containing copper sulfate solution, which was immersed in an unglazed earthenware container crammed with vitriol and a zinc electrode [37]. Due to the use of liquid electrolyte, these wet cells faced leakage and spillage problems. Researchers used glass jars to hold their components to overcome these problems, which made them fragile and potentially dangerous. Due to all these reasons, the cells in those days were not suitable for portable applications. At the end of the 19th century, scientists invented dry cell batteries where the liquid electrolyte was replaced with a paste of liquid electrolyte, which made the device available for portable applications [38]. Batteries are mainly classified into primary (nonrechargeable) and secondary (rechargeable) batteries. Primary batteries are designed in such a way that they can be used until they become exhausted, at which point they are discarded. The chemical reactions of these primary batteries are nonreversible and these batteries cannot be recharged once exhausted. When all the reactants within the battery are completely used, the battery does not produce any current and is useless. The main advantage of these batteries is that they produce current immediately. Therefore such types of batteries are generally used in portable devices that require low current. In general, these batteries have higher energy densities compared to that of rechargeable batteries. Disposable batteries cannot be used under high-drain applications with loads up to 75 Ω. Zinc—carbon batteries and alkaline batteries, which are disposable batteries, are examples of primary batteries. Secondary batteries are rechargeable. The main advantage is that these batteries regenerate the original chemical reactants and can be used again multiple times [38]. These batteries are charged by applying current. During the discharge process, the battery reverses the chemical reactions. The lead—acid battery, the oldest rechargeable battery consisting of liquid electrolyte in an unsealed container, was used in earlier days in automotive and boating applications. The disadvantage of this battery is its handling due to its heavy weight. However, the low manufacturing cost and high current made it popular at the time. Batteries used in mobile phones and MP3 players are the most common examples of these batteries. The miniaturization of these batteries was again a task for scientists to overcome and nowadays hearing aids and wristwatches are some examples of the miniaturization of secondary batteries. Depending on the application, secondary batteries are categorized into four types:

1. Sealed maintenance-free batteries, which are generally used in uninterruptible power supply (UPS) applications and are available from 12 V. These batteries offer reliable, consistent, and low-maintenance power, and are often subjected to deep cycle applications.
2. Lead—acid batteries are most popular and rechargeable batteries. Lead—acid batteries are available in several different configurations such as small sealed cells having a capacity in the range 1 Ah. One of the main applications of lead—acid batteries is in the automotive industry as they are primarily used as starting, lighting, and ignition batteries. Other applications of lead—acid batteries include energy storage, emergency power, electric vehicles (even hybrid vehicles), communication systems, emergency lighting systems, etc. The wide range of applications of lead—acid batteries is a result of their wide voltage ranges, different shapes and sizes, low cost, and relatively easy maintenance. When compared to other secondary battery technologies, lead—acid batteries are the least expensive option for any application and provide very good performance. The electrical efficiency of lead—acid batteries is between 75% and 80%. This efficiency makes them suitable for energy storage (UPS) and electric vehicles.
3. LIBs became popular during the last couple of decades. More than 50% of the consumer market has adopted the use of LIBs. Particularly, laptops, mobile phones, cameras, etc. are the largest applications of LIBs. These batteries are used for higher capacities and relatively lower self-discharge applications due to their high energy density potential.
4. Nickel—cadmium batteries have many advantages over the other three types of secondary rechargeable batteries. They can be recharged many times and possess more electrical- and physical-withstanding capacity. They provide constant potential during

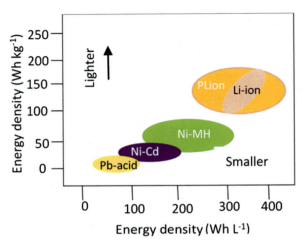

FIGURE 12.1 Commercial evolution of rechargeable batteries to higher energy density [39].

the discharge process. This battery contains nickel oxide as a cathode, a cadmium compound as an anode, and potash solution as the electrolyte. Historical development of the commercialization of rechargeable batteries is represented in Fig. 12.1. The energy density per unit volume and weight for the foremost common rechargeable battery systems can be seen in the figure. According to James et al., the development of batteries progressed to more energy dense systems with the use of less harmful chemicals with greater cycling efficiency [39].

12.3.1 Parts of a battery

The negative electrode in a battery is known as the anode. At this electrode, a reduction reaction takes place during charging and vice versa. The oxidation number of the reacting element increases during the discharging process and this number decreases during the charging process. The positive electrode of a battery, where a reduction reaction takes place during discharging and an oxidation reaction takes place during charging, is known as the cathode. The exact opposite of anode kinematics occurs in the cathode, i.e., the oxidation number of the reacting element increases during the charging process and this number decreases during the discharging process. The electrolyte is sandwiched between the cathode and anode terminals, which permits the flow of electrical charge between the two electrodes. Current density is a factor that determines the suitable candidate material for electrodes. C-rate is another term for charging current used in the literature. According to Odne et al. "1C refers to a charging time of 1 hour from 0% state of charge (SoC) to 100% SoC, which is a full cycle and 2C refers to half an hour charging of a full cycle." SoC is simply the maximum possible amount of charge that is shifted between the two electrodes. Another term often used for charging current is the depth of discharge (DoD). It has been observed that a battery loses its capacity with time [40]. Therefore the term state of health (SoH) has been introduced in some research papers in place of DoD. SoH refers to how much charge in coulombs is available for use in the battery at a given C-rate relative to a new battery [40].

12.3.2 Working of a battery

The stored energy within the battery is converted into electricity, which flows through the battery. The electrodes in a battery contain atoms of conducting materials. In an alkaline battery, the anode and cathode are made up of zinc and manganese dioxide, respectively. The electrolyte, which is sandwiched in between electrodes, contains ions. The series of electrochemical reactions takes place when electrolyte ions meet the electrode atoms, commonly referred to as oxidation—reduction (redox) reactions. In a battery, the cathode accepts electrons from the anode, referred to as the oxidant, and the anode is known as the reducer due to its electron-losing property. Finally, these reactions end in the flow of ions due to the freeing of electrons from the atoms of the electrode as well as between the anode and the cathode. These free electrons are accommodated inside the anode. Due to this reaction, the two electrodes gain different charges. The anode now becomes positively charged by realizing electrons and the cathode becomes negatively charged by accepting electrons. This difference causes the movement of electrons toward the charged cathode. When the electric current reenters the battery, the circuit of the battery is completed, which is shown in Fig. 12.2.

12.3.3 Theoretical potential, capacity, and energy of batteries

The energy stored within the electrodes of the battery decides the amount of electrical energy per mass delivered through the battery. The change in Gibbs free energy (ΔG) takes place due to the redox reaction within the battery. High energy conversion efficiency is a most important criterion for any battery, which governs chemical-to-electrical conversion within the cells. However, it has been reported that due to the polarization effect of the electrodes, the available energy is lower than the stored chemical energy [40]. Activation

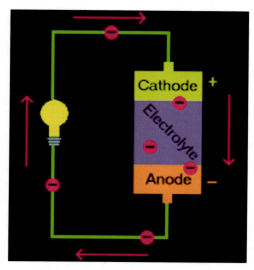

FIGURE 12.2 Working of a battery.

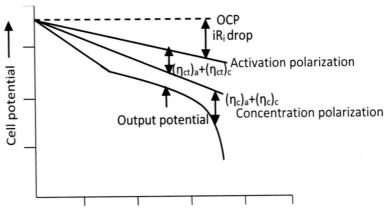

FIGURE 12.3 Cell potential as a function of current [39].

polarization and concentration polarization are mainly considered as two types of polarization within any cell or battery. The first polarization is due to electrochemical reaction within the cell and the second is due to concentration gradients between the reactants and products at the electrode surface. The values of both polarizations can be calculated after measuring the electrochemical parameters and mass transfer data. The internal resistance of the cell is another key factor that affects the overall performance of the battery, and is generally referred to as ohmic polarization. Its magnitude is proportional to the current delivered and causes a voltage drop during operation [40]. Cell potential as a function of current is shown in Fig. 12.3.

The main advantages of batteries are that they offer desirable cost-effective and compact solutions for storage purposes. They do not have moving parts and are not hazardous to the environment. They also possess high efficiency, sufficient cycle life, and durable shelf and service life [39]. The batteries are the best solution for delivering stored chemical energy into electrical energy without harmful emissions to the environment. Specific energy represents the amount of stored electrical energy within the battery and is a function of the cell voltage (V). Specific power represents the rate at which the energy can be extracted from a battery, which is the function of battery capacity (Ah/kg). The thermodynamic and kinetic constraints of the chemistry and materials used for a battery's components affect both parameters, i.e., specific energy and specific power. Battery life and safety are the other important factors on which researchers are still working. It has been noted that scientists are developing electrode materials that yield a high cell potential, which is related to high storage capacity. However, high cell potentials are corelated to high activity, which facilitates reaction with the electrolyte, and ultimately results in thermal runaway and fire, which are safety issues. Such types of issues have been reported in the case of LIBs using organic electrolytes [39,41].

12.4 Novel materials for batteries

Currently, research is focused on identifying a group of materials that can be used to improve the performance of batteries. Niobium tungsten oxides with a complex crystalline

12.4 Novel materials for batteries

structure have been used to improve the power of batteries by researchers from the University of Cambridge. It has been reported that the rate of movement of lithium ions is faster compared to that of typical electrode materials, which means faster charging for batteries. This provides valuable information to researchers to make next-generation batteries using unconventional materials [42]. It has been reported that it is difficult to make small batteries with nanoparticles. Use of nanoparticles for the fabrication of small batteries is challenging because all parts need to be assembled tightly together [42]. The niobium tungsten oxides were used by Griffith et al. to fabricate a small battery. It consisted of a rigid, open structure that did not trap the inserted lithium, and the particle size was also larger than other electrode materials. Most scientists have not used this material due to its complex atomic arrangements. Due to the porous structure of this material, Li ion can easily pass through it more quickly. Rechargeable batteries suffer from two problems: (1) batteries do not provide enough capacity (run time) and (2) they take longer to recharge. A research team at Cornell University addressed these problems by using a block copolymer-derived 3D interpenetrating gyroidal nanohybrid, which allowed the battery to self-assemble with a unique, higher-capacity internal structure than standard batteries and enabled far-faster recharging [43]. In contrast to having the battery's electrodes on either side of an electrolyte separator, this method intertwines the battery's internal parts during self-assembly of the 3D gyroidal structure. The structure has thousands of nanoscale pores that are filled with components that are useful for energy storage and delivery. The 3D architecture provides the facility for folding and integrating all nanoscaled electrochemical components into interpenetrating networks, which in turn decreases the footprint area to a substantial level. Thus this 3D architecture represents an atom-efficient approach. In nanohybrids, the cathode is made up of composite nanonetworks (sulfur/poly(3,4-ethylenedioxythiophene)) and a carbon anode, which are separated by a pinhole-free polymer electrolyte layer (PPO). This PPO is wrapped around the interpenetrating gyroidal electrodes. It has been shown that this type of nano-3D electrochemical energy storage device has a stable open-circuit voltage and a discharge plateau at around 2.7 V [2]. Conversion-based materials such as transition-metal oxides and spinel-metal oxides have attracted a great deal of attention as suitable anode electrode materials. Moreover, the theoretical capacities of these materials are greater compared to that of commercial graphite [44]. The electrochemical performance of binary transition-metal oxides possessing a spinel structure is comparatively good. AB_2O_4 spine-structured oxides containing two transition metals, A and B, have been introduced by researchers as novel materials. Hu et al. [45] developed $CoMn_2O_4$ with a hierarchical mesoporous structure delivering a high discharge capacity of 894 mAh/g at 500 mA/g after 100 cycles. Won et al. [46] developed a $ZnFe_2O_4$ yolk–shell with a specific capacity of 862 mAh/g at 500 mA/g. The specific capacity of 690 mAh/g at $C/10$ has been reported for $ZnMn_2O_4$ material prepared through the coprecipitation method [47]. A relatively high reversible capacity of 550 mAh/g run at 5 A/g has been reported by Bai et al. for $ZnCo_2O_4$ for 2000 cycles [48]. Li et al. used $NiCo_2O_4$ synthesized through a facile solvothermal process, and subsequent pyrolysis exhibited a specific capacity of 705 mAh/g even after 500 cycles with a current density of 800 mA/g [49]. The study of CoV_2O_4 (CVO) was extensively done by Lu et al. [44]. In this study, CVO was used with different binders. It was observed that electrochemical performance was improved due to the use of a carboxymethyl cellulose + styrene-butadiene rubber (CMC + SBR) binder. Improved electrochemical performance was reported due to the strong hydrogen bonding of carboxyl and hydroxyl groups in CMC with the active and conductive materials and Cu

III. Electrochemical energy conversion and storage

foil. The addition of SBR increased the adhesion of the electrode onto the Cu foil and it also generated better elasticity, which improved the rapid intercalation/deintercalation of Li^+ ions. However, the smaller ohmic and faradaic resistance was obtained for a CVO-CMC-SBR electrode compared to a CVO-P electrode [44]. Another commercialized novel anode that has been studied is the spinel-phase $Li_4Ti_5O_{12}$, which has high working potential that neglects the effect of occurrence of Li dendrite and SEI. Another advantage of using this material is that it possesses the zero strain effect, which provides a long lifespan with low theoretical capacity, high voltage, and lower energy density [50,51]. Ti-based anodes like $Zn_2Ti_3O_8$ and $Li_2MTi_3O_8$ (M = Co, Zn, Mg) have also been investigated and it was shown that these two spinel anodes work below 1 V with a capacity around 300 mAh g [52–54]. Li_2TiSiO_5, a layered structure, has also been studied and exhibited high capacity (>240 mAh g) with a different cut-off voltage below 1 V (~0.28 V) [55,56]. Insertion and conversion are two reactions found during charging and discharging processes in this material. Though this material improved cycle life and rate performance, the poor conductivity of this material makes carbon coating inevitable. Na is a low-cost source that has been used in place of Li in TiO_2-SiO_2Li_2O systems [50]. Among the numerous TiO_2-SiO_2-Na_2O ternary compounds systems, Na_2TiSiO_5 has been investigated as a potential candidate anode material [57]. Three polymorphs have been reported for this oxide. Li_3VO_4 has also been extensively studied due to its long cycling time and high reversible capacity [58]. Jiang et al. for the first time prepared an Li_3VO_4 microsphere as a positive electrode, which has a hollow lantern 3D structure and fairly open framework, which allow ions to be inserted reversibly [58]. Cyclic voltammograms of Li_3VO_4 at different scans have been studied for a better understanding. It has been observed that deoxidization peaks slowly moved to a lower potential when the oxidation peaks moved to higher potential due to enhancing the polarization effect of the battery [59]. Jiang et al. successfully prepared a highly open framework of $Mo_{2.5 + y}VO_{9 + z}$ as a positive electrode with $AlCl_3$/[BMIm]Cl electrolyte [60]. This material possesses a 3D framework constructed by MO_6 octahedral and pentagonal $[(Mo)Mo_5O_{27}]$ units sharing corners with each other. It has been observed that the batteries are discharged at different rates from 2 to 100 mA/g with a cut-off potential at 0.02 V versus Al/Al^{3+}. A capacity of 340 mAh/g at a discharge rate of 2 mA/g has been reported as the optimum in all Al-based batteries. Similarly, after insertion, Al species in $Mo_{2.5 + y}VO_{9 + z}$ structures provide a large storage capacity at a fast discharge rate [59]. Recently, olivine-structured lithium transition-metal phosphate $LiMPO_4$ (M = Fe, Mn, Co, Ni) has been studied as a novel potential cathode material. It has been investigated that the voltages of $LiCoPO_4$ and $LiNiPO_4$ materials were 4.9 and 5.1 V, respectively [61]. The electrochemical activity in $LiMnPO_4$ cathode material has been limited by synthesizing it with nanomaterial synthesis methods, which are generally designed for liquid-phase routes to avoid grain growth and agglomeration by leading to Mn_{21} disorder on the Li_1 sites in $LiMnPO_4$. A novel graphene-modified $LiMnPO_4$ composite prepared by Jiang et al. using the spray-drying process showed improved cathode material for batteries [61]. $MnWO_4$ has also been considered as a novel electrode material for batteries by other researchers [62–64]. The electrochemical properties of $MnWO_4$ have also been studied in detail and it was shown that they had slow fading capacity. To solve this slow fading capacity problem, some researchers confined the electrode material in a carbon matrix, such as porous carbon, ordered mesoporous carbon, carbon nanotubes, carbon nanofibers, and graphene [65–68]. Due to the confining conductive medium of these materials, they have shown improved cyclic stability, and structural stability improvement with carbon doping acted as a buffer to relieve volume expansion [69]. $ZnWO_4$ cuboids distributed on a reduced graphene oxide (rGO) surface were successfully synthesized by Wang et al. using the hydrothermal method with a subsequent annealing process in air. A specific capacity of more than 477.3 mAh/g at a current density of 100 mA/g after 40 cycles has been observed in ZnWO4/ rGO hybrids [70]. Therefore it has been concluded that the electrochemical performance for tungstate as electrode materials can be prepared with carbon doping. Gao et al. synthesized $MnWO_4$ composite using the facial hydrothermal method with glucose, which is a carbon source. A composite, which is a mixture of two materials or phases, has also attracted attention. It has been shown that carbon-doped $MnWO_4$ composites have better cycling performance and superior rate capability compared to pure $MnWO_4$ [71].

12.5 Lithium-ion batteries

Lithium-ion secondary batteries have gained increasing importance due to the variety of battery chemistries and were found to be superior compared to any other batteries available in the market. Remarkable progress has been achieved since 1980. Subsequently, Prof. John B. Goodenough and Rachid Yazami discovered the $LiCoO_2$ cathode and the graphite anode for LIBs, respectively [72]. And finally in 1985 the research team of Akira Yoshino developed the first LIB prototype. Moreover, a quite stable and rechargeable version of the LIB was established in 1991. The lithium-ion polymer battery surfaced in 1997. The advantages of the lithium-ion polymer battery are that it can hold electrolyte in a solid polymer composite, and the electrodes and separators are laminated with each other. These batteries can be encased in a flexible wrapping so that they can be shaped to fit a particular device. The energy density has been reported higher than that of LIBs. Due to its ability to fit a particular shape, the lithium-ion polymer battery has been used in portable electronics such as mobile phones [72].

12.5.1 Basics

Li^+ ions move in between the positive and negative electrode during the discharging and charging processes in LIBs. Fig. 12.4 represents the basic principles of an LIB. During discharging at the anode, Li ions move toward the electrolyte due to electrochemical oxidization. At the same time, electrons move through the external circuit and enter the cathode. The Li ions travel through the electrolyte to compensate for the electrons moving through the external circuit resulting in intercalation of Li ions toward the cathode. Exactly the reverse process of discharging occurs during charging. In this process, Li ions toggle back and forth between anode and cathode. Basically, the reactions on cathode and anode are described using two half reactions as follows:

$$\text{At cathode}: \ aA + ne^- + nLi^+ \rightarrow cC \tag{12.1}$$

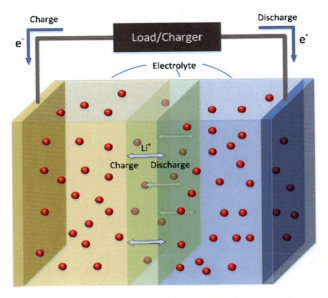

FIGURE 12.4 Schematic of an Li-ion battery [73].

$$\text{At anode}: bB \rightarrow dD + ne^- + nLi^+ \tag{12.2}$$

$$\text{Overall cell reaction}: aA + bB \leftrightarrow cC + dD \tag{12.3}$$

12.5.2 Potential

Each of the electrochemical reactions is related to a standard electrode potential, E_0, which can be calculated from Gibbs free energy (ΔG). The basic thermodynamic equations for the calculation of ΔG are given as:

$$\Delta G = \Delta H^0 - T\Delta S^0 \tag{12.4}$$

where ΔH = enthalpy

T = absolute temperature
ΔS^0 = entropy

If the released Gibbs energy is transformed to electrical work, then:

$$\Delta G = W = -nFE^0 \tag{12.5}$$

where ΔG = Gibbs free energy

n = number of electrons transferred during the process
F = Faraday's constant

Under standard conditions:

$$E^0 = -\Delta G/nF \tag{12.6}$$

12.5.3 Open-circuit voltage

Voltage is one of the simplest properties to calculate from first principles for a possible electrode material, since it relates to the reaction energy of the cathode material with lithium through the Nernst equation. The design of an LIB system also requires careful selection of electrode materials to obtain a high operating voltage (V_{oc}). Fig. 12.5 represents a schematic of an open-circuit energy diagram with an aqueous electrolyte [74]. The bandgap (E_g) in between the lowest unoccupied molecular orbital (LUMO) and the highest occupied molecular orbital (HOMO) of the electrolyte material is also known as the "window" for that electrolyte. μ_A and μ_C are the electrochemical potentials of anode and cathode, respectively. The anode

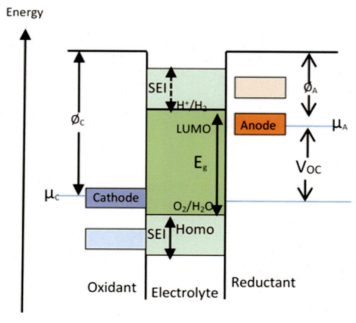

FIGURE 12.5 Schematic of an open-circuit energy diagram of an aqueous electrolyte, Φ_A and Φ_C [74]. *HOMO*, highest occupied molecular orbital; *LUMO*, lowest unoccupied molecular orbital; *SEI*, solid-electrolyte-interphase.

with electrochemical potentials above the LUMO will reduce the electrolyte until a passivation layer stops creating a barrier to electron transfer. Similarly, the cathode with a μ_C below the HOMO will oxidize the electrolyte until a passivation layer blocks electron transfer. Due to this mechanism, both electrodes become thermodynamically stable within the window of the electrolyte, which constrains the V_{oc} of a battery as represented by Eq. (12.7):

$$V_{oc} = (\mu_A - \mu_C)l_e \leq E_g \tag{12.7}$$

In this formula, e is the electron charge magnitude. The SEI layer provides kinetic stability for larger V_{oc} when $eV_{oc} - E_g$ is small [74].

12.5.4 Discharging

During the discharging process, chemical energy stored in a cell is converted into electrical energy that can be withdrawn into a load.

12.5.5 Charging

Charging is the operation in which energy is stored in a secondary cell using an electric current. The battery is restored to its original charged condition through charging.

12.5.6 Overcharging

A battery may explode or leak after attempting to charge beyond its capacity. It may also cause harm to the charger or device.

12.5.7 Electrical conductivity

Electrical conductivity is a measure of the ease with which a material conducts current.

12.5.8 Short circuit

A short circuit is an abnormal connection between two nodes of an electrical circuit, which results in overcurrent limited by the Thévenin equivalent resistance network. This is a circuit that permits a current in an unintended path where no or very small electrical impedance is encountered. It may cause circuit damage, overheating, or explosion.

12.5.9 Theoretical specific capacity

Theoretical specific capacity is an important parameter to evaluate active materials. It can be calculated from the following equation:

$$Q_{tsc} = \frac{n \times F}{3600.M} \tag{12.8}$$

where n is the number of moles of electrons transferred within an electrochemical reaction, F is the Faraday constant, and M is the relative molecular mass of active materials.

12.5.10 Specific capacity

Specific charge capacity (Q_c) or specific discharge capacity (Q_d) can be calculated based on the total amount of electrons transferred:

$$Q_c/Q_d = I \times t/m \tag{12.9}$$

where I is the current density (A), t is the time (h), and m is the mass of active materials (g). The unit of specific capacity is mAh/g or Ah/kg.

12.5.11 Energy density

Energy density is the amount of energy stored in a region of space per unit volume or mass. It is usually desirable that the energy density stored in an LIB system is as high as possible. The unit of energy density is Wh/kg, which is calculated by:

$$Energy\ density = \frac{E \times Q}{1000} \tag{12.10}$$

where E is the voltage (V) and Q is the specific capacity (Ah/kg).

12.5.12 Charge/discharge rate

The term charge/discharge rate, which is also known as the C-rate, is employed to discover how fast lithium ions can be transferred to a cell. It is an expression of the speed with which a battery is being charged/discharged during a particular time.

12.5.13 Irreversible capacity loss

Irreversible capacity results from irreversible lithium reactions, which do not result in insertion or extraction from the active materials. It equals the difference between the charge and discharge capacity for the nth cycle:

$$Irreversible\ capacity\ loss = n\text{th}\ Q_c - n\text{th}\ Q_d \tag{12.11}$$

12.5.14 Capacity retention

Capacity retention, which is always used to evaluate cycling stability, is the ratio of discharge capacity to initial discharge capacity for the nth cycle.

12.5.15 Coulombic efficiency

Coulombic efficiency is defined as the ratio of the output of charge to the input of charge of a battery. Internal resistance of a cell determines the coulombic efficiency of a battery or cell.

12.5.16 Elevated temperature

Chemical reactions happen far more readily at high temperatures than at low temperatures. Furthermore, the active materials are more porous and therefore the internal resistance is less at higher temperatures.

Other injurious effects include the destructive action of hot acid on the wooden separators used in batteries. Greater expansion of active material will also occur, and this expansion has not been found to be uniform over the surface of the plates. Therefore charging temperature is more important for LIBs than the operating limits. The chemical reaction in LIBs can perform well at elevated temperatures (>45°C). It has been observed that the performance of a battery is degraded during the charging and discharging process [73].

12.5.17 Construction

The LIB is composed of three main parts, i.e., anode, cathode, and electrolyte. Generally, the negative electrode is made of carbon material, the positive electrode is a metal oxide, and lithium salt in an organic solvent is used as electrolyte [75]. According to a study, the positive electrode can also be formed using a layered oxide (such as lithium cobalt oxide), a polyanion (such as lithium iron phosphate), or a spinel (such as lithium manganese oxide) [75]. Recently, graphene as an electrode material has been studied extensively. Graphene possesses 2D and 3D structures used for lithium batteries. The energy density, voltage, life, and safety of an LIB can also be changed using the appropriate choice of materials. A nonaqueous electrolyte is generally used in LIBs because pure lithium is highly reactive. It reacts vigorously with water and forms LiOH and hydrogen gas. Thus a sealed container is always used for the transportation of LIBs.

12.5.18 Self-discharge

Batteries are automatically self-discharged if not in use or not delivering current. The self-discharge rate of Li ions is stated by manufacturers at around 1.5%−2% per month. It has been studied that the self-discharge rate increases with temperature. The self-discharge rate drops somewhat at intermediate states of charge and does not increase monotonically [76]. Self-discharge rates increase with increase in battery age.

12.5.19 Battery life

The life of an LIB is one of the most important parameters that governs the number of full charge/discharge cycles in terms of capacity loss or impedance rise. Battery cycle life changes due to different parameters, which include stress factors such as temperature, discharge current, charge current, and state of charge ranges. In real day-to-day life applications such as smartphones it has been observed that batteries are not fully charged and discharged. Hence, battery life cannot be completely defined through full discharge cycles. Therefore some researchers use the term cumulative discharge, defined as the total amount of charge delivered by the battery during its full life cycle. This is simply "… the summation of the partial cycles as fractions of a full charge-discharge cycle" [77].

12.5.20 Degradation

It has been observed that batteries degrade gradually due to the chemical and mechanical changes to the electrodes, which in turn reduce battery capacity. Batteries are degraded through a variety of chemical, mechanical, electrical, and thermal failure mechanisms. SEI growth, mechanical cracking of SEI layer and electrode particles, lithium plating, and thermal decomposition of electrolyte are the main mechanisms leading to degradation in LIBs [78]. It has been reported that degradation depends on the temperature and minimum degradation observed at 25°C. Researchers have concluded that the formation of the SEI layer improves the performance of the battery but leads to thermal degradation. This SEI layer is composed of electrolyte–carbonate reduction products, which have an ionic conductor but an electronic insulator. This layer forms on both electrodes and has the capacity to determine many battery parameters. However, degradation of the device through several reactions may occur due to outside operation parameters. LIB capacity fades over thousands of cycles due to the formation of an SEI layer on the negative electrode by slow electrochemical processes. SEI leads to the consumption of lithium ions and in turn reduces the charge and discharge efficiency of a battery.

12.5.21 Damage and overloading

Damage to LIBs may take place for several reasons, such breakage of the battery or high electrical load withdrawn through the battery without overcharge protection. The battery may explode due to an external short circuit. Thermal runaway and cell rupture also occur due to overcharging [79]. In extreme cases, this can lead to leakage, explosion, or fire. To tackle all these problems, LIB packs contain fail-safe circuitry [79].

12.6 Prior state-of-the-art of Li-ion batteries

Many varieties of rechargeable LIBs are available nowadays in world markets. Among all the available batteries, the LIB is considered superior due to its better characteristics and performance. Researchers are now focusing their attention on the study of positive environmental impacts and the recycling potential of LIBs. From prior state-of-the-art, it is very important to know detailed information regarding LIB chemistries and key properties. A comparison of LIBs with other electric energy storage (EES) technologies explains the advantages of LIBs over EES devices. LIBs have many applications such as portable electronic devices and road transport. Scientists are now studying how LIBs can be used effectively in power supply systems. The state-of-the-art also provides information about the manufacturing process of LIBs from their raw materials to the formation of LIBs. Energy

12.6.1 Comparison of LIBs with other batteries

Lead−acid batteries, nickel−metal hydride (Ni−MH) batteries, and nickel−cadmium (Ni−Cd) batteries are some types of batteries available in the market. Mahammad Hannan et al. explained in detail the comparison between LIBs and other types of batteries [80]. According to them, high energy efficiency and power density are two main aspects of LIBs, which allow them to be designed lighter and smaller in size. The other advantages of LIBs include a wide temperature range of operation, long cycle life, rapid charge capability, low self-discharge rate and charge, energy, and voltage efficiency [80,81]. Due to these advantages, LIBs are considered superior compared to commercial batteries and their usage in powering medical instruments and portable devices [1].

12.6.2 Comparison between Li-ion battery types

Lithium cobalt oxide provides a high specific energy so is a better choice for laptops and mobile phones compared to other batteries. Lithium manganese oxide has moderate specific power, specific energy, and level of safety in comparison to other LIBs. In olivine-type $LiFePO_4$ material, a plateau at 3.4 V has been reported as lithium has been intercalated/deintercalated from the octahedral sites [82,83]. $LiFePO_4$ has good safety and cycling performance but has toxicity and low voltage, capacity, and energy problems. Fig. 12.6 compares the performance and lifecycles of various lithium battery types in terms of different parameters for electric vehicles [84]. It has been reported that lithium-titanate is a better choice because it provides safety, good performance, stable lifecycle, and economy but has low capacity and power. Lithium nickel manganese/cobalt has moderate performance compared to other batteries. The lifecycle and safety features are considered to be more significant than capacity in the application of LIBs for electric vehicles.

12.6.3 Li ion-based chemistries

Among all the metals available for battery chemistry, lithium has been considered a superior material due to its light and electropositive nature, which in turn provides higher potential for energy storage. The main challenge with LIBs is use of Li ions, which are highly reactive in nature. This problem has been resolved by using material that donates lithium ions instead of using metallic lithium [83]. LIB chemistries have been compared with six dimensions such as specific energy, specific power, safety, performance, lifetime, and cost. It has been inferred that LIB chemistries are superior to other battery chemistries in all six dimensional views [84]. Nowadays, Li−air batteries, which contain Li_2O_2 (Li−air), have been considered an emerging technology. The energy density of this new technology is close to the energy density of fossil fuels, making it candidate material for electric vehicles [85]. The Li−air battery consists of an anode made of porous carbon and a cathode with lithium metal. In this case, Li directly reacts with oxygen and delivers a high capacity around 1200 mAh g

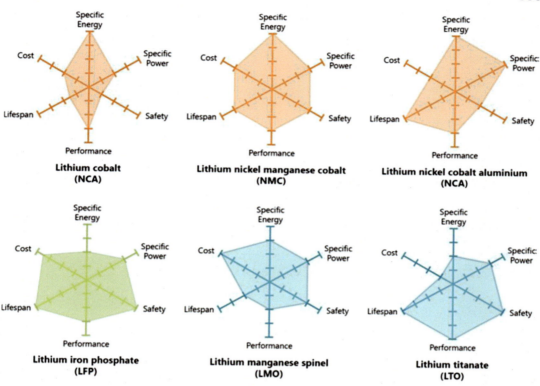

FIGURE 12.6 Comparison of Li-ion batteries for electric vehicles [84].

[83]. This value for Li–air batteries is much higher compared to that of available LIB technologies. Lithium–sulfur batteries have also been considered to be another suitable candidate material for high-energy batteries. In this case, Li metal acts as an anode and solid sulfur (not gaseous like oxygen) acts as a cathode so that it can produce solid Li_2S instead of solid Li_2O_2.

12.6.4 LIB industry and market trends

Production of the LIB industry consists of raw materials, cell components, cells, battery packs, and applications [83]. Many different types of sectors are involved during the production of LIBs such as the mining industry, which provide raw materials, the inorganic chemical industry, which is useful for the supply of cathode materials, the organic chemical industry, and the polymer chemical industry, which provides electrolyte, binder, and separator, respectively. The metal industry supplies electrode foils. Due to changing lifestyles and a shift to smart electronic devices, there is more demand for LIBs. More advanced applications of LIBs include renewable energy sector and stationary energy storage purposes. The LIB market comprises automotive, industrial, consumer electronics, medical, aerospace, and defense sectors. A survey revealed that the largest market for LIBs is the consumer electronics category.

12.6.5 Critical raw material

The sustainable growth of the LIB industry depends on a secure supply of raw materials and proper end-of-life management for products. Very limited attention has been paid to critical raw materials. Therefore a critical raw material evaluation model has been developed by Song et al. This model provides information about critical issues related to the supply risk and economic importance of different materials useful for the production of LIBs [86]. They used an integrating trade-linked model for dynamic materials, performed flow analyses for the relevant critical materials, and optimized subsequent materials flow analysis. A material is considered critical when very few countries assure its supply [83]. According to Song et al., the LIBs market will not be saturated before 2025. Therefore it is most important to find effective methods for utilizing the growing amount of waste materials and to provide a resource supplement for this industry. The lithium-richest countries are Argentina, Bolivia, Chile, China, USA, Australia, Canada, Russia, Congo, and Serbia [82]. It has been reported that global lithium consumption in 2016 was 37,800 t for batteries (39%), 30% for ceramics and glass, 8% for lubricating greases, 5% for polymer production, 5% for continuous casting mold flux powders, and 13% for other uses. In recent years it has been observed that lithium consumption has increased significantly for batteries [87].

12.6.6 LIB key properties

Durability, specific energy, power, and safety are the most important properties of an LIB. A battery degrades depending on its use in all conditions. The rapid degradation of LIBs takes place due to low or high operation temperatures and overcharge or deep discharge conditions. Cycle aging of batteries takes place due to the charging and discharging process. Battery aging reduces capacity and power loss due to the loss of cyclable lithium and electrode materials. Side reactions cause loss of cyclable lithium, and loss of electrode materials is due to dissolution, structural degradation, and particle isolation [84]. The durability of any LIB depends on the number of full cycles. The specific energy of an LIB depends on the types of material used for electrodes and material morphology such as nano- or microstructures. Specific energy is a most important criterion in mobile phone applications and ranges from 90 to 250 Wh kg. The maximum power-providing capacity of LIBs depends on many factors such as voltage, density of Li ions, SEI, diffusion coefficient of the electrodes, and conductivity. Sometimes LIBs have safety issues because they contain lithium, oxygen, and flammable electrolyte. Safety is the most important issue in LIBs due to their applications in road transport, aviation, and portable electronics. Therefore this issue has been studied in detail by various researchers [88]. According to Wang et al. [88], when LIBs are excessively heated, thermal runaway is the most serious issue. Different safety issues, which include operating conditions, accidents, and aging mechanisms, also need to be handled carefully during the operation of LIBs depending on the application.

12.6.7 Environmental impact

The increasing demand of LIBs in many applications, particularly in mobile and stationary energy storage, has caught the attention of several researchers. These researchers were keen to know the environmental impacts of LIBs on production and the effects of industrial products

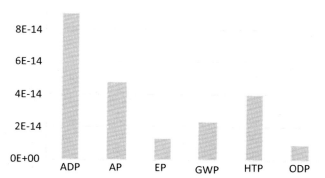

FIGURE 12.7 Normalized averaged environmental impact of a Li-ion battery [91]. *ADP*, abiotic depletion; *AP*, Acidification potential, *EP*, eutrophication potential; *GWP*, global warming potential; *HTP*, human toxicity; *ODP*, ozone depletion potential.

on the environment. Ellingsen et al. studied the lifecycle emissions of an LIB used in an electric vehicle [89]. According to them, approximately 170 kg CO_2 were emitted per kWh of battery capacity. A maximum 85% DoD of LIB emission of 2.2 kg CO_2-eq per100 km distance has been recorded. This has also been confirmed by other studies. A similar type of result has also been observed by Kim et al. for the mass production of the Ford Focus electric [90]. J.F. Peters et al. interpreted the data of the environmental impact of LIBs as shown in Fig. 12.7 [91].

According to them, a survey of average annual impacts in Europe proved that battery production causes more impacts in abiotic depletion, acidification potential, and human toxicity than that of greenhouse gas emissions. All of these are additional environmental impacts apart from cumulative energy demand and global warming potential, which suggests that there is a need for further research on assessing these impacts. The environmental impact of electronic vehicle LIBs has been studied by Notter et al. from production to use and disposal. He carried out an extensive comparison with an internal combustion engine vehicle [92]. This lifecycle assessment result concluded that the environmental impact caused by the battery was up to 15%, while there was only a 2.3% impact caused by the extraction of lithium [83].

12.7 Shortcomings of Li-ion batteries

The LIB is considered a new promising energy source. Though this battery captured a large market area due to its variety of applications, it has a number of shortcomings. The prominent drawback of the LIB is concerned with safety. Nevertheless, aging, transportation problems, storage problems, manufacturing costs, and discharge issues are also problems related to this battery. These problems are discussed in detail in the following subsections.

12.7.1 Safety

Li metal has an unstable nature during the charging process. LIBs have been found more sensitive to high temperatures, crashing, overcharging, overdischarging, and short circuits due to high energy density [93]. Moreover, frequent accidents occur in LIBs due to their components such as plastic packing, separators, and electrolytes, which are combustible in nature. Thermal hazard is another safety issue. The LIB is stable in nature in which lithium ions transfer between the two electrodes during charging and discharging processes where Li ion is regularly cycled many times. The original stable structure of the LIB can be damaged due to external factors such as physical, electrical, and thermal factors, as well as manufacturing defects. Some internal defects like a low-quality separator, material contaminant, and improperly arranged constituents also lead to thermally hazardous conditions [94]. The thermal hazard in LIB occurs due to several reasons such as the decomposition of electrode/electrolyte, the reaction between two electrodes, and the reaction between electrode and electrolyte. Thermal hazard conditions in a single battery are due to high temperature, ejection, combustion, and toxic gases during the thermal runaway process. The malfunctioning of LIBs has caused a number of incidents on airlines, which raises the safety issue. The Federal Aviation Administration reported 46 incidents in aircraft during 2017 due to malfunctioning LIBs [95]. A low-pressure environment exists on aircraft and in some high-altitude areas. Therefore there are many risks associated with both transportation and utilization in low-pressure environments. Fire also behaves differently at low pressure compared to normal conditions. The LIB is fragile in nature and therefore requires a protection circuit for safety.

12.7.2 Aging

The aging of LIBs is of more concern due to their applications in satellites, electric vehicles, and standby batteries because of their long battery life, which is more than 15 years. If an LIB is not in use, then it automatically fails after a number of years. There are two types of aging: one is during use and the other is in storage [96]. Aging during use, also known as cycling, is due to the degradation of active materials reversibility coming from phase transformations at the time of lithium insertion. Storage aging is due to the thermodynamic instability of materials, which results in side reactions; it also due to kinetically induced effects such as concentration gradients. The aging effect of an LIB can be reduced by storing it in a cool place at 40% charge. The recommend storage temperatures for LIBs have been reported at around 15°C (59°F). It has been reported by Feng Leng et al. that the degradation rates of all the components in an LIB increases with increase in temperature. It has been studied that operating temperature has less impact on the degradation rate of the charge transfer rate. The degradation rate of maximum charge storage operating temperature has been found to increase from 4.22% to 13.24% as temperature rises from 25 to 55°C [97].

12.7.3 Expense

An LIB is 40 % more expensive than a nickel-metal hydride battery. An LIB needs onboard computer circuitry, which also makes it expensive.

12.7.4 Deep discharge

The self-discharge rate of an LIB is low. There is no problem with the general integrity of this battery on partially discharging. However, an LIB becomes unusable on deep discharging or when the voltage drops below a particular level.

12.8 Lithium−sulfur batteries

Global warming is an increasingly pressing problem. To handle this problem, scientists around the world are trying to find new innovative devices that will help them to attenuate global warming. Some efforts such as reducing energy consumption and the search for renewable sources have been undertaken. The manufacture of Li−S batteries is a big step toward solving this. This battery was invented in 1962 and patented by Herbert and Ulam. Initially, the Li-S battery was made up of components such as lithium or lithium alloys as anode, sulfur as cathode, and an electrolyte composed of aliphatic saturated amines. After a few years, the introduction of organic solvents made this technology more advanced, delivering 2.35−2.5 V. At the end of the 1980s, a rechargeable lithium−sulfur battery was investigated. The lithium−sulfur battery has high specific energy and is lighter in weight due to the low and moderate atomic weight of lithium and sulfur, respectively [98]. Such a battery was used on the longest and highest-altitude solar-powered aeroplane flight [98]. The specific energy of this battery is about 500 Wh kg, which is better than that of LIBs [99]. Gibbs energy of the lithium−sulfur reaction has been reported to be five times more than the theoretical energy of an LIB. The polysulfide shuttle effect is considered to be the biggest problem in the Li-S system. This effect leads to low lifecycle of the battery due to leakage of active material from the cathode electrode. Fig. 12.8 represents the working principle and shuttle effect of a lithium−sulfur battery.

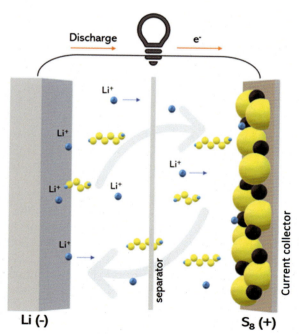

FIGURE 12.8 Working principle and shuttle effect of a lithium−sulfur battery.

The Li-S cell consists of lithium as anode material, nonaqueous liquid as electrolyte, and sulfur as cathode, which is dispersed in a carbon medium having a binder. In the first discharging process of a cell, molecules of elemental sulfur (S_8) generate a chain of lithium polysulfides by accepting the electrons. The phase diagram studies of James et al. showed that polysulfides with less than three sulfur atoms are insoluble in electrolyte [100]. Li_2S_3 was found to be the least soluble polysulfide. After the generation of insoluble and electronically nonconductive Li_2S_2, further sulfur reduction is stopped in the Li-S system. Furthermore, this polysulfide cannot be converted back into elemental S after recharging the first discharge. In the second and subsequent charging processes, the higher-order polysulfide (Li_2S_8) is generated at the sulfur cathode and through the latter stages of charging it diffuses to the lithium anode. At the anode surface, it reacts with Li to regenerate lower-order Li_2S_8. This Li_2S_8 diffuses back to the sulfur cathode to generate higher forms and thus leads to the formation of a shuttle mechanism [100]. The shuttle effect is considered to be the main reason for degradation in a lithium–sulfur battery. Moreover, the shuttle effect is responsible for the characteristic self-discharge of lithium–sulfur batteries due to the slow dissolution of Li_2S_8 [101]. The reactions during discharging processes are as follows:

At the anode (oxidation or loss of electrons):

$$2Li \rightarrow 2Li^+ + 2e^- \tag{12.12}$$

At the cathode (reduction or gaining electrons):

$$S + 2Li^+ + 2e^- \rightarrow Li_2S \tag{12.13}$$

Overall cell reaction (discharging process):

$$2Li + S \rightarrow Li_2S \tag{12.14}$$

However, the low conductivity and massive volume change of sulfur on discharging is the main problem related to lithium–sulfur batteries. Therefore a search for a suitable candidate material as cathode was the first challenge for researchers. A carbon/S cathode and lithium as an anode have been used as a suitable combination by most researchers [102]. The electronic conductivity of sulfur is 5×10^{-30} S/cm at 25°C. A carbon coating is used to increase its conductivity [103]. Recently, carbon nanofibers have been used, which provide a sufficient electron conduction path and structural integrity for Li-S batteries. Rapid degradation during Li–S battery design has been considered as a problem. When the sulfur absorbs lithium, volume expansion of the Li_xS composition takes place, which is 80% of the volume of the original S [104], leading large mechanical stresses on the cathode and in turn causing rapid degradation. This process reduces the contact between the carbon and therefore the sulfur, and prevents the flow of lithium ions to the carbon surface. A liquid organic electrolyte has been used in lithium sulfur batteries. The electrolyte is important because it is responsible for the shuttle effect by polysulfide dissolution and SEI stabilization at the anode surface. It has been observed that the organic carbonate electrolyte in LIBs is not compatible with the chemistry of lithium–sulfur batteries.

12.9 Extensive comparison

The reduction of CO_2 emissions is now gaining importance and because of this the demand for electric vehicles is increasing day by day. Therefore automobile manufacturers are now launching electric vehicles that are capable of competing with internal combustion engine vehicles in performance and economy. LIBs are generally used in electric vehicles, which reached to its saturation limit, i.e., 200–250 Wh/kg specific energy density [105]. In today's scenario, the Li-S battery is more promising than the LIB due to the positive consequences of replacing metals in the cathode of LIBs with sulfur. S is an electrochemically active material that is more abundant in nature and is also a valuable by-product, which can be recovered using the desulfurization processes of the oil and gas industries. In an LIB, lithium ions are inserted into the molecular structure of the carbon electrode through the discharge process. Li^+ ions then shuttle between the positive electrode and negative electrode. On the positive electrode, they are stored upon discharging and on the negative electrode, they are stored upon charging. Theoretical energy densities of LIBs have been reported at around 500 Wh/kg depending on the electrode materials used in the system. On the other hand, redox reaction in the lithium–sulfur system is different from the intercalation process of LIBs. The theoretical capacity of sulfur has been reported at around 1675 mAh/g [106]. When sulfur reacts with lithium (negative electrode) it forms Li_2S via the intermediates Li_2S_8, Li_2S_4, and Li_2S depending on the SoC, which gives high capacity to the anode, and in turn increases the energy density [106]. The theoretical energy density of the lithium–sulfur system has been reported at around 2500 Wh/kg, which is higher than the LIB [105]. Additionally, some main differences between both batteries are as follows:

1. The Li-S battery can work throughout all SoC windows from 0% to 100%, while Li-ion cells have to work by approximately 20% of SoC [106].
2. The volumetric energy density (Wh/L) and gravimetric energy density of the Li-S system are higher than in the LIB, which results in lower battery cost.
3. The Li-S battery is light weight compared to that of the LIB due to the use of sulfur.
4. The lighter weight of Li-S leads to increased mileage range of electric vehicles compared to the LIB.
5. Due to the low cost and light weight of the Li-S battery, researchers have been able to increase its lifespanLi-S.
6. The Li-S battery uses only elemental sulfur as cathode material, while different types of cathode materials such as lithium cobalt oxide, lithium manganese oxide, and lithium nickel manganese cobalt oxide have been used as cathode materials in LIBs [107].
7. The power limitations of LIBs are due to the diffusion of ions into the electrodes, while for Li-S batteries, power exhibits a high sensitivity to cycling parameters such as current profile or temperatures due to slow diffusion of species through the electrolyte.

Though the Li-S battery has many advantages over the LIB, the low conductivity of sulfur, which is round 5×10^{-30} S cm at 25°C, the solubility of sulfur and lithium polysulfides in the electrolytes, and the shuttle effect cause problems like self-discharge, short cycle life, and low coulombic efficiency, which affect the performance of the battery [105]. Increasing the

TABLE 12.1 Comparison of lithium-ion and lithium—sulfur batteries [105].

| Parameter | Li-ion battery | | | | Li—S battery |
	Nickel manganese cobalt oxide	Nickel cobalt aluminum oxide	Lithium cobalt oxide	Lithium iron phosphate	
Energy density	+	+	+	+ −	+ + +
Power density	+ +	+ +	+ +	+	+ −
Lifecycle	+ +	+ −	+	+ +	− −
Self-discharge	+ +	+ +	+ +	+ +	− −
Safety	+	−	−	+ −	+
Price	−	+ −	+ −	+	+ + +
Model	+ +	+ +	+	+ −	−
Environmental	−	−	− −	+ −	+ +

lifespan of an Li-S battery is the biggest challenge for scientists. If one can achieve 500 cycles for an Li—S battery, then this battery can be suitable for many applications.

According to Song et al., the voltage of a lithium—sulfur battery ranges between 2 and 7 V when fully charged and 1 and 5 V when depleted. Li battery provides voltage up to 4 and 15 Volts; that's why Li-S battery provides voltage output 50%. Therefore it was reported that an LIB is more advantageous than Li-S in terms of power density [108]. It has been reported that an Li-ion cell has 10 times more cell value, i.e., 20—50 Ah/cell compared to an Li—S cell. Wagner et al. reported that the decrease in capacity rate of Li-S batteries is 10—40 times less than LIBs [109]. A study by Benveniste et al. showed that lithium—sulfur batteries have 20% lower environmental impact compared to Li-ion cells. According to them, Li-S batteries are still going through testing phases and have few industrialized manufacturing processes [105]. Detailed analysis done by G. Benveniste et al. on the different key parameters suggests that Li-S batteries are clearly ahead compared to LIBs. Table 12.1 represents a comparison of both battery types depending on four key aspects: energy density, safety, price, and environmental impact [105]. From the table it is clear that Li-S batteries have received less interest in terms of life-cycle, self-discharge, power density, and modeling. LIBs are currently closer to their theoretical energy density limit than lithium sulfur. However, according to researchers, these four key parameters are not enough for Li-S batteries to be used instead of LIBs.

12.10 Current research on Li—S batteries

Lithium—sulfur batteries are considered to be a new and promising technology due to their high theoretical energy density of 2600 Wh/kg and high specific capacity at around 1675 mAh/g. These batteries also have greater energy storage ability through phase-transformation electrochemistry, which depends on a sulfur cathode and metallic lithium anode. It is well known that the conversion chemistry of lithium sulfur batteries provides significant advantages but it also has some drawbacks, which include the low conductivity of sulfur and lithium sulfide, the huge volumetric change from sulfur to lithium sulfide, and the shuttle effect [110]. Researchers have focused their attention on resolving all these issues related to Li—S cells. Most efforts have been focuseded on reduction of the shuttle effect and improvement in retention activity within the sulfur cathode. According to the studies of some researchers, discharge capacity, cyclability, and efficiency of Li— S batteries can be improved using nanostructured sulfur composites [111]. Moreover, other approaches have also been taken up, which involve novel cell configurations with trapping interlayers, Li/dissolved polysulfide cells, and efficient electrolytes. The following subsections illustrate these approaches in brief.

12.10.1 Sulfur cathode

The sulfur cathode has been considered the most prominent component in lithium—sulfur batteries because of the direct relation with the electrochemical performance of this battery. Numerous works have been carried out to minimize the shuttle effect, to increase electroconductivity, and limit the volume variation of sulfur [112]. Carbon materials have been extensively studied due to their excellent electrical conductivity. It has also been concluded that the introduction of conductive materials such as graphene can enhance the overall conductivity of Li-S cells. In this regard, researchers have used different forms of graphene such as carbon nanotubes, nitrogen-doped single-walled carbon nanohorns, and carbon nanofibers in the sulfur cathode [113]. Some researchers have used conductive carbons and conductive polymers in sulfur cathodes to form sulfur—carbon composites and sulfur-conductive polymer composites to enhance the electrical conductivity. However, a high initial discharge capacity has been reported after using sulfur-conductive polymer composites and polyacrylonitrile [114]. Different types of conductive/porous carbon materials and conductive polymers have been doped with sulfur to study their effect on electrical conductivity. Among all these materials, porous/conductive carbon has shown prominent results due to its porous structure and higher electrical conductivity compared to other materials. Papandrea et al. [115] prepared a lithium—sulfur battery using a freestanding 3D graphene framework (3DG—S90) with the help of a simple one-pot synthesis method, which showed improved performance. Hollow-structured carbon materials have also been studied due to their special shape and low carbon density. The internal void space within the hollow structured carbon materials accommodates high sulfur and in turn provide volume expansion. It has been concluded that by integrating this structure, the shuttle effect was prohibited by hosting the polysulfides with physical/chemical obstruction [116]. A novel polymer-encapsulated hollow sulfur nanosphere cathode material has been studied by Li et al. [117].

Although it has been reported that the shuttle effect in lithium—sulfur batteries can be minimized using composites, the occurrence of intrinsic material degeneration within the batteries cannot be solved completely by this approach. It has been reported that the conversion of sulfur into Li_2S results in the breakdown of material coatings during repetitive redox reaction. Therefore Li_2S has been considered a substitute for pure sulfur, which has a theoretical specific capacity around 1166 mAh/g [112]. Li_2S allows the use of lithium-free anodes, which mitigates the safety issue of the battery [118]. However, it has been studied that battery performance decreases due to the low conductivity of Li_2S. To tackle this problem, Cai et al.

III. Electrochemical energy conversion and storage

developed an $Li_2S–C$ composite by the ball-milling technique. This $Li_2S–C$ composite showed a starting discharge capacity of 1144 mAh/g [119].

12.10.2 Electrolyte

Electrolytes sandwiched in between two electrodes are used as an ion transport pathway between the electrodes. The ionic conductivity of liquid electrolytes is high, hence their use in batteries. The problem with liquid electrolyte in lithium–sulfur batteries is that the intermediate polysulfide dissolves in the liquid electrolyte that shuttles between the two electrodes. Furthermore, battery performance is also affected due to the solubility of polysulfides. It has been found that solid electrolytes in the form of either inorganic or polymer compounds have prevented polysulfide dissolution that occurred in liquid electrolytes [120]. In an Li-S battery, solid electrolytes act as separators and increase the utilization of both electrodes. Due to the advantage of reducing the solubility and shuttle effect of polysulfides of lithium sulfur batteries, polymer and solid-state electrolytes have been studied and used extensively. Long-chain poly(ethylene–methylene oxide) polymers have been used as polymer electrolytes in lithium–sulfur batteries. According to Shin et al. [121], the electrochemical performance of an Li–S system was improved after using poly(ethylene–methylene oxide):$10LiCF_3SO_3$ composite polymer electrolytes with titanium oxide additive because of increasing ionic conductivity and improved interfacial stability [121]. Hassoun et al. developed a solid-state lithium–sulfur battery employing a PEO-based gel-type polymer membrane [122].

12.10.3 Anode

An anode, which is an important component of the lithium–sulfur battery, determines the long-term cycle stability of a battery. For many years, metallic Li has been considered a suitable candidate material for LIBs due to low potential and high energy density. It has been mentioned in many research papers that coulombic efficiency reduces due to the consumption of metallic Li during the shuttle process. Metallic lithium is unstable in nature and when it reacts with organic electrolytes, it leads to the formation of an SEI layer, which raises a safety issue. In a further repeated cycling process, lithium dendrite formation took place due to the nonuniform growth of an SEI layer on the lithium interface leading to perforation in the separator, which finally resulted in a short circuit, explosion, or firing [112]. To solve this safety issue, coating a physical protection layer on the anode surface can isolate lithium from the electrolyte as well as from the mobile discharge products. Ma et al. [123] uniformly coated an Li_3N layer with a thickness of 200–300 nm to minimize the safety issue. However, the lithium metal with an optimal Al_2O_3 protection layer of 0.73 mg/cm^2 exhibited a capacity of 160 mA/g after 50 cycles [124]. Insertion of a functionalized interlayer between two electrodes is another way to protect the lithium anode. It is evident from the literature that most of the interlayers are made up of different forms of carbonaceous material.

Another suitable solution to solve this lithium anode issue is to replace the metallic Li electrode with another stable material, which can also avoid the corrosion issue. In this regard, researchers have found that silicon material is a better alternative because it possesses high theoretical specific capacity and high energy density [125]. Rapid prelithiation of Si nanowires was studied by Liu et al. [126]. Krause et al. [127] developed a novel method for the fabrication of a silicon nanowire/carbon anode showing high capacity for a lithium—sulfur battery. Hassoun's group studied tin as another option for the Li anode [128]. A lithium—sulfur cell was developed by Duan et al. using a carbonic ester liquid electrolyte and a lithium/Sn—C composite anode [129]. Carbon-based anode material is another anode candidate for Li—S batteries. Graphite materials with a layered structure have also been studied for the Li-ion intercalation/deintercalation process. However, one major drawback associated with graphite is its incompatibility with electrolyte material in Li—S batteries. Also, the use of graphite materials requires another electrolyte material that leads to surface stabilization with the graphite particles.

Tao Li et al. [112] summarized the different types of approaches that have been used for improvement in battery performance. According to them, a protective film on the anode surface is the best solution, which also minimizes the corrosion issue. Furthermore, insertion of an interlayer material between the separator and Li anode improves SEI stability and lifespan of the Li anode. Thus this anode structure design has been found to be a low cost, safe, and effective interlayer method to solve the anode issue for commercial production of lithium—sulfur batteries. The other approach is the replacement of the Li anode with other materials, which also minimizes some issues with lithium—sulfur batteries such as the shuttle effect, instability of lithium, and the capacity issue.

12.11 Applications

Battery technologies play vital roles for transforming society in a more sustainable way. Due to high energy storage systems, batteries are participating in an energy-ecological evolution through their expanded applications, which vary from portable electronics devices to electric vehicles. Application of lithium-based batteries in internal combustion engines reduces the consumption of fossil fuels and the emission of greenhouse gases in road transportation. Li-based batteries have many more applications in day-to-day life. These applications are as follows.

12.11.1 Electric vehicles and toys

By using lithium iron phosphate, different-sized batteries have been prepared. This type of battery has been used in large electric vehicles such as buses, electric cars, and hybrid vehicles as well as small electric vehicles like electric bikes, small cars, and forklifts. Such batteries have applications in remote-controlled toys and wind energy storage.

12.11.2 Small medical equipment and portable instruments

The LIB has a wide range of uses in small medical applications. Lightweight LIBs have been used in gadgets such as laptops, cell phones, camcorders, and iPods. Li-ion phosphate batteries are installed in cars such as the Mercedes Hybrid S-Class.

12.11.3 Emergency power backup or UPS

The lithium battery has the main advantage of being used for an emergency power backup that protects equipment from power loss. It is not a type of generator because it provides immediate power to run the equipment. It is mostly used in computers and communication and medical applications.

12.11.4 Solar power storage

Rechargeable LIBs are closely matched with solar panels. The LIB easily charged, which is a quality that maximizes potential solar power storage during the daytime. According to a recent survey, the use of solar power has drastically increased in an increasing number of countries. Li batteries are useful during the nighttime as well as when solar equipment needs repairing.

12.11.5 Surveillance or alarm systems in remote locations

Rechargeable Li batteries are suitable for remote monitoring applications due to their long life, small size, and less power-losing capacity through a self-discharging process when a system is not active. The self-discharging rate of Li batteries is lower than lead—acid batteries, which is useful in situations when they are not under continuous use. Lithium batteries are widely used in monitoring remote perimeters, a fleet of vehicles, and construction sites, for example, where a permanent alarm system has not been installed.

12.11.6 Personal freedom with mobility equipment

New technology has been developed in such a way that it has made daily life easier. Lightweight lithium batteries are considered suitable in devices for mobility equipment due to their smaller size, longer life, faster charging, and lower self-discharge capacity compared to that of any cell available in the market. Individuals are also using reliable mobility technology from electric wheelchairs to stair lifts, allowing less able individuals to live an independent life.

12.11.7 Portable power packs

Rechargeable LIBs are the most appropriate solution for powering phones and lightweight laptops and computers. Li batteries have the capability of tolerating movement and temperature changes and also maintaining their power delivery on usage. The lithium battery is a long-lasting and more efficient portable device. These batteries are more economical and durable in all conditions.

12.11.8 Aviation applications

Aircraft must be lightweight so that they need less energy to keep them airborne. Different companies are trying to develop Li batteries that are also lightweight for use in an aviation battery management system.

12.11.9 Marine applications

Lithium—sulfur batteries are advantageous in terms of specific energy, safety, and lower mass density. All these properties are useful for the powering of autonomous underwater vehicles. The National Oceanography Center has developed a pressure-tolerant lithium—sulfur battery capable of powering marine autonomous underwater vehicles to depths of over 6000 m. An Li—S battery can sustain extreme pressure at 4°C. Therefore it can be said that Li-S technology is the ultimate solution for use in deep-sea mining and offshore oil and gas applications.

12.11.10 Military applications

The components of a battery freeze at lower temperatures, thus preventing chemical reactions. It has been found that most batteries fail in such conditions. Therefore a battery having high energy density is required in these situations. Researchers are developing a lithium—sulfur battery that can be operated at a temperature of −60°C. Due to the light weight of a lithium—sulfur battery, mission costs can be reduced. OXIS Energy has developed a prototype, lightweight, portable battery for soldiers and IT equipment.

12.11.11 Aerospace applications

It has been reported that OXIS and NASA are researching the development of high specific energy lithium sulfur for lightweight applications such as drones and high-altitude vehicles. Several companies are also using these batteries in aerospace applications such as stratospheric balloons.

12.12 Future prospective

Li batteries are considered the best suitable technology in recent electrochemistry. These batteries dominate the electric automotive and renewable energy storage market. The main advantage of this device is that it has the potential to expand applications from medicine to robotics and space. New avenues can be opened in the future due to advancements in the technology of Li batteries. Over the past few years, researchers have been searching for technology that can make flexible and wearable electronic devices. The Li battery is the ultimate solution to their search due to its low power requirements. This battery can be used for powering the future generation of devices such as health tags, smart watches, and displays. Modern cars can now be built with new innovative concepts such as hybridization, which began with microhybrid vehicles. The new concept in electric vehicles means that cars can be boosted during acceleration and can be driven for short distances using a full-hybrid electric vehicle. Electric vehicles may be supported by battery power without an internal combustion engine. The voltage levels are different for different vehicle concepts depending on the battery pack design. Nowadays and in the future too, microhybrid vehicles can be built with batteries in combination with supercapacitors or high-powered Li-ion cells. Currently, vehicle-to-grid concepts are under development in which a battery is used for driving as well as for grid stability. If the progress of research is as fast as it was over the last 10 years, this field of technology advancement will produce batteries of high potential, which will require less time to charge than it takes to refuel a conventional car. Also, in line with this great technology is the need to reduce the effects of global warming and noise pollution. The lithium battery project has been intensively stimulated and accelerated in research and development. At the present stage, some projects for the development of basic technology of a new kind of battery are about to start in parallel with dispersed-type battery energy storage technology. Entry of a new kind of battery in the market in the near future is expected.

12.13 Conclusions

In this chapter, a comprehensive assessment of the developments in LIBs as well as in lithium—sulfur batteries was discussed in detail. Emphasis was given to the major historical progress and material developments of both batteries. Here, more attention was focused on recent developments from Li-ion cells to Li—S batteries. Moreover, different component materials used up to now were described in detail, which provided a clear understanding of all the reactions involved in the chemistry of both batteries. Applications of both Li and lithium—sulfur batteries were also explained in detail. All the achievements and efforts done by scientists in the development of rechargeable lithium—sulfur battery technology, which is a high energy density system, were also discussed. The time taken by lithium—sulfur batteries to be used in practical applications is lengthy compared to that of LIBs. The most challenging task associated with this battery is developing a battery pack with high energy density while maintaining high capacity with long-term cycle stability. Therefore, the efforts made by researchers to solve this issue are guided by using the concept of basic science. Efforts for the development of high-performance lithium—sulfur batteries were explained briefly. Developments in the materials used for cathode, anode, and electrolyte based on their electrochemistry, morphology, and engineering design to improve the cycle life of batteries were explained in detail. The parameters used for batteries, which include specific energy, safety, and cycle life, were also mentioned. Recent newly evaluated batteries meet all the criteria required for the automotive industry and other applications. One can switch to the miniaturization of these batteries by changing the materials used for cell components. Evolution of the LIB is in the hands of innovators who will give it the priority it deserves in the coming years.

References

[1] S.G. Liasi, S.M.T. Bathaee, Optimizing microgrid using demand response and electric vehicles connection to microgrid, in: Smart Grid Conference (SGC 2017), 2017, ISBN 978-1-5386-4279-5, pp. 1—7.
[2] T.M. Gur, Review of electrical energy storage technologies, materials and systems: challenges and prospects for large-scale grid storage, Energy Environ. Sci. 11 (2018) 2696—2767.
[3] O. Groger, H.A. Gesteiger, J.P. Suchsland, Review-electromobility: batteries or fuel cells, J. Electrochem. Soc. 162 (2015) A2605—A2622.
[4] M.N. Obrovac, V.L. Chevrier, Alloy negative electrodes for Li-ion batteries, Chem. Rev. 114 (2014) 1444—11502.
[5] C.K. Chan, H. Peng, G. Liu, K. Mcilwrath, X.F. Zhang, R.A. Huggins, Y. Cui, High performance lithium battery anodes using silicon nanowires, Nat. Nanotechnol. 3 (2008) 31—35.

References

367

[6] J.E. Trevey, A.F. Gross, J. Wang, P. Liu, J.J. Vajo, Stable cycling and excess capacity of a nanostructured Sn electrode based on $Sn(CH_3COO)_2$ confined within a nanoporous carbon scaffold, Nanotechnology 24 (42) (2013) 1–6.

[7] J. Muldoon, C.B. Bucur, T. Gregory, Quest for nonaqueous multivalent secondary batteries: magnesium and beyond, Chem. Rev. 114 (2014) 11683–11720.

[8] U. Kasavajjula, C. Wang, A.J. Appleby, Nano-and bulk-silicon-based insertion anodes for l-ion secondary lithium cells, J. Power Sources 163 (2007) 1003–1039.

[9] T. Morita, N. Takami, Nano Si cluster-SiOx-C composite material as high-capacity anode material for rechargeable lithium batteries, J. Electrochem. Soc. 153 (2006) A425–A430.

[10] C.M. Park, J.H. Kim, H. Kim, H.J. Sohn, Li-alloy based anode materials for Li secondary batteries, Chem. Soc. Rev. 39 (2010) 3115–3141.

[11] H. Groult, B. Kaplan, F. Lantelme, S. Komaba, N. Kumagai, H. Yashiro, T. Nakajima, B. Simon, A. Barhoun, Nanoparticles from electrolysis of molten carbonates and use as anode materials in lithium-ion batteries, Solid State Ionics 177 (2006) 869–875.

[12] T. Doi, Y. Iriyama, T. Abe, Z. Ogumi, Electrochemical insertion and extraction of lithium ion at uniform nano-sized $Li_{4/3}Ti_{5/3}O_4$ particles prepared by spray pyrolysis method, Chem. Mater. 17 (2005) 1580–1582.

[13] W. Weppner, R.A. Huggins, Electrochemical investigation of the chemical diffusion partial ionic conductivities, and other kinetic parameters in Li_3Sb and Li_3Bi, J. Solid State Chem. 22 (1977) 297–308.

[14] Z. Lei, Y. Zhang, X. Lei, Temperature uniformity of a heated lithium-ion battery cell in cold climate, Appl. Therm. Eng. 129 (2018) 148–154.

[15] M. Bakierska, M. Swietoslawski, K. Chudzik, M. Lis, M. Molenda, Enhancing the lithium ion diffusivity in $LiMn_2O_{4-y}S_y$ cathode materials through potassium doping, Solid State Ionics 317 (2018) 190–193.

[16] X. Xu, Z. Meng, X. Zhu, S. Zhang, W.Q. Han, Biomass carbon composited FeS_2 as cathode materials for high-rate rechargeable lithium-ion battery, J. Power Sources 380 (2018) 12–17.

[17] N. Rapulenyane, E. Ferg, H. Luo, High-performance $Li_{1.2}Mn_{0.6}Ni_{0.2}O_2$ cathode materials prepared through a facile one-pot co-precipitation process for lithium ion batteries, J. Alloys Compd. 762 (2018) 272–281.

[18] M. Thackeray, Lithium-ion batteries: an unexpected conductor, Nat. Mater. 1 (2002) 81–82.

[19] M.S. Whittingham, Ultimate limits of intercalation reactions for lithium batteries, Chem. Rev. 114 (2014) 11414–11443.

[20] E. Peled, The electrochemical behavior of alkali and alkaline earth metals in nonaqueous battery systems – the solid electrolyte interphase model, J. Electrochem. Soc. 126 (1979) 2047–2051.

[21] V. Thangadurai, W. Weppner, Recent progress in solid oxide and lithium ion conducting electrolytes research, Ionics 12 (2006) 81–92.

[22] V. Thangadurai, W. Weppner, Solid state lithium ion conductors: design considerations by thermodynamic approach, Ionics 8 (2002) 281–292.

[23] P. Knauth, Inorganic Li ion conductors: an overview, Solid State Ionics 180 (2009) 911–916.

[24] S. Stramare, V. Thangadurai, W. Weppner, Lithium lanthanum titanates a review, Chem. Mater. 15 (21) (2003) 3974–3990.

[25] R. Murugan, V. Thangadurai, W. Weppner, Lithium ion conductivity of $Li_{5+x}Ba_xLa_{3-x}Ta_2O_{12}$ ($x = 0-2$) with garnet-related structure in dependence of the barium content, Ionics 13 (2007) 195–203.

[26] Y.X. Gao, X.P. Wang, W.G. Wang, Q.F. Fang, Sol–gel synthesis and electrical properties of $Li_5La_3Ta_2O_{12}$ lithium ionic conductors, Solid State Ionics 181 (2010) 33–36.

[27] S. Hasegawa, N. Imanishi, T. Zhang, J. Xie, A. Hirano, Y. Takeda, O. Yamamoto, Study on lithium/air secondary batteries—stability of NASICON-type lithium ion conducting glass—ceramics with water, J. Power Sources 189 (2009) 371–377.

[28] Y. Shimonishi, T. Zhang, P. Johnson, N. Imanishi, A. Hirano, Y. Takeda, O. Yamamoto, N. Sammes, A study on lithium/air secondary batteries—stability of f NASICON-type glass ceramics in acid solutions, J. Power Sources 195 (18) (2009) 6187–6191.

[29] M. Barre, F. Berre, M.P. Crosnier-Lopez, C. Galven, O. Bohnke, J.L. Fourquet, The NASICON solid solution $Li_{1-x}La_x/3Zr_2(PO_4)_3$: optimization of the sintering process and ionic conductivity measurements, Ionics 15 (2009) 681–687.

[30] C. Yada, Y. Iriyama, T. Abe, K. Kikuchi, Z. Ogumi, A novel all-solid-state thin-film-type lithium-ion battery with in situ prepared positive and negative electrode materials, Electrochem. Commun. 11 (2009) 413–416.

[31] W.J. Fergus, Ceramic and polymeric solid electrolytes for lithium-ion batteries, J. Power Sources 195 (2010) 4554–4569.

III. Electrochemical energy conversion and storage

[32] J.M. Tarascon, A.S. Gozdz, C. Schmutz, F. Shokoohi, P.C. Warren, Performance of Bellcore's plastic rechargeable Li-ion batteries, Solid State Ionics 86–88 (1996) 49–54.

[33] N. Shukla, A.K. Thakur, Role of salt concentration on conductivity optimization and structural phase separation in a solid polymer electrolyte based on PMMA-LiClO$_4$, Ionics 15 (2009) 357–367.

[34] W. Pu, X. He, L. Wang, Z. Tian, C. Jiang, C. Wan, Ionics 14 (2008) 27–31.

[35] M. Bellis, Biography of Alessandro Volta, Stored Electricity and the First Battery, 2008.

[36] Battery History, Technology, Applications and Development, Mpower Solutions Ltd., 2007.

[37] G. Borvon, History of the Electrical Units, Association S-EAU-S, 2012.

[38] Columbia Dry Cell Battery, National Historic Chemical Landmarks, American Chemical Society, 2013.

[39] J.F. Rohan, M. Hasan, S. Patil, P.D. Casey, T. Clancy, Energy Storage: Battery Materials and Architectures at the Nanoscale, Chapter 6, 2014, pp. 107–138.

[40] O.S. Burheim, Engineering Energy Storage, 2017, pp. 111–145. https://www.sciencedirect.com/science/book/9780128141007 (Chapter 7), Secondary Batteries.

[41] T.M. Gur, Review of electrical energy storage technologies, materials and systems: challenges and prospects for large-scale grid storage, Energy Environ. Sci. 11 (2018) 2696–2767.

[42] K.J. Griffith, M.K. Wiaderek, G. Cibin, E.L. Marbella, P.C. Grey, Niobium tungsten oxides for high-rate lithium-ion energy storage, Nature 559 (7715) (2018) 556–563.

[43] J.G. Werner, G.G. Rodri, Calero, H.D. Abrun, U. Wiesner Werner, Block copolymer derived 3-D interpenetrating multifunctional gyroidalnanohybrids for electrical energy storage, Energy Environ. Sci. 11 (5) (2018) 1261–1270.

[44] J.S. Lu, I.V.B. Maggay, W.R. Liu, Chem. Commun. 54 (2018) 3094–3097.

[45] L. Hu, H. Zhong, X. Zheng, Y. Huang, P. Zhang, Q. Chen, CoMn$_2$O$_4$ spinel hierarchical microspheres assembled with porous nanosheets as stable anodes for Li ion batteries, Sci. Rep. 2 (986) (2012) 1–8.

[46] J.M. Won, S.H. Choi, Y.J. Hong, Y.N. Ko, Y.C. Kang, Electrochemical properties of yolk-shell structured ZnFe$_2$O$_4$ powders prepared by a simple spray drying process as anode material for lithium-ion battery, Sci. Rep. 4 (5857) (2014) 1–5.

[47] F.M. Courtel, H. Duncan, Y. AbuLebdeh, I.J. Davidson, High capacity anode materials for Li-ion batteries based on spinel metal oxides A Mn$_2$O$_4$(A= Co, Ni, and Zn), J. Mater. Chem. 21 (2011) 10206–10218.

[48] J. Bai, X. Li, G. Liu, Y. Qian, S. Xiong, Unusual formation of ZnCo$_2$O$_4$ 3D hierarchical twin microspheres as a high-rate and ultralong-life lithium-ion battery anode material, Adv. Funct. Mater. 24 (2014) 3012–3020.

[49] J. Li, S. Xiong, Y. Liu, Z. Ju, Y. Qian, High electrochemical performance ACS of monodisperse NiCO$_2$O$_4$ mesoporous microspheres as an anode material for Li-ion batteries, Appl. Mater. Interfaces 5 (2013) 981–988.

[50] D. He, T. Wu, B. Wang, Y. Yang, S. Zhao, J. Wang, H. Yu, Novel Na$_2$TiSiO$_5$ anode material for lithium ion batteries, Chem. Commun. 55 (2019) 2234–2237.

[51] B. Zhao, R. Ran, M. Liu, Z. Shao, A comprehensive review of Li$_4$Ti$_5$O$_{12}$-based electrodes for lithium-ion batteries: the latest advancements and future perspectives, Mater. Sci. Eng. 98 (2015) 1–71.

[52] Y.X. Xu, Z.S. Hong, L.C. Xia, J. Yang, M.D. Wei, One step sol–gel synthesis of Li$_2$ZnTi$_3$O$_8$/C nanocomposite with enhanced lithium-ion storage properties, Electrochim. Acta 88 (2013) 74–78.

[53] Z.S. Hong, M.D. Wei, Layered titanate nanostructures and their derivatives as negative electrode materials for lithium-ion batteries, J. Mater. Chem. 1 (2013) 4403–4414.

[54] Z.S. Hong, X.Z. Zheng, X. K Ding, L.L. Jiang, M.D. Wei, K.M. Wei, Complex spinel titanate nanowires for a high rate lithium-ion battery, Energy Environ. Sci. 4 (2011) 1886–1891.

[55] S. Wang, R. Wang, Y. Bian, D. Jin, Y. Zhang, L. Zhang, In-situ encapsulation of pseudocapacitive Li$_2$TiSiO$_5$ nanoparticles into fibrous carbon framework for ultrafast and stable lithium storage, Nanomater. Energy 55 (2019) 173–181.

[56] J.Y. Liu, Y. Liu, M.Y. Hou, Y.G. Wang, C.X. Wang, Y.Y. Xia, Li$_2$TiSiO$_5$ and expanded graphite nanocomposite anode material with improved rate performance for lithium-ion batteries, Electrochim. Acta 260 (2018) 695–702.

[57] G.W. Peng, H.S. Liu, FT-IR and XRD characterization of phase transformation of heat-treated synthetic natisite (Na$_2$TiOSiO$_4$) powder, Mater. Chem. Phys. 42 (1995) 264–275.

[58] J. Jiang, H. Li, J. Huang, K. Li, J. Zeng, Y. Yang, J. Li, Y. Wang, J. Wang, J. Zhao, Investigation of the reversible intercalation/deintercalationnovel Li$_3$VO$_4$@C microsphere composite cathode material for aluminum of Al ion batteries, ACS Appl. Mater. Interfaces 9 (2017) 28486–28494.

References

[59] Y. Ru, S. Zheng, H. Xue, H. Pang, Different positive electrode materials in organic and aqueous systems for aluminium ion batteries, J. Mater. Chem. 24 (2019) 14391–14418.

[60] W. Kaveevivitchai, A. Huq, S. Wang, M.J. Park, A. Manthiram, Rechargeable Aluminum-ion batteries based on an open-tunnel framework, Small 13 (2017) 1701296–1701306.

[61] Y. Jiang, R. Liu, W. Xu, Z. Jiao, M. Wu, A novel graphene modified $LiMnPO_4$ as a performance-improved cathode material for lithium-ion batteries, J. Mater. Res. 28 (18) (2013) 2584–2589.

[62] L. Zhang, C. Lu, Y. Wang, Y. Cheng, Hydrothermal synthesis and characterization of $MnWO_4$ nanoplates and their ionic conductivity, Mater. Chem. Phys. 103 (2007) 433–436.

[63] E. Zhang, Z. Xing, J. Wang, Z. Ju, Y. Qian, Enhanced energy storage and rate performance induced by dense nanocavities inside $MnWO_4$ nanobars, RSC Adv. 2 (2012) 6748–6751.

[64] H.W. Shim, A.H. Lim, J.C. Kim, G. Lee, D. Kim, Hydrothermal realization of a hierarchical, flowerlike $MnWO_4$@MWCNTs nanocomposite with enhanced reversible Li storage as a new anode material, Chem. Asian J. 8 (2013) 2851–2858.

[65] A. Varzi, C. Taubert, M. Mehrens, N. Kreis, W. Schutz, et al., Study of multi-walled carbon nanotubes for lithium-ion battery electrodes, J. Power Sources 196 (2011) 3303–3309.

[66] Y. Zhang, T. Chen, J. Wang, G. Min, L. Pan, Z. Song, Z. Sun, W. Zhou, J. Zing, The study of multi-walled carbon nanotubes with different diameter as anodes for lithium-ion batteries, Appl. Surf. Sci. 258 (2012) 4729–4732.

[67] H. Huang, S. Yin, T. Kerr, N. Tylor, L.F. Nazar, Nanostructured Composites: a high capacity, fast rate $Li_3V_2(PO_4)_3$/carbon cathode for rechargeable lithium batteries, Adv. Mater. 14 (2002) 1525–1528.

[68] H. Wang, L.F. Cui, Y. Yang, H. Casalongue, J. Robinson, Y. Liang, Y. Cui, H. Dai, Mn_3O_4–Graphene hybrid as a high-capacity anode material for lithium ion batteries, J. Am. Chem. Soc. 132 (40) (2010) 13978–13980.

[69] K. Novoselov, A. Geim, S. Morozov, D. Jiang, Y. Zhang, S.V. Dubonos, I.V. Grigorieva, A.A. Firsov, Electric field effect in atomically thin carbon films, Science 306 (2004) 666–669.

[70] X. Wang, B.L. Li, D.P. Liu, H. Xiang, $ZnWO_4$ nanocrystals/reduced graphene oxide hybrids: synthesis and their application for Li ion batteries, Sci. China Chem. 57 (2014) 122–126.

[71] G. Gao, W. Dang, H. Wu, G. Zhang, C. Feng, Synthesis of $MnWO_4$@C as novel anode material for lithium ion batteries, J. Mater. Sci. Mater. Electron. 29 (15) (2018) 12801–12812.

[72] X. Gao, Thesis University of Wolongong, Development of Novel Materials for Lithium Battery, 2014.

[73] J. Zhang, L. Zhang, F. Sun, Z. Wang, An overview on thermal safety issues of lithium-ion batteries for electric vehicle application, IEEE Access 6 (2018) 23848–23863.

[74] J.B. Goodenough, Y. Kim, Challenges for rechargeable Li batteries, Chem. Mater. 22 (2010) 587–603.

[75] M. Silberberg, Chemistry: The Molecular Nature of Matter and Change, fourth ed., McGraw-Hill Education, New York (NY), 2006, p. 935. ISBN 007721650.

[76] A.H. Zimmerman, Self-discharge losses in lithium-ion cells, IEEE Aero. Electron. Syst. Mag. 19 (2) (2004) 19–24.

[77] S. Saxena, C. Hendricks, M. Pecht, Cycle life testing and modeling of graphite/$LiCoO_2$ cells under different state of charge ranges, J. Power Sources 327 (2016) 394–400.

[78] C. Hendricks, N. Williard, S. Mathew, M. Pecht, A failure modes, mechanisms, and effects analysis (FMMEA) of lithium-ion batteries, J. Power Sources 327 (2016) 113–120.

[79] R. Spotnitz, J. Franklin, Abuse behavior of high-power lithium-ion cells, J. Power Sources 113 (1) (2003) 81–100.

[80] A.M. Hannan, M. Hoque, A. Hussain, Y. Yusof, P.J. Ker, Enabling the electric future of mobility: robotic automation for electric vehicle battery assembly, IEEE Access 7 (2019) 170961–170991.

[81] Q. Lai, H. Zhang, X. Li, L. Zhang, Y. Cheng, A novel single flow zincubromine battery with improved energy density, J. Power Source 235 (2013) 1–4.

[82] D. Son, E. Kim, T.-G. Kim, M.G. Kim, J. Cho, B. Parka, Nanoparticle iron-phosphate anode material for Li-ion battery, Appl. Phys. Lett. 85 (24) (2004) 5875–5877.

[83] G. Zubia, R. Dufo-Lopeza, M. Carvalhob, G. Pasaogluc, The lithium-ion battery: state of the art and future perspective, Renew. Sustain. Energy Rev. 89 (2018) 292–308.

[84] The Boston Consulting Group Inc., Batteries for Electric Cars: Challenges, Opportunities, and the Outlook to 2020, The Boston Consulting Group Inc., Boston, 2010, pp. 1–18. Tech. Rep.

[85] S. Vasquez, S. Lukic, E. Galvan, L. Franquelo, Energy storage systems for transport and grid applications, IEEE Trans. Ind. Electron. 57 (12) (2010) 3881–3895.

[86] J. Song, W. Yan, H. Cao, Q. Song, H. Ding, Z. Lv, Y. Zhang, Z. Sun, Material flow analysis on critical raw materials of lithium-ion batteries in China, J. Clean. Prod. 215 (2019) 570–581.

[87] A. Mancha, A look at some international lithium-ion battery recycling initiatives, J. Undergrad. Res. 9 (2016) 1–5.

[88] Y. Wang, B. Liu, Q. Li, S. Cartmell, S. Ferrara, Z.D. Deng, J. Xiao, Lithium and lithium ion batteries for applications in microelectronic devices: a review, J. Power Sources 286 (2015) 330–345.

[89] L. Ellingsen, G. Majeau-Bettez, B. Singh, A.K. Srivastava, L.O. Valoen, A.H. Stromman, Life-cycle assessment of a lithium-ion battery vehicle pack, J. Ind. Ecol. 18 (2014) 13–24.

[90] H.C. Kim, T.J. Wallington, R. Arsenault, C. Bae, S. Anh, J. Lee, Cradle-to-gate emissions from a commercial electric vehicle Li-ion battery: a comparative analysis, Environ. Sci. Technol. 50 (14) (2016) 7715–7722.

[91] F. Peters Jens, M. Baumann, B. Zimmermann, J. Braun, M. Weil, The environmental impact of Li-Ion batteries and the role of key parameters- a review, Renew. Sustain. Energy Rev. 67 (2017) 491–506.

[92] D.A. Notter, M. Gauch, R. Widmer, P.A. Wager, A. Stamp, R. Zah, H.J. Althaus, Contribution of Li-ion batteries to the environmental impact of electric vehicles, Environ. Sci. Technol. 44 (17) (2010) 6550–6556.

[93] D. Ouyang, M. Chen, Q. Huan, J. Weng, Z. Wang, J. Wang, A review on the thermal hazards of the lithium-ion battery and the corresponding countermeasures, Appl. Sci. 9 (12) (2019) 2483–2528.

[94] Y. Wu, S. Saxena, Y. Xing, Y. Wang, C. Li, W. Yung, M. Pecht, Analysis of manufacturing-induced defects and structural deformations in Lithium-ion batteries using computed tomography, Energies 11 (2018) 925–947.

[95] Werner, Patent, US-10-543,799 B1, 2020, pp. 1–15.

[96] M. Broussely, P. Biensan, F. Bonhomme, P. Blanchard, S. Herreyre, K. Nechev, R.J. Staniewicz, Main aging mechanisms in Li ion batteries, J. Power Sources 146 (2005) 90–96.

[97] F. Leng, C. Ming Tan, M. Pecht, Effect of temperature on the aging rate of li ion battery operating above room temperature, Sci. Rep. 5 (12967) (2015) 1–12.

[98] S.S. Zhang, Liquid electrolyte lithium/sulfur battery: fundamental chemistry, problems, and solutions, J. Power Sources 231 (2013) 153–162.

[99] R. Manthiram, F. Yongzhu, S. Yu-Sheng, Challenges and prospects of lithium–sulfur batteries, Acc. Chem. Res. 46 (2013) 1125–1134.

[100] R.J. Akridge, V.Y. Mikhaylik, N. White, Li/S fundamental chemistry and application to high-performance rechargeable batteries, Solid State Ionics 175 (2004) 243–245.

[101] A. Manthiram, F. Yongzhu, S.H. Chung, Z. Chenxi, S.Y. Sheng, Rechargeable lithium–sulfur batteries, Chem. Rev. 114 (23) (2014) 11751–11787.

[102] Y.J. Choi, K.W. Kim, Improvement of cycle property of sulfur electrode for lithium/sulfur battery, J. Alloys Compd. 449 (1–2) (2008) 313–316.

[103] J.A. Dean, Lange's Handbook of Chemistry, third ed., McGraw-Hill, New York, 1985, pp. 3–5.

[104] M.M. Islam, A. Ostadhossein, O. Borodin, A. Yeates, T. Tipton, W. William Hennig, G. Richard Kumar Nitin, Duin, C.T. Van Adri, ReaxFF molecular dynamics simulations on lithiated sulfur cathode materials, Phys. Chem. Chem. Phys. 17 (5) (2015) 3383–3393.

[105] G. Benveniste, H. Rallo, L. Canals Casals, A. Merino, B. Amante, Comparison of the state of lithium-sulphur and lithium-ion batteries applied to electromobility, J. Environ. Manag. 226 (2018) 1–12.

[106] A. Fotouhi, D. Auger, L.O. Neill, T. Cleaver, S. Walus, Lithium-Sulfur battery technology readiness and applications—a review, Energies 10 (12) (2017) 1–15.

[107] I. Buchmann, Batteries in a Portable World: A Handbook on Rechargeable Batteries for Non-engineers, fourth ed., 2016.

[108] M. Song, J. Cairns, Y. Zhang, Lithium/sulfur batteries with high specific energy: old challenges and new opportunities, RSC Publ. 5 (6) (2013) 2186–2204.

[109] N. Wagner, H. Eneli, M. Ballauff, K.A. mFriedrich, Correlation of capacity fading processes and electrochemical impedance spectra in lithium/sulfur cells, J. Power Sources 323 (2016) 107–114.

[110] L. Yan, M. Xiao, S. Wang, D. Han, Y. Meng, Edge sulfurized graphene nanoplatelets via vacuum mechanochemical reaction for lithium–sulfur batteries, J. Energy Chem. 26 (2017) 522–529.

[111] X. Ji, L.F. Nazar, Advances in Li–S batteries, J. Mater. Chem. 20 (2010) 9821–9826.

[112] T. Li, X. Bai, U. Gulzar, Y. Bai, C. Capiglia, W. Deng, X. Zhou, Z. Liu, Z. Feng, R. ProiettiZaccaria, Comprehensive understanding of lithium–sulfur battery technology, Adv. Funct. Mater. 29 (22) (2019).

[113] W.G. Chong, J.Q. Huang, Z.L. Xu, X. Qin, X. Wang, J.K. Kim, Lithium–sulfur battery cable made from ultralight, flexible graphene/carbon nanotube/sulfur compositefibers, Adv. Funct. Mater. 27 (2017) 1604815.

[114] J. Wang, J. Yang, J. Xie, N. Xu, A novel conductive polymer–sulfur composite cathode material for rechargeable batteries, lithium, Adv. Mater. 14 (2002) 963–965.

References

[115] B. Papandrea, X. Xu, Y. Xu, C.Y. Chen, Z. Lin, G. Wang, Y. Luo, M. Liu, Y. Huang, L. Mai, Three-dimensional graphene framework with ultra-high sulfur content for a robust lithium—sulfur battery, Nano Res. 9 (1) (2016) 240—248.

[116] Z. Li, H.B. Wu, X.W.D. Lou, Rational designs and engineering of hollow micro-/nanostructures as sulfur hosts for advanced lithium—sulfur batteries, Energy Environ. Sci. 9 (2016) 3061—3070.

[117] W. Li, G. Zheng, Y. Yang, Z.W. Seh, N. Liu, Y. Cui, High-performance hollow sulfur nanostructured battery cathode through a scalable room temperature, one-step, bottom-up approach, Proc. Natl. Acad. Sci. U.S.A. 110 (2013) 7148—7153.

[118] Y. Qiu, G. Rong, J. Yang, G. Li, S. Ma, X. Wang, Z. Pan, Y. Hou, M. Liu, F. Ye, Highly nitridated graphene cycles—Li_2S Cathodes with stable modulated, Adv. Energy Mater. 5 (2015) 1—8.

[119] K. Cai, M.K. Song, E.J. Cairns, Y. Zhang, Nanostructured Li_2S—C composites as cathode material for high-energy lithium/sulfur batteries, Nano Lett. 12 (2012) 6474—6479.

[120] X. Yao, N. Huang, F. Han, Q. Zhang, H. Wan, J.P. Mwizerwa, C. Wang, X. Xu, High-performance all-solid-state lithium—sulfur batteries enabled by amorphous sulfur-coated reduced graphene oxide cathodes, Adv. Energy Mater. 7 (2017) 1602923.

[121] J.H. Shin, K.W. Kim, H.J. Ahn, J.H. Ahn, Electrochemical properties and interfacial stability of (PEO) $10LiCF_3SO_3$—Ti_nO_{2n-1} composite polymer electrolytes for lithium/sulfur battery, Mater. Sci. Eng. B 95 (2) (2002) 148—156.

[122] J. Hassoun, B. Scrosati, Moving to a solid-state configuration: a valid approach to making lithium-sulfur batteries viable for practical applications, Adv. Mater. 22 (2010) 5198—5201.

[123] G. Ma, Z. Wen, M. Wu, C. Shen, Q. Wang, J. Jin, X. Wu, A lithium anode protection guided highly-stable lithium—sulfur battery, Chem. Commun. 50 (2014) 14209.

[124] H.K. Jing, L.L. Kong, S. Liu, G.R. Li, X.P. Gao, Protected lithium anode with porous Al_2O_3 layer for lithium—sulfur battery, J. Mater. Chem. 3 (2015) 12213—12219.

[125] S. Goriparti, E. Miele, F. De Angelis, E. Di Fabrizio, R.P. Zaccaria, C. Capiglia, Review on recent progress of nanostructured anode materials for Li-ion batteries, J. Power Sources 257 (2014) 421—443.

[126] N. Liu, N. Lue, M.T. HuMcDowell, A. Jackson, Y. Cui, Prelithiated silicon nanowires as an anode for lithium ion batteries, ACS Nano 5 (2011) 6487—6493.

[127] A. Krause, S. Dorfler, M. Piwko, F.M. Wisser, T. Jaumann, E. Ahrens, L. Giebeler, H. Althues, S. Schadlich, J. Grothe, High area capacity lithium-sulfur full-cell battery with prelitiathed silicon nanowire-carbon anodes for long cycling stability, Sci. Rep. 6 (27982) (2016) 1—12.

[128] J. Hassoun, B. Scrosati, A high-performance polymer tin sulfur lithium ion battery, Angew. Chem., Int. Ed. 49 (2010) 2371—2374.

[129] B. Duan, W. Wang, A. Wang, Z. Yu, H. Zhao, Y. Yang, A new lithium secondary battery system: the sulfur/lithium-ion battery, J. Mater. Chem. 2 (2014) 308—314.

CHAPTER
13

Glasses and glass-ceramics as sealants in solid oxide fuel cell applications

P.S. Anjana[1], M.S. Salinigopal[1,2], N. Gopakumar[2]

[1]Department of Physics, All Saints' College, University of Kerala, Trivandrum, Kerala, India;
[2]Department of Physics, Mahatma Gandhi College, Research Centre, University of Kerala, Trivandrum, Kerala, India

13.1 Introduction

Whether constrained by the world's limited hydrocarbon resources or by the need for emission reductions, the efficient production of electricity is a fundamental requirement for the modern world. Ever-increasing energy consumption, rising public awareness of environmental protection, and higher prices of fossil fuels have motivated many to look for alternative/renewable energy sources [1,2]. Fuel cells are the most efficient means to directly convert stored chemical energy into usable electrical energy (by electrochemical reaction), promising power generation with high efficiency and low environmental impact [3]. Friedrich Schönbein discovered the basic principle of the fuel cell in 1838. The first fuel cell was developed by Sir William Grove (in 1839) based on reversing the electrolysis of water [4]. The development of fuel cell technology is now grabbing the attention of the scientific world because it is considered to be a clean source of energy.

Different types of fuel cells are composed under the same basic principle and share some generic components. The classification of fuel cells depends on the electrolyte and fuel used, which in turn determine the electrode reactions and the type of ions that carry the current across the electrolyte. The most common classification of fuel cells includes the polymer electrolyte fuel cell (PEFC), alkaline fuel cell (AFC), phosphoric acid fuel cell (PAFC), direct methanol fuel cell (DMFC), molten carbonate fuel cell (MCFC), and solid oxide fuel cell (SOFC). Broadly, the nature of electrolyte dictates the operating temperature range of the fuel cell. The operating temperature and the life of a fuel cell will be decided by the physicochemical and thermomechanical properties of materials used as the cell components (i.e., electrodes, electrolytes, sealants, interconnects, etc.) [5]. The different categories of fuel cells are represented in Table 13.1 with their specifications.

Energy Materials
https://doi.org/10.1016/B978-0-12-823710-6.00012-1

Copyright © 2021 Elsevier Ltd. All rights reserved.

TABLE 13.1 Different categories of fuel cells [7–9].

	AFC	PEFC	DMFC	PAFC	MCFC	SOFC
Operating temperature (°C)	60–80	70–90	70–90	150–200	600–650	650–1000
Oxidant	O_2	Air	Air	Air	Air, CO_2	Air
Ionic conduction	OH^-	H^+	H^+	H^+	CO_3^{2-}	O_2^-
Fuel	H_2	H_2	CH_3OH	H_2, CH_4	H_2, CH_4, CO	H_2, CH_4, CO
Maximum electrical efficiency	60%	40%–55%	35%	40%	55%	65%
Power output	10–100 kW	1–250 kW	1–100 W	Up to 1 MW	Up to 100 MW	Up to 100 MW
Advantage	Cathode reaction faster in alkaline electrolyte, leads to high performance Low-cost components	Solid electrolyte reduces corrosion and electrolyte management problems Low temperature Quick startup	High energy density of fuel, easier storage	Higher efficiency with Combined Heat and Power (CHP) Increased tolerance to fuel impurities	High efficiency Fuel flexibility Can use a variety of catalysts	High efficiency Fuel flexibility Can use a variety of catalysts Solid electrolyte
Disadvantage	Sensitive to CO_2 in fuel and air Electrolyte management	Expensive catalysts Sensitive to fuel impurities Low-temperature heat	Efficiency is limited by permeation of methanol	Pt catalyst Long startup time Low current and power	High-temperature corrosion and breakdown of cell components Long startup time Low power density	High-temperature corrosion and breakdown of cell components High-temperature operation Requires long startup time

AFC, Alkaline fuel cell; DMFC, direct methanol fuel cell; MCFC, molten carbonate fuel cell; PAFC, phosphoric acid fuel cell; PEFC, polymer electrolyte fuel cell; SOFC, solid oxide fuel cell.

13.2 Solid oxide fuel cell

During the past few decades, significant efforts have been made in the development of fuel cell technology. Among the different categories of fuel cell, SOFC is the most efficient power-generating system based on electrochemistry principles and is currently in demand due to its better output with ease of handling, low emission, and simple manufacturing process. Hence, it is considered to be the most efficient and environmentally friendly power-generating device with ideal fuel flexibility [5,6]. SOFC technology is most suited to applications in the distributed generation (i.e., stationary power) market because of its high conversion efficiency, Which is beneficial when fuel costs are higher. SOFCs have a potential long-life expectancy of more than 40,000−80,000 h. Some fuel cell developers see SOFCs being used in motor vehicles [10]. SOFCs offer great promise to generate energy from hydrocarbons. They convert the fuel chemical energy directly into electrical energy through electrochemical reactions that are driven by the difference in the oxygen chemical potential between the anode and cathode of the cell [11].

13.3 Operation of the SOFC

Oxygen in the area of the cathode is converted into O^{2-} ions and is transported through an electrolyte based on stabilized zirconia to the anode. Oxygen ions are transferred to the anode, passing through the electrolyte during fuel oxidation. On the cathode, the air passes first through a heat exchanger with functions of preheating. Heat is generated by the reaction at the anode (Fig. 13.1). The reactions occurring at the cathode and anode are:

$$\frac{1}{2}O^{2-} + 2e^- \rightarrow O^{2-} \tag{13.1}$$

$$H_2 + O^{2-} \rightarrow H_2O + 2e^- \tag{13.2}$$

The hydrogen used by an SOFC is produced from natural gas through either internal or external steam reforming. The heat can be supplied by the overpotential loss. The chemical reactions contain the methane-reforming process and the water gas-shifting reaction, as follows:

$$\text{Methane reforming: } CH_4 + H_2O \rightarrow CO + 3H_2 \tag{13.3}$$

$$\text{Gas shifting: } CO + H_2O \rightarrow CO_2 + H_2 \tag{13.4}$$

SOFCs typically operate at 700−1000°C and, for this reason, they are more suitable for stationary applications with a long operating time horizon. Using sheets of electrolytes, it is possible to lower the operating temperature below 800°C, resulting in a lower resistance of the electrolyte and thus greater efficiency. The anode in some cases is made of porous ceramic nickel/zirconium, while the cathode can be made of lanthanum manganite. SOFCs are also more tolerant to sulfur, which generally poisons other types of cells [12].

Based on geometry, tubular and planar are the two most popular designs of SOFCs. For planar SOFCs, each cell is made into a flat disk, square, or rectangular plate. These cells are put in series and connected by the interconnect plates, as schematically shown in Fig. 13.2. For tubular SOFCs, usually, the electrode (either cathode or anode) is made into a long tube with a porous wall. Outside the electrode tube are the electrolyte and another electrode.

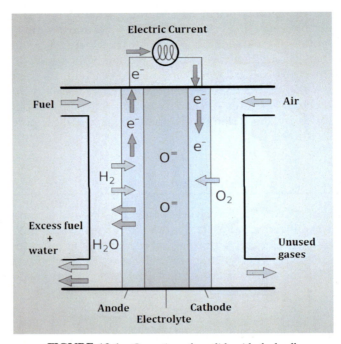

FIGURE 13.1 Operation of a solid oxide fuel cell.

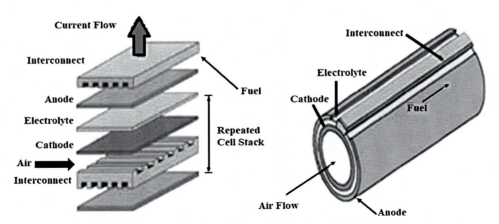

FIGURE 13.2 Planar and tubular solid oxide fuel cell designs.

III. Electrochemical energy conversion and storage

Cells are also connected in series through interconnects (Fig. 13.2). Among the SOFCs developed, the planar SOFC has lower manufacturing cost and higher power density compared to the tubular SOFC, but it requires hermetic sealants to maintain its gas tightness [13].

13.4 Components of the SOFC

The main functional components of SOFCs are cathode (air electrode), anode (fuel electrode), electrolyte, interconnect, and sealant (Fig. 13.3). The cathode is a thin porous layer on the electrolyte where oxygen reduction takes place. Due to the high operating temperature of the SOFC, noble metals or electronic conducting oxides can be used as cathode materials [14], generally made of lanthanum compounds. The most commonly used one is $LaMnO_3$ because it can easily acquire excess or deficiencies in either lanthanum or oxygen, making it easy to match the substrate crystalline structure. It offers excellent thermal expansion similar to zirconia electrolytes and provides good performance at operating temperatures. The choice of the electrode material depends on the target application [15]. Metals can be used as SOFC anode materials because of the reducing conditions of the fuel gas. Moreover, these metals must be nonoxidized since the composition of the fuel changes during the operation of the cell [16]. The ceramic (anode) layer must be very porous to allow the fuel to flow toward the electrolyte. The anode is commonly the thickest and strongest layer in each individual cell, because it has the smallest polarization losses and is often the layer that provides mechanical support [17]. The most commonly used anode is nickel oxide—yttrium-stabilized zirconia.

The electrolyte is a dense layer of oxygen ion-conducting ceramics. Popular electrolyte materials include yttrium-stabilized zirconia (YSZ) (often the 8% form, 8YSZ), doped cerium oxide, and doped bismuth oxide. YSZ has emerged as the most suitable electrolyte material.

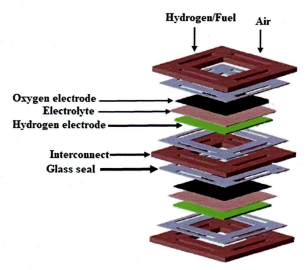

FIGURE 13.3 Components of a solid oxide fuel cell [31].

III. Electrochemical energy conversion and storage

Yttria serves the dual purpose of stabilizing zirconia into a cubic structure at high temperatures and also providing oxygen vacancies at the rate of one vacancy per mole of dopant. A typical dopant level is 10 mol% yttria. Interconnection serves as the electric contact to the air electrode and also protects the air electrode material from the reducing environment of the fuel on the fuel electrode side. The interconnector can be either a metallic or ceramic layer that sandwiches between each individual cell. Its purpose is to connect each cell in series, so that the electricity generated by each cell can be combined. The interconnector is exposed to both the oxidizing and reducing side of the cell at high temperatures and hence it must be extremely stable [17,18].

The sealant is a crucial material in the SOFC stack and maintains its operation and performance. It is needed to provide mechanical bonding of the components and to prevent mixing and leakage of fuel and oxidant within and outside the stack, respectively. Sealants need to have long-term stability, i.e., they should be able to survive even thousands of thermal cycles during cell operation. Due to the high operating temperature of SOFC, conventional sealing materials such as alloys, polymers, and organic adhesive cannot exist at the working temperature of SOFCs. Hence, different glasses and glass-ceramics are used as the sealants as they improve the lifespan of the SOFCs at high temperatures after many thermal cycles [19–21].

The three mainstream methods for sealing in SOFCs are compressive sealing, compliant sealing, and chemical bonding (rigid) sealing [22,23]. In compressive sealing a material is sandwiched between two sealing surfaces and is compressed by an externally applied load. In compressive seals at cell operating temperatures, an applied load complicates the cell design and increases fabrication cost. The primary advantage of compressive sealing is the tolerance to the mismatch of coefficient of thermal expansion (CTE). The disadvantages of the compressive sealing method are the lack of materials, which are not damaged at high temperature due to the chemical reactions occurring in the aggressive SOFC environment, and the need for complicated apparatus for mechanical loading [15,24]. Compliant seals are electrically conductive and do not bond with the other SOFC components. They are prone to oxidation and hydrogen embrittlement. Rigid seals have many advantages compared to compressive and compliant seals. The bonding seal approach relies on chemical bonding to provide hermetic sealing. The primary advantages of bonding seals are superior chemical stability under reactive atmospheres and no need for an external load frame for effective sealing. The main disadvantages of bonding seals are brittleness at low temperatures, which results in susceptibility to CTE mismatch, and poor long-term durability resulting from softening and crystallization of the glass phases. Mica and mica-based hybrid materials, metal brazes, and glass/glass-ceramic materials are used as compressive, compliant, and rigid seals, respectively [24–30].

13.5 SOFC component requirements

Each component of the SOFC serves several functions and must therefore meet certain requirements such as [14]:

- Proper stability (chemical, phase, morphological, and dimensional);
- Proper conductivity;

- Chemical compatibility with other components;
- Similar thermal expansion to avoid cracking during cell operation;
- Dense electrolyte to prevent gas mixing;
- Porous anode and cathode to allow gas transport to the reaction sites;
- High strength and toughness properties;
- Fabricability;
- Amenability to particular fabrication conditions;
- Compatibility at higher temperatures at which the ceramic structures are fabricated; and
- Low cost.

The aim of this review is to analyze available data on the reported sealants for SOFCs based on glass-ceramics. Particular attention is focused on their functional properties determining applicability under SOFC operating conditions, advantages and drawbacks with respect to other known systems, and the effects of various additives enabling sealant optimization. In light of the technological criteria discussed in the following sections, aluminosilicate-based materials constitute one of most promising families of rigid glass-ceramic sealants providing, in turn, serious advantages with respect to compressive (e.g., mica-based hybrids) and compliant (e.g., metal-brazes) seals. The chemical compatibility of sealing glass with metal interconnects is a critical issue in developing sealants for SOFCs [32–34].

13.6 Requirements of sealants

Seal optimization always needs a multifactor analysis. The general requirements for sealant materials are:

- Nearly zero gas permeability;
- Good adhesion to the solid electrolyte interconnects, electrodes, current collectors, and/ or other interfacing materials;
- Chemical inertness with respect to these materials under stack fabrication and operation conditions;
- CTEs ($9-13 \times 10^{-6}/°C$) compatible with those of the electrochemical cell constituents and other construction materials;
- High electrical resistivity ($>10^5 \ \Omega$ cm) under operating conditions to avoid short circuiting between different layers of the stack;
- Minimum volatilization and diffusion of the sealant components;
- No tendencies to bulk reduction, oxidation, hydration, carbonate formation, and reactions with other gaseous species such as SO_x and H_2S;
- Thermal and morphological stability at cell operation temperatures and during startup/ shutdown;
- Compatibility of the characteristic temperatures, primarily glass transition temperature (T_g), crystallization temperature (T_c), softening temperature (T_s), and maximum shrinkage temperature, with limitations arising from properties of the electrochemical device components and target operation regimes (i.e., no harmful reaction with joining components);

- Superior thermal shock resistance and high mechanical strength;
- Good sinterability, easy processing, and an absence of seal defects such as pores, bubbles, or microcracks;
- Stability with respect to local heating, high applied voltage, flame, carbon deposition, and other parasitic phenomena;
- Self-healing ability originating from viscous flow of the seal;
- Availability of seal components and low cost;
- During joining, the sealant has to show some viscous behavior to compensate for manufacturing tolerances. Viscosity must be $\approx 10^5$ Pa s at the joining temperature and $>10^9$ Pa S at the operating temperature;
- Sealants must provide tightness (hermeticity) to avoid leakage of reactant gases, which implies that a series of thermal, mechanical, and chemical requirements must be fulfilled;
- Must have good wetting behavior on both sealing surfaces;
- Should have deformability, but must be able to withstand a slight overpressure [35–38].

As mentioned earlier, an SOFC consists of different components such as cathode, anode, electrolyte (CTE in the range $10-11 \times 10^{-6}/°C$), interconnect (the mainly used one is Crofer22APU, with a CTE in the range $11-12 \times 10^{-6}/°C$), and sealants [39]. Glasses and glass-ceramics are the preferred SOFC seal materials. A wide range of material properties required for sealing can be achieved with a glass or glass-ceramic seal by suitable compositional design.

13.7 Methodology

Glass samples were prepared by the normal melt-quench technique [40]. The T_g, T_c, and peak crystallization temperature of glass samples were determined by differential thermal analysis (DTA). The nucleation temperature was fixed based on DTA data. Glass-ceramics were prepared by controlled heat treatment of the glasses. Temperature for the heat treatment varies according to the composition of the glasses [41]. The noncrystalline (amorphous) nature of the glasses and the phases obtained upon heat treatment (crystallization) in glass-ceramics were analyzed using X-ray diffraction (XRD). Density measurements were carried out using Archimedes' method [42]. Microstructural studies were done using scanning electron microscopy (SEM). Dilatometer measurements were performed on pressed samples after various thermal treatments to determine the thermal expansion behavior of the material. The temperatures corresponding to T_g and dilatometric softening temperature (T_{ds}) were extracted from the expansion curve. Joining ability has to be investigated under conditions similar to those employed in a stack with different interconnect materials [35].

The viscosity of glasses was also calculated from the dilatometric data using a method based on the Vogel–Fultcher–Tamman (VFT) equation as reported by Wang et al. Although this is not a very accurate method, it is very convenient for viscosity measurement. According to the simple liquid theory (Mott and Gurney) [43], there is a relationship between pseudo-critical temperature (T_k) and the absolute melting point (T_m):

$$\frac{T_k}{T_m} = \frac{2}{3} \tag{13.5}$$

Beaman [43,44] showed that this rule could be applied to glass/glass-ceramics and the terms T_k and T_m of Eq. (13.5) can be replaced with T_g and liquid temperature (T_l), respectively, of the glass. Thus Eq. (13.5) becomes:

$$\frac{T_g}{T_l} = \frac{2}{3} \tag{13.6}$$

Hence, T_l can be calculated using the value of T_g. The viscosity (η) values at T_g, dilatometric softening temperature (T_{ds}), and T_l are fixed and independent of materials, and are reported to be $10^{13.6}$, $10^{11.3}$, and 10^6 Pa s, respectively, at these temperatures. Then, according to the VFT equation [45,46]:

$$\log \eta = A + \frac{B}{T - T_0} \tag{13.7}$$

where A, B, and T_0 are all constants and T is the temperature at which the viscosity is measured. It is possible to determine all these constants by substituting values for T_g, T_d, and softening temperature (T_s) in Eq. (13.7). Thermal expansion coefficients of composites are very important in relation to dimensional stability and mechanical compatibility when used with other materials [47].

13.8 Sealant materials currently used

Glasses and glass-ceramics commonly used as sealing materials for SOFCs have to fulfill many requirements. Major attributes of glass-ceramics include increased refractory behavior and superior mechanical properties relative to glasses as well as ceramics. Undoubtedly, one of the major qualities, however, is an ability to tailor their thermal expansion characteristics. This makes glass-ceramics ideal candidates where compatible thermal expansions are necessary [18]. Glass and glass-ceramics, which have stable performance during high-temperature operation, are currently favorable candidates for sealing the components in SOFCs [48]. Additionally, glass and glass-ceramic seals are flexible in design, easy to fabricate, and cost competitive. Many glass and glass-ceramic systems have been investigated individually along with other additives as potential SOFC seals. However, the best results have been reported only for compositions based on silica. While the alkali silicate glasses tend to be very reactive toward other SOFC components [49], alkaline-earth aluminosilicate glasses have yielded promising results [50]. Based on performance, many authors [51−54] reported that boro-aluminosilicate-based sealing materials were found to be the best compared with other sealant materials. The properties of these glasses or glass-ceramics can be improved by introducing glass modifiers to the network. The most commonly used glass modifiers are alkaline-earth metals. This is because alkaline-earth metal ions have different chemical properties such as ionic radius, field strength, and electronegativity, which can strongly influence CTE, T_g and crystallization behavior, conductivity, and reactivity of the glasses [55].

Table 13.2 shows the glass and glass-ceramic compositions investigated for SOFCs. The density of G9 glasses increased with an increase in Bi_2O_3 content due to its highest density (8.9 g/cm^3) in comparison to other constituents of glasses. The molar volume of glasses

TABLE 13.2 Glass and glass-ceramic compositions investigated for solid oxide fuel cells.

Sample code	Composition														References
	BaO	CaO	MgO	SrO	B_2O_3	Al_2O_3	SiO_2	La_2O_3	ZnO	ZrO_2	NiO	Bi_2O_3	Y_2O_3	Others	
G18	35	15			10	5	35								[37]
G21	35	10			15	5	30								[37]
G23	32.5	12.5			12.5	2.5	35	5							[37]
G24	32.5	12.5			10	2.5	39	3.5							[37]
L2	35				16.7	10	33.3	5							[50]
Z2	35				16.7	10	33.3			5					[50]
N2	35				16.7	10	33.3			5					[50]
Mg1.5-55	27						55								[56]
G9		15.72	14.5	12.45	2	2.04	45.72	6.53			1				[57]
—	6.5	1.7			36.4	1	35.4		1					18	[58]
BAS	30				28.6	8.57	28.53	4.3							[58]
BCAS1	28.3	1.7			28.6	8.57	28.53	4.3							[59]
CAS	30				28.6	8.57	28.53	4.3							[60]
L09	50				30	5	15								[60]
L01	20				40	5	35								[60]
P00	12.62				16.67	7.59	63.12								[60]
G1	46.5				12.5	8.8	18.2	14							[35]
—	40				15	10	30	5							[61]
—				25	40	7	8	20							[62]
Zr-0		33.28	11.1	11.1	4.41	5.55	33.28	1.11				0.15			[63]
Ba32	32.88				14.91		52.19	20.13							[64]
14				24.56	40.29	6.92	8.11	6.98							[65]
10		21.63	17.27		2	2.18	48.93				1				[66]

SZS-1				51			40	9			[67]
SZS-4				51	10		30	9			[68]
BASP0	30			20	10	5	30	5			[69]
–		19.67	21.32		2	1.99	48.67	6.35			[70]
SLABS-3				10	30	15	30	15			[71]
–				25.7	12.8	13.1	44.3	4.1			[72]
Sr-0	1.36	21.09	16.84		2	2.13	48.77	6.81	1		[73]
N-25		15				10	45			30	[73]
SL				30	20		40	10			[74]
BL	30				20		40	10			[74]
7A	6.43	18.82	16.9			2.14	7.88	6.83	1		[75]
7(Sr)2B		18.83	16.92	4.35	2	2.14	47.92	6.84	1		[75]
–				30	10	15	30	15			[76]
Zn1.5-50	30						50	20			[76]
BAS-1	55					5	40				[77]
BMAS1	10			35		5	80				[77]
INA-1	30				10	0.62	40	20			[78]
–	40.78	4.36				0.83	47.41				[79]
INM1	30				7.5		40	20		2.5	[80]
CMAS		35		6.3		21.2	37.5				[81]

(*Continued*)

TABLE 13.2 Glass and glass-ceramic compositions investigated for solid oxide fuel cells.—cont'd

Sample code Composition	BaO	CaO	MgO	SrO	B_2O_3	Al_2O_3	SiO_2	La_2O_3	ZnO	ZrO_2	NiO	Bi_2O_3	Y_2O_3	Others	References
YSO-1		10		42.5	7.5		34						6		[61]
BBS	35				50		15								[82]
CBS		35			50		15								[82]
SBS				35	50		15								[82]
BABS	50				30	5	15								[83]
CABS		50			30	5	15								[83]
SABS				50	30	5	15								[83]
BNBS-5	50				30		15							5	[84]
BGBS-5	50				30		15							5	[84]
BLBS-5	50				30		15	5							[85]
BDBS-5	50				30		15							5	[85]

was observed to follow a similar trend as it increased with increasing Bi_2O_3 concentration in glasses. The dilatometric T_g decreased with increasing Bi_2O_3 content in glasses, while no significant impact of Bi_2O_3 concentration could be observed on the T_s of glasses [57]. This may be explained because Bi_2O_3 is an unconventional glass network former and increasing the Bi_2O_3/SiO_2 ratio in the glasses has been reported to increase the covalent character of Bi^{3+} atoms and decrease the covalency of Si^{4+} [65].

Ghosh et al. investigated barium aluminosilicate glass with the composition 46.5BaO-14.0La$_2$O$_3$-8.8Al$_2$O$_3$-12.5B$_2$O$_3$-18.2SiO$_2$ (G1) for its application as a sealant for SOFCs. Considering the G1 glass composition, it exhibited a higher CTE value of ~11.0 × 10^{-6} K^{-1}. The slope of the curve between T_g and T_d shows a dramatic increase in expansion just before the glass structure deforms by viscous flow. It should be noted that the difference in CTE between the G1 glass and Crofer22APU is also relatively less and lies below 10% mismatch [35]. Glass-ceramics, which can be prepared by the controlled crystallization of glasses, exhibit superior properties to glasses and can have various CTE values depending on the type of precipitated crystalline phases and their volume fraction in the glass matrix [50]. Glass-ceramics also show higher chemical stability than glasses, especially under SOFC operating conditions [86].

In the XRD pattern (Fig. 13.4) of the G1 glass heat treated at 800°C for different durations, the main crystalline phase precipitated is hexacelsian ($BaAl_2Si_2O_8$), and when it is heat treated for a longer period of time (10, 50, and 100 h), the hexacelsian phase begins to transform to the monoclinic celsian phase as the latter is thermodynamically stable below ~1590°C.

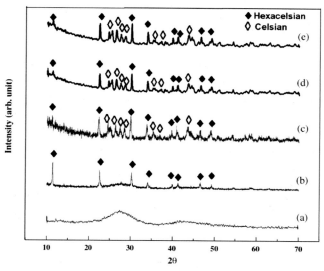

FIGURE 13.4 X-ray diffraction pattern of glass G1 heat treated at 800°C for different durations of time: (A) glass frit without heat treatment, (B) 3 h, (C) 10 h, (D) 50 h, and (E) 100 h [35]. *Copyright Elsevier (License Number: 4786330145295, March 2020).*

FIGURE 13.5 Microstructure of G1 glass-ceramic specimen heat treated at 800°C for (A) 10 h, (B) 50 h [35]. *Copyright Elsevier (License Number — 4786330145295, March 2020).*

Formation of the celsian phase is, however, undesirable for SOFC sealing applications as it has a very low CTE in between 30 and 1000°C. From the SEM micrographs (Fig. 13.5) of G1 glass heat treated at 800°C for different durations, it can be clearly confirmed that after heat treatment for longer durations (50 h), a large number of celsian crystals grow indicating enhanced ceramization of the parent glass [35].

Reddy et al. [63] reported the influence of zirconia on a series of melt-quenched alkaline-earth aluminosilicate glasses investigated for their application as sealants for SOFCs. Sintering precedes crystallization and well-sintered and mechanically strong glass-ceramics have been obtained in all zirconia-containing glasses. A good correlation can be observed in between the crystalline phases in glass-ceramics. Impedance spectroscopy analysis showed that all the investigated glass-ceramics exhibit good insulating properties. Smooth, nonporous, and adherent crack-free interfaces are observed between glass-ceramics and Crofer22APU after 500 h of prolonged heat treatment suggesting the long-term stability of zirconia-containing glasses as sealants in SOFCs [63].

Lahl et al. investigated the variation in crystallization kinetics of $AO-Al_2O_3-B_2O_3-SiO_2$ glass with the introduction of nucleating agents like TiO_2, ZrO_2, and Cr_2O_3 as the alkaline-earth metal changed from A = Ba, Ca, and Mg. Its activation energy of crystal growth increased significantly in response to the field strength [87]. The composition containing the combination of MgO/SrO-based glass samples doped by La, Al, and Y have shown higher crystallization temperatures compared to SrO-based glass samples. This may be attributed to the higher field strength of Mg^{2+} compared to Sr^{2+} cations. Thermal expansion of such glasses is higher and is comparable to the other components of SOFCs, which make it a suitable choice as sealant for SOFC applications [88]. Donald et al. reported that the use of calcium borosilicate and alkaline-earth lanthanum borate glass had good properties suitable for producing good-quality seals of Ti metal [89].

Recently, a number of alkaline-earth aluminosilicate glass sealants were synthesized for SOFCs. Alkaline-earth metals have different chemical properties such as field strength, ionic radius, and electronegativity that have a strong influence on T_g, T_s, CTE, crystallization behavior, electrical conductivity, as well as the reactivity of glasses. The most well-known alkaline-earth-based aluminosilicate systems that have been widely used for sealing applications are the BaO/SrO-Al$_2$O$_3$-SiO$_2$-, BaO/SrO-CaO-Al$_2$O$_3$-SiO$_2$-, and BaO/SrO-CaO/MgO-Al$_2$O$_3$-SiO$_2$-B$_2$O$_3$-based glass systems [90–97].

In one of our previous work, the CTE of glasses with the composition 35AO-50B$_2$O$_3$-15SiO$_2$ (A = Ba, Ca, Sr) falls in the range (8.18–10.5) $\times$ 10^{-6}/°C. The alkaline-earth metals in the borosilicate glass network changes the CTE. Here, the BaO (10.15 $\times$ 10^{-6}/°C)-based glass has the highest CTE compared to SrO (9.93 $\times$ 10^{-6}/°C) and CaO (8.18 $\times$ 10^{-6}/°C). This is due to the strong glass network formation and hence strong bonding in the BaO-based glass. This can be related to the ionic radius of the cations present in the systems, i.e., CTE is higher for glass having BaO, which can be attributed to the highest ionic radius of Ba^{2+} (1.35 Å) compared to that of Sr^{2+} (1.18 Å) and Ca^{2+} (1.00 Å) [82]. High ionic radius results in the creation of nonbridging oxygens (NBOs) in the network and the glass structure becomes looser, resulting in an increase of CTE [98]. Kaur et al. reported that in RO-Al$_2$O$_3$-B$_2$O$_3$-SiO$_2$ (A = Ba, Sr, Ca, Mg) systems, CTE is higher for Ba-based glasses and this can be assigned with the high ionic radius of Ba^{2+} (1.35 Å) [99].

35AO-50B$_2$O$_3$-15SiO$_2$ (A = Ba, Ca, Sr) glasses were not converted to glass-ceramics. During SOFC operation, glass is converted into glass-ceramics due to the formation of crystalline phases. In general, glass-ceramics consist of crystalline phases along with a glass matrix. The presence of a glass matrix in glass-ceramics is beneficial to heal the cracks during SOFC operation. Hence, glass-ceramic materials show superior properties and are considered ideal candidates for sealing applications. Glass-ceramics are more thermally stable than glasses. Hence, we mainly focus on glass-ceramic sealant materials. Al$_2$O$_3$ plays a vital role in the crystallization of glass networks. We try to develop glass-ceramics by adding Al$_2$O$_3$ to the alkaline-earth-based borosilicate glasses with different compositions. Glasses with the composition 50AO-5Al$_2$O$_3$-30B$_2$O$_3$-15SiO$_2$ (A = Ba, Sr, Ca) have been successfully synthesized. The CTE values of these systems lie in the range (10.15–11.40) $\times$ 10^{-6}/°C. Here, the BaO (11.4 $\times$ 10^{-6}/°C)-based glass has the highest CTE compared to SrO (10.32 $\times$ 10^{-6}/°C) and CaO (10.15 $\times$ 10^{-6}/°C) [83]. This is due to the strong glass network formation and hence strong bonding in the BaO-based glass compared to all other glass networks. In the present glass systems, CTE is higher for BABS glass due to the high ionic radius of Ba^{2+} (1.35 Å) compared to that of Sr^{2+} (1.18 Å) and Ca^{2+} (1.00 Å). An increased number of NBOs are created in glass networks having cations with high ionic radius. The creation of NBOs in a network results in the reduction of tightness of the glass structure, i.e., the glass structure becomes looser, which results in an increase of CTE [98].

At moderate temperatures, alkali oxides behave as modifiers in the glass matrix to achieve a homogeneous melt, to decrease the T_g, to adjust glass viscosity, and to improve the wettability of glasses. Barium-containing aluminosilicate glasses have the lowest T_g and T_s and the highest CTE values compared to other alkaline-earth-containing glasses, for instance, CaO- and SrO-containing glasses. This variation can be related to the basis of field strength values. This is because the force characteristics of Mg^{2+} cations are highest among alkali and alkaline-earth cations and lowest among network former ions. Hence, Mg^{2+} cations in the glass network can exist as both modifiers and network formers [98].

Studies on the interaction of glasses with a Ti-based alloy interconnect have been done. Using thermodynamic data on the free energy of reaction of Ti with a number of oxides, it is clear that some oxides should reduce the corrosion of Ti by molten glass. Many experiments confirm that the addition of Li$_2$O, BeO, CaO, SrO, BaO, and CeO$_2$ did indeed limit corrosion. Similarly, substitution of oxides in the glass by less reactive oxides, for example, replacement of Na$_2$O by B$_2$O$_3$, MnO, or ZnO, also minimize corrosion [99].

Glasses and glass-ceramics containing rare earth ions are suitable for SOFC applications due to significant physical and chemical properties such as high T_g and T_s, high microhardness, excellent chemical resistance and thermal resistance, etc., which are required for a sealing material [99]. Generally, in silicate systems, rare earth ions act as network modifiers. Rare earth oxides are added to improve the properties of glasses and glass-ceramics. Due to high cationic field strength and large coordination numbers, rare earth dopants can induce variation in thermal expansion, T_g and crystallization behavior of glasses and hence improved properties can be expected from glass-ceramics added with rare earth ions. However, to the best of our knowledge, there are no reports on the reaction between rare earth-added glass-ceramics and other components of SOFCs. So in our work we developed glass-ceramics with the composition 50BaO-(5-x)Al$_2$O$_3$-xR$_2$O$_3$-30B$_2$O$_3$-15SiO$_2$ ($x = 0, 1, 2, 3, 4, 5$, and R = Nd, Gd, La, Dy) [84,85].

Fig. 13.6 shows SEM micrographs of the interface between the BDBS-5 glass-ceramic/Crofer22APU and Crofer22APU/BDBS-5 glass-ceramic/Crofer22APU interface heat treated at 700°C for 10 h. A good adhesion between the 50BaO-5Dy$_2$O$_3$-30B$_2$O$_3$-15SiO$_2$ (BDBS-5) glass-ceramic sealant and metallic interconnect is observed. A continuous and defect/crack-free interface is present indicating good physical compatibility between the two materials. BDBS-5 glass-ceramic bonds well to the metallic part because no bubbles and cracks are observed at the interface. Although there are slight differences in CTE between the seals and adjoining components, the compressive stress loaded on them makes the interfaces maintain excellent contact. No significant diffusion of Ba into the Crofer22APU alloy or formation of Cr-rich reaction product is observed. This shows that a Cr-rich interface is not formed [100]. It was reported earlier that BaO-containing glasses interact chemically with chromia-forming alloys, e.g., BaCrO$_4$, which leads to Ba depletion in the glass-ceramics and to the

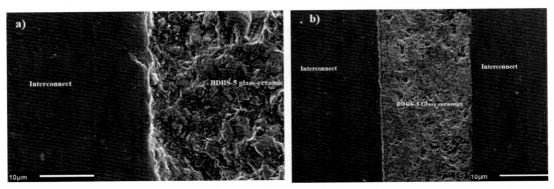

FIGURE 13.6 Scanning electron microscopy images of (A) interface between BDBS-5 glass-ceramics/Crofer22-APU, (B) Crofer22APU/BDBS-5 glass-ceramics/Crofer22APU interface heat treated at 700°C at 10 h [100].

separation of glass-ceramics from the alloy matrix due to thermal expansion mismatch [66]. In the present investigation, we did not observe any separation of glass-ceramics from the alloy matrix after heat treatment. This result indicates that presently prepared glass-ceramics can be a better alternative for sealant applications [100].

ZnO enhances sintering and crystallization processes over MgO and CaO in glass-ceramics, thus resulting in dense sintered glass powder compacts and mechanically strong glass-ceramics. In this study, it is clear that ZnO acts in a manner similar to B_2O_3 with regard to viscosity and thermal expansion [101]. La_2O_3 is known to control the viscosity and CTE of silicate glasses and glass-ceramics [102]. It is well known that, with the addition of B_2O_3 to the silicate glass network, viscosity and crystallization tendency of the glass matrix decreases. According to Sohn et al. [50], the B_2O_3/SiO_2 ratio plays an important role in tailoring the properties of the glass-ceramic sealants as the CTE of the glasses has been observed to increase with an increase in B_2O_3/SiO_2 ratio.

Glass-ceramics are formed by the controlled crystallization of glasses. The strength and properties of glass-ceramics mainly depend on the amount and nature of the crystalline phases formed in it. In the case of barium-containing glass-ceramics for SOFCs, crystallization increases thermal expansion. The CTE of $BaO-MgO-SiO_2$ and $BaO-ZnO-SiO_2$ increases with increasing BaO content for constant SiO_2 contents [103–106]. This increase in CTE is due to the formation of barium silicate ($BaSiO_3$) [107–110], which has a large CTE compared to enstatite ($MgSiO_3$). Fig. 13.7 represents the T_g and CTE of some of the silicate glass systems used as sealants in SOFCs.

Two sodium calcium aluminosilicate glass systems with compositions $45SiO_2$-$9Al_2O_3$-$10B_2O_3$-$18CaO$-$18Na_2O$ and $40SiO_2$-$9Al_2O_3$-$10B_2O_3$-$18CaO$-$23Na_2O$ have been developed

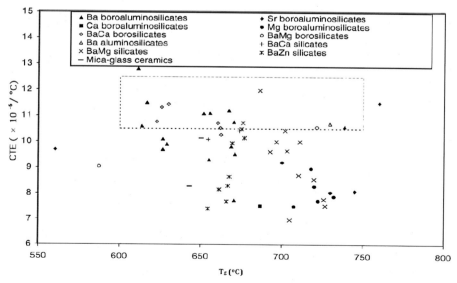

FIGURE 13.7 Glass transition temperatures (T_g) and coefficients of thermal expansion (CTE) for solid oxide fuel cell sealant materials [5]. *Copyright Elsevier (License Number: 4786310405564, 2020).*

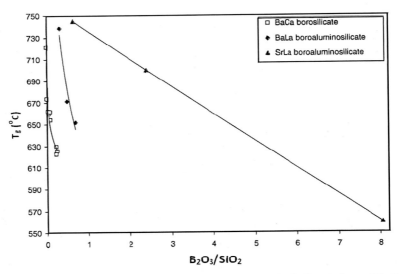

FIGURE 13.8 Effect of B_2O_3 on the glass transition temperatures (T_g) of silicate glasses [5]. *Copyright Elsevier (License Number: 4786310405564, 2020).*

and their suitability as sealants in SOFCs has been studied. These systems can be used as glassy seals and glass-ceramic seals. They show good bonding with the YSZ electrolyte and interconnect material in SOFCs. They have T_g values of 580 and 545°C, T_s values of 740 and 680°C, and CTE values of 11.2 and $12.5 \times 10^{-6}/°C$, which are suitable for sealants in SOFCs [111].

Alkaline-earth-based aluminosilicate glass-ceramics are the most commonly used sealant materials for SOFCs [5]. Boron oxide is an important additive to silicate glasses. Boron oxide is added to decrease the viscosity, T_g, and T_s of the glass matrix (Fig. 13.8) of SOFC sealants [112–115]. The amorphous structure is stabilized by boron oxide (Fig. 13.8) by the increase in activation energy for crystallization with increasing boron content (B/Al ratio) [5].

N.H. Menzler et al. investigated the interaction study of a sealant glass-ceramic (glass A with composition 40BaO-8CaO-2Al$_2$O$_3$-35SiO$_2$-15PbO) with metallic interconnect materials (Crofer22APU 1st and JS-3). Fig. 13.9A and B show the SEM cross-sections of samples of the combinations Crofer22APU 1st/glass A and JS-3/glass A (both glasses in a nonprecrystallized state). It is obvious from micrographs that even at the edges of the sealant, e.g., at the triple-phase boundaries (sealant, metal, atmosphere), no internal corrosion is detectable. Only the oxide layer formed on the steel is present. At the start of the test, the oxide layer formed is shown for both material combinations and is independent of the crystallization of the glass. After a short exposure time of about 1 h, the oxide layer formed is thin (1 μm) and consists of chromium, manganese, and oxygen. After prolonged duration of heat treatment (25 h), it does not increase in thickness. Separation into a two-layered structure is clear from Fig. 13.9. Exposure times of up to 500 h for specimens containing material combinations of a glass-ceramic sealant in the precrystallized or nonprecrystallized state with Crofer22APU 1st or JS-3 do not lead to corrosion or internal oxidation effects of the metal. Both metals

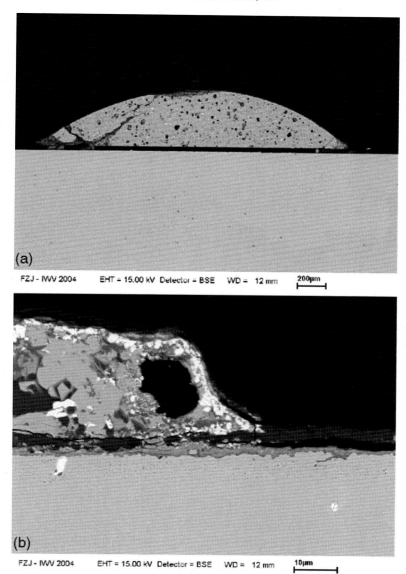

FIGURE 13.9 Scanning electron microscopy cross-sections of specimens annealed for 500 h under air: (A) Crofer22APU 1st/glass A (overview), (B) JS-3/glass A (detail from the triple-phase boundary); the gap between the steel and the glass results from sample preparation [116]. *Copyright Elsevier (License Number: 4786331316428, March 2020).*

interact with the glass-ceramic sealant to form a bonding layer. The formation of the oxidic double layer on the steel is influenced in the wetting zone of the glass-ceramic on the steel. However, no negative influence with respect to adhesion, cracking, enhanced interdiffusion, or double layer formation on the nonwetted steel surface is observed [116].

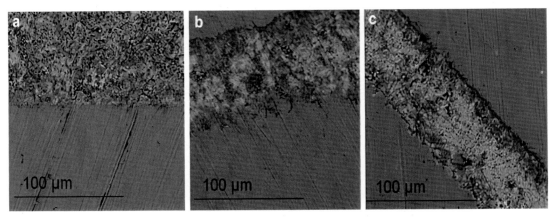

FIGURE 13.10 Microstructure at the interface of glass with Crofer22 APU: (A) BASP0, (B) BASP1, (C) BASP2 [68]. *Copyright Elsevier (License Number: 4786330639273, March 2020).*

Goel et al. [101] investigated the influence of ZnO on the structural and crystallization behavior of Di-Ca-Ts glasses and the properties of their sintered glass-ceramics have been investigated. Addition of ZnO promotes the phenomenon of liquid—liquid-phase separation in the glasses and leads to increasing connectivity of the silicate glass network.

Fig. 13.10 (a—c) shows the microstructures at the interface of BASP0, BASP1, BASP2 glasses ($30SiO_2$-$30BaO$-$20SrO$-$10B_2O_3$-$5La_2O_3$-$5Al_2O_3$-$0P_2O_5$ [BASP0], $28.8SiO_2$-$31.7BaO$-$19.2SrO$-$9.6B_2O_3$-$4.8La_2O_3$-$4.8Al_2O_3$-$0.9P_2O_5$ [BASP1], and $27.7SiO_2$-$33.3BaO$-$18.5SrO$-$9.2B_2O_3$-$4.6La_2O_3$-$4.6Al_2O_3$-$1.8P_2O_5$ [BASP2]) with Crofer22APU, respectively. SEM images reveal that glasses bonded well to the metallic part as no bubbles and cracks are observed at the interface. Across the boundary between BASP glass and Crofer22APU, a limited interdiffusion of Cr, Sr, and Ba is observed, which is required for good adhesion between metal and glass. The decrease in flow and sealing temperatures with the addition of $Ba_3(PO_4)_2$ in BASP glasses helps in forming good bonds with Crofer22APU at temperatures lower than 1000°C [68]. Batfalsky et al. reported that the BaO-containing glasses when subjected to interaction study reveal that BaO interacts chemically with chromia-forming alloys resulting in the formation of $BaCrO_4$, which leads to the separation of glass-ceramics from the alloy matrix due to thermal expansion mismatch [117]. This is due to the very high CTE of $BaCrO_4$. The $BaCrO_4$ phase has a very high CTE at $(16-18) \times 10^{-6}/°C$, which causes delamination of the glass-ceramics [118].

For glass systems having both B_2O_3 and SiO_2 as network formers, the properties are interpreted against the B_2O_3/SiO_2 ratio. By increasing the B_2O_3/SiO_2 ratio and decreasing SrO concentration in the glass matrix, both the T_g and T_s decrease. The content of the network modifier in the matrix also affects the properties as it controls glass structure. B_2O_3 is systematically substituted by SrO, hence the B_2O_3/SiO_2 ratio in the glass changes along with SrO content. For the SLABS series of glasses (SrO [x], La_2O_3 [15], Al_2O_3 [15], B_2O_3 [40 − x], and SiO_2 [30] [x = 10, 15, 20, 25, and 30 wt%]), with a decrease in B_2O_3/SiO_2 ratio and increase in SrO content in the glass the Si-O-Si/O-Si-O bending frequency and B-O-B stretching frequency increase monotonically. It is observed that if the ratio decreases from 1 to 0.333, the connectivity decreases with increased network breakage, which may be due to network modifiers having more impact over glass formers such as SiO_2 and B_2O_3 [70]. The addition of modifiers in silicate glasses creates more NBO that decreases the average symmetry of the Si—O bonds, which results in the increase of the CTE [119]. The CTE also increases if the molar free volume and bond bending of glass decrease [45]. For the SLABS series of glasses, increasing the concentration of SrO in the glass network results in an increasing network breakage, hence more NBOs are created, which also results in decreasing bond bending due to the rigidity of the structure as mentioned in the Fourier transform infrared results. Hence, the CTE of the glass samples increases with an increase in SrO content despite a lowering of the B_2O_3/SiO_2 ratio [70].

Thermal expansion of a series of SLABS glasses (SrO [x], La_2O_3 [15], Al_2O_3 [15], B_2O_3 [40 − x], and SiO_2 [30] [x = 10, 15, 20, 25, and 30 wt.%]) has been studied. The CTE of glasses calculated from the linear expansion data increases gradually from $8.29 \times 10^{-6}/°C$ for SLABS-3 to $9.72 \times 10^{-6}/°C$ for SLABS-7 (Fig. 13.11). The CTE of a seal glass mainly depends on the glass structure symmetry, nature of the bond bending, and molar free volume in a glass network [70]. For example, pure SiO_2 glass has a CTE of $0.6 \times 10^{-6}/°C$ due to its high symmetry and B_2O_3 has a CTE of $14.4 \times 10^{-6}/°C$ due to its low symmetry [45].

Lara et al. [106] reported on alkaline-earth-based silicate glasses with the composition RO-BaO-SiO_2 (RO = Mg, Ba, Zn) explaining their glass-forming ability, sinterability, and thermal and electrical properties. Generally, glass-ceramics show favorable chemical, thermal, dielectric, and biological properties superior to metals and various polymers [120]. Sun et al. investigated the alkaline-earth-based aluminoborosilicate system with the composition BaO-Al_2O_3-R_2O_3-B_2O_3-SiO_2 showing its suitability for sealants in a SOFC system with ZrO_2 electrolyte. Dependence of glass structure, wetting behavior, and crystallization with the

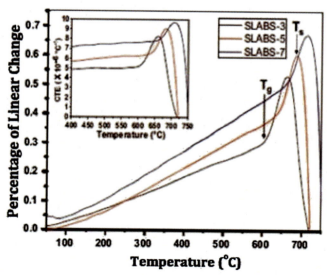

FIGURE 13.11 Dilatometer plots of SLABS series samples with coefficient of thermal expansion plots shown in the inset [70]. *Copyright Elsevier (License Number: 4786331046833, March 2020).*

addition of Al_2O_3 were also studied. The T_g of these systems lies in the range 600–625°C. The thermal stability values indicate that these systems are thermally stable. All the CTEs of the glasses are in the range $(9–12) \times 10^{-6}/°C$, which matches other components of SOFCs [121].

Reis et al. [122] evaluated the thermal properties of two glasses with the compositions $18.50SrO$-$19.23CaO$-$13.23ZnO$-$1.90B_2O_3$-$2.94Al_2O_3$-$42.20SiO_2$-$2.00TiO_2$ and $25.48SrO$-$25.48CaO$-$3.92ZnO$-$1.96B_2O_3$-$1.96Al_2O_3$-$39.20SiO_2$-$2.00TiO_2$, which matches the requirements of SOFC sealant materials. The T_g varies within the range (664–690)°C and CTE values lie in the range $(10–11) \times 10^{-6}/°C$. The thermal stability parameter (from DTA data) is higher than 100°C showing that the glass systems are thermally stable.

Ley et al. studied a set of glasses and glass-ceramics containing boron oxide as the primary glass former to obtain low T_g, SrO to increase the CTE, La_2O_3 to control the viscosity, Al_2O_3 to retard the formation of phases formed during crystallization, and SiO_2 to enlarge the glass-forming range with other constituents. The T_gs of these systems are in the range 500–800°C. The SrO-Al_2O_3-La_2O_3-B_2O_3-SiO_2 system possesses good CTE, which matches with the other components of SOFCs. The interaction studies indicate that it bonds well with the electrolyte, anode, and cathode of the SOFC [114].

Glass with the composition $55SiO_2$-$27BaO$-$18MgO$ has a T_g of 715°C and CTE of $8.1 \times 10^{-6}/°C$. After heat treatment, silicate-based crystalline phases are formed. This glass-ceramic exhibits suitable properties as a sealant material in SOFCs. The interaction study with YSZ electrolyte and aluminum-based interconnect material (Fe-Cr alloy) shows that the glass-ceramic bonded well to the electrolyte and the metallic part of the interconnect, as no bubbles and cracks are observed at the interface [123]. Zhang and Zou [124] investigated the properties of a glass-ceramic with the composition $30SiO_2$-$20BaO$-$20SrO$-$30B_2O_3$ as a sealant material in SOFCs. Its T_g and CTE are 650°C and $11.7 \times 10^{-6}/°C$, respectively. The thermal properties are well suited to the requirements of a sealing material.

A literature review showed that the silica extracted with other oxides from agricultural waste was exploited to form glasses and glass-ceramics [125–127]. G. Sharma et al. reported a study of sealant materials in SOFCs based on two samples such as silicate glass-ceramic synthesized from wheat straw ash (WSA) and a mineral oxide-derived (MOD) glass-ceramic with the composition CaO-MgO-Na_2O-K_2O-SiO_2. They studied its thermal, resistivity, hardness, and interaction properties to check its stability as sealant materials in SOFCs [127]. The thermal stability of these glasses is checked using thermogravimetric analysis (TGA). From TGA results, it is clear that these systems have good thermal stability with respect to temperature in the range 350–900°C. The CTE of the WSA-based glass is $10 \times 10^{-6}/°C$, which matches other components of SOFCs [5,7,87]. CTE increased after crystallization, which may have been due to the presence of crystalline phases and their bonding with glass matrix [127]. Glass sealant must be an insulator at the working temperature of SOFCs. The resistivity of WSA is $\sim 10^6 \ \Omega$ cm at 700°C, which is in the insulating range required for the sealing material in SOFCs. MOD-based glass-ceramics could not form an interface with the interconnect (Crofer22APU). However, in the case of the WSA sample, a semicoherent interface is formed, which strong, smooth, and adheres well to Crofer22APU even after several thermal cycles for different durations. This good adhesion may be due to the presence of a hydroxyl group in SiO_2 polyhedra. Microhardness increases at the interface of the WSA glass-ceramic and Crofer22APU in the range 237–384 HV. At the boundary of the glass-ceramic and Crofer22APU, microhardness is less than the glass-ceramic and higher than Crofer. This is due to the semicoherent bonding between glass-ceramic and Crofer22APU. Sharma et al. concluded that WSA glass-ceramic could be used as a sealant in SOFCs [127].

Kim et al. [128] reported a glass with the composition $18BaO-45B_2O_3-24SiO_2-6Al_2O_3-3.5CaO-3.5SrO$ as a suitable candidate for a sealant in SOFCs. This system showed good sealing properties even after exposure at 850°C for 2000 h. Qi et al. [129] reported the properties of $59.6SiO_2-11.0Al_2O_3-10.6ZrO_2-3.4CaO-15.4Na_2O$ glass as a sealant in SOFCs. Its CTE value fits with that of YSZ ceramics. The interaction study of this glass system with YSZ ceramics and Crofer22APU indicates that this glass system has high chemical stability when in contact with the YSZ ceramics and Crofer22APU for a long duration.

The effect of ZrO_2 substitution by Y_2O_3 on the physical and chemical properties of $59.6SiO_2-11.0Al_2O_3-10.6ZrO_2-3.4CaO-15.4Na_2O$ glass and its chemical stability in contact with an Fe-Ni-Co alloy was investigated by Krainova et al. [130]. The composition is chosen as $59.6SiO_2-11.0Al_2O_3-(10.6-x)ZrO_2-3.4CaO-15.4Na_2O-xY_2O_3$ ($x = 0$; 2; 4 wt%). Due to the specifics of the SOFC operation, the glass sealants should withstand a high temperature for a long duration. SOFC devices are assumed to be safely working at elevated temperatures across a range of 700−850°C. The sealant material should meet strong requirements so that the compatibility between all the elements to be joined could be achieved in terms of their thermal expansion behavior. From this study, it is clear that the change in the ZrO_2/Y_2O_3 ratio has a negligible effect on the CTE of the present system and it lies in the range (9.1−9.8) × $10^{-6}/°C$. The CTE value of this glass sealant is expected to be stable during prolonged SOFC operation. The interaction study indicates that no interaction was observed on the boundaries of YSZ/glass and glass/alloy. Moreover, the use of this noncrystallizing sealant allows the tightness of the junction to be maintained despite a sharp increase in the CTE of an Ni−Fe−Co alloy. Yttria has a dual nature and can act simultaneously as both a glass former and a modifier [131], which, in the case of glass sealants, has a positive effect on the glass properties. According to Singh [132], the introduction of small (up to 6 mol%) additives of Y_2O_3 suppresses the crystallization process and increases the characteristic temperatures. Similar results were obtained by Mahdy et al. [133].

Sealing materials for planar SOFCs have been reported by Monreal et al. [134]. Silicate-based glasses and glass-ceramics for sealing at the fuel side and calcium aluminate cement mixed with 20−30 vol% silicate-based glass at the air side are recommended. Gas-tight 40-cell stacks running for over 6000 h are achieved. Nevertheless, these joining materials are designed for the $LaCrO_3$ connector with a CTE of $10.4 × 10^{-6}/°C$ instead of a high chromium steel with a CTE of $(12−14)10^{-6}/°C$ [134]. Larsen and James [135] studied the chemical stability of $MgO-CaO-Cr_2O_3-Al_2O_3-B_2O_3-P_2O_5$ glass sealants. A significant volatilization of $0.2−1.4$ mg/(cm^2 h) is observed. Finite element analysis of the stress and strain tolerance in SOFC manifolds was reported by Schwab et al. [136]. Rupali et al. studied the properties of a series of $SiO_2-B_2O_3-CaO-A_2O_3$ (A = La, Y) glasses and glass-ceramics. During heat treatment the main phase formed is La_2SiO_5. The thermal expansion coefficient of glass ceramic is higher compared to parent glass. Hence, glass ceramics are better sealants with CTE values within the limit that is required as a sealant [137].

Larsen and Poulsen [138] also reported that that presence of Cr(VI) in the form of $CaCrO_4$ might cause pore formation inside the seal if severe reaction takes place at the interface. This may not only create leakage in the seal but will also reduce mechanical strength. In general, the long-term stability of the interconnect is a critical parameter for electrical performance [114]. Glass compositions based on B_2O_3 tend to exhibit excessive volatilization in the SOFC environment. P_2O_5-based glasses can be adjusted to minimize volatilization but their thermal expansion coefficients are too low and they have low mechanical strength [139]. To date, the best results have been obtained using compositions rich in silica. While alkali silicate glasses tend to be very reactive toward SOFC components [102], alkaline-earth aluminosilicate glasses have yielded promising results [140−143].

Imanaka et al. [143] examined the effect of ceramic additions containing Al, such as alumina, aluminum nitride, mullite, and spinel, into the borosilicate glass for the suppression of cristobalite precipitation. The results showed that mullite or aluminum nitride suppresses cristobalite formation more effectively than alumina or spinel. Although both follow a simple rule of mixtures, glass/mullite composites can be fabricated with lower dielectric constants than glass/alumina composites, while maintaining a thermal expansion coefficient close to Si.

Currently, alkaline-earth aluminosilicate-based glasses and glass-ceramics are being studied as suitable candidates for sealing material in SOFCs [144−148]. Chou et al. developed Sr-Ca-Ni-Y-B silicate glass with a focus on the effects of NiO on glass forming and thermal and mechanical properties for sealant in SOFCs. CTE values are in the range $(11.5−12.2) \times 10^{-6}/$ °C. XRD studies revealed the presence of an NiO phase instead of reaction products, which indicated the chemical compatibility of NiO with the glass [148]. Creep properties were investigated for SOFCs using the glass-ceramic $BaO-B_2O_3-Al_2O_3-SiO_2$ (GC-9) in variously aged conditions at 800°C. The crystallinity in GC-9 glass-ceramic is increased with thermal aging time. The values of creep stress exponent increase with an increase in the duration of thermal aging. The GC-9 system with a longer thermal aging and a greater extent of crystallization exhibits a higher creep stress exponent value [149].

Dai et al. developed Al_2O_3-based compressive tape seals containing 0%−30% of aluminum and evaluated their thermal-cycled leakage rates and applicability. With an increase in the content of aluminum, the porosity decreases its strength as a seal. This change is due to aluminum oxidation to Al_2O_3. As a result, sealing effectiveness and thermal cyclability are enhanced. The interaction of this material with other cell components is also observed. The seal interfaces bond with other components well without forming any cracks on the interface. This exposes the applicability of the compressive seal in intermediate temperature SOFCs [150,151].

The joining strength between GC-9 (0−40 mol% BaO, 0−15 mol% B_2O_3, 0−10 mol% Al_2O_3, 0−40 mol% SiO_2, 0−15 mol% CaO, 0−15 mol% La_2O_3, and 0−5 mol% ZrO_2) glass-ceramic and Crofer22APU was investigated by Lin et al. The oxidation treatment did not improve the shear or tensile strength. A longer aging time reduces the shear joint strength [152]. The bonding strength between a glass-ceramic sealant $SiO_2-Al_2O_3-CaO-Na_2O$ (SACN) and chromium-containing metallic interconnects (Crofer22APU and AISI 430) was investigated under tensile loading at room temperature. It was found that the fracture of the Crofer22-APU/SACN/Crofer22APU joint always occurred within the glass-ceramic layer and never at the interfaces in the joint [153].

The effect of crystallization on the high-temperature mechanical properties of GC-9 glass-ceramic (Table 13.2) was investigated for use in SOFCs. The T_g and T_s of the GC-9 glasses were determined as 668 and 745°C, respectively [154]. At a temperature below 650°C, the

III. Electrochemical energy conversion and storage

flexural strength and structural stiffness of the GC-9 glass were enhanced by the formation of crystalline phases in the glass matrix [155].

A barium-free glass-ceramic composition based on diopside has been designed and tested for use as a sealant in SOFCs. Its crystallization occurs predominantly from the bulk material. The glass-ceramic shows good densification after the sintering process. No elements from the sealants are present in the corroded region, and the possibility of interaction between the steel and the sealant is very low (i.e., chemically stable sealant) [156]. Chou et al. developed an alkali silicate glass (SCN-1 with composition as BaO [8.23 mol%] and CaO [3.34 mol%], alkalis of K_2O [10.0 mol%]) and Na_2O [7.3 mol%], Al_2O_3 [2.8 mol%], and some impurities [less than 1%] of Fe, Mg, and Ti with the balance of SiO_2), which was evaluated as a candidate sealant for SOFCs in terms of combined isothermal aging and deep thermal cycling at 700 and 750°C. The high-temperature leak test showed that all samples exhibited good combined stability with hermetic seal behavior. From the interfacial study, it was clear that the present glass systems were chemically compatible with YSZ electrolyte [157].

Nonalkaline glass-MgO glass-ceramic composites as sealing materials for SOFCs have been investigated. This study showed that the MgO additive in the glass matrix could improve physical properties of the composites. According to XRD analysis, no obvious reaction between MgO and the glass matrix was observed. This revealed that the glass-ceramic composites may have reasonable chemical stability for practical application. The CTE of glass-ceramic composite increased with the concentration of MgO and its value was 11×10^{-6}/°C for high concentrations of MgO [158,159].

Glasses with the composition xLa_2O_3-(10-x)Y_2O_3-15SrO-15CaO-10B_2O_3-10Al_2O_3-40SiO_2$ ($x = 2, 4, 6, 8$) showed suitable properties as sealants in SOFCs. In this study, Kaur et al. reported the influence of the La_2O_3/Y_2O_3 ratio on properties and interfacial interaction with Crofer22APU and YSZ. The values of T_g for the present glass samples were in the SOFC working temperature range. The crystallization peak directly depended on the heating rate and changed with the increase in heating rate [160]. As the La_2O_3/Y_2O_3 ratio increased, the density of the glass samples decreased. The values of CTE for all the glasses matched well with values of CTE of electrolyte and interconnect and were in an acceptable range $(10.3-12.5 \pm 0.1) \times 10^{-6}$/°C. The interaction study showed good bonding at the interface for the glass composition with $x = 2$ and an La_2O_3/Y_2O_3 ratio = 0.25. For all the other compositions, some cracks were seen on the glassy side. The composition with $x = 2$ and La_2O_3/Y_2O_3 ratio = 0.25 had good results exhibiting a clean and smooth interface with 8YSZ and Crofer22APU. These results revealed that higher Y_2O_3 content compared to La_2O_3 showed higher stability among all other compositions. Thus glass with $x = 2$ and La_2O_3/Y_2O_3 ratio = 0.25 may be selected for applications in SOFCs as sealants [161].

13.9 Conclusions

Increasing demand for carbon-free energy sources and the need for efficient energy utilization make finite fossil fuels more attractive and are considered an alternative energy resource. Due to advantages such as eco-friendliness, thermal and chemical stability, and long-term durability over other fuel cells, it is only in the last two decades that fuel cells have offered a realistic prospect of being commercially viable. Unlike other fuel cells, the SOFC is a solid-state device that operates at elevated temperatures and is a potential candidate for high-efficiency energy conversion. The development of suitable high-temperature sealants is one of the major issues facing the commercialization of SOFCs. The sealants are necessary to stop fuel and air mixing within a fuel cell and gas tightness between the anode and the cathode compartments at high operating temperature must be maintained (700–800°C). Sealants must be an electrical insulator and have high chemical stability with the stack components. A large number of materials and approaches have been investigated for the development of sealants for SOFC stacks. Glass or glass-ceramic materials have been shown to have high thermal and chemical stability as sealants, and can operate in fuel cells for a number of thermal cycles (more than 1000 h) with no significant degradation.

This chapter reviewed important glasses and glass-ceramic systems used as sealants with suitable requirements for their thermal, chemical, mechanical, and electrical properties. Requirements of properties for the use of glasses and glass-ceramics as sealants in SOFCs were also discussed, and sealant glass and glass-ceramic requirements and compositional dependence were described. Interfacial stability with interconnect alloys depends on many factors such as composition and sealing conditions. The development of new materials will certainly continue to make SOFCs increasingly affordable, efficient, and reliable. The thermal properties and thermochemical stability of different silicate glass compositions that crystallize to form stable pyro- and orthosilicate phases make them attractive candidates for SOFC applications. Promising compositions possess desirable thermal expansion values that are compatible with electrolyte (YSZ) and interconnect (Crofer22APU and JS-3) materials, are chemically stable during interaction with these materials, and remain stable under SOFC operational conditions.

Acknowledgments

The authors are thankful to the University Grants Commission (UGC), New Delhi, India, for financial support as a major research project.

References

[1] E.D. Wachsman, K.T. Lee, Lowering the temperature of solid oxide fuel cells, Science 334 (2011) 935–939.

[2] E.D. Wachsman, C.A. Marlowe, K.T. Lee, Role of solid oxide fuel cells in a balanced energy strategy, Energy Environ. Sci. 5 (2012) 5498–5509.

[3] I. Dincer, Hydrogen and fuel cell technologies for sustainable future, Jordan J. Mech. Ind. Eng. 2 (2008) 1–14.

[4] Fuel Cell Handbook, seventh ed., EG&G Technical Services, November 2004.

[5] J.W. Fergus, Sealants for solid oxide fuel cells, J. Power Sources 147 (2005) 46–57.

[6] D.J.L. Brett, A. Atkinson, N.P. Brandon, S.J. Skinner, Intermediate temperature solid oxide fuel cells, Chem. Soc. Rev. 37 (2008) 1568–1578.

[7] V.M. Janardhanan, O. Deutschmann, Review paper- modeling of solid oxide fuel cells, J. Phys. Chem. 221 (2007) 443–478.

[8] U.S. Department of Energy, Comparison of Fuel Cell Technologies, http://www1.eere.energy.gov/hydrogenandfuelcells/Fuelcellas/pdfs/fc_comparison.charts.pdf.

[9] A. Toleuova, V. Yufit, S. Simons, W.C. Maskell, D.J.L. Brett, Review A review of liquid metal anode, J. Electrochem. Sci. Eng. 3 (3) (2013) 91–105.

[10] A.B. Stambouli, E. Traversa, Solid oxide fuel cells (SOFCs): a review of an environmentally clean and efficient source of Energy, Renew. Sustain. Energy Rev. 6 (2002) 433–455.

References

[11] X. Zhang, S.H. Chan, G. Li, H.K. Ho, J. Li, Z. Feng, A review of integration strategies for solid oxide fuel cells, J. Power Sources 195 (2010) 685–702.

[12] L. Barelli, E. Barluzzi, G. Bidini, Review: diagnosis methodology and technique for solid oxide fuel cells: a review, Int. J. Hydrogen Energy 38 (2013) 5060–5074.

[13] A. Weber, E. Ivers-Tiffée, Materials and concepts for solid oxide fuel cells (SOFCs) in stationary and mobile applications, J. Power Sources 127 (2004) 273–283.

[14] Q.M. Nguyen, Ceramic fuel cells, J. Am. Ceram. Soc. 76 (1993) 563–588, https://doi.org/10.1111/j.1151-2916.1993.tb03645.x.

[15] Y.S. Chou, J.W. Stevenson, Compressive mica seals for solid oxide fuel cells, J. Mater. Eng. Perform. 15 (2006) 414–421.

[16] Fuel Cell Materials, Nextech Materials, August 2001.

[17] S.C. Singhal, Science and technology of solid oxide fuel cells, MRS Bull. (2000) 21.

[18] I.W. Donald, B.L. Metcalfe, L.A. Gerrard, Interfacial reactions in glass-ceramic-to- metal seals, J. Am. Ceram. Soc. 91 (2008) 715–720.

[19] A.A. Reddy, D.U. Tulyaganov, S. Kapoor, A. Goel, M.J. Pascual, V.V. Kharton, J.M.F. Ferreira, Study of melilite based glasses and glass-ceramics nucleated by Bi_2O_3 for functional applications, RSC Adv. 2 (2012) 10955–10967.

[20] V. Kumar, S. Sharma, O.P. Pandey, K. Singh, Thermal and physical properties of 30SrO -40SiO$_2$-20B$_2$O$_3$-10A$_2$O$_3$ (A= La,Y,Al) glasses and their chemical reaction with bismuth vanadate for SOFC, Solid State Ionics 181 (2010) 79–85.

[21] Y. Guo, R. Ran, Z. Shao, A novel way to improve performance of proton-conducting solid-oxide fuel cells through enhanced chemical interaction of anode components, Int. J. Hydrogen Energy 36 (2011) 1683–1691.

[22] Y.S. Chou, J.W. Stevenson, L.A. Chick, Ultra-low leak rate of hybrid compressive mica seals for solid oxide fuel cells, J. Power Sources 112 (2002) 130–136.

[23] S.P. Simner, J.W. Stevenson, Compressive mica seals for SOFC applications, J. Power Sources 102 (2001) 310–316.

[24] S. Ghosh, A.D. Sharma, P. Kundu, S. Mahanty, R.N. Basu, Development and characterizations of BaO–CaO–Al$_2$O$_3$-SiO$_2$ glass–ceramic sealants for intermediate temperature solid oxide fuel cell application, J. Non-Cryst. Solids 354 (2008) 4081–4088.

[25] N.P. Bansal, E.A. Gamble, Crystallization kinetics of a solid oxide fuel cell seal glass by differential thermal analysis, J. Power Sources 47 (2005) 107–115.

[26] J.M. Ralph, A.C. Schoeler, M. Krumpelt, Materials for lower temperature solid oxide fuel cells, J. Mater. Sci. 36 (2001) 1161–1172.

[27] EG & G Technical Services, Fuel Cell Handbook, seventh ed., US Department of Energy, Office of Fossil Energy, National Energy Technological Laboratory, Morgantown, West Virginia, 2004, pp. 7–49 (chapter 7).

[28] K.S. Weil, The state-of-the-art in sealing technology for solid oxide fuel cells, J. Miner. Met. Mater. Soc. 58 (2006) 37–44.

[29] R. Barfod, S. Koch, Y.L. Liu, P.H. Hendriksen, Long term tests of DK - SOFC cells, in: Proceeding of Electrochemical Society, 2003-07(SOFC VIII), 2003, pp. 1158–1166.

[30] M. Bram, S. Reckers, P. Drinovac, J. Monch, R.W. Steinbrech, H.P. Buchkremer, D. Stover, Deformation behavior and leakage tests of alternative sealing materials for SOFC stacks, J. Power Sources (2004) 138–141.

[31] http://www.alternative-energy-news.info/new-solid-oxide-fuel-cell-technology/.

[32] M.K. Mahapatra, K. Lu, Glass-based seals for solid oxide fuel and electrolyzer cells- a review, Mater. Sci. Eng. R 67 (2010) 65–85.

[33] M.K. Mahapatra, K. Lu, Seal glass for solid oxide fuel cell – review, J. Power Sources 195 (2010) 7129–7139.

[34] P.A. Lessing, A review of sealing technologies applicable to solid oxide electrolysis cell, J. Mater. Sci. 42 (2007) 3465–3476.

[35] S. Ghosh, P. Kundu, A.D. Sharma, R.N. Basu, H.S. Maiti, Microstructure and property evaluation of barium aluminosilicate glass–ceramic sealant for anode-supported solid oxide fuel cell, J. Eur. Ceram. Soc. 28 (2008) 69–76.

[36] D.U. Tulyaganov, A.A. Reddy, V.V. Kharton, J.M.F. Ferreira, Aluminosilicate-based sealants for SOFCs and other electrochemical applications - a brief review, J. Power Sources 242 (2013) 486–502.

[37] K.D. Meinhardt, D.S. Kim, Y.S. Chou, K.S. Weil, Synthesis and properties of a barium aluminosilicate solid oxide fuel cell glass–ceramic sealant, J. Power Sources 182 (2008) 188–196.

III. Electrochemical energy conversion and storage

[38] J. Milhans, D. Li, M. Khaleel, X. Sun, H. Garmestani, Short communication: statistical continuum mechanics analysis of effective elastic properties in solid oxide fuel cell glass ceramic seal material, J. Power Sources 195 (2010) 5726–5730.

[39] Karmakar, Glass and Glass-Ceramics for Solid Oxide Fuel cell(SOFC) Sealants, Functional Glasses and Glass-Ceramics, Elsevier, 2017, https://doi.org/10.1016/B978-0-12-805056-9.00008-8.

[40] S.R. Rejisha, P.S. Anjana, N. Gopakumar, N. Santha, Synthesis and characterization of strontium and barium bismuth borate glass-ceramics, J. Non-Cryst. Solids 388 (2014) 68–74.

[41] S.R. Rejisha, P.S. Anjana, N. Gopakumar, Effect of cerium (IV) oxide on the optical and dielectric properties of strontium bismuth borate glasses, J. Mater. Sci. Mater. Electron. 27 (2016) 5475–5482.

[42] S.R. Rejisha, P.S. Anjana, N. Gopakumar, Structural and dielectric studies of $MBi_2B_2O_7$ (M = Sr&Ba) glass-$Bi_{24}B_2O_{39}$ microcrystal composites, J. Non-Cryst. Solids 512 (2019) 189–196.

[43] R.G. Beaman, Relation between (apparent)second-order transition temperature and melting point, J. Polym. Sci. 9 (1952) 470–472.

[44] S. Sakka, J.D. Mackenzie, Relation between apparent glass transition temperature and liquid temperature for inorganic glasses, J. Non-Cryst. Solids 6 (1971) 145–162.

[45] M.B. Volf, Chemical Approach to Glass. Series on Glass Science and Technology, Elsevier, NewYork, 1984, p. 7.

[46] M.A. Matveev, G.M. Matveev, B.N. Frenkel, Calculation and Control of Chemical, Optical and Thermal Properties of Glass, 1975, pp. 14–31. Holon, Israel.

[47] N. Santha, T.K. Nideep, S.R. Rejisha, Synthesis and characterization of barium boro silicate glass Al_2O_3 composites, J. Mater. Sci. Mater. Electron. 23 (2012) 1435–1441.

[48] H. Nonnet, H. Khedim, F. Mear, Development and characterization of glass and glass ceramic sealants for solid oxide electrolyser cells, J. Aust. Ceram. Soc. 48 (2012) 205–210.

[49] R.N. Basu, G. Blass, H.P. Buchkremer, D. Stover, F. Tietz, E. Wessel, I.C. Vinke, Simplified processing of anode supported thin film planar solid oxide fuel cells, J. Eur. Ceram. Soc. 25 (2005) 463–471.

[50] S.B. Sohn, S.Y. Choi, Suitable glass ceramic sealants for planar solid oxide fuel cells, J. Am. Ceram. Soc. 87 (2004) 254–260.

[51] S.T. Reis, R.K. Brow, Designing sealing glasses for solid oxide fuel cells, J. Mater. Eng. Perform. 15 (2006) 410–413.

[52] K.A. Nielsen, M. Solvang, F.W. Poulsen, P.H. Larsen, Evaluation of sodium alumino silicate glass composite seal with magnesia filler, Ceram. Eng. Sci. Proc. 25 (3) (2004) 309–314.

[53] A. Flugel, M.D. Dolan, A.K. Varshneya, Y. Zheng, N. Coleman, M. Hall, D. Earl, S.T. Misture, Development of an improved devitrified fuel cell sealing glass1. Bulk properties, J. Electrochem. Soc. 154 (2007) B601–B608.

[54] L. Luo, Y. Lin, Z.Z. Huang, Y. Wu, L. Sun, L. Cheng, J. Shi, Application of BaO-CaO- Al_2O_3-B_2O_3-SiO_2 glass-ceramic seals in large size planar IT-SOFC, Ceram. Int. 41 (2005) 9239–9243.

[55] W.N. Liu, X. Sun, B. Koeppel, M. Khaleel, Experimental study of the aging and self healing of glass/ceramic sealant used in SOFCs, Int. J. Appl. Ceram. Technol. 7 (2010) 22–29.

[56] M.J. Pascual, A. Guillet, A. Duran, Optimization of glass-ceramic sealant compositions in the system MgO-BaO-SiO_2 for solid oxide fuel cells (SOFC), J. Power Sources 169 (2007) 40–46.

[57] A. Goel, M.J. Pascual, J.M.F. Ferreira, Stable glass-ceramic sealants for solid oxide fuel cells: influence of Bi_2O_3 doping, Int. J. Hydrogen Energy 35 (2010) 6911–6923.

[58] J.C. Lee, H.C. Kwon, Y.P. Kwon, J.H. Lee, S. Park, Porous ceramic fiber glass matrix composites for solid oxide fuel cell seals, Colloid. Surface. Physicochem. Eng. Aspect. 300 (2007) 150–153.

[59] P. Jinhua, S. Kening, Z. Naiqing, C. Xinbing, Z. Deru, Sealing glass of baruium-calcium aluminosilicate system for solid oxide fuel cells, J. Rare Earths 25 (2007) 434–438.

[60] S.F. Wang, Y.R. Wang, Y.F. Hsu, C.C. Chuang, Effect of additives on the thermal properties and sealing characteristic of BaO-Al_2O_3-B_2O_3-SiO_2 glass-ceramic for solid oxide fuel cell application, Int. J. Hydrogen Energy 34 (2009) 8235–8244.

[61] Y.S. Choua, J.W. Stevenson, R.N. Gow, Novel alkaline earth silicate sealing glass for SOFC, part I. The effect of nickel oxide on the thermal and mechanical properties, J. Power Sources 168 (2007) 426–433.

[62] Z. Yang, K.D. Meinhardt, J.W. Stevenson, Chemical compatibility of barium-calcium aluminosilicate-based sealing glasses with the ferritic stainless-steel interconnect in SOFCs, J. Electrochem. Soc. 150 (2003) A1095–A1101.

[63] A.A. Reddy, D.U. Tulyaganov, A. Goel, M. Sardo, P.V. Wiper, M.J. Pascual, V.V. Kharton, V.A. Kolotygin, E.V. Tsipis, L. Mafra, J.M.F. Ferreira, Melilite glass-ceramic sealants for solid oxide fuel cells:effects of ZrO_2 additions assessed by microscopy, diffraction and solid-state NMR, J. Mater. Chem. 1 (2013) 6471–6480.

References

401

[64] L. Rezazadeh, S. Baghshahi, A.N. Golikand, Z. Hamnabard, Structure, phase formation and wetting behavior of BaO-SiO$_2$−B$_2$O$_3$ based glass−ceramics as sealants for solid oxide fuel cells, Ionics 20 (2014) 55−64.

[65] V. Simon, M. Todea, A.F. Takács, M. Neumann, S. Simon, XPS study on silicate bismuth glasses and glass ceramics, Solid State Commun. 141 (2007) 42−47.

[66] A. Goel, D.U. Tulyaganov, V.V. Kharton, A.A. Yaremchenko, J.M.F. Ferreira, Short communication: electrical behavior of aluminosilicate glass ceramic sealants and their interaction with metallic solid oxide fuel cell interconnects, J. Power Sources 195 (2010) 522−526.

[67] B. Tiwari, A. Dixit, G.P. Kothiyal, Study of glasses/glass-ceramics in the SrO-ZnO-SiO$_2$ system as high temperature sealant for SOFC applications, Int. J. Hydrogen Energy 36 (2011) 15002−15008.

[68] K. Sharma, G.P. Kothiyal, L. Montagne, F.O. Mear, B. Revel, A new formulation of barium-strontium silicate glasses and glass-ceramics for high- temperature sealant, Int. J. Hydrogen Energy 37 (2012) 11360−11369.

[69] A.A. Reddy, D.U. Tulyaganov, A. Goel, M.J. Pascual, V.V. Kharton, E.V. Tsipis, J.M.F. Ferreira, Diopside − Mg orthosilicate and diopside - Ba disilicate glass-ceramics for sealing applications in SOFC: sintering and chemical interactions studies, Int. J. Hydrogen Energy 37 (2012) 12528−12539.

[70] P.K. Ojha, S.K. Rath, T.K. Chongdar, N.M. Gokhale, A.R. Kulkarni, Short communication: physical and thermal behaviour of Sr−La−Al−B−Si based SOFC glass sealant as function of SrO content and B$_2$O$_3$/SiO$_2$ ratio in the matrix, J. Power Sources 196 (2011) 4594−4598.

[71] P.K. Ojha, T.K. Chongdar, N.M. Gokhale, A.R. Kulkarni, Accelerated devitrification of a strontium lanthanum aluminoborosilicate based intermediate temperature solid oxide fuel cell glass sealant and its effect on thermophysical behaviour of the glass ceramics, J. Power Sources 22 (2013) 28−34.

[72] A.A. Reddy, D.U. Tulyaganov, M.J. Pascual, V.V. Kharton, E.V. Tsipis, V.A. Kolotygin, J.M.F. Ferreira, Diopside-Ba disilicate glass-ceramic sealants for SOFCs: enhanced adhesion and thermal stability by Sr for Ca substitution, Int. J. Hydrogen Energy 38 (7) (2013) 3073−3086.

[73] B. Kaur, K. Singh, O.P. Pandey, Microstructural study of Crofer22APU-glass interface for SOFC application, Int. J. Hydrogen Energy 37 (2012) 3839−3847.

[74] G. Kaur, O.P. Pandey, K. Singh, Interfacial study between high temperature SiO$_2$ − B$_2$O$_3$ −AO-La$_2$O$_3$ (A = Sr, Ba) glass seals, Crofer22APU for solid oxide fuel cell applications, Int. J. Hydrogen Energy 37 (2012) 6862−6874.

[75] A. Goel, D.U. Tulyaganov, V.V. Kharton, A.A. Yaremchenko, S. Eriksson, J.M.F. Ferreira, Optimization of La$_2$O$_3$ containing diopside based glass ceramic sealants for fuel cell applications, J. Power Sources 189 (2009) 1032−1043.

[76] C. Lara, M.J. Pascual, R. Keding, A. Duran, Electrical behaviour of glass−ceramics in the systems RO-BaO−SiO$_2$ (R = Mg, Zn) for sealing SOFCs, J. Power Sources 157 (2006) 377−384.

[77] J. Puig, F. Ansart, P. Lenormand, L. Antoine, J. Daily, Sol-gel synthesis and characterization of barium (magnesium) aluminosilicate glass sealants for solid oxide fuel cells, J. Non-Cryst. Solids 357 (2011) 3490−3494.

[78] A. Arora, K. Singh, O.P. Pandey, Thermal, structural and crystallization kinetics of SiO$_2$-BaO−ZnO−B$_2$O$_3$-Al$_2$O$_3$ glass samples as a sealant for SOFC, Int. J. Hydrogen Energy 36 (2011) 14948−14955.

[79] P. Namwong, N. Laorodphan, W. Thiemsorn, M. Jaimasith, A. Wannakon, T. Chairuangsri, A barium − calcium silicate glass for use as seals in planar SOFCs, Chiang Mai J. Sci. 37 (2010) 231−242.

[80] A. Arora, V. Kumar, K. Singh, O.P. Pandey, Structural, thermal and crystallization Kinetics of ZnO−BaO-SiO$_2$-B$_2$O$_3$-Mn$_2$O$_3$ based glass sealants for solid oxide fuel cells, Ceram. Int. 37 (2011) 2101−2107.

[81] A.A. Reddy, A. Goel, D.U. Tulyaganov, S. Kapoor, K. Pradeesh, M.J. Pascual, J.M.F. Ferreira, Study of calcium magnesium-aluminum silicate(CMAS) glass and glass-ceramic sealant for solid oxide fuel cells, J. Power Sources 231 (2013) 203−212.

[82] M.S. Salinigopal, N. Gopakumar, P.S. Anjana, Alkaline Earth based borosilicate glasses as sealants in solid oxide fuel cell applications, Siliconindia 12 (2020) 101−107.

[83] M.S. Salinigopal, N. Gopakumar, P.S. Anjana, B. Suresh Kumar, Structural, optical and dielectric properties of aluminoborosilicate glasses, J. Electron. Mater. 49 (1) (2020) 695−704.

[84] M.S. Salinigopal, N. Gopakumar, P.S. Anjana, Synthesis and characterization of 50BaO- (5-x)Al$_2$O$_3$-xR$_2$O-30B$_2$O$_3$-15SiO$_2$(R=Nd, Gd) glass-ceramics, J. Non-Cryst. Solids 535 (2020).

[85] M.S. Salinigopal, N. Gopakumar, P.S. Anjana, O. P. Pandey Studies of 50BaO-(5-x) Al$_2$O$_3$-xR$_2$O$_3$-30B$_2$O$_3$-15SiO$_2$ (R=La, Dy) Glass-Ceramics as Sealant in Solid Oxide Fuel Cell [To be Communicated].

[86] M.J. Hyatt, N.P. Bansal, Crystal growth kinetics in BaO−Al$_2$O$_3$−SiO$_2$ and SrO- Al$_2$O$_3$-2SiO$_2$ glasses, J. Mater. Sci. 31 (1996) 172−184.

III. Electrochemical energy conversion and storage

402   13. Glasses and glass-ceramics as sealants in solid oxide fuel cell applications

[87] N. Lahl, K. Sing, L. Singheiser, K. Hilpert, D. Bahadur, Crystallization kinetics in AO- Al_2O_3-SiO_2-B_2O_3 glasses (A = Ba, Ca, Mg), J. Mater. Sci. 35 (2000) 3089–3096.

[88] V. Kumar, A. Arora, O.P. Pandey, K. Singh, Studies on thermal and structural properties of glasses as sealants for solid oxide fuel cells, Int. J. Hydrogen Energy 33 (2008) 434–438.

[89] I.W. Donald, P.M. Mallinson, B.L. Metcalfe, L.A. Gerrard, J.A. Fernie, Recent developments in the preparation, characterization and applications of glass- and glass ceramic-to-metal seals and coatings, J. Mater. Sci. 46 (2011) 1975–2000.

[90] K. Lu, M.K. Mahapatra, Network structure and thermal stability study of high temperature seal glass, J. Appl. Phys. 104 (2008) 074910.

[91] M. Budd, Glass Ceramic Material and its Use as Means for Joining Different Types of Material and Support, European Patent, 1998, p. 0975554B1.

[92] M. Radovic, E. Lara-Curzio, R.M. Trejo, H. Wang, W.D. Porter, Thermo - physical properties of YSZ and Ni-YSZ as a function of temperature and porosity, Ceram. Eng. Sci. Proc. 27 (2006) 79–85.

[93] M. Yokoo, Y. Tabata, Y. Yoshida, K. Hayashi, Y. Nozaki, K. Nozawa, H. Arai, Highly efficient and durable anode-supported SOFC stack with internal manifold structure, J. Power Sources 178 (2008) 59–63.

[94] R.N. Singh, Sealing technology for solid oxide fuel cells (SOFC), Int. J. Appl. Ceram. Technol. 4 (2007) 134–144.

[95] J.F.B. Rasmussen, P.V. Hendriksen, A. Hagen, Study of internal and external leaks in tests of anode-supported SOFCs, Fuel Cell. 8 (2008) 385–393.

[96] R.E. Loehman, H.P. Dumm, H. Hofer, Evaluation of sealing glasses for solid oxide fuel cells, Ceram. Eng. Sci. Proc. 23 (2002) 699–705.

[97] T. Jin, K. Lu, Thermal stability of a new solid oxide fuel/electrolyzer cell seal glass, J. Power Sources 195 (2010) 195–203.

[98] M. Hubert, A.J. Faber, On the structural role of boron in borosilicate glasses, Eur. J. Glasses Sci. Technol. B Phys. Chem. Glasses 55 (3) (2014) 136–158.

[99] G. Kaur, O.P. Pandey, K. Singh, Self-healing behavior of barium lanthanum borosilicate glass and its reactivity with different electrolytes for SOFC applications, Int. J. Appl. Ceram. Technol. 11 (1) (2014) 136–145.

[100] M.S. Salinigopal, N. Gopakumar, P.S. Anjana, Interaction Studies of Rare Earth Added Alumino borosilicate Glass-Ceramics with Crofer22APU (To be Communicated).

[101] A. Goel, D.U. Tulyaganov, E.R. Shaaban, R.N. Basu, J.M.F. Ferreira, Influence of ZnO on the crystallization kinetics and properties of diopside-Ca-based glasses and glass- ceramics, J. Appl. Phys. 104 (2008), 043529-1–043529-11.

[102] S.B. Sohn, S.Y. Choi, G.H. Kim, H.S. Song, G.D. Kim, Stable sealing glass for planar solid oxide fuel cell, J. Non-Cryst. Solids 297 (2002) 103–112.

[103] R.H. Doremus, Glass Science, second ed., John Wiley & Sons, New York, 1994.

[104] P.W. McMillan, Glass–Ceramics, Academic Press, London, UK, 1979.

[105] W. Holand, G. Beall, Glass–Ceramics Technology, The American Ceramic Society, Westerville, OH, 2002.

[106] C. Lara, M.J. Pascual, A. Duran, Glass forming ability, sinter ability and thermal properties in the systems RO–BaO–SiO_2 (R = Mg, Zn), J. Non-Cryst. Solids 348 (2004) 149–155.

[107] Y.S. Touloukian (Ed.), Thermo-physical Properties of High Temperature Solid Materials Oxides and Their Solutions and Mixtures, vol. 4, The MacMillan Co., NewYork, NY, 1967.

[108] K.S. Weil, J.E. Deibler, J.S. Hardy, D.S. Kim, G.G. Xia, L.A. Chick, C.A. Coyle, Rapture testing as a tool for developing planar solid oxide fuel cell seal, J. Mater. Eng. Perform. 13 (2004) 316–326.

[109] N.H. Menzler, M. Bram, H.P. Buchkremer, D. Stoever, Development of a gas tight sealing material for ceramic components, J. Eur. Ceram. Soc. 23 (2002) 445–454.

[110] N.P. Bansal, M.J. Hyatt, C.H. Drummond III, Crystallization properties of Sr − Ba- alumino silicate glass-ceramic matrices, Ceram. Eng. Sci. Proc. 12 (1991) 1222–1234.

[111] F. Smeacetto, M. Salvo, M. Ferrairis, V. Casalegno, P. Asinari, Glass and composite seals for the joining of YSZ to metallic interconnect in solid oxide fuel cells, J. Eur. Ceram. Soc. 28 (2008) 611–616.

[112] P. Geasee, T. Schwickert, U. Diekmann, R. Conradt, J.G. Heinrich, F. Aldinger (Eds.), Ceramic Materials and Compounds for Engineering, Wiley-VCH Verlag GmbH, Weinheim, Germany, 2001, pp. 57–62.

[113] R. Zheng, S.R. Wang, H.W. Nie, T.L. Wen, SiO_2-CaO-B_2O_3-Al_2O_3ceramic glaze as sealant for planar IT-SOFC, J. Power Sources 128 (2004) 165–172.

[114] K.L. Ley, M. Krumpelt, R. Kumar, J.H. Meiser, I.D. Bloom, Glass ceramic sealants for solid oxide fuel cells: part I. Physical properties, J. Mater. Res. 11 (1996) 1489–1493.

III. Electrochemical energy conversion and storage

References

[115] K. Sharma, A. Dixit, A. Arvind, G.P. Kothiyal, Study of structural and thermal properties of barium strontium silicate glasses, BARC Newsl. (2011) 337. Founder's Day Special Issue.

[116] N.H. Menzler, D. Sebold, M. Zahid, S.M. Gross, T. Koppitz, Interaction of metallic SOFC interconnect materials with glass—ceramic sealant in various atmospheres, J. Power Sources 152 (2005) 156—167.

[117] P. Batfalsky, A.A.C. Haanappel, J. Malzbender, N.H. Menzler, V. Shemel, I.C. Vinke, Chemical interaction between glass ceramic sealant and interconnect steels in SOFC stacks, J. Power Sources 155 (2006) 128—137.

[118] G. Kaur, Intermediate Temperature Solid Oxide Fuel Cells: Electrolytes, Electrodes and Interconnects, Elsevier, 2020, ISBN 978-0-12-817445-6.

[119] J.E. Shelby, Introduction to Glass Science and Technology, The Royal Society of Chemistry, Cambridge, 2005.

[120] G.P. Kothiyal, A. Ananthanarayanan, G.K. Dey, Glass and Glass-Ceramics, Functional Materials, Elsevier, 2012, https://doi.org/10.1016/B978-0-12-385142-0.00009-X.

[121] T. Sun, H. Xiao, W. Guo, X. Hong, Effect of Al_2O_3 content on BaO— Al_2O_3—B_2O_3—SiO_2 glass sealant for solid oxide fuel cell, Ceram. Int. 36 (2) (2010) 821—826.

[122] S.T. Reis, M.J. Pascual, R.K. Brow, C.S. Ray, T. Zhang, Crystallization and processing of SOFC sealing glasses, J. Non-Cryst. Solids 356 (2010) 3009—3012.

[123] C. Lara, Glass—ceramic Sealants of the System RO—BaO—SiO2 (R = Zn, Mg) for Solid Oxide Fuel Cells (SOFC), PhD Thesis, Autonoma University of Madrid, 2006.

[124] T. Zhang, Q. Zou, Tuning the thermal properties of borosilicate glass ceramic seal for solid oxide fuel cells, J. Eur. Ceram. Soc. 32 (2012) 4009—4013.

[125] F. Naghizadeh, M.R. Abdul Kadir, A. Doostmohammadi, F. Roozbahani, N. Iqbal, M.M. Taheri, S.V. Naveen, T. Kamarul, Rice husk derived bioactive glass-ceramic as a functional bio-ceramic: synthesis, characterization and biological testing, J. Non-Cryst. Solids 427 (2015) 54—61.

[126] R. Yuvakkumar, V. Elango, V. Rajendran, N. Kannan, High-purity nano silica powder from rice husk using a simple chemical method, J. Exp. Nanosci. 9 (2012) 272—281.

[127] G. Sharma, K. Singh, Agro-waste ash and mineral oxides derived glass-ceramics and their interconnect study with Crofer2APU for SOFC application, Ceram. Int. 45 (2019) 20501—20508.

[128] C.W. Kim, R.K. Brow, Inventors; The Curators of the University of Missouri, Mo-Sci Corporation, Assignees, Viscous Sealing Glass Compositions for Solid Oxide Fuel Cells, May 7, 2015. United States Patent 20150125780.

[129] S. Qi, N.M. Porotnikova, M.V. Ananyev, High-temperature glassy-ceramic sealants SiO_2—Al_2O_3—BaO—MgO and SiO_2—Al_2O_3—ZrO2—CaO—Na_2O for solid oxide electro - chemical devices, Trans. Nonferrous Met. Soc. China 6 (2016) 2916—2924.

[130] D.A. Krainova, N.S. Saetova, A.V. Kuzmina, A.A. Raskovalova, V.A. Eremina, M.V. Ananyeva, R.S. Wilckens, Non-crystallising glass sealants for SOFC: Effect of Y_2O_3 addition, Ceram. Int. (2019), https://doi.org/10.1016/j.ceramint.2019.10.266.

[131] V. Kumar, O.P. Pandey, K. Singh, Effect of A_2O_3 (A= La, Y, Cr, Al) on thermal and crystallization kinetics of borosilicate glass sealants for solid oxide fuel cells, Ceram. Int. 36 (2010) 1621—1628.

[132] K. Singh, N. Gupta, O.P. Pandey, Effect of Y_2O_3 on the crystallization behavior of SiO_2 —MgO—B_2O_3—Al_2O_3 glasses, J. Mater. Sci. 42 (2007) 6426—6432.

[133] E.A. Mahdy, S. Ibrahim, Influence of Y_2O_3 on the structure and properties of calcium magnesium aluminosilicate glasses, J. Mol. Struct. 1027 (2012) 81—86.

[134] E. Monrea, R. Spaeh, D. Stolten, Sealing materials for SOFC manifolds, DVS 184 (1997) 217—219.

[135] P.H. Larsen, P.F. James, Chemical stability of MgO-CaO-Cr_2O_3-Al_2O_3-B_2O_3 phosphate glasses in solid oxide fuel cell environment, J. Mater. Sci. 33 (1998) 2499—2507.

[136] K. Schwab, H. Baier, W. Schaefer, D. Stolten, Thermomechanical analysis of the joining process of a planar ceramic high temperature fuel cell system, DVS 184 (1997) 76—79.

[137] R. Gupta, Study of Physical Structural and Thermal Properties of Calcium borosilicate Glass Sealants, Thesis, School of Physics and Material Science, Thapar University, Patiala, 2009.

[138] P.H. Larsen, F.W. Poulsen, R.W. Berg, The influence of SiO_2 addition to 2MgO- Al_2O_3 -3.3P_2O_5 glass, J. Non-Cryst. Solids 244 (1999) 16—24.

[139] D. Bahadur, N. Lahl, K. Singh, L. Singheiser, K. Hilpert, Influence of nucleating agents on chemical interaction of MgO-Al_2O_3-B_2O_3-SiO_2, J. Electrochem. Soc. 151 (2004) A558.

[140] R.A. Meyers, Encyclopedia of Physical Science and Technology, vol. 4, Academic Press Inc., 1987.

[141] S.P. Jiang, L. Christiansen, B. Hughan, K. Foger, Effect of glass sealant materials on microstructure and performance of Sr-doped $LaMnO_3$ cathodes, J. Mater. Sci. Lett. 20 (2001) 695—697.

13. Glasses and glass-ceramics as sealants in solid oxide fuel cell applications

[142] N. Gupta, K. Singh, O.P. Pandey, in: Proceedings of the National Conference on Materials and Related Technologies, September 19–20, 2003, p. 141.

[143] Y. Imanaka, S. Akoi, N. Kamehara, K. Niwa, Cristobalite phase formation in glass or ceramic composites, J. Am. Ceram. Soc. 78 (1995) 1265–1271.

[144] S. Taniguchi, M. Kadowaki, T. Yasuo, Y. Akiyama, Y. Miyake, K. Nishio, Improvement of thermal cycle characteristics of a planar type solid oxide fuel cell by using ceramic fiber as sealing material, J. Power Sources 90 (2000) 163–169.

[145] V.A.C. Haanappel, V. Shemet, I.C. Vinke, S.M. Gross, T.H. Koppitz, N.H. Menzler, M. Zahid, W.J. Quadakkers, Evaluation of the suitability of various glass sealant-alloy combinations under SOFC stack conditions, J. Mater. Sci. 40 (2005) 1583–1592.

[146] N. Lahl, D. Bahadur, K. Singh, L. Singheiser, K. Hilpert, Chemical interaction between aluminosilicate base sealants and the components on the anode side of solid oxide fuel cells, J. Electrochem. Soc. 149 (5) (2002) A607–A614.

[147] Z. Yang, J.W. Stevenson, K.D. Meinhardt, Chemical interactions of barium calcium aluminosilicate based sealing glasses with oxidation resistant alloys, Solid State Ionics 160 (2003) 213–225.

[148] Y.S. Chou, J.W. Stevenson, R.N. Gow, Novel alkaline earth silicate sealing glass for SOFC Part I. The effect of nickel oxide on the thermal and mechanical properties, J. Power Sources 168 (2007) 426–443.

[149] C.K. Lin, K.L. Lin, J.H. Yeh, W.H. Shiu, C.K. Liu, R.Y. Lee, Aging effects on high temperature creep properties of a solid oxide fuel cell glass-ceramic sealant, J. Power Sources 241 (2013) 12–19.

[150] Z. Dai, J. Pu, D. Yan, B. Chi, L. Jian, Thermal cycle stability of Al_2O_3 based Compressive seals for planar intermediate temperature solid oxide fuel cells, Int. J. Hydrogen Energy 36 (2011) 3131–3137.

[151] B. Hua, J. Pu, W. Goan, J. Zhang, F. Lu, J. Li, Cyclic oxidation of Mn-Co spinel coated SUS 430 alloy in the cathodic atmosphere of solid oxide fuel cells, J. Power Sources 185 (2008) 419–422.

[152] C.K. Lin, J.Y. Chen, J.W. Tian, L.K. Chiang, S.H. Wu, Joint strength of a solid oxide fuel cell glass-ceramic sealant with metallic interconnect, J. Power Sources 205 (2012) 307–317.

[153] F. Smeacetto, M. Salvo, M. Ferraris, V. Casalegno, P. Asinari, A. Chrysanthou, Characterization and performance of glass-ceramic sealant to join metallic interconnects to YSZ and anode supported electrolyte in planar SOFCs, J. Eur. Ceram. Soc. 28 (2009) 2521–2527.

[154] H.T. Chang, C.K. Lin, C.K. Liu, High-temperature mechanical properties of a glass sealant for solid oxide fuel cell, J. Power Sources 189 (2009) 1093–1099.

[155] H.T. Chang, C.K. Lin, C.K. Liu, Effects of crystallization on the high-temperature mechanical properties of a glass sealant for solid oxide fuel cell, J. Power Sources 195 (2010) 3159–3165.

[156] A.G. Sabato, G. Cempura, D. Montinaro, A. Chrysanthou, M. Salvo, E. Bernardo, M. Secco, F. Smeacetto, Glass-ceramic sealant for solid oxide fuel cells application: characterization and performance in dual atmosphere, J. Power Sources 328 (2016) 262–270.

[157] Y.S. Chou, E.C. Thomsen, J.P. Choi, J.W. Stevenson, Compliant alkali silicate sealing glass for solid oxide fuel cell applications: combined stability in iso thermal ageing and ermal cycling with YSZ coated ferritic stainless steels, J. Power Sources 197 (2012) 154–160.

[158] S. Sakuragi, Y. Funahashi, T. Suzuki, y. Fujishiro, M. Awano, Non-alkaline glass-MgO composites for SOFC sealant, J. Power Sources 185 (2008) 1311–1314.

[159] Y. Funahashi, T. Shimanmori, T. Suzuki, Y. Fujishiro, M. Awano, Fabrication and characterization of components for cube shaped micro tubular SOFC bundle, J. Power Sources 163 (2007) 731–736.

[160] G. Kaur, O.P. Pandey, K. Singh, Glass stability and effect of heat treatment duration on chemical interaction between calcium lanthanum borosilicate glass sealant and electrolyte, J. Electrochem. Soc. 159 (2012) F717–F724.

[161] M. Kaur, G. Kaur, V. Kumar, Impact of La_2O_3/Y_2O_3 ratio on crystallization kinetics of CaO based aluminoborosilicate glasses and compatibility with Crofer and 8YSZ for SOFCs, J. Non-Cryst. Solids 521 (2019) 119539.

Further reading

[1] https://en.wikipedia.org/wiki/Solid_oxide_fuel_cell.

[2] http://www.ceramicindustry.com/articles/86115-ceramic-energy-advances-in-sofc- materials and-manufacturing.

SECTION IV

Lighting and Light emitting diodes

CHAPTER 14

Role of rare-earth ions for energy-saving LED lighting devices

Nutan S. Satpute, S.J. Dhoble

Department of Physics, RTM Nagpur University, Nagpur, Maharashtra, India

14.1 Introduction

Most developed lighting systems today are simply light-emitting diodes (LEDs). Around 95% of the LEDs available in the market generate white light due to the combined effect of blue light produced by a semiconductor diode (GaN or InGaN) and broad yellow radiation given by $Y_3Al_5O_{12}:Ce^{3+}$ (YAG:Ce) phosphor. Phosphor has been an important part of white light-emitting LEDs (WLEDs) for some time but there is yet no alternative available. Low color rendering index (CRI < 80) and high correlated color temperature (CCT > 5000 K) are possible with a combination of YAG:Ce/blue LED, but additionally there is a requirement for the addition of red phosphors (fluoride or nitride), which are expensive and hygroscopic [1]. Rare-earth-doped inorganic phosphors are particularly used in LED lighting and display devices as well as wind turbines.

Various techniques and steps have been implemented to produce efficient blue light emission LEDs but currently research is focused on WLEDs, which is mainly concentrated on the development of rare-earth free phosphors for WLEDs [2,3]. This chapter comprises a broader range of luminescence phenomena and the mechanisms involved therein, phosphors for LED lighting, and the prospects for energy-saving technologies as well as applications.

14.2 Fundamentals

This section discusses: the luminescence phenomenon; solid-state luminescent materials, beginning with a discussion of the different types of luminescence and the different ways in which luminescence can be excited; characteristics of luminescence; various types of phosphors; synthesis and characterization techniques of phosphor material; and applications.

Energy Materials
https://doi.org/10.1016/B978-0-12-823710-6.00018-2

Copyright © 2021 Elsevier Ltd. All rights reserved.

14.2.1 Luminescence

Light is mainly generated by two important phenomena, namely incandescence (involving the process of heating materials) and luminescence (the process of spontaneous emission of light by a substance without heating). Cold body radiation or cold light is also called luminescence. In 1888, Eilhardt Wiedemann (a German physicist) was the first to propose the concept of luminescence. The Latin word *lumen* means light, and materials that possess this phenomenon are known as "luminescent materials." In the 17th century, Italian scientist Vincentinus Casciarolo of Bologna introduced the term phosphor [4,5]. Luminescence is possible through various ways and it has many categories depending on the mode of excitation, given as follows:

1. Photoluminescence: excitation is by electromagnetic radiation/photons.
2. Cathodoluminescence: excitation is by high-energy electrons (cathode rays).
3. Electroluminescence: electric spark is used to trigger light emission.
4. Radioluminescence: excitation is possible with high-energy X-rays or γ-rays.
5. Sonoluminescence: excitation is by ultrasonic waves.
6. Triboluminescence: mechanical treatment of the material produces light glow.
7. Chemiluminescence: light is emitted during chemical reactions.
8. Bioluminescence: a form of chemiluminescence from living organisms.
9. Thermoluminescence: thermally stimulated luminescence in which material is thermally activated after initial irradiation by some other means such as ultraviolet (UV) or X-rays.

14.2.1.1 General characteristics of luminescence

Fig. 14.1 depicts the phenomenon of luminescence, showing two possible ways of the return process to the ground energy level such as (1) the radiative process, and (2) the nonradiative process. Luminescence occurs basically due to the radiative process only, but can include the nonradiative process. The nonradiative process is not responsible for luminescence. However, it is inherent in the process of luminescence as some photons absorbed by the atoms of a material cause lattice vibrations, which transport energy in the form of heat. To obtain an efficient luminescent material, radiative transition should dominate the process of nonradiative transition. In practice, exciting radiation is not easily absorbed by the

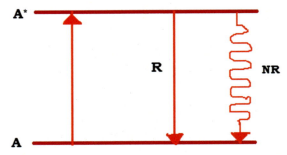

FIGURE 14.1 Energy-level scheme of the luminescent ion A. The * indicates the excited state, R the radiative return, and NR the nonradiative return to the ground state.

activators in luminescent material, which indicates that the situation is somewhat complicated as depicted in Fig. 14.1 [6–8].

The time duration of emission divides luminescence into subclasses:

1. **Fluorescence**: The process in which glow persists even after removing an excitation is called fluorescence. It is independent of the intensity of excitation and temperature. Its lifetime is generally less than 10^{-8} s.
2. **Phosphorescence**: The process in which the phenomenon of afterglow exists (decay is very slow with complex kinetics). It is often dependent on intensity of excitation and is strongly temperature dependent. Its lifetime is more than 10^{-8} s. Metastable states may cause a delay in the process of luminescence emission causing this effect. The defect centers, activators, impurities, and electron or hole traps present in the lattice are responsible for the creation of metastable states in the crystal structure, since thermal activation of the metastable activator or trap is essential for emission to occur.

14.2.1.2 Mechanism of luminescence

Fig. 14.2 illustrates the mechanism of luminescence with the help of a configurational coordinate diagram [9]. Configurational coordinates are a combination of specific normal coordinates, which are used to represent a large number of actual vibrational modes of the lattice. In Fig. 14.2, activator ion total energy E is plotted against distance R (where R is the distance between metal cations and anions of the host material, and R_o and R_1 are equilibrium distances of the ground-state activator and the excited-state activator, respectively). The vertical blue line represents the absorption (E_{ab}) and the vertical red line represents the emission (E_{emi}) transitions. From E_{ab} and E_{emi} transitions, it is observed that during the optical process, the activator remains at the same position as stated in the Franck–Condon

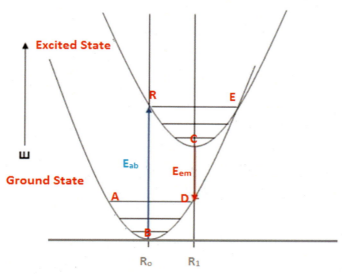

FIGURE 14.2 Configuration coordinate diagram illuminating the luminescence mechanism.

principle. This is because nuclei are much larger than electrons and the electronic transition takes place faster than the nuclei can respond. The process of optical absorption starts at R_o, i.e., at the ground-state equilibrium position. The probability that excited electrons will lose energy due to vibrations of the lattice is $10^{12}-10^{13}\,s^{-1}$, whereas the maximum probability of light emission is $10^9\,s^{-1}$. Due to this, before the process of luminescence, state B settles at equilibrium position C. Thus the emission and relaxation process from D $\rightarrow$ A helps to complete the cycle by releasing heat energy (nonradiative transition). Hence, Stokes shift occurs in the host lattice. The system is said to be in an equilibrium position at C; however, the higher-level excited curve is not at rest because of thermal energy of the system, which migrates over a small region around C. It covers an increasingly larger area with the increase in temperature of the system.

Emission occurs through point D by downward transition. The region around D is also covered at the ground state. At this point, a very rapid change of energy occurs. Thus, even for a small change in the value of configuration coordinates, a wider change of energies in the optical transition process occurs, which causes broad absorption as well as an emission band. Thus the change in temperature width of the bands also varies (at temperatures above and below room temperature).

Point E is the point of intersection of excited vibrational energy and ground-state energy levels, from where atoms can be raised due to energetic vibrations and can return to the ground state by releasing energy in the form of heat or infrared radiation. If the active centers remain at the excited state for relatively longer, then intense temperature quenching occurs. Due to this, decay time of the emission, which occurs in this temperature regime, decreases characteristically. If point E is at energy E_Q, which is just above the minimum energy level of the excited state, then the efficiency of luminescence η is given by:

$$\eta = \left[1 + C_{exp}\left(-\frac{E_q}{kT}\right)\right]^{-1} \tag{14.1}$$

where: C is a constant, k is the Boltzmann's constant, and T is the temperature (in K).

Therefore for fluorescence emission:

$$\frac{1}{t} = P_r + P_{nr} \tag{14.2}$$

where t is the mean lifetime, P_r is the probability of temperature-dependent radiative transition, and P_{nr} is the probability of nonradiative transitions increasing with temperature. The lifetime is given as if there were no nonradiative transitions at zero temperature:

$$\frac{1}{t_0} = P_r \tag{14.3}$$

$$\eta(T) = \frac{t(T)}{t(0)} = \frac{P_r}{(P_r + P_{nr})} \tag{14.4}$$

14.2.2 Phosphors—an overview

Luminescent materials are those materials that are able to emit light, particularly in the visible region (4000−7000 Å). So, many luminescent materials are available in nature but all can be used practically. Efficient luminescent material should be prepared in the laboratory with specific composition and should be reproducible and reliable. Such materials can be used for practical applications. Phosphors are generally solid materials existing in fine powder form. They give luminescence when excited with the help of a suitable frequency. Phosphor materials have an illustrious history and have been studied extensively since 1950. The first synthesized phosphor material was barium sulfide, but it has many inadequacies, for example, it exists in an impure form with poor efficiency, produces hydrogen sulfide, and also decomposes easily in air. In the literature, it is reported that phosphor material in pure form does not show good luminescence properties. To have good luminescence efficiency, phosphor should contain an activator metal. In sulfide phosphor form, it is reported that zinc and cadmium have great importance. Useful phosphor materials are generally made up of alkaline-earth sulfide phosphors, which contain an activator metal (some specific metal impurities) and rare-earth ions. Phosphors that are codoped with one or a combination of lanthanides and transition-metal ions have been studied by many researchers [10−18]. Impurities in the form of an activator can be introduced in phosphor material in the following ways: (1) impurities existing in very small concentrations in the host material itself and (2) impurities added in the host material at the cost of one of the constituents in a stoichiometric ratio. This deliberate addition of impurities generates certain localized prohibited energy levels.

Luminescence can be of two types depending on the energy levels involved in the process, namely characteristic luminescence and noncharacteristic luminescence. In the case of characteristic luminescence, activator atom energy levels or modified energy levels of the host lattice take part in the process of luminescence. Incident energy in the form of a quantum state is absorbed by the activator, which may be due to the transition of one of these electrons from one quantum state to another. This excited atom when returned to the ground state emits low-energy photons because some part of its energy is lost due to lattice interaction. But when a charge is transferred through the lattice, and it causes modification in the energy levels of the host atoms due to activator atoms, then noncharacteristic luminescence occurs [19,20].

In the crystal lattice, host crystal ions surround the activator ions and produce luminescent centers in the host materials. Luminescent centers in the host materials should be widely separated from each other. Excitation of phosphor and emission take place at these widely separated centers. If activator atom concentration in the host material is kept very low and its arrangement is kept regular, then high efficiency can be achieved. If the concentration of activator atoms increases beyond a certain limit, then it acts like a poison or a killer and hence reduces efficiency. Iron, cobalt, and nickel ions are some well-known killer atoms in the host crystal, which inhibit emission of light even though they are present in low concentrations. In sulfide phosphors, generally, coactivators (donors) are also observed as coimpurities, which is necessary for luminescence, but it does not have the same effect as that of activator (acceptor) atoms. When monovalent or trivalent impurities are added to the material, there is a lack of positive charge. This lack of positive charge can be balanced by adding suitable flux materials, for example, sodium thiosulfate. The addition of flux is useful to produce a solution and fix activators at appropriate places in the host material on firing [21]. A very small amount of flux added to the phosphor is sufficient to achieve luminescence efficiency; this flux facilitates the incorporation of activators in the host crystal, hence they are called coactivators. This process produces many luminescent centers and gives strong activation. To produce useful luminescent phosphor, one should take into account the following parameters: type or class of host crystal, chemical composition, crystal class, activator to be added (type and concentration), suitable coactivator, synthesis method depending on starting material, crystallization process temperature and duration, emission spectrum, and persistence. Alkaline-earth sulfides are widely used phosphors and by using various activators and coactivators it is possible to produce a variety of phosphors with different properties, which can be used for many applications. These sulfide materials have wide bandgaps; excited states of dopant atoms are quite properly distributed between the valence and conduction bands. Such host materials create a suitable environment around the luminescent centers [22].

14.2.2.1 Types of phosphor

Blue phosphor

Blue phosphor is an important candidate, which is used in the development of WLEDs. The activator Eu^{2+} ion is most commonly used as the blue light-emitting activator as it has a small Stokes shift and less decay time. It has broad near-UV excitation and emission bands. Some reported blue-emitting phosphors are $CaLaGa_3S_6O:Ce^{3+}$ [23], $LiSrPO_4:Eu^{2+}$ [24], $SrMg_2Al_{16}O_{27}:Eu^{2+}$ [25], $Sr_{3.5}Mg_{0.5}Si_3O_8Cl_4:Eu^{2+}$ and $Ba_{3.5}Mg_{0.5}Si_3O_8Cl_4:Eu^{2+}$ [26], $LiCaPO_4:Eu^{2+}$ [27], $Ba_3LaNa(PO_4)_3F:Eu^{2+}$ [28], $BaHfSi_3O_9:Eu^{2+}$ [29], $Ca_2LuScZrAl_2GeO_{12}:Ce^{3+}$ [30], $Ca_3Hf_2SiAl_2O_{12}:Ce^{3+}$ [31], $Ca_3Zr_2SiGa_2O_{12}:Ce^{3+}$ [32], $Na_3RbMg_7(PO_4)_6:Eu^{2+}$ [33], $Ca_2B_5O_9Br:Eu^{2+}$ [34], $Na_{3-2x}Sc_2(PO_4)_3:xEu^{2+}$ [35], and so on.

An efficient oxyhalide blue phosphor, $Sr_4OCl_6:Eu^{2+}$, has been developed by Gwak and coworkers with better photoluminescence emission and increased conversion efficiency [36]. A blue-emitting $RbBaPO_4:Eu^{2+}$ phosphor was developed by Song et al. by adopting the solid-state synthesis method. An intense purple/blue emission band maximum at 430 nm has been observed in the photoluminescence emission spectra of this phosphor. This material exhibits an orthorhombic structure on the basis of crystallographic information. Four O atom coordinates with one P atom in a rigid tetrahedral structure but only one Ba^{2+} site were observed in this phosphor. Chromaticity and efficacy of this synthesized phosphor were found to be more stable than that of the commercially used $Sr_3MgSi_2O_8:Eu^{2+}$ phosphor [37]. A novel blue phosphor, $SrLu_2O_4:Ce^{3+}$, matching well with a near-400 nm chip and showing high thermal stability, has been developed by Zang et al. [38]. Their study of the photoluminescence spectrum revealed the presence of a broad emission band at 460 nm with a bandwidth of nearly 90 nm. Internal quantum efficiency was found to be increased up to 76% by optimizing the Ce^{3+} concentration. They developed a series of WLEDs that were fabricated based on 405 nm chips coated with a blend of the new synthesized blue phosphor with the commercial yellow and red phosphors. It is reported that high CRIs (≥ 90) were achieved, while the CCT was tunable in the range of 3094−8990 K. These results suggest that $SrLu^2O^4:Ce^{3+}$ can be used as a blue-emitting phosphor in near-UV-based WLEDs.

III and IV group rare-earth-doped nitrides such as AlN, GaN, InGaN, and AlInGaN have been studied extensively because they show wide applications in blue−UV optoelectronic

14.2 Fundamentals

413

devices. Nitrides are generally preferred for blue light sources for two reasons: (1) large bandgap (incorporates the entire visible spectrum) and (2) strong chemical bonding (which is very stable and resistant to degradation under conditions of intense light illumination).

Red phosphors

Narrowband emission spectra are characteristic of red-emitting phosphors, which play an important role in the manufacturing of good-quality LEDs. Silicon nitride is found to be a suitable host material for red-emitting phosphor materials [39,40]. Eu^{3+}, Pr^{3+}, Sm^{3+}, and Eu^{2+} are rare-earth activators that are able to produce red color radiation in various host materials such as phosphate, nitride, sulfide, borate, vanadate, silicate, and tungstate [41].

Some reported red-emitting phosphors are $(Y_2MoO_4)_3$:Eu^{3+} [42], Ca_9Sc $(PO_4)_7$:Ce^{3+}, Mn^{2+} [43], (K_xNa_{1-x}) SiF_6:Mn^{2+} [44], $Cs_2WO_2F_4$:Mn^{4+} [45], Cs_2NbOF_5:Mn^{4+}, $BaNbOF_5$:Mn^{4+} [46—48], and so on.

A structural study of (K_xNa_{1-x}) SiF_6:Mn^{2+} phosphor identified various phases such as trigonal, orthorhombic, and cubic. Also, it is observed that emission intensity of the cubic form of this phosphor is five times greater than the trigonal form of the reported phosphor, which provides a relation between the structural and optical properties of matter [49]. Similarly, the bandgap of $NaEu(WO_4)_2$ (NaEuW) phosphor at different doses of high-energy electron beam irradiation is irradiated to increase the efficiency of red emissions as reported by Munirathnappa et al. [50]. Their study showed that electron beam irradiation had a great effect on crystal structure, as well as morphology, which in turn also affects the red fluorescence properties of NaEuW to generate efficient white light [50].

Yang et al. [51] synthesized red-emitting phosphor $Rb_5Nb_3OF_{18}$:Mn^{4+} by an ion-exchange method. Mn^{4+} in $Rb_5Nb_3OF_{18}$, which has an appropriate bandgap, experiences a strong crystal field and thus emits intense red color radiation at 633 nm and one broad excitation band at $\sim$470 nm. Yang et al. reported that LEDs coated with Mn^{4+}-doped $Rb_5Nb_3OF_{18}$ phosphor possess excellent luminescent properties, proving that the WLED could emit warm white light with high luminous efficiency (113.6 lm/W). Thus this Mn^{4+}-doped $Rb_5Nb_3OF_{18}$:Mn^{4+} was found to be a good red color candidate for WLEDs [51].

The photoluminescence properties and applications of $Na_2WO_2F_4$:Mn^{4+} were studied by Wang and coworkers [52]. An ultra-intense zero phonon line Mn^{4+} can be observed in such phosphor, which is beneficial to the color purity of red emission. Thus this reported phosphor has great potential for application in WLED lighting and displays.

Green phosphors

Green-emitting phosphor must have strong emission intensity for the production of white light. These phosphors have an emission peak at 510—550 nm The Tb^{3+} activator is a prominent green-emitting phosphor, which is efficient for excitation in the UV region [53]. Eu^{2+} and Ce^{3+} also produce green emissions. Eu^{2+} is generally used as a sensitizer for the transfer of energy with other activators such as Tb^{3+} and Ce^{3+}, and causes enhancement in emission intensity. Some reported green-emitting phosphors are strontium thiogallate $(SrGa_2S^{4+}MgGa_2O_4)$:Eu^{2+} prepared by the solid-state reaction method [54], $CaSc_2O_4$:Ce^{3+} [55], $BaSi_2O_5$:Eu^{2+} [56], $Sr_6BP_5O_{20}$:Eu^{2+} [57], Ca_3Y_2 $(Si_3O_9)_2$:Ce^{3+}, Tb^{3+} [58], $SrAl_2O_4$:Eu^{2+} [59], $Ca_2MgSi_2O_7$:Eu^{2+}, Mn^{2+} [60], and so on.

Green oxynitride phosphor β-SiAlON:Eu^{2+} has been synthesized by Hirosaki et al. [61]. This phosphor gives intense green emission at 538 nm. It has been reported that this

III. Electrochemical energy conversion and storage

phosphor has a broad excitation range from 300 to 420 nm, which makes it possible to emit strongly under near-UV or blue light excitation.

Green-emitting phosphor $BaLu_{1.94}Al_4SiO_{12}:0.06Ce^{3+}$ ($BLASO:0.06Ce^{3+}$) has been synthesized by the conventional solid-state reaction method [62]. When an LED device is fabricated by using $BLASO:0.06Ce^{3+}$ as the green phosphor, it shows good optical performance with the specification Ra = 88.5, CCT = 3562 K, and CIE 1931 chromaticity coordinates $x = 0.4041$, $y = 0.3951$. This shows that this phosphor can be used as a good candidate for developing thermally stable WLEDs with high CRI [62].

Yellow phosphors

Yellow phosphors have a garnet structure, and $A_3B_2C_3O_{12}$ doped with rare-earth ions (general formula) is widely useful for many applications. Commercially available WLEDs are designed with a combination of InGaN LED chips with Ce^{3+}-doped $Y_3Al_5O_{12}$ yellow-emitting phosphor, which possesses superoptical characteristics [63]. However, this phosphor has some disadvantages due to high thermal quenching. Thus there is a need to develop a novel yellow phosphor with low thermal quenching that generates warm white light.

Some reported yellow-emitting phosphors are $La_3Si_6N_{11}:Ce$ [64], $Ba_{0.93}Eu_{0.07}Al_2O_4$ [65], $Sr_3SiO_5:Ce^{3+}$, Li^+ phosphor [66], $SrAlSi_4N_3:Ce_3$ [67], and so on.

A new yellow-emitting $Sr_9MgLi(PO_4)_7:Eu^{2+}$ phosphor was developed and employed in WLEDs by Qiao et al. [68]. Developed $Sr_9MgLi(PO_4)_7$ is basically derived from the β-$Ca_3(PO_4)_2$-type mineral structure. On excitation with UV light (365 nm), this Eu^{2+}-doped phosphor shows a broad emission band (450 −700 nm).

A $Ba_2YAlO_5:Dy^{3+}$ yellow light-emitting phosphor was developed by Deng et al. [69]. On excitation at 357 nm, this phosphor showed a yellow emission at 580 nm, which is related to the $^4F_{9/2} \rightarrow {}^6H_{13/2}$ transition of Dy^{3+} ions. Concentration quenching occurred for a doping concentration above 2 mol%, which is due to the electric dipole−dipole interaction in this phosphor. It has been reported that Dy^{3+}-activated Ba_2YAlO_5 phosphors would be a promising yellow emission phosphor for WLED applications.

Further phosphors can be classified as upconversion and downconversion phosphors also.

Upconversion phosphors   When two or more photons are absorbed sequentially by atoms and the atoms reemit energy in the form of photons of a shorter wavelength than the incident wavelength, then that process is termed upconversion [70]. Nicolaas Bloembergen et al. were the first to propose this phenomenon, revealing the photon-producing ability of lanthanide ions. This phenomenon has tremendous importance in various applications, for example, energy conversion and medicine. Two low-energy photons successively absorbed by the single ion are excited to a higher energy level, then a single high-energy photon is generated when this high-energy ion makes a downward transition. The process in which energy transfer takes place between two rare-earth ions (e.g., Yb^{3+}, Er^{3+}) with the release of high-energy photons is known as energy transfer upconversion. It shows an anti-Stokes-type emission, in which infrared radiation is converted into visible radiation [71−75]. Generally, rare-earth elements from the d-block or f-block show the upconversion phenomenon. Examples are Ln^{3+}, Mo^{3+}, Ti^{2+}, Re^{4+}, Os^{4+}, Ni^{2+}, and so on. Fig. 14.3A shows the upconversion phosphor mechanism when sunlight is incident on the solar cell.

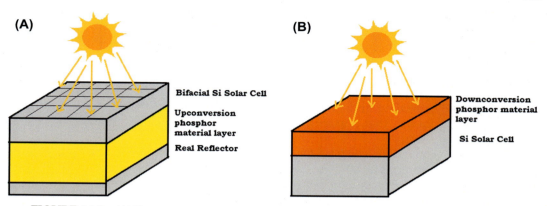

FIGURE 14.3 (A) Upconversion phosphor mechanism. (B) Downconversion phosphor mechanism.

Downconversion phosphors When a high-energy photon is absorbed by the photovoltaic cell atoms it then reemits two or more lower-energy photons; this process is termed downconversion. These photons have sufficient energy so that they can be easily absorbed by the photovoltaic cell with negligible thermalization loss. In this process, the energy loss caused by the thermalization of hot charge carriers after the absorption of high-energy photons may be reduced if the downconverted photons are absorbed by solar cells. This process is also referred to as quantum cutting [76–80]. Therefore near-infrared quantum cutting is a promising option to enhance the efficiency of Si solar cells. Generally, rare-earth elements from the d-block or f-block show downconversion phenomenon, for example, Eu^{2+}, Eu^{3+}, Tb^{3+}, Yb^{3+}, Dy^{3+}, and so on. Fig. 14.3B shows the downconversion phosphor mechanism when sunlight is incident on the solar cell.

14.2.3 Synthesis and characterization of phosphor

Phosphors can be synthesized by using many methods, but one should finalize the method of synthesis depending upon the final requirement. Most phosphor material compositions contain more than four elements, which makes it difficult to obtain a uniform single-phase material. Phosphor material performance is greatly affected by atomic structure, defects in the material, its interfaces, microstructure, composition, and many other factors, which can be controlled by thermodynamics and kinetics of the synthesis method. Generally, adopted synthesis methods are combustion synthesis, solid-state reaction, sol-gel/Pechini method, coprecipitation, solvothermal method, sonochemical synthesis, and microwave-assisted techniques.

14.2.3.1 Solid-state diffusion

Solid-state diffusion is also called solid-state reaction, which is a widely used method for the synthesis of phosphor. It is used to prepare oxide, oxyfluoride, oxychloride, as well as oxynitride materials. In this method, solid precursors are weighed in a proper stoichiometric ratio, then brought into close contact by mixing and grinding, preferably in acetone, followed by heat treatment at a particular high temperature (ranging from 500 to 2000°C) to facilitate diffusion of atoms and ions in the chemical reaction. Grain boundary contacts also affect the diffusion of atoms. Impurities and defects present in the material also affect the transportation of atoms or ions across grain boundaries. These grain boundaries sometimes stop the movement of charge carriers and thus further progress ceases. Thus to start the further movement of atoms or ions across the grain boundary, a fresh surface must be created in the material by repeated annealing and grinding of material [81,82]. In the solid-state diffusion method, it is well known that impurity phases are generated, which may be responsible for poor luminous efficacy.

Solid-state diffusion can also be used for the synthesis of polycrystalline ceramic materials. In this case, the procedure is as follows: (1) a medium-range temperature 200—450°C is used to heat the mixture of the precursor to remove unwanted residual gases, (2) the mixture is ground and reheated at a temperature preferably near to melting point (1000—1500°C) of the sample, and (3) the mixture is repeatedly mixed and heated until a homogeneous product formation is complete to obtain the required crystallinity.

14.2.3.2 Combustion synthesis

The combustion synthesis method is simply a wet-chemical method in which after synthesis, calcination and reheating are not necessary. It is a popular method because it is relatively simple, saves energy, and is cost effective. To produce multi-ingredient phosphor rapidly, this method is very effective. It is an exothermic reaction in which the emission of heat along with light occurs. For the combustion process, fuel and oxidizer are required; metal nitrate acts as an oxidizing reactant and urea acts as a reducing reactant. The synthesis process involves weighing solid precursors, then grinding and mixing generally in an aqueous solution of distilled water in which the ingredients are dissolved. Then, the aqueous solution is ignited at low temperature (around 500°C), which allows the highly exothermic and oxidative reaction to occur by generating a light flame [83,84]. During production of the flame, the temperature of the system is raised to nearly 1200°C and the reaction time is usually less than 10 min. This short duration at high temperature is sometimes responsible for the formation of nanocrystalline powder. However, after synthesis, agglomeration may occur within a few days, which can be crushed into fine powder. Fuel and oxidizer must be used in appropriate amounts to generate the maximum temperature at the time of synthesis.

14.2.3.3 Sol-gel synthesis

The sol-gel method is one of the best methods for the synthesis of phosphor material with nano/microsized particles with uniform distribution. Precursor (in the form of metal nitrates) after proper weighing in a stoichiometric ratio is mixed in distilled water by adding a surfactant such as polyethylene glycol (PEG) and a chelating agent like citric acid. The obtained solution is then ready for the stirring process at around 80°C. After stirring for a certain time, the formation of sol takes place due to condensation, which has a polymeric network, and is then converted into gel. The process of gelation is due to the sudden loss of fluidity of the solution, which forms an elastic solid gel [85]. The produced gel is then kept in the furnace for heat treatment to get rid of some of the organic content from the phosphor. Sometimes, annealing of the phosphor at higher temperature (<500°C) is required to remove residue of the PEG. The size of particles and morphology of the final product can be controlled by maintaining the proportion of chelating agent, surfactant, and precursor materials. When rare-earth ions are acquired in the form of oxides, it is necessary to convert them to nitrate form by dissolving them in concentrated HNO_3, heating and stirring the solution, and adding a few drops of distilled water. The sol-gel technique is an excellent method for producing a high purity and quality product, but it is quite a lengthy procedure. This technique is also termed the Pechini method. The only difference is that the Pechini method was developed for materials in which material processing lacks hydrolysis equilibria and involves citrate gel formation, which uses a combustion step to remove the solvent. However, the sol-gel technique involves controlled hydrolysis to form a gel that is dried to obtain the final product in xerogel, aerogel, ambigel, or cryogel form depending on the process involved for drying and removing residue from the solvent.

14.2.3.4 Coprecipitation method

Precipitation is the formation of a solid in a solution during a chemical reaction. In the coprecipitation technique, required amounts of chemical precursors along with trace amounts of an activator are added to form the desired compound. When the chemical reaction occurs, the solid formed (called the phase) sinks to the bottom of the solution precipitate. This can occur when an insoluble substance, the precipitate, is formed in the solution due to a reaction or when a compound has supersaturated the solution. The solute phase sinks to the bottom of the solution (though it will float if it is less dense than the solvent or forms a suspension). Precipitation reactions can be used for making pigments, removing salts from water, and performing qualitative chemical analysis. The precipitated compound (after appropriate washing and drying of the product) is sintered at high temperature to facilitate diffusion of the activator and formation of an appropriate number of defects in the crystalline materials [86].

14.2.3.5 Solvothermal synthesis

The solvothermal synthesis method is takes place in a closed system in which high temperature and pressure are maintained to ignite a chemical reaction so that the decomposition of precursor occurs to derive the required product directly from the formed solution. This method can be used for both aqueous and nonaqueous solvents. Solvent materials used in these techniques are ethanol, CO_2, water, HCl, ammonia, methanol, hydrofluoric acid, etc. When the solvent is water, then this method can be termed hydrothermal. The solvothermal process involves mixing the precursor in a proper solvent and then sealing it in an autoclave, which is then heated at a temperature greater than the boiling point of the preferred solvent. The autoclave acts as a closed system in which the temperature and pressure of the solution are elevated; this subsequently helps to crystallize the dissolved solvent material. A high-purity and homogeneous product containing nanosized particles, which are distributed periodically, can be controlled by proper selection of the precursor materials in a stoichiometric ratio and maintaining a proper temperature and pressure at the time of the reaction conditions. Postsynthesis annealing is not required in this method. Even though the product is formed at relatively low temperatures, phosphors are developed with high crystallinity. Reaction temperature, choice of solvent, precursor material, and pH affect the size, morphology, and phases formed in phosphor [87]. The only disadvantage with this method is that the product cannot be developed in bulk amounts, i.e., mass production is not possible.

14.2.4 Characterization for phosphor

14.2.4.1 X-ray diffraction techniques

X-ray diffraction (XRD) helps to find the geometry or shape of a molecule using X-rays. The elastic scattering phenomenon of X-rays from the atoms of material has a long range order. In XRD, collimated X-ray (wavelength $\lambda \sim 0.5-2\text{Å}$) radiation is allowed to fall on a sample in powder form and is diffracted by the crystalline phases in the sample. Bragg's law ($\lambda = 2d \sin \theta$, where d is the spacing between atomic planes in the crystalline phase) helps to understand the crystalline phases present in the sample. The intensity of the diffracted X-rays is measured as a function of the diffraction angle 2θ and the specimen's orientation. The obtained pattern from XRD helps to identify the various crystalline phases and to study their internal structural properties thoroughly. An XRD pattern reveals information about the various phases present in the crystalline composition under study and also gives information about particle/crystal size, crystallinity of the sample, solid solution, stress, and texture [88]. XRD also helps to determine concentration profiles, film thicknesses, defects, and atomic arrangements in amorphous as well as crystalline materials. XRD is a nondestructive technique, which makes it very useful in many studies. Almost all the characteristics of a material can be studied with the help of XRD. Fig. 14.4 shows a basic functional block diagram of an X-ray diffractometer. The observed XRD pattern of the specimen sample is then compared with Joint Committee on Powder Diffraction Spectra data to confirm the formation of crystalline phases in the sample under study and also the purity of the sample.

14.2.4.2 Scanning electron microscope

A scanning electron microscope (SEM) is a nondestructive technique because the secondary electrons generated do not affect the interatomic structure of the material. It is used to

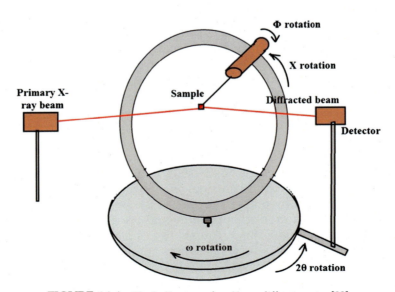

FIGURE 14.4 Block diagram of an X-ray diffractometer [88].

study the surface morphology of samples at very high magnification. Thus the sample under study can be used repeatedly for further characterization. A block diagram of an SEM is shown in Fig. 14.5.

Fundamental principles of SEM

In SEM, electrons are accelerated with the help of a high-voltage power supply and allowed to fall on the sample. These energetic electrons (energy ranges from a few to 50 keV) fall on the material and energy is dissipated due to the many signals produced during the electron—material interaction, when the incident electrons are deaccelerated in the solid sample. An electron beam is focused on the sample with the help of a condensing lens assembly, so that it targets a very acute area of the sample. This electron beam travels through the objective lens and is deflected by the pairs of coils, which are arranged linearly over a targeted area of the sample, as shown in Fig. 14.5. When this electron beam strikes the surface of the sample specimen, it is scattered by atoms of the sample in all directions. Through this electron—material interaction, interaction volume is created inside the sample of size about 5 μm. During this interaction, secondary electrons are also generated. which are sensed by a detector to form an image. Fig. 14.6 shows SEM images of $LiAlO_2$ phosphor synthesized via the solution combustion method [90]. From this SEM micrograph, it is seen

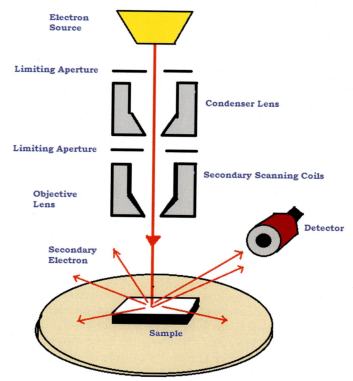

FIGURE 14.5 Schematic diagram of a scanning electron microscope [89].

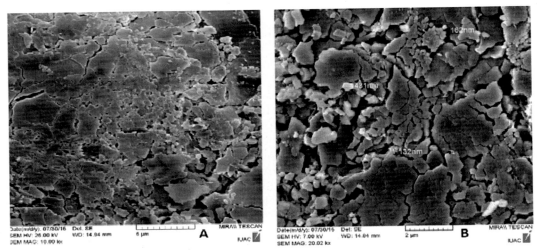

FIGURE 14.6 Scanning electron microscope images of LiAlO$_2$ phosphor synthesized via the solution combustion method. *Reproduced with permission from K.K. Gupta, R.M. Kadam, N.S. Dhoble, S.P. Locha, S.J. Dhoble, A comparative investigation of Ce^{3+}/Dy^{3+} and Eu^{2+} doped LiAlO$_2$ phosphors for high dose radiation dosimetry: explanation of defect recombination mechanism using PL, TL and EPR study, J. Lumin. 188 (2017) 81−95. Copyright 2017 Elsevier.*

that the particles have an irregular shape, and the multigrains agglomerate forming clusters with an average size in the order of about 1−5 μm. This SEM micrograph of the prepared sample was produced using a JEOL-JSM 6380A SEM instrument.

As mentioned earlier, SEM can produce a variety type of signals, such as secondary electrons, characteristic X-rays, specimen currents, backscattered electrons, cathode luminescence, and transmitted electrons. On the basis of energies, the secondary and backscattered electrons can be separated, which are used to produce images of samples to show morphology and topography and also reveal contrasts in the composition containing multiphase samples.

14.2.4.3 Transmission electron microscopy

In 1932, Knoll and Ruska developed the first transmission electron microscope. TEM is a microscopic technique where a beam of electrons is allowed to transmit through a very thin sample where electrons interact and pass through it. The image is formed through electron−material interactions of the sample. It is then magnified and detected by a sensor, such as a charge-coupled device (CCD). Finally, it is displayed on a fluorescent screen.

Construction and operation

Fig. 14.7 shows a block diagram of a transmission electron microscope with its different working components containing mainly a vacuum system, a series of electromagnetic lenses, and electrostatic plates. In TEM, a defocused beam of electrons is bombarded on a sample, which passes through the sample and then forms an image. The condenser lenses produce the illuminating electron beam. A lens system, which is arranged below the sample, contains an assembly of objective lens, intermediate lens, and projector lens, which work together to form the image. With this arrangement, two different modes are possible: the imaging or the diffraction mode. They are

14.2 Fundamentals

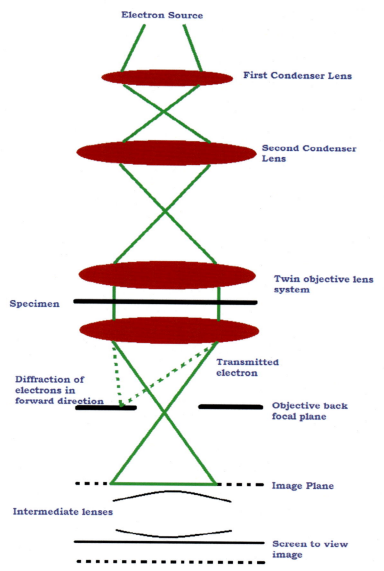

FIGURE 14.7 Functional block diagram of a transmission electron microscope [89].

selected by the excitation of the intermediate lens used to magnify the first intermediate image (imaging mode) or the diffraction pattern (diffraction mode) formed by the objective lens. The first intermediate image or the diffraction pattern is then further magnified by the projector lens. Finally, a CCD camera detects the final image and it is displayed on a screen. Contrast in the final TEM image occurs due to the scattering of the incident electrons by the sample specimen. Fig. 14.8 shows a TEM image of ZnO-TiO_2:Dy^{3+} powder phosphor [91].

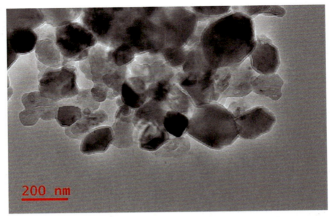

FIGURE 14.8 Transmission electron microscope image of ZnO-TiO$_2$:Dy^{3+} powder phosphor. *Reproduced with permission from S.J. Mofokeng, V. Kumar, R.E. Kroon, O.M. Ntwaeaborwa, Structure and optical properties of Dy^{3+} activated sol-gel ZnO-TiO$_2$ nanocomposites, J. Alloys Compd. 711 (2017) 121–131. Copyright 2017 Elsevier.*

14.2.4.4 Photoluminescence

The spectrofluorophotometer irradiates a sample with excitation light and measures the fluorescence emitted from the irradiated sample to perform a qualitative or quantitative analysis. A typical configuration of the spectrofluorophotometer is schematically described in Fig. 14.9A) taking the RF-5301 PC instrument [92] as an example. The basic components of the spectrofluorophotometer are as follows:

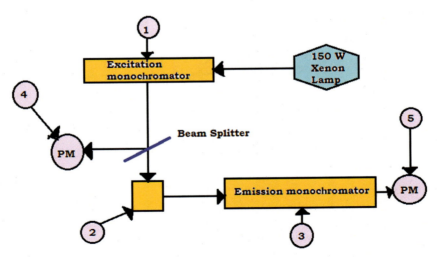

FIGURE 14.9A Configuration of an RF-5301 PC [89]. *PM*, Photomultiplier.

1. An excitation monochromator is used to isolate a required wavelength from the light coming from a xenon lamp to obtain excitation light. The excitation monochromator has a diffraction grating with a larger aperture to collect a large amount of light, which affects the sensitivity of the spectrofluorophotometer.
2. A cell holder is used to hold a cell, which is filled with the sample under study.
3. An emission monochromator is used to selectively collect fluorescence produced by the sample material. A photomultiplier tube connected to it measures the intensity of the fluorescence. Diffraction grating aperture size is the same as that of the excitation monochromator, which collects the greatest possible amount of light.
4. A monitor side photomultiplier tube is used for monitoring. High emission intensity and an uninterrupted radiation spectrum are the characteristics of xenon lamps used on spectrofluorophotometers. However, their tendency to emit unstable light will result in greater signal noise if no countermeasure is incorporated. Thus nonuniformity in the radiation spectrum of the xenon lamp and in the spectral sensitivity characteristics of the photomultiplier tube may cause distortion in the spectrum. Monitoring the function of the photomultiplier tube helps to overcome these factors; a portion of excitation light is monitored and feeds the resultant signal back to the photomultiplier tube.
5. Fluorescence side photomultiplier tube.

Optical system of spectrofluorophotometer

The optical system of the RF-5301 PC instrument is shown in Fig. 14.9B. (1) A uniquely designed xenon lamp (150 W) is used as the light source. It contains ozone, which decomposes by means of the heat produced by the lamp. The bright spot from the xenon lamp is magnified and converged by using an ellipsoidal mirror (2) and then further converged on the inlet slit toward the excitation site (3) by the concave mirror (4). A portion of the light isolated by the concave grating (5) passes through the outlet slit, travels through the condenser lens (11), and illuminates the sample cell. To achieve light-source compensation, a portion of the excitation light is reflected by the beam splitter quartz plate (6) and directed to the Teflon reflector plate 1 (7). The diffusely reflected light from the reflector plate 1 (7) then passes through the aperture for light quantity balancing (21) and illuminates the Teflon reflector plate 2 (8). Reflected by the reflector plate 2 (8), the diffuse light is attenuated to a specific ratio by the optical attenuator (9) and then reaches the photomultiplier for monitoring (10). The fluorescence occurring on the cell is directed through the lens (13) to the emission monochromator that comprises the slit assembly (14) and the concave grating (15). Then, the isolated light is introduced through the concave mirror (16) into the photomultiplier for photometry (17) and the resultant electrical signal is fed to the preamplifier.

The procedure for measurement of the excitation and emission spectra to record the photoluminescence spectra, sample under investigation, it can be either in powder form i.e., solid state or the powder dissolvent in appropriate solvent i.e., solvated state. The sample cell consists of a round sample holder, quartz disc (window), and a threaded cap. The quartz disc is fixed into the sample holder and the powder sample is spread on it. Then, the threaded cap is fitted to hold the powder sample. The metal frame is put on the sample cell containing the sample so that the front protrusion of the cell fits into the metal frame aperture. While analyzing the sample, optical axis is aligned along the centerline of the powder surface (quartz window). First, the excitation spectra are recorded by setting the emission

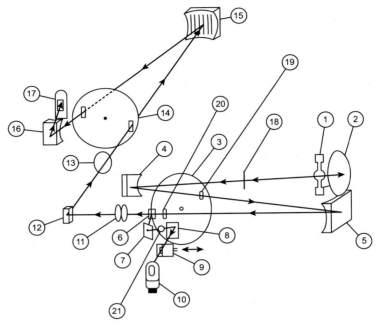

FIGURE 14.9B Typical optical system configuration of spectrofluorophotometer [92].

wavelength at zero order and keeping other parameters as specified in the manual. The excitation bands are identified from these spectra and the emission spectra are scanned for identified excitation wavelengths. It is necessary to know the approximate nature of excitation and emission spectra. While doing so the direct scattered light may superimpose on the excitation spectrum, so it is necessary to select a particular band in the emission for scanning the excitation. Therefore for proper excitation wavelengths, emission is set at the position as identified from the earlier emission spectrum. Again, the same procedure is followed for identifying correct excitation positions and emission is recorded for each excitation band separately.

In ordinary measurements, a spectrum is affected by wavelength characteristics of the analysis system (monochromator, photomultiplier, etc.). To measure a spectrum, correction is performed using Rhodamine B as a standard. Similarly, emission spectrum is corrected by using the diffuser and attenuator mentioned in the instrument manual. Both spectra are correctable in the range of 220–600 nm. The sample whose emission wavelengths are within 220–600 nm is scanned in the correct spectrum mode and the samples whose wavelengths are beyond 600 nm are scanned in ordinary mode. Emission spectra are recorded with excitation band pass 5 nm and emission band pass 1.5 nm, while the excitation spectra are recorded with excitation band pass 1.5 nm and emission band pass 5 nm.

14.2.5 Importance of rare-earth elements

Rare-earth ions (lanthanides) are very important in much of modern optical technology as an active constituent of materials. They are widely used in many applications. Many of today's cutting-edge optical technologies and future innovations rely on their unique properties.

Rare-earth ions are used as luminescent center/activator ions in numerous host lattices. The luminescent materials are used for various purposes such as making lamps [93—95] and in cathode ray tubes and TV screens [96], laser crystals [97], dosimetry of ionizing radiation paints [98], and chemical and biochemical analysis and medical diagnosis.

In many of these applications, inorganic solids doped with rare-earth impurities are used. To understand how the rare-earth impurities make various applications possible, it is necessary to know the luminescence characteristics of these materials. Basically, there are four important parameters, namely excitation type and spectrum, relaxation to emitting state and decay time, emission intensity, and emission spectrum, which determine the utility of rare-earth-doped phosphors. All these parameters may further depend on concentration and temperature. This dependence is equally important in the context of phosphors in various applications.

For the development of highly efficient nanoparticles for various applications, rare-earth ions are used as luminescent centers/activator ions in numerous host lattices. Europium (Eu atomic number 63) was discovered in the spectral line. The discovery of europium is generally credited to Demarcay, who separated the rare earth in reasonably pure form in 1901. Europium is prepared by mixing Eu_2O_3 with a 10% excess of lanthanum metal. Europium ignites in air at about 150—180°C, and is about as hard as lead and is quite ductile. Europium is a reactive rare-earth metal and is quickly oxidized in air. The electronic configuration is [Xe] $4f^7 6s^2$. Dysprosium (atomic number 66) was discovered in 1886 by Lecoq de Boisbaudran. Dysprosium occurs along with other rare-earth elements in a variety of minerals such as xenotime, fergusonite, gadolinite, euxenite, polycrase, and blomstrandime. Monzaite and bastanasite are the most important sources. Dysprosium is relatively stable in air temperatures and dissolves in mineral acids. It is very reactive and has been used for laser materials.

To understand the luminescent properties of rare-earth ions, it is necessary to know their key energy levels. Energy-level diagrams of rare-earth ions doped in various solids have been well studied [99—107]. The energy levels may be divided into three categories: those corresponding to $4f_n$ configuration and $4f_{n-1}5d$ configuration, and those corresponding to charge transfer involving neighboring ions.

In spite of their resemblance to energy levels of free rare-earth ions and rare-earth ions in solids, there is an important difference in emission properties. In solids, the emission of rare-earth ions is observed at different spectral positions compared to absorption. This difference is described by the Stokes shift. The shift for transitions within a 4f shell results from the absorption and emission that takes place between different levels. Usually, absorption corresponds to the transition from the ground state to higher excited states. Electrons in the higher excited state then lose energy to lattice until the state lying just below the previous excited states is available. When the difference between the excited states is large, the energy corresponding to this transition cannot be transferred to the lattice and is dissipated in the form of an emission. The emission then corresponds to the transition from the intermediate state to the ground state.

Rare-earth ions are usually trivalent. Ions corresponding to configuration $4f^0$ (La^{3+}), $4f^7$ (Gd^{3+}), and $4f^{14}$ (Lu^{3+}) are stable. Rare-earth elements next to these three tend to exchange electrons and acquire this stable configuration. Thus adjacent to La, Ce can exist in tetravalent form, adjacent to Gd, Eu^{2+} and Tb^{4+} are stable, and adjacent to Lu, Yb^{2+} is stable. Besides these, the elements that are next to adjacent exist in an ionic state, which is near to the stable, completely filled shell or filled shell configuration. Thus we have tetravalent Pr and divalent Sm as stable ions and divalent Tm has been found in some solids. Summarizing, besides the usual trivalent state for all rare-earth ions, tetravalent Ce, Tb, and Pr and divalent Eu, Sm, and Yb are not uncommon.

The property of rare-earth ions that sets them apart from other transitional elements is that their 4f electrons remain highly localized to the ion and their optical transitions maintain much of an atomic-like character even when the ion is an element of a crystalline solid. This atomic-like behavior of rare-earth ions' 4f electrons is in sharp contrast to the transition metals' d electrons, whose behavior is strongly affected by the presence of the host lattice and may show significant delocalization and mixing with the electronic states of other ions in the lattice.

The actinides' 5f electrons provide an intermediate case whose properties may vary between these extremes depending on the nature of the material and ion. This characteristic of rare-earth ions arises from the unique situation in which the lowest-energy electrons are not spatially the outermost electrons of the ion, and thus have a limited direct interaction with the ion's environment. The "shielding" of the 4f electrons from the environment by the outer filled shells of 5p and 5s electrons prevents the 4f electrons from directly participating in bonding and allows them to maintain much of the character exhibited by a free ion. This nonbonding characteristic of 4f electrons is responsible for the well-known chemical similarity of the different rare-earth ions. Since transitions between the electronic states of the "shielded" 4f electrons give rise to the ion's optical transitions, materials that contain rare-earth ions exhibit unique optical properties [108].

Rare-earth ions have the following luminescent characteristics. Basically, there are four important parameters, namely excitation type and spectrum, relaxation to emitting state and decay time, emission intensity, and emission spectrum, which determine the utility of rare-earth doped phosphors. All these parameters may further depend on concentration and temperature.

Apart from the applied aspects, rare-earth emission is of interest because of academic reasons as well. The first observation of rare-earth spectra was by Becquerel [109], who became interested in the extreme sharpness of the lines. These are otherwise generally observable only in the spectra of free atoms and molecules. It was hoped that the rare-earth spectra would reveal as much about solids as the free ion spectra have about the atom. These aspects are discussed in detail by Huffner [105].

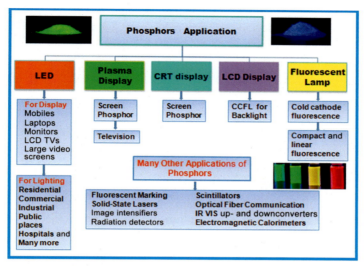

FIGURE 14.10 The range of phosphors in various applications. *CCFL*, Cold-cathode fluorescent lamp; *CRT*, cathode ray tube; *IR Vis*, infrared visual; *LCD*, liquid crystal display; *LED*, light-emitting diode.

14.3 Applications of various phosphor materials

Technologically, phosphor materials are of great use as they are utilized in a wide range of applications. Fig. 14.10 shows the application range of phosphors in different fields. From the varied research into phosphors, a variety of materials have been developed for different applications such as display devices [110,111], optical communication amplifiers [112–114], fluorescent lamps [115], cathode ray tubes [116], and lasers [117]. Phosphors have also been used in applications such as X-ray detector systems [118] and scintillators [119]. Phosphor research has been extended into areas such as solar cells [120–122] and WLEDs [123,124] over the last three decades. Nowadays, phosphor has shown rapid growth in the field of illumination as solid-state lighting, and in the coming decades it will replace all existing sources of illumination.

14.4 Phosphor for LED lighting

14.4.1 Basic requirements for phosphor in solid-state lighting

Phosphors can be used for solid-state applications, especially for the production of white light. A few of the important properties of phosphor are as follows:

1. The absorption capacity of phosphor in the emission region of LED must be high. Also, its excitation maxima must lie in the pumping region of LEDs.

2. Developed phosphors should possess an energy transfer mechanism, so that its emission may produce white light radiation having promising CRI and CCT.
3. LEDs made from phosphor should have strong electrical to optical conversion, leading to high quantum efficiency.
4. Phosphors should be chemically and thermally stable.
5. Phosphors should be free from photobleaching.
6. To reduce thermal quenching, phosphors should be capable of operating at elevated temperatures.
7. The manufacturing process of phosphors must be cost effective, comparatively simple, and capable of mass production.
8. Handling, use, and disposal of the phosphor material should be safe. Thus phosphor materials should be nontoxic and environmentally friendly.

In the field of general illumination, new methods should be generated to enhance the efficacy of white light, e.g., WLEDs. Solid-state lighting applications include the development of simple LEDs, organic LEDs, or light-emitting polymers. The evolution of solid-state lighting has an illustrious history. In 1907, H.R. Round [125] observed a green light/glow from an SiC diode junction when a current flowed through it. This was termed electroluminescence. A similar observation was made by Losev [126]. These inventions remained unnoticed until the invention of the first practical LED in 1962 by Nick Holonyak [127]. Thereafter, LEDs were popularized in different applications such as signage, and became available in different colors such as green, amber, and red. Later in 1994, Nakamura [128] made a significant development, i.e., the fabrication of a blue LED, which was made by using double-heterostructural InGaN/GaN blue LED chips, and was the first LED to generate white light from semiconductor material. It was commercialized by the Nichia Corporation in 1996 [129]. Solid-state lighting devices have very important characteristics such as high light efficiency, low energy consumption, and long service lifetime compared to other conventional light sources. Thus solid-state lighting has become an extremely strong field and provided a number of possible alternatives to replace existing lighting technologies. Unlike other traditional light sources, LEDs are not a typical source of white light, but they do generate single-color (monochromatic) radiation. Thus they are of great interest for applications that involve colored lights such as traffic lights, display boards, and exit signs. Basically, semiconductor materials are the basis of LEDs, which are from deliberately doped impure elements to form a p-n junction. When a suitable voltage is provided between the end points of a diode, i.e., across the junction, it emits a colored light. The bandgap of a diode material affects or decides the color of emitted radiation from an LED. But to make it a source of general white light, a device is required to produce white light from monochromatic sources.

To obtain white light from a semiconducting material, it should be doped to produce luminescence in the host matrix. Many researchers are working on LED materials to develop WLEDs with single-phase host materials, which operate under UV or near-UV excitation.

As Dy^{3+} has a characteristic emission in the blue and yellow regions, it is generally preferred for producing white light emission. To study the effect of different host lattices, A.K. Bedyal adopted the combustion synthesis method to prepare [130] orthophosphate phosphors by doping Dy^{3+}. Emission spectra of this phosphor show characteristic 4f-4f transition of the Dy^{3+} ions. It is reported that with respect to host material, the luminescence intensity of Dy^{3+} ions changes, which is evident from photoluminescence results. Changing of host also causes changes in luminescence intensity between the blue-to-yellow emission band and stark splitting. This result shows that Dy^{3+} ions were present in a different symmetrical environment in different host lattices. The symmetry of the orthophosphate lattices containing potassium is lower than the symmetry of the lattice orthophosphate lattices containing sodium. The orthophosphate ($K_3Y(PO_4)_2:Dy^{3+}$, $K_3La(PO_4)2:Dy^{3+}$, $Na_3Y(PO_4)_2:Dy^{3+}$) intensities were found to be greater by 75% than that of the original intensities at 423 K. The CIE having coordinates very close to the white light coordinates, and the other optic properties of the phosphors, proves that these presented phosphors would be good candidates for single-phosphor-based WLEDs.

The transition $^4F_{9/2} \rightarrow {}^6H_{15/2}$ corresponds to the blue region (470–500 nm) and $^4F_{9/2} \rightarrow {}^6H_{13/2}$ corresponds to the yellow region (570–590 nm) of Dy^{3+} emissions. The yellow emission in Dy^{3+} is hypersensitive. Tuning of the ratio of yellow emission to blue emission is essential to get a white light with the desired color temperature.

When Dy^{3+} ions are doped in a host matrix, they may occupy a site so that inversion symmetry is produced within the crystal; the blue color emission is then emitted dominantly. However, if it occupies a site so that noninversion symmetry forms within the matrix, then yellow color emission is emitted dominantly.

Nair and Dhoble synthesized single-phase white light-emitting $Ca_3Mg_3(PO_4)_4:Ce^{3+}$, Dy^{3+}, Li^+ phosphors by the urea-assisted combustion method [131]. They reported that energy transfer successfully occurred from Ce^{3+} to Dy^{3+} in the $Ca_3Mg_3(PO_4)_4$ host lattice, which is observed in the spectral overlap of the excitation spectrum of Dy^{3+} ions and the emission spectrum of Ce^{3+} ions as shown in Fig. 14.11, which occurred due to electric dipole–dipole

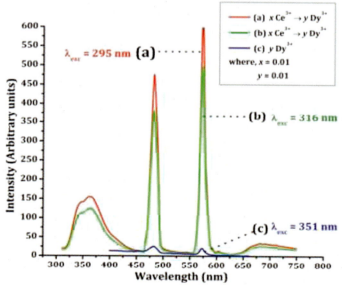

FIGURE 14.11 Comparison of the photoluminescence emission spectra of (A) $Ca_3Mg_3(PO_4)_4:Ce^{3+}$, Dy^{3+}, Li^+ phosphors at 295 nm excitation wavelength, (B) $Ca_3Mg_3(PO_4)_4:Ce^{3+}$, Dy^{3+}, Li^+ phosphors at 316 nm excitation wavelength, and (C) $Ca_3Mg_3(PO_4)_4:Dy^{3+}$, Li^+ phosphors at 351 nm excitation wavelength. *Reproduced with permission from G.B. Nair, S.J. Dhoble, White light emission through efficient energy transfer from Ce^{3+} to Dy^{3+} ions in $Ca_3Mg_3(PO_4)_4$ matrix aided by Li^+ charge compensator, J. Lumin. 192 (2017) 1157–1166. Copyright © 2017 Elsevier.*

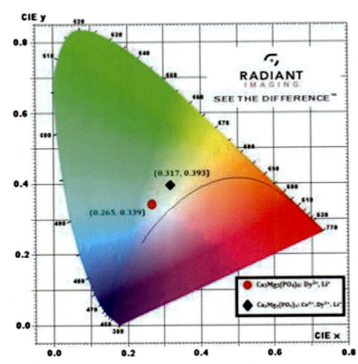

FIGURE 14.12 CIE chromaticity coordinates of $Ca_3Mg_3(PO_4)_4$:Dy^{3+}, Li^+ and $Ca_3Mg_3(PO_4)_4$:Ce^{3+}, Dy^{3+}, Li^+ phosphors. *Reproduced with permission from G.B. Nair, S.J. Dhoble, White light emission through efficient energy transfer from Ce^{3+} to Dy^{3+} ions in $Ca_3Mg_3(PO_4)_4$ matrix aided by Li^+ charge compensator, J. Lumin. 192 (2017) 1157–1166. Copyright © 2017 Elsevier.*

interaction. At 295 and 316 nm excitation, the Dy^{3+} emission peaks at 482 and 572 nm were enhanced significantly and a 10-fold increase was recorded in their intensity (Fig. 14.12).

Nair and Dhoble also reported the CIE coordinates of $Ca_{2.96}Mg_3(PO_4)_4$:$0.01Ce^{3+}$, $0.01Dy^{3+}$, $0.02Li^+$ phosphor (as shown in Fig. 14.11), which were located at (0.317,0.393) in the white region with a correlated color temperature of 4739 K, which shows that these phosphors have good potential to be used as a UV-convertible white light-emitting phosphor for LEDs.

So many Dy^{3+}-doped phosphors were reported that can be used as a dopant in the host matrix to obtain white light. They are as follows: $NaLa(PO_3)_4$:Dy^{3+} [132], $Ca_8MgBi(PO_4)_7$:Dy^{3+} [133], $CaZr_4(PO_4)_6$:Dy^{3+} [134,135], $Ca_3B_2O_6$:Dy^{3+} [136], $Sr_3Gd(PO_4)_3$:Dy^{3+} [137], etc. However, transitions from Dy^{3+} ions are generally due to the 4f-4f transitions. These transitions are the forbidden transitions by selection rules. Generally, luminescence produced by these Dy^{3+}-doped phosphors shows less efficiency. To enhance this efficacy, the host matrix must be sensitized with sensitizer so that energy transfer to the Dy^{3+} ions occurs [131].

It is also possible to produce white light by doping single europium in Eu^{2+} or Eu^{3+} ion form in a host. Red emission is the characteristic of Eu^{3+} ions, but in some rare cases they may give narrowband emission peaks, which extend from the blue to the red region of the visible

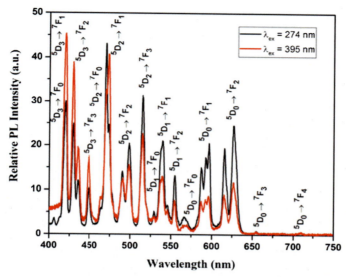

FIGURE 14.13 Emission spectra of $Ba_5Zn_4Y_{7.92}Eu_{0.08}O_{21}$ nanophosphor, excited at 274 and 395 nm, respectively. *PL, Photoluminescence. Reproduced with permission from M. Dalal, V.B. Taxak, J. Dalal, A. Khatkar, S. Chahar, R. Devi, S.P. Khatkar, Crystal structure and Judd-Ofelt properties of a novel color tunable blue-white-red $Ba_5Zn_4Y_8O_{21}:Eu^{3+}$ nanophosphor for near-ultraviolet based WLEDs, J. Alloys Compd. 698 (2017) 662–672. Copyright © 2016 Elsevier.*

light spectrum and hence are useful for white light production [138,139]. The emission spectra of $Ba_5Zn_4Y_{7.92}O_{21}:0.08Eu^{3+}$ nanophosphor at 274 and 395 nm excitation are shown in Fig. 14.13 [138].

Multiphonon relaxation and cross-relaxation occurring between energy levels must be lowered to achieve special tuning of emission bands of Eu^{3+} ions, which may be possible by choosing host materials that have lower vibrational energies (phonon frequencies) and as far as possible lower concentration of Eu^{3+}.

Europium-doped $Eu^{3+}:BaNb_2O_6$ phosphors were prepared by the solid-state reaction method [140]. Formation of the pure orthorhombic phase for $BaNb_2O_6$ at 1200°C by calcination has been confirmed by XRD. The photoluminescence properties of these phosphors under near-UV/blue excitations (at 394 and 466 nm) reveals the presence of intense red emission at 611 nm, which is due to $^5D_0 \rightarrow {}^7F_2$ transitions of Eu^{3+} as shown in Fig. 14.14.

The emission spectrum helps CIE coordinates that are falling in the pure red region of the CIE 1931 diagram as shown in Fig. 14.15. Interestingly, these coordinates are found to be very near to the commercial red phosphor ($Y_2O_2S:Eu^{3+}$). Reported results show that $Eu^{3+}:BaNb_2O_6$ phosphors are found to be good materials to use in WLEDs.

Yu et al. [141] synthesized Eu-activated CaF_2 single crystals by the Bridgman–Stockbarge method and the dependence of photoluminescence properties of $Eu:CaF_2$ crystals on the concentration of EuF_3 was studied in the UV-Vis region. On excitation with 398 nm, $Eu:CaF_2$ single crystals show both characteristic f–f transitions of Eu^{3+} ions and 5d $\rightarrow$ 4f transition emission of Eu^{2+} ions in the emission spectrum. It has been reported that the emission spectrum of $Eu:CaF_2$ single crystal is affected by the atomic concentration ratios of Eu^{2+}/Eu^{3+},

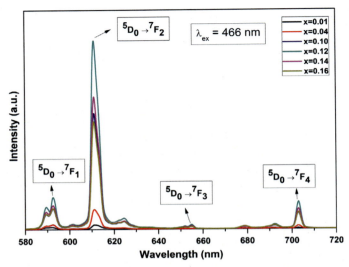

FIGURE 14.14 Room temperature photoluminescence emission spectra ($\lambda_{ex} = 466$ nm) of BaNb$_2$O$_6$:Eu^{3+} at different concentrations. *Reproduced with permission from A.K. Vishwakarma, K. Jha, M. Jayasimhadri, A.S. Rao, K. Jang, B. Sivaiaha, D. Haranath, Red light emitting BaNb$_2$O$_6$:Eu^{3+} phosphor for solid state lighting applications, J. Alloys Compd. 622 (2015) 97–101. Copyright © 2014 Elsevier.*

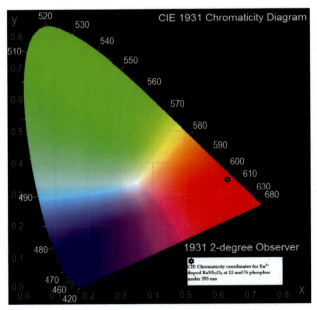

FIGURE 14.15 CIE chromaticity diagram for 12 mol% Eu^{3+}-doped BaNb$_2$O$_6$ phosphor. *Reproduced with permission from A.K. Vishwakarma, K. Jha, M. Jayasimhadri, A.S. Rao, K. Jang, B. Sivaiaha, D. Haranath, Red light emitting BaNb$_2$O$_6$:Eu^{3+} phosphor for solid state lighting applications, J. Alloys Compd. 622 (2015) 97–101. Copyright © 2014 Elsevier.*

local lattice environment of Eu ions, and energy transfer from Eu^{2+} ions to Eu^{3+} ions. The corresponding luminescence color can be tuned from blue, pink, orange/red to orange. In addition, 0.6% $Eu:CaF_2$ single crystal gives the emission of warm white light at an excitation wavelength around 322 nm, which makes it a potential candidate for applications in the areas of UV WLEDs.

Multicolor emission has been reported in $YGa_{1.5}Al_{1.5}(BO_3)_4(YGAB):Tb^{3+}, Eu^{3+}$ [142]. It has been reported that the energy transfer mechanism is greatly affected by distribution of the isolated sites in the host matrix, which also influences the luminescence properties. Paired ions are distributed at a larger distance and hence concentration quenching is found to be less, which affects the interaction between the luminescence centers. In Tb^{3+}/Eu^{3+} codoped systems, the color can be tuned by transferring energy from Tb^{3+} to Eu^{3+} with the help of a single excitation wavelength.

Wei et al. [143] reported the luminescence properties of $Li_3BaSrLn_3(WO_4)_8:Eu^{3+}$ phosphors emitting red light and having good thermal stability. The CIE diagram shows the color coordinate values (0.670,0.330). The detailed study of $Li_3BaSrLn_3(WO_4)_8:Eu^{3+}$ at 90% phosphor revealed no concentration quenching. Thus cross-relaxation of the ions and the nonradiative energy transfer process make it potential material for near-UV-excited WLEDs.

$Eu^{2+}/Tb^{3+}/Eu^{3+}$ codoped glass G-5Tb4Eu shows good thermal stability and also at 423 K possesses good emission intensity, which is about 86% of its initial intensity at 298 K. These excellent properties make these glasses a potential candidate for designing and fabricating WLED devices [144].

$5d \rightarrow 4f$ transition of Eu^{2+} ions causes a broadband emission. As the outermost electrons of Eu^{2+} ions are present in the 5d state, the coordination environment around Eu^{2+} ions is greatly dependent on the type of host. In some hosts, Eu^{2+} ions are capable of covering broad emission band, for example $Ba_{0.97}Sr_{0.99}Mg(PO_4)_2:0.04Eu^{2+}$ phosphor has given full color emission band on excitation with 350 nm, at 447 and 536 nm [145].

Appropriate selection of codopant makes it possible to generate white light in almost every type of host material. It has been reported that Mn^{2+} can be used as codopant to support Eu^{2+} ions to produce white light [146–150]. Mn^{2+} ions generally give red emission, which makes them ideal to pair with the blue emission from Eu^{2+} ions. It is well known that Mn^{2+} ions do not show strong red luminescence when they are singly doped. The $^4T_1-^6A_1$ transition of Mn^{2+} is forbidden and inefficient to produce luminescence unless it is sensitized with codopants like Eu^{2+} or Ce^{3+} ions.

Eu^{2+} and Eu^{3+} ions simultaneously existing in same host lattice can be used for color emission tuning and are able to produce white light with desired chromaticity coordinates [151]. Similar types of host materials are: $LiMgPO_4$ [152], $KCaBO_3$ [153], $\beta-Ca_2SiO_4$ [154], $Sr_5(PO_4)_3Cl$ [155], and $Ca_3Y_2Si_3O_{12}$ [156].

The pair of $Ce^{3+}-Eu^{2+}$ ions is an ideal combination to give the luminescence effect, which helps to tune the color from yellow to blue to give white light in many host matrices that are affected by the distribution of ions in crystal symmetry [157–160]. The compositional map given for phosphor $Li_2(Sr_{1-x-y}Eu_xCe_y)SiO_4$ with Ce^{3+} and Eu^{2+} doping at various concentrations in Fig. 14.16 gives a better understanding of the color-tuning concept in phosphors [161].

There are several reported results for color tuning to obtain Ce^{3+} along with Eu^{2+}, Tb^{3+}, Eu^{3+}, etc. [162]. However, it is observed and reported that in most cases, there is no overlap

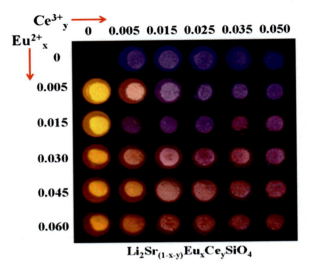

FIGURE 14.16 Composition map and luminescence photograph of the Li$_2$(Sr$_{1-x-y}$Eu$_x$Ce$_y$)SiO$_4$ phosphor with varying Ce^{3+} and Eu^{2+} concentrations. *Reproduced with permission from L. Chen, A. Luo, Y. Zhang, F. Liu, Y.Jiang, Q. Xu, X. Chen, Q. Hu, S.F. Chen, K.J. Chen, H.C. Kuo, Optimization of the single-phased white phosphor of Li$_2$SrSiO$_4$: Eu^{2+}, Ce^{3+} for light-emitting diodes by using the combinatorial approach assisted with the Taguchi method, ACS Comb. Sci. 14 (2012) 636–644. Copyright © 2012, American Chemical Society.*

in between their excitation spectra and the emission spectra of InGaN or GaN chips. Thus for pc-LED application, the use of Ce-doped phosphors is not included [163,164]. However, in certain cases like Y$_2$SiO$_5$:Ce^{3+}, Tb^{3+}, Eu^{3+} [165] the required results for LEDs are given. The combination of Ce^{3+}, Tb^{3+}, and Eu^{3+} ions is used together to tune the color emission from blue → white → orange/red, so that it can be very useful when used in silicate, borate, and nitride host matrices [165,166]. The excitation energy is generally utilized by Ce^{3+} ions; then, part of this energy is transferred to both Tb^{3+} and Eu^{3+} ions through nonradiative transition. A proposed energy-level model for energy transfer from Ce^{3+} → Tb^{3+} → Eu^{3+} in a GdBO$_3$ host can be understood from Fig. 14.17 [166].

Solid-state reaction under a weak reductive atmosphere has been used to synthesize Ca$_8$MgLa(PO$_4$)$_7$:Ce^{3+}, Mn^{2+} [167]. XRD and photoluminescence were used for the reported phosphor. The study revealed that the red emission at 640 nm originated from the $^4T_1(^4G)$ → $^6A_1(^6S)$ transition of Mn^{2+}. This red emission increased by a factor of 6.4 with maximum concentration of Ce^{3+} as codoped element. The electric dipole–dipole mechanism is responsible for the energy transfer from Ce^{3+} to Mn^{2+}. As this phosphors are reported to show, broadband UV absorption of Ce^{3+} and has good color coordinates, hence can be used as a promising candidate for red emission in UV-based WLEDs.

Jiang et al. reported a series of synthesized Y$_3$Al$_5$O$_{12}$:Mn^{4+} phosphors with different doping concentrations. Again, Zn^{2+} is added to optimize the value to these series. Tellurite glass is used as a suitable host to disperse YAG:Mn^{4+}/Zn^{2+} phosphors [168].

LED fabrication was done with a combination of Mn^{4+}/Zn^{2+}:YAG inside the matrix of tellurite glass.

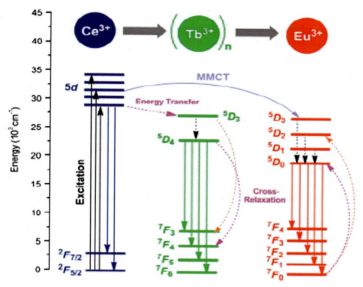

FIGURE 14.17 Proposed energy-level model for energy transfer from $Ce^{3+} \rightarrow Tb^{3+} \rightarrow Eu^{3+}$ in a $GdBO_3$ host. *MMCT, Metal-to-metal charge transfer. Reproduced with permission from J. Zhou, Z. Xia, Luminescence color tuning of Ce^{3+}, Tb^{3+} and Eu^{3+} codoped and tri-doped $BaY_2Si_3O_{10}$ phosphors via energy transfer, J. Mater. Chem. C 3 (2015) 7552–7560. Copyright © 2015 Royal Society of Chemistry publications.*

Fluoride and oxide phosphors doped with tetravalent manganese ions (Mn^{4+}) possess good optical characteristics: low photon absorption, narrow emission, high color purity, etc.

Fluoride-based phosphors with Mn^{4+} were studied for possible applications as promising phosphors for LED applications [169–176] such as $A_2MF_6:Mn^{4+}$ (where A is an alkali metal ion such as Na^+, K^+, Rb^+, Cs^+ and M = Si, Ge, Sn, Zr, Ti); $BMF_6:Mn^{4+}$ (where B is an alkaline-earth metal ion such as Mg^{2+}, Ca^{2+}, Ba^{2+}, Zn^{2+} and M = Si, Ge, Ti); $A_3MF_6:Mn^{4+}$ (where A = Na^+, K^+ and M = Al, Ga); $A_2XF_6:Mn^{4+}$ (where A is an alkali metal ion such as K^+, Rb^+, Cs^+, NH_4^+ and X = Si, Ge, Zr, Al, Ti); and $BSiF_6:Mn^{4+}$ (where B is divalent Ba^{2+} or Zn^{2+}).

14.5 Prospects for energy-saving technology

14.5.1 Need for energy-saving technology

Today, the whole world is facing an energy crisis. Thus the demand for electricity is growing rapidly due to the high requirement of industries in developed and developing countries as well as population growth in developing countries. Due to this excessive energy consumption, greenhouse gas emissions are rising quickly and becoming one of the great problems for everyone. Major devices are the lighting devices that consume energy and are utilized for indoor and outdoor applications. As per the global database, lighting utilizes around 650 Mt of energy and in turn emits 1900 Mt of CO_2. In comparison with the amount of energy (650 Mt) generated in a power plant, the amount of CO_2 generated by lighting energy is nearly 70% higher than that generated in a power plant and three times higher compared to emissions from aircraft [177]. Thus lighting utilizes around 15% of total world electricity consumption and causes 5% of global greenhouse gas emissions. Detailed analysis shows that the commercial sector utilizes 43%, the residential sector 31%, the industrial area 18%, and other sectors around 8% of global lighting [177]. Thus there is a need to develop energy-efficient and -saving technologies to overcome these problems and this task is being undertaken worldwide. European countries are adopting new policies related to energy saving and developing devices as well as technologies that will enhance the efficiency of lighting devices for the future. Thus energy-saving LEDs are promising devices for the replacement of conventional lighting components used in many areas such as industry, power plants, public places, commercial buildings, residential areas, etc. These energy-saving solid-state devices, i.e., LEDs, need to be studied thoroughly in different ways to make them ideal candidates for energy-saving devices. Doing so will greatly improve energy efficiency, mitigating impacts of global warming and reducing energy dependence.

As a source of light, the evolution of LEDs has shown remarkable progress in the last two - decades, which includes improvement in several areas such as efficacy, which is measured in output lumens/input watts. This efficacy enhancement is possible by advances in LED chip technology. LED products are more advantageous because they consume less power and have longer lifecycles (up to 10 years) compared to incandescent bulbs.

Because LEDs are basically semiconductor devices, it is possible to integrate additional electronics into bulbs, which makes them multifunctional, e.g., they can be used as occupancy and daylight sensors connected to a network interface. Due to this multifunctional integration, certain actions are possible such as automatic dimming of lights if natural light is already present in an area; these devices can also sense human activity and accordingly adjust lighting intensity. Using such multifunctioning technology helps to achieve sufficient high intensity light output, and saves energy as well as money.

In most cities, half of electricity generation is utilized in street light illumination, public places, hospitals, commercial areas, and for industry, especially during night periods. For such places, advanced LED lighting technology is useful because LEDs also help to lower electricity consumption and hence cost. LEDs used in parking areas, garages, staircases, and street walkways can be designed in such a manner that they can be automatically dimmed or turned on when sensors detect movement in an unoccupied area. This type of automation is an especially attractive and cost-saving option for many metro cities. LED bulbs are chemically and thermally stable, thus they are less prone to cracks and fractures and less susceptible to damage from external temperatures. Hence, the robustness and durability of LEDs is greater than conventional bulbs. Moreover, they are easy to replace, as LEDs lose power incrementally. LEDs are mercury-free devices and hence they are eco-friendly.

14.6 Conclusion and future outlook

Advances in solid-state white lighting technologies have witnessed the explosive development of phosphor materials (downconversion luminescent materials). A large amount of evidence has demonstrated the revolutionary role of emerging nitride phosphors in producing superior WLEDs for lighting and display applications. Structural and compositional versatility, together with unique local coordination environments, enables nitride materials to have remarkable luminescent properties such as abundant emission colors, controllable photoluminescence spectra, high conversion efficiency, and small thermal quenching/degradation. Here, we summarized the state-of-the-art progress of this novel family of luminescent materials and discussed the topics of materials discovery, crystal chemistry, structure-related luminescence, temperature-dependent luminescence, and spectral tailoring. Various nitride phosphors and their possible applications in solid-state lighting for general illumination, backlighting, and laser-driven lighting were also reviewed. Also, the challenges and outlook for these types of promising downconversion materials were highlighted.

The phenomenon of luminescence and its fundamentals, the importance of rare-earth elements in luminescence, phosphors and their background, current various methods of synthesis, essential characterization for luminescence study, different types of phosphors for LEDs in various host materials that are available for producing white light with several properties such as CRI, efficacy, and CCT, and the prospects of solid-state lighting were also summarized in this chapter. We discussed the synthesis and characterization of various rare-earth-doped phosphors such as dysprosium Dy^{3+}, europium (Eu^{2+} and Eu^{3+}), manganese Mn^{4+}, cerium Ce^{3+}, terbium Tb^{3+}, and others, which are utilized in the lighting industry for the production of white light. The emission color of rare-earth ions greatly affects by the presence of surrounding environments, which includes coordination with other atoms, symmetry, covalence, strength of crystal field, bond length between successive atoms, size of the field, etc. in which ions are embedded. This environment makes it possible to tune the emission color by varying the concentration of doping ions. An attempt was made to discuss various phosphors emitting different colors, and their applications were reviewed. The properties of phosphors play an important role in determining their various characteristics such as the efficacy of luminosity, CRI, CCT, lifespan, etc., which are greatly affected by certain synthetic conditions and precision at the time of LED device fabrication. Also, the functioning of LEDs also depends on the method of synthesis, the make and purity of the precursor used for the synthesis of phosphors, and climatic conditions at the time of synthesis.

LEDs are considered a growing eco-friendly source of light because they have so many advantages over conventional incandescent and fluorescent light sources. LEDs are capable of operating at low voltage, help to control the level of emissions (CO_2) in the environment, and are nontoxic, which makes them a safe alternative for the replacement of existing commercial light source technologies. Hence, they are essential components to handle economic as well as environmental sustainability. For the development of such a useful and eco-friendly technology, phosphors are of great interest, which are needed to be synthesized with unique luminescent properties such as broad excitation spectrum or tunable emission wavelength. The future will find LEDs in a wide range of applications such as solid-state lighting, optoelectronic devices, display devices, plant growth, tissue engineering, health and medical, biological imaging, security, optical storage, etc.

References

[1] A. Piquette, W. Bergbauer, B. Galler, K.C. Mishra, On choosing phosphors for near-UV and blue LEDsfor white ligh, ECS J. Solid State Sci. Technol. 5 (2016) R3146—R3159.

[2] M.S. Wang, G.C. Guo, Inorganic—organic hybrid white light phosphors, Chem. Commun. 52 (2016) 13194—13204.

[3] T. Mondal, S. Mondal, S. Bose, D. Sengupta, U.K. Ghorai, S.K. Saha, Pure white light emission from a rare earth-free intrinsic metal—organic framework and its application in a WLED, J. Mater. Chem. C 6 (2018) 614—621.

[4] S. Shionoya, W. Yen, Phosphor Handbook, CRC Press, Boca Raton, 1999.

[5] M.H.V. Werts, Making sense of lanthanide luminescence, Sci. Prog. 88 (2) (2005) 101—131.

[6] D.R. Vij, S. Shionoya, Luminescence of Solids, Plenum Press, New York, 1998.

[7] S.W.S. Mckeever, Thermoluminescence of Solids, Cambridge University, Cambridge, 1985.

[8] L. Li, S. Zhou, S. Zhang, Investigation on charge transfer bands of Ce^{4+} in Sr_2CeO_4 blue phosphor, Chem. Phys. Lett. 453 (2008) 283—289.

[9] J. Rumble, Handbook of Chemistry and Physics, ninety ninth ed., CRC Press, Boca Raton, FL, 2018, ISBN 9781138561632, p. 36.

[10] D. Curie, Luminescence in Crystals, John Wiley Sons Inc., New York, 1963.

[11] S. Asano, N. Yamashita, Effect of magnetic field on luminescence of MgS, CaS and CaO phosphors activated with Bi^{3+} ions, Phys. Lett. 86A (1981) 191—193.

[12] K. Kato, F. Okamoto, Preparation and cathodoluminescence of CaS: Eu and $Ca_{1-x}Sr_xS$: Eu phosphors, Jpn. J. Appl. Phys. 22 (1983) 76—83.

[13] V. Shanker, S. Tanaka, M. Shiiki, H. Kobayashi, H. Sasakura, Electroluminescence in thin-film CaS:Ce, Appl. Phys. 45 (1984) 960—961.

[14] K. Tanaka, S. Mikami, T. Ogwa, K. Taniguchi, M. Yoshida, S. Nakajima, High brightness red electroluminescence in CaS:Eu thin films, Appl. Phys. Lett. 48 (1986) 1730—1732.

[15] S. Tanaka, Color electroluminescence in alkaline-earth sulfide thin-films, J. Lumin. 40&41 (1988) 20—23.

[16] P.K. Patil, R.D. Lawanger-Pawar, Electroluminescence and electrical properties of. CaS: Zn phosphors, Indian J. Phys. 64A (1990) 121—127.

[17] R. Morimo, J.C. Brickerton, S. Clough, I.V.F. Viney, B. Ray, Preparation of CaS:Ce and SrS:Ce thin films by a spin-coating method and their characterization, J. Matter. Sci. Lett. 16 (1997) 1882—1884.

[18] A. Tiwari, S.J. Dhoble, Tunable lanthanide/transition metal ion-doped novel phosphors for possible application in w-LEDs: a review, Luminescence 35 (2020) 4—33.

[19] S. Asano, N. Yamashita, Y. Nakao, Luminescence of the Pb^{2+}-ion dimer center in CaS and CaSe phosphors, Basic Solid State Phys. 89 (1978) 663—667.

[20] H. Fujita, K. Kato, F. Okamoto, Photoluminescence of CaS: Ce, Na phosphor, Jpn. J. Appl. Phys. 19 (1980) 1063.

[21] L. Marton (Ed.), Methods of Experimental Physics - Solid State of Physics, Academic Press, New York, 1959, p. 6.

[22] R.P. Rao, The preparation and thermoluminescence of alkaline earth sulphide phosphors, J. Mater. Sci. 21 (1986) 3357—3386.

[23] R.J. Yu, J. Wang, M. Zhang, J.H. Zhang, H.B. Yuan, Q. Su, A new blue-emitting phosphor of Ce^{3+}-activated $CaLaGa_3S_6O$ for white-light-emitting diodes, Chem. Phys. Lett. 453 (2008) 197—201.

[24] Z.C. Wu, J. Shi, J. Wang, M.L. Gong, Q. Su, A novel blue-emitting phosphor $LiSrPO_4:Eu^{2+}$ for white LEDs, J. Solid State Chem. 179 (2006) 2356—2360.

[25] V.B. Pawade, S.J. Dhoble, Novel blue-emitting $SrMg_2Al_{16}O_{27}:Eu^{2+}$ phosphor for solid-state lighting, Luminescence 26 (2011) 722—727.

[26] N.S. Dhoble, V.B. Pawade, S.J. Dhoble, Combustion synthesis of $X_{3.5}Mg_{0.5}Si_3O_8C_{l4}$ ($X^{3.5}$ = Sr,Ba):Eu^{2+} blue emitting phosphors, Adv. Mater. Lett. 2 (2011) 327—330.

[27] M. Kim, M. Kobayashi, H. Kato, M. Kakihana, A highly luminous LiCaPO:Eu phosphor synthesized by a solution method employing a water-soluble phosphate ester, Optic Photon. J. 3 (2013) 13—18.

[28] M. Jiao, N. Guo, W. Lü, Y. Jia, W. Lv, Q. Zhao, B. Shao, H. You, Tunable blue-green-emitting $Ba_3LaNa(PO_4)_3F:Eu^{2+}$, Tb^{3+} phosphor with energy transfer for near-UV white LEDs, Inorg. Chem. 52 (2013) 10340—10346.

[29] S. Xin, G. Zhu, Enhanced luminescence and abnormal thermal quenching behaviour investigation of $BaHfSi_3O_9:Eu^{2+}$ blue phosphor co-doped with $La^{3+}—Sc^{3+}$ ion pairs, RSC Adv. 6 (2016) 41755—41760.

References

[30] Y. Zheng, W. Zhuang, X. Xing, J. Zhong, R. Liu, Y. Li, Y. Liu, Y. Hu, Synthesis, structure and luminescent properties of a new blue-green-emitting garnet phosphor $Ca_2LuScZrAl_2GeO_{12}:Ce^{3+}$, RSC Adv. 6 (2016) 68852−68859.

[31] X. Ding, G. Zhu, W. Geng, Q. Wang, Y. Wang, Highly efficient cyan-emitting garnet $Ca_3Hf_2SiAl_2O_{12}: xCe^{3+}$ phosphor for solid state white lighting, CrystEngComm 17 (2015) 3235−3242.

[32] J. Zhong, W. Zhuang, X. Xing, R. Liu, Y. Li, Y. Zheng, Y. Hu, H. Xu, Synthesis, structure and luminescence properties of new blue-green-emitting garnet-type $Ca_3Zr_2SiGa_2O_{12}: Ce^{3+}$ phosphor for near-UV pumped white-LEDs, RSC Adv. 6 (2015) 2155−2161.

[33] D. Böhnisch, J. Rosenboom, A.G. Fuente, W. Urland, T. Jüstel, F. Baur, On a blue emitting phosphor $Na_3RbMg_7(PO_4)_6:Eu^{2+}$ showing ultra high thermal stability, J. Mater. Chem. C 7 (2019) 6012−6021.

[34] Z. Xia, H. Du, J. Sun, D. Chen, X. Wang, A novel blue-emitting $Ca_2B_5O_9$ Br:Eu $^{2+}$ phosphor prepared by a microwave calcination route, Mater. Chem. Phys. 119 (2010) 7−10.

[35] Y.H. Kim, P. Arunkumar, B.Y. Kim, S. Unithrattil, E. Kim, S.H. Moon, J.Y. Hyun, K.H. Kim, D. Lee, J.S. Lee, W.B. Im, A zero-thermal-quenching phosphor, Nat. Mater. 16 (2017) 543−550.

[36] S.J. Gwak, P. Arunkumar, W.B. Im, A new blue-emitting oxohalide phosphor $Sr_4OCl_6: Eu^{2+}$ for thermally stable, efficient white-light-emitting devices under near-UV, J. Phys. Chem. C 118 (2014) 2686−2692.

[37] H.J. Song, D.K. Yim, H.S. Roh, I.S. Cho, S.J. Kim, Y.H. Jin, H.W. Shim, D.W. Kim, K.S. Hong, $RbBaPO_4:Eu^{2+}$: a new alternative blue phosphor for UV-based white light-emitting diodes, J. Mater. Chem. C 1 (2013) 500−505.

[38] S. Zhang, Z. Hao, L. Zhang, G.H. Pan, H. Wu, X. Zhang, Y. Luo, L. Zhang, H. Zhao, J. Zhang, Efficient blue-emitting phosphor $SrLu_2O_4:Ce^{3+}$ with high thermal stability for near ultraviolet (~ 400 nm) LED-chip based white LEDs, Sci. Rep. 8 (2018) 10463.

[39] Y.Q. Li, A.C.A. Delsing, G. de With, H.T. Hintzen, Luminescence properties of Eu^{2+}-activated alkaline-earth silicon-oxynitride $MSi_2O_{2-\delta}N_{2+2/3\delta}$ (M = Ca, Sr, Ba): A promising class of novel LED conversion phosphors, Chem. Mater. 17 (2005) 3242−3248.

[40] K. Uheda, N. Hirosaki, Y. Yamamoto, A. Naito, T. Nakajima, H. Yamamoto, Luminescence properties of a red phosphor, $CaAlSiN_3:Eu^{2+}$, for white light-emitting diodes, Electrochem. Solid State Lett. 9 (2006) H22−H25.

[41] R. Le Toquin, A.K. Cheetham, Red-emitting cerium-based phosphor materials for solid-state lighting applications, Chem. Phys. Lett. 423 (2006) 352−356.

[42] Y. Tian, X. Qi, X. Wu, R. Hua, B. Chen, Luminescent properties of $Y_2(MoO_4)_3:Eu^{3+}$ red phosphors with flower-like shape prepared via co-precipitation method, J. Phys. Chem. C 113 (2009) 10767−10772.

[43] H.F. Li, Y.L. Jia, T.F. Ma, R. Pang, W.Z. Sun, D. Li, J.P. Fu, S. Zhang, L.H. Jiang, C.Y. Li, $Ca_9Sc(PO_4)_7 :Ce^{3+}$,Mn^{2+} $^-$ a red-emitting phosphor based on energy transfer, Eur. J. Inorg. Chem. 6 (2016) 867−873.

[44] F. Tang, Z.C. Su, H. Ye, S.J. Xu, G. Wang, W. Guo, Y. Cao, W. Gao, X. Pan, Boosting phonon-induced luminescence in red fluoride phosphors via composition-driven structural transformations, J. Mater. Chem. C 5 (2017) 12105−12111.

[45] P.Q. Cai, L. Qin, C.L. Chen, J. Wang, H.J. Seo, Luminescence, energy transfer and optical thermometry of a novel narrow red emitting phosphor: $Cs_2WO_2F_4:Mn^{4+}$, Dalton Trans. 46 (2017) 14331−14340.

[46] Q. Wang, Z.Y. Yang, H.Y. Wang, Z.K. Chen, H.L. Yang, J. Yang, Z.L. Wang, Novel Mn^{4+}-activated oxyfluoride $Cs_2NbOF_5:Mn^{4+}$ red phosphor for warm white light-emitting diodes, Opt. Mater. 85 (2018) 96−99.

[47] H. Ming, J.F. Zhang, L.L. Liu, J.Q. Peng, F. Du, X.Y. Ye, Y.M. Yang, H.P. Nie, A novel $Cs_2NbOF_5:Mn^{4+}$ oxyfluoride red phosphor for light-emitting diode devices, Dalton Trans. 47 (2018) 16048−16056.

[48] X.L. Dong, Y.X. Pan, D. Li, H.Z. Lian, J. Lin, A novel red phosphor of Mn^{4+} ion-doped oxyfluoroniobate $BaNbOF_5$ for warm WLED applications, Cryst. Eng. Commun. 20 (2018) 5641−5646.

[49] R. Kasa, S. Adachi, Red and deep red emissions from cubic $K_2SiF_6:Mn^{4+}$ and hexagonal K_2MnF_6 synthesized in $HF/KMnO_4/KHF_2/Si$ solutions, J. Electrochem. Soc. 159 (2012) J89.

[50] A.K. Munirathnappa, N. Sundaram, J. Dwivedi, V.C. Petwal, Enhanced red luminescence and improved crystallinity in $NaEu(WO_4)_2$ phosphor: an electron beam irradiation study, New J. Chem. 42 (4) (2018) 2726−2732.

[51] Z. Yang, Z. Yang, Q. Wei, Q. Zhou, Z. Wang, Luminescence of red-emitting phosphor $Rb_5Nb_3OF_{18}:Mn^{4+}$ for warm white light-emitting diodes, J. Lumin. 210 (2019) 408−412.

[52] T. Hu, H. Lin, Y. Cheng, Q.M. Huang, J. Xu, Y. Gao, J.M. Wang, Y.S. Wang, A highly-distorted octahedron with a C_{2v} group symmetry inducing an ultra-intense zero phonon line in Mn^{4+}-activated oxyfluoride $Na_2WO_2F_4$, J. Mater. Chem. C 5 (2017) 10524−10532.

[53] R.C. Ropp, The Chemistry of Artificial Lighting Devices, Elsevier, Amsterdam, 1993, p. 486.

[54] M. Nazarov, D.Y. Noh, C.C. Byeon, H. Kim, Efficient multiphase green phosphor based on strontium thiogallate, J. Appl. Phys. 105 (2009) 073518−073520.

[55] Y. Shimomura, T. Kurushima, N. Kijima, Photoluminescence and crystal structure of green-emitting phosphor $CaSc_2O_4:Ce^3$, J. Electrochem. Soc. 154 (2007) J234.

[56] Q. Zhang, Q. Wang, X. Wang, X. Ding, Y. Wang, Luminescence properties of Eu^{2+}-doped $BaSi_2O_5$ as an efficient green phosphor for light-emitting devices and wide color gamut field emission displays, New J. Chem. 40 (2016) 8549–8555, https://doi.org/10.1039/C6NJ01831A.

[57] M. Zhang, J. Wang, W. Ding, Q. Zhang, Q. Su, A novel white light-emitting diode (w-LED) fabricated with $Sr_6BP_5O_{20}:Eu^{2+}$ phosphor, Appl. Phys. B 86 (2007) 647–651.

[58] Y.C. Chiu, W.R. Liu, Y.T. Yeh, S.M. Jang, T.M. Chen, Luminescent properties and energy transfer of green-emitting $Ca_3Y_2(Si_3O_9)_2:Ce^{3+},Tb^3$ phosphor, J. Electrochem. Soc. 156 (2009) 221.

[59] K.Y. Jung, H.W. Lee, H.K. Jung, Luminescent properties of (Sr, Zn) Al_2O_4: $Eu^{2+}B^{3+}$ particles as a potential green phosphor for UV LEDs, Chem. Mater. 18 (2006) 2249–2255.

[60] C.K. Chang, T.M. Chen, White light generation under violet-blue excitation from tunable green-to-red emitting $Ca_2MgSi_2O_7:Eu,Mn$ through energy transfer, Appl. Phys. Lett. 90 (2007) 161901.

[61] N. Hirosaki, R.J. Xie, K. Kimoto, Characterization and properties of green-emitting β-SiAlON:Eu^{2+} powder phosphors for white light-emitting diodes, Appl. Phys. Lett. 86 (2005) 211905–211909.

[62] Y. Qiang, Z. Pan, X. Ye, M. Liang, J. Xu, J. Huang, W. You, H. Yuan, Ce^{3+} doped $BaLu_2Al_4SiO_{12}$: a promising green-emitting phosphor for white LEDs, J. Lumin. 203 (2018) 609–615.

[63] B. Liu, C. Shi, Potential white-light long-lasting phosphor: Dy^{3+}-doped aluminate, Appl. Phys. Lett. 86 (19) (2009) 191111.

[64] T. Seto, N. Kijima, N. Hirosaki, A new yellow phosphor $La_3Si_6N_{11}:Ce^{3+}$ for white LEDs, ECS Trans. 25 (2009) 247–252.

[65] X. Li, J.D. Budai, F. Liu, J.Y. Howe, J. Zhang, X.-J. Wang, Z. Gu, C. Sun, R.S. Meltzer, Z. Pan, New yellow $Ba_{0.93}Eu_{0.07}Al_2O_4$ phosphor for warm-white light-emitting diodes through single-emitting-center conversion, Light Sci. Appl. 2 (2013) 50.

[66] H.S. Jang, D.Y. Jeon, Yellow-emitting $Sr3SiO5:Ce^{3+}$, Li^+phosphor for white-light-emitting diodes and yellow-light-emitting diodes, Appl. Phys. Lett. 90 (2007) 041906.

[67] C. Hecht, F. Stadler, P.J.S. Schmidt, J.S. Gunne, V. Baumann, W. Schnick, $SrAlSi_4N_7:Eu^{2+}$ – a nitridoalumosilicate phosphor for warm white light (pc)LEDs with edge-sharing tetrahedra, Chem. Mater. 21 (2009) 1595–1601.

[68] J. Qiao, Z. Xia, Z. Zhang, B. Hu, Q. Liu, Near UV-pumped yellow-emitting $Sr_9MgLi(PO_4)_7:Eu^{2+}$ phosphor for white-light LEDs, Sci. China Mater. 61 (2018) 985–992.

[69] B. Deng, J. Chen, C. Zhou, H. Liu, A novel UV pumped yellow-emitting phosphor $Ba_2YAlO_5:Dy^{3+}$ for white light-emitting diodes, E3S Web Conf. 79 (2019) 03005.

[70] V. Mahalingam, F. Mangiarini, F. Vetrone, V. Venkatramu, M. Bettinelli, A. Speghini, J.A. Capobianco, Bright white upconversion emission from $Tm^{3+}/Yb^{3+}/Er^{3+}$-doped $Lu_3Ga_5O_{12}$ nanocrystals, J. Phys. Chem. C 112 (2008) 17745–17749.

[71] B. Zhou, B. Shi, D. Jin, X. Liu, Controlling upconversion nanocrystals for emerging applications, Nat. Nanotechnol. 10 (2015) 924–936.

[72] M. Haase, H. Schäfer, Upconverting nanoparticles, Angew. Chem. Int. Ed. 50 (2011) 5808–5829.

[73] F. Auzel, Upconversion and anti-Stokes processes with f and d ions in solids, Chem. Rev. 104 (2004) 139–174.

[74] D.R. Gamelin, H.U. Güdel, Design of luminescent inorganic materials: new photophysical processes studied by optical spectroscopy, Acc. Chem. Res. 33 (4) (2000) 235–242.

[75] Z. Huang, X. Li, M. Mahboub, K.M. Hanson, V.M. Nichols, H. Le, M.L. Tang, C.J. Bardeen, Hybrid molecule–nanocrystal photon upconversion across the visible and near-infrared, Nano Lett. 15 (8) (2015) 5552–5557.

[76] T. Trupke, M.A. Green, P. Wurfel, Improving solar cell efficiencies by down-conversion of high-energy photons, J. Appl. Phys. 92 (2002) 1668–1674.

[77] B.S. Richards, Enhancing the performance of silicon solar cells via the application of passive luminescence conversion layers, Sol. Energy Mater. Sol. Cell. 90 (2006) 2329–2337.

[78] B.S. Richards, Luminescent layers for enhanced silicon solar cell performance: down-conversion, Sol. Energy Mater. Sol. Cell. 90 (2006) 1189–1207.

[79] R.T. Wegh, H. Donker, K.D. Oskam, A. Meijerink, Visible quantum cutting in $LiGdF_4:Eu^{3+}$ through downconversion, Science 283 (1999) 663–666.

[80] G.H. Dieke, H.M. Crosswhite, The spectra of the doubly and triply ionized rare earths, Appl. Optic. 2 (1963) 675–686.

References

[81] G.M. Chow, L.K. Kurihara, Synthesis and processing of nanostructured powders and films, Nanostruct. Mater. Process. Prop. Appl. (2006) 784.

[82] G.B. Nair, S.J. Dhoble, White light emitting $MZr_4(PO_4)_6:Dy^{3+}$ (M = Ca, Sr, Ba) phosphors for WLEDs, J. Fluoresc. 27 (2017) 575−585.

[83] G.B. Nair, S.J. Dhoble, Orange light-emitting $Ca_3Mg_3(PO_4)_4:Sm^{3+}$ phosphors, Luminescence 32 (2017) 125−128.

[84] S.A. Pardhi, G.B. Nair, R. Sharma, S.J. Dhoble, Investigation of thermoluminescence and electron-vibrational interaction parameters in $SrAl_2O_4:Eu^{2+}$, Dy^{3+} phosphors, J. Lumin. 187 (2017) 492−498.

[85] C.J. Brinker, G.W. Scherer, Sol-Gel Science: The Physics and Chemistry of Sol-Gel Processing, Academic Press INC, UK. London, 1990.

[86] T. Rivera, J. Roman, I. Azorin, R. Sos, J. Guzman, K. Serrano, M. Garcia, G. Alarcon, Preparation of CaSO$_4$:Dy by precipitation method to gamma radiation dosimetry, Appl. Radiat. Isot. 68 (2010) 623−625.

[87] Q.Y. Zhang, X.Y. Huang, Recent progress in quantum cutting phosphors, Prog. Mater. Sci. 55 (2010) 353−427, https://doi.org/10.1016/j.pmatsci.2009.10.001.

[88] A. Bishnoi, S. Kumar, N. Joshi, Chapter 9 - Wide-Angle X-ray Diffraction (WXRD): Technique for Characterization of Nanomaterials and Polymer Nanocomposites, 2017.

[89] V.B. Pawade, S.J. Dhoble, Phosphors for Energy Saving and Conversion Technology, CRC Press, September 17, 2018, ISBN 978-1-138-59817-1, p. 238. Taylor & Francis Group; International Standard Book Number-13.

[90] K.K. Gupta, R.M. Kadam, N.S. Dhoble, S.P. Locha, S.J. Dhoble, A comparative investigation of Ce^{3+}/Dy^{3+} and Eu^{2+} doped $LiAlO_2$ phosphors for high dose radiation dosimetry: explanation of defect recombination mechanism using PL, TL and EPR study, J. Lumin. 188 (2017) 81−95.

[91] S.J. Mofokeng, V. Kumar, R.E. Kroon, O.M. Ntwaeaborwa, Structure and optical properties of Dy^{3+} activated sol-gel $ZnO\text{-}TiO_2$ nanocomposites, J. Alloys Compd. 711 (2017) 121−131.

[92] K.N. Shinde, S.J. Dhoble, H.C. Swart, K. Park, Phosphate Phosphors for Solid-State Lighting. Springer-Verlag Berlin Heidelberg, in: Springer Series in Materials Science, vol. 174, 2012, pp. 79−99.

[93] T. Welker, Recent developments on phosphors for fluorescent lamps and cathode-ray tubes, J. Lumin. 48−49 (1991) 49−56.

[94] B.M.J. Smets, Phosphors based on rare-earths, a new era in fluorescent lighting, Mater. Chem. Phys. 16 (1987) 283−299.

[95] B.M.J. Smets, Adv. Non Radiative Proc. In Solids. Ed. Di. Bartolo, Plenum Press, N.Y., 1991, p. 375.

[96] R. Raue, A.T. Vink, T. Welker, Phosphor screens in cathode-ray tubes for projection television, Philips Tech. Rev. 44 (1989) 335−347.

[97] A. Penzkofer, Solid state lasers, Prog. Quant. Electron. 12 (1988) 291−428.

[98] W.H. Byler, J.J. Maffis, in: T.C. Paton (Ed.), Pigment Handbook, Wiley, NY, 1988, p. 881.

[99] G.H. Dieke, in: H. Crosswhite (Ed.), Spectra & Energy Levels of RE Ions in Crystals, John Wiley and Sons, N.Y., 1968.

[100] C.A. Morrison, R.P. Levitt, in: K.A. Schneider, L. Eyring (Eds.), Handbook of Phys. & Chem. of Rare Earths, vol. 5, North-Holland, N.Y., 1982, p. 461.

[101] B.R. Judd, Operator Techniques in Atomic Spectra, Mc Graw-Hill, N.Y., 1963.

[102] B.R. Judd, in: K.A. Schneidner, L. Eyring (Eds.), Handbook of Phys. & Chem. of Rare Earths, vol. 11, North-Holland, N.Y., 1988, p. 81.

[103] B.G. Wyborne, Spectroscopic Properties of RE, Wiley, N.Y., 1965.

[104] R.D. Peacock, The intensities of lanthanide f—f transitions, Struct. Bond 22 (1975) 83−122.

[105] S. Huffner, Optical Spectra of Transparent RE Compounds, Acad. Press., N.Y., 1978.

[106] E.Y. Wong, O.M. Stafsudd, D. R Johnston, Optical absorption spectrum of Nd^{3+}-doped LaF_3 single crystal: the evidence of a hidden selection rule, Phys. Rev. 131 (1963) 990−992.

[107] W. T Carnall, G.L. Goodman, K. Rajnak, R.S. Rana, A systematic analysis of the spectra of the lanthanides doped into single crystal LaF_3, J. Chem. Phys. 90 (1989) 3443−3457.

[108] C.W. Thiel, Y. Sun, R.L. Cone, Progress in relating rare-earth ion 4f and 5d energy levels to host bands in optical materials for hole burning, quantum information and phosphors, J. Mod. Optic. 49 (2002) 2399−2411.

[109] J. Becquerel, Influence de variation s de temperature sur l'absorption dans les corps solides, Radium 4 (1907) 328−339.

[110] A. Pandey, V.K. Rai, $Pr^{3+}-Yb^{3+}$ codoped Y_2O_3 phosphor for display devices, Mater. Res. Bull. 57 (2014) 156−161.

III. Electrochemical energy conversion and storage

[111] Y. Xu, J. Chen, H. Zhang, H. Wei, L. Zhou, Z. Wang, Y. Pan, X. Su, A. Zhang, J. Fu, White-light-emitting flexible display devices based on double network hydrogels crosslinked by YAG:Ce phosphors, J. Mater. Chem. C 8 (2020) 247–252.

[112] A. Bastos, B. McKenna, M. Lima, P.S. André, L.D. Carlos, R.C. Evans, R.A.S. Ferreira, Flexible optical amplifier for visible-light communications based on organic–inorganic hybrids, ACS Omega 3 (10) (2018) 13772–13781.

[113] S. Mei, X. Liu, W. Zhang, R. Liu, L. Zheng, R. Guo, P. Tian, High-bandwidth white-light system combining a micro-LED with perovskite quantum dots for visible light communication, ACS Appl. Mater. Interfaces 10 (2018) 5641–5648.

[114] J. Zmojda, M. Kochanowicz, P. Miluski, D. Dorosz, Side-detecting optical fiber doped with Tb^{3+} for ultraviolet sensor application, Fibers 2 (2) (2014) 150–157.

[115] A. Poddar, S.C. Gedam, S.J. Dhoble, Luminescence of Eu and Ce in $K_3Ca_2(SO_4)_3$ F fluoride material, Luminescence 30 (2015) 914–917.

[116] L. Ozawa, M. Itoh, Cathode ray tube phosphors, Chem. Rev. 103 (2003) 3835–3856.

[117] C.T. Yejh, Y. Chou, K.S. Yang, S.K. Wu, C.C. Wang, Luminescence material characterizations on laser-phosphor lighting techniques, Optic Express 27 (2019) 7226–7236.

[118] I. Kandarakis, D. Cavouras, G.S. Panayiotakis, D. Triantis, C.D. Nomicos, Europium-activated phosphors for use in X-ray detectors of medical imaging systems, Eur. Radiol. 8 (1998) 313–318.

[119] G. Kertzscher, S. Beddar, Inorganic scintillation detectors based on Eu-activated phosphors for ^{192}Ir brachy therapy, Phys. Med. Biol. 62 (2017) 5046–5075.

[120] H. Lin, S. Zhou, H. Teng, Y. Li, W. Li, X. Hou, T. Jia, Near infrared quantum cutting in heavy Yb doped $Ce_{0.03}Yb_3xY_{(2.97 - 3x)}$ Al_5O_{12} transparent ceramics for crystalline silicon solar cells, J. Appl. Phys. 107 (2010) 043107–043111.

[121] K.L. Rik, V. Deun, Enhancing the energy transfer from Mn^{4+} to Yb^{3+} via a Nd^{3+} bridge role in$Ca_3La_2W_2O_{12}$:Mn^{4+},Nd^{3+},Yb^{3+} phosphors for spectral conversion of c-Si solar cells, Dyes Pigments 162 (2018) 1–32.

[122] S.G. Pérez, J. Sanchizb, V.D. Rodríguez, D. Cañadillas-Ramallo, J.G. Platas, D. Borchert, B.G. Díaz, C.H. Rodríguez, R.G. Lemus, Highly luminescent film as enhancer of photovoltaic devices, J. Lumin. 201 (2018) 148–155.

[123] Y.N. Ahn, K.D. Kim, G. Anoop, G.S. Kim, J.S. Yoo, Design of highly efficient phosphor-converted white light-emitting diodes with color rendering indices (R1 – R15) $\geq$ 95 for artificial lighting, Sci. Rep. 9 (2019) 16848.

[124] A.R. Kadam, S.J. Dhoble, Synthesis and luminescence study of Eu^{3+}-doped $SrYAl_3O_7$ phosphor, Luminescence 34 (2019) 846–853.

[125] H.J. Round, A note on carborundum, Electr. World 19 (1907) 309.

[126] O.V. Lossev, Luminous carborundum detector and detection with crystals, Wireless Telegrap, Telephon 26 (1924) 403.

[127] N. Holonyak, S.F. Bevacqua, Light emission from Ga(As1-XPX) junctions, Appl. Phys. Lett. 1 (1962) 82–83.

[128] S. Nakamura, T. Mukai, M. Senoh, Candela class high-brightness InGaN/AlGaN double heter-ostructure blue light-emitting diodes, Appl. Phys. Lett. 64 (1994) 1687–1689.

[129] S. Peralta, H. Ruda, Applications for advanced solid-state lamps, IEEE Ind. Appl. Mag. 4 (1998) 31–42.

[130] A.K. Bedyal, A.K. Kunti, V. Kumar, H.C. Swart, Effects of cationic substitution on the luminescence behavior of Dy^{3+} doped orthophosphate phosphor, J. Alloys Compd. 806 (2019) 1127–1137.

[131] G.B. Nair GB, S.J. Dhoble, White light emission through efficient energy transfer from Ce^{3+} to Dy^{3+} ions in $Ca_3Mg_3(PO_4)_4$ matrix aided by Li^+ charge compensator, J. Lumin. 192 (2017) 1157–1166.

[132] F. Liu, Q. Liu, Y. Fang, N. Zhang, B. Yang, G. Zhao, White light emission from $NaLa(PO_3)_4$: Dy^{3+} single-phase phosphors for light-emitting diodes, Ceram. Int. 41 (2015) 1917–1920.

[133] Z.W. Zhang, A.J. Song, M.Z. Ma, X.Y. Zhang, Y. Yue, R.P. Liu, A novel white emission in $Ca_8MgBi(PO_4)_7$: Dy^{3+} single-color phosphor, J. Alloys Compd. 601 (2014) 231–233.

[134] W. Geng, G. Zhu, Y. Shi, Y. Wang, Luminescent characteristics of Dy^{3+} doped calcium zirconium phosphate $CaZr_4(PO_4)_6$ (CZP) phosphor for warm-white LEDs, J. Lumin. 155 (2014) 205–209.

[135] Z. Zhang, L. Liu, X. Zhang, J. Zhang, W. Zhang, D. Wang, Preparation and investigation of $CaZr_4(PO_4)_6$:Dy^{3+} single-phase full-color phosphor, Spectrochim. Acta Part A : Mol Biomol Spectrosc 137 (2015) 1–6.

[136] X.Y. Sun, J.C. Zhang, X.G. Liu, L.W. Lin, Enhanced luminescence of novel $Ca_3B_2O_6$:Dy^{3+} phosphors by Li^+-codoping for LED applications, Ceram. Int. 38 (2012) 1065–1070.

References

[137] Q. Xu, J. Sun, D. Cui, Q. Di, J. Zeng, Synthesis and luminescence properties of novel $Sr_3Gd(PO_4)_3:Dy^{3+}$ phosphor, J. Lumin. 158 (2015) 301–305.

[138] M. Dalal, V.B. Taxak, J. Dalal, A. Khatkar, S. Chahar, R. Devi, S.P. Khatkar, Crystal structure and Judd-Ofelt properties of a novel color tunable blue-white-red $Ba_5Zn_4Y_8O_{21}:Eu^{3+}$ nanophosphor for near-ultraviolet based WLEDs, J. Alloys Compd. 698 (2017) 662–672.

[139] L. Chunxia, C. Zang, Z. Hou, L. Wang, Z. Quan, H. Lian, J. Lin, β-$NaYF_4$ and β-$NaYF4:Eu^{3+}$ microstructures: morphology control and tunable luminescence properties, J. Phys. Chem. C 113 (2009) 2332–2339.

[140] A.K. Vishwakarma, K. Jha, M. Jayasimhadri, A.S. Rao, K. Jang, B. Sivaiaha, D. Haranath, Red light emitting $BaNb_2O_6:Eu^{3+}$phosphor for solid state lighting applications, J. Alloys Compd. 622 (2015) 97–101.

[141] H. Yu, B. Zhang, X. Chen, X. Qian, D. Jiang, Q. Wu, J. Wang, J. Xu, L. Su, Color-tunable visible photoluminescence of $Eu:CaF_2$ single crystals: variations of valence state and local lattice environment of Eu ions, Optic Express 27 (2019) 523–532.

[142] J. Li, Z. Zhang, X. Li, Y. Xu, Y. Ai, J. Yan, J. Shi, M. Wu, Luminescence properties and energy transfer of $YGa_{1.5}Al_{1.5}(BO_3)_4:Tb^{3+},Eu^{3+}$ as a multi-colour emitting phosphor for WLEDs, J. Mater. Chem. C 5 (2017) 6294–6299.

[143] B. Wei, Z. Liu, C. Xie, S. Yang, W. Tang, A. Gu, W.T. Wong, K.L. Wong, Fast synthesis of red $Li_3BaSrLn_3(WO_4)_8:Eu^{3+}$ phosphors for white LEDs under near-UV excitation by a microwave-assisted solid state reaction method and photoluminescence studies, J. Mater. Chem. C 3 (2015) 12322–12327.

[144] Y. Ma, M. Fei, W. Zhang, L. Teng, F. Hu, R. Wei, H. Guo, Energy transfer and tunable luminescent properties in $Eu^{2+}/Tb^{3+}/Eu^{3+}$co-doped oxyfluoride alumino silicate glass, J. Lumin. 219 (2020) 116966.

[145] Z.C. Wu, J. Liu, W.G. Hou, J. Xu, M.L. Gong, A new single-host white-light-emitting $BaSrMg(PO4)2:Eu2+$ phosphor for white-light-emitting diodes, J. Alloys Compd. 498 (2010) 139–142.

[146] Y. Liu, A. Lan, Y. Jin, G. Chen, X. Zhang, $Sr_3Bi(PO_4)_3:Eu^{2+}$, Mn^{2+}: single-phase and color-tunable phosphors for white-light LEDs, Opt. Mater. 40 (2015) 122–126.

[147] X.M. Zhang, W.L. Li, H.J. Seo, Luminescence and energy transfer in Eu^{2+}, Mn^{2+} co-doped $Li_4SrCa(SiO_4)_2$ for white light-emitting-diodes, Phys. Lett. A 373 (2009) 3486–3489.

[148] W. Wu, Z. Xia, Synthesis and color-tunable luminescence properties of Eu^{2+} and Mn^{2+}-activated $Ca_3Mg_3(PO_4)_4$ phosphor for solid state lighting, RSC Adv. 3 (2013) 6051–6057.

[149] C. Shen, Y. Yang, S. Jin, H. Feng, Synthesis and luminous characteristics of $Ba_2MgSi_{2-x}Al_xO_7:0.1Eu^{2+},0.1Mn^{2+}$ phosphor for WLED, Optik 121 (2010) 29–32.

[150] N. Guo, H. You, C. Jia, R. Ouyang, D. Wu, A Eu^{2+} and Mn^{2+}-coactivated fluoro-apatite-structure $Ca_6Y_2Na_2(PO_4)_6F_2$ as a standard white-emitting phosphor via energy transfer, Dalton Trans. 43 (2014) 12373–12379.

[151] Z. Wang, X. Hou, Y. Liu, Z. Hui, Z. Huang, M. Fang, X. Wu, Luminescence properties and energy transfer behavior of colour-tunable white-emitting $Sr_4Al_{14}O_{25}$ phosphors with co-doping of Eu^{2+}, Eu^{3+} and Mn^{4+}, RSC Adv. 7 (2017) 52995–53001.

[152] A. Baran, S. Mahlik, M. Grinberg, P. Cai, S.I. Kim, H.J. Seo, Luminescence properties of different Eu sites in $LiMgPO_4:Eu^{2+}$, Eu^{3+}, J. Phys. Condens. Matter. 26 (2014) 385401.

[153] A.A. Reddy, S. Das, S. Ahmad, S.S. Babu, J.M.F. Ferreira, G.V. Prakash, Influence of the annealing temperatures on the photoluminescence of $KCaBO_3:Eu^{3+}$ phosphor, RSC Adv. 2 (2012) 8768.

[154] A. Baran, J. Barzowska, M. Grinberg, S. Mahlik, K. Szczodrowski, Y. Zorenko, Binding energies of Eu^{2+} and Eu^{3+} ions in β-Ca_2SiO_4 doped with europium, Opt. Mater. 35 (2013) 2107–2114.

[155] J. Chen, Y. Liang, Y. Zhu, S. Liu, H. Li, W. Lei, Abnormal reduction of Eu^{3+} to Eu^{2+} in $Sr_5(PO_4)_3Cl$: Eu phosphor and its enhanced red emission by the charge compensation, J. Lumin. 214 (2019) 116569.

[156] A. Dobrowolska, E. Zych, Spectroscopic characterization of $Ca_3Y_2Si_3O_{12}:Eu^{2+}$, Eu^{3+} powders in VUV-UV–vis region, J. Phys. Chem. C 116 (2012) 25493–25503.

[157] Q. Wang, D. Deng, Y. Hua, L. Huang, H. Wang, S. Zhao, G. Jia, C. Li, S. Xu, Potential tunable white-emitting phosphor $LiSr_4(BO_3)_3:Ce^{3+}$, Eu^{2+} for ultraviolet light-emitting diodes, J. Lumin. 132 (2012) 434–438.

[158] F. Xiao, Y.N. Xue, Q.Y. Zhang, $Ca_2BO_3Cl:Ce^{3+}$, Eu^{2+}: a potential tunable yellow–white–blue-emitting phosphors for white light-emitting diodes, Phys. B Condens. Matter. 404 (2009) 3743–3747.

[159] Z. Zheng, T. Wanjun, Tunable luminescence and energy transfer of $Ce^{3+}/Eu^{2+}/Mn^{2+}$-tridoped $Sr_8MgLa(PO_4)_7$ phosphor for white light LEDs, J. Alloys Compd. 663 (2016) 731–737.

[160] Y. Song, G. Jia, M. Yang, Y. Huang, H. You, H. Zhang, $Sr_3Al_2O_5Cl_2:Ce^{3+}$, Eu^{2+}: a potential tunable yellow-to-white-emitting phosphor for ultraviolet light emitting diodes, Appl. Phys. Lett. 94 (2009) 091902.

III. Electrochemical energy conversion and storage

[161] L. Chen, A. Luo, Y. Zhang, F. Liu, Y. Jiang, Q. Xu, X. Chen, Q. Hu, S.F. Chen, K.J. Chen, H.C. Kuo, Optimization of the single-phased white phosphor of Li_2SrSiO_4: Eu^{2+}, Ce^{3+} for light-emitting diodes by using the combinatorial approach assisted with the Taguchi method, ACS Comb. Sci. 14 (2012) 636−644.

[162] C. Lv, X. Min, S. Li, Z. Huang, Y. Liu, X. Wu, M. Fang, Luminescence properties of emission tunable single-phased phosphor $La_7O_6(BO_3)(PO_4)_2$:Ce^{3+}, Tb^{3+}, Eu^{3+}, Mater. Res. Bull. 97 (2018) 506−511.

[163] M. Zhao, Z. Zhao, L. Yang, L. Dong, A. Xia, S. Chang, Y. Wei, Z. Leu, The generation of energy transfer from Ce^{3+} to Eu^{3+} in $LaPO_4$:Ce^{3+}/Tb^{3+}/Eu^{3+} phosphors, J. Lumin. 194 (2018) 297−302.

[164] X. Zhang, X. Fu, J. Song, M. Gong, Luminescent properties and energy transfer studies of color-tunable $LuBO_3$:Ce^{3+}/Tb^{3+}/Eu^{3+} phosphors, Mater. Res. Bull. 80 (2016) 177−185.

[165] X. Zhang, L. Zhou, Q. Pang, J. Shi, M. Gong, Tunable luminescence and $Ce^{3+} \rightarrow Tb^{3+} \rightarrow Eu^{3+}$ energy transfer of broadband-excited and narrow line red emitting Y_2SiO_5:Ce^{3+}, Tb^{3+}, Eu^{3+} phosphor, J. Phys. Chem. C 118 (2014) 7591−7598.

[166] J. Zhou, Z. Xia, Luminescence color tuning of Ce^{3+}, Tb^{3+} and Eu^{3+} codoped and tri-doped $BaY2Si3O10$ phosphors via energy transfer, J. Mater. Chem. C 3 (2015) 7552−7560.

[167] Y.N. Xue, F. Xiao, Q.Y. Zhang, A red-emitting $Ca_8MgLa(PO_4)_7$:Ce^{3+},Mn^{2+} phosphor for UV-based white LEDs application, Spectrochim. Acta Mol. Biomol. Spectrosc. 78 (2011) 1445−1448.

[168] J. Jiang, Y. Cheng, W. Chen, Z. Liu, T. Xu, M. He, L. Zhou, R. Yuan, W. Xiang, X. Liang, Mater. Res. Bull. 105 (2018) 277.

[169] X. Jiang, Y. Pan, S. Huang, X. Chen, J. Wang, G. Liu, Hydrothermal synthesis and photoluminescence properties of red phosphor $BaSiF_6$:Mn^{4+} for LED applications, J. Mater. Chem. C 2 (2014) 2301−2306.

[170] Z. Wang, Y. Liu, Y. Zhou, Q. Zhou, H. Tan, Q. Zhang, J. Peng, Red-emitting phosphors Na_2XF_6:Mn^{4+} (X = Si, Ge, Ti) with high colour-purity for warm white-light-emitting diodes, RSC Adv. 5 (2015) 58136−58140.

[171] L.L. Wei, C.C. Lin, Y.Y. Wang, M.H. Fang, H. Jiao, R.S. Liu, Photoluminescent evolution induced by structural transformation through thermal treating in the red narrow-band phosphor K_2GeF_6:Mn^{4+}, ACS Appl. Mater. Interfaces 7 (2015) 10656−10659.

[172] Z. Wang, Y. Zhou, Z. Yang, Y. Liu, H. Yang, H. Tan, Q. Zhang, Q. Zhou, Synthesis of $K_{2X}F_6$:Mn^{4+} (X = Ti, Si and Ge) red phosphors for white LED applications with low-concentration of HF, Opt. Mater. 49 (2015) 235−240.

[173] E.H. Song, J.Q. Wang, S. Ye, X.F. Jiang, M.Y. Peng, Q.Y. Zhang, Room-temperature synthesis and warm-white LED applications of Mn^{4+} ion doped fluoroaluminate red phosphor Na_3AlF_6:Mn^{4+}, J. Mater. Chem. C 4 (2016) 2480−2487.

[174] Q. Zhou, H. Tan, Y. Zhou, Q. Zhang, Z. Wang, J. Yan, M. Wu, Optical performance of Mn^{4+} in a new hexa-coordinated fluorozirconate complex of Cs_2ZrF_6, J. Mater. Chem. C 4 (2016) 7443−7448.

[175] R. Hoshino, S. Adachi, Light-Induced degradation in red-emitting $ZnSiF_6.6H_2O$:Mn^{4+} hydrate phosphor, ECS J. Solid State Sci. Technol. 3 (2014) R144.

[176] M.H. Fang, H.D. Nguyen, C.C. Lin, R.S. Liu, Preparation of a novel red Rb_2SiF_6:Mn^{4+} phosphor with high thermal stability through a simple one-step approach, J. Mater. Chem. C 3 (2015) 7277−7280.

[177] Light's Labour's Lost: Policies for Energy-Efficient Lighting; 2005 Electricity Consumption Estimated from IEA, World Energy Outlook, Paris, France, 2006.

CHAPTER

15

Synthesis and luminescence study of silicate-based phosphors for energy-saving light-emitting diodes

Nilesh Ugemuge[1], Yatish R. Parauha[2], S.J. Dhoble[2]

[1]Department of Physics, Anand Niketan College, Warora, Chandrapur, India; [2]Department of Physics, RTM Nagpur University, Nagpur, Maharashtra, India

15.1 Introduction

A luminescent material, also known as a phosphor, is a solid material that converts certain types of energy into electromagnetic radiation over and above thermal radiation [1]. It consists of a host lattice and a luminescent center, often called the activator. The activator absorbs the excitation energy, rising to an excited state, and then returns to the ground state by emission of radiation. The nature of both the activator ion and the lattice site determines the absorption and emission characteristics. Phosphors are highly pure inorganic materials consisting of a small amount of activator ion and an inert host matrix. Different host lattices may be used like oxides, sulfides, silicates, aluminates, borates, aluminosilicates, halides, nitrides, oxynitrides, selenides, stannates, phosphates, titanates, etc. for exploring phosphor materials [2,3]. The function of a host lattice is to absorb the incident radiation and transfer it to the activator ion. Rare-earth ions are widely studied for selecting a suitable activator ion in combination with a host lattice. Synthetic methods used for material exploration also depend on the spectroscopic properties of these activator ions. The most commonly used activator ions are europium, terbium, and cerium. Red phosphor materials are prepared by doping europium (III) ions that emit strong orange/red light. Green-emitting phosphor materials are prepared by doping terbium (III) ions as an activator. Europium (II) and cerium (III) ions are the blue-emitting ones used for synthesizing blue phosphors. Europium (II) also emits green light but that depends on the type of host lattice used for material synthesis. In silicate phosphors, the emission and excitation peaks can be studied with the help of energy-level diagrams of doped rare earths. The vast majority of accessible phosphor materials have a place with certain host lattices like oxides, silicates, aluminates, sulfides, oxysulfides, nitrides,

Energy Materials
https://doi.org/10.1016/B978-0-12-823710-6.00017-0

445

Copyright © 2021 Elsevier Ltd. All rights reserved.

and so on. Silicate matrices have attracted in a great deal of attention and are important in developing effective luminescent materials as they exhibit excellent photoluminescence (PL) properties, good chemical stability, and relative ease of preparation compared to others. Silicate-based materials can be investigated for their general synthetic condition, and their composition, shape, and size can be effortlessly controlled to improve the PL execution of the phosphor materials. In recent years, much consideration has been centered around luminescent materials, which are based on silicate because of their broad applications in various fields. Low cost, high color rendering index (CRI) and stability, and small color aberration moisture-resistant properties have prompted scientists to explore and characterize silicate-based luminescent materials [4,5]. When rare-earth ions are doped into the silicate host, the properties of the materials are changed due to the occupation of metal ion sites by doped ion in the host lattice. Therefore the properties of doped ions should be kept in mind during the design and development of advanced silicate phosphor materials having specific practical approaches in white light-emitting diodes (WLEDs). For research and development of WLEDs, highly efficient and stable red silicate materials having low color-related temperature and high CRI are required [6].

Disilicates and trisilicates of rare earths have been widely studied and special attention has been given to the silicate host lattices because of their potential applications in the fields of nuclear medical diagnostics, high-energy physics, positron emission tomography, plasma display panels, flat panel displays, ionizing radiation detection, solid-state lighting, and excitation with gamma rays and thermal neutrons. When silicate materials are doped with lanthanide ions, they show excellent luminescence response, which can be used in lasers and several display devices, for example, $Y_2Si_2O_7$ material when doped with cerium (III) ion finds its practical applicability in cathode ray tubes (CRTs) [7].

Disilicates of europium, terbium, and erbium (III) all exhibit a high degree of polymorphism reflecting unstable ($4f^5$, $4f^6$, $4f^{10}$, and $4f^{11}$) electronic configuration of these ions in this state. The crystal chemistry of these rare-earth disilicates is found to be quite complicated because the phase of these materials changes with temperature. Over the years, many efforts have been made to develop new silicate materials having excellent luminescence properties. Europium (III)-doped $M_2Si_2O_7$ (M = Y, Gd, and La) is a very efficient phosphor material and exhibits the phenomenon of polymorphism. Also, different synthetic procedures have been used for preparing silicate-type luminescent materials like solid-state reaction [8,9], spray pyrolysis [10], the solution combustion method [11], the hydrothermal method [12], etc. However, the sol-gel technique is an attractive and potential alternative for the synthesis of silicate phosphors because it is much simpler and has high chemical homogeneity, low processing temperature, controls the size and morphology of the particles, and has good optics transparency [13]. Currently, there is growing interest in preparing disilicates and trisilicates by solid-state reaction and the sol-gel process at low temperature [14]. In addition, using the proper synthesis technique is an important consideration in the development of phosphors [2].

15.2 Approach for energy-saving and eco-friendly lighting

In the age of digitization, demand for energy increases day by day. So, energy saving plays a vital role in the present scenario. The proposed work intends to energy-efficient lighting systems. In the age of digitization, the insufficiency of energy is regularly experienced when electrical power created on Earth is wastefully expended for lighting applications [15]. Calamitous conditions arise if energy is not used prudently and shrewdly for lighting. The ever-rising interest in energy may cause a host of changes on the environment and this should be supported before time runs out. The root of this issue might be the wastefulness shown by gadgets that consume high quantities of energy for a low yield. With regard to lighting, the key is the proper selection of high-productivity phosphors capable of giving good visible light emission. In the last few years, LEDs have attracted a lot of attention and popularity. It is foreseen as the ultimate lamp for the near future, outperforming all other lighting sources that have demonstrated their wastefulness when converting electrical energy into light [16,17]. In this era, LEDs are the best choice for lighting technology. LEDs are the most recent contestants in the field of lighting; they have many benefits and are promising candidates for the future's low-cost lighting. LEDs can transmit light of an intended color without the utilization of color channels, which conventional lighting techniques require. This proves to be more efficient and can lower the initial costs of incandescent and fluorescent lamps. Incandescent and fluorescent sources often require an external reflector to collect light and direct it in a usable manner. However, LEDs can be designed to focus their light.

15.2.1 Light sources and luminaires

15.2.1.1 Light sources

Currently, various lighting technologies are available in the lighting industry. These are available in the market with different colors, voltages, lumen output, shapes, sizes, and cost. In general, lighting devices can be classified into three categories: (1) incandescent lamps, (2) fluorescent lamps, and (3) solid-state lamps.

Incandescent lamps

Incandescent lamps are a widely used, well-known, and simple lighting technology. These types of lamp work on the principle of incandescence, which produces light by heat. When the electric current is passed through the thin metal filament, the filament is heated, glows, and produces light. Low output is the one big drawbacks of this lighting technology. According to the literature, incandescent lamps convert only 15%—30% of electrical energy into light; the remaining 70%—85% of energy is dissipated as heat. This means that the efficiency of these lamps is very low [18,19]. In the current scenario, various lighting devices are available in the market, which are highly efficient and environmentally friendly.

Fluorescent lamps

A fluorescent lamp is also known as a discharge lamp. It is a type of electrical lamp that uses ultraviolet (UV) light emitted by mercury vapor to excite a phosphor, which emits visible light. Fluorescent lights are normally used to provide lighting for business premises, dwellings, and retail units. The sizes, light colors, and wattage of fluorescent tubes fluctuate fundamentally. The most widely recognized trait of fluorescent lighting is that the lights are mounted inside an apparatus, regularly containing somewhere in the region of one to eight lights, with the installations themselves mounted in, or dangling from, a building's ceiling. There are a few issues with this type of bright light. In spite of the fact that these lights are

IV. Lighting and light emitting diodes

low cost, maintenance and performance issues regularly emerge after some time. Fluorescent lighting is intended to work for a specific time, and it is well known that fluorescent tubes tend to have shorter lifetimes. The more often they are actuated, the shorter the lifetime of the tube. Fluorescent lights are additionally susceptible to temperature variances and fail to perform in colder temperatures.

Light-emitting diodes

Solid-state lighting as LEDs is a developing innovation with potential to enormously surpass the effectiveness of traditional lamp-based lighting frameworks. While energy efficiency is the essential inspiration behind solid-state lighting, LEDs are additionally foreseen to have altogether new functionalities to lighting frameworks, enormously improving the way in which we utilize light. LEDs have already supplanted conventional lights in various lighting frameworks, including traffic lights, signs, and displays. A significant number of these applications require monochrome light, and the narrowband emission properties of LEDs present a reasonable favorable position over separated light methodologies. Nonetheless, the best effect of solid-state lighting will probably be to brighten applications that demand a high-quality white light source. Combined with the requirement for high effectiveness, the stringent color necessities of white light create new problems for LED innovations, a significant number of which presently do not last. A significant test for LEDs is conveying notable effectiveness at the present densities and temperatures, which are important for high-control activity. LED-based lighting frameworks should likewise provide productive and adaptable methods for dealing with the extraction and distribution of light. Because LEDs have altogether distinctive form factors and dielectric properties compared to conventional lights, new systems for proficiently extracting light from the LED chip as well as controlling the light emission properties are being investigated [20,21].

LEDs are one of the best applications of solid-state lighting, because they have superior lighting output. LEDs can revolutionize the lighting and display industries because of several advantages such as small size, durability, long lifetime, fast switching, and high luminescence efficiency [22—24]. Generally, three methods are known for the generation of white light by LEDs and phosphors: (1) a combination of individual monochromatic LED lamps that emit red, green, and blue colors; this method has several difficulties like disordered electrics, mismatched aging properties, and high cost, (2) red, green, and blue phosphors are excited by a near-UV LED chip, and (3) yellow-emitting phosphors are excited by blue LEDs. Methods (2) and (3) are very good for academic research and practical applications. The blue InGaN LED chip-excited yellow color-emitting $Y_3Al_5O_{12}:Ce^{3+}$ phosphor is used commercially; it was founded by the Nichia Corporation in 1996 [25—28].

15.2.1.2 Luminaires

The fundamental capacity of the luminaire is to direct light by utilizing reflective and protecting materials. Many lighting upgrade projects consist of replacing at least one of these parts to improve apparatus proficiency. Then again, users may consider replacing the whole luminaire with one that will effectively give a suitable amount of light. Luminaires can be of the conventional type, for example, a recessed or surface-mounted incandescent, fluorescent, or other electric-emission luminaire. Luminaires can likewise be of the nontraditional type, for example, fiber optics with the light source at one location and the fiber center or "light

15.2.2 Conventional lighting versus energy-efficient lighting

The present energy-efficient bulbs are accessible in a wide scope of colors and light levels. While the underlying cost of energy-efficient bulbs is regularly higher than customary incandescent bulbs, the more up-to-date bulbs are more economical. Many of the current bulbs last longer than traditional bulbs, so they will not have to be changed as frequently (Table 15.1).

15.2.3 Energy consumption in new arrangement

Energy is one of the significant contributors to financial improvement of any nation. In the case of developing nations, energy generation may be quite basic but the ever-increasing desire for energy will need enormous efforts to meet the demand. Financial development is attractive for developing nations, and energy is fundamental for economic development. However, the connection between economic development and the need to expand energy is not a direct one. For instance, under current conditions, a 6% expansion in India's gross domestic product would force an expanded interest of 9% on its energy segment [29]. The fundamental point of energy security for a country is to decrease its reliance on imported

TABLE 15.1 Comparison between incandescent lighting and light-emitting diode lamps [Include references].

Topic	Incandescent lighting	LED	Best lighting system
CCT	A wide range of CCT values. There are three primary options: soft white (2700 −3000 K), cool white (3500−4100 K), and daylight (5000−6500 K)	Wide CCT values generally in the range of 2200−6000 K	—
CRI	Excellent CRI values in the range of 95−100	Very broad spectrum of CRI values in the range of 65−95	Incandescent lamp
Efficiency	Poor efficiency	Highly efficient, generally 50 lm/watt	LED
Directionality	All incandescent lights emit light omnidirectionally	Emits light at 180 degrees	LED
Lifetime	Poor lifetime of any bulb (roughly 1200 h)	Long lifetime generally in the range from 25,000 to 200,000 h	LED
Cost	Cost is very low but it has the highest energy costs	High initial costs and low lifetime costs	LED
Size	Available in all sizes	Extremely small size	LED
Warranty	No warranty	5−10 years	LED

CCT, Correlated color temperature; *CRI*, color rendering index; *LED*, light-emitting diode.

IV. Lighting and light emitting diodes

energy hotspots for its financial development. Energy protection and energy efficiency are independent but related ideas. Energy protection is accomplished when the development of energy utilization is decreased, estimated in physical terms. Energy conservation can, accordingly, be the consequence of a few procedures or advancements, for example, profitability increment or mechanical advancement. Then again, energy proficiency is accomplished when energy power in a particular item, procedure, or area of creation or utilization is diminished without influencing yield, utilization, or comfort levels. Advancement of energy productivity will add to energy preservation and is hence a basic piece of energy protection limited by time strategies. In the current scenario, the inadequacy of energy is all the more regularly felt when electrical power produced on Earth is wastefully expended on lighting applications [30]. Disastrous conditions will prevail if energy is not used prudently for lighting. The ever-rising interest in energy may cause a large number of changes on Earth and this should be addressed before time runs out. The root of this issue might be the wastefulness shown by gadgets that consume high amounts of energy for a low yield. With regard to lighting, the key is the proper selection of high-proficiency phosphors capable of giving good light emission. LEDs have attracted a lot of attention and popularity and are predicted to be the definitive source of light, outperforming all other traditional sources that have demonstrated their wastefulness in their transformation of electrical energy into light [16,17]. LEDs are the most recent contestants in the field of lighting; they have many benefits and are promising candidates for the future's low-cost lighting. Aside from energy productivity, it is similarly essential to guarantee the environmentally friendly nature of the lighting innovation. If every one of these conditions is to be fulfilled, there is no other choice than solid-state lighting. Phosphor-converted light-emitting diodes (pc-LEDs) have favorable advantages over other conventional sources of lighting, especially in terms of luminous efficiency, low power consumption, durability, eco-friendliness, and long operational lifetime [31–34].

15.3 Synthesis methods for alkaline-earth silicate-based phosphor

In this work, we explain the different synthetic procedures and characterization techniques used to develop silicate-based phosphor materials, including their future prospective. There are many different methods to synthesize inorganic phosphors. The most popular method is solid-state synthesis. However, the sol-gel/Pechini, coprecipitation, hydrothermal, spray pyrolysis, and combustion synthesis methods are also often-used methods. The selection of a synthesis method depends on end use.

15.3.1 Solid-state method

The solid-state reaction method is a very popular and extensive method for preparing microcrystalline phosphor. It is also called the solid-state diffusion method. Generally in this method we use a precursor in the form of oxide and carbonate. The solid-state method is the best method for solid-state lighting because prepared microsized phosphor is very helpful in coatings. In this method, chemical reactance does not require any solvent for the preparation of our samples. With the help of the solid-state reaction method, we have prepared our sample material in powder form without any impurities. The advantage of this method is that no waste materials are produced during the synthesis process and thereafter. So we can say that this method is also eco-friendly [35]. However, the solid-state reaction method is lengthy, requires a high temperature, and is a high-power consumption method. Another limitation is that the size of crystallites cannot be controlled by changing the synthesis condition. The solid-reaction method is easily explained step by step with the help of Fig. 15.1.

The synthesis of phosphors through the solid-state reaction method uses the following steps:

1. All precursors are weighed in a stoichiometric ratio and put in a crucible.
2. After all the precursors are weighed, they are ground for about 30 min by using a mortar and pestle. Grinding is necessary to obtain a homogeneous powder of reactance.
3. The crucible is then put in a furnace for 4–8 h at 400–600°C for preheating.
4. The sample is ground again and returned to the furnace at a higher temperature (800–1500°C) for 2448 h.
5. Finally, the product is ready and cooled to room temperature. The heating and mixing is repeated until a homogeneous product formation is complete, further obtained powder sample is used for characterization.

15.3.2 Coprecipitation method

The coprecipitation method is very effective for the synthesis of phosphor. It is a simple method and does not require any costly equipment, stringent reaction conditions, or complex

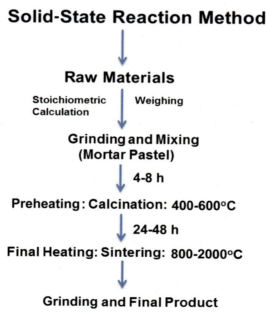

FIGURE 15.1 Sample preparation diagram of the solid-state reaction method.

procedures. This method is an excellent choice for research scholars and students because it provides higher purity and better stoichiometric control. Through the coprecipitation method we prepared phosphor in microsized and submicrosized particles. The main disadvantages of this method are high cost, low final output, and length of time. This method always requires distilled water during synthesis [36,37]. In this method, all precursors are weighed in a stoichiometric ratio and placed in a beaker. When chemical reaction occurs, the solid formed sinks to the bottom of the solution precipitate, and is transformed into a solution due to the reaction. Good mixing results in a more homogeneous product, particularly in the case of coprecipitation. The precipitated compound is sintered at high temperature to facilitate diffusion of the activator and formation of an appropriate number of defects in the crystallite materials.

Several silicate-based luminescent materials have been prepared by the coprecipitation process. $Sr_2MgSi_2O_7:Eu^{2+}$, Dy^{3+} materials were prepared by this synthetic procedure. Preparation was done using aminopropyltriethoxysilane to carry out the precipitation of metal nitrates [38]. The coprecipitation process gave rise to $La_{10}Si_6O_{27}$ and (Y,Gd) $SiO_5:Tb^{3+}$ green-emitting materials [39]. An $La_{9.33+x}Si_6O_{26+1.5x}$ (where $x = 0-0.67$) composition was also investigated by this method using NH_3 as a precipitant [40].

Silicate materials are prepared by keeping in mind the composition of final products. T.T.H. Tam et al. synthesized Eu^{2+}-doped $Ba_2MgSi_2O_7$ silicate materials using the procedure given in Ref. [41]. A stoichiometric amount of metal nitrates was dissolved in distilled water and tetraethyl orthosilicate along with an activator oxide in ethanol and nitric acid solution. The aqueous form of all these initial precursors was stirred for 1 h until the formation of homogeneous solution. Next, the precursors were allowed to mix and stirred again continuously for 3 h. Consecutively, a stoichiometric amount of ammonium hydroxide solution was added to carry out the precipitation process. The resulting solution was then continuously stirred at 100°C until a white gel was formed. The gel was then allowed to dry at 200°C. Furthermore, the sample was calcined at high temperature for some time. Grinding of the powder was done with the help of agate mortar. If the doping of Eu^{2+} ion occurred, then the as-prepared powder underwent lengthy ion reduction using an H_2/N_2 (10%) mixture at high temperature. In the coprecipitation process, pH is the most important factor and must be carefully controlled for single-phase product formation. Several other factors also influence the coprecipitation process such as concentration of starting reactants, precipitant concentration, separation of solid precipitant from filtrate, and drying and calcination of the powder form of materials.

In this modified technique, $(C_2H_5O)_4Si$ powder was dissolved in C_2H_5OH with drops of nitric acid and then stirred for about half an hour [42]. Furthermore, a stoichiometric amount of activator and metal nitrates dissolved in N,N-dimethyl formamide (DMF) solvent was slightly heated for complete dissolution of the precursors. After the appearance of a transparent solution, this mixture was allowed to cool at room temperature and then silica solution was added to the mixture gently with continuous stirring. Subsequently, sodium hydroxide solution was added to the mixture to carry out the precipitation process. The solution was stirred again and then filtered; the precipitate obtained was washed with C_2H_5OH. The precipitate was allowed to dry at 80°C. Finally, the sample was calcined to a high temperature. If Eu^{2+} ion was used as a dopant, then a reductive atmosphere of H_2 and N_2 was used. The sample was then allowed to cool.

In the modified coprecipitation process, DMF was used as a solvent because initial precursors like metal nitrates and chlorides can be suitably dissolved in this solvent. It was used to avoid the agglomeration of silicate particles by reducing the rate of precipitation reactions. The coprecipitation method requires a very low reaction temperature, gives rise to uniformly distributed nanosized materials, and is suitable for the large-scale production of phosphor materials. However, there is wide agglomeration, poor grain size distribution, and morphological study of nanoparticles is needed using this method.

Currently, a modified coprecipitation technique is used to overcome these limitations. In this method, stabilizing agents are used to achieve better particle properties of the luminescent materials. To implement a reaction using the modern coprecipitation technique, inert atmospheric conditions are employed to carry out the reduction process.

Since the precursors should dissolve but the precipitate should not dissolve in the solvent, the solubility of both precipitates and precursors in the solvent should be confirmed before starting the procedure. The precipitates are separated by filtering or centrifuging and are then heat treated to decompose the resulting hydroxide or carbonate. The powder is then annealed at high temperature to crystallize the powders. Because the use of expensive, high-purity, rare-earth fluorides as starting materials is not required (as for solid-state reaction), this is the preferred method for producing complex fluorides. Smaller particle sizes and a homogeneous distribution of the activator ions can be achieved using this technique [43]. In addition, the particles prepared by this method have a uniform, narrow size distribution, as opposed to powders prepared by solid-state reaction synthesis. The particle size and morphology are dependent on the pH, precursor and solvent choice, precipitating agent, stirring rate, and the order of addition of the compounds. The main drawbacks of this method are the difficulty in preparing chloride-based phosphors due to the dissolution of precipitates (metal chlorides) in the solvent, and the additional high-temperature annealing step required for the production of oxides [2].

15.3.3 Spray pyrolysis method

Spray pyrolysis is a technique used to synthesize a wide variety of high-purity, homogeneous ceramic powders with very uniform size and spherical morphology [2]. Spray pyrolysis is one of the most common, inexpensive, simple, and quite versatile methods used for synthesizing submicron and spherical-shaped luminescent materials [44], multicomponent oxides, chlorosilicates, metals, catalysts, and nanopowders [45]. Spray pyrolysis is a continuous and single-step process that gives rise to materials having finite size and uniform shape of particles in nonaggregated form. Spray pyrolysis may be applied in the form of aerosol spray pyrolysis or flame spray pyrolysis for producing phosphor materials. Flame spray pyrolysis is the modified form of ultrasonic spray pyrolysis.

The process of spray pyrolysis used for material synthesis is completed in four steps. Drop formation occurs from the precursor in first step, which on evaporation resulted in shrinkage of drop size. Now, precursors are transformed into oxides that result in the formation of solid products in the last step [38]. The experimental setup involves an ultrasonic generator to produce droplets, filter precipitator, and burner. The precursor solution is prepared by dissolving metal nitrates, activator ions, and silica content in deionized water. For aerosol formation of the precursor, the initial solution is allowed to set on an ultrasonic vibrator. Now, O_2 and fuel gas (CH_4 diluted with N_2) are provided via an outer and inner circular coaxial tube for the oxidation of precursor aerosols. A vacuum pump is also used to withdraw the aerosol. Finally, a silicon fiber filter is used for collecting the particles [40]. Later on, changes are made during the processing of silicate formation, where, prior to filtration, the precursor aerosols are preheated in the furnace at low temperature for solvent evaporation and then introduced in the furnace at high temperature to decompose the precursors. There is good control of the chemical composition and morphology of products [46], which are highly pure and homogeneous. However, materials synthesized using the large-scale spray pyrolysis technique are hollow and porous, which is a major drawback. This hollowness or porosity has strong effects on the thermal stability and morphological characteristics of the phosphor materials. To overcome the limitation of hollowness or porosity, a modified spray pyrolysis technique is proposed where artificial colloidal solution is employed for material synthesis [47].

Different silicate-based reactions were experimented on by the spray pyrolysis process. $Y_2Si_2O_7$:Eu/SiO_2 core−shell luminescent materials were synthesized using the flame spray pyrolysis technique. In these silicate materials, the phosphor core present at the center was surrounded by a silica coating. This coating helped in stabilizing the surface of particles and resulted in luminescence enhancement of the core materials [48]. Mn-doped Zn_2SiO_4 materials were synthesized by this technique using a filter expansion aerosol generator [49]. $Ca_8Mg(SiO_4)_4Cl_2$ luminescent materials doped with europium were successfully synthesized using the pyrolysis technique from the colloidal solution having NH_4Cl. Europium-doped barium strontium silicate materials were also explored using the pyrolysis process in the presence or absence of ammonium chloride as a flux [50]. This technique gave rise to Y_2SiO_5:Tb^{3+} green-emitting materials having NH_4F as a flux [51]. Eu^{2+}-doped strontium silicate phosphors were also prepared using this technique and the characteristics of the powder were studied with or without the addition of a flux like NH_4Cl [52]. Eu^{3+}-doped $La_2Si_2O_7$ materials were explored using the spray pyrolysis process and their PL characteristics were also studied. Mechanical activation effects on yttrium disilicate formation were studied in 2001 by G. Tzvetkov and N. Minkova.

15.3.4 Hydrothermal method

The hydrothermal method is a process that uses high temperature and pressure to precipitate materials directly from a solution. This method makes use of the increased solubility of almost all inorganic substances in water at elevated temperatures and pressures and subsequent crystallization of the dissolved material from the solution [53]. The term hydrothermal refers to any heterogeneous chemical reaction in the presence of a solvent (whether aqueous or nonaqueous) above room temperature and at a pressure greater than 1 atm in a closed system [54]. Hydrothermal processing is used to synthesize powders, single crystals, fibers, and ceramic materials [55]. It is a self-purifying technique that requires low processing temperature and is used to explore nanomaterials at high vapor pressures and temperatures.

Nowadays, the hydrothermal method has been coupled with various other processes, including ultrasound-assisted, electrochemical-assisted, surfactant-assisted, and microwave-assisted hydrothermal processing to control the reaction medium and improve the material characteristics [56]. Today, modified microwave-assisted hydrothermal synthesis is the trend for preparing these phosphor materials, which reduces the processing time and energy required for carrying out the reaction. This method also has a potential to improve material characteristics like morphology and the crystallite size of particles.

The main advantage of the technique is that despite the low synthetic temperatures, crystalline powders are obtained without the need for further heat treatment. By changing the pH, precursor selection, and reaction temperature, the size, phase, and morphology of the powders can be modified [57]. In addition, special sealed vessels, which can be operated at very high temperature and pressure (usually around 500°C and 340 atm), are required [58]. Furthermore, due to the small volume of the reaction vessels, mass production of phosphors is difficult.

15.3.5 Combustion method or self-propagating low-temperature synthesis

Combustion is a simple, rapid, versatile, low-cost, and effective method for producing a wide range of technologically useful oxides, ceramics, catalysts, and nanosized phosphor materials [59]. This is one of the most suitable and scaled-up methods for producing highly pure and homogeneous products having the desired properties of phosphor materials. Combustion is a promising powder preparation technique because it provides a high-temperature environment favorable for material synthesis. Different oxides like $Y_2O_3:Eu^{2+}$, $Y_2SiO_5:Ce^{3+}$, $Y_3Al_5O_{12}:Tb^{3+}$, $BaMgAl_{10}O_{17}:Eu^{2+}$, etc. have been successfully prepared by this technique [60].

Phosphor materials were synthesized by the combustion of aqueous solutions having stoichiometric compositions of starting reactants like metal nitrates and fuels. This process involves highly exothermic reaction between metal nitrates (oxidizers) and fuels like urea, glycine, carbohydrazide, oxalyl dihydrazide, citric acid, malonic acid dihydrazide, and tetraformal trisazine, which play the role of reducers. A semisolid paste of this redox mixture when introduced in a muffle furnace at high temperature results in a voluminous and fluffy powder in large amounts. Chemical reaction between reactants starts at 500°C), the mixture is ignited vigorously producing flame, and the process is completed within a few minutes [61].

A number of gases evolved during combustion such as CO_2, O_2, N_2, NO_2, H_2O, NH_3, etc. Different fuels were used for producing oxide materials but easy availability and high exothermicity of the urea made it a first choice for producing a large variety of alumina-based materials [62]. By variation of fuel and oxidizer flow rates, flame temperature and particle residence time can be easily controlled. Factors affecting combustion synthesis are stoichiometric ratio, particle shape, size and processing of reactant particles, ignition technique, adiabatic temperature, etc. This method is comparatively simple and low cost. Due to a very short reaction time, the process completes within 10 min and results in the formation of highly pure final products. However, the combustion process has a number of complications during the synthesis of nanophosphor materials because there is the requirement for additional heating in this process for exploring the pure-phase nanosized materials. This method is not environmentally friendly because harmful gases like CO_2, N_2, N_2O, etc. are also evolved during the processing of these materials. The combustion procedure is difficult to control due to the very short reaction time. Postprocessing steps like grinding or milling are also required, which may introduce impurities in the final products.

IV. Lighting and light emitting diodes

Combustion synthesis is a popular method to rapidly produce multiconstituent oxides [63]. The products are generally chemically homogeneous, have fewer impurities, and have higher surface areas compared to solid-state-synthesized powders. Due to the short duration at high temperatures, nanocrystalline powders are produced. However, they are agglomerated into hard, submicron-sized powders that are not easily separated. To improve the purity, homogeneity, and crystallinity of complex oxide powders, Garcia et al. [56] developed a new method using hydrazine as a fuel in a high-pressure reaction vessel. This method is superior because of: (1) increased personal safety due to protection by the vessel; (2) total recuperation of the product; (3) more precise control of reaction parameters as it is possible to measure the temperature, pressure, and gas flow rates; and (4) a controlled atmosphere (reactive or inert). By this method, a variety of luminescent materials such as silicates, phosphates, borates, oxyfluorides, oxychlorides, and aluminates have been simply and expeditiously prepared [2].

15.3.6 Sol-gel method

The sol-gel method is one of the most well-establish methods for the preparation of nanophosphors. This method has potential control of the whole reaction during synthesis. Generally in this method, we use nitrides, sulfates, and chlorides to form precursors. The sol-gel method can be classified in two ways: aqueous sol-gel and - nonaqueous sol-gel methods.

- **Aqueous sol-gel method**: In this method, we use water as a solvent medium.
- **Nonaqueous sol-gel method**: In this method we use organic solvent as a solvent medium.

The method proceeds through a series of chemical processes, including hydrolysis, gelation, drying, and thermal treatment. The process involves the formation of sol through hydrolysis and polymerization reactions and subsequent attainment of a rigid porous gel. The final product is then obtained by removal of the solvent and residuals from the pores of the gel by aging drying and annealing [64–66].

There has been an intense level of research activity on the development of luminescent materials by the sol-gel method [67]. Silica gel and powders can be easily prepared from rice husk ash either by applying thermal treatment ($500-1400°C$) or by leaching due to acidic or basic solutions followed by neutralization with an acid [68]. Sol-gel is an effective, versatile, and low-cost synthesis technique used for the preparation of nanoscale materials of excellent brightness and chromaticity [69,70]. This synthetic route is very simple and is used for large-scale production of silicate-based phosphor materials having high luminous efficiency. Several other luminescent materials based on different host lattices like oxides, halides, borates, phosphates, tungstates, molybdates, sulfides, oxysulfides, etc. have also been synthesized using sol-gel technology. Rare-earth-doped lutetium pyrosilicate ($Lu_2Si_2O_7$)-nanosized materials were also synthesized by this technique [71]. Eu^{3+}-doped $Y_2Si_2O_7$ phosphors were prepared using this synthetic procedure [72]. An Eu^{3+}-doped M_2SiO_4 type of silicate-based nanomaterial was also successfully synthesized using the sol-gel process [73]. Now, the method used for preparing a series of europium (III)-doped $M_2Si_2O_7$ (M = Y, Gd, and La) materials is explained briefly here. A comparative study of different conventional techniques applied for the various material syntheses is also compiled in Table 15.2.

IV. Lighting and light emitting diodes

TABLE 15.2 Comparative analysis of the synthesis method, which is used for phosphor synthesis [48].

Synthesis method	Solid-state reaction	Sol-gel/pechini	Co precipitation	Hydrothermal	Combustion	Spray pyrolysis
Particle size	>5 μm	10 nm−2 μm	10 nm−1 μm	10 nm−1 μm	500 nm−2 μm	100 nm−2 μm
Particle size distribution	Narrow−broad	Narrow	Narrow	Narrow	Medium	Very narrow
Morphological control	Poor−good	Medium	Very good	Good	Poor	Very good
Purity	Poor−very good	Good	Medium	Medium−good	Medium−good	Medium−good
Cost	Low	Medium	Medium	Medium−high	Low−medium	Medium
Synthesis time	Short−long	Medium	Medium	Very long	Short	Short
Type of material synthesis	Aluminates, silicates, aluminosilicates, nitrides, sulfides, and a number of metal oxides	Magnetite and magnetite nanoparticles	Magnetite and magnetite nanoparticles	Submicron to nanosized materials, powders, single crystals, fibers, and ceramic materials	Oxides, ceramics, catalysts, and nanosized phosphor materials	Multicomponent oxides, chlorosilicates, metals, catalysts, nanopowders, and submicron- and spherical-shaped materials
Reaction conditions	Reaction is carried out at high temperature (1200−1600°C)	Highly pure and uniform nanomaterials achievable at low processing temperature	Coprecipitation method requires very low reaction temperature	Product formation occurs at low-temperature reaction conditions	Exothermic reaction occurs between metal nitrates and organic fuels at a temperature of ~500°C in ~5−10 min	Process involves drop formation from the precursors and its shrinkage due toevaporation. Finally, precursor transformation to solid products.
Advantages	Environmentally friendly method Easy availability of starting materials Simplicity of the method Thermodynamically stable Product formation	Simple and low-cost method High chemical homogeneity Controls the size and morphology of the particles Has good optics transparency	Uniformly distributed nanosized materials obtained	Reheating is not required, which prevents the particles from agglomeration Highly pure and homogeneous product formation	Scaled-up method for producing highly pure and homogeneous products having desired properties	There is good control of the chemical composition and morphology of the products formed

(Continued)

TABLE 15.2 Comparative analysis of the synthesis method, which is used for phosphor synthesis [48].—cont'd

Synthesis method	Solid-state reaction	Sol-gel/pechini	Co precipitation	Hydrothermal	Combustion	Spray pyrolysis
Disadvantages	Long reaction time High-temperature processing conditions are required, which enhance the particle size of materials Repeated milling and washing with chemicals tends to degrade the luminescence property of materials	Cost of precursors, shrinkage of a wet gel upon drying, and difficulty avoiding residual porosity and OH groups	Wide agglomeration, poor grain size distribution, and morphological study of nanoparticles	Requirement of expensive autoclaves and long processing time	Scaled-up method for producing highly pure and homogeneous products having desired properties	Requires additional heating for exploring pure-phase nanosized materials. Not eco-friendly.
Limitations	Extensive grinding and milling required	Requires a soluble precursor Carbon contamination Difficult to obtain nitrides, sulfides, and other nonoxide materials	Requires a soluble precursor Difficult to obtain nitrides	Requires a soluble precursor Special equipment needed	Requires a soluble precursor Carbon contamination Difficult to obtain nitrides	Requires a soluble precursor Difficult to obtain nitrides and other nonoxide materials

Metal oxides, in their appropriate molar ratios, were dissolved in a nitric acid solution (which was less costly than buying metal nitrate powders). Citric acid, a chelating agent, was then added along with ethylene glycol and polyethylene glycol (cross-linking agents). The solution was stirred at various temperatures, usually below 100°C, which produced a transparent gel. The gel was calcined at low temperature (~ 350°C) to remove organics and residual water, resulting in a nanocrystalline powder. The particle size and morphology can be controlled by adjusting the solvent and heating conditions, or by changing the precursors. A variety of stoichiometric, single-phase materials can be simply prepared by this method without the need for extensive grinding and milling. However, the phosphors can become contaminated with carbon due to the use of organic reagents and thus a preannealing step in air may be needed. Also, a reducing atmosphere may be required to obtain Eu^{2+}- and Ce^{3+}-activated phosphors [2].

15.3.7 Microwave method

The performance of rare-earth-doped inorganic phosphor depends on the composition and crystallinity of the host compound. Microwave heating is one of the fastest synthesis techniques used for preparing nanoparticles. Utilization of microwaves in the synthesis of materials is gaining significance. The microwave-assisted synthesis method is commonly a lot quicker, cleaner, and more prudent than traditional techniques. An assortment of materials, for example, carbides, nitrides, complex oxides, silicides, zeolites, apatite, polymers, metals, halides, sulfides, phosphates, and so on have been synthesized utilizing the microwave synthesis method [74].

Pedroso et al. [75] reported modified microwave-assisted solid-state synthesis for the preparation of Pr^{3+}-, Eu^{3+}-, or Tb^{3+}-doped Lu_2O_3-persistent luminescence materials. In the reported work, 15 g of granular activated carbon was placed in an alumina crucible, which was used as the microwave susceptor and CO was used as a reducing gas. The prepared materials mixture was placed in a second crucible, which was partially surrounded by activated carbon for efficient heat transfer to the reagents. The larger crucible was packed with alumina discs to avoid exposure to O_2. Both crucibles were placed inside a cavity cut into a block of aluminosilicate thermal insulation brick. The top and bottom of the block were closed with two flat lids of the same material. A schematic of a microwave-assisted solid-state synthesis device and crucible setup is depicted in Fig. 15.2 [75].

Work was also reported on material synthesis using the microwave method. C_3S or Ca_3SiO_5 material was prepared using the microwave technique [75]. Europium (III)-doped Sr_2SiO_4 materials were also explored by the microwave sintering technique [76]. Microwave heating is a rapid synthetic method that controls reaction parameters, the nucleation process, and has excellent morphological control of nanoporous materials. This method also reduces the processing time of chemical reactions and suppresses the formation of side products. However, the equipment used (dedicated microwave reactor) in this technique is not economically beneficial.

IV. Lighting and light emitting diodes

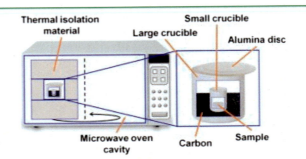

FIGURE 15.2 Microwave-assisted solid-state synthesis sdevice (*left*) and crucible setup (*right*). Reprinted with permission obtained from the American Chemical Society (2016)C.C.S. Pedroso, J.M. Carvalho, L.C.V. Rodrigues, J. Hölsä, H.F. Brito, Rapid and energy-saving microwave-assisted solid-state synthesis of Pr^{3+}-, Eu^{3+}-, or Tb^{3+}-doped Lu_2O_3 persistent luminescence materials, ACS Appl. Mater. Interfaces 8 (2016) 19593–19604. https://doi.org/10.1021/acsami.6b04683.

15.3.8 Liquid-phase precursor method

Liquid-phase synthesis is another method for the synthesis of silicate-based phosphor materials. This method may give a better level of transformation (~90%) compared to 20% for the high-pressure process. Solvents are broadly utilized as reaction media in liquid-phase synthesis. The variety of solvents can drastically change the response rate and system. This process is referred to as solvation impact, which can be quantitatively surveyed through solvation free energy. Compared to vapor-phase synthesis methods, significantly less attention has been given to understanding the variables that influence presentation during liquid-phase synthesis. Green-emitting phosphor $(Ba_{1-x},Sr_{1-x})SiO_4:2xEu^{2+}$, $x = 0.03$, 0.05, 0.1, and 0.15, was synthesized through the liquid-phase precursor process [77]. The liquid-phase precursor method is reported to result in phosphors with markedly increased emission intensities compared to other synthesis methods.

15.3.9 Micro-pulling method

The micro-pulling-down (μ-PD) technique is a helpful method to search for new materials. This technique can prepare single-crystalline materials rapidly. The synthesized materials by micro-pulling-down (μ-PD) technique are of adequate measurements for all the fundamental characterization by optical and luminescence techniques, including photo- and radioluminescence, decay kinetics, and light yield measurements. It is possible to grow shaped and/or device-sized crystals from the liquid by utilizing a solitary advanced procedure. Ongoing improvements to the μ-PD strategy make the nature of μ-PD precious crystals similar to those prepared by the Czochralski, Bridgeman, or other traditional development systems [78]. The μ-PD method employs an iridium crucible and after heating was optimized for Li_4SiO_4 crystal growth taking into account Li evaporation. To grow high-quality crack-free single crystals, the heating power was increased to establish a milder temperature gradient, thicker meniscus, and smaller crystal diameter, resulting in smaller stresses in the as-grown crystals. The crystals were grown by the μ-PD method with radiofrequency inductive heating. An iridium crucible with a die of 5 mm in diameter was used. Growth was encouraged under

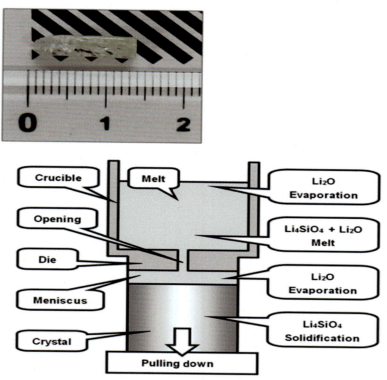

FIGURE 15.3 Schematics of the micro-pulling-down method adapted to grow Li$_4$SiO$_4$ single crystals employing self-flux growth with excessive Li$_2$CO$_3$. *Reprinted with permission obtained from Elsevier publication J. Pejchal, V. Babin, A. Beitlerova, S. Kurosawa, Y. Yokota, A. Yoshikawa, M. Nikl, Improvement of the growth of Li 4 SiO 4 single crystals for neutron detection and their scintillation and luminescence properties, J. Cryst. Growth 3 (2016) 1–8. https://doi.org/10.1016/j.jcrysgro.2016.02.008.*

an argon atmosphere using iridium wire as a seed (Fig. 15.3). The pulling speed was around 0.05–0.07 mm/min. The starting materials were prepared by mixing 4 N purity Li$_2$CO$_3$, glassy SiO$_2$ powder, and corresponding oxides of the doping ions (Ti$_2$O$_3$, Cr$_2$O$_3$, or Al$_2$O$_3$) in a mortar in a ratio corresponding to the stoichiometric Li$_4$SiO$_4$ composition. The mass balance during the crystal growth was kept in a way that the pulling speed was experimentally adapted to the Li evaporation rate, so that the melt composition remained constant and stable crystal growth was enabled [79].

15.4 Review of recently reported silicate-based phosphors

At present, a number of silicate-based phosphors have been reported for various applications such as WLEDs, through-lens detectors, solar cells, display applications, scintillators, etc. This section is divided into two parts. In the first part, we report a detailed study of earlier-reported silicate-based phosphors; in the second part, we list some silicate-based phosphors assisted by different synthesis techniques, which may be applicable for WLEDs.

15.4.1 Detailed study of earlier-reported silicate-based phosphors

15.4.1.1 Eu²⁺-activated barium silicate phosphors

Som et al. [80] reported Eu^{2+}-activated dual light-emitting barium silicate phosphors for the generation of warm WLEDs. In the present work, Som et al. synthesized $Ba_{1.95}Eu_{0.05}MgSi_2O_7$ phosphors by the solid-state reaction method. The XRD pattern of synthesized $Ba_2MgSi_2O_7$ phosphor was well matched with standard data and monoclinic crystal structure. Fig. 15.4A shows PL emission and excitation spectrum of the synthesized $Ba_{1.95}Eu_{0.05}MgSi_2O_7$

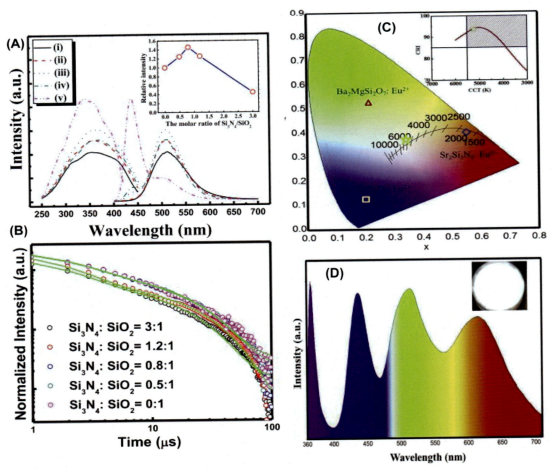

FIGURE 15.4 Characterization results of Eu^{2+} ion-activated $Ba_2MgSi_2O_7$ phosphor with various molar ratios of $Si_3N_4:SiO_2$: (A) photoluminescence (PL) excitation and emission spectra. (B) PL decay curve. (C) CIE coordinates of the synthesized phosphors. (D) Electroluminescence spectra of the optimized fabricated white light-emitting diode. *Reprinted with permission obtained from Elsevier publication S. Som, P.H. Kuo, C.H. Lu, Facile synthesis of Eu^{2+} activated dual light emitting barium silicate phosphors for warm-white light emitting diodes, J. Alloys Compd. 790 (2019) 1060–1064. https://doi.org/10.1016/j.jallcom.2019.02.322.*

15.4 Review of recently reported silicate-based phosphors

phosphors with various molar ratios of $Si_3N_4:SiO_2$. The PL emission spectra of prepared phosphors at an $Si_3N_4:SiO_2$ ratio of 0:1 revealed a broad emission band, which emitted blue and green emissions, and the center point of the band was observed at 513 nm due to the $4f_6^5d_1-4f_7$ transition of Eu^{2+} ions. When the molar ratio of $Si_3N_4:SiO_2$ was increased to 0.8:1, emission intensity also increased. The excitation spectra of synthesized phosphor for an $Si_3N_4:SiO_2$ ratio of 0:1 revealed a broad excitation band in the range of 250—450 nm. This excitation spectrum was observed due to 4f—5d transition for Eu^{2+}. Fig. 15.4B shows a PL decay curve of Eu^{2+} ion-activated $Ba_2MgSi_2O_7$ phosphor at 1250°C with various molar ratios of $Si_3N_4:SiO_2$. In the present study, synthesized Eu^{2+} ion-activated $Ba_2MgSi_2O_7$ phosphor with an $Si_3N_4:SiO_2$ ratio of 0.8:1 at 1200°C revealed blue and green emissions under near-UV light with excellent color stability. The WLEDs were fabricated with the prepared blue/green phosphors and commercial red phosphors. In the present case, it is obvious from the figure that white light with correlated color temperature (CCT) < 5500 K and CRI > 90 (blue-shaded zone) can be essentially accomplished. An arrangement of WLEDs was then fabricated by covering as-arranged blue/green and commercial red phosphors on an InGaN UV-LED chip to confirm the simulation results. The WLEDs with a chromaticity coordinate at (0.34, 0.37), a CRI of 94, and a CCT of 5213 K were obtained and appear as a star point in Fig. 15.4C. The electroluminescence spectra of the fabricated WLEDs are shown in Fig. 15.4D. The inset image of Fig. 15.4D presents the fabricated WLED with high brightness. In the current research, rather than two unique phosphors, a solitary blue/green light-producing phosphor was consolidated with commercial red phosphors to accomplish proficient, warm white light with high CRI in WLEDs [80].

15.4.1.2 Near-UV-converting cyan-emitting $RbBaScSi_3O_9:Eu^{2+}$ phosphor

Ray et al. [81] successfully synthesized $RbBaScSi_3O_9:xEu^{2+}$ ($x = 0.2, 0.5, 2, 5, 7$) phosphor, which was confirmed by XRD pattern. Rietveld refinement was also performed, and showed a crystalline form of the synthesized phosphor with a single phase. $RbBaScSi_3O_9:Eu^{2+}$ phosphor has a monoclinic phase with a P21/n space group. The electronic structure of synthesized $RbBaScSi_3O_9:xEu^{2+}$ phosphor was also studied by the authors. The luminescence properties of the synthesized phosphor were also investigated in detail. The PL excitation and emission spectra are revealed in Fig. 15.5A and B. The PL excitation spectra revealed strong absorption at 365 nm in the near-UV region and peak excitation peak at 425 nm in the blue region, which arises due to the $4f_7 \to 4f_6^5d_1$ transitions of Eu^{2+} ions doped in the $RbBaScSi_3O_9$ host. These excitation wavelengths are well matched with commercial near-UV (360—400 nm) LED chips, which are applicable in WLEDs. Under 365 nm excitation, PL emission spectra were investigated for various concentrations of Eu^{2+} ions. The PL emission spectra depict broad emission spectra at 492 nm. In the PL emission spectra, two negligible traces of Eu^{3+} emission peaks can be observed at 589 and 612 nm corresponding to the $^5D_0 \to {}^7F_1$ and $^5D_0 \to {}^7F_2$ transitions of Eu^{3+}, respectively. Emission spectra depict the highest emission intensity at 2 mol% of europium ions. Subsequently, PL emission intensity decreased due to the concentration-quenching mechanism. By using the PL results, the authors also calculated critical distance (R_c) and type of interaction. According to Ray et al., the critical distance of $RbBaScSi_3O_9:2mol\%Eu^{2+}$ phosphor was found to be 27.21 Å and electric dipole—dipole transition was responsible for energy transfer. The authors also reported 79% internal quantum efficiency of $RbBaScSi_3O_9:2mol\%Eu^{2+}$ phosphor in the near-UV

IV. Lighting and light emitting diodes

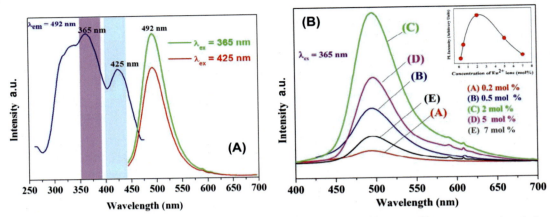

FIGURE 15.5 (A) Photoluminescence (PL) excitation spectrum of RbBaScSi$_3$O$_9$:Eu^{2+} phosphor and emission spectra, which were monitored under excitation at 365 and 492 nm. (B) PL emission spectra of RbBaScSi$_3$O$_9$:xEu^{2+} (x = 0.2, 0.5, 2, 5, 7) phosphor under 365 nm excitation and concentration-quenching spectra. *Reprinted with permission obtained from Elsevier publication S. Ray, P. Tadge, S.J. Dhoble, G.B. Nair, A. Singh, A.K. Singh, M. Rai, T.M. Chen, V. Rajput, Structural and spectroscopic characterizations of a new near-UV-converting cyan-emitting RbBaScSi$_3$O$_9$:Eu^{2+} phosphor with robust thermal performance, J. Alloys Compd. 713 (2017) 138–147. https://doi.org/10.1016/j.jallcom.2017.03.366.*

region. The synthesized phosphor showed very low thermal quenching (89.28%) relative to the PL intensity at 150°C with respect to that measured at room temperature. The chromaticity coordinates were located at (x, y) = (0.121, 0.307) giving a clear indication that the light emitted by the phosphor was cyan in color. All these results indicate that synthesized phosphor has great potential for industrial applications of this particular phosphor [81].

15.4.1.3 Ce^{3+}/Tb^{3+} codoped CaLa$_4$Si$_3$O$_{13}$ phosphors

In the present work, Khan et al. [82] reported that Ce^{3+}/Tb^{3+} co-doped CaLa$_4$Si$_3$O$_{13}$ phosphors were successfully synthesized by the solid-state reaction method. The authors reported that synthesized phosphors have a hexagonal apatite structure with a P6$_3$/m space group. The luminescence properties of synthesized phosphor were investigated by PL characterization. The PL emission spectra of CaLa$_4$Si$_3$O$_{13}$:Ce^{3+} phosphors show broad emission spectra in the range of 370–550 nm, which was monitored under a 355 nm excitation wavelength, as shown in Fig. 15.6A. The emission spectra show highest intensity at Ce = 0.06. The PL emission and excitation spectra also observed for CaLa$_4$Si$_3$O$_{13}$:Ce^{3+} phosphors are depicted in Fig. 15.6B, which shows PL emission peaks at 415, 440, 490, 552, 594, and 625 nm, ascribed to the transitions due to $^5D_3 \rightarrow {}^7F_j$ (J = 5, 4) and $^5D_4 \rightarrow {}^7F_J$ (J = 6, 5, 4, 3) under 237 nm excitation. Subsequently, the energy-transfer mechanism of Ce^{3+}/Tb^{3+} codoped CaLa$_4$Si$_3$O$_{13}$ phosphors was studied in detailed. The PL excitation and emission spectra of Ce^{3+}/Tb^{3+} codoped CaLa$_4$Si$_3$O$_{13}$ phosphors are shown in Fig. 15.6C. The emission spectra show blue and green emission under 355 nm excitation. The PL studies of Ce^{3+}/Tb^{3+} codoped CaLa$_4$Si$_3$O$_{13}$ phosphors reveal an energy-transfer mechanism Ce$^{3+} \rightarrow$ Tb^{3+}. Khan et al. reported 93% efficiency of energy transfer. Fluorescence decay time curves of CaLa$_4$Si$_3$O$_{13}$:0.06Ce^{3+}/yTb^{3+} phosphors were also investigated. Fig. 15.6D shows a CIE

15.4 Review of recently reported silicate-based phosphors

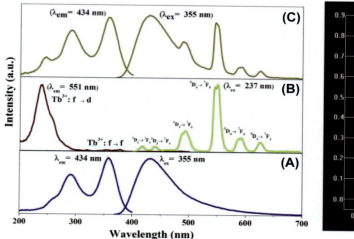
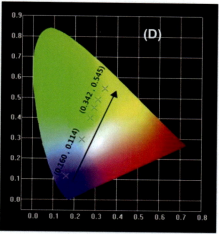

FIGURE 15.6 Photoluminescence (PL) emission and excitation spectra of (A) CaLa$_4$Si$_3$O$_{13}$:0.06Ce^{3+}, (B) CaLa$_4$Si$_3$O$_{13}$:0.4Tb^{3+}, and (C) CaLa$_4$Si$_3$O$_{13}$:0.06Ce^{3+}/0.2Tb^{3+} phosphors. (D) CIE chromaticity diagram of CaLa$_4$Si$_3$O$_{13}$:0.06Ce^{3+}/yTb^{3+} phosphors (y = 0, 0.1, 0.2, 0.3, 0.4, and 0.5; ultraviolet excitation at 293 nm). *Reprinted with permission obtained from Elsevier publication S.A. Khan, W. Ji, L. Hao, X. Xu, S. Agathopoulos, N.Z. Khan, Synthesis and characterization of Ce^{3+}/Tb^{3+} co–doped CaLa$_4$Si$_3$O$_{13}$ phosphors for application in white LED, Opt. Mater. 72 (2017) 637–643. https://doi.org/10.1016/j.optmat.2017.06.034.*

chromaticity diagram of CaLa$_4$Si$_3$O$_{13}$:0.06Ce^{3+}/yTb^{3+} phosphors (y = 0, 0.1, 0.2, 0.3, 0.4, and 0.5; UV excitation at 293 nm). The CIE chromaticity diagram shows color-tuning characteristics from the blue region (0.160, 0.114) to the green region (0.342, 0.545), on account of the characteristic blue emission of Ce^{3+} ions and the green emission of Tb^{3+} ions, respectively. All these results and experimentations revealed that synthesized CaLa$_4$Si$_3$O$_{13}$:Ce^{3+}/Tb^{3+} phosphor has great applicability as WLED phosphors [82].

15.4.1.4 BaLu$_2$Si$_3$O$_{10}$:Eu^{3+} red-emitting phosphors

Annadurai et al. [83] synthesized a series of BaLu$_{2-x}$Eu$_x$Si$_3$O$_{10}$ (x = 0.1, 0.5, 0.6, 0.7, 0.9, and 1.0) phosphors by the solid-state method. The synthesized phosphors show a monoclinic crystal structure with a P2$_1$/m space group. The PL excitation and emission spectra of the synthesized phosphors are shown in Fig. 15.7A. The PL emission spectra show several sharp emission peaks at 579, 591, 610, 656, and 703 nm under excitation at 395 nm, which can be assigned to the transitions from $^5D_0$ to ^7F_J (J = 0, 1, 2, 3, and 4) of Eu^{3+} ions, respectively, as shown in Fig. 15.7B. Critical distance (R_c) was also calculated by using the Blasse formula. It was found to be 8.35 Å, which is higher than 5 Å and confirms electric multipolar interaction is responsible for energy transfer between Eu^{3+} ions and host. The fluorescence decay curve indicates the lifetime of synthesized phosphor. The lifetime values of BaLu$_{2-x}$Eu$_x$Si$_3$O$_{10}$ phosphors were found to be 2.024, 1.957, 1.914, 1.865, 1.722, and 1.538 ms for x = 0.1, 0.5, 0.6, 0.7, 0.9, and 1.0, respectively. The synthesized phosphor showed good thermal stability. The quantum efficiency of synthesized phosphor was found to be 83.2%. Fig. 15.7C shows the CIE chromaticity diagram of BaLu$_{1.3}$Eu$_{0.7}$Si$_3$O$_{10}$ phosphor with CIE coordinates (x = 0.6541,

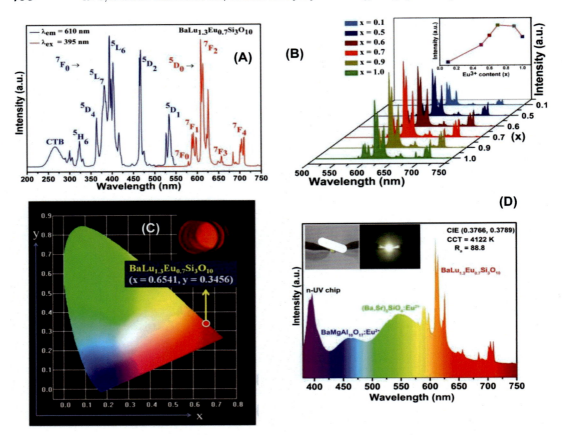

FIGURE 15.7 (A) Photoluminescence (PL) excitation and emission spectra of BaLu$_{1.3}$Eu$_{0.7}$Si$_3$O$_{10}$ phosphor. (B) PL spectra for BaLu$_{2-x}$Eu$_x$Si$_3$O$_{10}$ (x = 0.1, 0.5, 0.6, 0.7, 0.9, and 1.0) phosphors (λ_{ex} = 395 nm). (C) CIE chromaticity diagram for BaLu$_{1.3}$Eu$_{0.7}$Si$_3$O$_{10}$ phosphor. (D) Electroluminescence spectrum of the fabricated white light-emitting diode device. *Reprinted with permission obtained from Elsevier publication G. Annadurai, B. Devakumar, L. Sun, H. Guo, S. Wang, X. Huang, Crystal structure, photoluminescence properties and thermal stability of BaLu$_2$Si$_3$O$_{10}$:Eu^{3+} red-emitting phosphors with high color purity for near-UV-excited white LEDs, J. Lumin. 215 (2019) 116623. https://doi.org/10.1016/j.jlumin.2019.116623.*

y = 0.3456) and 97.8% color purity. The fabricated WLED device shown in Fig. 15.7D can emit bright white light with good CIE color coordinates (0.3766, 0.3789), low CCT (4122 K), and high R_a (88.8). These results further certified that the red-emitting BaLu$_{1.3}$Eu$_{0.7}$Si$_3$O$_{10}$ phosphor has great potential in the application of near-UV chip-based WLEDs [83].

15.4.1.5 Tm^{3+} and Dy^{3+} ion-codoped Ba$_3$La$_6$(SiO$_4$)$_6$ phosphors

Patnam et al. reported near-UV light-excited Tm^{3+} and Dy^{3+} ion-codoped barium lanthanum silica oxide phosphors for white light applications [84]. In the present work, the authors successfully synthesized Tm^{3+} and Dy^{3+} ion-codoped Ba$_3$La$_6$(SiO$_4$)$_6$ (BLSO) phosphor via the citrate-based sol-gel method. The synthesized phosphor was characterized by

15.4 Review of recently reported silicate-based phosphors

various techniques such as XRD, scanning electron microscopy (SEM), Fourier transform infrared, and PL techniques. The XRD patterns confirmed their phase purity and revealed a hexagonal crystal structure with a P63/m (176) space group. The SEM image of synthesized phosphor revealed particle size, and an energy-dispersive X-ray spectroscopy (EDS) study showed elemental composition. The EDS spectra showed elements of barium (Ba), lanthanum (La), silicate (Si), and oxygen (O) of the BLSO phosphors. The PL study was also carried out in detail. First, the authors studied PL excitation and emission spectra of Tm^{3+} and Dy^{3+} ion-doped BLSO phosphor separately, then energy-transfer and color-tuning studies were carried out. Fig. 15.8A and B shows PL excitation and emission spectra of BLSO:$0.1Dy^{3+}/0.05Tm^{3+}$ phosphor. PL excitation spectra show several excitation peaks monitored under 453 and 573 nm emission wavelength. In this figure we can clearly see that the excitation peaks

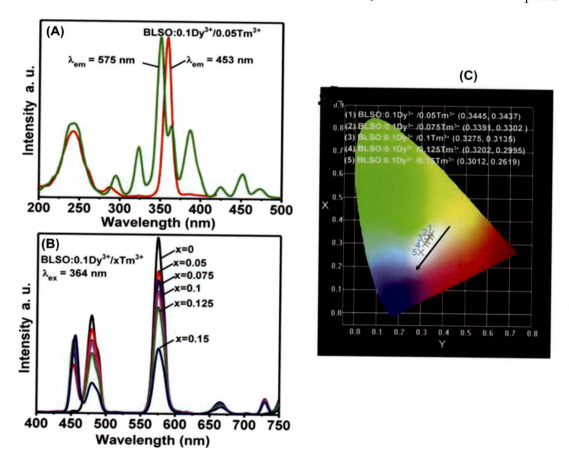

FIGURE 15.8 (A) and (B) Photoluminescence excitation and emission of $Ba_3La_6(SiO_4)_6$ (BLSO)$0.1Dy^{3+}/xTm^{3+}$ ($x = 0$, 0.05, 0.75, 0.1, 0.125, and 0.15) phosphors. (C) CIE chromaticity coordinate of BLSO$0.1Dy^{3+}/xTm^{3+}$ ($x = 0$, 0.05, 0.75, 0.1, 0.125, and 0.15) phosphors. *Reprinted with permission obtained from Elsevier publication H. Patnam, S.K. Hussain, L.K. Bharat, J.S. Yu, Near-ultraviolet excited Tm^{3+} and Dy^{3+} ions co-doped barium lanthanum silica oxide phosphors for white-light applications, J. Alloys Compd. 780 (2019) 846–855. https://doi.org/10.1016/j.jallcom.2018.11.321.*

overlapped at the 364 nm wavelength. Under 364 nm excitation, the PL emission spectra of the BLSO:0.1Dy^{3+}/xTm^{3+} (x = 0.05, 0.075, 0.1, 0.125, and 0.15) phosphors shows several emission bands at around 453 nm ($^1D_2 \rightarrow {}^3F_4$), 479 nm ($^4F_{9/2} \rightarrow {}^6H_{15/2}$), 575 nm ($^4F_{9/2} \rightarrow {}^6H_{13/2}$). Emission spectra (Fig. 15.8B) depicted that the concentration of Tm^{3+} ions increased, then the emission intensity of the 450 nm emission peak gradually increased and the intensity of the 479 nm emission peak drastically decreased. All these results imply that an efficient energy transfer occurred in the Tm^{3+} and Dy^{3+} ions. The fluorescence decay and thermal study also analyzed by the authors showed excellent results for white light generation. Fig. 15.8C shows the CIE chromaticity coordinates of synthesized phosphor (0.3391, 0.3302) under near-UV excitation, which is very close to standard white light (0.33, 0.33). These results show that synthesized phosphors can be efficiently used as white-emitting phosphors in solid-state lighting applications [84].

15.4.1.6 Eu^{2+}-doped Ca$_8$Sc$_2$(PO$_4$)$_{6-y}$(SiO$_4$)$_{1+y}$ phosphor

In the present work, Ruan [85] investigated Eu^{2+}-activated Ca$_8$Sc$_2$(PO$_4$)$_6$(SiO$_4$) phosphor. The authors successfully synthesized by the high-temperature solid-state method. The structural and luminescence properties were investigated in detail by using X-ray powder diffraction, Rietveld refinement, diffuse reflection spectra, emission excitation spectra, decay curves, and temperature dependence spectra. The PL emission and excitation spectra are depicted in Fig. 15.9A. The PL emission spectra of Eu^{2+}-doped Ca$_8$Sc$_2$(PO$_4$)$_6$(SiO$_4$) phosphor show a broad emission spectra center at 365 nm. Under a 365 nm excitation wavelength, broad emission spectra are observed at 470 nm, which is ascribed due to 4f–5d transition of Eu^{2+} ions. The insert image of Fig. 15.9A shows concentration-quenching spectra, which confirm Ca$_{7.96}$Sc2(PO$_4$)$_{6-y}$(SiO$_4$)$_{1+y}$:0.04Eu^{2+} phosphors showing the highest emission intensity. The

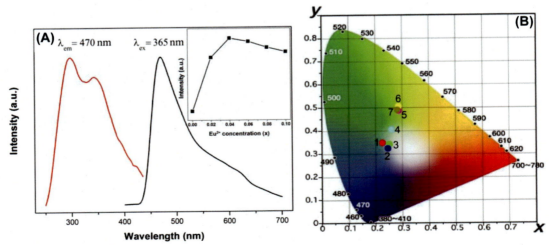

FIGURE 15.9 (A) Photoluminescence excitation and emission spectra of Ca$_{8-x}$Sc$_2$(PO$_4$)$_6$(SiO$_4$):xEu^{2+} (x = 0.04) phosphors. (B) CIE chromaticity diagram of Ca$_{8-x}$Sc$_2$(PO$_4$)$_6$(SiO$_4$):xEu^{2+} (x = 0, 0.02, 0.04, 0.06, 0.08, and 0.1) phosphors under 365 nm excitation. *Reprinted with permission obtained from Elsevier publication F. Ruan, D. Deng, H. Yu, M. Wu, B. Chen, S. Xu, R. Lei, Color-tunable properties based on complex anion substitution in Eu^{2+} doped Ca$_8$Sc$_2$(PO$_4$)$_{6-y}$(SiO$_4$)$_{1+y}$ phosphor, Ceram. Int. 44 (2018) 15432–15439. https://doi.org/10.1016/j.ceramint.2018.05.197.*

critical energy-transfer distance was calculated to be 30.58 Å by concentration quenching, which was related to dipole—dipole interaction. Average lifetime of synthesized phosphor was also characterized. Fig. 15.9B shows the CIE chromaticity coordinates of $Ca_{7.96}Sc2(PO_4)_{6-y}(SiO_4)_{1+y}:0.04Eu^{2+}$ phosphors. The CIE coordinates varied significantly from (0.2266, 0.3475) to (0.2824, 0.4904) with the emission color changing from greenish blue to green. All these obtained results demonstrated that synthesized phosphor has great potential; it can be used as a color-tunable phosphor for WLED applications [85].

15.4.2 List of some silicate-based phosphors assisted by different synthesis techniques

Until now, a number of silicate-based phosphor materials have been prepared by using various synthesis routes. Some of the phosphor materials are discussed in the foregoing section and some of the silicate-based phosphor materials are listed in Table 15.3, which may be applicable to WLEDs.

15.5 Applications of silicate-based phosphor

15.5.1 Light-emitting diodes

Commercial LED enterprises have come a long way in producing white light. LEDs were a profoundly encouraging contender in the development of a quality lighting system for the lighting industry. A few critical surveys have been published that have concentrated on recent advances in the revelation and basic structure of LED phosphors. It has been a troublesome task to imitate the full range of natural light because of challenges in developing phosphors that have expansive emission spectra. New phosphors are being persistently explored to achieve a high CRI and better color consistency [107]. Solid-state lighting is emerging as an alternative to preexisting lighting technologies and is highly desirable. This type of lighting (LEDs) involves semiconductor LEDs, organic LEDs, and polymer organic LEDs as the source of illumination. As a new solid-state light source, LEDs are considered the next generation of lighting devices because of their long lifetime, energy-saving, reliability, safety, and environmentally friendly characteristics. To fulfill the need for LED applications, silicate phosphors have been designed and developed over the past few years. In the current scenario, LEDs are used in security alarms, electronic calculators, mobile phones, traffic signals, digital computers, home lighting, factory lighting, street lighting, digital multimeters, remote control systems such TVs or LCD remotes, digital watches, automotive heat lamps, aviation lights for aircraft warning signals, etc.

15.5.2 Display devices

Significant secrecy has surrounded the class of gadgets known as panel displays. For example, industry in general has consistently shown a keen interest in definitive "color picture-on-the-wall"-type TVs, while the military has been searching for sizable level flat display boards for combat information centers, air-traffic control boards, radar plotting

TABLE 15.3 List of reported silicate-based phosphors.

Phosphor material (composition) with activator	Synthesis method	Excitation wavelength (nm)	Emission wavelength (nm)	Emission of color	CIE chromaticity coordinate /CCT/color purity	References
$Ca_3Y_2Si_3O_{12}:Sm^{3+}$	Citric-based sol-gel synthesis	402	562, 600, 650	Orange/red color	(0.592, 0.402)	[86]
$CaZrSi_2O_7:Eu^{2+}$	Solid-state reaction method	363, 410	512	Green color	(0.216, 0.618)	[87]
$Sr_3SiO_5:Ce^{3+}/Dy^{3+}$	Citric-based sol-gel synthesis	360	434, 572	Blue and yellow color	[$1Ce^{3+}/1Dy^{3+}$ (0.1923,0.1255)], [$2Ce^{3+}/2Dy^{3+}$ (0.1998, 0.1270)], [$3Ce^{3+}/3Dy^{3+}$ (0.2108, 0.1438)]	[88]
$K_4CaSi_3O_9:Eu^{3+}$	Solid-state reaction method	310, 323, 365, 383, 416	578, 592, 616, 654, 707	Yellow, orange, red	(0.64, 0.33)	[89]
(a) $Sr_2MgSi_2O_7:Ce^{2+}$ (b) $Sr_2MgSi_2O_7:Tb^{3+}$	Solid-state reaction method	(a) 348 (b) 243	(a) 405 (b) 543	(a) Violet/blue (b) Green color	(a) Not available (b) Green emission (0.2806, 0.6761)	[90]
$Rb_3YSi_2O_7:Eu$	–	450	622	Red color	CIE (0.653, 0.346), R_a (93), CCT (4013 K)	[91]
$K_2ZrSi_2O_7:Eu^{2+}$	Solid-state reaction method	350	462	Blue color	–	[92]
$Y_2SiO_5:Eu^{3+}$	Combustion route	397, 464	507, 591, 599, 612, 655, 689, 700		CIE (0.5866, 0.4026), CCT = 2018.5 K	[93]
$KBaScSi_2O_7:Eu^{2+}$	Solid-state reaction method	350	488	Blue color	CIE (0.3898, 0.3692), R_a = 88.6 CCT (3770)	[94]
$BaSiF_6:Mn^{4+}$	Hydrothermal process	363 and 467	635	Red color	–	[95]

15.5 Applications of silicate-based phosphor

Material	Synthesis method	Excitation (nm)	Emission (nm)	Color	Properties	Ref.
$Ca_2SiO_4:Eu^{3+}$	Flux fusion reaction method	325, 385, 400, 470, 511, 539	578, 587, 592, 597, 612, 618, 626, 656, 705, 710	Red color	—	[96]
$Ba_5Si_2O_6Cl_6:Eu^{2+}$	Solid-state reaction method	400	510	Green color	CIE (0.324, 0.338), R_a = 88.45, CCT = 5890 K	[97]
$Ba_4La_6(SiO_4)O:Eu^{3+}$	Citrate sol-gel process	283, 288, 363, 382, 394, 414, 464, 533	579, 586, 591, 594, 598, 613, 650, 702	Red color	CIE (0.654, 0.344)	[98]
$Ba_2Gd_2Si_4O_{13}:Eu^{3+}$	(a) Microwave synthesis (b) Solid-state method	393, 464	591, 612, 700	Red color	CIE (x = 0.65, y = 0.35)	[99]
$Ba_2CaZn_2Si_6O_{17}:Eu^{2+}$	Solid-state method	340	467, 527	Blue/green color	CIE (0.259, 0.463)	[100]
$SrHfSi_2O_7:Eu^{2+}$	Solid-state method	352	456	Blue color	CIE (0.162, 0.171), CCT (4531 K), CRI (90.2)	[101]
$(BaSr)_2SiO_4:Eu^{2+}$	Citrate sol-gel method	370	505	Green color	CIE (0.275, 0.523), (0.266, 0.554), (0.261, 0.589)	[102]
$RbLi(Li_3SiO_4)_2:Eu^{2+}$	Solid-state reaction method	460	530, 540	Green color	CIE (0.3182, 0.3275) luminous efficacy of 97.28 lm/W, CCT (6221 K)	[103]
$Na_2Ca_4Mg_2SiO_{15}:Eu^{2+}, Mn^{2+}$	Solid-state reaction method	330	480, 645	Blue and red color	CIE (0.305, 0.319)	[104]
$Y_2MoSiO_8:Eu^{3+}$	Solid-state reaction method	319, 362, 382, 395, 415, 465	581, 595, 618, 653, 706	Orange/red color	CIE (0.65, 0.34)	[105]
$Li_2SrSiO_4:Eu^{3+}, Sm^{3+}$	Solid-state reaction method	Eu^{3+} (242, 396) Sm^{3+} (405)	Eu^{3+} (579, 588, 591, 612, 620, 650, 680, 700) Sm^{3+} (560, 570, 582, 598, 643, 653)	Orange/red emission	CIE (0.570,0.366) to (0.640,0.330)	[106]

CCT, Correlated color temperature; CRI, color rendering index.

displays, and displays utilized for reading PC data. In many of these applications, quick writing, bright display withholding, or capacity are sought. A display device is an output device used for presenting information in visual form. RE^{3+}-doped (Ln = Eu, Ce, Dy, and Sm) $BaCa_2Si_3O_9$ and $K_2CaSi_3O_9$ silicate materials have been used in the field of lighting and display devices [108]. Europium (III)-doped Y_2SiO_5 material has been found to be suitable for the lamp and display industries because it exhibits high luminescence efficiency and good stability under conditions of high irradiance with an electronic beam [109]. $Ba_2Gd_2Si_4O_{13}$:Ce^{3+}, Tb^{3+} and rare-earth-doped $Ca_3Y_2Si_3O_{12}$ materials could be used in display device applications.

15.5.3 Plasma display panels

Plasma displays most notably use the properties of plasma as a light source. The main advantage of plasma display technology is that you can produce a very wide screen using extremely thin materials. And because each pixel is lit individually, the image is very bright and looks good from almost every angle. Plasma display panels (PDPs) are viewed as a promising innovation for large, flat TV displays. In this manner, many major electronics organizations are directing their focus on PDPs, and large-scale manufacturing has already begun. The principle of light generation in a PDP is described in several textbooks [110]. $(Y,Gd,Eu)BO_3$ is generally utilized as a red phosphor for PDPs [110−112] because of its high emission intensity under excitation of vacuum ultraviolet light at 147 and 172 nm. However, the main emanation line of the phosphor is at 593 nm (in the orange/red region), which prompts a low color purity [113].

15.5.4 Field emission display

A field emission display (FED) is a vacuum electron device with a flat panel display innovation that utilizes large areas of field electron emission sources to give electrons that strike a colored phosphor to generate a color picture. Display electrons from a large number of microtips pass through gates and light up pixels on a screen. FEDs are a type of matrix display with quick reaction times, wide response times, high brightness, high resolution, high contrast levels, high radiant productivity, and a wide color range. They are self-emissive, distortion-free displays and therefore a backlighting system and thin-film transistor active matrix are not required, which reduces the unpredictability of the presentation [114]. This innovation offers cold cathode-type surface emanation and consumes little energy. It shares the benefits of CRTs by bundling the focal points of LCDs and other level broad innovations. However, FEDs require high vacuum levels to work, which is time consuming and difficult to accomplish; they also require maintenance. Another genuine obstacle to the advancement of full-color FEDs is the absence of appropriate blue luminescent materials with high luminance, high productivity, great chromaticity, low saturation, and great aging properties. Intense electron bombardment of the phosphor layer will likewise cause gas discharge during use, which is harmful [115].

IV. Lighting and light emitting diodes

15.5.5 Cathode ray tubes

A CRT is a presentation screen that produces pictures as a video signal. It is a sort of vacuum tube that display pictures when electron beams from an electron gun strike a luminous surface. In other words, the CRT produces beams, accelerates them at high speed, and deflects them to make pictures on a phosphor screen. Electronic presentation gadgets were initially made with CRTs, because they were the most established and least expensive electronic presentation innovation; they were derived from vacuum tubes utilized for TV. CRTs work at any resolution, geometry, and aspect ratio without the need to rescale the picture. They work on the principle of cathodoluminescence. This technology offers a large viewing angle, high brightness, high resolution, and good color gamut [116].

15.5.6 Scintillators

Scintillators are materials that can alter high-energy radiation, for example, X- or γ-rays to a near-visible or visible light. They are generally utilized as detectors in clinical diagnostics, high-energy material science, and geophysical investigation. Scintillators can be gaseous, liquid or solid, organic or inorganic (glass, single crystal, ceramics). Detectors dependent on scintillators are basically made from a scintillator material and a photodetector that can be either a photomultiplier tube or a photodiode. The role of the photodetector is to change the light coming from the scintillator into an electrical signal. In designing and assessing radiation detectors, it is essential to decide precisely the division of scintillator-emitted light gathered by the optical sensor [117].

15.5.7 Solar cell

A solar cell is also called a photovoltaic cell. It is an electrical device that directly converts sunlight energy into electrical energy. It works on the principle of photovoltaic effect, which is a physical and chemical phenomenon. The demand for electric power, which is a fundamental need for life, increases exponentially with the ever-growing world population. Based on materials utilization, innovation, and their efficiencies, accessible sunlight-based cells are characterized into three ages. The first is silicon sun-based cells called first-generation photovoltaic devices. Second-generation solar cells comprise CdTe, CIGSe, and CZTSe, etc., based on thin-film technology. Dye-sensitized and perovskite solar cells are categorized into third-generation photovoltaic cells [118]. There is huge research and development going on in the field of solar cell innovation. It is a difficult undertaking for the photovoltaic community to plan productive and cost-effective conversion of sunlight into power.

15.6 Conclusions

The rising development of optoelectronic display applications has fueled an increasing requirement for novel techniques for the preparation of materials having excellent optical as well as electronic characteristics. Performance in certain aspects of designed materials has even exceeded that of other existing materials. We anticipate that the new developments, which may shape the potential of optoelectronic knowledge, will be on the rise with the help of silicate materials. To accomplish that purpose, a series of red, blue, green, and orange phosphors has been designed and successfully synthesized using different synthesis methods. XRD patterns show structures that depend on the synthesis and annealing temperature used for the preparation of phosphors. The effect of temperature on PL properties of silicates has been studied and discussed successfully. Transmission electron microscopy results show particle size. Hence, silicate phosphors demonstrate a practical approach to different fields of study and industry, including lighting, lasers, display panels, scintillators, nuclear medical diagnostic systems, etc. The host lattices of these phosphor materials would be a better choice for research and would help in exploring new luminescent materials for further photonic applications. An additional direction for photonic applications is the development of silicate materials having better particle properties and chromaticity coordinates that can be integrated with UV LED chips to generate light.

15.7 Future scope

This chapter was dedicated to silicate-based inorganic phosphors and their potential for solid-state lighting. Our objective was to provide an idea of the basics, extensions, and varieties of this research field. We have attempted to highlight the current main avenues of research and to reveal the most significant subjects, particularly those requiring significant advancements or breakthroughs. As we see it, the investigation of inorganic materials for pc-LEDs is an interdisciplinary field that requires capability and imaginative research in various regions, for example, luminescence and optical spectroscopy, inorganic chemistry, solid-state physics, and materials science. The future of silicate-based inorganic phosphors lies in an effective approach to find and grow new silicate-based phosphors for pc-LEDs by altering the structure and luminescence tuning of existing phosphor materials. This idea consists of different techniques, for example, control of the doping level, cationic substitution, anionic substitution, cationic—anionic substitution, crystal-site engineering, and mixing of nanophases etc. These strategies were effectively utilized in the fitting of novel silicate-based phosphors for WLEDs, as has appeared in recent papers. Recently, the focus on the tunable luminescence of phosphors has become a research hotspot. This methodology is by all accounts promising for significant advancements in the development of luminescent materials.

Acknowledgments

One of the authors, Yatish R. Parauha, is thankful to the Department of Science and Technology (DST), India, for financial support through INSPIRE fellowship (INSPIRE Code — IF180284). One of the authors, SJD, is thankful to DST, India, for financial assistance (Nano-Mission Project Ref. No. DST/NM/NS/2018/38(G), dated January 16, 2019) and to R.T.M. Nagpur University, Nagpur, for constant encouragement.

References

[1] S.K. Misra, S.I. Andronenko, A variable-temperature X-band EPR study of the Gd3+ ion in a La2Si2O7 crystal characterized by monoclinic site symmetry, Appl. Magn. Reson. 32 (2007) 377–384, https://doi.org/10.1007/s00723-007-0020-5.

[2] G. Blasse, B.C. Grabmaier, G. Blasse, B.C. Grabmaier, A general introduction to luminescent materials, Lumin. Mater. (1994) 1–9, https://doi.org/10.1007/978-3-642-79017-1_1.

[3] J. Grandgirard, D. Poinsot, L. Krespi, J.P. Nénon, A.M. Cortesero, Costs of secondary parasitism in the facultative hyperparasitoid *Pachycrepoideus dubius*: does host size matter? Entomol. Exp. Appl. 103 (2002) 239–248, https://doi.org/10.1023/A.

[4] D. Singh, V. Tanwar, S. Bhagwan, V. Nishal, S. Sheoran, S. Kadyan, A.P. Samantilleke, P.S. Kadyan, Synthesis and optical characterization of europium doped MY 2 O 4 (M = Mg, Ca, and Sr) nanophosphors for solid state lightning applications, Indian J. Mater. Sci. 2015 (2015) 1–8, https://doi.org/10.1155/2015/845065.

[5] A.J. Fernández-Carrión, M. Ocaña, P. Florian, J. García-Sevillano, E. Cantelar, A.N. Fitch, M.R. Suchomel, A.I. Becerro, Crystal structure and luminescent properties of Eu^{3+}-doped A-$La_2Si_2O_7$ tetragonal phase stabilized by spray pyrolysis synthesis, J. Phys. Chem. C 117 (2013) 20876–20886, https://doi.org/10.1021/jp407172z.

[6] A.M. Gilmore, Luminescence: The Instrumental Key to the Future of Nanotechnology, CRC Press, United State, 2013.

[7] A. Jose, M. Ocan, J. Garc, E. Cantelar, A.I. Becerro, New single-phase , white-light-emitting phosphors based on δ - Gd 2 Si 2 O 7 for solid-state lighting, J. Phys. Chem. C 118 (31) (2014) 18035–18043.

[8] P.A. Trusty, K.C. Chan, C.B. Ponton, Synthesis of sinteractive single-phase microstructure yttrium disilicate precursor powder using hydrothermal processing, J. Mater. Res. 13 (1998) 3135–3143, https://doi.org/10.1557/JMR.1998.0426.

[9] S. Shinde, S. Pitale, S.G. Singh, M. Ghosh, B. Tiwari, S. Sen, S.C. Gadkari, S.K. Gupta, Structural and luminescence properties of $Gd_2Si_2O_7$:Ce prepared by solution combustion followed by heat treatment, J. Alloys Compd. 630 (2015) 68–73, https://doi.org/10.1016/j.jallcom.2014.12.223.

[10] M. Díaz, C. Pecharromán, F. Del Monte, J. Sanz, J.E. Iglesias, J.S. José, C. Yamagata, S. Mello-Castanho, Synthesis, thermal evolution, and luminescence properties of yttrium disilicate host matrix, Chem. Mater. 17 (2005) 1774–1782, https://doi.org/10.1021/cm047957b.

[11] Y. Li, B.G. You, W. Zhao, W.P. Zhang, M. Yin, Synthesis and luminescent properties of nano-scale $Y_2Si_2O_7$:Re3+ (Re=Eu, Tb) phosphors via sol-gel method, Chinese J. Chem. Phys. 21 (2008) 376, https://doi.org/10.1088/1674-0068/21/04/376-380.

[12] D. Hreniak, W. Strek, A. Opalińska, M. Nyk, M. Wołcyrz, W. Lojkowski, J. Misiewicz, Luminescence properties of Tb-doped yttrium disilicate prepared by the sol-gel method, J. Sol. Gel Sci. Technol. 32 (2004) 195–200, https://doi.org/10.1007/s10971-004-5788-2.

[13] H. Terraschke, C. Wickleder, UV, blue, green, yellow, red, and small: newest developments on Eu^{2+}-doped nanophosphors, Chem. Rev. 115 (2015) 11352–11378, https://doi.org/10.1021/acs.chemrev.5b00223.

[14] C.D. Walters, UC San Diego electronic theses and dissertations essays in applied microeconomics, Peer Rev. (2017). https://escholarship.org/uc/item/0xp4r2hb.

[15] J. Mckittrick, L.E. Shea-rohwer, Review: down conversion materials for solid-state lighting, J. Am. Ceram. Soc. 26 (2014) 1–26, https://doi.org/10.1111/jace.12943.

[16] J. Hobohm, O. Krüger, S. Basu, K. Kuchta, S. Van Wasen, C. Adam, Chemosphere recycling oriented comparison of mercury distribution in new and spent fluorescent lamps and their potential risk, Chemosphere 169 (2017) 618–626, https://doi.org/10.1016/j.chemosphere.2016.11.104.

[17] N. Holonyak, N. Holonyak, Is the light emitting diode (LED) an ultimate lamp? J. Phys. 68 (2013) 864, https://doi.org/10.1119/1.1301966.

[18] C.-H. Chung, K.-S. Yang, K.-H. Chien, M.-S. Jeng, M.-T. Lee, Heat transfer characteristics in high power LED packaging, Smart Sci. 2 (2014) 1–6, https://doi.org/10.1080/23080477.2014.11665596.

[19] C.J. Humphreys, USE solid-state lighting, MRS Bull. 33 (2008) 459–471.

[20] M.H. Crawford, LEDs for solid-state lighting: performance challenges and recent advances, IEEE J. Sel. Top. Quantum Electron. 15 (2009) 1028–1040, https://doi.org/10.1109/JSTQE.2009.2013476.

[21] J.J. Wierer, D.A. Steigerwald, M.R. Krames, J.J. O'Shea, M.J. Ludowise, G. Christenson, Y.C. Shen, C. Lowery, P.S. Martin, S. Subramanya, W. Götz, N.F. Gardner, R.S. Kern, S.A. Stockman, High-power AlGaInN flip-chip light-emitting diodes, Appl. Phys. Lett. 78 (2001) 3379–3381, https://doi.org/10.1063/1.1374499.

[22] S. Ye, F. Xiao, Y.X. Pan, Y.Y. Ma, Q.Y. Zhang, Phosphors in phosphor-converted white light-emitting diodes: recent advances in materials, techniques and properties, Mater. Sci. Eng. R Rep. 71 (2010) 1–34, https://doi.org/10.1016/j.mser.2010.07.001.

[23] Z. Xia, Q. Liu, Progress in discovery and structural design of color conversion phosphors for LEDs, Prog. Mater. Sci. 84 (2016) 59–117, https://doi.org/10.1016/j.pmatsci.2016.09.007.

[24] K.F. Hsu, C.W. Lin, J.M. Hwang, Selecting conversion phosphors for white light-emitting diodes package by generalized reduced gradient method in dispensing application, in: Proc. Int. Conf. Numer. Simul. Optoelectron. Devices, NUSOD. 2015-May, 2015, pp. 91–92, https://doi.org/10.1109/NUSOD.2015.7292837.

[25] B. Shao, J. Huo, H. You, Prevailing strategies to tune emission color of lanthanide-activated phosphors for WLED applications, Adv. Opt. Mater. 7 (2019) 1–23, https://doi.org/10.1002/adom.201900319.

[26] H. Zhu, X. Liu, R. Fu, Y. Shi, H. Wang, Q. He, X. Song, Luminous efficiency enhancement of WLEDs via patterned RGB phosphor arrays, J. Lumin. 211 (2019) 1–7, https://doi.org/10.1016/j.jlumin.2019.02.063.

[27] T. Walther, E.S. Fry, Optics in Remote Sensing, 2016, https://doi.org/10.1007/978-3-319-31903-2_9.

[28] H. Zuo, Y. Liu, J. Li, X. Shi, S. Ma, M. Zhao, Enhancement of red emission in $KLa(MoO4)2:Eu3+$, $Bi3+$ phosphor for WLEDs, Ceram. Int. 41 (2015) 14834–14838, https://doi.org/10.1016/j.ceramint.2015.08.006.

[29] S. Energy, Energy scenario, Energy (2001) 1–36.

[30] J. McKittrick, L.E. Shea-Rohwer, Review: down conversion materials for solid-state lighting, J. Am. Ceram. Soc. 97 (5) (2014) 1327–1352, https://doi.org/10.1111/jace.12943.

[31] C.M. Mehare, Y.R. Parauha, N.S. Dhoble, C. Ghanty, S.J. Dhoble, Synthesis of novel Eu^{2+} activated $K_3Ca_2(SO_4)_3F$ down-conversion phosphor for near UV excited white light emitting diode, J. Mol. Struct. 1212 (2020) 127957, https://doi.org/10.1016/j.molstruc.2020.127957.

[32] M. Cui, J. Wang, M. Shang, J. Li, Q. Wei, P. Dang, H.S. Jang, J. Lin, Full visible light emission in $Eu2+,Mn2+$-doped $Ca9LiY0.667(PO4)7$ phosphors based on multiple crystal lattice substitution and energy transfer for warm white LEDs with high colour-rendering, J. Mater. Chem. C 7 (2019) 3644–3655, https://doi.org/10.1039/c9tc00109c.

[33] M. Zhao, H. Liao, M.S. Molokeev, Y. Zhou, Q. Zhang, Q. Liu, Z. Xia, Emerging ultra-narrow-band cyan-emitting phosphor for white LEDs with enhanced color rendition, Light Sci. Appl. 8 (2019), https://doi.org/10.1038/s41377-019-0148-8.

[34] J. Xiang, J. Zheng, Z. Zhou, H. Suo, X. Zhao, X. Zhou, N. Zhang, M.S. Molokeev, C. Guo, Enhancement of red emission and site analysis in Eu^{2+} doped new-type structure $Ba_3CaK(PO_4)_3$ for plant growth white LEDs, Chem. Eng. J. 356 (2019) 236–244, https://doi.org/10.1016/j.cej.2018.09.036.

[35] G.B. Nair, S.J. Dhoble, Highly enterprising calcium zirconium phosphate $[CaZr_4(PO_4)_6:Dy^{3+}, Ce^{3+}]$ phosphor for white light emission, RSC Adv. 5 (2015) 49235–49247, https://doi.org/10.1039/c5ra07306e.

[36] G. Xu, X. Zhang, W. He, H. Liu, H. Li, R.I. Boughton, Preparation of highly dispersed YAG nano-sized powder by co-precipitation method, Mater. Lett. 60 (2006) 962–965, https://doi.org/10.1016/j.matlet.2005.10.052.

[37] A.B. Gaikwad, S.C. Navale, V. Samuel, A.V. Murugan, V. Ravi, A co-precipitation technique to prepare $BiNbO_4$, $MgTiO_3$ and $Mg_4Ta_2O_9$ powders, Mater. Res. Bull. 41 (2006) 347–353, https://doi.org/10.1016/j.materresbull.2005.08.010.

[38] T.T.H. Tam, N.V. Du, N.D.T. Kien, C.X. Thang, N.D. Cuong, P.T. Huy, N.D. Chien, D.H. Nguyen, Co-precipitation synthesis and optical properties of green-emitting $Ba_2MgSi_2O_7:Eu^{2+}$ phosphor, J. Lumin. 147 (2014) 358–362, https://doi.org/10.1016/j.jlumin.2013.11.066.

[39] K.Y. Jung, K.H. Han, Y.C. Kang, H.K. Jung, Preparation of $CaMgSi_2O_6:Eu$ blue phosphor particles by spray pyrolysis and its VUV characteristics, Mater. Chem. Phys. 98 (2006) 330–336, https://doi.org/10.1016/j.matchemphys.2005.09.032.

[40] H. Hasegawa, T. Ueda, T. Yokomori, $Y_2Si_2O_7:Eu/SiO_2$ core shell phosphor particles prepared by flame spray pyrolysis, Proc. Combust. Inst. 34 (2013) 2155–2162, https://doi.org/10.1016/j.proci.2012.07.067.

[41] H.S. Kang, Y.C. Kang, K.Y. Jung, S. Bin Park, Eu-doped barium strontium silicate phosphor particles prepared from spray solution containing NH_4Cl flux by spray pyrolysis, Mater. Sci. Eng. B Solid State Mater. Adv. Technol. 121 (2005) 81–85, https://doi.org/10.1016/j.mseb.2005.03.013.

[42] H.J. Lee, S.K. Hong, D.S. Jung, S.H. Ju, H.Y. Koo, Y.C. Kang, The characteristics of X1 type $Y_2SiO_5:Tb$ phosphor particles prepared by high temperature spray pyrolysis, Ceram. Int. 32 (2006) 865–870, https://doi.org/10.1016/j.ceramint.2005.07.001.

References

[43] Q.Y. Zhang, X.Y. Huang, Recent progress in quantum cutting phosphors, Prog. Mater. Sci. 55 (2010) 353–427, https://doi.org/10.1016/j.pmatsci.2009.10.001.

[44] S.H. Lee, H. Young, Koo, Y.C. Kang, Characteristics of α'- and β-Sr_2SiO_4:Eu^{2+} phosphor powders prepared by spray pyrolysis, Ceram. Int. 36 (2010) 1233–1238, https://doi.org/10.1016/j.ceramint.2010.01.007.

[45] W.N. Wang, I.W. Lenggoro, Y. Terashi, T.O. Kim, K. Okuyama, One-step synthesis of titanium oxide nanoparticles by spray pyrolysis of organic precursors, Mater. Sci. Eng. B Solid State Mater. Adv. Technol. 123 (2005) 194–202, https://doi.org/10.1016/j.mseb.2005.08.006.

[46] W.J. Dawson, Hydrothermal synthesis of advanced ceramic powders, Am. Ceram. Soc. Bull. 67 (1988) 1673–1678, https://doi.org/10.4028/www.scientific.net/ast.45.184.

[47] A.I. Becerro, M. Naranjo, A.C. Perdigón, J.M. Trillo, Hydrothermal chemistry of silicates: low-temperature synthesis of y-yttrium disilicate, J. Am. Ceram. Soc. 86 (2003) 1592–1594, https://doi.org/10.1111/j.1151-2916.2003.tb03519.x.

[48] H.S. Kang, Y.C. Kang, H.D. Park, Y.G. Shul, Green-emitting yttrium silicate phosphor particles prepared by large scale ultrasonic spray pyrolysis, Korean J. Chem. Eng. 20 (2003) 930–933, https://doi.org/10.1007/BF02697301.

[49] S. Komarneni, Nanophase materials by hydrothermal, microwave-hydrothermal and microwave solvo hydrothermal, Curr. Sci. 85 (12) (2003) 1730–1734.

[50] S. Komarneni, H. Katsuki, Nanophase materials by a novel microwave-hydrothermal process, Pure Appl. Chem. 74 (2002) 1537–1543, https://doi.org/10.1351/pac200274091537.

[51] D. Polsongkram, P. Chamninok, S. Pukird, L. Chow, O. Lupan, G. Chai, H. Khallaf, S. Park, A. Schulte, Effect of synthesis conditions on the growth of ZnO nanorods via hydrothermal method, Phys. B Condens. Matter. 403 (2008) 3713–3717, https://doi.org/10.1016/j.physb.2008.06.020.

[52] J. Ortiz-Landeros, C. Gómez-Yáñez, R. López-Juárez, I. Dávalos-Velasco, H. Pfeiffer, Synthesis of advanced ceramics by hydrothermal crystallization and modified related methods, J. Adv. Ceram. 1 (2012) 204–220, https://doi.org/10.1007/s40145-012-0022-0.

[53] X.P. Chen, X.Y. Huang, Q.Y. Zhang, Concentration-dependent near-infrared quantum cutting in $NaYF_4$: Pr^{3+}, Yb^{3+} phosphor, J. Appl. Phys. 106 (2009) 3–7, https://doi.org/10.1063/1.3224906.

[54] K.C. Patil, M.S. Hegde, T. Rattan, S.T. Aruna, Chemistry of Nanocrystalline Oxide Materials, 2008, https://doi.org/10.1142/6754.

[55] S.M. Mohun, M. Wilkinson, The eye of the caecilian Rhinatrema bivittatum (Amphibia: Gymnophiona: Rhinatrematidae), Acta Zool. 96 (2015) 147–153, https://doi.org/10.1111/azo.12061.

[56] H. Peelaers, D. Steiauf, J.B. Varley, A. Janotti, C.G. de Walle, (In x Ga 1- x) 2 O 3 alloys for transparent electronics, Phys. Rev. B. 92 (2015) 85206.

[57] S. Kundu, K. Wang, H. Liang, Size-controlled synthesis and self-assembly of silver nanoparticles within a minute using microwave irradiation, J. Phys. Chem. C. 113 (2009) 134–141, https://doi.org/10.1021/jp808292s.

[58] R.J. Xie, N. Hirosaki, Y. Li, T. Takeda, Rare-earth activated nitride phosphors: synthesis, luminescence and applications, Materials 3 (2010) 3777–3793, https://doi.org/10.3390/ma3063777.

[59] S. Lu, J. Zhang, J. Zhang, Synthesis and luminescence properties of Eu^{3+}-doped silicate nanomaterial, Phys. Procedia 13 (2011) 62–65, https://doi.org/10.1016/j.phpro.2011.02.015.

[60] S. Horikoshi, N. Serpone, Microwaves in Nanoparticle Synthesis: Fundamentals and Applications, 2013, https://doi.org/10.1002/9783527648122.

[61] H. Li, D.K. Agrawal, J. Cheng, M.R. Silsbee, Formation and hydration of C3S prepared by microwave and conventional sintering, Cem. Concr. Res. 29 (1999) 1611–1617, https://doi.org/10.1016/S0008-8846(99)00145-3.

[62] G. Castro, R. Salvini, F.A.A.M.N. Soares, A.A. Nierenberg, G.S. Sachs, B. Lafer, R.S. Dias, Applying association rules to study bipolar disorder and premenstrual dysphoric disorder comorbidity, in: Can. Conf. Electr. Comput. Eng. 2018-May, 2018, pp. 1–4, https://doi.org/10.1109/CCECE.2018.8447747.

[63] E.J. Bosze, J. McKittrick, G.A. Hirata, Investigation of the physical properties of a blue-emitting phosphor produced using a rapid exothermic reaction, Mater. Sci. Eng. B Solid State Mater. Adv. Technol. 97 (2003) 265–274, https://doi.org/10.1016/S0921-5107(02)00598-6.

[64] B. Wang, H. Lin, J. Xu, H. Chen, Y. Wang, $CaMg_2Al_{16}O_{27}$:Mn^{4+}-based red phosphor: a potential color converter for high-powered warm W-led, ACS Appl. Mater. Interfaces 6 (2014) 22905–22913, https://doi.org/10.1021/am507316b.

IV. Lighting and light emitting diodes

[65] N. Bajpai, A. Tiwari, S.A. Khan, R.S. Kher, N. Bramhe, S.J. Dhoble, Effects of rare earth ions (Tb, Ce, Eu, Dy) on the thermoluminescence characteristics of sol-gel derived and γ-irradiated SiO_2 nanoparticles, Luminescence 29 (2014) 669–673, https://doi.org/10.1002/bio.2604.

[66] T. Peng, L. Huajun, H. Yang, C. Yan, Synthesis of $SrAl_2O_4$:Eu, Dy phosphor nanometer powders by sol-gel processes and its optical properties, Mater. Chem. Phys. 85 (2004) 68–72, https://doi.org/10.1016/j.matchemphys.2003.12.001.

[67] S. Yao, Y. Li, L. Xue, Y. Yan, A promising blue-green emitting phosphor for white light-emitting diodes prepared by sol-gel method, J. Alloys Compd. 491 (2010) 264–267, https://doi.org/10.1016/j.jallcom.2009.10.113.

[68] S.E. Park, J.S. Chang, K.H. Young, S.K. Dae, H.J. Sung, S.H. Jin, Supramolecular interactions and morphology control in microwave synthesis of nanoporous materials, Catal. Surv. 8 (2004) 91–110, https://doi.org/10.1023/B:CATS.0000026990.25778.a8.

[69] O.H.Y. Zalloum, M. Flynn, T. Roschuk, J. Wojcik, E. Irving, P. Mascher, Laser photoluminescence spectrometer based on charge-coupled device detection for silicon-based photonics, Rev. Sci. Instrum. 77 (2006), https://doi.org/10.1063/1.2173030.

[70] A. Singha, P. Dhar, A. Roy, A nondestructive tool for nanomaterials: Raman and photoluminescence spectroscopy, Am. J. Phys. 73 (2005) 224–233, https://doi.org/10.1119/1.1819933.

[71] M.P. Saradhi, U.V. Varadaraju, Photoluminescence Studies on Eu 2 + -Activated Li 2 SrSiO 4 s a Potential Orange-Yellow Phosphor for Solid-State Lighting, vol. 150, 2006, pp. 5267–5272.

[72] Y. Deng, W. Song, W. Dong, R. Dai, Z. Wang, Z. Zhang, Z. Ding, White light emission of Eu^{3+}/Ag co-doped $Y_2Si_2O_7$, J. Rare Earths 32 (2014) 779–786, https://doi.org/10.1016/S1002-0721(14)60140-X.

[73] D. Singh, S. Sheoran, S. Bhagwan, S. Kadyan, Optical characteristics of sol-gel derived M_3SiO_5:Eu3+ (M = Sr, Ca and Mg) nanophosphors for display device technology, Cogent Phys. 0 (2016), https://doi.org/10.1080/23311940.2016.1262573.

[74] D. Singh, V. Tanwar, A. Simantilke, P.S. Kadyan, I. Singh, Luminescent characterization of Eu^{2+} doped $BaMAl_{10}O_{17}$ (M = Ca/Mg or both) blue nanophosphors for white light emitting applications, J. Mater. Sci. Mater. Electron. 26 (2015) 9977–9984, https://doi.org/10.1007/s10854-015-3676-x.

[75] C.C.S. Pedroso, J.M. Carvalho, L.C.V. Rodrigues, J. Hölsä, H.F. Brito, Rapid and energy-saving microwave-assisted solid-state synthesis of Pr^{3+}-, Eu^{3+}-, or Tb^{3+}-doped Lu_2O_3 persistent luminescence materials, ACS Appl. Mater. Interfaces 8 (2016) 19593–19604, https://doi.org/10.1021/acsami.6b04683.

[76] H.Y. Chen, M.H. Weng, S.J. Chang, R.Y. Yang, Preparation of Sr_2SiO_4:Eu^{3+} phosphors by microwave-assisted sintering and their luminescent properties, Ceram. Int. 38 (2012) 125–130, https://doi.org/10.1016/j.ceramint.2011.06.044.

[77] U. Bin Humayoun, Y.H. Song, M. Lee, K. Masato, H. Abe, K. Toda, Y. Sato, T. Masaki, D.H. Yoon, Synthesis of high intensity green emitting $(Ba,Sr)SiO_4$:Eu^{2+} phosphors through cellulose assisted liquid phase precursor process, Opt. Mater. 51 (2016) 110–114, https://doi.org/10.1016/j.optmat.2015.11.036.

[78] A. Yoshikawa, M. Nikl, G. Boulon, T. Fukuda, Challenge and study for developing of novel single crystalline optical materials using micro-pulling-down method, Opt. Mater. 30 (2007) 6–10, https://doi.org/10.1016/j.optmat.2006.10.030.

[79] J. Pejchal, V. Babin, A. Beitlerova, S. Kurosawa, Y. Yokota, A. Yoshikawa, M. Nikl, Improvement of the growth of Li 4 SiO 4 single crystals for neutron detection and their scintillation and luminescence properties, J. Cryst. Growth. 3 (2016) 1–8, https://doi.org/10.1016/j.jcrysgro.2016.02.008.

[80] S. Som, P.H. Kuo, C.H. Lu, Facile synthesis of Eu^{2+} activated dual light emitting barium silicate phosphors for warm-white light emitting diodes, J. Alloys Compd. 790 (2019) 1060–1064, https://doi.org/10.1016/j.jallcom.2019.02.322.

[81] S. Ray, P. Tadge, S.J. Dhoble, G.B. Nair, A. Singh, A.K. Singh, M. Rai, T.M. Chen, V. Rajput, Structural and spectroscopic characterizations of a new near-UV-converting cyan-emitting $RbBaScSi_3O_9$:Eu^{2+} phosphor with robust thermal performance, J. Alloys Compd. 713 (2017) 138–147, https://doi.org/10.1016/j.jallcom.2017.03.366.

[82] S.A. Khan, W. Ji, L. Hao, X. Xu, S. Agathopoulos, N.Z. Khan, Synthesis and characterization of Ce^{3+}/Tb^{3+} co–doped $CaLa_4Si_3O_{13}$ phosphors for application in white LED, Opt. Mater. 72 (2017) 637–643, https://doi.org/10.1016/j.optmat.2017.06.034.

[83] G. Annadurai, B. Devakumar, L. Sun, H. Guo, S. Wang, X. Huang, Crystal structure, photoluminescence properties and thermal stability of $BaLu_2Si_3O_{10}$:Eu^{3+} red-emitting phosphors with high color purity for near-UV-excited white LEDs, J. Lumin. 215 (2019) 116623, https://doi.org/10.1016/j.jlumin.2019.116623.

References

479

[84] H. Patnam, S.K. Hussain, L.K. Bharat, J.S. Yu, Near-ultraviolet excited Tm^{3+} and Dy^{3+} ions co-doped barium lanthanum silica oxide phosphors for white-light applications, J. Alloys Compd. 780 (2019) 846−855, https://doi.org/10.1016/j.jallcom.2018.11.321.

[85] F. Ruan, D. Deng, H. Yu, M. Wu, B. Chen, S. Xu, R. Lei, Color-tunable properties based on complex anion substitution in Eu^{2+} doped $Ca_8Sc_2(PO_4)_{6-y}(SiO_4)_{1+y}$ phosphor, Ceram. Int. 44 (2018) 15432−15439, https://doi.org/10.1016/j.ceramint.2018.05.197.

[86] V.R. Bandi, B.K. Grandhe, K. Jang, H.S. Lee, S.S. Yi, J.H. Jeong, Citric based sol-gel synthesis and photoluminescence properties of un-doped and Sm^{3+} doped $Ca_3Y_2Si_3O_{12}$ phosphors, Ceram. Int. 37 (2011) 2001−2005, https://doi.org/10.1016/j.ceramint.2011.02.026.

[87] V.R. Bandi, B.K. Grandhe, K. Jang, S.S. Kim, D.S. Shin, Y.I. Lee, J.M. Lim, T. Song, Luminescent properties of a new green emitting Eu^{2+} doped $CaZrSi_2O_7$ phosphor for WLED applications, J. Lumin. 131 (2011) 2414−2418, https://doi.org/10.1016/j.jlumin.2011.06.004.

[88] K.G. Krishna, L.K. Bharat, J.S. Yu, Energy transfer and luminescence properties of Ce^{3+}/Dy^{3+} co-doped Sr_3SiO_5 phosphors for WLEDs, Ceram. Int. 43 (2017) 2586−2591, https://doi.org/10.1016/j.ceramint.2016.11.065.

[89] L. Zhao, F. Fan, X. Chen, Y. Wang, Y. Li, B. Deng, Luminescence and thermal-quenching properties of silicate-based red-emitting $K_4CaSi_3O_9$:Eu^{3+} phosphor, J. Mater. Sci. Mater. Electron. 29 (2018) 5975−5981, https://doi.org/10.1007/s10854-018-8571-9.

[90] Y. Wang, Y. Chen, Q. Sun, B. Yan, Synthesis, structure, and photoluminescence properties of Ce^{3+} and Tb^{3+} doped alkaline-earth silicate $Sr_2MgSi_2O_7$ phosphors for WLEDs, J. Mater. Res. 32 (2017) 547−556, https://doi.org/10.1557/jmr.2016.524.

[91] J. Qiao, L. Ning, M.S. Molokeev, Y. Chuang, Q. Zhang, K.R. Poeppelmeier, Z. Xia, Site-selective occupancy of Eu 2+ toward blue-light-excited red emission in a Rb 3 YSi 2 O 7:Eu phosphor, Angew. Chem. 2−8 (2019), https://doi.org/10.1002/ange.201905787.

[92] Z. Tang, D. Wang, W.U. Khan, S. Du, X. Wang, Y. Wang, Novel zirconium silicate phosphor $K_2ZrSi_2O_7$:Eu^{2+} for white light-emitting diodes and field emission displays, J. Mater. Chem. C 4 (2016) 5307−5313, https://doi.org/10.1039/c6tc01449f.

[93] G. Ramakrishna, H. Nagabhushana, B.D. Prasad, Y.S. Vidya, S.C. Sharma, K.S. Anantharaju, S.C. Prashantha, N. Choudhary, Spectroscopic properties of red emitting Eu^{3+} doped Y_2SiO_5 nanophosphors for WLED's on the basis of Judd−Ofelt analysis: *Calotropis gigantea* latex mediated synthesis, J. Lumin. 181 (2017) 153−163, https://doi.org/10.1016/j.jlumin.2016.08.050.

[94] J. Zhong, J. Li, M. Liu, K. Wang, Y. Zhu, X. Li, Z. Ji, D. Chen, Novel cyan-emitting $KBaScSi_2O_7$:Eu^{2+} phosphors with ultrahigh quantum efficiency and excellent thermal stability for WLEDs, J. Am. Ceram. Soc. 102 (2019) 7376−7385, https://doi.org/10.1111/jace.16644.

[95] H. Jia, Y. Duan, J. Gao, X. Yuan, H. Wang, G. Li, Enhanced red emission for $BaSiF_6$:Mn^{4+} hexagonal nanorod phosphor via using spherical silica, Mater. Lett. 223 (2018) 163−165, https://doi.org/10.1016/j.matlet.2018.03.152.

[96] Q. Yu, Y. Liu, S. Wu, X. Lü, X. Huang, X. Li, Luminescent properties of Ca_2SiO_4:Eu^{3+} red phosphor for trichromatic white light emitting diodes, J. Rare Earths 26 (2008) 783−786, https://doi.org/10.1016/S1002-0721(09)60005-3.

[97] S. Dutta, T.M. Chen, An efficient green-emitting $Ba_5Si_2O_6Cl_6$:Eu^{2+} phosphor for white-light LED application, RSC Adv. 7 (2017) 40914−40921, https://doi.org/10.1039/c7ra07221j.

[98] G.S.R. Raju, E. Pavitra, S.K. Hussain, D. Balaji, J.S. Yu, Eu^{3+} ion concentration induced 3D luminescence properties of novel red-emitting $Ba_4La_6(SiO_4)O$:Eu^{3+} oxyapatite phosphors for versatile applications, J. Mater. Chem. C 4 (2016) 1039−1050, https://doi.org/10.1039/c5tc03919c.

[99] T. shuai Lü, X. hui Xu, D. jian Wang, L. Sun, J. bei Qiu, Fabrication, structure and photoluminescence properties of Eu^{3+}-activated red-emitting $Ba_2Gd_2Si_4O_{13}$ phosphors for solid-state lighting, Optoelectron. Lett. 10 (2014) 106−110, https://doi.org/10.1007/s11801-014-3210-z.

[100] G. Annadurai, S.M.M. Kennedy, V. Sivakumar, Luminescence properties of a novel green emitting $Ba_2CaZn_2Si_6O_{17}$:Eu^{2+} phosphor for white light - emitting diodes applications, Superlattices Microstruct. 93 (2016) 57−66, https://doi.org/10.1016/j.spmi.2016.02.045.

[101] Q. Zhang, X. Ding, Y. Wang, Novel highly efficient blue-emitting $SrHfSi_2O_7$:Eu^{2+} phosphor: a potential color converter for WLEDs and FEDs, Dye. Pigm. 163 (2019) 168−175, https://doi.org/10.1016/j.dyepig.2018.11.052.

IV. Lighting and light emitting diodes

[102] E. Pavitra, G. Seeta Rama Raju, J. hyun Oh, L. Krishna Bharat, J.H. Jeong, Y.S. Huh, J.S. Yu, $(BaSr)_2SiO_4:Eu^{2+}$ nanorods with enhanced luminescence properties as green-emitting phosphors for white LED applications, Dyes Pigm. 142 (2017) 447–456, https://doi.org/10.1016/j.dyepig.2017.03.065.

[103] M. Zhao, H. Liao, L. Ning, Q. Zhang, Q. Liu, Z. Xia, Next-generation narrow-band green-emitting $RbLi(Li_3SiO_4)_2:Eu^{2+}$ phosphor for backlight display application, Adv. Mater. 30 (2018) 1–7, https://doi.org/10.1002/adma.201802489.

[104] W. Lv, Y. Jia, Q. Zhao, M. Jiao, B. Shao, W. Lü, H. You, Color tuning and energy transfer investigation in $Na_2Ca 4Mg_2Si_4O_{15}:Eu^{2+}, Mn^{2+}$ phosphor and its potential application for UV-excited UV-WLEDs, RSC Adv. 4 (2014) 7588–7593, https://doi.org/10.1039/c3ra46474a.

[105] G. Dong, J. Zhao, M. Li, L. Guan, X. Li, A novel red $Y_2MoSiO_8:Eu^{3+}$ phosphor with high thermal stability for white LEDs, Ceram. Int. 45 (2019) 2653–2656, https://doi.org/10.1016/j.ceramint.2018.10.050.

[106] H. Yu, W.W. Zi, S. Lan, H.F. Zou, S.C. Gan, X.C. Xu, G.Y. Hong, Photoluminescence characteristics of novel red emitting phosphor $Li_2SrSiO_4:Eu^{3+}, Sm^{3+}$ for white light emitting diodes, Mater. Res. Innov. 16 (2012) 298–302, https://doi.org/10.1179/1433075X12Y.0000000013.

[107] A. Tiwari, S.J. Dhoble, Tunable lanthanide/transition metal ion-doped novel phosphors for possible application in w-LEDs: a review, Luminescence 35 (2020) 4–33, https://doi.org/10.1002/bio.3712.

[108] S.J. Dhoble, S.K. Raut, N.S. Dhoble, Synthesis and photoluminescence characteristics of rare earth activated some silicate phosphors for LED and display devices, J. Mater. Sci. Mater. Electron. 5 (2015) 178–182.

[109] G. Tzevekov, N. Minkova, Effects of mechanochemical treatment on yttrium oxyapatite formation, J. Mater. Synth. Process. 9 (2001) 125–130, https://doi.org/10.1023/A:1013293329770.

[110] T. Jüstel, H. Nikol, Optimization of luminescent materials for plasma display panels, Adv. Mater. 12 (2000) 527–530, https://doi.org/10.1002/(sici)1521-4095(200004)12:7<527::aid-adma527>3.3.co;2-#.

[111] W. Zhuang, X. Cui, Z. Yu, C. Zhao, H. He, X. Huang, 41.2: influence of doping on the crystal structures and luminescent properties of aluminate phosphors for plasma display panel, SID Symp. Dig. Tech. Pap. 34 (2003) 1223, https://doi.org/10.1889/1.1832507.

[112] E.M.J. Weber, a V. Dotsenko, L.B. Glebov, V. a Tsekhomsky, Handbook of Optical Laser and Optical Science and Technology Series Physics and Chemistry of Photochromic Glasses, 2003, https://doi.org/10.1109/MEI.2004.1342445.

[113] C.R. Ronda, Recent achievements in research on phosphors for lamps and displays, J. Lumin. 72–74 (1997) 49–54, https://doi.org/10.1016/S0022-2313(96)00374-2.

[114] R. Hischier, Life cycle assessment study of a field emission display television device, Int. J. Life Cycle Assess. 20 (2015) 61–73, https://doi.org/10.1007/s11367-014-0806-2.

[115] S. Itoh, M. Tanaka, T. Tonegawa, M. Taniguchi, K. Tamura, Y. Marushima, Y. Fujimura, M. Namikawa, Y. Naito, F. Kataoka, K. Nawamaki, Y. Kubo, T. Niiyama, Y. Takeya, K. Deguchi, S. Kawata, Y. Sato, T. Yamaura, Development of field emission displays (FEDs), J. Vac. Sci. Technol. B 3 (2006) 1821–1824, https://doi.org/10.1116/1.1691409. IDW '06 - Proc. 13th Int. Disp. Work.

[116] N.T. Kalyani, H. Swart, S.J. Dhoble, History of organic light-emitting diode displays, in: Princ. Appl. Org. Light Emit. Diodes, 2017, pp. 205–225, https://doi.org/10.1016/b978-0-08-101213-0.00008-4.

[117] E. Roncali, M.A. Mosleh-Shirazi, A. Badano, Modelling the transport of optical photons in scintillation detectors for diagnostic and radiotherapy imaging, Phys. Med. Biol. 62 (2017) R207–R235, https://doi.org/10.1088/1361-6560/aa8b31.

[118] R. Rajeswari, N. Islavath, M. Raghavender, L. Giribabu, Recent progress and emerging applications of rare earth doped phosphor materials for dye-sensitized and perovskite solar cells: a review, Chem. Rec. 20 (2020) 65–88, https://doi.org/10.1002/tcr.201900008.

CHAPTER 16

Investigations of energy-efficient RE(TTA)₃dmphen complexes dispersed in polymer matrices for solid-state lighting applications

Akhilesh Ugale[1], N. Thejo Kalyani[2], S.J. Dhoble[3]

[1]Department of Applied Physics, G.H. Raisoni Academy of Engineering and Technology, Nagpur, Maharashtra, India; [2]Department of Applied Physics, Laxminarayan Institute of Technology, Nagpur, Maharashtra, India; [3]Department of Physics, RTM Nagpur University, Nagpur, Maharashtra, India

16.1 Introduction

In this contemporary era, the climate of the earth has been greatly impacted and is in a state of distress. Extreme development in all fields in the last few decades has resulted in a drop in the availability of natural resources. It has also shown numerous adverse effects indicating the negligence of human beings toward nature while achieving personal comfort. A serious and increasing crisis has been observed particularly in the field of energy production and its use, which needs an immediate healing action to preserve the global environment. But development demands a huge requirement of electrical energy, which ultimately results in global warming. Because of this threat, all developed and developing countries are becoming increasingly concerned about sustainable development. Sustainability has three dimensions, generally known as the three pillars of sustainability, namely: environment, society, and economy (Fig. 16.1), which are interrelated and interlinked with each other. The field of lighting consumes almost one-third of total energy production [1]. Hence, fair and wise utilization of electrical energy is only possible with energy-efficient, optoelectronic, solid-state light-emitting devices. These devices are energy-efficient and eco-friendly sources, which significantly reduce the level of CO_2 emission, thereby leading to positive impacts on climate change. With the invention of inorganic light-emitting diode (LED) sources, this consumption

Energy Materials
https://doi.org/10.1016/B978-0-12-823710-6.00016-9

FIGURE 16.1 Sustainability dimensions: three pillars of sustainability.

has decreased to some extent [2]. Implementation of organic light-emitting diode (OLED) sources could further curtail energy consumption, leading to sustainable development.

Keeping this in view, we propose the basic concepts of solid-state lighting (SSL) and highlight the synthesis and characterization of energy-efficient, rare earth-based RE(TTA)$_3$dmphen (RE = Sm/Eu$_{0.5}$Sm$_{0.5}$, TTA = thenoyltrifluoroacetone, dmphen = 4,7-dimethyl-1,10-phenanthroline) complexes dispersed in polymer matrices for SSL applications.

16.2 Solid-state lighting

Lighting can be considered as the application of light to attain a realistic optical effect. To fulfill this requirement, artificial lighting technology evolved to provide light especially at nighttime, and as a consequence a huge amount of energy is invested to generate electricity worldwide [3]. Despite many failures, Thomas Edison finally succeeded with his discovery of the incandescent bulb. Since then, it has been the most common way of lighting. But this kind of lighting source consumes more energy and hence the search for energy-efficient sources began. Later, this form of lighting became less popular with the evolution of fluorescent lamps and compact fluorescent lamps (CFLs). CFL technology uses harmful mercury for excitation and these lamps are nondisposable. The most efficient way of producing light artificially is through solid-state devices, which include LEDs, OLEDs, and polymer LEDs as sources of illumination [4,5]; hence, the name SSL. SSL is seen to be a better solution to the energy crises by reducing the threat of global warming. Remarkable research is ongoing to develop such efficient sources and energy-saving technology, which offers many benefits like improved lifetime, less power consumption, flat-panel display, and eco-friendliness compared to previous lighting technologies that produce cool white light. The appearance of color in light is evaluated on the basis of characteristics like correlated color temperature (CCT), color-rendering index (CRI), and Commission Internationale de l'Eclairage (CIE) coordinates [6].

16.2.1 Correlated color temperature

The CCT of white light represents an equivalence to the radiated heat flux from incandescent lighting and is measured in kelvin (K). CCT has a wide range from 2700 to 6500 K where a temperature above 5000 K refers to cool colors and that from 2700 to 3000 K refers to warm colors.

16.2.2 Color-rendering index

CRI is the capability of white light to represent the real colors of objects for human visual perception. Values of CRI on an internationally accepted scale of 1−100 indicate possible luminous efficacy.

16.2.3 CIE coordinates

The CIE system is used to represent the composition of a colors in terms of chromaticity coordinates x and y on a CIE diagram. In this system, mathematical calculations are used to express a color in terms of three primaries called tristimulus values (X, Y, Z) whose sum $X + Y + Z = 1$. The ratios $\frac{X}{1}$, $\frac{Y}{1}$ and $\frac{Z}{1}$ are called chromaticity coordinates [7].

16.3 Requisite of solid-state lighting

It is a fact, universally recognized, that in this digital era, there are still billions of people who make use of fuel-based lighting like kerosene lamp sources (which are health and even climate hazardous), due possibly to the unavailability of or their inability to pay for alternatives. For this reason, LEDs/OLEDs have been promoted for their use in SSL applications, either indoor or outdoor, possibly powered by solar panels, which curtails the problem of energy consumption. Hence, the entire world is looking forward to the implementation of novel energy-saving and eco-friendly technologies for lighting. OLED sources play a vital role in generating cool white light and have a potential for replacing existing light technology due to the features listed in Fig. 16.2 [8].

16.4 Eco-friendly and energy-efficient organic complexes

Many eco-friendly and energy-efficient complexes have been reported. However, with improved characterization and fabrication techniques, researchers are now developing novel materials for improved lifetime and quantum yield. The state-of-the-art of the past decade is shown in Table 16.1.

IV. Lighting and light emitting diodes

FIGURE 16.2 Advantages of solid-state lighting.

TABLE 16.1 State-of-the-art of light-emitting complexes for solid-state lighting in the past decade.

S. No.	Complex	Structure	Emission wavelength (nm)	References
1	Eu(TPBDTFA)₃Phen		612	[9]
2	Eu(TTA)₃Phen		611	[3]
3	Eu(DBM)₃Cl-MIP		611	[10]
4	Eu(TTA)₃bipy		614	[11]

16.4 Eco-friendly and energy-efficient organic complexes

485

TABLE 16.1 State-of-the-art of light-emitting complexes for solid-state lighting in the past decade.—cont'd

S. No.	Complex	Structure	Emission wavelength (nm)	References
5	Eu(dmh)$_3$phen		613	[12]
6	Eu$_x$RE$_{(1-x)}$(TTA)$_3$Phen		611	[13]
7	Sm(HTH)$_3$phen		611 648	[14]
8	Sm(hfa)$_3$(phen)$_2$		643	[15]
9	Sm(hfa)$_3$(phen)$_2$MeOH		564 598 645	[16]
10	Sm(hfa)$_3$(phen)$_2$		562 592 643	[17]
11	Eu$_{0.5}$Sm$_{0.5}$(TTA)$_3$dpphen		613	[18]

(*Continued*)

IV. Lighting and light emitting diodes

486 16. Investigations of energy-efficient RE(TTA)$_3$dmphen complexes dispersed in polymer matrices for solid-state lighting applications

TABLE 16.1 State-of-the-art of light-emitting complexes for solid-state lighting in the past decade.—cont'd

S. No.	Complex	Structure	Emission wavelength (nm)	References
12	Eu(TTA)$_3$dpphen		613	[19]
13	Eu(TTA)$_3$tppo		616	[20]
14	Sm(TTA)$_3$tppo		566 600 645	[20]
15	Eu$_{0.5}$Sm$_{0.5}$(TTA)$_3$tppo		616	[20]
16	Y$_{(1-x)}$Eu$_x$(DBM)$_3$(Phen)		611	[21]

16.5 Role of rare earths in SSL

Rare earth ions are the family of 17 elements of group III and period 6 with atomic number 21, 39, and from 57 to 71 in the periodic table. Considerable improvement in the field of luminescent materials has been made with the introduction of these rare earth ions as luminescent species. An important breakthrough was the use of Eu^{3+}-activated materials as the red component for color-television screens. A couple of years later, the use of rare earth-activated phosphors yielded considerable improvements in luminescent lamps, X-ray intensifying screens, and optoelectronic display devices. Even more recently, Philips introduced energy-saving SL lamps (solid-state lamps) [22].

The characteristics of rare earth ions are attributable to the 4f shell, which is not completely filled. The electron of this shell is screened by those in the outer shell and as a result they give rise to a number of discrete energy levels. These special structural properties enable rare earth ions to produce a strong and sharp emission with good efficiency. Because of these characteristics, rare earths along with proper chelating compounds have been extensively used as emissive materials in SSL devices. One of the most widely used classes of chelating ligands, which can form a stable complex with rare earth metal ions, are β-diketonates. Many complexes have been synthesized so far with these combinations, which can emit red, green, and blue color independently, and are ultimately suitable for white light emission. Out of 17 rare earth elements, europium (Eu) and samarium (Sm) are widely used as red−orange light emission centers. Work was therefore undertaken and efforts have been made to understand the luminescence in europium- and samarium-based phosphor. Also, major focus has been given to polymer matrices of red light-emitting europium and samarium complexes doped in polymethylmethacrylate (PMMA) and polystyrene (PS) at different weight percentages because they can be drawn into thin films, which can be used to make large-size flat panels for SSL using more economical solution techniques. Table 16.2 reflects the Eu and Sm fact box.

TABLE 16.2 Physical and chemical properties of europium and samarium—fact box [23].

	Europium	Samarium
Symbol	Eu	Sm
Atomic number	63	62
Discoveror	Eugène-Antole Demarçay	Paul-Émile Lecoq de Boisbaudran
Year of discovery	1901	1879
Electronic configuration	[Xe] $4f^7\ 6s^2$	[Xe] $4f^6\ 6s^2$
Appearance	Silvery white with a pale yellow tint; but rarely seen without oxide discoloration	A silvery-white metal
Element category	Lanthanide	Lanthanide
Period no.	6	6
Block	f	f
Phase at room temperature	Solid	Solid
Element classification	Metal	Metal
Crystal structure	Body-centered cubic	Body-centered cubic
Atomic weight (g)	151.964	150.36
Melting point (°C)	822	1072

(*Continued*)

488 16. Investigations of energy-efficient RE(TTA)₃dmphen complexes dispersed in polymer matrices for solid-state lighting applications

TABLE 16.2 Physical and chemical properties of europium and samarium—fact box [23].—cont'd

	Europium	Samarium
Boiling point (°C)	529	1794
Density (g/cm^3)	5.24	7.52
Molar mass (cm^3)	28.97	—
Atomic radius (pm)	180	180
Isotopes	^{151}Eu and ^{153}Eu	^{144}Sm, ^{147}Sm, ^{148}Sm, ^{149}Sm, ^{150}Sm, ^{152}Sm, ^{154}Sm
Relative atomic mass (u)	151.964	150.36
Covalent radius (A^0)	1.98	1.85
Estimated abundance	2.0 mg/kg (crustal) 1.3×10^{-7} mg/L (oceanic)	0.3 mg/kg (crustal)
Reactivity with air	High	Less
Ignition temperature in air (°C)	150–180	150
Ionization energy (eV)	5.670	—
Oxidation states	+3, +2, +1 (a mildly basic oxide)	+3, +2
Ionization energies (kJ/mol)	547.1, 1085, 2404	544.534, 1068.09, 2257.8, 3994.5
Electronegativity	Pauling scale: 1.2	Pauling scale: 1.7
Heat of fusion (kJ/mol)	9.21	8.62
Heat of vaporization (kJ/mol)	176	192
Molar heat capacity (J/mol K)	27.66	29.54

16.6 Experimental

All the chemicals used during the synthesis were of analytical (analytical reagent) grade purchased from Sigma-Aldrich. All the reactions were carried out at room temperature and in an inert atmosphere.

16.6.1 Reagents and solvents

The precursors used for the synthesis of Sm(TTA)₃dmphen and Eu₀.₅Sm₀.₅(TTA)₃dmphen include: europium chloride hexahydrate (EuCl$_3$.6H$_2$O) purity >99%, M.wt = 366.41 g/mol, samarium chloride hexahydrate (SmCl$_3$.6H$_2$O) purity >99%, M.wt = 364.81 g/mol, 2-thenyoltrifluoroacetone (TTA = C$_8$H$_5$F$_3$N$_2$O$_2$S) purity 99%, M.wt = 222.19 g/mol, melting point 40–45°C, dmphen = 4,7-dimethyl-1,10-phenanthroline, ethanol absolute (C$_2$H$_5$OH) purity 99.98%, M.wt = 46.07 g/mol, chloroform (CHCl$_3$) purity 99.7%, M.wt = 119.38 g/mol, dichloromethane (CH$_2$Cl$_2$) M.wt = 84.93 g/mol, purity 99.5%, toluene (C$_6$H$_5$CH$_3$),

IV. Lighting and light emitting diodes

M.wt = 92.14 g/mol, purity 99.9%, acetic acid (CH_3COOH) purity 99.5%, M.wt = 60.05, formic acid (HCO_2H) M.wt = 46.03 g/mol, purity 99.9%, potassium hydroxide (KOH), and double distilled water.

16.6.2 Structure

Chemical structure of the synthesized complexes Eu(TTA)$_3$dmphen, Sm(TTA)$_3$dmphen, and Eu$_{0.5}$Sm$_{0.5}$(TTA)$_3$dmphen are depicted in Fig. 16.3a and b, respectively.

16.6.3 Synthesis procedure

The Sm(TTA)$_3$dmphen and Eu$_{0.5}$Sm$_{0.5}$(TTA)$_3$dmphen hybrid organic complex was synthesized by the solution technique at room temperature according to the scheme outlined in Fig. 16.4. Pinkish precipitate was obtained with good quantitative yield (79%). This precipitate was thoroughly washed with ethanol and then with double distilled water twice. The precipitate was collected by filter paper and then dried at 80°C for 1 h to remove the residual moisture. When exposed to a UV source, orange–red light emission was observed from the synthesized complexes. Another complex, namely Eu$_{0.5}$Sm$_{0.5}$(TTA)$_3$dmphen, was synthesized by following the same procedure and replacing SmCl$_3$.6H$_2$O with EuCl$_3$.6H$_2$O and SmCl$_3$.6H$_2$O. Fig. 16.5a and b shows the synthesized complexes under (A) normal light and (B) UV light.

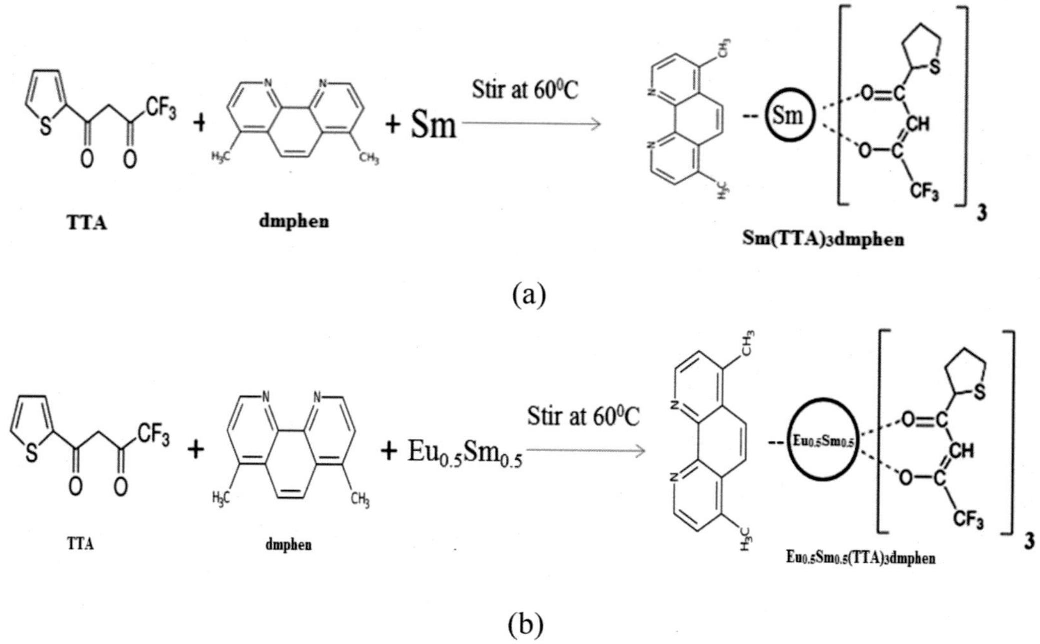

FIGURE 16.3   Synthesis scheme of (a) Sm(TTA)$_3$dmphen and (b) Eu$_{0.5}$Sm$_{0.5}$(TTA)$_3$dmphen.

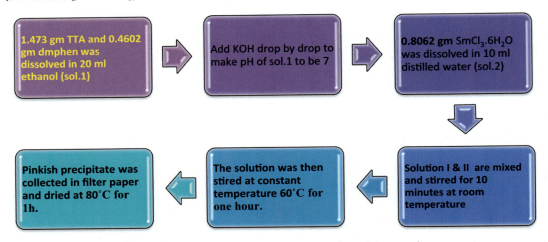

FIGURE 16.4 Outline of the synthesis procedure of the complexes.

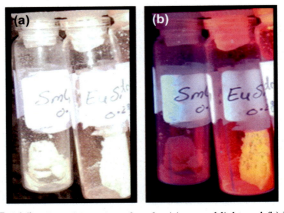

FIGURE 16.5 Complexes viewed under (a) normal light and (b) UV light.

16.6.4 Preparation of blended thin films

Along with the synthesized RE(TTA)$_3$dmphen complexes, the reagents and solvents used for preparing the blended thin films are:

1. Polymethylmetacrylate (PMMA)

Molecular formula $(C_5O_2H_8)_n$, melting point = 180°C, Qualigens Fine Chemicals, density = 1.19 g/cc, refractive index = 1.492.

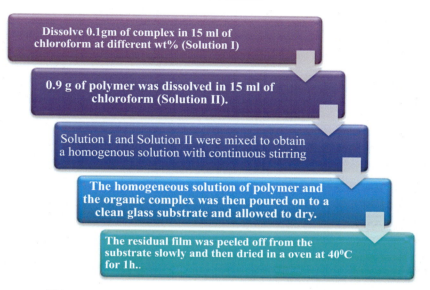

FIGURE 16.6 Steps involved in the preparation of blended films.

2. Polystyrene (PS)

Molecular formula $(C_8H_8)_n$, melting point = 237.5°C, Qualigens Fine Chemicals, density = 1.05 g/cc, refractive index = 1.519.

3. Dichloromethane

Molecular formula CH_2Cl_2, M.wt = 84.93 g/mol, Sigma-Aldrich, purity 99.8%, density 1.3 g/cm^3, refractive index = 1.4244.

4. Formic acid

Molecular formula CH_2O_2, M.wt = 46.03 g/mol, SD Fine Chemicals, purity 99%.

16.6.4.1 Preparation of blended films

The synthesized RE(TTA)$_3$dmphen complexes were then dispersed in the polymer matrices at different weight percentages using a molecular doping method by wt%. The steps involved in the preparation of the blended thin films are shown in the flowchart of Fig. 16.6.

16.7 Result and discussion

The photophysical properties of polymer matrices of RE(TTA)$_3$dmphen doped thin films were characterized by UV-Vis optical absorption spectra and photoluminescence (PL) spectra on an HR 4C 4568 UV-Vis spectrophotometer and Humamatsu F-4500 spectrofluorometer, respectively, at 5 and 10 wt%. The chromatic coordinates (x, y) were calculated using 1931 CIE color calculator program radiant imaging.

492 16. Investigations of energy-efficient RE(TTA)₃dmphen complexes dispersed in polymer matrices for solid-state lighting applications

TABLE 16.3 Summary of photoluminescence spectra of RE(TTA)$_3$dmphen in the polymer matrix.

Parameter	Sm(TTA)$_3$dmphen	Eu$_{0.5}$Sm$_{0.5}$(TTA)$_3$ dmphen
λ_{exc} in PMMA 10% (nm)	385	379
λ_{emi} in PMMA 10% (nm)	646	613
λ_{exc} in PMMA 5% (nm)	372	379
λ_{emi} in PMMA 5% (nm)	613	613
λ_{exc} in PS 10% (nm)	377	379
λ_{emi} in PS 10% (nm)	655	624
λ_{exc} in PS 5% (nm)	350	349
λ_{emi} in PS 5% (nm)	614	613

PMMA, Polymethylmethacrylate; *PS*, polystyrene.

16.7.1 Characterization of blended films in the solid state

PL analysis was carried out for RE(TTA)$_3$dmphen complexes doped in PMMA and PS at 5 and 10 wt% in the solid state. The excitation and emission peaks in molecularly doped RE(TTA)$_3$dmphenthin films were determined in the solid state with the help of these PL spectra.

16.7.1.1 Photoluminescence of RE(TTA)$_3$dmphen blended films in PMMA and PS

The excitation spectra of RE(TTA)$_3$dmphen blended thin films showed almost the same characteristics for all the polymer matrices except with a wt% of 5% in PS. For all wt%, they showed excitation around 380 nm and for the 5% in PS it was found to be around 350 nm, which may be due to the dispersion effect of the complex into the polymer. Sm(TTA)$_3$dmphen blended films showed the characteristic emission of Sm^{3+} ions with good intensity. The emission spectra of Eu$_{0.5}$Sm$_{0.5}$(TTA)$_3$dmphen blended films showed the characteristic emission of Eu^{3+} ions at 613 nm and a very low-intensity emission peak of Sm^{3+} at 624 nm, indicating the dominance of Eu^{3+} characteristic emission over Sm^{3+} in terms of intensity. The excitation and emission values of RE(TTA)$_3$dmphen are summarized in Table 16.3. The PL spectra of the RE(TTA)$_3$dmphen complexes in the polymer matrix are shown in Fig. 16.7a and b.

16.7.2 UV-Vis absorption spectra of blended thin films in various organic solvents

The electronic structure of the synthesized organic luminescent complex can be seen from the UV-Vis absorption spectra. A wide range of absorption spectra of blended thin films of synthesized RE(TTA)$_3$dmphen complexes in basic (dichloromethane) and acidic (formic acid) media at different wt% were studied. Absorption spectra were analyzed with UV-Vis apparatus with a PC interface by first calibrating solvent in a cuvette to obtain a baseline, then by running a cuvette filled with each of the solutions made from the complexes

IV. Lighting and light emitting diodes

16.7 Result and discussion 493

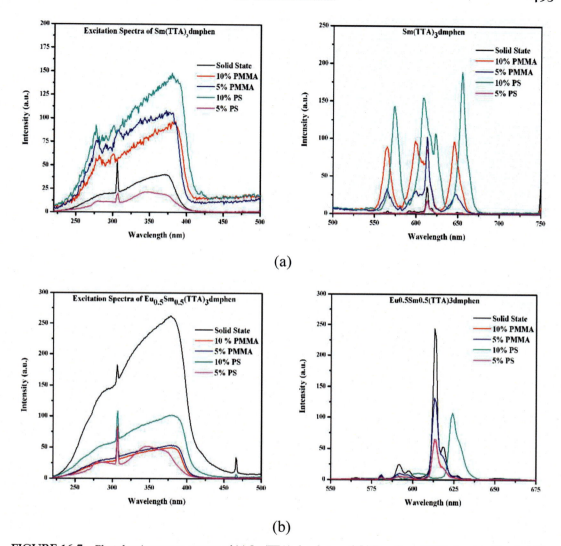

FIGURE 16.7 Photoluminescence spectra of (a) Sm(TTA)$_3$dmphen and (b) Eu$_{0.5}$Sm$_{0.5}$(TTA)$_3$dmphen blended thin films in polymethylmethacrylate (PMMA) and polystyrene (PS).

dissolved in solvent. Later, 5 and 10 wt%, 0.05 g thin film was dissolved in 5 mL of dichloromethane and formic acid, taken individually. The absorption spectra showed two peaks corresponding to $\pi \rightarrow \pi^*$ and $n \rightarrow \pi^*$ optical transitions in the range of 240–275 and 370–390 nm, respectively, in both solvents. However, a bathochromic shift (a shift toward a higher wavelength) in the peak position was observed in the solvated films in formic acid compared to that in dichloromethane. The optical densities of all the wt% of the films solvated in dichloromethane were found to be almost the same, whereas they decreased by a small value for the acidic medium. The UV-Vis absorption spectra of the Sm(TTA)$_3$dmphen blended films solvated in dichloromethane showed a peak at 251 nm for $\pi \rightarrow \pi^*$ optical transition and a second peak at 380, 375, 385, and 380 nm due to $n \rightarrow \pi^*$ optical transition for 10% PMMA, 5% PMMA, 10% PS, and 5% PS, respectively. However, the solvated films of $Eu_{0.5}Sm_{0.5}(TTA)_3$dmphen showed two peaks at 246 and 384 nm for $\pi \rightarrow \pi^*$ and $n \rightarrow \pi^*$ optical transitions. The UV-Vis analysis of the RE(TTA)$_3$dmphen blended thin films in dichloromethane and formic acid is shown in Fig. 16.8a–d.

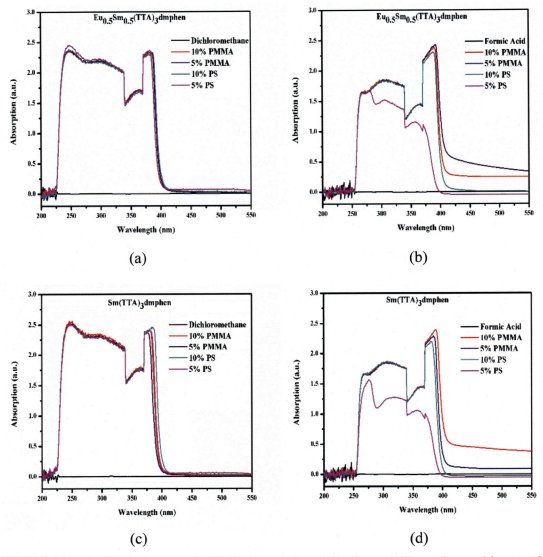

FIGURE 16.8 (a–d) UV-Vis spectra of RE(TTA)$_3$dmphen blended thin films in dichloromethane and formic acid. *PMMA*, Polymethylmethacrylate; *PS*, polystyrene.

16.7.2.1 Determination of the optical energy gap of RE(TTA)₃dmphen thin films in PMMA/PS

Optical excitation of electrons across the bandgap produced an abrupt increase in absorptivity at the wavelength corresponding to the gap energy. This transition in the absorption spectrum is known as the optical absorption edge. The procedure described by Morita et al. was employed to measure the optical energy gap of the blended thin films [24]. The bandgap of material can be determined from the absorption edge of the UV-Vis spectra. The optical energy gap of RE(TTA)₃dmphen blended thin films in PMMA and PS and solvated in dichloromethane and formic acid is shown in Fig. 16.9a–d. The graphs indicating the optical energy bandgap for RE(TTA)₃dmphen blended films in PMMA and PS in dichloromethane and formic acid are shown in Table 16.4.

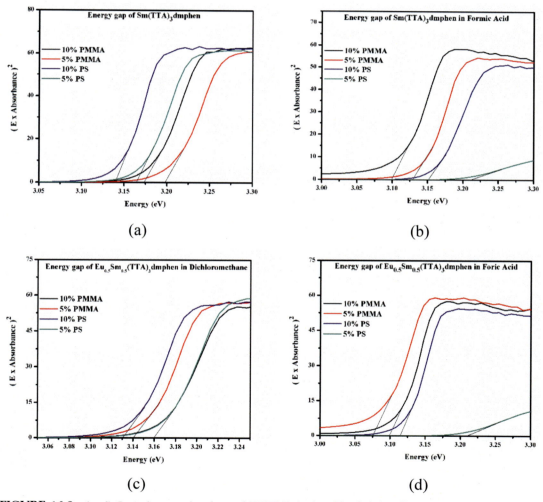

FIGURE 16.9 (a–d) Optical energy bandgap of RE(TTA)₃dpphen blended thin films in dichloromethane and formic acid.

TABLE 16.4 Optical energy gap values of RE(TTA)$_3$dmphen complex blended films in dichloromethane and formic acid.

Parameter	Energy gap in dichloromethane (eV)	Energy gap in formic acid (eV)
Sm(TTA)$_3$dmphen 10% PMMA	3.17	3.10
Sm(TTA)$_3$dmphen 5% PMMA	3.20	3.13
Sm(TTA)$_3$dmphen 10% PS	3.14	3.15
Sm(TTA)$_3$dmphen 5% PS	3.16	3.21
Eu$_{0.5}$Sm$_{0.5}$(TTA)$_3$dmphen 10% PMMA	3.16	3.10
Eu$_{0.5}$Sm$_{0.5}$(TTA)$_3$dmphen 5% PMMA	3.14	3.07
Eu$_{0.5}$Sm$_{0.5}$(TTA)$_3$dmphen 10% PS	3.13	3.11
Eu$_{0.5}$Sm$_{0.5}$(TTA)$_3$dmphen 5% PS	3.16	3.21

PMMA, Polymethylmethacrylate; PS, polystyrene.

16.7.3 Photoluminescence spectra of thin films in organic solvents

Blended films of all the synthesized complexes of RE(TTA)$_3$dmphen were solvated in basic and acidic organic solvents, namely dichloromethane and formic acid, respectively, at a molar concentration of 10^{-3} mol/L and their PL spectra were noted.

16.7.3.1 Photoluminescence spectra of RE(TTA)$_3$dmphen thin films solvated in dichloromethane

A homogeneous mix obtained by dissolving 0.05 g of film of the RE(TTA)$_3$dmphen complexes in PMMA and PS in 5 mL of dichloromethane and formic acid was used to evaluate the PL spectra. The result of PL characterization is shown in Table 16.5. From the recorded values, it is observed that the emission wavelength for Eu$_{0.5}$Sm$_{0.5}$(TTA)$_3$dmphen showed strong emission at 614 nm due to $^5D_0 \rightarrow {}^7F_2$ transition, and other transitions were found

TABLE 16.5 Photoluminescence parameters of RE(TTA)$_3$dmphen blended thin films in dichloromethane.

Parameter in dichloromethane	λ_{ext} (nm)	λ_{emi} (nm)
Sm(TTA)$_3$dmphen 10% PMMA	391	614
Sm(TTA)$_3$dmphen 5% PMMA	396	614
Sm(TTA)$_3$dmphen 10% PS	396	613
Sm(TTA)$_3$dmphen 5% PS	393	614
Eu$_{0.5}$Sm$_{0.5}$(TTA)$_3$dmphen 10% PMMA	392	614
Eu$_{0.5}$Sm$_{0.5}$(TTA)$_3$dmphen 5% PMMA	395	614
Eu$_{0.5}$Sm$_{0.5}$(TTA)$_3$dmphen 10% PS	396	614
Eu$_{0.5}$Sm$_{0.5}$(TTA)$_3$dmphen 5% PS	392	614

PMMA, Polymethylmethacrylate; PS, polystyrene.

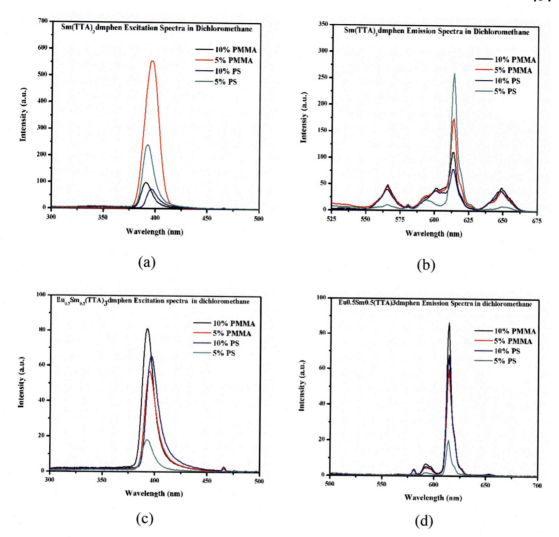

FIGURE 16.10 (a–d) Photoluminescence spectra of RE(TTA)₃dmphen blended thin films in dichloromethane. *PMMA*, Polymethylmethacrylate; *PS*, polystyrene.

to be very weak. On the other hand, in the case of the Sm(TTA)₃dmphen complex blended and solvated films, three characteristic peaks were recorded at 566, 614, and 649 nm, of which the emission intensity at 614 nm was found to be high, as shown in Fig. 16.10.

16.7.3.2 Photoluminescence spectra of RE(TTA)₃dmphen thin films solvated in formic acid

The PL parameters of RE(TTA)3dmphen complex blended thin films in PMMA and PS solvated in formic acid are shown in Table 16.6, and the PL spectra of these films are depicted in Fig. 16.11a–d.

TABLE 16.6 Photoluminescence parameters of RE(TTA)₃dmphen complex blended thin films in formic acid.

Parameter in formic acid	λ_{ext} (nm)	λ_{emi} (nm)
Sm(TTA)₃dmphen 10% PMMA	398	448
Sm(TTA)₃dmphen 5% PMMA	399	448
Sm(TTA)₃dmphen 10% PS	400	450
Sm(TTA)₃dmphen 5% PS	400	453
Eu₀.₅Sm₀.₅(TTA)₃dmphen 10% PMMA	418	462
Eu₀.₅Sm₀.₅(TTA)₃dmphen 5% PMMA	418	470
Eu₀.₅Sm₀.₅(TTA)₃dmphen 10% PS	420	451
Eu₀.₅Sm₀.₅(TTA)₃dmphen 5% PS	420	455

PMMA, Polymethylmethacrylate; *PS*, polystyrene.

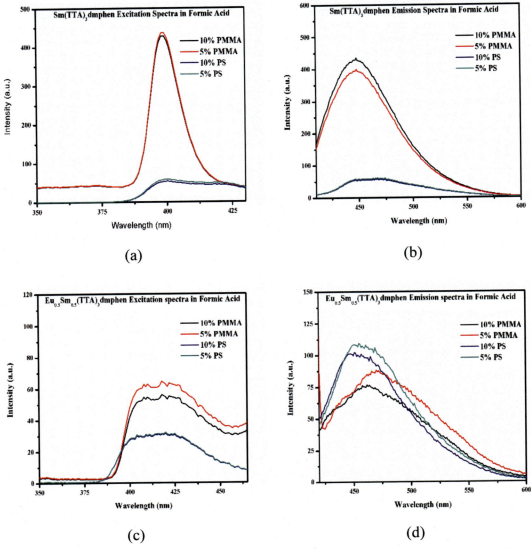

FIGURE 16.11 (a–d) Photoluminescence spectra of RE(TTA)₃dmphen blended thin films in formic acid. *PMMA*, Polymethylmethacrylate; *PS*, polystyrene.

16.7.3.3 CIE coordinates for blended films of RE(TTA)$_3$L in PMMA and PS

The CIE 1931 system was used to explore the photometric characteristics of the PMMA and PS blended films of synthesized complexes on the basis of chromatic coordinates. The determined CIE coordinates of RE(TTA)$_3$dmphen are shown in Table 16.7 and the CIE diagrams are shown in Fig. 16.12a and b.

16.7.3.4 CIE coordinates for solvated blended films of RE(TTA)$_3$dmphen of PMMA and PS in dichloromethane and formic acid

Blended thin films weighing 0.05 g were dissolved in 5 mL of basic (dichloromethane) and acidic (formic acid) media and their PL spectra were determined [25]. From the PL emission spectra, CIE coordinates were obtained. It can be seen that the coordinates for the blended films solvated in dichloromethane are in the red region of the spectrum and are shown within circles/ovals, and coordinates for the solvated films in formic acid are shown within the rectangle shape in Fig. 16.13.

TABLE 16.7 CIE coordinates of the blended films of RE(TTA)$_3$dmphen.

S. No.	Environment	Solvent	CIE coordinates		Correlated color temperature (K)
			x	y	
Sm(TTA)$_3$dmphen					
1.	Thin film	10% PMMA	0.4562	0.3426	2128
		5% PMMA	0.4216	0.3094	2383
		10% PS	0.5030	0.3484	1772
		5% PS	0.4545	0.3081	1926
Eu$_{0.5}$Sm$_{0.5}$(TTA)$_3$dmphen					
2.	Thin film	Chloroform	0.6129	0.3336	—
		Dichloromethane	0.6114	0.3315	—
		Toluene	0.6426	0.3177	—
		Formic acid	0.5673	0.3267	—

PMMA, Polymethylmethacrylate; *PS*, polystyrene.

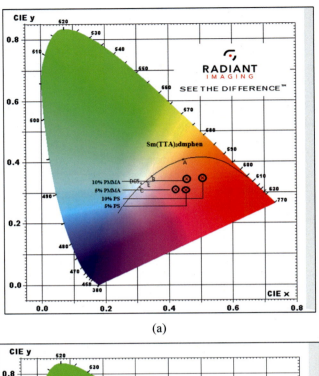

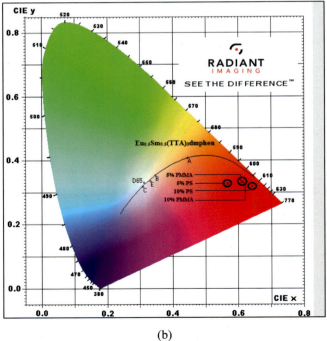

FIGURE 16.12 (a and b) CIE 1931 diagram for RE(TTA)$_3$dmphen blended films. *PMMA*, Polymethylmethacrylate; *PS*, polystyrene.

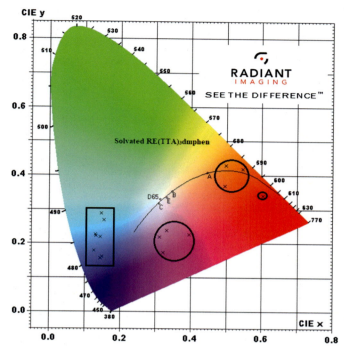

FIGURE 16.13 CIE 1931 diagram for RE(TTA)$_3$L blended films solvated in dichloromethane and formic acid.

16.8 Conclusions

We successfully prepared thin films of the previously synthesized RE(TTA)$_3$dmphen complexes in polymers PMMA and PS at different weight percentages (5% and 10%). The synthesized thin films were characterized by PL in the solid state and solvated in dichloromethane and formic acid. UV-Vis absorption characteristics of the same films in the solvated state were determined and the optical energy gap of all these complexes was calculated. In the PL spectra of the films in the solid state, emission in the orange—red region of the visible spectrum was observed for all the films. The intensity of emission in polymers PMMA and PS was found to be greater compared to the pure complex in most of the complexes, which may be caused by the interaction between chelate and the polymers. UV-Vis characterization showed two absorption peaks for all the complexes in basic as well as acidic media. The peaks corresponded to the characteristic $\pi \rightarrow \pi^*$ and $n \rightarrow \pi^*$ optical transitions observed in the chelating ligands. Bathochromic shift of the peak was observed for the complexes from basic to acidic media. The values of the optical energy gap from all the blended thin films were noted in the range of 3.09—3.25 eV. These values were found to be in good agreement with the values of energy gap obtained when the complex was solvated in different basic and acidic organic solvents as well as in the solid state, a major requirement of materials during the fabrication of OLEDs. The PL spectra showed the emission wavelength in the orange -red region of the visible spectrum for films solvated in dichloromethane, whereas the emission wavelength was in the blue region of the visible spectrum for films solvated in formic acid. These results when plotted on a CIE chromaticity diagram prove the color tunability of the complexes in different media. Thus synthesized complexes can be blended in polymers to obtain flexible films for the fabrication of OLEDs with sustained results.

References

[1] J.M. Phillips, et al., Research challenges to ultra-efficient inorganic solid-state lighting, Laser Photon 4 (2007) 307—333.

[2] N.T. Kalyani, S.J. Dhoble, Novel materials for fabrication and encapsulation of OLEDs, Renew. Sustain. Energy Rev. 44 (2015) 319—347.

[3] N.T. Kalyani, S.J. Dhoble, Organic light emitting diodes: energy saving lighting technology-a review, Renew. Sustain. Energy Rev. 16 (2012) 2696—2723.

[4] E.F. Schubert, Light-Emitting Diodes, Cambridge University Press, 2006.

[5] M.R. Krames, O.B. Shchekin, R. Mueller-Mach, G.O. Mueller, L. Zhou, G. Harbers, M.G. Craford, J. Disp. Technol. 3 (2) (2007) 160—175.

[6] Q.Y. Zhang, K. Pita, W. Ye, W.X. Que, Influence of annealing atmosphere and temperature onphotoluminescence of Tb^{3+} or Eu^{3+} activated zinc silicate thin film phosphors via sol-gel method, Chem. Phys. Lett. 351 (2002) 163—170.

[7] N. Thejokalyani, S.J. Dhoble, Importance of eco-friendly OLED lighting, Defect Diffusion Forum 357 (2014) 1—27.

[8] D. Chitnis, N. Thejokalyani, S.J. Dhoble, Escalating opportunities in the field of lighting, Renew. Sustain. Energy Rev. 64 (2016) 727—748.

[9] N. Xiang, Y. Xu, Z. Wang, X. Wang, L.M. Leung, J. Wang, Q. Su, M. Gong, Spectrochim. Acta, A 69 (2008) 1150—1153.

[10] J.-F. Lee, H.-L. Chen, et al., Photosensized controlling benzyl methacrylate based matrix enhanced Eu^{3+} narrow band emission for fluorescence applications, Int. J. Mol. Sci. 13 (2012) 3718—3737.

[11] D. Chitnis, N. Thejokalyani, S.J. Dhoble, Exploration of spectroscopic properties of solvated tris (thenoyltrifluoroacetone)(2,2'-bipyridine) europium(III) red hybrid organic complex for solution processed OLEDs and displays, J. Lumin. 185 (2017) 61—71.

[12] D. Chitnis, N.T. kalyani, S.J. Dhoble, Assessment of spectroscopic parameters of solvated Eu(dmh)3phen organometallic complex in various basic and acidic solvent, J. Biol. Chem. Lumin. 5 (2018) 968—980.

[13] N. Thejokalyani, R.G. Atram, S.J. Dhoble, Synthesis and characterization of $Y_{(1-x)}Eu_x(DBM)_3(phen)$ organic luminescent thin films, J. Biol. Chem. Lumin. 29 (2014) 674—678.

[14] Y. Zheng, L. Fu, et al., Electroluminescence based on a β-diketonate ternary samarium complex, J. Mater. Chem. 12 (2002) 919—923.

[15] Y. Hasegawa, S. Tsuruoka, et al., Enhanced deep red luminescence of tris(hexafluoroacetylacetonato)samarium(II) complex with phenanthroline in solution by control of ligand coordination, J. Phys. Chem. 112 (2008) 803—807.

[16] Z. Kin, H. Kajii, et al., Optical and electroluminescent properties of samarium complex-based organic light-emitting diodes, Thin Solid Films 516 (2008) 2735—2738.

[17] Y. Hasegawa, S. Tsuruoka, et al., Emission properties of Sm(III) complex having ten-coordination structure, Thin Solid Films 516 (2008) 2704—2707.

[18] A. Ugale, N.T. Kalyani, S.J. Dhoble, Colour tunable emission from $Eu_{0.5}Sm_{0.5}(TTA)_3$dpphen β-diketonate hybrid organic complex in various organic solvents, Optik 06 (2018) 171—182.

[19] A. Ugale, N. Kalyani, S.J. Dhoble, Reddish orange to blue tunable emission from rare earth β-diketonateEu(TTA)3dpphen complex for solid state lighting applications, Mater. Sci. Energy Technol. 2 (2019) 57—66.

[20] A. Ugale, N. T. Kalyani, S.J. Dhoble, Investigations on RE(TTA)3tppo (RE = $Eu/Sm/Eu_{0.5}$ $Sm_{0.5}$) rare earth b-diketonate complexes for OLEDs and solid state lighting,

[21] N.T. kalyani, R.G. Atram, S.J. Dhoble, Synthesis and characterization of $Y_{(1-x)}Eu_x(DBM)_3(phen)$ and $La_{(1-x)}Eu_x(DBM)_3(phen)$ complexes for solid state lighting, Int. J. Lumin. Appl. 6 (2) (2016) 139—145.

References

503

[22] A. Ugale, N.T. Kalyani, S.J. Dhoble, Potential of europium and samarium β -diketonates as red light emitters in organic light-emitting diodes, in: Lanthanide-Based Multifunctional Materials, 01/2018, ISBN 9780128138403, pp. 59—97, https://doi.org/10.1016/B978-0-12-813840-3.00002-8.

[23] http://www.rsc.org/periodictable/element/63/europium, https:www.livescience.com/38186-europium.html, http://www.rsc.org/periodic table/element/62/samarium.

[24] S. Morita, T. Akashi, A. Fujii, M. Yoshida, Y. Ohmori, K. Yoshimoto, T. Kawai, A.A. Zakhidrov, S.B. Lee, K. Yoshino, Synth. Met. 69 (1995) 433.

[25] H.F. Brito, O.L. Malta, J.F. Menezes, Luminescent properties of diketonates of trivalent europium with dimethyl sulfoxide, J. Alloys Compd. 303 (2000) 336.

IV. Lighting and light emitting diodes

CHAPTER

17

Synthesis and characterization of energy-efficient Mq_2 (M = Zn, Cd, Ca, and Sr) organometallic complexes for OLED display applications

Prajakta P. Varghe[1], N. Thejo Kalyani[2], P.G. Shende[1], S.J. Dhoble[3]

[1]Department of Surface Coating Technology, Laxminarayan Institute of Technology, Nagpur, Maharashtra, India; [2]Department of Applied Physics, Laxminarayan Institute of Technology, Nagpur, Maharashtra, India; [3]Department of Physics, RTM Nagpur University, Nagpur, Maharashtra, India

17.1 Introduction

In this world of escalating prospects, employing energy-efficient displays with the ultimate objective of curtailing energy consumption without compromising the quality of the image is a gigantic challenge for researchers and scientists exploring this field [1]. Designing environmentally friendly and energy-efficient displays could be achieved by implementing technological changeovers and process novelty. Moreover, these innovations need to be evaluated for various parameters such as environmental impact, lifetime, cost, and other engineering performance aspects [2]. Display technology is presently in a state of adaptation as bulky cathode ray tubes are being replaced by flexible flat-panel displays and organic light-emitting diode (OLED) technology because of the desire for up-to-the-minute products such as televisions, solid-state display panels, laptops, computers, mobiles, etc.

Despite being a relatively new technology, their versatile attributes such as slimness, fast response time, low power consumption, self-illuminating mechanism, exceptional brightness,

Energy Materials
https://doi.org/10.1016/B978-0-12-823710-6.00003-0

Copyright © 2021 Elsevier Ltd. All rights reserved.

FIGURE 17.1 Organic light-emitting diode displays—superior attributes.

high contrast, rich colors, deep blacks, nontoxicity, recyclability, wide viewing angle, etc. [3,4] are revolutionizing the field of displays. The superior OLED display attributes are depicted in Fig. 17.1. Color from any display device, say a flat-panel solid-state lighting display, television, desktop, or laptop display screen, is generated from primary colors of light (RGB) and the diverse colors we perceive are due to different permutations and intensities of these primary colors [5]. Each pixel on these display panels is composed of three small dots of light-emitting phosphors (RGB taken separately) surrounded by a black mask, as shown in Fig. 17.2. Such pixels are arranged in rows and columns, depending on the size of the display.

Any acceptable display device should satisfy certain attributes such as (1) high resolution, (2) high luminance, (3) large color gamut, (4) high brightness and contrast, (5) light weight, (6) slender thickness, (7) fast switching response, (8) low operating voltage, and (9) good lifetime. Among all the display devices invented so far, OLED displays are lighter, slimmer, and satisfy all the foregoing attributes.

FIGURE 17.2 Single RGB pixel.

17.1.1 OLED displays: a brief background

In this digital era, display is an imperative interface of communication [6]. It is an output device, which gives optical depiction of the information on the screen when energized with electricity. Displays can be categorized as active and passive display devices. Active displays are self-emissive; they operate at low voltage with higher lifetime as they do not require any backlight. Conversely, passive devices are nonemissive devices; they operate at high voltage with less lifetime as they require backlight. Contemporarily thin and flat-panel displays with light-emitting diodes (LEDs)/OLEDs are intended to be energy efficient and eco-friendly. However, LED displays are useless due to limited flexibility and high processing cost. Hence, OLEDs have been exhaustively explored as a subsequent invention in energy-efficient display applications to attain environmental benefits. They employ simpler carbon-based materials, which are eco-friendly and have an energy-saving emissive layer. They are made by placing a stack of thin organic films, which includes a hole injection layer, a hole transport layer (HTL), an emissive layer (EML), and an electron transport layer (ETL), between two conductors, one a metallic cathode and the other a transparent (typically glass) anode [7]. These displays have the advantage of printing on any medium, leading to the possibility of flexible plastic display devices, which can be folded or rolled. However, to protect them from dirt and moisture, they are encapsulated, which adds to the cost [8].

17.1.2 Working mechanism of displays

OLEDs work much the same as LED displays, but instead of using n-type and p-type semiconductor layers, they employ organic molecules to generate electron and hole pairs called excitons. When a suitable voltage is applied to these devices, excitons form and recombine in the EML by emitting a photon of characteristic wavelength by a process known as electroluminescence.

17.2 Role of quinoline in display technology

Quinoline is a small molecule; hence, emissive materials have many advantages such as:

- They can be sublimed easily
- Ease of handling
- Ability to form uniform thin films
- Good carrier electron transporter
- High fluorescent yield
- High thermal stability and glass transition temperature
- High efficiency of photoluminescence
- High luminescence quantum yield
- High stability even at high operating voltages
- Cover almost the entire range of the visible region

Generally, thermal vibration of the chromophore leads to loss of energy by non-radiative pathway. However, this can be avoided by improving the rigidity of the chromophore by chelating a suitable chromophore to an appropriate metal ion, which improves the emission efficiency as the degree of freedom of the thermal vibrations of the chromophore is suppressed. However, metal ions for electroluminscent chelates are constrained to those metals that do not demonstrate d—d transitions because they may interfere with the luminescence of the ligand [9,10].

Luminescent metal chelates contain one or more organic chromophores and supporting metal substituent ligands attached to the metal center through a donor acceptor bond or covalent bond. Because metal chelate (Fig. 17.3A) compounds are the supporting part of chromophore (Fig. 17.3B) small molecules, they can be readily sublimed.

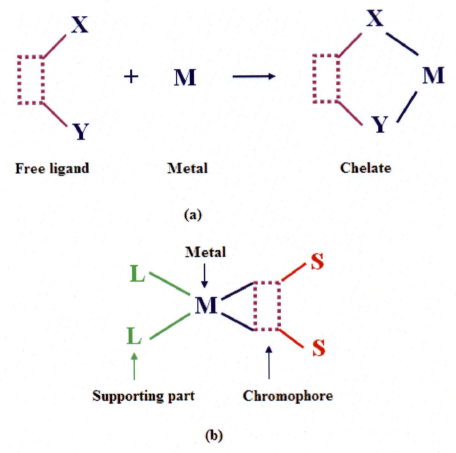

FIGURE 17.3 Illustration of (A) metal chelates and (B) small molecules.

17.3 Quinoline complexes: prior state of the art

Fluorescent coordination compounds, mainly ligand-based, have caught the attention of global researchers and scientists due to their appealing applications as emissive material in

OLEDs [11]. Particularly, small molecule metal chelates of 8-hydroxyquinoline-based Zn, Ca, Sr, Cd, Mg, and Ba have been used in OLED devices as the ELT and HTL. They can also be used as an EML, ETL, and HTL in OLEDs because they almost cover all the primary colors (RGB) of the visible region [12–14]. The literature speaks volumes about the nature of metal ions tuning emission wavelength and stability [15,16]. It was also found that with the increase in atomic number (Z) of the metal ion, the fluorescence efficiency of the complex was suppressed due to the heavy atom effect (increase in intersystem crossing) [17,18]. Prior state of the art reports the usage of iridium and platinum [19–21]. Hence, we propose the synthesis and evaluation of the molecular structure and optical properties of some most interesting organometallic complexes—Znq_2, Caq_2, Srq_2, and Cdq_2—with nitrates as divalent metal with two 8-hydroxyquinoline molecules in the solid state and solvated state.

17.4 Synthesis of green phosphor for displays

An environmentally friendly technique, namely the precipitation method, has been adopted for the synthesis of energy-efficient Mq_2 (M = Zn, Cd, Ca, and Sr) organometallic complexes for display devices.

17.5 Experimental

The raw materials employed were of analytical grade (analytical reagent) and hence analysis of the raw materials was not executed.

17.5.1 Reagents and solvents

The reagents and solvents employed in the present study included 8-hydroxyquinoline [$C_{27}H_{18}AlN_3O_3$] (Aldrich Chemicals) of molecular weight 145.16 g and a melting point ranging between 192 and 194°C, zinc nitrate [$Zn(NO_3)_2$] (Aldrich Chemicals) of molecular weight 189.36 g and a melting point of 110°C, cadmium nitrate [$Cd(NO_3)_2.4H_2O$] (Aldrich Chemicals) of molecular weight 236.43 g and a melting point of 350°C, calcium nitrate [$Ca(NO_3)_2.2H_2O$] (Aldrich Chemicals) of molecular weight 164.09 g and a melting point of 42.7°C, strontium nitrate [$Sr(NO_3)_2$] (Aldrich Chemicals) of molecular weight 211.63 g and a melting point of 570°C, acetic acid [CH_3COOH] (Alfa Aesar Chemicals) of molecular weight 60.05 g and a melting point ranging between 16 and 17°C, which was purchased along with ammonia solution [NH_4OH] (Aldrich Chemicals) of molecular weight 35.04 g and a melting point of 91.5°C, and double distilled water.

17.5.2 Structure and synthesis of Mq_2

Znq_2 was synthesized at room temperature by a simple precipitation method as described by the literature [22]. The complete synthesis scheme is depicted in Fig. 17.4.

Other complexes, namely Cdq_2, Caq_2, and Srq_2, were synthesized by following the same procedure. The structures of Mq_2 (M = Zn, Ca, Sr, and Cd) organometallic complexes are shown in Fig. 17.5.

IV. Lighting and light emitting diodes

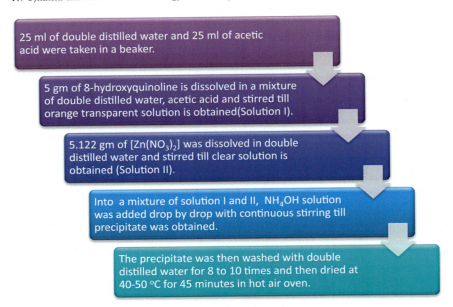

FIGURE 17.4 Znq₂ synthesis scheme: photographic illustration.

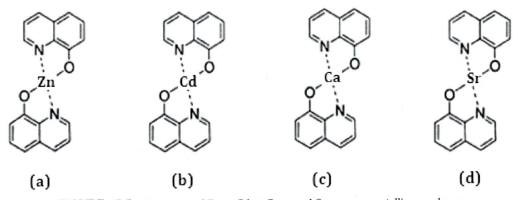

FIGURE 17.5 Structures of Znq₂, Cdq₂, Caq₂, and Srq₂ organometallic complexes.

17.6 Results and discussion

The Fourier transform infrared (FT-IR) spectrum was produced on a SHIMADZU Model 8101A infrared spectrophotometer. The optical absorption spectra of the synthesized complexes were carried out on an Analytic Specrod-50 spectrophotometer and photoluminescence (PL) spectra were obtained on an HITACHI F-4000 spectrofluorometer. 1931 Commission Internationale de l'Eclairage (CIE) system software was used to explore CIE coordinates of the synthesized complexes. All the measurements were carried out at room temperature.

17.6.1 Structural analysis

The synthesized complexes were characterized by FT-IR, recorded in the range of 4000–400 cm^{-1}, and carried out on a SHIMADZU Model 8101A infrared spectrophotometer using the KBr pellet technique [23].

17.6.1.1 Fourier transform infrared spectroscopy

FT-IR is generally used to confirm the molecular structure and quality of a sample under investigation. Fig. 17.6A shows vibrations at 1602.85, 1575.84, and 1327.03 cm^{-1}, consigned to

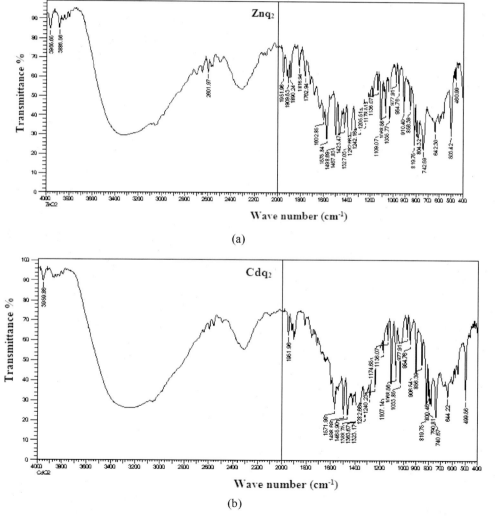

FIGURE 17.6 Fourier transform infrared spectra of Mq$_2$ (M = Zn, Cd, Ca, and Sr) organometallic complexes.

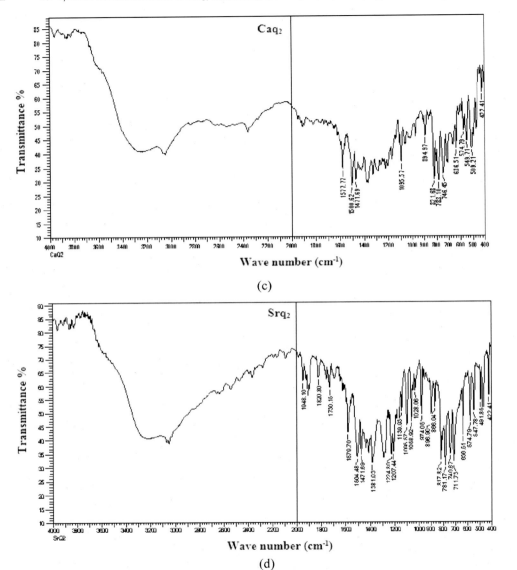

FIGURE 17.6 cont'd

the quinoline group of Znq$_2$. The bands at 1498.69 and 1467.83 cm^{-1} can be accredited to pyridyl and phenyl groups. The peaks at 742 and 642 cm^{-1} are allied with in-plane ring deformation [24], while the absorption peaks at 2852 and 2923 cm^{-1} can be allocated to the stretching modes of methylene groups in Znq$_2$. Similarly, for Cdq$_2$, IR bands at high wave number (3959.86–1951 cm^{-1}) are due to localized hydrogen stretching vibrations. Absorption between 1571.99 and 1033.85 cm^{-1} corresponds to heavy atom (Cd) in-plane and bending vibrations. Absorption below $1000 \, cm^{-1}$ is a trait of out-of-plane modes. The bands at $1571.99 \, cm^{-1}$ are consigned to C—C stretching vibration from the quinoline ligand. The bands at 1500 and $1465.90 \, cm^{-1}$ correspond to C—C/C—N stretching. A band at $1100 \, cm^{-1}$ may be the result of C—O stretching vibration at the C—O—M site. The spectrum reveals a sharp absorption band near $1110 \, cm^{-1}$. The extreme out-of-plane absorption in this region was found at $740.67 \, cm^{-1}$, as shown in Fig. 17.6B. For the Caq$_2$ complex, C—N vibrations (1577.77, $1284 \, cm^{-1}$), C—O stretching vibrations ($1284 \, cm^{-1}$), C—C vibrations (1471.69, $787 \, cm^{-1}$), C—N vibrations ($1577.77 \, cm^{-1}$), C—C—H bending/stretching ($1284 \, cm^{-1}$), C—H vibrations (1205, $746.45 \, cm^{-1}$), Ca—O stretching vibration, and Ca—N stretching vibration ($509.21 - 422.41 \, cm^{-1}$) were observed. The vibrations at $1095.57 \, cm^{-1}$ were designated to the quinoline group of Caq$_2$, as shown in Fig. 17.6C.

IR absorption bands of Srq$_2$ at high wave number ($1948.1 - 1820.80 \, cm^{-1}$) may be attributed to localized hydrogen stretching vibrations. The vibrations at 1579.70, 1471.69, and $1381.03 \, cm^{-1}$ can be designated to the quinoline group and the bands at 1471.69 and $1381.03 \, cm^{-1}$ belong to both the pyridyl and phenyl groups. The peaks at 740.67 and $636 \, cm^{-1}$ are coupled with in-plane ring deformation, as shown in Fig. 17.6D.

17.6.2 Photophysical analysis

Photophysical analysis of the synthesized complexes was accomplished by UV-Vis absorption spectra in acetic acid solution (10^{-3} M), carried out on an Analytic Specrod-50 spectrophotometer. Photoluminescent spectra on an HITACHI F-4000 spectrofluorometer and CIE coordinates were typically characterized by CIE system software.

17.6.2.1 Absorption spectra

Absorption spectra of Znq$_2$, Cdq$_2$, Caq$_2$, and Srq$_2$ organometallic complexes in acetic acid (acidic medium) at 10^{-3} M reveal two absorption peaks of Znq$_2$ at 399 and 296 nm, attributed to $n \rightarrow \pi^*$ and $\pi \rightarrow \pi^*$ transition of the aromatic ring. Similarly, the maximum absorption and shoulder peaks of Caq$_2$, Cdq$_2$, and Srq$_2$ were observed at 302 and 397, 305 and 408, and 319 and 410 nm, respectively, as shown in Fig. 17.7.

The optical energy bandgap of Znq$_2$, Caq$_2$, Srq$_2$, and Cdq$_2$ complexes at 10^{-3} (mol/L), determined by the literature method [25,26], is shown in Fig. 17.8. The energy gap of Znq$_2$ was found to be 2.78 eV, while for Cdq$_2$, Caq$_2$, and Srq$_2$ complexes, it was found to be 2.70, 2.67, and 2.6 eV, respectively. The UV-Vis absorption parameters of Mq$_2$ complexes in acidic acid are shown in Table 17.1.

17.6.2.2 Photoluminescence spectra

Fig. 17.9 illustrates the excitation spectra of Mq$_2$ (M = Zn, Cd, Ca, and Sr) organometallic complexes between 250 and 500 nm. All the complexes exhibit similar spectra with an excitation peak and a shoulder toward the lower wavelength. Such similarity can be attributed to the presence of 8-hydroxyquinoline molecules, a major contributor in all the complexes. Znq$_2$ portrays an excitation peak at 436 nm with a shoulder at 384 nm. Similarly, Cdq$_2$, Caq$_2$, and Srq$_2$ illustrate excitation peaks at 436, 427, and 385 nm with a shoulder at 383, 389, and 332 nm, respectively.

IV. Lighting and light emitting diodes

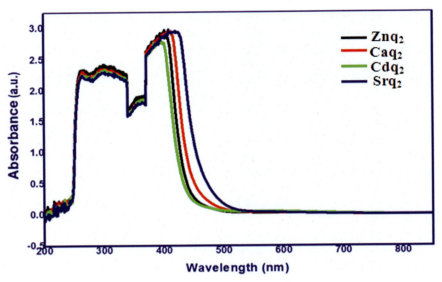

FIGURE 17.7 Absorption spectra of Mq$_2$ (M = Zn, Cd, Ca, and Sr) organometallic complexes in acetic acid solution (10^{-3} M).

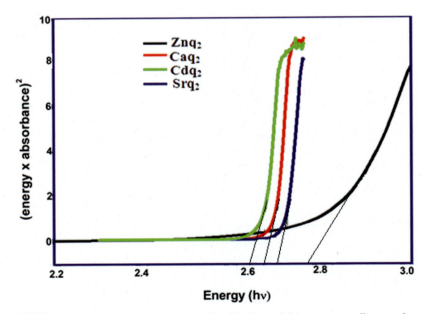

FIGURE 17.8 Energy gap of Mq$_2$ (M = Zn, Cd, Ca, and Sr) organometallic complexes.

17.6 Results and discussion

TABLE 17.1 UV-Vis optical absorption parameters of Mq_2 complexes in acidic acid.

Complex	Molar ratio (mol/L)	Band	λ_{max} (nm)	Energy gap (eV)
Znq_2	10^{-3}	I	296	2.78
		II	399	
Caq_2		I	305	2.67
		II	408	
Cdq_2		I	302	2.70
		II	397	
Srq_2		I	319	2.60
		II	410	

The emission spectra of hydroxyquinoline metal complexes Znq_2, Cdq_2, Caq_2, and Srq_2 are shown in Fig. 17.10. Znq_2 illustrates its emission peak at 499 nm ($\lambda_{ext} = \lambda_{ext} = 436$ nm) [27]. When Cdq_2 was excited at 383 nm, the emission wavelength was found at 491 nm, while Caq_2 displayed an emission peak at 490 nm under an excitation wavelength of 389 nm.

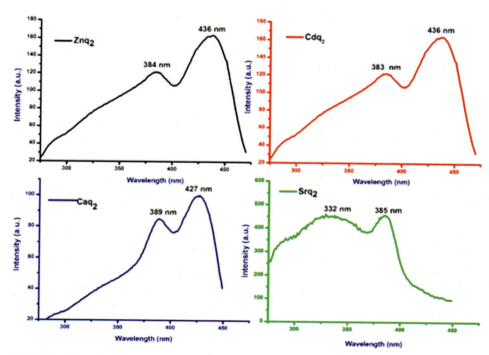

FIGURE 17.9 The excitation and emission spectra of Mq_2 (M = Zn, Cd, Ca, and Sr) complexes.

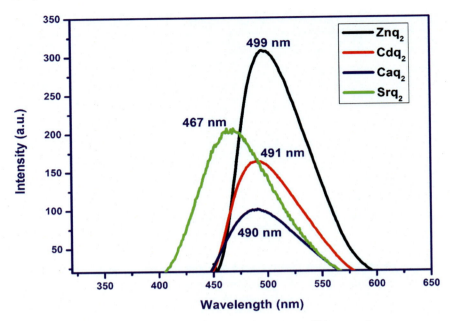

FIGURE 17.10 The photoluminescence spectra of Mq2 complexes.

Similarly, when Srq2 was excited at 385 nm, $\lambda_{ext}\lambda_{ext}$ peaked at 467 nm. All these complexes have emission in the green/bluish-green/blue region of visible radiation. The blue/greenish-blue emission peaks in the range of 490—499 nm arise from the singly ionized oxygen vacancy due to the recombination of a photogenerated hole with a singly ionized electron in the valance band [28,29]. The PL emission of Srq2 was blue shifted by 23—32 nm from the PL emission of Znq2, which may be attributed to delocalization of $\pi \rightarrow \pi^*$ transition of the excited state [30,31].

17.6.3 Photometric properties

Any emission color can be assigned with a pair of chromaticity coordinates (x, y) [32] on the CIE chromaticity diagram. For Znq2, it was found to be (0.1805, 0.5026), corresponding to the near-green region as shown in Fig. 17.11. Similarly, the chromaticity coordinates for Caq2 and Cdq2 were found to be (0.1629, 0.3834) and (0.15, 0.1939), which fall in the bluish-green region and for Srq2, it was found to be (0.1689, 0.4253), respectively, corresponding to the blue region. Correlated color temperature (CCT) is the temperature of a black body whose chromaticity closely resembles that of a light source. The synthesized Mq2 complexes also show CCT values in between the standard range. A summary of PL spectra and CIE coordinates of metal complexes (Mq2: M = Cd, Ca, Sr, and Zn) is displayed in Table 17.2.

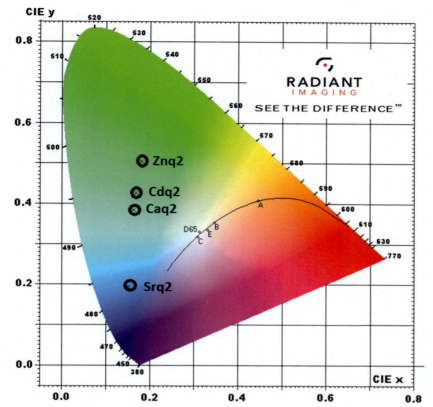

FIGURE 17.11 CIE 1931 (x,y) diagram showing emission color coordinates of Mq$_2$ complexes.

17.7 Superiority: OLED displays

- **Lambertian emitters**

OLED pixels are Lambertian emitters—the displays that appear equally bright from viewing angles up to 170 degrees. This compares with the narrower viewing angles of liquid crystal display (LCD) technology whereby the brightness of the light source typically falls by half outside of a 90–100 degrees viewing angle (although some disingenuous TV manufacturers claim differently).

- **Power consumption**

An OLED display does not rely on backlight to illuminate the screen. It emits visible light in low ambient light conditions without compromising the contrast ratio; hence, it eliminates the backlight, which reduces the power consumption typically by up to 60%–80% as compared to devices that rely on backlighting.

TABLE 17.2 Summary of photoluminescence spectra and CIE coordinates of metal complexes.

S. No.	Complex	Under natural light	Under UV light	$\lambda_{exc(nm)}$	$\lambda_{emi(nm)}$	(x, y)	Correlated color temperature (K)
1.	Znq_2			384 436	499	(0.1805, 0.5026)	9593
2.	Cdq_2			383 436	491	(0.1689, 0.4253)	12,247
3.	Caq_2			389 427	490	(0.1629, 0.3834)	15,608
4.	Srq_2			332 467	467	(0.150, 0.1939)	–

- **Response time**

Theoretically, OLEDs are capable of rapidly changing their response times for leading consumer TVs that respond in 0.01 ms and can refresh a screen at 200 Hz or 100 kHz. LCDs are cited in the 4–8 ms range.

- **Thinner and lighter displays**

OLED technology provides thinner, flexible, and lighter displays because of the elimination of backlight, and because of their relatively simple construction.

- **Wide viewing angle**

OLEDs offer drastic viewing angles—up to 84 degrees with no luminance degradation.

17.8 Limitations

There are some disadvantages to OLED displays:

- **Limited lifetime**

The organic materials can be easily subjected to chemical processes such as oxidation and gradually lose their light-emitting properties. Particularly, blue OLEDs offer a lifetime of around 5000 h, which is much lower compared to their red and green counterparts. The technology is immature and manufacturers are working toward improving OLED lifetime. An alternative approach is to add a light blue subpixel to the standard three-color RGB matrix.

- **Water intrusion**

The fact that the organic materials can be easily damaged by water intrusion into the displays, demanding relatively exclusive sealing and development of the technology, is restrained by expensive techniques.

- **Cost**

The cost of OLED displays is relatively high due to complicated fabrication techniques and the necessity for encapsulation.

17.9 Applications

OLED flat-panel displays are mainly used in digital devices such as television screens, desktop monitors, laptop screens, solid-state lighting panels, mobile screen displays, digital cameras, and many more.

17.10 Conclusions

Green/bluish-green/blue light-emitting 8-hydroxyquinoline derivative metal complexes (Mq_2: where M = Cd, Ca, Sr, and Zn) were synthesized successfully by the precipitation method at room temperature. FT-IR spectra of all Mq_2 metal complexes were found to be nearly the same, while metal stretching vibrations were found to be different for different metal derivatives. Absorption peaks of Znq_2 were registered at 399 nm with a shoulder at 296 nm, attributed to $n \rightarrow \pi^*$ and $\pi \rightarrow \pi^*$ transition of the aromatic ring. Similarly, the maximum absorption and shoulder peaks of Caq_2, Cdq_2, and Srq_2 were observed at 302 and 397, 305 and 408, and 319 and 410 nm, respectively. The energy gap of Znq_2 was found to be 2.78 eV, while for Cdq_2, Caq_2, and Srq_2 complexes, it was found to be 2.70, 2.67, and 2.6 eV, respectively. PL spectra of these metal complexes offer tunable emission wavelength ranging inbetween blue to green color (467–499 nm). The chromaticity coordinates (x, y) for Znq_2 were found to be (0.1805, 0.5026), while for Caq_2 and Cdq_2, they were found to be (0.1629, 0.3834) and (0.15, 0.1939). For Srq_2, CIE coordinates were found to be (0.1689, 0.4253), respectively, which corresponds to the blue region. The tunability of light-emitting 8-hydroxyquinoline derivative metals finds applications in tunable emissions, demonstrating their wide and versatile applications in electroluminescent devices.

17.11 Future perspectives

With the ultimate objective of potential applications, numerous design strategies like amalgamation of novel materials, sensible framework for device configuration, and expansion of superior and low-price tag fabrication techniques have been put forward to accomplish elevated efficiency and high-quality color stability lighting. Realistic applications, which encourage specialists in this field to expand innovative directions for upcoming research progress of OLEDs, are at the forefront.

References

[1] I. Overland, Subsidies for fossil fuels and climate change: a comparative perspective, Int. J. Environ. Stud. 67 (2010) 303–317.

[2] Organic light-emitting diode, http://en.wikipedia.org/wiki/Oled, Wikipedia.org.

[3] N. Thejo Kalyani, H. Swart, S.J. Dhoble, Organic light emitting diodes: the future lighting sources, in: Principles and Applications of Organic Light Emitting Diodes, Elsevier: Imprint Wood head Publishing Series in Electronic and Optical Materials, United Kingdom, 2017, ISBN 978-0-08-101213-0, pp. 141–170. Ch. 6.

[4] N. Thejo Kalyani, H. Swart, S.J. Dhoble, History of organic light emitting diode displays, in: Principles and Applications of Organic Light Emitting Diodes, Elsevier: Imprint Woodhead Publishing Series in Electronic and optical materials, United Kingdom, 2017, ISBN 978-0-08-101213-0, pp. 205–226. Ch. 8.

[5] A. Joseph, in: Castellano (Ed.), Handbook of Display Technology, Gulf Professional Publishing, 1992, ISBN 0-12-163420-5, p. 9.

[6] K.S. Chen, C.H. Wang, H.T. Chen, A MAIC approach to TFT-LCD panel quality improvement, Microelectron. Reliab. 46 (2006) 1189–1198.

[7] C.W. Tang, S.A. Van Slyke, C.H. Chen, Electroluminescence of doped organic thin films, J. Appl. Phys. 65 (1989) 3610–3616.

[8] N. ThejoKalyani, S.J. Dhoble, Novel materials for fabrication and encapsulation of OLEDs, Renew. Sustain. Energy Rev. 44 (2015) 319–347.

[9] Y. Hamada, T. Sano, M. Fujita, T. Fujii, Y. Nishio, K. Shibata, Organic light emitting diodes 3- or 5-hydroxyflavone-metal complexes, J. Appl. Phys. 324A (1993) 514.

[10] W.Y. Yang, Luminescent Zinc and Lanthanide Complexes Based on 2,2′-DipyridylamineDerivatives, Department of Chemistry Queen's University, Kingston, Ontario, Canada, August 2000, pp. 22–30.

References

[11] H. Friend, R. Gymer, A. Holmes, J. Burroughes, R. Marks, C. Taliani, D. Bradley, D. Santos, J. Bredas, M. Logdlund, W.R. Salaneck, Electroluminescence in conjugated polymers, Nature 397 (6715) (1999) 121.

[12] M. Brinkmann, G. Gadret, M. Muccini, C. Taliani, N. Masciocchi, Sironi, Correlation between molecular packing and optical properties in different crystalline polymorphs and amorphous thin films of *mer-* Tris (8-hydroxyquinoline)aluminium(III), J. Am. Chem. Soc. 122 (21) (2000) 5147−5157.

[13] A. Meyers, M. Weck, Design and synthesis of Alq_3-functionalized polymers, Macromolecules 36 (6) (2003) 1766−1768.

[14] M. Cankurtaran, H. Ç elik, E. Tiras, A. Bayrakli, N. Balkan, Hot electron energy relaxation via acoustic phonon emission, Phys. Status Solidi B 207 (1998) 139.

[15] Y. Harima, K. Takeda, K. Yamashita, Molecular solid of zinc tetraphenylporphyrin as a model organic semiconductor with a well-defined depletion layer, J. Phys. Chem. 56 (9) (1995) 1223−1229.

[16] M. Brinkmann, G. Gadret, M. Muccini, C. Taliani, N. Masciocchi, Sironi, Single-crystal structure determination of new polymorph (Alq_3) of the electroluminescence OLED material tris(8-hydroxyquinoline)aluminium(Alq_3), J. Am. Chem. Soc. 122 (2000) 21.

[17] H.P. Santos, E.S. Gomes, M.V. dos Santos, K.A. D'Oliveira, New class of organic hole-transporting materials based on xanthene derivatives for organic electronic applications, Inorg. Chim. Acta. 484 (2019) 0020−1693.

[18] X. Xiong, F. Song, J. Wang, Y. Zhang, Y. Xue, L. Sun, N. Jiang, P. Gao, L. Tian, X. Peng, Thermally activated delayed fluorescence of fluorescein derivative for time resolved and confocal fluorescence imaging, J. Am. Chem. Soc. 136 (2014) 9590.

[19] W. Huang, Highly crystalline multimetallic nanoframes with three dimensional electrocatalytic surfaces, Adv. Sci. 4 (2017) 1600−1666.ta.

[20] M. Amati, F. Lelj, Luminescent compounds fac- and mer- aluminium tris(quinolin-8-olate), J. Phys. Chem. 40 (2003) 2560.

[21] S. Tălu, S. Solaymani, M. Bramowicz, N. Naseri, S. Kulesza, A. Ghaderi, Anal. Microstructure and micromorphology of Cu/Co nanoparticles: surface texture analysis, Inside Chem. (2016) 27228.

[22] P. Shinde, S. Pandharipande, N. Thejokalyani, S.J. Dhoble, Exploration of photophysical properties of green light emitting bis(8-hydroxyquinoline) zinc (Znq_2) metal chelate under various environments, Optik 162 (2018) 151−160.

[23] D. Chitnis, N. Thejokalyani, S.J. Dhoble, Exploration of spectroscopic properties of solvated tris (thenoyltrifluoroacetonate)(2,2'-bipyridine)europium(III)red hybrid organic complex for solution processed OLEDs and displays, J. Lumin. 185 (2017) 61−71.

[24] F.E. Lytle, D.R. Storey, M.E. Juricich, Spectrochim. Acta 29 A (1973).

[25] N. Khotele, N.T. Kalyani, S.J. Dhoble, Synthesis and characterization of green-light emitting $(pbi-Cl)_2Ir(acac)$ complex, J. Phys. Chem. Solid. 130 (2019) 19−31.

[26] A. Ugale, N.T. Kalyani, S.J. Dhoble, Reddish orange to blue tunable emission from rare earth β-diketonate Eu(TTA)3 dpphen complex for solid state lighting applications, Mater. Sci. Energy Technol. 2 (2019) 57−66.

[27] L.S. Amrut, S.J. Satish, P. Ramchandara, Functional nanomaterials synthesis and characterization, Adv. Appl. Sci. Res. 1 (2010) 36.

[28] C. Michael, G. Jurgen, M. Wolfgang, H. Harald, B. Wolfgang, Preparation and characterisation of blue luminescent Tris(8-hydroxyquinoline) aluminium(Alq_3), Adv. Funct. Mater. 13 (2003) 108.

[29] V.B. Devi, P. Arulmozhichelvan, P. Murugakoothan, Synthesis and characterisation of znq_2 and znq_2: CTAB particles for optical applications, Bull. Mater. Sci. 40 (6) (2017) 1049−1053.

[30] F. Michelotti, F.M. Bertolotti, E. Cianci, V. Foglietti, Synth., Alq_3/PVK hetrojunction electroluminescent devices, Metals 105 (2000) 111−112.

[31] K. Binnemans, P. Lenaerts, K. Driesen, C. Gorller-Walrand, A luminescent tris(2-thenoyltrifluroacetonato)europium(III) complex covalently linked to a 1,10- phenanthroline -functionalised sol-gel glass, J. Mater. Chem. 14 (2004) 191−195.

[32] C. Calculator, A Software from Radiant Imaging. Version 2, 2007.

SECTION V

Practical concerns and beyond

CHAPTER

18

Spectral response and quantum efficiency evaluation of solar cells: a review

M.Z. Farah Khaleda[1], B. Vengadaesvaran[1], N.A. Rahim[1,2]

[1]Higher Institution Centre of Excellence (HICoE), UM Power Energy Dedicated Advanced Center (UMPEDAC), Level 4, Wisma R&D, University of Malaya, Jalan Pantai Baharu, Kuala Lumpur, Malaysia; [2]Renewable Energy Research Group, King Abdulaziz University, Jeddah, Makkah Province, Saudi Arabia

Nomenclature

AM Air mass
ARC Antireflection coating
c-Si Crystalline silicon
DSSC Dye-sensitized solar cell
EQE External quantum efficiency
EVA Ethylene vinyl acetate
FF Fill factor
I Output current
IQE Internal quantum efficiency
IR Infrared
J Current density
LSC Luminescent solar concentrator
NCA Nanocone array
PSC Perovskite solar cell
PECVD Plasma-enhanced chemical vapor deposition
PV Photovoltaic
QE quantum efficiency
SiPV Silicon-based PV
SR Spectral response
STPV Solar thermophotovoltaic
UV Ultraviolet

Energy Materials
https://doi.org/10.1016/B978-0-12-823710-6.00014-5

Copyright © 2021 Elsevier Ltd. All rights reserved.

Subscripts

oc Open circuit
p P-polarization
ref Reference cell
s S-polarization
SC Solar cell
sc Short circuit
test Testing device

Greek symbols

λ Wavelength
$\mathcal{E}$ Dielectric constant
μ Relative permittivity
$\mathcal{T}$ Transmission
$\mathcal{P}$ Reflectance
θ_i Angle of incidence
θ_r Angle of refraction

18.1 Introduction

The pressing need for carbon-free energy and high energy demand has paved the way for the diffusion of renewable technologies. It is said that solar energy is the most abundant renewable energy available, and the cleanest. While conventional sources of energy such as petroleum, coal, and natural gas are being depleted, scientists are searching for solutions to meet the increasing energy demands, and solar energy appears to be a promising way due to its infinite supply. Currently, there are two main techniques that generate electricity from solar energy, namely solar thermal and photovoltaic (PV). Solar cells, due to their simplicity of operation and maintenance, have attracted a lot of research to improve the performance and operability of solar panels. Various approaches have been tried to improve the conversion efficiency and reduce the manufacturing cost. To date, the highest reported efficiency is 46%, achieved through multijunction silicon-based solar cells [1]. Over the past few years, power conversion efficiency has been reported to be steadily improving, as shown as Fig. 18.1 [2–4].

Besides its manufacturing and installation cost [5], there are various factors such as shading, availability of sunlight, heat, humidity [6], and others that affect its efficiency, but the main focus in this chapter will be on its spectral response (SR) and quantum efficiency (QE). SR is a cornerstone that affects the performance of solar cells as is measured from a solar cell itself [7], and from there, QE can be calculated. Spectral responsivity is also known as spectral sensitivity and has units of amperes per watts [8]; it is related to the detection of light for a device, with a function of wavelength or frequency of the light. It is said that the higher the QE of a solar cell, the higher its efficiency. There are several solutions that could tackle problems related to SR and QE such as using solar concentrators, encapsulation of solar cells, and foreign materials such as nanocrystals and luminescence materials, which will be further discussed in this chapter.

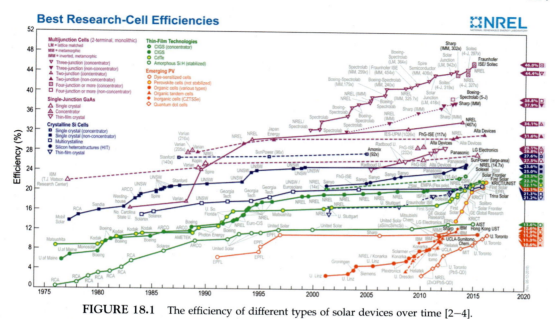

FIGURE 18.1 The efficiency of different types of solar devices over time [2–4].

18.1.1 Structure of the chapter

This chapter is a meticulous review of SR and QE evaluation of solar cells. The content of the study is organized in such a way that it is suitable for academic as well as industrial purposes. Starting with an introduction to solar energy in Section 18.1, the chapter continues with crucial characteristics of solar cells that affect their performance which is explained in Section 18.2 of the chapter. Research methodology is adopted for review in Section 18.3. Section 18.4 describes the main focus of the chapter, which is the SR of solar cells and its related measurement quantities. This part is key to understanding the recent advancements in solar technologies that are mainly involved with spectral modification, which is presented in Section 18.5. Next, challenges and difficulties faced by researchers in their efforts to search for the solution to solar technology problems are mentioned in Section 18.6. This review chapter ends concisely with conclusions and future suggestions for the improvisation of solar technologies along with recent trends in the topic in Section 18.7.

18.2 Factors affecting the performance of solar cells

Solar irradiance, which is defined as the radiance flux received from the sun in the form of electromagnetic radiation in a wavelength by surface per unit area of the cell (W/m^2), is one of the crucial factors determining the performance of solar cells. It is a parameter that projects power density at a given wavelength. The angle of the sun, the clouds in hazy weather, as well as air pollution affect irradiance levels. The spectral distribution of radiance that arrives on Earth's surface is another important parameter that influences the efficiency of PV cells [9]. Previous research showed that irradiance varies throughout a certain duration of time [10,11]. Fig. 18.2 shows $I-V$ characteristics for a clean PV at different solar irradiance levels by using a variable resistor [12]. The graph shows that the output voltage increases with increasing solar irradiance. The result shows that at a maximum solar intensity of 773 Wm^{-2}, the output voltage obtained is 1138 mA.

Temperature has an impact on all solar cell module parameters, such as short-circuit current (I_{sc}), open-circuit voltage (V_{oc}), efficiency, and many others [13,14]. Different from irradiance, I_{sc} and V_{oc} increase when irradiance increases, where temperature has the opposite concept because it is a function of irradiance. When the temperature of a cell increases, V_{oc} decreases and I_{sc} increases under standard test conditions (under a global AM1.5 spectrum, 25°C temperature, and 1000 Wm^{-2}) [15,16]. The concept states that as the temperature of the solar cell's material increases, resistance increases, and electrons move through it slowly. Consequently, the open-circuit voltage decreases because fewer electrons can move from the valence band to the conduction band. This leads to another situation where the quality of solar panels is a significant factor. High-quality solar panels are designed to withstand extreme heat. Lower-quality panels lose efficiency and produce less energy [17]. Fig. 18.3 shows the power curve ($P-V$) of a PV panel at constant 1000 Wm^{-2} solar irradiance with different PV temperatures [17]. By analyzing the figure, as the panel temperature increases,

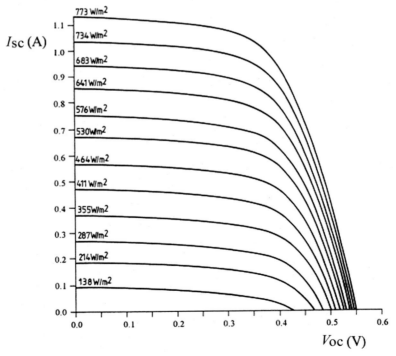

FIGURE 18.2 The $I-V$ curve of a photovoltaic cell at different solar intensities [12].

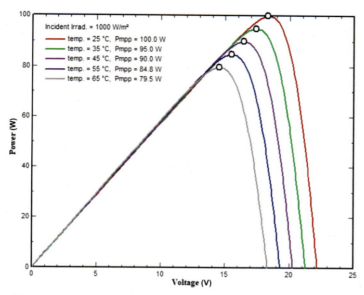

FIGURE 18.3   P–V curve of a photovoltaic (PV) panel at constant solar irradiance with distinct PV panel temperatures [17].

the output power decreases gradually. The graph shows that by increasing the temperature of the PV panel by 10°C, about 5 W or 5% of output power is decreased. Observing the graph, the minimum output power obtained is 79.5 W at a temperature of 65°C, while the maximum power reached is 100 W when the temperature is 25°C.

The bandgap of a cell's material affects the efficiency of a solar cell significantly. Material with a large bandgap such as GaN will result in insufficient energy absorption. On the other hand, materials with a narrow bandgap such as PbS and Ge will result in insufficient photovoltage. If we use low-bandgap polymers, the light absorbed can be extended into the red and infrared (IR) region, causing the cell to harvest more solar energy leading to power conversion efficiency [18–20]. According to Wang et al. [19], this weakens the charge for the driving force in V_{oc}, since the V_{oc} of the bulk heterojunction of solar cells is governed by the donor's highest occupied molecular orbital level and acceptor's.

Lowest unoccupied molecular orbital level [21–23]. Fig. 18.4 shows the performance of solar cells as a function of bandgap with 1000 W/m² irradiation and a spectrum of AM1.5. The lines demonstrate the empirical trends as well as physical limits of these parameters. From the figure, for a-Si:H solar cells, the performance is lower than the fundamental conversion efficiency limits. Even though V_{oc} and J_{sc} could potentially improve, the fill factor contributes the most in terms of the difference between the device's performance and potential [24].

The design and insulation of PV systems should also be taken into consideration for optimum power output. Shaded solar panels contribute to nonuniform solar radiation; hence, they produce less electricity [25]. Shading issues due to trees and the size, height, and proximity of surrounding barriers can be minimized or eliminated with proper and suitable system design [26]. Besides, an unclean or soiled solar panel also produces less electricity. The term

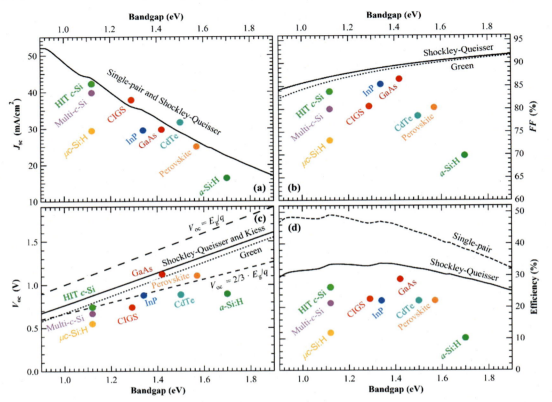

FIGURE 18.4 The performance and theoretical limits of single-junction solar cells of distinct classes: (A) short-circuit current density, (B) fill factor (FF), (C) open-circuit voltage, and (D) conversion efficiency [24].

"soiling" refers to foreign substances that settle on the surface of solar panels such as dust, dirt, and other debris [27,28]. This stops sunlight from reaching the solar system properly, affects its overall performance, and hence its efficiency cannot be maximized [29]. In places with frequent rain, soiling does not have much of an affect on output as the soiling tends to be washed away naturally. Regions that experience a long period of dry and hot weather like California as well as the south of Navarre (Spain) are more likely to experience increased soiling compared to other countries with high humidity, especially during the summer [27,30]. Soiling can also occur near construction sites or beside beaches, where dust can be rapidly collected. Cleaning systems regularly may be the solution to this problem but it is quite costly and time consuming. Self-cleaning systems have been proposed as a possible solution [31].

Wind speed is one of the major factors that also significantly affects the performance of solar cells. In a study by Goossens et al. [12], wind velocity had an impact on cells' performance. High winds affect the sedimentological structure of dust coatings on the surface of solar cells, which in turn create a higher transmittance of light for coatings. Wind speed affects the temperature of the PV cell [15,32]. Mekhilef et al. [15] stated that increasing wind velocity reduces heat and humidity on the surface of the PV cell, which leads to an increase in overall efficiency of a PV system.

Photoaging is influenced by several factors such as air, solar ultraviolet (UV) radiation, and other pollutants, where parameters such as temperature, mechanical stresses, as well as the presence of organic solvents and water accelerates the process [33]. Photodegradation is a type of deterioration that involves chemical reactions caused by exposure to UV rays [34]. This phenomenon impacts on the mechanical properties of some materials such as polystyrene [34], polycarbonate [35], and several other polymers, and causes the materials to become useless over a certain period of time. Photodegradation happens when there is activation of polymer macromolecules upon photon absorption by the polymers. Fig. 18.5 shows a basic illustration of the formation of microcracks and weak centers caused by UV radiation and mechanical stress.

Photodegradation is a serious aspect that needs to be taken into account to bring the performance of polymer-based solar cells to another level. Generally, it disturbs the mechanical, physical, and chemical properties of most polymers [34,36,37]. For solar cells such as thin-film and organic types, photodegradation needs to be avoided to prolong the lifetime of the cells. Therefore there are several methods that can be followed to overcome these drawbacks, for example, by mixing certain polymers with fullerenes, which has proven to show positive results for plastic solar cells [38]. The analysis indicates that the rate of degradation of conjugated polymers, which is high under the influence of light and oxygen, is remarkably decreased by adding fullerenes as used in plastic solar cells.

Photodegradation can be considered the reason why various solutions such as nanofilters and antireflection coatings (ARCs) for improving the efficiency of solar cells have ended in failure. This is because that particular phenomenon gives negative impact on barrier and optical properties of many coatings that could damage the target applications [39]. The presence

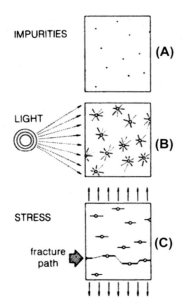

FIGURE 18.5 (A) Initial chemical heterogeneities. (B) Subsequent effects of ultraviolet radiation. (C) Mechanical stress on the formation of weak centers and microcracks [33].

of certain impurities also affects the rate of degradation. Most pure polymers do not absorb light at wavelengths of more than 200 nm; however, photodegradation of polymers can even occur at wavelengths of more than 300 nm. This means that the polymers contain impurities that are most likely to contain conjugated double bonds or carbonyl groups, which are responsible for the absorption of radiation over 220 nm [33].

The durability of materials is the key element in the progression of solar cells other than primary type (crystalline) applications specifically for those involving polymeric materials. Researchers have developed many ways for the photostabilization of polymers. One exceptional work is the use of well-selected stabilizers. Preventive stabilizers such as UV absorbers function to absorb and transform the detrimental radiation into harmless thermal energy [40]. Fig. 18.6 shows that in sample A, a yellow coloration started to appear after applying UV radiation for 500 h and reached the highest intensity after 800 h. On the other hand, in sample B, which contains organic UV absorbers (Tinuvin 1130), faded yellowing is spotted, showing a low intensity of coloration after 800 h of UV radiation. This proves that Tinuvin 1130 works as a very effective protection against UV light in sample B [41].

All of these factors must be taken into account when fabricating systems for improved solar technologies. These might be the core reasons why efficiencies could not be standardized sufficiently over time. In our opinion, to eliminate factors contributing to the inefficiency of solar devices, it is also important to study other effects of combining different methods into one system as these might affect the lifetime of the devices. In other words, the efficiencies of solar cells could be vastly improved with newly invented systems, but the cells could not be used for a long period of time. While increasing the efficiency of those cells, researchers should design systems that can adapt to the degradation of the cells' materials at the same time.

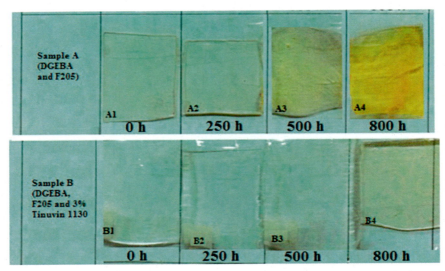

FIGURE 18.6 Optical image of coloration of samples A and B after 0, 250, 500, and 800 h of ultraviolet radiation exposure [41]. *DGEBA*, Diglycidyl ether of bisphenol A.

18.3 Research methodology

The scope of research in solar cells is very wide and researchers around the globe are working toward improvements in efficiency. New developments for solar technologies have been remarkably increased due to higher efficiencies achieved each year. The focus of research for renewable energy has slowly been shifted to solar cells, specifically in their SR. Scientists are finding ways to allow only spectrums of certain wavelengths (which are needed by solar devices) to have access to solar cells. There are numerous creative ways that have been introduced such as using the solar filtration method, solar concentrators, and even a combination of these methods with various materials.

One particular topic in solar cells that is quite basic yet crucial is focusing on their SR measurements. Scientists came to realize the importance of this focus area, as the number of publications has increased from 2012 (the year when this topic became popular) to 2019/2020. Graphical details of year-wise publications from well-known search engines, Scopus and Web of Science, are displayed in Figs. 18.7 and 18.8, respectively, by using the keywords "Spectral response of solar cell." From about 2562 papers that were published (by using Scopus and Web of Science), 142 relevant papers were narrowed down for inclusion in this review chapter.

From the Scopus search engine, which is popular for research purposes, it was found that publications related to the SR of solar cells with different fabrication methods, techniques, and materials had been increasing steadily from 2012 to 2016, and started to decrease from 2017 onward, as shown in Fig. 18.7 (searched by using the keywords "Spectral response of solar cell" on the Scopus database). However, from 2019 to March 2020, the number of studies in this area increased with a total of 154 publications.

In addition, the Web of Science database also showed similar patterns, confirming a growing interest in this matter by using the same keywords, "Spectral response of solar cell." The number of publications in this topic on the Web of Science from 2019 to March 2020 was 192, from a total of 1410 publications since 2012 (Fig. 18.8).

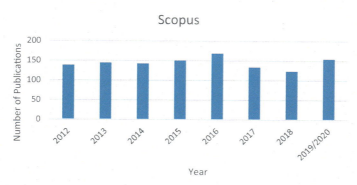

FIGURE 18.7 Year-wise publications on the scopus database for the spectral response of solar cells.

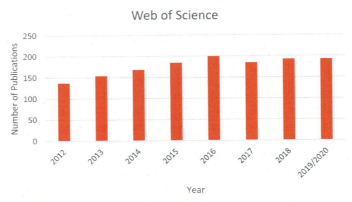

FIGURE 18.8 Year-wise publications on the web of science database for the spectral response of solar cells.

18.4 Spectral response and measurement quantities

SR of a solar cell is a crucial criterion that determines its performance. Only certain wavelengths in the spectrum contribute to the efficiency of the cell. There are many aspects that need to be taken into account in the determination of the spectral distribution of total radiation. In this section, important measurement quantities for SR are outlined for the purpose of calculating the properties of materials such as external quantum efficiency (EQE), internal quantum efficiency (IQE), reflectance, transmittance, and other related and practical parameters.

18.4.1 Light conversion process

A PV solar cell works like a semiconductor diode where the semiconductor's material absorbs incident photons and turns them into electron–hole pairs or excitons [42]. Those excitons exploit the PV effect, which exists at semiconductor junctions [43,44]. With the correct wavelength of an incident photon with energy more than the bandgap's energy (E_{gap}), an electron from the valence band can be excited to the conduction band, leaving a hole in the valence band. For that stage, due to uneven distribution of charge carriers [45], the process leads to the formation of forward bias by in-built voltage that sweeps away these carriers from across the p-n junction. Fig. 18.9 shows some basic features of PV energy conversion.

The check valve as shown in Fig. 18.9 is to prevent backflow of excited electrons [46]. In addition, no photons with energy below E_{gap} can excite the electrons to the conduction band. Nevertheless, due to thermalization, excess energy will be lost rapidly during the process [42]. Eventually, at some point, some electrons from the conduction band will lose energy and return to recombine with a hole in the valence band, a process known as recombination.

18.4.2 Source spectrum

The solar spectrum has a crucial influence on PV performance in terms of temporal and geographic conditions [47]. The right amount of light source is important to ensure that

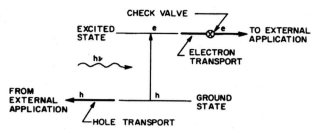

FIGURE 18.9 Basic characteristic of photovoltaic energy conversion [46].

optimum conversion efficiency is achieved. This might relate to the distribution of solar irradiance with average photon energy at the site that affects the amount of electrical energy lost to the spectral mismatch of a device over its lifetime [48]. This means different sites or places have different solar spectra or solar irradiances, which, in turn, give different effects for a solar cell's performance. It is, however, difficult to decouple several variables such as temperature, average photon energy, irradiance, light-induced degradation, and annealing cycles for spectral matching studies [48,49]. Nonetheless, as reported by several researchers, for outdoor applications or standard test conditions, the standard global AM1.5G spectrum is used [48,50–53]. A device such as a solar spectral irradiance meter is also useful to measure direct normal irradiance in several wavelength bands [47]. Combining with certain models, the resulting spectral irradiance distribution as well as the key atmospheric transmittances under all sky conditions can be obtained.

On the other hand, there is no standardized light source for indoor applications as mentioned by Minnaert et al. [50]. This is the reason why Minnaert et al. [50] designed an experiment and proposed to characterize indoor light sources in different categories. As such, different types of solar technologies respond to those fluctuations differently. It is said that multijunction solar devices are more influenced by the change in solar spectral irradiance compared to single junction devices [47].

The zero air mass (AM) radiation spectrum of spectral distribution is referred to as AM0, where the AM is the measure of the absorption in the atmosphere that affects the intensity of solar radiation and its spectral content that reaches the surface of Earth. AM calculation is given by:

$$AM = 1/\cos\theta \tag{18.1}$$

18.4.3 Spectral response

SR is defined as the proportion of current that is generated by the cell to the incident power on the surface of the cell [54]. It is often measured in amperes per watt. The basic formula of SR, in terms of cell current $I_{sc}(\lambda)$ divided by light intensity $I_{light}(\lambda)$ under monochromatic illumination [54–56], is:

$$SR(\lambda) = \frac{I(\lambda)}{P(\lambda)} = \frac{I_{SC}(\lambda)}{I_{light}(\lambda)} \tag{18.2}$$

Fig. 18.10 shows an example of a normalized SR for round-robin intercomparison for the international characterization of a dye-sensitized solar cell (DSSC) [57]. A few SRs for crystalline silicon-based PV (SiPV) were obtained and used for confirmation of spectral mismatch between the bias light and the AM1.5 global standard as the dominant cause for SR inconsistency of the nonlinearly responsive samples of DSSC. In this case, the participants involved are the Research Centre for New Generation Photovoltaics, Korea Institute of Energy Research, and Korea and Kanagawa Academy of Science and Technology. According to Fig. 18.10, the three participants managed to obtain good agreement in crystalline SiPV-1 for achieving a major response in the visible region.

18.4.4 External quantum efficiency

EQE is the number of electrons supplied to the external circuit for each photon incident on the device. For an ideal solar cell, each photon would produce an electron–hole pair. Consequently, all these carriers will go toward the depletion region and then be separated and collected. Nevertheless, those photons that do not have more energy than the bandgap energy, E_g, could not be photocarriers and might not contribute to photocurrents even with sufficient energy [58–60]. EQE can directly be determined by SR measurement with an instrument called a quantum efficiency tester, as shown in Fig. 18.11. It is a crucial indoor/outdoor calibration device where the characterization for indoor current–voltage can be matched spectrally to the standard spectrum of AM1.5 [61].

Fig. 18.12 shows the EQE graph for a monocrystalline-Si (m-Si) solar cell in the wavelength range 350–1200 nm at room temperature [50]. As can be seen from Fig. 18.12, for m-Si and polycrystalline (p-Si) silicon solar cells, EQE increases as wavelength increases. The peak for m-Si is higher than the peak for p-Si. It is then seen to slowly decrease for both types,

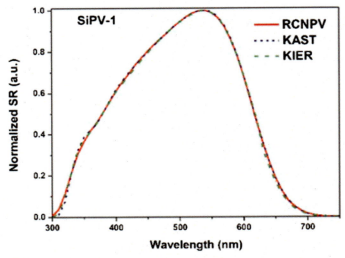

FIGURE 18.10 Spectral responsivity spectra of an SiPV-1 cell [57]. *KAST*, Korea and Kanagawa Academy of Science and Technology; *KIER*, Korea Institute of Energy Research; *RCNPV*, Research Centre for New Generation Photovoltaics.

FIGURE 18.11 Photographic view of a quantum efficiency tester in the UM Power Energy-Dedicated Advanced Centre.

where m-Si starts to decrease rapidly after around a wavelength of 1100 nm, where the same phenomenon happens to p-Si at a wavelength of 1000 nm. The speedy decrease is perhaps due to the optical losses and recombination that occur due to the effect of transmission and reflection [58,60]. The amorphous silicon solar cell (a-Si) has a lower peak compared to the other types and the graph decreases at a very much lower wavelength as well, which is around 600 nm.

Where, m-Si is monocrystalline silicon, p-Si is multicrystalline silicon, t-Si is thin film transfer silicon, GaAs is gallium arsenide, CIGS is coper indium gallium selenide, CdTe is

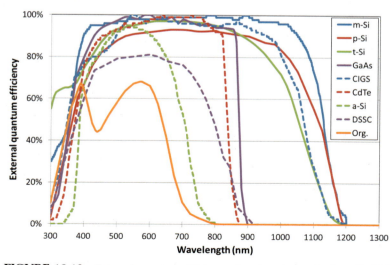

FIGURE 18.12 External quantum efficiency of various types of solar cells [50].

cadmium telluride, a-Si is amorphous silicon, DSSC is dye sensitized solar cell, and Org. is organic solar cell type DTDCTP:C70. The QE at a particular wavelength can be given as [62]:

$$QE(\lambda) = \frac{1}{q} \frac{hc}{\lambda} \frac{I_{SC}(\lambda)}{P(\lambda)} \tag{18.3}$$

where $I_{sc}(\lambda)$ is the short-circuit current, $P(\lambda)$ is the output light power for a silicon solar cell at varying wavelengths, λ is the photon wavelength, q is the electronic charge, h is Planck's constant, and c is the speed of light. By placing a reference cell, absolute spectral response (ASR_{ref}), near the testing device, incident light intensity can be measured. The equation is as follows [63]:

$$EQE(\lambda) = \frac{A_{ref}dI_{SC}(\lambda)}{A_{test}dI_{SC,ref}(\lambda)} \frac{hc}{e\lambda} ASR_{ref}(\lambda) \tag{18.4}$$

where A_{ref} and A_{test} are the reference cell and the testing device areas, respectively, $dI_{sc}(\lambda)$ is the short-circuit current values of the test device, $dI_{sc,ref}(\lambda)$ is the reference cell, both measured at the wavelength value λ, and e is the electron charge. It is reported that the addition of a backside diffusive reflector (white, 97% reflection) increases efficiency by about 32%, where EQE and short-circuit current of a solar device and the luminescent solar concentrator (LSC) were significantly affected [61]. This shows that EQE differs with modifications made to solar modules. Fig. 18.13 shows an example of an experimental setup made at the European Solar Test Installation with a 300 W steady-state xenon lamp as source light [61].

18.4.5 Internal quantum efficiency

Similar to EQE, the IQE of a solar cell is the ratio between the carrier's numbers collected and all photons absorbed by the solar cell at a certain wavelength [64]. The only difference is that EQE focuses on incident photons on the solar cell, while IQE calculates photons absorbed by the cell without considering light transmitted through and reflected from it [64]. Thus the value IQE is always greater than EQE. In other words, according to Yang et al. [56], IQE is the number of minority carriers that contribute to the short-circuit current over the number of photons that enter the cell. The general equation for IQE is [65]:

$$IQE(\lambda) = \frac{EQE(\lambda)}{1 - Transmission - Reflectance} = \frac{EQE(\lambda)}{1 - \tau(\lambda) - \rho(\lambda)} \tag{18.5}$$

Therefore IQE can be derived from Eq. (18.5) showing its relation with SR as follows [56]:

$$IQE(\lambda) = \frac{1}{1 - \rho(\lambda)} \cdot \frac{I_{SC}(\lambda)/e}{I_{light}(\lambda)/\left(\frac{hc}{\lambda}\right)} \tag{18.6}$$

$$= \frac{1}{1 - \rho(\lambda)} \cdot \frac{hc}{e\lambda} \cdot SR(\lambda) \tag{18.7}$$

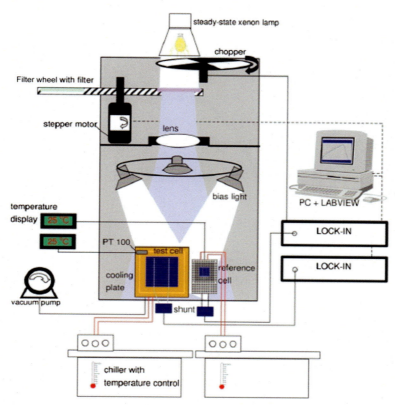

FIGURE 18.13 External quantum efficiency measurement setup at the European Solar Test Installation [61].

From Eq. 18.6, it can be concluded that internal quantum efficiency is proportional to spectral response whereby IQE increases as SR increases. An example of IQE graph is as shown in Fig. 18.14, a graph of internal quantum efficiency versus wavelength for AlGaInP solar cells. According to Yukun Sun et al. [66], in about all aspects of 2.09 and 2.19 eV AlGaInP solar cells, an improvement was observed after rapid thermal annealing (RTA). As can be seen from the graph, compared to 2.02 eV AlGaInP (orange), the inferior IQE across all wavelengths of as-grown 2.09 and 2.19 eV AlGaInP cells (dashed green and blue, respectively) shows that AlGaInP with wider optimized bandgap energy (Eg) suffers from lower low minority carrier diffusion lengths ($L_p = 70$ nm and $L_n = 450$ nm) in both the emitter and base regions, respectively. With the RTA, the enhancement in the wider-Eg cells is only significant at wavelengths close to the band edge (solid green and blue curves) [66].

18.4.6 Spectral mismatch

Spectral mismatch, M, is a parameter that is employed to rectify the measured values of short-circuit current and it can be obtained by the use of an arbitrary test spectrum [67]. Spectral mismatch is very useful for solar simulator's spectral classification for a specific device,

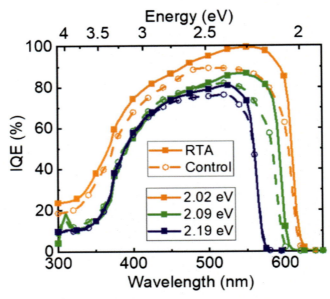

FIGURE 18.14   IQE of 2.02–2.19 eV AlGaInP solar cells [66].

which involves a reference cell to measure the intensity of a light source and standard spectrum of interest [68,69]. It can be used to calculate the J_{sc} of a solar device as well as the cell's current, I_{sc}, under a standard AM1.5G light source. The performance of a solar cell or PV device should be determined with respect to the standard test conditions considering its area, temperature, intensity, and spectrum [66].

By using a calibrated reference cell along with spectral mismatch correction factor, the error in the current of the device can be reduced to less than 2%, as stated by Emery et al. [69], leading to an increase in efficiency. Often, the respective SRs of a reference cell and solar cell do not match, causing a phenomenon called measurement mismatch error [70]. This measurement mismatch error can be quantified by the spectral mismatch factor [71]. The spectral mismatch factor, M, is shown in Eq. (18.8) [57,69,70]:

$$M = \frac{\int E(\lambda) \cdot S_t(\lambda) \cdot d\lambda}{\int E_o(\lambda) \cdot S_t(\lambda) \cdot d\lambda} \cdot \frac{\int E_o(\lambda) \cdot S_r(\lambda) \cdot d\lambda}{\int E(\lambda) \cdot S_r(\lambda) \cdot d\lambda} \qquad (18.8)$$

where $E(\lambda)$ is the solar simulator's spectral distribution, $E_o(\lambda)$ is the AM1.5 reference spectral distribution, $S_t(\lambda)$ is the test device SR, and $S_r(\lambda)$ is the reference device SR. A study conducted by Müllejans et al. [72] showed the importance of spectral mismatch in the calibration of reference devices using the sun's rays; in this case, c-Si reference devices were involved. The measurements could further be improved by estimating the uncertainties for the spectral mismatch correction factor of a device [73]. Spectral mismatch might affect the short-circuit current's estimation and occurs when using a reference cell with an SR that is different from the SR obtained from the device under test. According to Eke et al. [74], the value of spectral mismatch depends on spectral irradiance distribution between solar simulator and reference spectrum AM1.5G, as shown in Fig. 18.15.

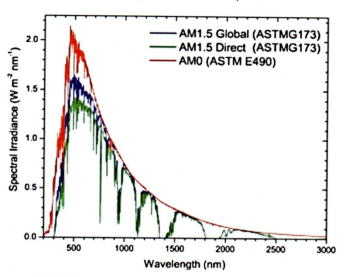

FIGURE 18.15 Standard reference solar spectra [74].

18.4.7 Reflectance and transmittance

Reflection is a process whereby a portion of the radiant flux from incident rays on a surface is returned to the same hemisphere that contains incident radiation [75]. Hence, spectral reflectance can be defined as the ratio of reflected radiant flux over incident radiant flux [75]. It is a fraction of light reflected from a surface as a function of wavelength. An ARC or film can play an important role in reducing the reflection of sunlight from solar cells as well as increasing the light coupling into the active region of PV devices, leading to increased efficiency [76]. Reflectance can be measured with instruments such as a reflectance spectrometer. However, effects such as particle size, scatter, and multicollinearity affect the performance in reflectance spectrometry diffusion [77]. Besides, a goniophotometer can also be used to measure reflectance as a function of illumination angle and viewing angle, where those angles can be varied independently [78]. Fig. 18.16 shows specular reflectance, $R_{specular, solar}$, at near-normal incidence (15 degrees) for five solar reflective materials. From the figure, aluminized aluminum sheet, which is denoted as AlSheet, has the lowest reflectance of 0.860, while silvered sheet, AgSheet#2, achieved the highest reflectance of 0.948 in the wavelength range between 0 and 2500 nm [79].

Transmittance is defined as the ratio of solar energy falling on a cell that is transmitted through it. In the case where all light passes through a cell without any absorption, it is said that the absorbance is zero and the transmittance is 100%. Besides reflectance, light intensity can be multiplied by transmittance at a given wavelength to determine radiant energy [75,78]. A filter with spectrally varying optical properties has better transmittance compared to normal clear glass, which has flat transmittance [80]. Fig. 18.17 indicates the optical transmission curve of $NiCo_2O_4$ films under various gas atmospheres.

From Fig. 18.17, in the wavelength range of 600−2400 nm, the highest optical transmittance of light is displayed by the $NiCo_2O_4$ thin-film deposition in the oxygen ambient

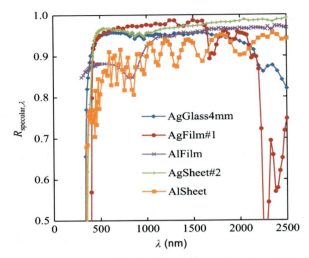

FIGURE 18.16 Specular reflectance spectra of back-silvered glass, silvered and aluminized polymer films, and silvered and aluminized aluminum sheets [79].

atmosphere (O_2), which is approximately 90%. This proves that the oxygen ambient atmosphere leads to a remarkable improvement in the IR $NiCo_2O_4$ film's transmittance. According to the law of conservation of energy, the sum of reflection coefficient, R, and transmission coefficient, T, is [82,83]:

$$R + T = 1 \tag{18.9}$$

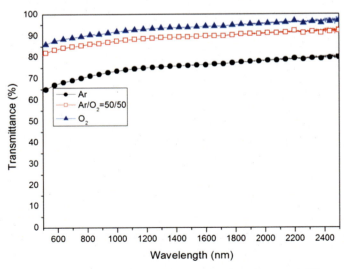

FIGURE 18.17 Optical transmittance for $NiCo_2O_4$ films under the influence of different gases at 300°C [81].

18.4 Spectral response and measurement quantities

543

Depending on the irradiance's polarization, whether it is perpendicular or parallel to the plane, values of R and T can be determined [82]. S-polarization happens when the vector of an electric field is perpendicular to the plane that contains incident irradiance, both refracted and reflected. In this case, the reflection coefficient is denoted as R_s. On the other hand, for the case where the vector of the electric field is in line with the incident plane, it is called p-polarization, where the reflection coefficient is denoted as R_p. By using the law of conservation of energy, their respective transmission coefficients can be calculated [83]:

$$R_s + T_s = 1 \tag{18.10}$$

$$R_p + T_p = 1 \tag{18.11}$$

where T_s and T_p are transmission coefficients for s-polarization and p-polarization, respectively. From Fresnel's equations, R_s and R_p can be determined as shown in Eqs. (18.12)–(18.15) [84]. They are often used to determine the values of reflectance and transmittance at the interface between two mediums with different refractive indexes at various incident angles [83,85]:

$$R_s = \left[\frac{\sin(\theta_t - \theta_i)}{\sin(\theta_t + \theta_i)} \right] = \left[\frac{n_1 \cos(\theta_i) - n_2 \cos(\theta_t)}{n_1 \cos(\theta_i) + n_2 \cos(\theta_t)} \right]^2 \tag{18.12}$$

$$R_s = \left[\frac{n_1 \cos(\theta_i) - n_2 \sqrt{1 - \left(\frac{n_1}{n_2} \sin \theta_i\right)^2}}{n_1 \cos(\theta_i) + n_2 \sqrt{1 - \left(\frac{n_1}{n_2} \sin \theta_i\right)^2}} \right]^2 \tag{18.13}$$

$$R_p = \left[\frac{\tan(\theta_t - \theta_i)}{\tan(\theta_t + \theta_i)} \right] = \left[\frac{n_1 \cos(\theta_t) - n_2 \cos(\theta_i)}{n_1 \cos(\theta_t) + n_2 \cos(\theta_i)} \right]^2 \tag{18.14}$$

$$R_p = \left[\frac{n_1 \sqrt{1 - \left(\frac{n_1}{n_2} \sin \theta_i\right)^2} - n_2 \cos(\theta_i)}{n_1 \sqrt{1 - \left(\frac{n_1}{n_2} \sin \theta_i\right)^2} + n_2 \cos(\theta_i)} \right]^2 \tag{18.15}$$

where n_1 and n_2 are mediums with different refractive indexes and θ_i and θ_t are the angle of incidence (toward the normal of the surface) and the angle of refraction (the outgoing beam), respectively [83].

V. Practical concerns and beyond

18.5 Recent advancements

In this modern era, solar cell technology is evolving and developing at an astounding rate. Advancements in these solar cell devices affect many aspects of today's society. The prime improvement of solar cells can be seen from their efficiency as well as their construction. Today, there are various types of solar cells with improved efficiency such as solar cells with foreign particles, transparent solar cells, encapsulation of solar cells, and many more. Each of these solar cells has its own uniqueness and advantages.

To create the perfect solar cell, certain variables that can potentially affect the output should be considered [86]. Solar radiation contains a few wavelengths that are related to energy frequencies, where every one of the wavelengths affects the solar cells in different ways. To create a PV effect, the material must absorb photons from sunlight to create excess carriers. Solar irradiance is found in abundance in the blue and UV regions; hence, an increased SR in those areas is needed to improve efficiency [86]. Spectral management of solar radiation offers the ability to match the spectral content of an energy source with the sensitivity of a solar device [87]. This not only increases the output performance of solar cells but also minimizes wavelengths that cause degradation to solar cells. Fig. 18.18 shows an example of a popular approach in spectral management by using a refractive-waveguide concentrator. From the figure, by using a certain set of prisms in specific geometrical arrangement, collimated light is refracted onto a focal point or line [87].

18.5.1 Solar filter

One of the methods to improve the performance of solar cells is by using a filter [89]. Solar filters have caught the attention of many scientists. Astronomically, a solar filter functions to block off some sunlight that can cause harm to the eyes. Depending on the materials used, solar filters can eliminate certain unwanted wavelengths that degrade solar cells. Long-term exposure to UV rays might induce photocatalytic activity in some materials, which leads to the degradation of solar devices. Park et al. [90] stated that UV excitation can largely be prevented by the use of a fluorine-doped tin oxide-conducting glass support as well as TiO_2 or SnO_2 layers to filter out UV photons.

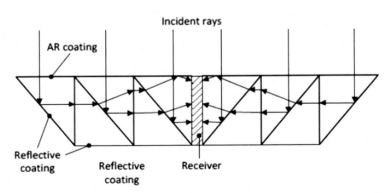

FIGURE 18.18 Refractive approach in spectral management [87,88]. *AR*, Antireflection.

Research was conducted by Kazem et al. [91] to find the best wavelength or color for a filter that gives the best performance of solar cells. The result of the study proved that the sunlight spectrum colors affected the output of solar PV cells. One way to determine the total irradiance of artificial light from a solar simulator on different types of solar filters is by normalization of the chronoamperometry measurement, as shown in Fig. 18.19 [92]. From the graph, the UV-B/C blocking filter that cuts below the 350 nm and above the 500 nm wavelength shows an improved normalized current density compared to open spectrum. This indicates that the light portion in the 350—500 nm range is indeed effective in photocurrent production [92].

Nanoparticles can also be used for achieving selective spectral filtration for solar cells, which is cost effective. In a study by Otanicar et al. [93], efficient solar energy conversion can be achieved with a hybrid system composed of selective solar filters by using a suspended nanoparticle fluid to absorb non-PV photons directly. In the study, indium tin oxide and gold nanoparticles were chosen for a well-blended atomic mixture. Nanoparticles in the fluid enable the absorption and transmission of specific wavelengths. Its flexibility is one of advantages of using nanoparticles, where it can be grown into distinct shapes, materials, and sizes, which makes it convenient to tune and scale up the spectral properties to massive volume production.

An optical filter is proven to be useful in solar thermophotovoltaic (STPV) systems [94]. Referring to Fig. 18.20, in the system the absorption and emission of solar radiation as thermal radiation take place before the illumination of solar cells. The thermophotovoltaic enables the utilization of selective filters as well as sub-banding gap photon reflection to the emitter to increase efficiency [95]. Mbakop et al. [94] used the emitter and spectral filter of a 1D photonic

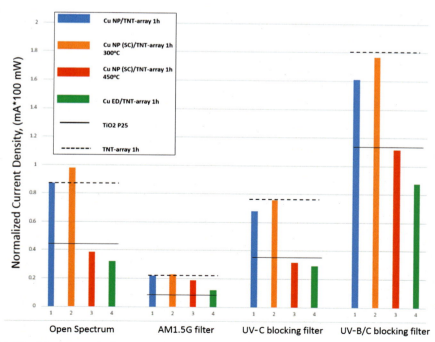

FIGURE 18.19 Measurement of normalized chronoamperometry for different filters to solar simulator light [92].

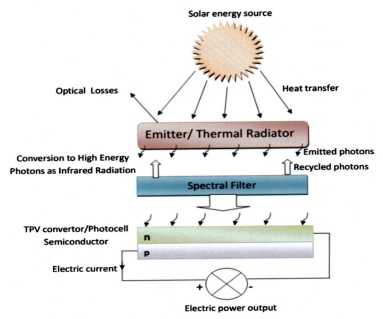

FIGURE 18.20 An illustration of the basic concept of the solar thermophotovoltaic [94]. *TPV*, Thermophotovoltaic.

crystal, which were applied to an STPV system to study its thermal efficiencies. A TiO_2/SiO_2 photonic crystal filter was chosen to make the filter structure match well with the spectral distribution of the emitter with high temperature within the corresponding transmission band.

18.5.2 Solar concentrator

Another method for collecting and concentrating solar energy is by using solar concentrators [96]. They harvest sunlight over large areas and focus it on PVs or solar cells [97]. In spite of the fact that many types of solar concentrators have been invented since the evolution of the solar cell, this type of solar technology has been overlooked until recently. Researchers have started developing new materials for enhancing solar concentrators again. One type of new solar concentrator is the organic solar concentrator [98]. Not only does it portray high QE of more than 50%, but the cost of PV power can also be reduced with this device.

Besides, scientists are exploiting LSCs to replace active PV materials by using inexpensive luminophores [99–102]. The concentrators utilize fluorescent materials to absorb incident light and release lower-energy photons into confined modes within a polymer waveguide. Consequently, they can be directed to an attached solar cell through total internal reflection. The use of nanocrystals has also been introduced for LSCs in solar cell advancement research [97]. Reabsorption losses of various types of nanocrystals can be quantified by using a 1D liquid waveguide with translatable excitation, as shown in Fig. 18.21, to measure their luminescence spectra as a function of photon transport distance, *L*.

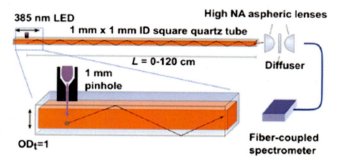

FIGURE 18.21 Schematic illustration of a proposed setup for the measurement of excitation distance-dependent photoluminescence spectra [97]. *ID*, Inside diameter; *LED*, light-emitting diode; *NA*, numerical aperture; OD_t, outside diameter.

A recent work proposed that a solar cell's temperature can be decreased by combining a solar concentrator and a filter. A solar concentrator was used to optimize solar radiation and a long-wavelength cut filter was used to reduce the cell's temperature. Through this study, Ahmad et al. [103] proved that the lifetime of a solar cell can be extended with an increase of 1.9×10^5 h with a Fresnel lens optical concentration system, which can reduce the solar cell's temperature. The transmittance curve for the system in Fig. 18.22 shows that more than 90% transmittance was obtained in a wavelength range of 400–1300 nm. The range of the long-length of over 1300 nm was successfully reduced.

Recently, nanotechnology was introduced in the concentrator–filtration combination method, whereby a nanoparticle fluid filter, which consisted of gold with nanosized particles and indium tin oxide nanocrystal, was fabricated and the system was combined with a solar concentrator [104]. Filter efficiency was 62% for a thin-film solar cell (GaAs) and 56% for a crystalline silicon solar cell (c-Si). Installation of filter fluid, which was placed directly in front of the PV receiver, re-collected the light that went through the filter and reflected it off the PV cell. A schematic diagram of the system is shown in Fig. 18.23.

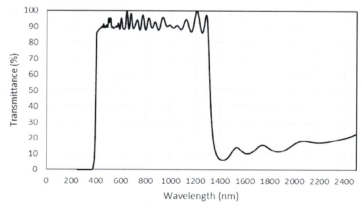

FIGURE 18.22 Transmittance curve as a function of wavelength for the solar cell system [103].

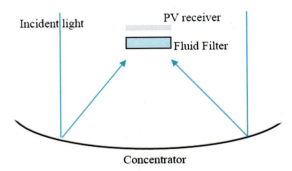

FIGURE 18.23 A concentrator system with a photovoltaic (PV) receiver and fluid filter in the direction of the concentrated light [104].

18.5.3 Antireflection coatings

A popular alternative to improve solar technology performance is the use of an ARC. This is a coating that is used for reducing reflection from a surface [105]. It is said to be efficiency improving since less light is lost. It is an old technology but is still undergoing development to match the efficiency target. Some of the useful techniques used in fabricating ARCs are sputtering [106], plasma-enhanced chemical vapor deposition (PECVD) [107], dip coating [108], spin coating [109], spray pyrolysis [108], and liquid-phase deposition [110].

According to Jiang Y. et al. [111], in the attempt to increase the performance of colorful perovskite solar cells, it was proved that power conversion efficiency managed to reach 13%−15% by modifying the device's structure with the use of the transparent conducting polymer PEDOT:PSS^T as the top electrode, as shown in Fig. 18.24. The conducting polymer serves as an ARC, where its thickness variation leads to the suppression of reflectance at certain wavelengths, instead of a whole visible spectral wavelength [111].

Uzum et al. [108] stated that the key factor to enhance the efficiency of silicon solar cells is to reduce optical losses, and the absorption properties of the solar cell must be improved. Therefore Uzum et al. [108] introduced an alternative for ARC layers for silicon solar cells by using TiO_2^- compact as well as ZrO_2^- polymer composite coating layers. It was proved that reflectance further decreased with a ZrO_2^- polymer composite/TiO_2^- compact layer compared to a single TiO_2^- compact film in a range of spectra around 550−1050 nm. Fig. 18.25 shows the reflectance curve for both cases. From the graph, the average reflectance was lowered with the addition of ZrO_2^- polymer composite to the TiO_2^- compact film with 5%−12% reflectance gain in a wavelength range of 300−450 nm. A steady low reflectance can also be found in a wavelength range of around 700−1000 nm.

In addition, an improved version of ARCs on crystalline silicon solar cells has also been developed by implementing argon plasma treatment of silicon nitride (SiN), which results in lower reflectance [112]. By using the PECVD technique, hydrogenated silicon nitride (a-SiN:H) was deposited on a silicon substrate with a mixture of ammonia (NH_3) and hydrogen (H_2) gases, and subsequently with argon plasma treatment. The EQE from this study showed a relative increase of 2.72% and 4.46% in current density and conversion efficiency, respectively.

18.5 Recent advancements

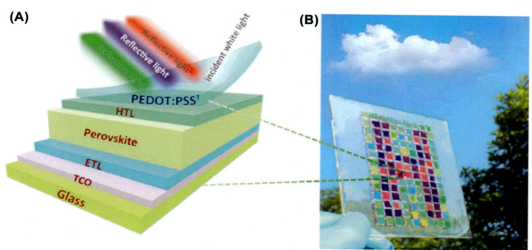

FIGURE 18.24 (A) The cell's architecture layer with conducting polymer, (poly(3,4-ethylenedioxythiophene):poly(styrenesulfonate), PEDOT:PSS^T as the top electrode and (B) a photographic colored image of the letter H collected by colorful perovskite solar cells, where the size of each pixel substrate is approximately 5 × 5 mm^2 [111]. HTL is the hole transport layer, ETL is the electron transport layer and TCO stands for transparent conductive oxides.

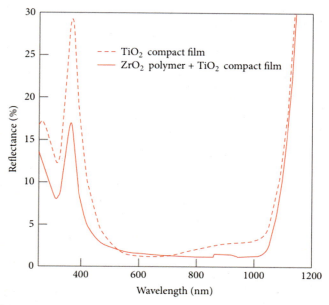

FIGURE 18.25 Reflectivity comparison of textured silicon wafers with single TiO$_2^-$ compact film and ZrO$_2^-$ polymer composite/TiO$_2^-$ compact layer [108].

V. Practical concerns and beyond

Like in many other methods, nanotechnology is also suitable for improving the performance of ARCs. In a work by Elshorbagy et al. [113], short-circuit current was enhanced by 15.2% with a reference flat solar cell, and a lower reflectivity was achieved with a particular design of ARC in the form of Si_3N_4 subwavelength nanostructured dielectric layers. This study utilized the light-trapping mechanism in the amorphous silicon hydrogenated (a-Si:H) solar cells by using the Si_3N_4 nanostructured layers that trap light toward the a-Si:H active layer. From this research, it was found that the method of preparation was simple and cost effective [113]. Duan et al. [114] also conducted similar research on utilizing a light-trapping mechanism by using silver nanoparticles, and managed to enhance the light absorption of c-Si cells by 26%.

For a practical antireflective application, the ultralow refractive index thin films must have good transparency and be economical with high durability against harsh temperature and weather conditions. Karthik et al. [115] introduced the application of a facile synthesis of ink-bottle mesoporous MgF_2 nanoparticles with excellent crystallinity and dispersible properties in ARCs. With the coatings, an outstanding result was achieved, whereby almost 100% transmittance was observed in the wavelength range of 615—660 nm (visible range) as well as average transmittance of 99% and 97% in the wavelength range of 400—800 nm (visible range) and 300—1500 nm (active solar range), respectively.

18.5.4 Solar cell encapsulation

Solar cell encapsulation is also a promising method to further boost solar PV's output and avoid degradation. It is said that without encapsulation, devices will commonly exhibit a formidable degradation under hours of continuous illumination, whereas encapsulated devices will have a longer lifetime [116]. This method is specifically beneficial to organic solar cells that have high degradation rate and cells that are exposed to high humidity conditions. Basically, complete solar cell encapsulants should consist of an adhesive layer, a weather-resistant layer, and a pottant [117]. The pottant acts as an insulator that also protects the fragile electrical as well as mechanical elements from external forces in the environment. A weather-resistant layer is needed to prevent solar cells from being exposed to rain, dust, and other debris, while an adhesive layer is needed to connect the solid outer weather-resistant layer to the shock-proof pottant layer.

Rizzo et al. [118] further investigated the effect of temperature, moisture, as well as humidity on the degradation of perovskite solar cells (PSCs) and applied different sealing techniques while encapsulating the devices to analyze any differences during the experiment. Two different glass-to-glass sealing methods were used, without sealing, glass sealing perovskite solar cell (GS-PSC) and with sealing, edge sealing perovskite solar cell (ES-PSC), to study the intrinsic stability of PSCs. Fig. 18.26 shows two encapsulation strategies with and without the addition of a light-curable glue. It is proven that the one with light-curable glue edge sealing managed to reduce moisture percolation as well as mechanisms of oxidation at a stress temperature of below 80°C. Consequently, this resulted in a prolonged lifetime of the cell [118].

Ethylene vinyl acetate (EVA) has been reported to be a popular choice for the majority of solar module manufacturers and will probably dominate encapsulation materials in future years [119,120]. The stability of EVA has improved remarkably with the addition of UV

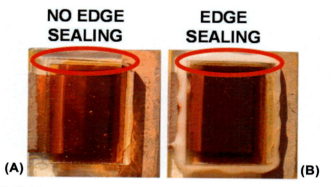

FIGURE 18.26   Images of samples: (A) nonsealed edge and (B) sealed edge [118].

absorbing agents [121,122]. The result of a study by Witteck et al. [120] showed that UV transparent encapsulation polymers might produce UV unstable modules; hence, stable UV passivation layers are proposed to further improve the performance of UV encapsulation materials for solar modules.

From a study by Dennler et al. [123], the extended lifetime of solar cells was achieved by using a new type of transparent, flexible, ultrahigh barrier material produced by PECVD to encapsulate the cells. According to Bryant et al. [124], degradation caused by light and oxygen was the main reason for the low operational stability of PSCs that were exposed in ambient environments. It was found that those without encapsulated cells went through a speedy degradation on timescales of minutes to several hours when they were exposed to light and dry air. A successful encapsulation as well as effective barrier layers will contribute to long-term stability [124].

Nanotechnology has also been implemented in an attempt to make improved encapsulations, whereby nanocone arrays (NCAs) are fabricated as ARCs on the encapsulated cover glass, and this increases the power conversion efficiency of solar cells [125]. Fig. 18.27 shows that over a broad wavelength of 330–1800 nm, NCA ARC glass exhibited the highest transmittance spectrum due to its conical nanostructures with small periods of <200 nm.

18.5.5 Spectral splitting

The efficiency of a single-bandgap solar cell is restricted by the inability to efficiently convert photon energy into a broad range of solar spectra. One suggested way to overcome this problem is to split the wide solar spectrum into ranges of smaller energies [126]. Then, utilization of each energy range can take place on the appropriately tuned bandgap of the PV cell. The concept of spectral splitting was introduced in the 1960s [127,128]. It is a technique of separating sunlight spectrally into various wavelength bands with the use of a spectral beam splitter and targeting each band at the most efficient receiver. Distribution of incident angles on a wavelength's interference filter can be created by concentrating spectral splitting systems, which results in diverged transmission or reflection characteristics [129]. Theoretically, photons with energies that are near to the bandgap of solar cells are more

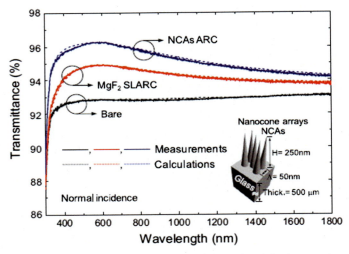

FIGURE 18.27 A graph of measured (*solid lines*) and calculated (*dotted lines*) transmittance spectra at normal incidence of bare MgF$_2$ single-layer antireflection coating (SLARC) and nanocone array (NCA) antireflection coating (ARC) [125].

efficient in producing electricity and the nonutilized energy is often dissipated as heat [130]. This problem led to the discovery of the spectral beam approach.

There have been few attempts at introducing methods for spectral splitting. Among primary technology for spectral splitting are the use of a thin-film interference filter and a selective absorbing or transmitting filter, as shown in Fig. 18.28 with small graphs portraying the spectrum range present at each stage [129]. For the filter's composition, a number of thin layers ranging from a few nanometers to hundreds of nanometers of high refractive index contrast nonabsorbing dielectric materials are deposited on a transparent substrate. The

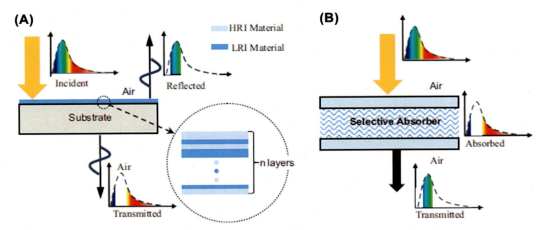

FIGURE 18.28 Mechanisms of spectral splitting by using (A) a thin-film interference filter with layers of high refractive index (HRI) and low refractive index (LRI) materials and (B) a selective absorber (solid or liquid) [129].

V. Practical concerns and beyond

center of the research by Macleod et al. is the filter that functions like an edge filter, band stop, or band pass.

It is known that UV and IR parts of the spectrum contribute to the heating of the solar cell; hence, while utilizing the visible region SR, spectral beam splitting could eliminate the excessive heating problem. In a study by Kandilli et al. [143], concentrated solar irradiance can be divided into "light" and "heat" parts by implementing the spectral beam splitting approach with a system assembled by a cold mirror, fiber-optic bundle, paraboloidal dish, dual axes tracking system, and Stirling engine on a lighting-power generation combined system. The setup of the experiment is as shown in Fig. 18.29. The cold mirror is the core element to experimentally separate the full solar spectra into distinct wavelengths. An average lighting efficiency of 14% ± 0.03 was achieved [143]. Bicer et al. [131] also conducted similar work on a PV panel by utilizing cold mirrors in a spectral splitting system and increased the power output from 3.50 to 6.75 W.

Sarswat et al. [126] also proved that energy conversion efficiency of solar cells (GaAs and c-Si solar cells) can be improved by 10% with the use of a prism—cylindrical lens combination method for spectral splitting of white light. To achieve the subdivision of a broad spectrum into several energy ranges, the setup is rather simple, and consists of a dispersive equilateral prism, a cylindrical lens, sample solar cells mounted on a scanning stage, and a light source, as shown in Fig. 18.30. For this experiment, the prism and the lens were the key components, and their positions were chosen as the setup to maximize dispersion. When the white light is

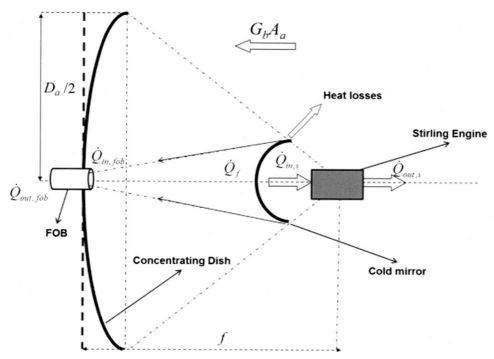

FIGURE 18.29 Illustration of the lighting-power generation combined system setup [143]. FOB is a fibre optic bundle.

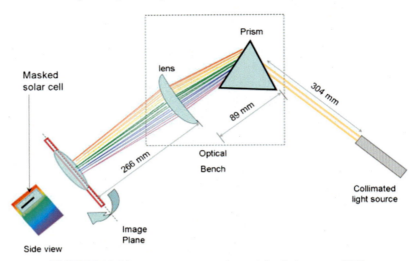

FIGURE 18.30 Schematic view of spectral splitting setup [126].

emitted from the collimated fiber source, it is dispersed by the prism. Every color in the dispersed beam is then presented to a laterally separated line by the lens [126].

A remarkable energy conversion efficiency of 23.4% was also achieved by implementing a tandem system in spectral splitting with the use of a high bandgap PSC [132]. In this work, a $CH_3NH_3PbBr_3$ solar cell was coupled with a 22.7% of an efficient silicon passivated emitter rear locally diffused solar cell to produce a positive result, which makes this as a promising method to be further improvised in the future. The $CH_3NH_3PbBr_3$ creates a high-voltage output without excessive light diversion from the Si cell that has an outstanding SR to a broad range of wavelengths.

18.5.6 Tandem system

Another concept for improving solar devices is the "tandem junction." This is achieved by placing a wide band-gap "top cell" on top of a silicon cell, the "bottom cell" [133]. Generally, tandem cells can be fabricated in either 2-terminal or 4-terminal configurations, representing distinct extremums in the optical and electronic coupling space [134]. The difference between these two terminals is that the 2-terminal design is monolithic and a series of interconnected cells is needed, whereas the 4-terminal design consists of cells that are stacked mechanically and the design allows those cells to be connected independently. Both designs have certain drawbacks that could cause efficiency losses of about 10%—11% [135,136]. Therefore a 3-terminal configuration was introduced with the possibility to select the advantages of both standard configurations and avoid any disadvantages [137,138].

With the tandem system approach, the efficiency of silicon solar cells can increase from 25.6% to above 30% [139,140]. An example of the device architecture of the planar heterojunction of the solar cell using the tandem approach is shown in Fig. 18.31 by using a scanning electron microscope. McMeekin et al. [133] stated that a metal halide perovskite PV cell

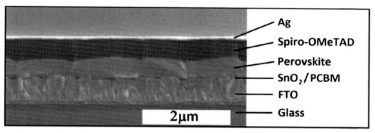

FIGURE 18.31 Scanning electron microscope image of the cross-section of a planar heterojunction of a solar cell system [133]. Ag is a silver electrode, Spiro-OMeTAD is 2,2',7, 7'-tetrakis(N,N-di-p-methoxyphenylamine)-9,9'- spirobifluorene, SnO_2/PCBM is Tin(IV) oxide/phenyl-C60-butyric acid methyl ester (PC60BM) electron-selective layer, and FTO is Fluorine-doped Tin Oxide.

has the potential to increase the efficiency of commercial silicon PV cells by about 20%–30% when a tandem architecture is introduced into the system. Nonetheless, this system faces many difficulties, e.g., finding the ideal wide band-gap materials that offer stability, cost effectiveness, and high performance for crystalline and thin-film solar technologies [133].

The improvement in efficiency by using tandem systems has also been proven in a similar work by Bush et al. [141], where mechanically stacked perovskite/silicon tandem performance was studied, as shown in Fig. 18.32. From the $J-V$ curve, the efficiency of m-Si alone is 17%. However, with the semitransparent PSC stacked on top of the m-Si cell, the efficiency increased to 18% (12.3% + 5.7%) in the tandem, with a J_{sc} of 13.3 mA/cm² from the bottom filtered Si cell.

Remarkable recent work by Luis M. et al. [142] showed that power conversion efficiency could be raised from 34% to 45% by using tandem systems. Luis M. et al. [142] promoted the use of this system for spectral stability by using voltage-matched tandem solar cells, taking into account their traditional series-connected counterparts. The study demonstrated how

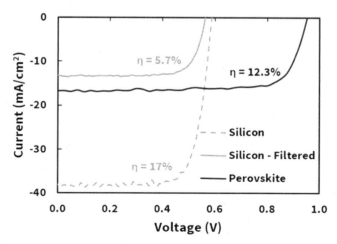

FIGURE 18.32 $J-V$ curve of the tandem with maximum power calculated from the addition of silicon cells and perovskite [141].

singlet fission can be used to produce simple voltage-matched tandems, as shown in Fig. 18.33A. The tandem cell demonstrates an efficient photocurrent addition, leading to EQE exceeding 100% at pentacene's main absorption peak, as shown in Fig. 18.33B.

Some advancements that have been made by researchers have obviously shown numerous improved results; however, increased efficiencies are yet to compete with the conventional energy source that is being currently exploited worldwide. In the authors' opinion, to replace the use of fossil fuels, which are continuously depleting, as well as focusing on increasing the efficiencies of solar technologies, the costs for constructing contemporary solar engineering systems should also be considered since large-scale highly efficient solar devices often require a lot of maintenance and space. It is, however, a difficult challenge to achieve both efficiency and economy at the same time, but if it can be accomplished, it will be *the* news sensation in the field, and a major energy shift toward solar energy will become a reality.

18.6 Challenges and limitations

Despite all the progress and impressive achievements that have been made in solar facilities, there are indeed numerous obstacles that should be overcome to upgrade this technology to an optimum standard. From this review, one of the great challenges faced by many in this industry is the imbalance between efficiency and cost. IQE can often be close to 100% in many studies by using high-quality materials and the performance of solar devices improves with good structures, careful installations, and proper maintenance, but they can be very expensive [142]. In particular, tandem solar cells are proven to attain better power conversion efficiencies but they require high precision of fabrication with materials and deposition techniques that are expensive. Consequently, generating electricity by using current solar panels can be double or triple the cost of using fossil fuels power plants. On top of that, some solar-based technologies like solar panels are not space efficient and their performance can only be improved with larger surface areas for the cells to achieve superior spectral sensitivity. Another problem found in solar research is the transient lifetime of solar gadgets. It is known that an ideal SR could not be achieved at long wavelengths because semiconductors

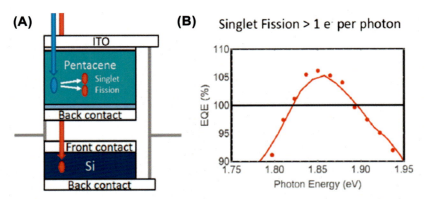

FIGURE 18.33 (A) A mechanism of the parallel tandem cell and (B) external quantum efficiency (EQE) graph of the tandem [142]. *ITO*, Indium tin oxide.

in solar cells are unable to absorb photon energies below the bandgap. Hence, while using numerous methods such as solar concentrators and multijunction solar cells to obtain a better SR, degradation appears to be faster than normal. Presumably, more intruding particles might be developed with more sunlight striking the cells. Above all, researchers face difficulties in drawing standardized conclusions to their studies because the smallest shift in solar irradiance affects the findings and data collected, as well as the overall quantum efficiency. Instruments that are in use for solar research at the moment could not compromise with a great deal of uncertainty. For example, a high efficiency solar panel might not adapt to climate extremities, leading to the short lifetime of the solar device. As the consequence to the matter, more budget will be used to replace the solar panels with the new ones, on top of the cost of regular maintenance. Besides, to precisely forecast definite trends in solar energy, decades of related data obtained from satellites are required, which is still not enough as of today since satellites are built not long ago to collect data from the ancient time, those days before satellites exist.

18.7 Conclusions, future recommendations, and publication trends

The use of solar technologies has been proven to bring positive outcomes in the renewable energy sector. To ensure a well-rounded system of solar devices for commercialization, many efforts have been implemented in various creative ways. Nonetheless, to catch up with the efficiencies of conventional energy sources that are being widely used today, further improvisations are needed. This section sums up the overall review study and future recommendations that can be made toward the betterment of solar technologies and publication trends on advancements made on solar devices.

18.7.1 Conclusions

It is clear that solar technology is a promising solution over other renewable energies to satisfy the world's energy deficit issues. However, the main concern regarding solar devices is their efficiencies or performance. On top of that, solar cells face further problems such as degradation of materials, where the lifespan of the cell is short and difficult to maintain. The output of a PV is still quite far behind the energy obtained from the burning of fossil fuels. There have been a variety of methods introduced to improvise solar cell's SR and QEs for the progression of solar devices. In this chapter, among the modern advancements suggested in the literature were solar filtering, solar concentrators, encapsulation methods, ARCs, spectral splitting, and tandem systems.

From the review, solar filtering with different materials gives similar positive results in terms of UV filtration. There are color filters, UV-B/C blocking filters, nanoparticles such as indium tin oxide and gold, as well as optical filters, all of which can lead to an increase in efficiency in absorption enhancement of photons in certain wavelengths. Solar concentrators using LSCs have managed to improve the performance of solar cells by using inexpensive materials such as nanocrystals and luminophores. A combination of methods was also achieved by a few researchers, such as combining a concentrator and a filter, whereby solar radiation can be optimized while reducing the cell's temperature. Among the results, when combined with solar concentrators, more than 90% of transmittance was obtained by using a long wavelength cut filter, and about 56%–62% of filter efficiencies were achieved by using a nanoparticle fluid filter in separate events.

For ARCs, among effective methods to increase the efficiency of solar cells are transparent conducting polymer layers, which lead to the suppression of reflectance at certain wavelengths, using a combination of ZrO_2^- polymer composite/TiO_2^- compact layer. In this case, the average reflectance was found to be reduced, and using the PECVD technique on SiN solar cells a relative increase of 2.72% for current density with 4.46% conversion efficiency was achieved. The use of nanotechnology is also practical in ARCs, whereas I_{sc} can be increased by 15.2% with Si_3N_4 subwavelength nanostructured dielectric layers. Light absorption can be enhanced by 26% using silver nanoparticles and nearly 100% transmittance can be achieved with facile synthesis of ink-bottle mesoporous MgF_2 nanoparticles. On the other hand, the solar cell encapsulation method with different strategies, the sealing technique and PECVD, gives similar positive outcomes in terms of efficiency and lifetime of solar cells. Materials such as EVA and NCA further improve the performance of those cells.

Spectral splitting has many methods and each method has its own core function. To eliminate unwanted spectra, components such as thin-film interference filter, selective absorber, and cold mirrors were used in separate studies. A combination of techniques was also applied for achieving an increment in efficiency. It has been proven that conversion efficiency can be improved by 10% with a prism—cylindrical lens combination and 23.4% of energy conversion efficiency was obtained with the application of a tandem system in the spectral splitting method. The tandem system enables the combination of advantages of individual components. Different architectures of tandem systems result in distinct outcomes. From the studies, the efficiency of solar cells with tandem systems can increase by up to 30%–45%.

Therefore, from the design and engineering of a PV to the materials involved in its creation, everything was studied in detail by researchers to produce a clean and cost-effective solar technology that can compete with the current main source of energy usage. Knowing solar studies in depth is the main focus to find problems related to solar cells and to overcome the obstacles that prevent their efficiency from increasing. Foremost is the capability to match the spectral wavelength according to the receiver's sensitivity. By using these techniques, solar technologies have shown vast improvements compared to the first batch of solar cells in the 1950s.

18.7.2 Future recommendations

A significant positive aspect of solar technologies is that they have been improving remarkably over time. With some modifications to the structures and materials, it is believed that the performance of solar PVs will go beyond the current achievable efficiency. SR of a solar cell is a crucial feature in improving devices. All the recent advancements in the industry with various methods have their own uniqueness, advantages, and setbacks. Some can obtain a very impressive performance but degrade quickly. Some solar cells can last for a long period of time but the output might not be competitive enough to be used widely in the market. Hence, the combination of several methods with properly designed architectures is suggested to further enhance efficiency and sustain the cells. For example, the use of solar concentrators with nanoparticle-coated ARCs can be used with the spectral splitting method in one system. Because light utilization can be executed by the concentrators, spectral splitting can be used to repel unwanted wavelengths that can contribute to the heating of the sensitive receiver, and the cell's temperature could be reduced with the ARC. Besides, in solar filtering, the combination of materials can contribute to the cell's improvisation. The integration between materials such as photoluminescence materials and nanomaterials, for example, might be good news for the solar industries. Photoluminescence materials that reemit photons from ambient light into dark areas have the potential to sustain light in cells for a certain period of time, and nanomaterials are well known for their flexibility and enhancement characteristics. Titanium dioxide, most commonly known as titania (TiO_2), is a well-known material that can eliminate the UV spectrum, an unwanted spectrum that potentially contributes to the heating of the cell. TiO_2 has different structures: anatase and rutile. A 100% anatase TiO_2 structure gives full transparency. It can be assumed that the lifetime of a cell can be increased and more solar spectrum can be absorbed by integrating anatase TiO_2 in solar filtration or ARC methods. With the right developments and engineering designs for solar cells, these suggestions might raise solar technologies to a whole new level.

18.7.3 Publication trends

To date, the performance of solar technologies has indeed improved; nonetheless, the efficiency of these technologies has not yet been fully utilized. Exposed to sunlight, normal solar cells tend to absorb light from various wavelengths; however, only certain wavelengths are needed for solar cells to work. Some unwanted spectra contribute to the inefficiency of solar cells apart from reducing their lifetime. Based on this drawback, researchers have been working for years to overcome problems related to the SR and QE of solar facilities.

Figs. 18.34 and 18.35 show the data regarding research publications on solutions to reduce unwanted spectra that are contributing to the degradation of solar cells (updated on March 8,

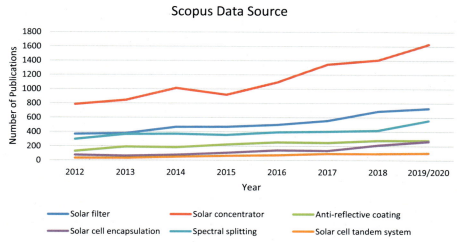

FIGURE 18.34 Recent trend on number of publications on method used to improve solar cell efficiency, which is taken from the Scopus data source.

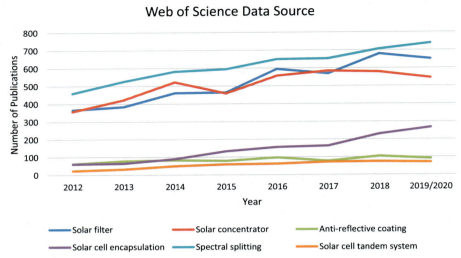

FIGURE 18.35 Recent trend on number of publications on method used to improve solar cell efficiency, which is taken from the web of science data source.

2020). Also, those articles propose several creative and innovative ways to maximize the performance of solar cells. Fig. 18.34 shows the data of a number of publications from 2012 to 2019/2020 found on Scopus. According to the trend, solar concentrator is the most popular choice of all methods with a total of 9030 articles published (searched by keyword: solar concentrator; from Scopus). On the other hand, tandem system of solar cells has the least number of publications with 528 articles published (searched by keyword: solar cell tandem system; from Scopus). With a difference of 8502 published papers from solar concentrator, researchers in this field are most likely to fail to improvise by using this method for the betterment of solar cell efficiency.

Fig. 18.35 shows a trend of publication counts for number of methods proposed for solar cell improvisation for 9 years from the Web of Science. In the Web of Science, tandem system of solar cell and antireflection coating have the least number of publications from 2012 to March 2020 with 442 (searched by keyword: solar cell tandem system; from Web of Science) and 671 (searched by keyword: antireflection coating; from Web of Science) papers published, respectively. Moreover, unlike results obtained from Scopus, the Web of Science data source shows that spectral splitting is the most favorable choice of method among scientists in solar research with 4910 publications (searched by keyword: spectral splitting; from Web of Science). Although it has a comparably lower number of articles compared to solar concentrator searched through Scopus, spectral splitting has a more stable, increasing trend until 2019/2020, which has the potential to develop its scope for further publications in future years.

Acknowledgments

The authors would like to express their gratitude for the assistance of UM Power Energy Dedicated Advanced Centre (UMPEDAC) and the Higher Institution Centre of Excellence (HICoE) Program Research Grant, UMPEDAC — 2016 (MOHE HICOE — UMPEDAC).

References

[1] M.A. Green, Y. Hishikawa, W. Warta, E.D. Dunlop, D.H. Levi, J. Hohl-Ebinger, et al., Solar cell efficiency tables (version 50), Prog. Photovoltaics Res. Appl. 25 (2017) 668–676.

[2] M.S. Ahmad, A. Pandey, N.A. Rahim, Advancements in the development of TiO 2 photoanodes and its fabrication methods for dye sensitized solar cell (DSSC) applications. A review, Renew. Sustain. Energy Rev. 77 (2017) 89–108.

[3] M.A. Green, K. Emery, Y. Hishikawa, W. Warta, E.D. Dunlop, Solar cell efficiency tables (version 45), Prog. Photovoltaics Res. Appl. 23 (2015) 1–9.

[4] A. Pandey, V. Tyagi, A. Jeyraj, L. Selvaraj, N. Rahim, S. Tyagi, Recent advances in solar photovoltaic systems for emerging trends and advanced applications, Renew. Sustain. Energy Rev. 53 (2016) 859–884.

[5] H. H Aberlin, H. Eppel, Photovoltaics, Wiley, 2012.

[6] I. Visoly-Fisher, K.D. Dobson, J. Nair, E. Bezalel, G. Hodes, D. Cahen, Factors affecting the stability of CdTe/CdS solar cells deduced from stress tests at elevated temperature, Adv. Funct. Mater. 13 (2003) 289–299.

[7] B. Chambers, C. Backus, Measurement of the spectral response of silicon solar cells under highly illuminated conditions, in: Photovoltaic Solar Energy Conference, Springer, 1981, pp. 418–422.

[8] A. Stephen, Survey of Instrumentation and Measurement, John Willey & Sons, New York, 2001.

[9] M. Iqbal, An Introduction to Solar Radiation, Elsevier, 2012.

[10] R.C. Willson, S. Gulkis, M. Janssen, H. Hudson, G. Chapman, Observations of solar irradiance variability, Science 211 (1981) 700–702.

[11] M.A. Green, Solar Cells: Operating Principles, Technology, and System Applications, 1982.

[12] D. Goossens, E. Van Kerschaever, Aeolian dust deposition on photovoltaic solar cells: the effects of wind velocity and airborne dust concentration on cell performance, Sol. Energy 66 (1999) 277–289.

[13] A. Luque, S. Hegedus, Handbook of Photovoltaic Science and Technology, Wiley, Chichester, 2003.

[14] I.B. Karki, Effect of temperature on the IV characteristics of a polycrystalline solar cell, J. Nepal Paediatr. Soc. 3 (2016) 35–40.

[15] S. Mekhilef, R. Saidur, M. Kamalisarvestani, Effect of dust, humidity and air velocity on efficiency of photovoltaic cells, Renew. Sustain. Energy Rev. 16 (2012) 2920–2925.

[16] A. Virtuani, E. Lotter, M. Powalla, Influence of the light source on the low-irradiance performance of Cu (In, Ga) Se 2 solar cells, Sol. Energy Mater. Sol. Cells 90 (2006) 2141–2149.

[17] A. Amelia, Y. Irwan, W. Leow, M. Irwanto, I. Safwati, M. Zhafarina, Investigation of the effect temperature on photovoltaic (PV) panel output performance, Int. J. Adv. Sci. Eng. Inf. Technol. 6 (2016).

[18] E. Perzon, X. Wang, F. Zhang, W. Mammo, J.L. Delgado, P. de la Cruz, et al., Design, synthesis and properties of low band gap polyfluorenes for photovoltaic devices, Synth. Met. 154 (2005) 53–56.

[19] E.P.X. Wang, J.L. Delgado, P. De La Cruz, F. Zhang, F. Langa, M.R. Andersson, O. Inganas, Appl. Phys. Lett. 85 (2004) 5081.

[20] M.N.K. Colladet, L. Goris, L. Lutsen, D. Vanderzande, Thin Solid Films 451 (2004) 7.

[21] E.P. Xiangjun Wang, F. Oswald, F. Langa, S. Edmassie, M.R. Andersson, O. Inganas, Adv. Funct. Mater. 15 (2005) 1665–1670.

[22] A.C.C.J. Brabec, D. Meissner, N.S. Sariciftci, T. Fromhertz, M.T. Rispens, L. Sanchez, J.C. Hummelen, Adv. Funct. Mater. 11 (2001) 374.

[23] M.S.A. Gadisa, M.R. Andersson, O. Inganas, Appl. Phys. Lett. 84 (2004) 374.

[24] M. Stuckelberger, R. Biron, N. Wyrsch, F.-J. Haug, C. Ballif, Review: progress in solar cells from hydrogenated amorphous silicon, Renew. Sustain. Energy Rev. 76 (2017) 1497–1523.

[25] S.R. Chowdhury, H. Saha, Maximum power point tracking of partially shaded solar photovoltaic arrays, Sol. Energy Mater. Sol. Cells 94 (2010) 1441–1447.

[26] H. Patel, V. Agarwal, MATLAB-based modeling to study the effects of partial shading on PV array characteristics, IEEE Trans. Energy Convers. 23 (2008) 302–310.

[27] M. García, L. Marroyo, E. Lorenzo, M. Pérez, Soiling and other optical losses in solar-tracking PV plants in navarra, Prog. Photovoltaics Res. Appl. 19 (2011) 211–217.

[28] A.M. Pavan, A. Mellit, D. De Pieri, The effect of soiling on energy production for large-scale photovoltaic plants, Sol. Energy 85 (2011) 1128–1136.

[29] J. Zorrilla-Casanova, M. Piliougine, J. Carretero, P. Bernaola-Galván, P. Carpena, L. Mora-López, et al., Losses produced by soiling in the incoming radiation to photovoltaic modules, Prog. Photovoltaics Res. Appl. 21 (2013) 790–796.

[30] J.R. Caron, B. Littmann, Direct monitoring of energy lost due to soiling on first solar modules in California, in: Photovoltaic Specialists Conference (PVSC), vol. 2, IEEE, 2012, pp. 1–5. IEEE 38th.

[31] L.K. Verma, M. Sakhuja, J. Son, A.J. Danner, H. Yang, H.C. Zeng, et al., Self-cleaning and antireflective packaging glass for solar modules, Renew. Energy 36 (2011) 2489–2493.

[32] S. Nann, K. Emery, Spectral effects on PV-device rating, Sol. Energy Mater. Sol. Cell. 27 (1992) 189–216.

[33] J.F. Rabek, Polymer Photodegradation: Mechanisms and Experimental Methods, Springer Science & Business Media, 2012.

[34] E. Yousif, R. Haddad, Photodegradation and photostabilization of polymers, especially polystyrene: review, SpringerPlus 2 (2013) 398.

[35] C. Pryde, Photoaging of Polycarbonate: Effects of Selected Variables on Degradation Pathways, ACS Publications, 1985.

[36] A. Ghasemi-Kahrizsangi, J. Neshati, H. Shariatpanahi, E. Akbarinezhad, Improving the UV degradation resistance of epoxy coatings using modified carbon black nanoparticles, Prog. Org. Coating 85 (2015) 199–207.

[37] Z. Zhang, S. Wang, J. Zhang, Large stabilizing effect of titanium dioxide on photodegradation of PVC/α-methylstyrene-acrylonitrile copolymer/impact modifier-matrix composites, Polym. Compos. 35 (2014) 2365–2375.

[38] H. Neugebauer, C. Brabec, J.C. Hummelen, N.S. Sariciftci, Stability and photodegradation mechanisms of conjugated polymer/fullerene plastic solar cells, Sol. Energy Mater. Sol. Cell. 61 (2000) 35–42.

[39] I. Topolniak, A. Chapel, J. Gaume, P.-O. Bussiere, G. Chadeyron, J.-L. Gardette, et al., Applications of polymer nanocomposites as encapsulants for solar cells and LEDs: impact of photodegradation on barrier and optical properties, Polym. Degrad. Stabil. 145 (2017) 52–59.

[40] J. Pospíšil, J. Pilař, N.C. Billingham, A. Marek, Z. Horák, S. Nešpůrek, Factors affecting accelerated testing of polymer photostability, Polym. Degrad. Stabil. 91 (2006) 417–422.

[41] S. Nikafshar, O. Zabihi, M. Ahmadi, A. Mirmohseni, M. Taseidifar, M. Naebe, The effects of UV light on the chemical and mechanical properties of a transparent epoxy-diamine system in the presence of an organic UV absorber, Materials 10 (2017) 180.

[42] A. Shah, P. Torres, R. Tscharner, N. Wyrsch, H. Keppner, Photovoltaic technology: the case for thin-film solar cells, Science 285 (1999) 692–698.

[43] K. Zweibel, P. Singh, Harnessing solar power: the photovoltaics challenge, Am. J. Phys. 61 (1993) 286–287.

[44] G.P. Smestad, Education and solar conversion:: demonstrating electron transfer, Sol. Energy Mater. Sol. Cells 55 (1998) 157–178.

[45] R.C. Neville, Solar Energy Conversion: The Solar Cell, Elsevier, 1995.

[46] S. Fonash, Solar Cell Device Physics, Elsevier, 2012.

[47] V. Tatsiankou, K. Hinzer, J. Haysom, H. Schriemer, K. Emery, R. Beal, Design principles and field performance of a solar spectral irradiance meter, Sol. Energy 133 (2016) 94–102.

[48] S. Reynolds, V. Smirnov, Modelling performance of two-and four-terminal thin-film silicon tandem solar cells under varying spectral conditions, Energy Procedia 84 (2015) 251–260.

[49] A. Virtuani, L. Fanni, Seasonal power fluctuations of amorphous silicon thin-film solar modules: distinguishing between different contributions, Prog. Photovoltaics Res. Appl. 22 (2014) 208–217.

[50] B. Minnaert, P. Veelaert, A proposal for typical artificial light sources for the characterization of indoor photovoltaic applications, Energies 7 (2014) 1500–1516.

[51] J. Peet, J.Y. Kim, N.E. Coates, W.L. Ma, D. Moses, A.J. Heeger, et al., Efficiency enhancement in low-bandgap polymer solar cells by processing with alkane dithiols, Nat. Mater. 6 (2007) 497.

[52] Y. Liang, Z. Xu, J. Xia, S.T. Tsai, Y. Wu, G. Li, et al., For the bright future—bulk heterojunction polymer solar cells with power conversion efficiency of 7.4%, Adv. Mater. (2010) 22.

[53] Z. He, C. Zhong, S. Su, M. Xu, H. Wu, Y. Cao, Enhanced power-conversion efficiency in polymer solar cells using an inverted device structure, Nat. Photon. 6 (2012) 591.

[54] T. Mueller, Heterojunction Solar Cells (a-Si/c-Si): Investigations on PECV Deposited Hydrogenated Silicon Alloys for Use as High-Quality Surface Passivation and Emitter/BSF, Logos Verlag Berlin GmbH, 2009.

[55] H. Mackel, A. Cuevas, Spectral Response of the Photoconductance: A New Technique for Solar Cell Characterization, 2001.

[56] W. Yang, Z. Ma, X. Tang, C. Feng, W. Zhao, P. Shi, Internal quantum efficiency for solar cells, Sol. Energy 82 (2008) 106–110.

[57] C.-Y. Chen, S.K. Ahn, D. Aoki, J. Kokubo, K.H. Yoon, H. Saito, et al., International round-robin intercomparison of dye-sensitized and crystalline silicon solar cells, J. Power Sources 340 (2017) 309–318.

V. Practical concerns and beyond

References

[58] S. Chander, A. Purohit, A. Nehra, S. Nehra, M. Dhaka, A study on spectral response and external quantum efficiency of mono-crystalline silicon solar cell, Int. J. Renew. Energy Resour. 5 (2015) 41–44.

[59] C.S. Solanki, B. Arora, J. Vasi, M. Patil, Solar Photovoltaics, PHI publication, 2013.

[60] L. Fang, L. Danos, T. Markvart, Solar cell as a waveguide: quantum efficiency of ultrathin crystalline silicon solar cell, in: 28th European Photovoltaic Solar Energy Conference and Exhibition, 2013, pp. 31–33.

[61] M. Pravettoni, D.J. Farrell, A.J. Chatten, R. Bose, R.P. Kenny, K.W. Barnham, External Quantum Efficiency Measurements of Luminescent Solar Concentrators: A Study of the Impact of Backside Reflector Size and Shape, 2009, p. 332789268.

[62] S.R. Wenham, Applied Photovoltaics, Routledge, 2012.

[63] W. Weber, J. Lambe, Luminescent greenhouse collector for solar radiation, Appl. Optic. 15 (1976) 2299.

[64] F.C. Krebs, Polymeric Solar Cells: Materials, Design, Manufacture, DEStech Publications, Inc., 2010.

[65] M. Law, M.C. Beard, S. Choi, J.M. Luther, M.C. Hanna, A.J. Nozik, Determining the internal quantum efficiency of PbSe nanocrystal solar cells with the aid of an optical model, Nano Lett. 8 (2008) 3904–3910.

[66] Y. Sun, et al., 2.0–2.2 eV AlGaInP solar cells grown by molecular beam epitaxy, Sol. Energy Mater. Sol. Cells 219 (2021) 110774.

[67] D.L. King, J.A. Kratochvil, W.E. Boyson, Measuring solar spectral and angle-of-incidence effects on photovoltaic modules and solar irradiance sensors, in: Photovoltaic Specialists Conference, 1997, Conference Record of the Twenty-Sixth IEEE, IEEE, 1997, pp. 1113–1116.

[68] K. Emery, Solar simulators and I–V measurement methods, Sol. Cells 18 (1986) 251–260.

[69] K. Emery, C. Osterwald, T. Cannon, D. Myers, J. Burdick, T. Glatfelter, et al., Methods for measuring solar cell efficiency independent of reference cell or light source, in: Proc 18th IEEE PV Spec Conf, 1985, pp. 21–25.

[70] H. Field, K. Emery, An uncertainty analysis of the spectral correction factor, in: Photovoltaic Specialists Conference, 1993, Conference Record of the Twenty Third IEEE, IEEE, 1993, pp. 1180–1187.

[71] H. Curtis, Errors in Short Circuit Current Measurements Due to Spectral Mismatch between Sunlight and Solar ulators, 1976.

[72] H. Müllejans, A. Ioannides, R. Kenny, W. Zaaiman, H.A. Ossenbrink, E.D. Dunlop, Spectral mismatch in calibration of photovoltaic reference devices by global sunlight method, Meas. Sci. Technol. 16 (2005) 1250.

[73] M.A. Haridy, Uncertainty estimation of spectral mismatch correction factor for incandescent lamps, Int. J. Curr. Res. Aca. Rev. 3 (2015) 262–273.

[74] R. Eke, T.R. Betts, Spectral irradiance effects on the outdoor performance of photovoltaic modules, Renew. Sustain. Energy Rev. 69 (2017) 429–434.

[75] J.M. Palmer, The measurement of transmission, absorption, emission, and reflection, Handb. Optics 2 (1995) 251–255.

[76] Y.-J. Lee, D.S. Ruby, D.W. Peters, B.B. McKenzie, J.W. Hsu, ZnO nanostructures as efficient antireflection layers in solar cells, Nano Lett. 8 (2008) 1501–1505.

[77] R. Barnes, M.S. Dhanoa, S.J. Lister, Standard normal variate transformation and de-trending of near-infrared diffuse reflectance spectra, Appl. Spectrosc. 43 (1989) 772–777.

[78] J.T. Woolley, Reflectance and transmittance of light by leaves, Plant Physiol. 47 (1971) 656–662.

[79] P. Good, T. Cooper, M. Querci, N. Wiik, G. Ambrosetti, A. Steinfeld, Spectral reflectance, transmittance, and angular scattering of materials for solar concentrators, Sol. Energy Mater. Sol. Cell. 144 (2016) 509–522.

[80] C.A. Gueymard, Spectral effects on the transmittance, solar heat gain, and performance rating of glazing systems, Sol. Energy 83 (2009) 940–953.

[81] S.Y. Tsai, C.T. Ni, K.Z. Fung, Characterization of infrared transmittance in mixed transition metal oxides for solar cells application, Ceram. Int. 43 (2017) S460–S463.

[82] M. Born, E. Wolf, Principles of Optics: Electromagnetic Theory of Propagation, Interference and Diffraction of Light, Elsevier, 2013.

[83] A. Reinders, P. Verlinden, A. Freundlich, Photovoltaic Solar Energy: From Fundamentals to Applications, John Wiley & Sons, 2017.

[84] G.R. Valenzuela, Theories for the interaction of electromagnetic and oceanic waves—a review, Boundary-Layer Meteorol. 13 (1978) 61–85.

[85] E. Hecht, Optics, 4th International edition, Addison-Wesley, San Francisco, 2002, p. 3.

[86] H. Tada, J. Carter Jr., B. Anspaugh, R. Downing, Solar Cell Radiation Handbook, 1982.

[87] C. Stanley, A. Mojiri, G. Rosengarten, Spectral light management for solar energy conversion systems, Nanophotonics 5 (2016) 161–179.

V. Practical concerns and beyond

[88] J.-h. Huang, W.-C. Fei, W.-C. Hsu, J.-c. Tsai, Solar concentrator constructed with a circular prism array, Appl. Optic. 49 (2010) 4472–4478.

[89] O. Gunawan, B. Lei, Solar Cell Characterization System With an Automated Continuous Neutral Density Filter, 2017. Google Patents.

[90] N.-G. Park, M. Grätzel, T. Miyasaka, K. Zhu, K. Emery, Towards stable and commercially available perovskite solar cells, Nat. Energy 1 (2016) 16152.

[91] H.A. Kazem, M.T. Chaichan, The impact of using solar colored filters to cover the PV panel on its outcomes, Bull. J. 2 (2016) 464–469.

[92] F. Tavella, C. Ampelli, L. Frusteri, F. Frusteri, S. Perathoner, G. Centi, Development of photoanodes for photoelectrocatalytic solar cells based on copper-based nanoparticles on titania thin films of vertically aligned nanotubes, Catal. Today (2017).

[93] T. Otanicar, D. DeJarnette, N. Brekke, E. Tunkara, K. Roberts, P. Harikumar, Full spectrum collection of concentrated solar energy using PV coupled with selective filtration utilizing nanoparticles, MRS Adv. 1 (2016) 2935–2940.

[94] F.K. Mbakop, N. Djongyang, G.W. Ejuh, D. Raïdandi, P. Woafo, Transmission of light through an optical filter of a one-dimensional photonic crystal: application to the solar thermophotovoltaic system, Phys. B Condens. Matter 516 (2017) 92–99.

[95] B.I. Ismail, Power generation using nonconventional renewable geothermal & alternative clean energy technologies, in: Planet Earth 2011-Global Warming Challenges and Opportunities for Policy and Practice, InTech, 2011.

[96] R. Winston, Principles of solar concentrators of a novel design, Sol. Energy 16 (1974) 89–95.

[97] L.R. Bradshaw, K.E. Knowles, S. McDowall, D.R. Gamelin, Nanocrystals for luminescent solar concentrators, Nano Lett. 15 (2015) 1315–1323.

[98] M.J. Currie, J.K. Mapel, T.D. Heidel, S. Goffri, M.A. Baldo, High-efficiency organic solar concentrators for photovoltaics, Science 321 (2008) 226–228.

[99] L. Xu, Y. Yao, N.D. Bronstein, L. Li, A.P. Alivisatos, R.G. Nuzzo, Enhanced photon collection in luminescent solar concentrators with distributed Bragg reflectors, ACS Photonics 3 (2016) 278–285.

[100] J. Yoon, L. Li, A.V. Semichaevsky, J.H. Ryu, H.T. Johnson, R.G. Nuzzo, et al., Flexible concentrator photovoltaics based on microscale silicon solar cells embedded in luminescent waveguides, Nat. Commun. 2 (2011) 343.

[101] X. Sheng, L. Shen, T. Kim, L. Li, X. Wang, R. Dowdy, et al., Doubling the power output of bifacial thin-film GaAs solar cells by embedding them in luminescent waveguides, Adv. Energy Mater. 3 (2013) 991–996.

[102] N.D. Bronstein, L. Li, L. Xu, Y. Yao, V.E. Ferry, A.P. Alivisatos, et al., Luminescent solar concentration with semiconductor nanorods and transfer-printed micro-silicon solar cells, ACS Nano 8 (2013) 44–53.

[103] N. Ahmad, Y. Ota, K. Nishioka, Temperature reduction of solar cells in a concentrator photovoltaic system using a long wavelength cut filter, Jpn. J. Appl. Phys. 56 (2017) 032301.

[104] D. DeJarnette, E. Tunkara, N. Brekke, T. Otanicar, K. Roberts, B. Gao, et al., Nanoparticle enhanced spectral filtration of insolation from trough concentrators, Sol. Energy Mater. Sol. Cell. 149 (2016) 145–153.

[105] Three Layer Anti-Reflection Coating, 1965. Google Patents.

[106] K. Ali, S.A. Khan, M.M. Jafri, Effect of double layer (SiO 2/TiO 2) anti-reflective coating on silicon solar cells, Int. J. Electrochem. Sci. 9 (2014) 7865–7874.

[107] Y. Wan, K.R. McIntosh, A.F. Thomson, Characterisation and optimisation of PECVD SiNx as an antireflection coating and passivation layer for silicon solar cells, AIP Adv. 3 (2013) 032113.

[108] A. Uzum, M. Kuriyama, H. Kanda, Y. Kimura, K. Tanimoto, H. Fukui, et al., Sprayed and spin-coated multilayer antireflection coating films for nonvacuum processed crystalline silicon solar cells, Int. J. Photoenergy (2017) 2017.

[109] S.-Y. Lien, D.-S. Wuu, W.-C. Yeh, J.-C. Liu, Tri-layer antireflection coatings (SiO 2/SiO 2–TiO 2/TiO 2) for silicon solar cells using a sol–gel technique, Sol. Energy Mater. Sol. Cell. 90 (2006) 2710–2719.

[110] J.-J. Huang, Y.-T. Lee, Self-cleaning and antireflection properties of titanium oxide film by liquid phase deposition, Surf. Coating. Technol. 231 (2013) 257–260.

[111] Y. Jiang, B. Luo, F. Jiang, F. Jiang, C. Fuentes-Hernandez, T. Liu, et al., Efficient colorful perovskite solar cells using a top polymer electrode simultaneously as spectrally selective antireflection coating, Nano Lett. 16 (2016) 7829–7835.

[112] H. Ghosh, S. Mitra, H. Saha, S.K. Datta, C. Banerjee, Argon plasma treatment of silicon nitride (SiN) for improved antireflection coating on c-Si solar cells, Mater. Sci. Eng. B 215 (2017) 29–36.

References

[113] M. Elshorbagy, K. Abdel-Hady, H. Kamal, J. Alda, Broadband anti-reflection coating using dielectric Si 3 N 4 nanostructures. Application to amorphous-Si-H solar cells, Optic Commun. 390 (2017) 130−136.

[114] Z. Duan, M. Li, T. Mwenya, Y. Li, D. Song, Morphology optimization of silver nanoparticles used to improve the light absorption in thin-film silicon solar cells, Plasmonics (2017) 1−7.

[115] D. Karthik, S. Pendse, S. Sakthivel, E. Ramasamy, S.V. Joshi, High performance broad band antireflective coatings using a facile synthesis of ink-bottle mesoporous MgF 2 nanoparticles for solar applications, Sol. Energy Mater. Sol. Cell. 159 (2017) 204−211.

[116] Q. Dong, F. Liu, M.K. Wong, H.W. Tam, A.B. Djurišić, A. Ng, et al., Encapsulation of perovskite solar cells for high humidity conditions, ChemSusChem 9 (2016) 2597−2603.

[117] A. Gupta, J.D. Ingham, A.H. Yavrouian, Solar Cell Encapsulation, 1983. Google Patents.

[118] A. Rizzo, L. Ortolan, S. Murrone, L. Torto, M. Barbato, N. Wrachien, et al., Effects of thermal stress on hybrid perovskite solar cells with different encapsulation techniques, in: Reliability Physics Symposium (IRPS), 2017 IEEE International, IEEE, 2017. PV-1−PV-6.

[119] A. Metz, G. Demenik, A. Richter, T. Vlasenko, I. Buchovskaya, M. Zwegers, et al., International Technology Roadmap for Photovoltaic, ITRPV, Frankfurt, 2015, p. 1.

[120] R. Witteck, B. Veith-Wolf, H. Schulte-Huxel, A. Morlier, M.R. Vogt, M. Köntges, et al., UV-induced degradation of PERC solar modules with UV-transparent encapsulation materials, Prog. Photovoltaics Res. Appl. 25 (2017) 409−416.

[121] A. Jentsch, K.-J. Eichhorn, B. Voit, Influence of typical stabilizers on the aging behavior of EVA foils for photovoltaic applications during artificial UV-weathering, Polym. Test. 44 (2015) 242−247.

[122] G. Griffini, S. Turri, Polymeric materials for long-term durability of photovoltaic systems, J. Appl. Polym. Sci. 133 (2016).

[123] G. Dennler, C. Lungenschmied, H. Neugebauer, N. Saricftci, M. Latreche, G. Czeremuszkin, et al., A new encapsulation solution for flexible organic solar cells, Thin Solid Films 511 (2006) 349−353.

[124] D. Bryant, N. Aristidou, S. Pont, I. Sanchez-Molina, T. Chotchunangatchaval, S. Wheeler, et al., Light and oxygen induced degradation limits the operational stability of methylammonium lead triiodide perovskite solar cells, Energy Environ. Sci. 9 (2016) 1655−1660.

[125] J.W. Leem, J.S. Yu, J. Heo, W.-K. Park, J.-H. Park, W.J. Cho, et al., Nanostructured encapsulation coverglasses with wide-angle broadband antireflection and self-cleaning properties for III−V multi-junction solar cell applications, Sol. Energy Mater. Sol. Cells 120 (2014) 555−560.

[126] P.K. Sarswat, M. Jagannathan, M.L. Free, Performance of photovoltaic cells in different segments of spatial-spectral distributions, Vacuum (2017).

[127] R.E. Nelson, A brief history of thermophotovoltaic development, Semicond. Sci. Technol. 18 (2003) S141.

[128] R.I. Rabady, Optimized spectral splitting in thermo-photovoltaic system for maximum conversion efficiency, Energy 119 (2017) 852−859.

[129] A. Mojiri, R. Taylor, E. Thomsen, G. Rosengarten, Spectral beam splitting for efficient conversion of solar energy—a review, Renew. Sustain. Energy Rev. 28 (2013) 654−663.

[130] X. Ju, C. Xu, X. Han, X. Du, G. Wei, Y. Yang, A review of the concentrated photovoltaic/thermal (CPVT) hybrid solar systems based on the spectral beam splitting technology, Appl. Energy 187 (2017) 534−563.

[131] Y. Bicer, A.F.V. Sprotte, I. Dincer, Concentrated solar light splitting using cold mirrors for photovoltaics and photonic hydrogen production applications, Appl. Energy 197 (2017) 169−182.

[132] R. Sheng, A.W. Ho-Baillie, S. Huang, M. Keevers, X. Hao, L. Jiang, et al., Four-terminal tandem solar cells using $CH_3NH_3PbBr_3$ by spectrum splitting, J. Phys. Chem. Lett. 6 (2015) 3931−3934.

[133] D.P. McMeekin, G. Sadoughi, W. Rehman, G.E. Eperon, M. Saliba, M.T. Hörantner, et al., A mixed-cation lead mixed-halide perovskite absorber for tandem solar cells, Science 351 (2016) 151−155.

[134] G.W. Adhyaksa, E. Johlin, E.C. Garnett, Nanoscale back contact perovskite solar cell design for improved tandem efficiency, Nano Lett. 17 (2017) 5206−5212.

[135] J.P. Mailoa, M. Lee, I.M. Peters, T. Buonassisi, A. Panchula, D.N. Weiss, Energy-yield prediction for II−VI-based thin-film tandem solar cells, Energy Environ. Sci. 9 (2016) 2644−2653.

[136] M.W. Rowell, M.D. McGehee, Transparent electrode requirements for thin film solar cell modules, Energy Environ. Sci. 4 (2011) 131−134.

[137] A. Martí, A. Luque, Three-terminal heterojunction bipolar transistor solar cell for high-efficiency photovoltaic conversion, Nat. Commun. 6 (2015).

V. Practical concerns and beyond

[138] D. Bahro, M. Koppitz, A. Mertens, K. Glaser, J. Mescher, A. Colsmann, Understanding the external quantum efficiency of organic Homo-tandem solar cells utilizing a three-terminal device architecture, Adv. Energy Mater. 5 (2015).

[139] V. Sivaram, S.D. Stranks, H.J. Snaith, Outshining silicon, Sci. Am. 313 (2015) 54–59.

[140] M.A. Green, K. Emery, Y. Hishikawa, W. Warta, E.D. Dunlop, Solar cell efficiency tables (version 46), Prog. Photovoltaics Res. Appl. 23 (2015) 805–812.

[141] K.A. Bush, C.D. Bailie, Y. Chen, A.R. Bowring, W. Wang, W. Ma, et al., Thermal and environmental stability of semi-transparent perovskite solar cells for tandems enabled by a solution-processed nanoparticle buffer layer and sputtered ITO electrode, Adv. Mater. 28 (2016) 3937–3943.

[142] L.M. Pazos-Outón, J.M. Lee, M.H. Futscher, A. Kirch, M. Tabachnyk, R.H. Friend, et al., A silicon–singlet fission tandem solar cell exceeding 100% external quantum efficiency with high spectral stability, ACS Energy Lett. 2 (2017) 476–480.

[143] C. Kandili, K. Ulgan, et al., Review and modelling the systems of transmission concentrated solar energy via optical fibres, Renew. Sust. Energy Rev. 13 (1) (2009) 67–84. https://doi.org/10.1016/j.rser.2007.05.005.

CHAPTER

19

Energy materials: Applications and propelling opportunities

N. Thejo Kalyani[1], S.J. Dhoble[2]

[1]Department of Applied Physics, Laxminarayan Institute of Technology, Nagpur, Maharashtra, India; [2]Department of Physics, RTM Nagpur University, Nagpur, Maharashtra, India

19.1 Introduction

Energy is essential to propel vehicles, drive machinery in industry, generate electricity, provide light and heat, etc. However, the energy crisis is a harsh reality [1]. This rapidly escalating energy catastrophe has propelled industrialists, researchers, academicians, and scientists to formulate detailed endeavors for the advancement of novel materials and technologies that are proficient in generating energy and curtailing power demands. Exploiting any source of energy upsets the environment in some way or another. In any given situation, the source of energy chosen depends on factors such as ease of methods of extraction, economics, efficiency of the accessible technology, and the ecological harm that may be incurred. Emerging novel materials will overcome these problems to a certain extent. These innovative materials, when scaled up, will offer new solutions to the daunting energy crisis challenges and hopefully will transition to a sustainable energy system before it is too late [2,3]. Scaling up these materials provides positive synergy between these materials for the harvesting, storage, and saving of energy, and will strive to shorten the distance between research and applications. Hence, an attempt has been made to provide an overview of current progress, energy-related applications, and propelling opportunities of these energy materials. The basic requirements of energy material technologies for sustainable development are depicted in Fig. 19.1.

19.2 Overview of novel energy materials for energy-related applications

Materials are the key to improvised performance in numerous technologies. As worldwide energy demand is expected to increase continuously at a faster rate, it is critical to improve

Energy Materials
https://doi.org/10.1016/B978-0-12-823710-6.00011-X

Copyright © 2021 Elsevier Ltd. All rights reserved.

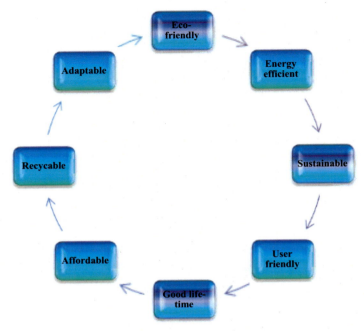

FIGURE 19.1 Energy material technologies: basic requirements for commercialization.

technologies for sustainable production, conversion, storage, as well as energy. To accelerate the exploration of a wide spectrum of energy materials, discovery and use of new, high-performance, low-cost, clean energy materials is essential. Development of the next generation of energy technology faces the challenges of finding and integrating new materials at a faster rate. Even with the current state-of-the-art technology, the innovation process for the translation of new material from land to market is a big challenge. Cross-cutting is needed by technology to accelerate the exploration, discovery, and integration of clean energy materials that expand the opportunities for speeding up the transition to a low carbon economy. Novel energy materials for energy-related applications include three active research fields, namely (1) materials for energy harvesting/conversion, (2) materials for energy storage, and (3) materials for energy saving [4]. A partial list of novel energy material technologies for energy-related applications is shown in Fig. 19.2.

19.2.1 Materials for energy harvesting and conversion

Many new-fangled materials as well as improvised technologies have been explored to achieve fruitful energy harvesting and conversion. This encompasses a variety of materials and designs to convert heat energy to electrical energy by means of a thermoelectric generator, solar energy into electricity by the photovoltaic effect, and fuel cells that convert the chemical energy of a fuel (generally H_2) and an oxidizing agent (O_2) into electricity through a pair of redox reactions.

19.2 Overview of novel energy materials for energy-related applications   569

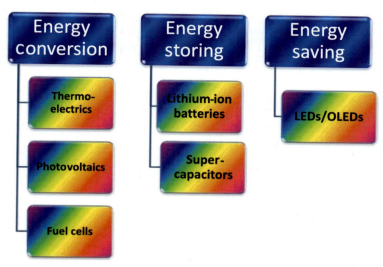

FIGURE 19.2 Novel energy material technologies for energy-related applications. *LEDs*, Light-emitting diodes; *OLEDs*, organic light-emitting diodes.

19.2.1.1 Thermoelectrics

The materials that proffer a sustainable pathway to convert heat energy to electrical energy or refrigeration directly from electrical energy (based on either the Seebeck or Peltier effects) are known as thermoelectrics. The Peltier effect has applications in cooling, while the Seebeck effect is used in thermoelectric generators for power generation from heat [5], as depicted in Fig. 19.3. The advantages of thermoelectrics include its: (1) eco-friendly nature, (2) scalability (the device can be applied to any size of heat source), (3) economy, and (4) energy recyclability (waste heat to electricity). However, technology progression is very slow because energy conversion efficiency and coefficient of performance are very low [6]. However, only 34% of the supplied heat energy is converted into useful energy and the rest is lost unutilized.

Materials selection criteria for thermoelectrics includes: (1) high figure of merit (requires low electrical resistivity (ρ) and thermal conductivity (κ)) (because the value of these parameters depends on temperature, the figure of limit is temperature dependent), (2) power factor (a parameter that defines the ability of material to produce electric power), which depends on electrical conductivity σ and the Seebeck coefficient (S), given by power factor $= \sigma S^2$, and (3) efficiency of a thermoelectric generator (η, the ratio of the energy provided to the load to the heat energy absorbed at a hot junction). Popularly used thermoelectric materials include bismuth chalcogenides (Bi_2Te_3, Bi_2se_3) and lead telluride (PBTe). Effective thermal energy to electrical energy conversion is possible with nanostructured compounds.

19.2.1.2 Photovoltaics

Photovoltaic devices have the potential to generate sustainable energy. Their efficiency of conversion depends purely on the wavelength of incident light that the photovoltaic material

V. Practical concerns and beyond

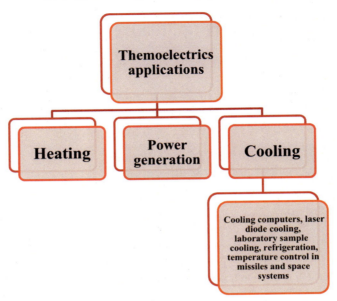

FIGURE 19.3 A few applications of thermoelectrics.

absorbs. This $\eta-\lambda$ relationship is characterized by spectral response. These materials have unique properties that can be tailored to meet the requirements set by the spectral distribution, angle of incidence, and intensity of the electromagnetic radiation prevailing in our natural surroundings [7,8]. The solar spectrum ranges between the ultraviolet and infrared regions, which encompasses the visible region also. However, only a small fraction of incident is converted into electrical energy depending on the nature of the absorbing material, and the rest is left unutilized. To harness this unharnessed energy, spectral conversion techniques such as upconversion or downconversion phosphors can be employed; this further improves the conversion efficiency of solar cells [9]. Versatile applications of solar cells are depicted in Fig. 19.4. However, this harnessing is only possible in the presence of the sun. Against this backdrop, the need to develop technologies that can store this converted energy that can be used effectively at night or when the sky is cloudy is prerequisite. Hence, the use of nano-optimized cells, polymeric dyes, quantum dots, thin films, multiple junctions, antireflection coatings, and ultrathin wafer semiconductor technology has gained momentum.

19.2.1.3 *Fuel cells*

Fuel cells are electrochemical cells with a pair of electrodes (anode and cathode) and an electrolyte that converts the chemical energy of a chemical reaction between a fuel and an oxidizing agent directly into electricity through a pair of redox reactions [10]:

$$\textbf{Fuel + Oxidizing agent = Electricity + Water vapor}$$

Common fuels include hydrogen, methanol, ethanol, and ammonia [11], while a common oxidizing agent is generally oxygen. Other popular energy sources and fuels for fuel cells

19.2 Overview of novel energy materials for energy-related applications

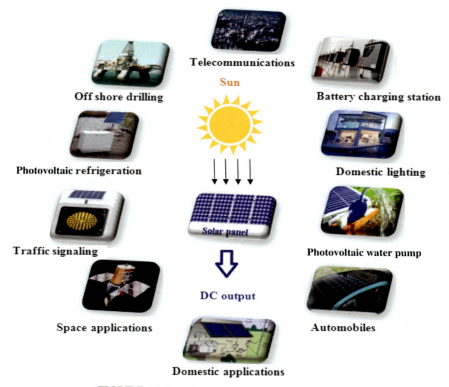

FIGURE 19.4 The many applications of solar cells.

include: (1) conventional fuels like natural gas, propane, diesel, and other hydrocarbons, (2) biomass like methane and methanol, and (3) renewable resources like solar, wind, and ocean energy. These are primarily used in three versatile areas, as described in Fig. 19.5.

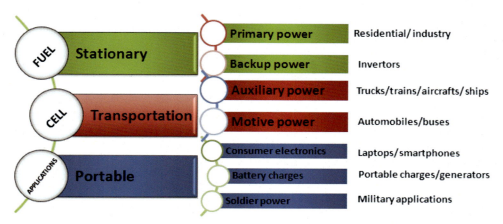

FIGURE 19.5 Applications of fuel cells.

The efficiency of fuel cells can be enhanced to a certain extent by employing nano-optimized membranes. However, these cells are sensitive to fuel impurities, generate low current and power, and employ expensive catalysts.

19.2.2 Materials for energy storage

These materials have the potential to store energy, as in batteries and supercapacitors. Also, chemical fuels such as hydrogen can also be stored; however, this is beyond the topic of discussion.

19.2.2.1 Lithium-ion batteries

Renewable energy storage has been ubiquitously utilized in the modernized world, such as portable electronics, electrical vehicles, and grid-scale energy storage. In the existing technologies, as far as energy storage is concerned, Li-ion batteries have gained massive momentum due to their potential to hold the greatest energy density [12]. Still, novel materials and structures with high specific capacity and high rate performance are needed. They offer portable electricity, powering electronic gadgets such as mobile phones, laptops, and tablets. Recently, Li—sulfur and Li—air batteries have emerged to supersede Li-ion batteries. Nanostructured flexible electrodes in optimized Li-ion batteries can further improve their performance.

19.2.2.2 Supercapacitors

A supercapacitor is an electrical energy storage device with the ability to store capacitance much higher than normal capacitors, but with lower voltage limits. Its possesses two electrodes, separated by a membrane, that have the ability to store electrical energy using reversible adsorption of ions from an electrolyte onto two porous electrodes to form an electric double layer at an electrode/electrolyte interface. The superior materials generally employed for these capacitors are carbon, graphite, conducting polymers, and metal oxides. They are potentially important components of a future energy storage system as they combine the high-energy/lower-pow characteristics of a battery with the high-power/low-energy characteristics of a supercapacitor [13,14]. With the invention of supercapacitors that offer good electrical conductivity, fabrication scalability, cycling stability, and mechanical robustness, the specific capacitance energy storage process has become simpler and easier. Applications of energy storage devices are portrayed in Fig. 19.6.

19.2.3 Materials for energy saving

Energy-saving materials are generally direct bandgap semiconducting materials used to create light through novel electro-optical devices like light-emitting diodes (LEDs), organic light-emitting diodes (OLEDs), and flat-panel display devices, which substantially reduce the consumption of electricity.

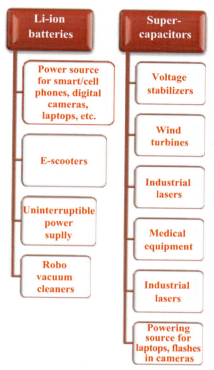

FIGURE 19.6 Applications of lithium-ion batteries and supercapacitors.

19.2.3.1 Optoelectronic materials

Optoelectronic devices are those devices in which either of the following occurs: incident light creates electrons and holes in materials, thereby generating current, as in solar cells, or injects electron–holes into a device and recombining; as a result, photons are generated as in LEDs/OLEDs, lasers, etc. In this chapter we will deal with the latter case. In such devices when light of energy E is less than the energy bandgap (E_g) of the material, no optical absorption takes place and the semiconductor is said to be transparent [15]. On the other hand, when light of energy E is greater than or equal to the energy bandgap (E_g) of the material, optical absorption takes place, covalent bonds break, electrons are created at the bottom of the conduction band, holes are created at the top of valence band, and excess energy is lost by the process of thermalization where the heat is transferred to the lattice; such semiconductors are opaque [16]. Due to excess energy, the electron may jump into higher energy levels above the conduction band leaving behind a hole. However, the electron returns to the bottom of the conduction band by liberating heat energy. The electron may also jump to the bottom of the conduction band from energy levels deep in the valence band, leaving behind a hole. Here, the hole loses energy and moves to the top of the valence band by the process of thermalization. LEDs/OLEDs/lasers work on the same principle, popularly known as electroluminescence, where the electrical energy is converted into light energy;

these devices employ direct bandgap semiconductors because (1) they do not require a photon to generate electro—hole pairs, so the electro-optical devices are very efficient, and (2) the top of the valence band and bottom of the conduction band occur at the same values of momentum. In these material-based devices, electrons and holes recombine at the interface of the junction and create photons through recombination [17,18]. On the other hand, in indirect bandgap semiconductors, a phonon is emitted. The performance of these devices is evaluated by quantum efficiency and power efficiency. Different metrics for quantum efficiency include internal and external quantum efficiencies.

Internal quantum efficiency:

$$\text{Internal quantum efficiency } (\eta_{\text{int}}) = \frac{\text{No. of photons generated internally}}{\text{No. of carriers}}$$

$$= \frac{\text{Radiative recombination rate}}{\text{Total recombination rate}} = \frac{R_r}{R_r + R_{nr}}$$

where R_r = radiative recombination and R_{nr} is nonradiative recombination. For better efficiency, $R_r >> R_{nr}$.

External quantum efficiency:

External quantum efficiency depends on internal and optical quantum efficiencies, given by:

$$\eta_{\text{ext}} = \eta_{\text{int}} \times \eta_{\text{opt}}$$

It further depends on device geometry, reflectance, transmission, and absorbance of different layers of the device [19].

Power efficiency:

$$\text{Power efficiency } (\eta_P) = \frac{\text{Optical power output}}{\text{Input power}}$$

$$= \frac{\text{No. of photons} \times \text{Energy}}{\text{Input power}} = \frac{\rho_{Ph} \times E}{P_{in}} = \frac{\rho_{Ph} \times h\nu}{VI}$$

19.2.3.1.1 LEDs/OLEDs

LEDs/OLEDs are luminescent devices where optical emission is due to electronic excitation (creation of electrons and holes). LEDs/OLEDs employ organic/inorganic semiconductor devices with bandgaps in the range of 1.7—3 eV. Visible LEDs at the top of this range are called infrared LEDs, and those at the bottom are called ultraviolet LEDs. In OLEDs, the organic emissive layer is sandwiched between the electrodes. Here, the valence band is referred to as the highest occupied molecular orbit and the conduction band corresponds to the lowest unoccupied molecular orbit [20].

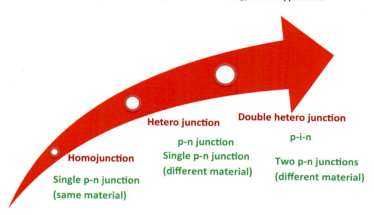

FIGURE 19.7 Different generations of p-n junction devices used in photovoltaics and light-emitting diodes.

LEDs For simple LEDs, a heavily doped forward bias homojunction (both p- and n-types are of the same material) is used. In such biasing conditions, the potential barrier (V_0) decreases and minority charge carriers can be easily injected on both sides. This is an example of injection electroluminescence because electrons and holes are injected for light generation. It is a spontaneous process and hence inefficient. The efficiency of LEDs can be maximized by generating the number of electron—hole pairs in the depletion region; this can be achieved by heterojunction (a junction between two different materials) and double heterojunction (a p-type semiconductor intrinsic n-type semiconductor) structures. By using double heterojunction structures, we can create potential wells and trap the electrons and holes in the intrinsic region, the recombination of which produces light [21]. Different generations of p-n junction devices used in photovoltaics and LEDs are depicted in Fig. 19.7.

OLEDs Light sources made from organic materials offer immense potential for a wide range of applications such as large space lighting, back lighting, and advertising displays only because they can be spread onto a large substrate by blending them in a polymer matrix. OLEDs have a lightweight source, which can potentially be made of any size, and a range of substrates [22,23].

In spite of the exceptional properties of organic materials for usage in display technologies, their potential is not well realized due to early degradation, laborious manufacturing processes, and encapsulation requirements. A driving voltage below 5 V is needed to match well with the distinctive incorporated electronics utilized for passive matrix displays. Voltage drop can be partially attributed to the high resistivity and poor conductivity encountered by organic devices. The merits, perils, and challenges of OLEDs [24,25] are demonstrated in Fig. 19.8. Application lists of LEDs and OLEDs are almost the same, and include ambient lighting, aviation lighting, automobile headlamps, general lighting, camera flashes, LED strip lighting, horticulture growth lights, medical devices, displays, high-definition televisions, solid-state lighting, computer monitors, backlighting, driver information systems, smartcards, interior car lighting, airport runway lighting, and advertising panels [26,27]. The diverse applications of OLEDs are detailed in Fig. 19.9.

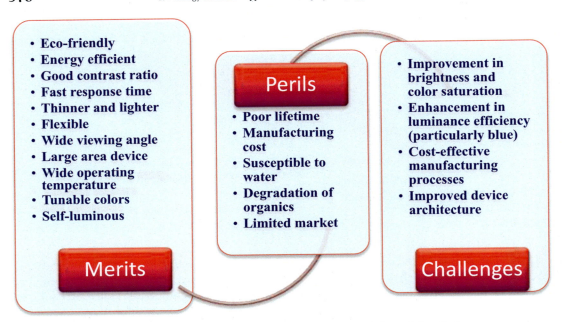

FIGURE 19.8 Merits, perils, and challenges of organic light-emitting diodes.

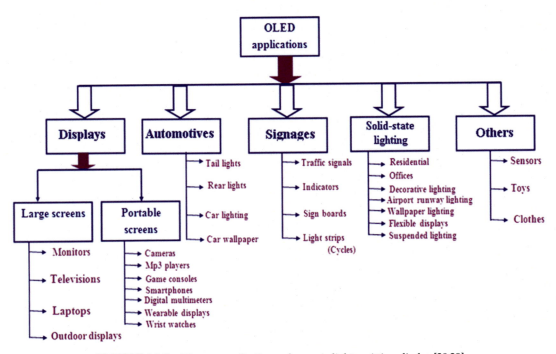

FIGURE 19.9 Diverse applications of organic light-emitting diodes [28,29].

19.3 Research area connections

Energy applications employing a wide-ranging energy landscape have been developed because of the efforts from researchers in understanding basic fundamental research, starting from a theoretical understanding of novel energy materials, their synthesis and characterization with improvised device anatomy, and leading to fabrication technologies pertaining to energy applications, to commercialize end products at more affordable prices. The research areas involving these energy materials include thermoelectric materials, fuel cells, rechargeable battery materials, photovoltaic materials, semiconducting materials, and a wide range of materials for impending research into existing and future energy technologies [30,31]. This is an incredibly dynamic area of research with real significance to global challenges, particularly the need to generate energy more sustainably and cost effectively to meet carbon reduction targets and aid economic growth. Today's brilliant technologies in energy generation have evoked new-fangled materials that can power independent devices by harnessing energy from the environment through various renewable energy resources such as solar energy, wind energy, geothermal energy, etc. Noteworthy improvements have also been made to a variety of energy storage technologies in terms of lifetime, compactness, compatibility, and energy density through efficient structures and nanosized memory storage devices. Drastic approaches to reduce power consumption have been made by semiconducting devices, which are now playing a major role in many areas of information and communication technology. For instance, solid-state lighting through LEDs/OLEDs curtails the consumption of energy by many orders of magnitude from compact fluorescent technology. The emergence of these energy materials has revolutionized versatile research in various fields of energy-related applications with greater impact on the future scaling up of devices based on these materials.

19.4 Opportunities and daunting challenges

Global monetary and ecological aspects are creating havoc for the resourceful generation, storing, and saving of energy with no major effect on greenhouse gases and other pollutants in the environment. Harvesting solar energy, the most abundant resource, in a more effective manner is a great challenge. The first solar cell system was built in the 1950s by the Department of Energy in the United States, whose efficiency was only 4%; later, with the tireless effort of many researchers, efficiency rose to around 23%, and now research will enhance efficiency even further. After more than 70 years since the harnessing of solar energy, efficiency has not appreciably increased because (1) in solar radiation, frequency is different for different colors and hence the energy of photons is also different. There exist less energetic as well as highly energetic photons, depending on the frequency. Only a part of energy ($E = E_g$) is sufficient in ejecting an electron—hole pair, while the remaining energy is wasted; on the other hand, less energetic photons are not capable of ejecting electron—hole pairs. Hence, many solar cells are interconnected (known as a solar panel) to achieve maximum conversion and hence efficiency. (2) Solar intensity is reduced by the atmosphere due to various optical phenomena such as scattering, reflection, refraction, etc. (3) Because the efficiency of a solar cell is reduced, a number of photovoltaic arrays are required, which would lead to higher cost. These challenges remain in spite of tremendous efforts. However, scientists have developed a few approaches that have enhanced efficiency to certain extent, e.g., (1) novel materials like perovskites, graphene, and quantum dots for fabricating solar cells, (2) antireflection coatings for solar cells so that the maximum amount of light can be absorbed by them (thicknesses of about 1700 A^0 ($\approx \frac{\lambda}{4}$) or 8500 A^0 ($\approx \frac{5\lambda}{4}$) to prevent the scattering of incident radiation), (3) thin surface layers of the cell (p-type) to enable maximum light to penetrate through the junction because single-crystal Si is not transparent to visible light, (4) focusing of considerable solar light onto the cell using mirrors, (5) flow of electrons from one side of the cell to the other side through an external circuit (to minimize losses, the cell is covered with a metallic contact grid that shortens the distance that electrons should travel while covering a small part of the cell surface, although some photons are blocked by the grid, which cannot be too small or else its own resistance would be too high), (6) heterojunction structures, (7) upconversion and downconversion phosphor, (8) tandem structures, (9) selection of appropriate bandgap materials, (10) dye-sensitized solar cells [32], (11) tandem structures, to name a few. The invention of organic solar cells, infrared plastic solar cells, and traditional silicon-based solar cells is challenging the photovoltaic landscape. Energy transmission and storage systems should be revised to augment energy density storage with respect to existing devices. There are hazards associated with H_2 fuel cells, and safety is a great challenge.

As far as energy-saving devices are concerned, LEDs/OLEDs play a vital role. With the invention of LEDs and OLEDs for solid-state lighting, traditional lamps like filament bulbs and compact fluorescent lamps are now out of date. The essential requirements of LEDs include (1) significant light must be emitted at the output, (2) light must be emitted from one side of the device, (3) almost all light should emerge from the device, and (4) encapsulation of the device is essential to protect it from moisture and contamination. Furthermore, the efficiency of LEDs/OLEDs can be enhanced by maximizing the radiative transitions and suppressing nonradiative transitions. The use of organic-printed electronic applications is now becoming popular. However, we need to explore unconventional energy sources like differences in soil pH, the movement of trees, and other possibilities of energy-harvesting mechanisms very closely.

19.5 Conclusions

An overview of current progress, energy-related applications, and propelling opportunities of energy-related demonstration, relevant technologies, and device architectures reveals the sense of emergency in developing more and more effective and efficient energy materials for future applications. As the scope of energy materials is comprehensive, efforts have been made to focus mainly on thermoelectrics, photovoltaics, Li-ion batteries, supercapacitors, fuel cells, and optoelectronic devices such as LEDs/OLEDs. In most of these areas, intensified progress in existing technologies is not enough to deal with the imperative concerns of robustness, efficiency, and cost. Innovative avenues in the field of materials research are urgently needed to propose and integrate these materials for energy applications for attaining desired objectives. Optimistically, for the future, other innovative inventions, which may replace existing materials, will always be welcomed.

19.6 Future outlook

In an age when the requirement for traditionally exploited natural resources is outpacing supply, conventional industrial practices are contributing to undesirable climatic change and developing regions are competing for a greater share of finite fuel stocks; the search for innovative ways to meet this need becomes more urgent than ever. Endorsing the idea of popularizing the use of energy materials will be the viable option. If improper handling and effective use of these materials are repeatedly ignored, attempts to balance the environment against our battle for a better lifestyle will not succeed.

References

[1] M. Song, S. Wang, H. Yu, L. Ynag, J. Wu, To reduce energy consumption and to maintain rapid economic growth: analysis of the condition in China based on expended IPAT model, Renew. Sustain. Energy Rev. 15 (5) (2011) 5129–5134.

[2] L.D. Taylor, The demand for electricity: a survey, Bell J. Econ. 6 (1) (1975) 75–110.

[3] W. Paw, S.D. Cummings, M.A. Mansour, D.K. Geiger, R. Eisenberg, Clean energy from solar fuels essay in chemistry for changing times, Coord. Chem. Rev. 171 (17) (1998) 125–135.

[4] J.P. Agarwal, Recent trends in high-energy materials, Prog. Energy Combust. Sci. 24 (1) (1998) 1–30.

[5] G.S. Nolas, J. Sharp, J. Goldsmid, Thermoelectrics: Basic Principles and New Materials Developments, Springer Science & Business Media, Technology & Engineering, 2013, pp. 1–293.

[6] C.B. Vining, An inconvenient truth about thermoelectric, Nat. Mater. 8 (2009) 83–85.

[7] C.G. Granqvist, Transparent conductors as solar energy materials: a panoramic review, Sol. Energy Mater. Sol. Cell. 91 (17) (2007) 1529–1598.

[8] C.G. Granqvist, Solar energy materials, Adv. Mater. 15 (21) (2003) 1789–1803.

[9] Y. Li, W. Shan, S. Cheng, M. Ko, Intematix Corp USA, Wavelength-converting Phosphors for Enhancing the Efficiency of a Photovoltaic Device. Patent: WO 2007/133344.

[10] W. Vielstich, A. Lamm, H. Gasteiger, Handbook of Fuel Cells. Fundamentals, Technology, Applications, United Kingdom: N. p.Web, 2003, pp. 1–2690.

[11] K. Kordesch, G. Simader, Fuel Cells and Their Applications, Germany: N. p.Web, 1996, ISBN 3-527-28579-2, pp. 1–389.

[12] J. Wan, A.F. Kaplan, J. Zheng, et al., Two dimensional silicon nanowalls for lithium ion batteries, J. Mater. Chem. 2 (17) (2014) 6051–6057.

[13] M. Jayalakshmi, K. Balasubramanian, Simple capacitors to super capacitors - an overview, Int. J. Electrochem. Sci. 3 (2008) 1196–1217.

[14] P. Simon, Y. Gogotsi, B. Dunn, Where do batteries end and super capacitors begin? Science 343 (6176) (2014) 1210–1211.

[15] J.R. Sanderocock, Light scattering from surface acoustic phonons in metals and semiconductors, Solid State Commun. 26 (8) (1978) 547–551.

[16] A. Pinczuk, E. Burstein, Fundamentals of inelastic light scattering in semiconductors and insulators, in: M. Cardona (Ed.), Light Scattering in Solids I. Topics in Applied Physics, vol. 8, Springer, Berlin, Heidelberg, 1983.

[17] G.W. Mudd, M.R. Molas, X. Chen, V. Zólyomi, K. Nogajewski, Z.R. Kudrynskyi, Z.D. Kovalyuk, G. Yusa, O. Makarovsky, L. Eaves, M. Potemski, V.I. Fal'ko, A. Patanè, The direct-to-indirect band gap crossover in two-dimensional van der Waals Indium Selenide crystals, Sci. Rep. 6 (2016) 1–10.

580 19. Energy materials: Applications and propelling opportunities

[18] R. Soniya, N. Thejokalyani, S.J. Dhoble, Ch. 2, in: D. Rivera (Ed.), Organic Light Emitting Diodes(OLED) Materials, Technology and Advantages, NOVA Science Publishers, Newyork, 2016, ISBN 978-1-63484001-9, pp. 41−92.

[19] N. ThejoKalyani, S.J. Dhoble, Novel materials for fabrication and encapsulation of OLEDs, Renew. Sustain. Energy Rev. 44 (2015) 319−347.

[20] N. Thejokalyani, S.J. Dhoble, Novel approaches for energy efficient solid state lighting by RGB organic light emitting diodes − a review, Renew. Sustain. Energy Rev. 32 (2014) 448−467.

[21] T. Georgiou, R. Jalil, B. Belle, et al., Vertical field-effect transistor based on graphene−WS2 heterostructures for flexible and transparent electronics, Nat. Nanotechnol. 8 (2013) 100−103.

[22] M.R. Krames, O.B. Shchekin, R. Mueller-Mach, G.O. Mueller, L. Zhou, G. Harbers, Status and future of high-power light-emitting diodes for solid-state lighting, J. Disp. Technol. 3 (2007) 160−175.

[23] E.F. Schubert, J.K. Kim, H. Luo, J.Q. Xi, Solid-state lighting—a benevolent technology, Rep. Prog. Phys. 69 (2006) 3069.

[24] www.flatpanelshd.com.

[25] www.cdtltd.co.uk.

[26] G.E. Jabbour, Screen printing for the fabrication of organic light emitting devices, IEEE 7 (2001) 5.

[27] D.W. Lin, J.K. Huang, C.Y. Lee, R.W. Chan, Y.P. Lan, C.C. Lin, K.Y. Lee, C.H. Lin, P.T. Lee, G.C. Chi, H.C. Kuo, Enhanced light output power and growth mechanism of GaN-based light emitting diodes grown on cone-shaped SiO_2 patterned template, J. Disp. Technol. 9 (2013) 285−291.

[28] D. Chitnis, N. Thejokalyani, S.J. Dhoble, Escalating opportunities in the field of lighting, Renew. Sustain. Energy Rev. 64 (2016) 727−748.

[29] J.A. Castellano (Ed.), Handbook of Display Technology, Gulf Professional Publishing, 1992, ISBN 0-12-163420-5, p. 9.

[30] N. Thejo Kalyani, S.J. Dhoble, Empowering the future with organic solar cell devices, Nanomater. Green Energy (2018) 325−350, https://doi.org/10.1016/B978-0-12-813731-4.00010-2.21, 9780128137314.

[31] I. Hadjipaschalis, A. Poullikkas, V. Efthimiou, Overview of current and future energy storage technologies for electric power applications, Renew. Sustain. Energy Rev. 13 (6−7) (2009) 1513−1522.

[32] M. Law, L.E. Greene, J.C. Johnson, R. Saykally, P. Yang, Nano wire dye-sensitized solar cells, Nat. Mater. 4 (6) (2005) 455−459.

CHAPTER

20

Sustainability, recycling, and lifetime issues of energy materials

N. Thejo Kalyani[1], S.J. Dhoble[2], B. Vengadaesvaran[3], Abdul Kariem Arof[4]

[1]Department of Applied Physics, Laxminarayan Institute of Technology, Nagpur, Maharashtra, India; [2]Department of Physics, RTM Nagpur University, Nagpur, Maharashtra, India; [3]Higher Institution Centre of Excellence (HICoE), UM Power Energy Dedicated Advanced Center (UMPEDAC), Level 4, Wisma R&D, University of Malaya, Jalan Pantai Baharu, Kuala Lumpur, Malaysia; [4]Centre for Ionics University of Malaya, Department of Physics, Faculty of Science, University of Malaya Kuala Lumpur, Malaysia

20.1 Introduction

In today's technology-driven world, energy is an elementary constituent in human life because humans are becoming more dependent on industrial, agricultural, health, domestic, and many other sectors, which need external sources of energy [1]. Hence, energy conversion and harvesting, storing, and saving are the three pillars that ensure the steady supply of available energy from a global perspective. Without discarding existing energy sources, amendments to eradicate or lessen their environmental impact are the need of the hour [2]. Also, new-fangled renewable sources need to be encouraged. Hence, research into energy materials has gained importance all over the world. It encompasses a wide area of study ranging from the basic sciences to engineering applications. To meet the pace of today's energy requirement, the materials chosen in this area should possess various attributes, as shown in Fig. 20.1.

However, the main concerns with these energy materials are issues of sustainability, recycling, and lifetime, which play a crucial role in dealing with the problem of a growing demand for energy (Fig. 20.2). This chapter mainly deals with these three issues in connection with energy-converting, -harvesting, -storing, and -saving materials, and the reasons and possible solutions to cope with these issues.

Energy Materials
https://doi.org/10.1016/B978-0-12-823710-6.00015-7

Copyright © 2021 Elsevier Ltd. All rights reserved.

FIGURE 20.1 Requirements of today's energy materials.

FIGURE 20.2 Energy materials challenges.

20.2 Sustainability, recycling, and lifetime issues

Energy and environment represent the two top-most challenges for modern society. With the current situation, things will get worse rather than better. Revolutionary transformative materials that enable advanced technologies need to be designed and developed. Hence, it is the need of the hour to accelerate deep insight into sustainability, recycling, and lifetime issues for a better tomorrow.

20.2.1 Sustainability

Sustainability is the continuity of benefits for today's generation so that they can be enjoyed by the generations of tomorrow. An example of today's benefits is energy. For energy to be enjoyed by future generations, it has to be sustained. To sustain energy, it must have the ability to be renewed and last for generations to come. Wind, water, solar, and geothermal are natural sources of sustainable energy. Wind power is obtained from electricity generation by conversion of kinetic energy into mechanical energy via wind turbines. Wind is free and abundant. Energy from wind decreases greenhouse gas (GHG) emissions and saves electricity costs. Energy from water in motion can produce water power, better known as hydropower. Likewise, the sun is abundant and free. Energy from the sun does not produce pollution or emit GHGs. Both hydropower and solar energy need little maintenance. Geothermal energy originates from Earth's internal energy sources. Other than electricity generation, it can also be used in pumps and for heating systems. Bioenergy is the energy created from biological masses such as wood, plants, food crops, algae, agricultural products, garbage, and so on. Biomass can be processed into biofuel (bioethanol and biodiesel) and converted to bioproducts (bioplastics, biocomposites, biofoams, biochemicals) as well as be transformed to electrical energy directly. To sustain energy, the extraction rate should be lower or equal to the natural replenishment rate. Sustainable development aims at (1) conservation of natural resources by maintaining the balance between ecology and development, (2) economical feasibility, and (3) safeguarding the health and lifestyle of the population, as shown in Fig. 20.3. It also aims at renewable resources, which have numerous benefits, so that the usage of nonrenewable sources can be curtailed. One can also use eco-friendly lightbulbs in homes and sustainable reprocessed timber and bricks in buildings. Although this seems like a small gesture, each action is supportive and contributes toward sustainability.

Sustainable development benefits people by crafting peace, prosperity, and global partnership to protecti the natural resources for future generations. Governments should also play their part through legal assistance in promoting sustainability by conserving natural resources and reducing developments that bring adverse effects on the environment. However,

FIGURE 20.3 Sustainability: salient features.

currently, we mostly rely on nonrenewable resources like oil, gas, and coal rather than resources based on solar, wind, and water. Overpopulation is a rising concern since most resources like fossil fuels are finite. Therefore sustainability is the best option. One aspect of sustainability is recycling, which helps protect the environment. Another aspect of sustainability is reducing energy consumption. One can install photovoltaic panels in households to reduce electricity usage or even take small actions such as turning off taps to save water when brushing teeth, watering plants by using rainwater, etc. In addition, one can also use natural fertilizers or compost instead of chemicals in gardening and agriculture. Leading a sustainable lifestyle cannot resolve all environment-related issues but it can assist in decreasing the consumption of resources.

20.2.2 Recycling

Recycling is a valuable alternative; it is the reuse of a raw material that has already been made into a product. Recycling reduces the need to extract even more raw material from natural resources. Aluminum cans, for example, can be recycled. Bauxite is the main ore for aluminum. Recycling aluminum cans will reduce the need to process aluminum from bauxite and in this way help to economize and conserve energy. The most common recycled articles include cardboard, food boxes, mail, glass and plastic bottles with caps, jars and jugs (glass and plastic), food and beverage cans, and paper to name a few. Recycling 1 kg of aluminum can saves more than 160 MJ, which is about the equivalent of 4655 L of gasoline (Fig. 20.4).

Recycling 1000 kg of paper (and thus helping to save our forests) can conserve energy from about 1464 L of gasoline. Recycled aluminum or paper can also be used to make new

FIGURE 20.4   Recycling: advantages.

products. Recycling can be seen to benefit society, help the economy, and protect the environment. It is to be realized that to effectively reduce waste, it should not be created in the first place. Therefore to minimize waste, the source of the waste has to be reduced. This can be achieved if some of the waste materials can be reused or recycled. Recycling is a very practical way to salvage natural resources and to care for the environment. Recycling also saves money. A new product involves numerous materials and energy, including raw materials extraction, product fabrication, and transportation to the acquisition place.

Production of photovoltaic panels does not create waste or emit chemical products and radiation during electricity generation. However, when solar panels are no longer usable, they can pose environmental issues related to improper waste disposal. Fortunately, photovoltaic panels have a long lifespan of 25 years [3]. Recycling solar panel technology is relatively insignificant and underdeveloped compared to photovoltaic-making technology itself. Strachala and coworkers [4] predicted around 100,000—545,000 tons of photovoltaic waste in Czech Republic by the end of 2025. The recycling process of first-generation silicon-based solar panels is more complex than second-generation thin-film-based photovoltaic panels since the former requires the dismantling of individual parts before being subjected to recycling. Furthermore, it is not profitable to recycle silicon modules, whereas thin-film-based panels contain cadmium and lead that are hazardous. Strachala et al. [4] estimated that about 13,567 tons of silicon, 425 tons of silver, 51,500 tons of aluminum, and 351,500 tons of glass can be conserved by recycling. It has been reported that in silicon photovoltaic modules, the most costly material is silver followed by aluminum, glass, and silicon and these metals are also highly impactful in terms of global warming potential (GWP) [5]. On the other hand, Yi et al. [6] reported that recycling of silicon-based solar modules could save around 207,000 tons per annum of silicon by 2050. Theoretically, it is said that about 30,000 tons of silicon can be reused by 2030 [7].

Three common approaches to recycling photovoltaic modules are chemical, thermal, and mechanical routes. Thermal treatment is more economical and takes less time than the chemical process but its drawbacks are release of gases when ethylene-vinyl acetate copolymer decomposes, and the possibility of damaging the silicon wafer [4]. Nonetheless, the mechanical route is efficient and beneficial to the environment as well as effectively reduces raw material extraction even though it is less cost effective than the chemical process [4]. Usually, a combination of both mechanical and chemical processes is used to recycle photovoltaic modules. Similar to solar panels, the lifetime of wind turbines is about 25 years and most of their constituents can also be recycled and reused [8]. Nevertheless, wind turbine recycling faces two major issues: i.e., operational cost and lack of market demand for the recycled materials [8]. Moreover, the recycling and recovery of these materials is not 100%.

To strive for a sustainable condition for recycling, the waste formed after recycling must not be more detrimental to the environment than the disposal of the objects after end of lifespan. Also, the recycled materials should not be less efficient than new raw materials. For instance, recycling of aluminum alloys involves a reforging process, which uses toxic and harmful chemical substances and a high amount of electricity. As a result, the waste and pollution arising from recycling could be damaging to Mother Nature compared to just disposing of the chemically stable aluminum alloys. With the depletion of forests around the world, reuse and recycling of forestry products like paper and cardboard is advantageous to the environment. Also, it is less energy consuming to make paper items using recycled materials. Conversely, the quality of products made from recycled materials is usually lower than that of those items when they were first used. Furthermore, the recycling process is deemed unsustainable when particular objects are transported aboard to be recycled for the sake of cheaper operational cost. Nevertheless, recycling can cut down the demand for raw materials, particularly in the case of recycled plastic-derived products that contain ingredients such as oil, which take a long time to redevelop. Hence, it is better to reuse products multiple times before recycling by following the 3Rs concept: reduce, reuse, and recycle.

Waste from electrical and electronic equipment, better known as e-waste, is harmful to human health and the environment when disposed through landfill and incineration. E-waste must be disposed of separately from other wastes like food waste, plastics, papers, aluminum cans, etc. Examples of e-waste include televisions, DVD players and stereos, rice cookers, refrigerators, baking ovens, toasters, fans, air-conditioners, hair dryers, washing machines, printers, computers, tablets, iPads, and cell phones along with their batteries. It should be noted that the lifetime of this electrical and electronic equipment is relatively shorter than photovoltaic modules and wind turbines as already stated. Hence, recycling of e-waste is essential even though recycling facilities may be lacking especially in developing countries. Malaysia is one popular destination for e-waste imported from abroad and is also a top plastic dumping location. The quantity of harmful e-waste from electrical and electronic production is estimated to be beyond 130,000 tons annually in Malaysia [9]. Furthermore, e-waste recycling systems are more sophisticated and expensive than conventional waste recycling facilities.

20.2.3 Lifetime issues

Whether it is an energy-conversion, -harvesting, -storing, or -saving device, its lifetime plays a crucial role in the economic prospective. For devices that are powered by renewable sources of energy like solar cells and solar electric vehicles, lifetime is not a big deal as they can be economically recharged without any effort. However, if the device lifetime is to be enhanced further, the recharging cycle can be minimized as well recycling can be thought as a rare case. As far as storage devices are concerned, the duration of charging and discharging cycles decides the lifetime of the device. Hence, batteries with low charging duration and high discharging duration are highly preferred. In the case of energy-saving materials like phosphor for fluorescent and solid-state devices like light-emitting diodes (LEDs) and organic light-emitting diodes (OLEDs), the lifetime of the light-emitting device depends on the characteristic features of phosphor. Hence, a phosphor with a considerable lifetime is highly appreciated [10].

20.3 Energy-harvesting and -conversion materials

The "energy-harvesting" concept is not new. The first hydroelectric plant that combined water and gravity to drive an electric-powered turbine was designed in 1882 and generated a relatively large-scale "green" and sustainable source of electricity. However, because this form of energy source relies heavily on natural resources such as wind and water, a complex and costly transmission network is needed. It significantly decreases the amount of output power as the transmission loss rises with size. Even so, only a few milliwatts of power are needed in many instances to turn on a wireless sensor node. Therefore a much smaller-scale solution is required. A solar cell that transforms energy from sun directly into electrical energy is truly a renewable energy source. A solar cell is one of the energy-harvesting materials that is a promising alternative to operate small electrical components. Furthermore, solar energy is considered an important alternative resource for the world's transition from fossil fuels to zero-carbon output in producing energy by 2050 [11]. To date, there are three generations of solar cells as deliberated next.

20.3.1 First-generation solar cells

Solar cells from the first generation are prepared from silicon. They are also recognized as conventional, traditional, and wafer-based solar cells. Currently, first-generation photovoltaic cells based on silicon can be grouped into four types, i.e., monocrystalline silicon, polycrystalline silicon, amorphous silicon, and hybrid silicon cells.

20.3.1.1 Monocrystalline silicon cells

Monocrystalline silicon is the most common and efficient silicon-based material employed in photovoltaic cell production. This element is often referred to as single-crystal silicon. It consists of silicon, where the entire solid's crystal lattice is continuous, unbroken to its edges, and free from grain limits. Monocrystalline silicon can be treated as an intrinsic semiconductor consisting only of excessively pure silicon. It can also be a p-type and n-type silicon by doping with other elements. In the production of solar cells, monocrystalline silicon is sliced from large single crystals and meticulously grown in a highly controlled environment. The cells are usually a few centimeters thick and arranged in a grid to form a panel. Monocrystalline silicon cells can yield higher efficiencies of up to 24.4% [12].

20.3.1.2 Polycrystalline silicon cells

Polycrystalline silicon, known as multicrystalline silicon, is a high-purity silicon used as the base material in solar cells. It is made by a chemical purification process from metallurgical-grade silicon. The polycrystalline structure results from molten silicon in which flat thin films have been drawn. Such a polycrystalline structure is called ribbon silicon. Solar cells based on polycrystalline silicon are simpler to produce since they do not require a tight atmosphere (controlled atmosphere/vacuum) compared to monocrystalline silicon solar cells, thus making them less expensive. The cells are carefully cooled and solidified from large blocks of molten silicon. The efficiency of polycrystalline silicon cells is slightly less (up to 22.9%) than that of monocrystalline silicon [13].

20.3.1.3 Amorphous silicon cells

Amorphous silicon is a noncrystalline type of silicon used as a semiconductor material for energy-harvesting cells. It is coated in thin film on a number of flexible substrates such as glass, metal, and plastic. The efficiency of solar cells constructed from this type of silicon is relatively low (about 10%) and is ideal for use in small electronic devices. Despite its low efficiency, amorphous silicon cells are one of the cleanest photovoltaic technologies since they do not use harmful heavy metals such as cadmium or lead [14].

V. Practical concerns and beyond

588   20. Sustainability, recycling, and lifetime issues of energy materials

20.3.1.4 Hybrid silicon cells

Solar cells formed by a combination of silicon with various materials are known as hybrid silicon cells. This approach was developed to achieve a more efficient, longer lifetime, and low-cost photovoltaic cell. Typically, the amorphous silicon layer is placed on a single crystal wafer. This results in good performance especially in indirect light conditions but tends to lose efficiency as the temperature rises [15].

20.3.2 Second-generation solar cells

Basically, second-generation photovoltaic cells are denoted as thin-film-based solar cells because they are prepared from layers of semiconducting materials with thicknesses of several micrometers. Such a solar cell is very inexpensive since thin-film technology uses much less active material in a cell and has a very low production cost. Basically, there are three types of base energy-harvesting materials used in commercial second-generation photovoltaic cells, namely amorphous silicon and two nonsilicon materials, i.e., cadmium telluride (CdTe) and copper indium gallium selenide (CIGS).

20.3.2.1 Amorphous silicon thin film

Amorphous silicon thin-film-based solar cells have a p-i-n (or n-i-p) form duality structure in which the p- and n-layers are primarily used to induce an internal electric field (i-layer). Such a solar cell is widely produced by the plasma-enhanced chemical vapor deposition method to place a very thin layer of amorphous silicon (1 μm) on a transparent conducting oxide substrate. Sputtering and hot wire chemical vapor deposition techniques are other means to prepare the silicon on the substrate.

20.3.2.2 Cadmium telluride thin film

CdTe in thin-film form is used in solar cells to absorb and transform visible light into electricity. It is an active material that can surpass the cost/watt of crystalline silicon. Even so, cadmium is highly toxic and not environmentally friendly, and the limited supply of tellurium (telluride in the form of an anion) makes this type of base material a limiting factor for industrial technology.

20.3.2.3 Copper indium gallium selenide thin film

Photovoltaic cells containing CIGS are made by placing a thin layer of copper, indium, gallium, and selenium on the front and back conducting substrate. CIGS has a direct bandgap that is ideal for use as material for energy harvesting in solar cells. Solar cells with CIGS thin film can produce a solar power efficiency up to 21.7% [16].

20.3.3 Third-generation solar cells

Solar cells from the third generation are capable of overcoming the optimum theoretical efficiency of a single-junction photovoltaic cell. The novel devices comprise dye-sensitized solar cell (DSSC), quantum dot-sensitized solar cell (QDSSC), perovskite solar cell (PSC), and organic solar cell (OSC).

V. Practical concerns and beyond

20.3.3.1 Dye-sensitized solar cells

DSSCs are economical cells that can fit in the thin-film-based group of photovoltaic cells. They are also branded as Grätzel cells, and were coinvented by Brian O'Regan and Michael Grätzel [17]. The DSSC's principle differs from first- and second-generation solar cells. A typical DSSC consists of a redox electrolyte placed between the dye/semiconductor oxide photoanode and the counter electrode. The dyes are adsorbed onto the semiconductor oxide layer, which is coated on the transparent conducting side of the substrate. When visible light is emitted into the DSSC, the dyes excite from the highest occupied molecular orbital (HOMO) to the lowest unoccupied molecular orbital (LUMO). Subsequently, the electrons are injected into the semiconductor's conduction band and flow out to the external circuit to do their work. The electrons then accumulate at the counter electrode and are ready to combine with triiodide ions from the electrolyte to become iodide ions. The iodide ions are then oxidized and the electrons are released and recombine with the ionized dyes. This process is repeated in the DSSC system.

Dyes

In DSSC, dyes are the most important energy-harvesting material responsible for capturing visible light. As dye plays a key role in DSSC, much effort is made on its synthesis and the exploration of novel dyes. The ideal sensitizer must be stable and capable of adhering to the semiconductor oxide surface. The dye should be able to absorb light at all wavelengths below 920 nm, thereby achieving a distance of light entering Earth's surface and optimizing efficiency. The ruthenium complex is a synthetic dye commonly used as a photosensitizer in DSSC applications. This type of dye is preferred because of its favorable photoelectrochemical properties, which offer high oxidized stabilization [18]. Some examples of ruthenium-based dyes include K19, K73, N3, N621, N719, N712, N945, Z907, Z910, Z955, HRS-1, and Black. Among all the dyes mentioned, DSSCs based on ruthenium(II)—polypyridyl complexes as materials for energy harvesting currently recorded light-to-electricity conversion efficiencies of more than 11% under 1 sun illumination [19]. The ability of these dye materials to absorb electromagnetic radiation from the visible to the near-infrared wavelengths results in high power conversion efficiency when applied in DSSCs. The attachment of carboxylate groups with the bipyridyl moiety lowers the energy of the ligand π^* orbital, thus more light is absorbed within the range of the visible wavelength. Through close analysis of HOMO and LUMO energy levels, the absorption spectra of ruthenium—polypyridyl systems can be modified to achieve the desired energy gap suitable for DSSCs. In DSSC, dyes are excited from HOMO to LUMO when illuminated by light. The electrons from metal-based HOMO are transferred to ligand-based LUMO and result in a metal-to-ligand charge transfer transition. Next, these electrons are injected into the semiconductor's conduction band through the dye's ligand anchored to it.

Semiconductor oxides

A photoelectrode in a DSSC consists of semiconductor oxide materials coated on a transparent conducting substrate. The semiconductor oxide should have a large surface area to allow more dyes to anchor on it. The electrons generated by the dyes are injected into and flow through the semiconductor oxide materials to the current collector and then to the external circuit. It can be seen that the semiconductor oxide acts as a medium for dye loading and electron transport. Therefore a semiconductor oxide with high dye loading and fast charge transport is of concern to DSSCs. The conduction band energy level position of the semiconductor oxide materials is also an important parameter in DSSCs as it determines the electron injection rate from photogenerated dyes. Some examples of semiconducting metal oxides are titanium dioxide (TiO_2), tin(IV) oxide (SnO_2), zinc oxide (ZnO), niobium pentoxide (NbO_5), and strontium titanate ($SrTiO_3$). Among all these types, TiO_2 and ZnO are widely used in DSSC applications due to their large bandgap as well as good morphology. TiO_2 is an n-type semiconductor having a 3.2 eV energy bandgap [20]. Its surface area can reach 200 m^2/g, which is particularly suitable for dyes to adsorb [21]. DSSCs fabricated with TiO_2-based photoanodes typically show open-circuit voltage (V_{oc}) between 500 and 700 mV [22]. ZnO semiconductor oxide is a suitable alternative to TiO_2 as an efficient photocatalyst. It has a 3.37 eV energy bandgap of as well as excellent electron mobility (200−300 cm^2/Vs at 80 K) [23]. In addition, higher surface-to-volume ratio and chemical stability of ZnO allow more dyes to anchor with it.

Electrolytes

Another important component in a DSSC is the electrolyte, which has the function of transferring electrons from counter electrodes to ionized dyes [24]. The ideal electrolyte is one with properties of high conductivity (and thus high dielectric constant), high boiling temperature, low viscosity, and low vapor pressure. Simple processing, durability, and environmental sustainability are additional factors for electrolytes for industrial manufacturing. In DSSCs, the electrolyte must have a redox mediator to enable oxidation and reduction processes for the charge transfer process to take place. Several types of redox couple are iodide/triiodide (I^-/I_3^-), copper(I)/copper(II) (Cu^I/Cu^{II}), cobalt(II)/cobalt(III) (Co^{II}/Co^{III}), ferrocene/ferrocenium (Fe/Fc^+), and nickel(III)/nickel(IV) (Ni^{III}/Ni^{IV}). An iodide/triiodide couple is commonly used in electrolytes for DSSC applications due to its advantages of strong solubility, propensity not absorb too much light, satisfactory redox potential, and ability to regenerate dye in a short time. Redox electrolyte affects the V_{oc} of DSSCs. This is because the V_{oc} of a cell is governed by the difference between the semiconductor's Fermi level and the electrolyte's redox potential. Since redox electrolyte consists of free cations resulting from salt dissociation, the Fermi level of a semiconductor oxide changes dependent on the quantity of cations absorbed onto the semiconductor's surface. The more cations, the more the Fermi level shifts toward the valence band. In addition, the redox electrolyte can influence the short-circuit current density (J_{sc}) of DSSCs. The J_{sc} of DSSCs is the total dynamic electrons induced from the dye's excited state to the semiconductor oxide conduction band and from the counter electrode to the ionized dyes. A large energy bandgap between the dye's excited state to the semiconductor oxide's conduction band influenced by free cation absorption on the photoanode surface induces high dynamic electrons. The high conductivity as well as high dielectric properties of redox electrolytes have further enhanced the J_{sc} of DSSC by increasing the transfer rate of electrons from the counter electrode to the ionized dyes. Studies on redox electrolytes are still being performed actively to find the best medium for DSSCs with excellent properties as well as great DSSC stability.

Counter electrode

The counter electrode is another key element of DSSCs. It accumulates electrons from the external circuit and catalyzes the redox electrolyte's reduction, which has a major effect on power conversion efficiency, durable stability, and device cost. Some types of counter electrodes used in DSSCs are platinum (Pt), carbon, metals and alloys, composites, polymers, and transition metal compounds of oxide, nitride, carbide, and sulfide [25]. So far, Pt is the most favored active material as counter electrode for DSSCs. There are several methods for coating Pt on a transparent conducting substrate. These include sputtering, electrochemical deposition, spray pyrolysis, thermal decomposition, thermal vapor deposition, and electrochemical reduction processes. Pt nanoparticles have outstanding properties, including high surface area, good transmittance, exceptional charge transfer mobility, high electrical conductivity, and slow corrosion rate. Pt counter electrodes with 3D nanostructures such as nanowires, nanotubes, nanoflowers, nanocups, and multipods of high surface areas improve their functions. A large catalytic surface area results in a more efficient electrochemical reduction due to high dynamic charge transfer at the electrolyte/counter electrolyte surface.

20.3.3.2 Quantum dot-sensitized solar cells

QDSSCs are third-generation solar cells that use quantum dots as energy harvesters due to their optoelectronic properties suitable for photovoltaic reactions. They have optical bandgaps that can be adjusted to various levels of energy that match the solar spectrum for efficient light absorption by altering the size of the dots. The highest power conversion efficiency recorded by a QDSSC was 16.6% as reported by the National Renewable Energy Laboratory [26]. The fabrication and mechanism of a QDSSC are very similar to that of a DSSC. What distinguishes them is the energy harvester material where the dye in a DSSC is replaced with a quantum point to form a QDSSC. The ideal quantum dot for solar cell applications should have a higher conduction band relative to semiconductor oxide for efficient electron injection. In addition, it should have large absorption of the solar spectrum as well as good stability to electrolytes, heat, and light. Cadmium chalcogenide (CdX, $X =$ Te, Se, or S) is commonly used as the base harvesting material in QDSSCs due to its high photoemission performance, easy fabrication, and low cost. There are several techniques to deposit quantum dots onto the semiconductor oxide/transparent conducting glass such as chemical bath deposition, spin casting, hot injection, and spray methods.

20.3.3.3 Perovskite solar cells

PSCs employ perovskite materials as a photosensitizer. Due to the high efficiency of PSCs based on single-junction architecture reported at 25.2% [26], these materials have attracted considerable attention. Perovskite compounds generally have an ABX_3 crystal structure as a chemical formula [27]. Here, A is an organic cation such as ions of methyl ammonium, whereas B represents metal cations consisting of elements like Pb and Sn. Meanwhile, X denotes halide anions such as Cl^-, Br^-, and I^- that are attached to both cations. It is reported that cations with radii of 1.60—2.50 Å are used to shape perovskite structures [28]. Basically, perovskite was coated on a semiconductor oxide placed on a transparent conducting glass to act as a photoanode. This photoanode then clamps the carrier transport layer (most commonly spiro-OMeTAD) together with the counter electrode (e.g., gold, platinum, or carbon). Methylammonium lead trihalide ($CH_3NH_3PbX_3$, $X = I$, Br, or Cl) is widely used as energy-harvesting material in PSCs due to its direct optical bandgap between 1.55 and 2.3 eV. The perovskite optical absorption spectrum depends on the halide element. In addition, this type of perovskite offers high dielectric constant, long diffusion length, fast charge separation reaction, long lifetime separation of charge carriers, and long transport distance of electrons and holes. An inexpensive and simple PSC fabrication process using methylammonium lead trihalide makes this type of perovskite a promising material as a light-harvesting active layer. Perovskite that has properties close to methylammonium lead trihalide is formamidinium lead trihalide ($H_2NCHNH_2PbX_3$; $X = I$, Br, or Cl), which has an energy bandgap of 1.48–2.2 eV, therefore it is able to show high efficiency when used as a light-harvesting material in solar cells [29].

20.3.3.4 Organic solar cells

OSCs are among the new technologies for photovoltaics, and use conductive organic polymers to absorb sunlight and transport charges to generate electricity through photovoltaic reaction. Organic polymers are used in OSCs because of their advantages in being able to change the energy gap by changing the length and polymer functional group. The high optical absorption coefficient of organic polymer makes this material selectable, thus more light is absorbed and therefore delivers high efficiency solar cells. It is reported that OSCs are able to produce efficiencies of up to 16.54% [30]. Since organic polymers are inexpensive, the production of this type of solar cell is cheap. Comparison with first- and second-generation solar cells reveals that OSCs are lightweight, flexible, customizable, low cost, and eco-friendly.

20.4 Energy-storing materials

Current renewable energy sources have created a viable means to store the generated energy system. Energy storage technologies are the key components to offset this issue by creating a sustainable energy system. Rechargeable lithium-ion batteries (LIBs) have been deployed to power small-size applications like electric vehicles as well as electric grids. This is because of their high power density, high energy efficiency, and environmental friendliness. According to the Chinese Energy Store Alliance, data showed that LIBs had the major proportion (exceeding 86%) of cumulative installed capacity in 2018. It is predicted that the demand for LIB will cost $99.98 billion and shipment volume will be 439.32 GWh by 2035. Such a massive demand will trigger a substantial consumption of material resources for their manufacture. The emergence of electric vehicles has driven the search for next-generation rechargeable batteries to meet the strong demand for high-energy density batteries. This assumes the current generation of electric vehicles will be required for recycling process of spent LIBs. Most of the spent LIBs are compounds containing heavy metal elements, namely cobalt (Co) and nickel (Ni), as well as toxic and flammable organic electrolytes. However, among these raw materials, Li and Co are in demand due to scarcity. It was reported in 2018 that the cost of Co was 5 and 10 times higher than Ni and Mn, respectively. Besides, the export price of the Li source, Li_2CO_3, was typically as high as $12,514 per ton in 2019. Considering the fact that Earth's crust can only supply 62 million tons of Li and 145 million tons of Co, the recovery process of metals from spent LIBs is an important route to sustain the environment for future generations. Inspired by this, researchers have been devoted to designing a low-cost and highly efficient pollution-free recycling process. However, the various kinds of rechargeable batteries that have been developed over the past decade may result in many challenges that will hinder the recycling process of spent LIBs because these batteries are made from many different elements. Hence, the sustainability of battery technology will be reviewed in this chapter from the perspective of life-cycle analysis (LCA), strategic resources, and economic analysis.

20.4.1 Life-cycle analysis

20.4.1.1 LCA of current LIBs

The life-cycle of a battery primarily consists of three stages, as shown in Fig. 20.5, namely (1) production use, (2) disposal, and (3) recycling. The battery production stage has presented growing concerns for researchers due to its significant impact on the environment. This stage is then followed by the use stage. To address the impact of battery production on the environment, a few issues must be pointed out, for example, (1) raw materials processing and battery manufacturing, (2) the production of battery components, and (3) the exploration of new alternative cathode and anode materials all contribute to environmental decline.

Previous studies have performed LCA on continuously produced batteries tested in a Ford Focus electric vehicle, in which the functional unit had 1 kWh of battery energy capacity. It was found that 45% of GHG emissions was caused by battery manufacturing. The enclosure and production of battery components such as cathodes, anodes, electrolytes, separators, and pouch and coin cell materials account for ∼40% of GHG emissions. Additionally, transportation, pack manufacturing, battery management system, thermal management, and the electrical system contribute less than 20% to GHG emissions [31]. Similar to GHG emissions, battery manufacturing, the production of battery components, and battery enclosures are the largest contributors to pollutant emissions, at 82%−92%

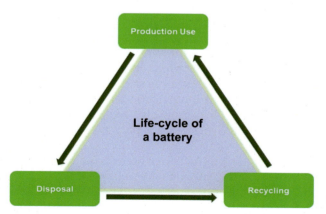

FIGURE 20.5 Life-cycle analysis of rechargeable batteries.

594　　　20. Sustainability, recycling, and lifetime issues of energy materials

depending on the pollutant. Pollutants such as carbon monoxide and particulate matter account for the largest proportion, which is contributed to by battery enclosures due to the production of steel used in the enclosures. Furthermore, the production of battery components causes high emissions of SO_2, which is similar to copper smelting [32,33]. Peters et al. [34] conducted a study to identify the impact of battery production on GWP, which is based on 1 kg of battery. Fifty to sixty percent of GWP is generated by component packaging, energy (electricity and heat), and battery management system (BMS). In addition, GWP is highly affected by the production of cathodes and cathode binders. Other items that also contribute to GWP are anode, anode binder, electrolyte, and separator. There was another study done by Yu et al. [35], which revealed that different cathode materials could also affect environmental sensitivity. For example, $Li_{0.98}Mn_{0.02}PO_4/C$ as electrodes in batteries have a higher index value than pristine $LiFePO_4/C$ batteries. This indicates that a higher index value signifies a larger impact on the environment because of the complexity of the doping process of $Li_{0.98}Mn_{0.02}PO_4$. Additionally, $LiFePO_4$ cathode materials have a higher index value compared to $LiMn_2O_4$ because Mn is less environmentally damaging than Fe. Additionally, $FeF_3(H_2O)_3/C$ batteries have only a small impact on the environment, due to their facile and environmentally benign synthesis route and because fewer raw materials are required. This shows that environmental sensitivity is highly affected by Li-containing compounds compared to Fe-based compounds. Thus it is important to explore more facile synthesis routes and alternative electrode materials to reduce the dependency of heavy metal elements. Subsequent research done by Wu et al. [36] presented a cradle-to-gate LCA of batteries with lithium metal, silicon nanowire, and graphite anodes. It was revealed that lithium metal as the anode is more environmentally benign than the other two anodes, based on a 1 kWh storage capacity. Batteries with silicon nanowires have higher specific energy, and also have greater environmental impact than graphite anodes, because of high upstream emission of silicon nanowire preparation.

20.4.1.2 LCA of next-generation batteries

From life cycle assesment (LCA) analysis, battery manufacturing give high impact to enviromental impact at high risk. Thus it is crucial to analyze the prospect of environmental impact on the production of next-generation batteries. However, there is a lack of data since these batteries have not been successfully commercialized yet, although a number of researchers have started doing LCA. Sodium-ion batteries (NIBs) are preferred over LIBs, because sodium is plentiful in nature, and the batteries are environmentally benign and economic. In 2016, Peters et al. [37] discussed a potential LCA for the production of NIBs with layered transition metal oxide as a cathode material and hard carbon as an anode material. It should be noted that the anode material is a key factor, which contributes to the depletion of fossil fuels, GWP, and eutrophication of water resources since hard carbon is prepared from the pyrolysis of sugar at extreme temperatures. Based on 1 kWh of storage capacity, a sodium-based battery has a lower environmental impact compared to lithium-based batteries. Additionally, this environmental impact can be further reduced by the production of hard carbon and reduction of nickel content in the transition metal elements of the cathode material.

Lithium—sulfur and lithium—air have emerged as competent candidates because of their theoretically high energy densities, and are potential targets for batteries in electric vehicles. Li-Air battery gives higher impact in all environmental problem such as greenhouse effect

V. Practical concerns and beyond

and etc. as compared to LiBs. This is believed to originate from power losses of the battery during operation, which results in the need for additional energy for their transportation out of the battery. A study has revealed that by recycling Li—air batteries, ~10%—30% of environmental impact due to production-related issues may be reduced. However, no recycling efforts have been considered so far. The future recycling process of Li—air batteries needs to be developed now because the study confirmed that lithium—air batteries may have 4—9 times less impact on the climate compared to current LIBs [38]. Yuan et al. [39] analyzed an LCA model of Li—S batteries, based on lab experimentation, industrial battery production, and literature data. Life-cycle energy consumption is largely dominated by battery phase use at 70%, with 22% accounting for battery production use and the balance being battery materials processing. Same trend is observed as battery use phase contributes for 70% of life cycle GWP and depletion of fossil fuels impacts while battery production only costs 28%. Compared with a lithium—nickel—cobalt manganese oxide (NCM)/graphite battery, the Li—S battery is produced with more environmentally benign technology, resulting in a reduction of 9%—90% of environmental impact in most categories. Yuan et al. also suggested that the environmental impact of an Li—S battery can be reduced by introducing a binder-free electrode and improving the decay rate of the battery.

In another study, Arvidsson et al. [40] analyzed the cradle-to-gate LCA of Li—S over cradle-to-grave due to a lack of data of the life and usage conditions of Li—S batteries. It was found that the energy used is largely contributed by the production of lithium bis(trifluoromethanesulfonyl)imide electrolyte salt and climate change is due to electricity use during the battery manufacturing process. Thus four strategies are proposed, namely (1) control of the use of electricity, (2) use of renewable sources to power electricity and heat, (3) improvement in the specific energy of the Li—S battery, and (4) deployment of carbon black as a cathode. These efforts could be potentially reduce energy and climate change impact by 54% and 93%, respectively. Other next-generation batteries, for example, potassium ion batteries, aluminum-ion batteries, and magnesium ion batteries, have yet to be investigated using LCA for environmental effects. Assuming their active material composition is similar to LIBs, reducing or replacing the heavy metal element with more environmentally friendly elements for cathode and anode material will play a significant role in minimizing their impact on the environment. In addition, other strategies that should be taken into consideration are increasing the energy efficiencies of material and battery production processes and reducing the dependency of electricity and electronic components. Realizing that the production of cathode binder may also have a high impact on the environment, water-based binders for cathode material should be utilized , as proposed by Notter et al. [41], to reduce GHG emissions.

20.4.2 Economic analysis of resources

The increasing demands for achieving high energy densities have encouraged the development of next-generation batteries in electric vehicles. However, a few factors should be reviewed to determine the feasibility of material resources. These are geographical concentration of manufacturing, element reservation, and product capacities in material processing. As Li is a must-have element in LIBs, Li resources have been extracted from mineral and brine deposits. However, Li resources are not plentiful in nature because the average level in Earth's crust is limited only to 20 ppm [42]. Since next-generation batteries are not ready for commercialization yet, the exploration of Li resources is still ongoing and there are 62 million tons of Li resources worldwide, according to the United States Geological Survey Report in 2019 [43].

NIBs are well known for their low cost, low environmental impact, and abundance in nature with even geographic distribution [44]. Because of these advantages, they are readily available in the form of minerals and brines such as Na_2SO_4, Na_2CO_3, and $NaCl$. It should be noted that synthetic soda ash (Na_2CO_3) can be produced from salt and limestone, as both products are never-ending resources. However, the process is costly and creates additional products. The waste products are usually discharged into waterways, which destroy coral reefs, seagrass, and seaweed. Potassium can be extracted in the form of potassium oxide (K_2O) or potash such as K_2NO_3, KSO_4, KCl, and potassic fertilizers. The amount of potash is estimated to range from 160 to 250 billion tons worldwide, which is somewhat lower than sodium resources. Canada has 60 billion tons of potash and the United States has 6 billion tons, enough to last for several thousand years [45]. Aluminum reserves are calculated to be 55–57 billion tons and manganese resources are thought to be 12 billion tons, which is larger than the annual consumption (16 million tons) [46].

Besides Li, Na, and K elements, some transition metal elements are also used in the production of electrode materials for batteries, for example, cobalt, nickel, and manganese. In 2019, Ni and Co reserves were identified at 25 million and 89 million tons, respectively, and can commonly be found in manganese crusts and nodules on the seabed. Over 120 million tons of Co is largely found on the ocean floor of the Pacific, Indian, and Atlantic oceans. In Earth's crust, cobalt deposits are mostly identified in Congo and Australia, which are the world's biggest cobalt-producing countries. Manganese reserves are not evenly distributed but more than 80% is located in South Africa [47]. USGS's Assessment of Undiscovered Copper Resources of the World in 2019 has revised identified Cu deposits at 2.1 billion tons and unexplored Cu deposits are expected to be 3.5 billion tons [48]. For the production of anode materials, there are more than 800 million tons of graphite deposits in the world. China leads the world in graphite production, with 70% global production capacity. Brazil, Madagascar, and Canada are also leading graphite exploration countries producing flake graphite; meanwhile, Sri Lanka is the major exporter of lump graphite. Manufacturers are primarily concerned about the increasing demand for flake graphite, and this could push global natural graphite production up by 7.6% in 2021 to 1.0255 million tons [49,50].

It is revealed that NCM111 and $LiCoO_2$ use more Li resources during production compared to other cathode materials. Modifying the cathode chemistries will exponentially increase the demand for cobalt compared to Li by 2030. Since cobalt reserves are limited, battery producers have made a transition from producing NCM111 into NCM811 and NCM622 cathode materials and the high-energy densities are still achievable. The only major influence on demand for Li is mainly due to the battery design factor. Thus it is necessary to explore new alternatives or solutions for Co- and Li-containing materials for batteries. For example, decreasing the nickel content in NCM-based cathode materials (NCM111 to NCM622) can save up to 7% of the costs.

20.5 Energy-saving materials

Energy-saving materials play a crucial role in the field of lighting as this field contributes toward 33% of energy consumption. Artificial lighting technology has a narration of thousands of years, and continues to advance even today. The evolution of artificial lighting started with fire, then kerosene lamps, halogen lamps, traditional incandescent bulbs, followed by fluorescent lamps and now solid-state lighting by LEDs [51]. These solid-state devices revolutionized the field of lighting as they are energy efficient and eco-friendly. They have already opened a doorway to transform the way we light the world. The growing significance of energy-saving and eco-friendly concepts has also revolutionized LEDs and OLEDs as highly proficient and feasible substitutes for traditional lighting technologies. However, there are certain concerns, particularly in OLEDs, like degradation of organic material, low quantum yields, and lifetime. If and when these barriers are overcome, they will have enduring ecological and cost-effective benefits. The prior state-of-the-art reveals that solid-state lighting technology can change the way we see the world.

20.6 Possible solutions

The correlation between materials research and energy technologies speaks volumes as it is now a comprehensive quest worldwide. However, sustainability, recyclability, and lifetime issues are at the forefront of global materials research. Because they are operated by clean and renewable energy source, solar cells are expanding their limits to infinity. However, the reality of their disposal is a big concern. If there is no effective and efficient recycling process, the toxic materials employed in framing these solar cells will be a major concern. Hence, regular landfills are unsuitable for their disposal because they may be breached and toxic materials could percolate into the soil, polluting the water table. Instead of disposing of solar panels, they could be recycled; however, the cost of the material recovered after the recycling mechanism is much less than the costs incurred for recycling. One way to solve this issue is by dismantling the solar cell/module and segregating its components, e.g., glass, which is the top-most layer of a solar cell, can be reused or recycled separately, metal components can be used for remolding cell frames, and the remaining miscellaneous materials can be thermally processed to shape them into new materials. The search for innovative mechanisms for recycling these solar cells is under way (Fig. 20.6).

As far as storage devices are concerned, the production of LIBs is surging every year with the increasing demand for electric vehicles. This development trend will eventually lead to an excess of spent Li batteries and Li scarcity. Many quantum leaps in battery technologies have been achieved to solve these problems. Thus some of these issues, with their tentative solutions, are reviewed here. In the recycling process, a secure and systematic route should be considered when dismantling spent LIBs to prevent any potential hazards in the future. In addition, LIB packs should be redesigned, making them more serviceable and easier to dismantle for the recycling process. A facile recycling process for discarded LIBs should be extremely effective to recover different components due to the complex structure of LIBs. From an economic perspective, low cost, high added value, and high purity of the product are the criteria needed to increase the economic efficiencies.

FIGURE 20.6 Possible solutions-before disposal.

Note that different types of batteries require different types of recycling technologies, thus proper recycling technologies should be proposed to achieve the maximum value of recycled products. To decrease environmental impacts, manufacturing processes must be highly energy efficient. As the production of cathodes and cathode binders contributes to increased GWP emissions, environmentally friendly materials are preferred. Reducing or replacing the heavy metal element with more environmentally benign elements for cathode and anode materials will play a significant role in minimizing their impact on the environment.

In the case of energy-saving materials, fluorescent lamps, which use mercury, are very toxic and disposal is a big concern. This problem of disposal can be prolonged with the use of solid-state light-emitting devices because their life-time is greater than traditional light-emitting devices. However, after use, when the necessity for disposal arises, electronic components and other materials that can be reused or remanufactured are separated and can be recycled because they are nonhazardous.

20.7 Conclusions

Despite these technological accomplishments, a large part of the world lives without adequate energy. Moreover, contemporary dependence on fossil fuels can cause significant damage to global resources, leading to GHG emissions that lead to climate change. Hence, renewable energy sources that are clean and eco-friendly come into play. Substantial improvements have been made in the field of solar cells as far as their conversion efficiency is concerned, just by selecting novel designs and materials for their fabrication. The converted energy will be more useful if the withdrawn energy can be effectively stored by suitable means and saved by an appropriate fixture or device. Regarding the LCA of batteries, a few issues are of major concern: (1) raw materials processing and battery manufacturing, (2) the production of battery components, and (3) the exploration of new alternative cathode and anode materials all contribute to worsening environmental decline. Next-generation batteries are more environmentally benign and this would help to reduce the environmental burden. The development of future rechargeable batteries has been accompanied by rapid growth in the global usage of electric vehicles. The limited resources of Li and Co have modified cathode chemistries and battery design to ensure the longevity of remaining resources for the future. In this chapter we proposed some potential solutions for improving the energy conversion factor, battery sustainability, and energy-saving technologies. However, subsequent research, development, and exploration are required to boost the efficiency, economy, environmental effects, and resources of energy materials development and recycling.

References

[1] N. Thejo Kalyani, S.J. Dhoble, Organic light emitting diodes: novel energy saving lighting technology-a review, Renew. Sustain. Energy Rev. 16 (2012) 2696–2723.

[2] D. Chitnis, N. Thejo Kalyani, S.J. Dhoble, Lighting up lives with eco-friendly and energy efficient OLEDs for solid state lighting, Bio Nano Front., Mater. Sci. 2 (2012) 346–348.

[3] J.-K. Choi, V. Fthenakis, Crystalline silicon photovoltaic recycling planning: macro and micro perspectives, J. Clean. Prod. 66 (2014) 443–449.

[4] D. Strachala, J. Hylský, J. Vaněk, G. Fafilek, K. Jandová, Methods for recycling photovoltaic modules and their impact on environment and raw material extraction, Acta Montan. Slovaca 22 (2017) 257–269.

[5] T. Maani, I. Celik, M.J. Heben, R.J. Ellingson, D. Apul, Environmental impacts of recycling crystalline silicon (c-SI) and cadmium telluride (CDTE) solar panels, Sci. Total Environ. 735 (2020) 138827.

[6] Y.K. Yi, H.S. Kim, T. Tran, S.K. Hong, M.J. Kim, Recovering valuable metals from recycled photovoltaic modules, J. Air Waste Manag. Assoc. 64 (2014) 797–807.

[7] M.M. Lunardi, J.P. Alvarez-Gaitan, J.I. Bilbao, R. Corkish, Chapter 2: a review of recycling processes for photovoltaic modules, in: B. Zaidi (Ed.), Book: Solar Panels and Photovoltaic Materials, 2018, pp. 9–27.

[8] J.P. Jensen, K. Skelton, Wind turbine blade recycling: experiences, challenges and possibilities in a circular economy, Renew. Sustain. Energy Rev. 97 (2018) 165–176.

[9] H. Ismail, M.M. Hanafiah, Discovering opportunities to meet the challenges of an effective waste electrical and electronic equipment recycling system in Malaysia, J. Clean. Prod. 238 (2019) 117927 (16 pp.).

[10] N. Thejokalyani, S.J. Dhoble, Novel approaches for energy efficient solid state lighting by RGB organic light emitting diodes – a review, Renew. Sustain. Energy Rev. 32 (2014) 448–467.

[11] R. Khan, Y.I. Go, Assessment of Malaysia's large-scale solar projects: power system analysis for solar PV grid integration, Glob. Challenges 4 (2020) 1900060 (17 pp.).

[12] J. Zhao, A. Wang, M.A. Green, F. Ferrazza, 19.8% efficient "honeycomb" textured multicrystalline and 24.4% monocrystalline silicon solar cells, Appl. Phys. Lett. 73 (14) (1998) 1991–1993.

[13] T. Kato, J.L. Wu, Y. Hirai, H. Sugimoto, V. Bermudez, Record efficiency for thin-film polycrystalline solar cells up to 22.9% achieved by Cs-treated Cu (In, Ga)(Se, S) 2, IEEE J. Photovoltaics 9 (1) (2018) 325–330.

[14] D. Derkacs, S.H. Lim, P. Matheu, W. Mar, E.T. Yu, Improved performance of amorphous silicon solar cells via scattering from surface plasmonpolaritons in nearby metallic nanoparticles, Appl. Phys. Lett. 89 (9) (2006) 093103.

[15] S. Jeong, E.C. Garnett, S. Wang, Z. Yu, S. Fan, M.L. Brongersma, M.D. McGehee, Y. Cui, Hybrid silicon nanocone—polymer solar cells, Nano Lett. 12 (6) (2012) 2971–2976.

[16] P. Jackson, D. Hariskos, R. Wuerz, O. Kiowski, A. Bauer, T.M. Friedlmeier, M. Powalla, Properties of Cu (In, Ga) Se$_2$ solar cells with new record efficiencies up to 21.7%, Phys. Status Solidi Rapid Res. Lett. 9 (1) (2015) 28–31.

[17] B. O'regan, M. Grätzel, A low-cost, high-efficiency solar cell based on dye-sensitized colloidal TiO$_2$ films, Nature 353 (6346) (1991) 737–740.

[18] Y. Qin, Q. Peng, Ruthenium sensitizers and their applications in dye-sensitized solar cells, Int. J. Photoenergy 2012 (2012), 291579 (21 pp.).

V. Practical concerns and beyond

600　　20. Sustainability, recycling, and lifetime issues of energy materials

[19] M.K. Nazeeruddin, P. Pechy, T. Renouard, S.M. Zakeeruddin, R. Humphry-Baker, P. Comte, P. Liska, L. Cevey, E. Costa, V. Shklover, L. Spiccia, Engineering of efficient panchromatic sensitizers for nanocrystalline TiO_2-based solar cells, J. Am. Chem. Soc. 123 (8) (2001) 1613−1624.

[20] N. Kamarulzaman, M.F. Kasim, R. Rusdi, Band gap narrowing and widening of ZnO nanostructures and doped materials, Nanoscale Res. Lett. 10 (1) (2015) 346 (12 pp.).

[21] Y. Zhou, L. Zhang, S. Tao, Porous TiO_2 with large surface area is an efficient catalyst carrier for the recovery of wastewater containing an ultrahigh concentration of dye, RSC Adv. 8 (2018) 3433−3442.

[22] F.I. Chowdhury, M.H. Buraidah, A.K. Arof, B.E. Mellander, I.M. Noor, Impact of tetrabutylammonium, iodide and triiodide ions conductivity in polyacrylonitrile based electrolyte on DSSC performance, Sol. Energy 196 (2020) 379−388.

[23] A.K. Chandiran, M. Abdi-Jalebi, M.K. Nazeeruddin, M. Grätzel, Analysis of electron transfer properties of ZnO and TiO_2photoanodes for dye-sensitized solar cells, ACS Nano 8 (3) (2014) 2261−2268.

[24] A.K. Arof, I.M. Noor, M.H. Buraidah, T.M.W.J. Bandara, M.A. Careem, I. Albinsson, B.E. Mellander, Polyacrylonitrile gel polymer electrolyte based dye sensitized solar cells for a prototype solar panel, Electrochim. Acta 251 (2017) 223−234.

[25] J. Wu, Z. Lan, J. Lin, M. Huang, Y. Huang, L. Fan, G. Luo, Y. Lin, Y. Xie, Y. Wei, Counter electrodes in dye-sensitized solar cells, Chem. Soc. Rev. 46 (19) (2017) 5975−6023.

[26] https://www.nrel.gov/pv/assets/pdfs/best-research-cell-efficiencies.20190802.pdf.

[27] L. Shamsuddin, I.M. Noor, I. Albinsson, B.E. Mellander, A.K. Arof, Perovskite solar cells using polymer electrolytes, Mol. Cryst. Liq. Cryst. 655 (1) (2017) 181−194.

[28] N.G. Park, Perovskite solar cells: an emerging photovoltaic technology, Mater. Today 18 (2) (2015) 65−72.

[29] G.E. Eperon, S.D. Stranks, C. Menelaou, M.B. Johnston, L.M. Herz, H.J. Snaith, Formamidinium lead trihalide: a broadly tunable perovskite for efficient planar heterojunction solar cells, Energy Environ. Sci. 7 (3) (2014) 982−988.

[30] S. Liu, J. Yuan, W. Deng, M. Luo, Y. Xie, Q. Liang, Y. Zou, Z. He, H. Wu, Y. Cao, High-efficiency organic solar cells with low non-radiative recombination loss and low energetic disorder, Nat. Photon. 14 (5) (2020) 300−305.

[31] H.C. Kim, T.J. Wallington, R. Arsenault, C. Bae, S. Ahn, J. Lee, Cradle-to-gate emissions from a commercial electric vehicle Li-ion battery: a comparative analysis, Environ. Sci. Technol. 50 (14) (2016) 7715−7722.

[32] B.P. Weidema, C. Bauer, R. Hischier, C. Mutel, T. Nemecek, J. Reinhard, C. Vadenbo, G. Wernet, Overview and Methodology: Data Quality Guideline for the Ecoinvent Database Version 3, 2013.

[33] V. Fthenakis, W. Wang, H.C. Kim, Life cycle inventory analysis of the production of metals used in photovoltaics, Renew. Sustain. Energy Rev. 13 (3) (2009) 493−517.

[34] J.F. Peters, M. Weil, Providing a common base for life cycle assessments of Li-Ion batteries, J. Clean. Prod. 171 (2018) 704−713.

[35] Y. Gong, Y. Yu, K. Huang, J. Hu, C. Li, Evaluation of lithium-ion batteries through the simultaneous consideration of environmental, economic and electrochemical performance indicators, J. Clean. Prod. 170 (2018) 915−923.

[36] Z. Wu, D. Kong, Comparative life cycle assessment of lithium-ion batteries with lithium metal, silicon nanowire, and graphite anodes, Clean Technol. Environ. Policy 20 (6) (2018) 1233−1244.

[37] J. Peters, D. Buchholz, S. Passerini, M. Weil, Life cycle assessment of sodium-ion batteries, Energy Environ. Sci. 9 (5) (2016) 1744−1751.

[38] M. Zackrisson, K. Fransson, J. Hildenbrand, G. Lampic, C. O'Dwyer, Life cycle assessment of lithium-air battery cells, J. Clean. Prod. 135 (2016) 299−311.

[39] Y. Deng, J. Li, T. Li, X. Gao, C. Yuan, Life cycle assessment of lithium sulfur battery for electric vehicles, J. Power Sources 34 (3) (2017) 284−295.

[40] R. Arvidsson, M. Janssen, M. Svanström, P. Johansson, B.A. Sandén, Energy use and climate change improvements of Li/S batteries based on life cycle assessment, J. Power Sources 383 (2018) 87−92.

[41] D.A. Notter, M. Gauch, R. Widmer, P. Wager, A. Stamp, R. Zah, H.-J. Althaus, Contribution of Li-Ion Batteries to the Environmental Impact of Electric Vehicles, ACS Publications, 2010.

[42] N. Yabuuchi, K. Kubota, M. Dahbi, S. Komaba, Research development on sodium-ion batteries, Chem. Rev. 114 (23) (2014) 11636−11682.

[43] H. Wang, Y. Zhong, B. Du, Y. Zhao, M. Wang, Recovery of both magnesium and lithium from high Mg/Li ratio brines using a novel process, Hydrometallurgy 175 (2018) 102−108.

V. Practical concerns and beyond

References 601

[44] S.-W. Kim, D.-H. Seo, X. Ma, G. Ceder, K. Kang, Electrode materials for rechargeable sodium-ion batteries: potential alternatives to current lithium-ion batteries, Adv. Energy Mater. 2 (7) (2012) 710–721.

[45] T. Roberts, W. Stewart, Inorganic phosphorus and potassium production and reserves, Better Crops 86 (2) (2002) 6–7.

[46] P.E. Sorensen, W.J. Mead, A cost-benefit analysis of ocean mineral resource development: the case of manganese nodules, Am. J. Agric. Econ. 50 (5) (1968) 1611–1620.

[47] J.B. Nicolas, P.W.S. Edward, W. Herve, Manganese deposits of Africa, Int. Union Geol. Sci. 39 (2) (2016) 285–317.

[48] J.M. Hammarstrom, M.L. Zientek, H.L. Parks, C.L. Dicken, Assessment of Undiscovered Copper Resources of the World, 2015, 2328-0328; US Geological Survey: 2019.

[49] Z. Babakhanova, M. Aripova, N. Khodjaev, R. Khamidov, The Study of the Mineralogical Composition and Flotation Parameters of Quartz-Graphitized Schist, 2016.

[50] U.S. Geological Survey, Mineral Commodity Summaries, U.S. Geological Survey (2019) 200.

[51] N. ThejoKalyani, S.J. Dhoble, Novel materials for fabrication and encapsulation of OLEDs, Renew. Sustain. Energy Rev. 44 (2015) 319–347.

V. Practical concerns and beyond

Index

'*Note*: Page numbers followed by "f" indicate figures and "t" indicate tables.'

A

Absorption layer, 260–261
Active displays, 507
Advanced colorful approach, 193–194, 195t–197t, 198f, 198t, 199f
Aerospace applications, 365
Air mass (AM), 535
  1.5 global, 185
Alkaline-earth aluminosilicate glass sealants, 386–387
Alkaline-earth-based aluminosilicate glass-ceramics, 390
Alkaline-earth-based silicate glasses, 393–394
Alkaline-earth silicate-based phosphor, 450
  combustion method, 455–456
  coprecipitation method, 451–453
  hydrothermal method, 454–455
  liquid-phase precursor method, 460
  micro-pulling-down (μ-PD) technique, 460–461, 461f
  microwave method, 459
  self-propagating low-temperature synthesis, 455–456
  sol-gel method, 456–459
  solid-state method, 450–451
  spray pyrolysis method, 453–454
Alkaline-earth sulfides, 411–412
Alkaline fuel cell (AFC), 373
Alkali oxides, 387
Amorphous silicon cells, 587
Amorphous silicon thin film, 588
Amorphous Si solar cell technologies, 54–55
  amorphous silicon, 54
  A-Si/mc-Si tandem cell, 54
  A-Si/pc-Si tandem cell, 55
  silicon, 54
Analytical techniques, 34–37
  current–voltage characteristics, 34, 35f
  current–voltage curve under illuminated conditions, 35
  dark current characteristics, 35–36
  PCVD analysis technique, 37, 38f
  photocurrent decay (PCD), 37
  spectral response solar cell, 36, 37f

Analytic Specrod-50 spectrophotometer, 510
Anode materials, 336–337
Antifog coating system
  polydimethylsiloxane (PDMS). *See* Polydimethylsiloxane (PDMS)
  Sylgard resin. *See* Sylgard resin
Antireflection coatings (ARCs), 531–532, 548–550, 549f, 558
Aqueous lithium—air batteries (LAB), 316–317, 316f
Aqueous sol-gel method, 456
Atmospheric pressure chemical vapor deposition (APCVD), 50, 51f
Atomic force microscopy (AFM), 70, 124, 266–267, 267f
Aviation applications, 364

B

Back surface illuminated solar cell spectral response, 122
Barium-free glass-ceramic composition, 397
Batteries, 15
Bauxite, 584
Biochemicals, 583
Biocomposites, 583
Bioenergy, 583
Biofoams, 583
Biofuel, 583
Bioluminescence, 408
Biomass, 583
Bioplastics, 583
Blue phosphors, 412–413, 445–446
Boltzmann's constant, 410
Bridgman—Stockbarge method, 431–433

C

Cadmium telluride, 44
  thin film, 588
Calcium titanium oxide ($CaTO_3$), 250–251
Capacitance measurements, 76
Capacity retention, 349
Carbon-based materials, 150–154
Carbon black, 150–151

603

604 INDEX

Carbon-free energy, 526
Carbon nanotubes, 153—154
Carboxymethyl cellulose (CMC), 318
Cathode materials, 24, 337
Cathode ray tubes (CRTs), 446, 473
Cathodoluminescence, 408
Cell parameters, upper limits of
    efficiency, 29
    fill factor (FF), 28—29
    open-circuit voltage (VOC), 28
    short-circuit current (I_{SC}), 28
Chalcopyrite (CIGS) solar cell technology, 55
Characteristic luminescence, 411
Charge-coupled device (CCD), 420
Charge/discharge rate, 349
Charging process, 348
Chemical method, 170—174, 170f
Chemical vapor deposition (CVD), 47—48
Chemiluminescence, 408
Chromophore, 507—508
Citric acid, 459
Closed-paced oxygen arrays, 299—301
Closed space sublimation, 51, 52f
Cold body radiation, 408
Cold light, 408
Color-rendering index (CRI), 483
Combustion method, 64, 455—456
Combustion synthesis, 64, 416
Commercialization, 194
Commission Internationale de l'Eclairage (CIE)
    coordinates, 483, 499, 501f
Compact fluorescent lamps (CFLs), 16, 482
Conductive polymer-based materials, 154—156
Configurational coordinates, 409—410, 409f
Conformational change, 78
Conventional lighting, 449, 449t
Copper indium gallium selenide, 45
    thin film, 588
Coprecipitation method, 63, 417, 451—453
Correlated color temperature (CCT), 483, 516
Coulombic efficiency, 350
Counter electrode (CE), 141, 149—158, 149f, 184—185,
    193, 217—218, 217f
    carbon-based materials, 150—154
    carbon black, 150—151
    carbon nanotubes, 153—154
    conductive polymer-based materials, 154—156
    graphene, 152—153
    inorganic compound-based materials, 156—158
    mesoporous carbon, 151—152
    metal nanoparticle materials, 150
    organic materials, 158
    platinum-based materials, 149

    Pt-free composites, 156
    transition-metal carbides, 157
    transition-metal oxide, 157—158
Cross-relaxation, 431
Crystalline silicon solar cells
    biased p-n junction, 108
    2D/3D screen-printed metallization effect
      characterization, 132
      experimental procedure, 126, 127f
      industrial crystalline silicon solar cells, 131—137
      metallization, 132—137, 134f—136f, 137t
      NaOH—NaOCl-textured crystalline silicon solar
        cells, 125—131
      solar cells efficiency, 126—131, 128f—131f
    efficient solar cells, 111—117
      electrical losses, 112—115, 115f
      high efficiency, 111, 112f
      high-performance solar cells method, 117
      optical losses, 115—116
      structure, 117
    fabrication process, 118—121
      manufacture technology, 118—119
      silicon solar cell technology, 119—121
    illuminated infinite p-n junction, 108—109
    large area crystalline silicon solar cells
      atomic force microscopy, 124
      back surface illuminated solar cell spectral
        response, 122
      current—voltage characteristics, 121—122
      dark current characteristic measurement, 122
      front-illuminated solar cell spectral response, 122
      illuminated I—V curve measurement, 122
      KLA-tencor development series, stylus profilers, 124
      PCVD measurement technique, 122—123
      photocurrent decay (PCD), 123—124, 123f
      recombination velocity at the surface, 124
      saturation current at the emitter, 124
      scanning electron microscope, 124
      spectral response of solar cells, 122
      Zeta instruments, 124—125
    Ohmic resistance, 110
    open-circuit voltage, 108
    photocurrents, 109—110
    physics of, 107—111
    real solar cells, 109—111, 110f
    series resistance, 110
    shunt resistance, 110
    solar array, 111, 111f
    space charge region, 108
    two-diode model, 110
μc-Si:H deposition conditions, 56
Current—voltage characteristics, 71—72, 121—122, 185,
    186f

D

Dark current characteristic measurement, 122
2D/3D screen-printed metallization effect
  characterization, 132
  experimental procedure, 126, 127f
  industrial crystalline silicon solar cells, 131–137
  metallization, 132–137, 134f–136f, 137t
  NaOH–NaOCl-textured crystalline silicon solar cells, 125–131
  solar cells efficiency, 126–131, 128f–131f
Deep discharge, 356
Deintercalation, 287–288
Device
  construction, 180–181, 181f
  evaluation, 267–269, 269f, 270t
  workings, 181–183, 181f–182f
Differential scanning calorimetry (DSC), 288
Diffusion technologies, 41
Dip coating, 160–161, 162f, 162t
Direct methanol fuel cell (DMFC), 373
Discharge lamp, 447–448. *See also* Fluorescent lamps
Disilicates, 446
Display devices, 469–472
Display technology, 505. *See also* Organic
      light-emitting diode (OLED)
  green phosphors, 509
  light-emitting diodes (LEDs), 507
  Mq_2, 509
  photometric properties, 516, 517f, 518t
  photophysical analysis
    absorption spectra, 513, 514f, 515t
    photoluminescence (PL) spectra, 513–516, 515f–516f
  quinoline, 507–508
  reagents and solvents, 509
  structural analysis, 511–513, 511f
  Znq_2, 509, 510f
Doctor-blade method, 160, 160f, 161t, 219
Downconversion phosphors, 415, 415f
Drop coating, 162, 163t
Dyes, 216
Dye-sensitized solar cell (DSSC), 536
  advanced colorful approach, 193–194, 195t–197t, 198f, 198t, 199f
  anatomy, 142–143
  commercialization, 194
  construction and working, 180–183
  counter electrodes, 193, 217–218, 217f, 591
  device construction, 180–181, 181f
  device workings, 181–183, 181f–182f
  dyes, 216, 589
  electrolytes, 216–217, 590
  fluorine tin oxide (FTO) glass substrate, 215–216

  future prospects, 194
  highest occupied molecular orbital (HOMO), 216
  hole conductor, 192–193
  indium tin oxide (ITO), 215–216
  $J-V$ characteristics, 238–239, 238f–239f, 239t
  light harvester, 191
  limitations, 194
  lowest unoccupied molecular orbital (LUMO), 216
  materials as electrodes
    counter electrode, 149–158
    photoanode, 143–148, 144f
  materials requirement for, 183–185
    counter electrode, 184–185
    hole conductor, 184
    metal oxide semiconductor, 183–184
    photosensitizer, 184
    sealing material, 185
    substrate (TCO), 183
  molecular photovoltaic (PV) materials, 180
  nanooxidic photoanode materials, 218–228
  nonrenewable energy sources, 213
  organic photovoltaics, 140
  performance of, 236–241
  photoanode, 215–216
    1D construction for, 229–235
    materials, 190–191
  second-generation solar cells, 140
  semiconductor oxides, 589–590
  solar cell performance, 185–190
    air mass 1.5 global, 185
    current–voltage characteristics, 185, 186f
    efficiency, 188
    electrochemical impedance spectroscopy (EIS), 189
    fill factor, 187–188
    Mott–Schottky analysis, 188–189
    output parameters, 185–188
    photocurrent, 186
    photovoltage, 187
    quantum efficiency, 189–190
    series and shunt resistance, 188
  solar cells, 139–140
  solar power technology, 214–215
  solar radiation, 214
  synthesis, 158–174
    chemical method, 170–174, 170f
    deposition method and procedure, 159
    dip coating, 160–161, 162f, 162t
    doctor blade, 160, 160f, 161t
    drop coating, 162, 163t
    electrospray deposition, 165–166, 165f
    gas-phase deposition method, 166–170
    inkjet printing, 165
    liquid-phase precursor. *See* Liquid-phase precursor

606

INDEX

Dye-sensitized solar cell (DSSC) (*Continued*)
    physical method, 159–170, 159f
      screen printing, 162–164, 164t
      spin coating, 162, 163f, 163t
      TiO_2 paste procedure, 158
    thin-film solar cells, 140
    TiO_2/ZnO nanocomposite photoanode, 237–238, 237f, 238t
    traditional p-n junction solar cells, 183, 183t
    working principle, 141–142, 142f
Dysprosium, 425

E

Eco-friendly lighting, 446–450
Efficient solar cells, crystalline silicon solar cells, 111–117
    electrical losses, 112–115, 115f
    high efficiency, 111, 112f
    high-performance solar cells method, 117
    optical losses, 115–116
    structure, 117
Electrical characterization, 71–76
    capacitance measurements, 76
    current–voltage characteristics, 71–72
    electron impedance spectroscopy, 75, 75f
    Hall effect, 72–73, 74f
    quantum efficiency (QE), 74–75
Electrical conductivity, 348
Electrical losses
    back surface field (BSF), 112
    emitter, photocurrent/saturation current from, 113
    emitter, resistance in, 115, 115f
    grid finger, contact resistance, 114
    Ohmic metal–semiconductor contact, 113
    recombination losses, 112
    Schottky contact, 113
    semiconductors, Ohmic resistance losses in, 115
Electric energy storage (EES) technologies, 351–352
Electric vehicles (EVs), 314, 363
Electrochemical impedance spectroscopy (EIS), 189
Electroluminescence, 408, 428, 507, 573–574
Electrolyte, 216–217
Electromagnetic converters, 5
Electromagnetic technologies, 5–6
Electron impedance spectroscopy (EIS), 75, 75f
Electron transport layer (ETL), 19–20, 507
Electrospinning method, 229–230
Electrospray deposition, 165–166, 165f
Electrostatic converters, 5
Elevated temperature, 350
Emergency power backup, 364
Emitter diffusion processes, 41
Energy consumption, 449–450

Energy conversion, 7, 568–572, 569f, 581, 586–592
Energy density, 349
Energy-efficient lighting, 449, 449t
Energy harvesting, 4, 4f, 568–572, 569f, 581, 586–592
    sources, 5–6, 5f–6f
Energy materials
    characterization techniques
      electrical characterization, 71–76
      lifetime assessment, 78
      optical properties, 76–78
      structural evaluation, 65–67
      surface evaluation, 67–71
    energy conversion, 7
    energy harvesting, 4, 4f
      sources, 5–6, 5f–6f
    energy saving, 6f, 8, 9t–11t
    energy storage, 7–8
    failure analysis/damage evaluation, 79–80
    failure parameters, 79
    fundamental physics, 8–21, 12f
      batteries, 15
      compact fluorescent lamps (CFLs), 16
      external quantum efficiency, 21
      fuel cell, 12–14, 14f
      internal quantum efficiency, 20
      light-emitting diodes, 17–19, 18f, 19t
      optoelectronic light-emitting devices, 16
      organic light-emitting diodes, 19–20
      photovoltaics cell, 14, 15f
      power efficiency, 21, 22f, 22t
      supercapacitor, 15–16, 16f
      thermoelectrics, 12, 13f
      white light approaches, 17
    future, 23–24
    potential applications, 23, 23f
    pressure to electrical energy, 7
    solar to electric energy, 7
    synthesis techniques, 62–65
      combustion method, 64
      coprecipitation method, 63
      hydrothermal synthesis, 65
      Pechini/citrate gel methods, 65
      sol-gel method, 63
      solid-state diffusion method, 62–63
      spray pyrolysis, 64
    thermal to electric energy, 7
Energy protection, 449–450
Energy-related applications, 567–568
    energy harvesting and conversion, 568–572, 569f
Energy saving, 6f, 8, 9t–11t
    lighting, 446–450
    materials, 597
    materials for, 572

optoelectronic, 573–575
technology, 435–436
Energy storage, 7–8
materials for, 572, 592–593
economic analysis, 595–596
life-cycle analysis (LCA), 593–595
lithium-ion batteries, 572
supercapacitors, 572, 573f
Energy transfer upconversion, 414
Ethylammonium (EA), 256
Ethylene carbonate (EC), 314–315
Ethylene vinyl acetate (EVA), 550–551
Europium (Eu), 425
chemical properties, 487, 487t–488t
physical properties, 487, 487t–488t
Evaporation, 48, 48f
E-waste, 586
External quantum efficiency (EQE), 21, 74–75,
536–538, 537f, 539f, 574

F

Failure analysis/damage evaluation, 79–80
Failure parameters, 79
Field emission display (FED), 472
Field emission scanning electron microscopy
(FESEM), 86–87, 90, 91f
Fill factor, 187–188
First-generation solar cells
amorphous silicon, 587
hybrid silicon, 588
monocrystalline silicon, 587
polycrystalline silicon, 587
Fluorescence, 409
spectroscopy, 76–77
Fluorescent lamps, 447–448
Fluorine tin oxide (FTO) glass substrate, 215–216
Formamidinium (FA), 256
Fourier transform infrared (FT-IR) spectroscopy,
86–87, 92–93, 93f, 511–513, 511f
Franck–Condon principle, 409–410
Front-illuminated solar cell spectral response, 122
Fuel cells, 12–14, 14f, 570–572
alkaline fuel cell (AFC), 373
applications of, 570–571, 571f
direct methanol fuel cell (DMFC), 373
Molten carbonate fuel cell (MCFC), 373
phosphoric acid fuel cell (PAFC), 373
polymer electrolyte fuel cell (PEFC), 373
solid oxide fuel cell (SOFC). *See* Solid oxide
fuel cell (SOFC)
types, 373, 374t

G

Gallium arsenide, 45, 45f
Gaseous phosphorous oxychloride ($POCl_3$), 41
Gas-flow sputtering, 50
Gas-phase deposition method
evaporation, 167–170, 168f
electron beam evaporation, 168, 169f
sputtering, 168–170
thermal evaporation, 167–168, 168f
physical vapor deposition, 166–167, 166f
arc deposition, 167
pulse laser deposition (PLD), 166–167, 167f
vacuum deposition, 166
Gel polymer electrolytes (GPEs), 313–314, 319–322,
321f, 321t
Geothermal energy, 583
Global warming potential (GWP), 585
Graphene, 152–153
Greenhouse gas (GHG) emissions, 583
Green phosphors, 413–414, 445–446
display technology, 509

H

Hall effect, 72–73, 74f
High-efficiency silicon solar cell technologies, 40–43
antireflection technologies, 43
auxiliary technologies, 42
oxidation technologies, 41–42
p–n Junction production technologies, 41
solar cell metallization technique, 43
Higher energy production, hot conditions, 44
Higher hydrocarboxylic acid, 65
Highest occupied molecular orbital (HOMO), 216
High open-circuit voltage, 33–34
back surface field (BSF), 33–34
requirements for, 33–34
High refractive index (HRI), 552–553, 552f
High short-circuit current, 30–33
collection probability, 30
generation rate, 30
junction depth, 30
optical losses minimization, 30–32
orientation, 30
recombination minimization, 32–33
requirements, 30
High-target utilization sputtering, 50
HITACHI F-4000 spectrofluorometer, 510
Hole conductor, 184, 192–193
Hole transport layer (HTL), 19–20, 254, 255t, 261,
507
Host lattices, 445–446

608 INDEX

Hot wire chemical vapor deposition (HWCVD), 51, 52f, 58–59
Hybrid silicon cells, 588
Hydrofluoric acid (HF), 288
Hydropower, 583
Hydrothermal method, 454–455
Hydrothermal synthesis, 65

I

Illuminated infinite p-n junction, 108–109
  efficiency, 109
  fill factor (FF), 109
  I–V (current–voltage) characteristics, 108
  open-circuit voltage, 109
  short-circuit current, 109
Illuminated I–V curve measurement, 122
Incandescence, 408
Incandescent lamps, 447
  vs. light-emitting diodes (LEDs), 449, 449t
Indium tin oxide (ITO), 215–216
Inorganic compound-based materials, 156–158
Inorganic perovskite, 254–256
Intercalation compounds, 24
Internal conversion/nonradiative relaxation, 78
Internal quantum efficiency (IQE), 20, 74–75, 538–539, 540f, 574
Intersystem crossing, 78
Ion-assisted deposition (IAD), 50
Ion-beam-assisted deposition, 52, 53f
Ion-beam sputtering (IBS), 49
Ion transfer medium, 313f
Irreversible capacity loss (IRC), 299–301

J

J–V characteristics, 238–239, 238f–239f, 239t

K

Kinetic Monte Carlo simulations, 288
KLA-tencor development series, stylus profilers, 124

L

Lambertian emitters, 517
Large area crystalline silicon solar cells
  atomic force microscopy, 124
  back surface illuminated solar cell spectral response, 122
  current–voltage characteristics, 121–122
  dark current characteristic measurement, 122
  front-illuminated solar cell spectral response, 122
  illuminated I–V curve measurement, 122
  KLA-tencor development series, stylus profilers, 124
  PCVD measurement technique, 122–123
  photocurrent decay (PCD), 123–124, 123f

recombination velocity at the surface, 124
  saturation current at the emitter, 124
  scanning electron microscope, 124
  spectral response of solar cells, 122
  Zeta instruments, 124–125
Large-grain thin-film crystalline Si solar cells, 58, 59f
Lead zirconate titanate, 5–6
Life-cycle analysis (LCA)
  lithium-ion batteries (LIBs), 593–594, 593f
  next-generation batteries, 594–595
Lifetime assessment, 78
  minority carrier lifetime measurement, 78–79, 79f
Lifetime issues, 586
Light conversion process, 534, 535f
Light-emitting diodes (LED), 17–19, 18f, 19t, 407, 446–447, 574–575, 575f, 586
  applications, 448
  fabrication, 434
  flexibility, 507
  vs. incandescent lamps, 449, 449t
  luminaires, 448–449
  phosphors. *See also* Phosphors
  processing cost, 507
Light harvester, 191
Light sources
  fluorescent lamps, 447–448
  incandescent lamps, 447
  light-emitting diodes (LEDs), 448
Light trapping, 52–53
Liquid-phase precursor, 460
  chemical bath deposition, 171
  chemical vapor deposition, 173–174
  electrochemical deposition, 172–173
  hydrothermal, 172
  sol-gel, 170–171
  solvothermal, 171–172
  spray pyrolysis, 171
  template method, 171
Lithium–air batteries (LABs)
  anion trapping agent, 326–327, 327t
  aqueous, 316–317, 316f
  Coulomb's law, 325f
  electric vehicles (EVs), 314
  electrolytes, 318–322
    gel polymer electrolytes, 321f, 321t, 10011s0050
    lithium salts, 320t
    solid polymer electrolytes (SPEs), 318–319, 319f
  gel polymer electrolytes (GPEs), 313–314
  ion transfer medium, 313f
  lithium-ion batteries (LIBs), 313–314
  nonaqueous, 314–315, 315f
  PAN, 323
  PMMA, 324

poly(vinylidene fluoride) (PVDF), 323, 323t
polymer host, 322–324, 322f
PVDF-HFP, 323
salt, 324–326, 324t
solid-state, 317–318, 317f
solvents, 328–329, 328f
types, 314–318
Lithium bis(oxalato)borate (LiBOB), 324–325
Lithium bis(trifluoromethanesulfonyl)imide
    (LiTFSI), 324–325
Lithium cobalt oxide, 287–291, 287f
Lithium-intercalated compounds, cathode materials
  cathode materials for, 286t
  ideal cathode materials, 286f
  layered structures
    hydrofluoric acid (HF), 288
    Kinetic Monte Carlo simulations, 288
    lithium cobalt oxide, 287–291, 287f
    lithium manganese oxide, 292–293, 293f
    lithium nickel cobalt aluminum, 297–299, 298f
    lithium nickel cobalt manganese oxide,
        295–297, 296f
    lithium nickel manganese oxide,
        293–295, 294f–295f
    lithium nickel oxide, 291–292, 291f
    lithium-rich and manganese-rich oxide,
        299–301, 299f–300f
    mechanothermal process, 288–290
  spinel structures, 301–304
    lithium manganese oxide, 301–303, 302f
    lithium nickel manganese oxide, 303–304, 303f
  transition metal (TM) elements, 285
Lithium-ion batteries (LIBs), 313–314,
        572, 593–594, 593f
  aging, 356
  applications, 363–365
    aerospace applications, 365
    aviation applications, 364
    electric vehicles and toys, 363
    emergency power backup, 364
    marine applications, 365
    military applications, 365
    mobility equipment, 364
    portable instruments, 363
    portable power packs, 364
    small medical equipment, 363
    solar power storage, 364
    surveillance/alarm systems, 364
    UPS, 364
  basics, 345–346
  batteries, 338–342
    advantages of, 342
    capacity, 341–342

energy, 341–342
life, 350–351
novel materials, 342–345
parts of, 340
theoretical potential, 341–342
types, 338–340
working of, 341, 341f
capacity retention, 349
charge/discharge rate, 349
charging process, 348
chemistry, 352–353
construction, 350
Coulombic efficiency, 350
damage, 351
deep discharge, 356
degradation, 351
discharging process, 348
electrical conductivity, 348
electric energy storage (EES) technologies, 351–352
elevated temperature, 350
energy density, 349
energy storage materials, 348–349
  anode materials, 336–337
  cathode material, 337
  electrolyte materials, 338
  electrolytic capacitors, 336
  graphite, 336
  solid-electrolyte-interphase (SEI) layer, 336
  supercapacitors, 336
environmental impact, 354–355
expense, 356
future prospective, 365–366
industry, 353
irreversible capacity loss, 349
lithium–sulfur batteries, 357–360, 357f
market trends, 353
open-circuit voltage, 347–348, 347f
other batteries vs., 352
overcharging process, 348
overloading, 351
potential, 346–347
properties, 354
raw material, 354
research, 360–363
  anode, 362–363
  electrolytes, 362
  sulfur cathode, 361–362
safety, 355–356
schematic of, 346f
self-discharge, 350
short circuit, 348
specific capacity, 349
state-of-the-art of, 351–355

610 INDEX

Lithium-ion batteries (LIBs) (*Continued*)
  theoretical specific capacity, 348–349
Lithium manganese oxide, 292–293, 293f, 301–303, 302f
Lithium nickel cobalt aluminum, 297–299, 298f
Lithium nickel cobalt manganese oxide, 295–297, 296f
Lithium nickel manganese oxide, 293–295, 294f–295f, 303–304, 303f
Lithium nickel oxide, 291–292, 291f
Lithium niobate ($LiNbO_3$), 7
Lithium-rich and manganese-rich oxide, 299–301, 299f–300f
Lithium salts, 320t
Lithium–sulfur batteries, 357–360, 357f
Lowest unoccupied molecular orbital (LUMO), 216
Low-pressure chemical vapor deposition (LPCVD), 50, 50f
Low refractive index (LRI), 552–553, 552f
Luminaires, 448–449
Luminescence, 434f
  characteristic luminescence, 411
  characteristics of, 408–409, 408f
  mechanism of, 409–410, 409f
  mode of excitation, 408
  noncharacteristic luminescence, 411
  phosphors, 411. *See also* Phosphors
Luminescent solar concentrator (LSC), 538

M

Marine applications, 365
Mechanical energy, 5
Mechanothermal process, 288–290
Mesoporous carbon, 151–152
Mesoporous perovskite solar cells, 257–258, 258f
Metal contacts, 18–19
Metal electrode, 261–262
Metallurgical-grade silicon, 56–57
Metal nanoparticle materials, 150
Metal oxides, 459
  semiconductor, 183–184
Metal oxide semiconductor field effect transistor (MOSFET), 71–72
Microcrystalline Si thin-film technology, 55–56
Micro-pulling-down (µ-PD) technique, 460–461, 461f
Microstructure of films, 56
Microwave method, 459, 460f
Military applications, 365
Minority carrier lifetime measurement, 78–79, 79f
Mobility equipment, 364
Molecular photovoltaic (PV) materials, 180
Molten carbonate fuel cell (MCFC), 373
Monocrystalline silicon cells, 587

Mott–Schottky analysis, 188–189
Multicrystalline silicon, 587
Multiphonon relaxation, 431

N

Nanocone arrays (NCAs), 551
Nanomaterials, 21–23
Nanooxidic photoanode materials, 218–228. *See* Nanooxidic photoanode materials
  doctor-blade method, 219
  SnO_2, 225–227, 225t, 226f, 227t, 228f
  titanium (IV) oxide (TIO_2), 218–228, 218t–219t, 220f, 222f, 224f
  ZnO, 218–228, 221t, 222f, 224f
Nanoparticles, 545
Nanostructured anodes, 23
Nanostructured flexible electrodes, 21–23
Nanotechnology, 547, 551
Neutron diffraction (ND), 294
Next-generation batteries, 594–595
Nickel cobalt aluminum (NCA), 297
Nonalkaline glass-MgO glass-ceramic composites, 397
Nonaqueous LAB, 314–315, 315f
Nonaqueous sol-gel method, 456
Noncharacteristic luminescence, 411
Nonradiative process, 408–409
Nonrenewable energy sources, 213

O

Optical filter, 545–546
Optical losses
  antireflection process, 116
  nonabsorbed light losses, 116
  shadowing contact fingers, 116
  solar cell, 115–116
  textured surfaces, 116
Optical properties
  fluorescence spectroscopy, 76–77
  polarized light microscopy, 77
  spatial and time-resolved photoluminescence, 77–78
Optimized diffusion conditions, 58
Optoelectronic light-emitting devices, 16
Optoelectronic materials, 573–574
  external quantum efficiency (EQE), 574
  internal quantum efficiency (IQE), 574
  light-emitting diodes (LEDs), 574–575
  organic light-emitting diode (OLED), 574–575
  power efficiency, 574
Organic light-emitting diode (OLED), 19–20, 481–482, 505, 506f, 574–575, 576f, 586
  active displays, 507
  applications, 519

INDEX

611

attributes, 506
Lambertian emitters, 517
limitations
   cost, 519
   lifetime, 519
   water intrusion, 519
mechanism, 507
passive displays, 507
power consumption, 517
response time, 519
single RGB pixel, 505–506, 506f
thinner and lighter displays, 519
wide viewing angle, 519
Organic materials, 158
Organic photovoltaics, 140
Organic semiconductors, 19–20
Organic solar cells (OSCs), 592
Organometal perovskite, 256
Output parameters, 185–188
Overpopulation, 583–584
Oxidizing agent, 570–571

P

PAN, 323
Passive displays, 507
PCVD measurement technique, 122–123
Pechini/citrate gel methods, 65
Perovskite solar cells (PVSCs), 550, 591–592
  calcium titanium oxide (CaTO₃), 250–251
  cations, 250–251, 250f
  challenges, 270
  device anatomy, 257–259, 257f
  device evaluation, 267–269, 269f, 270t
  device operation, 262–263
  ethylammonium (EA), 256
  fabrication approach, 257–259
  formamidinium (FA), 256
  future outlook, 271–272
  general characterization, 265–267
    atomic force microscopy (AFM), 266–267, 267f
    scanning electron microscopy (SEM), 265–266, 265f
    transmission electron microscopy (TEM), 266, 266f
    X-ray diffraction spectrometer, 265
  hole transport layers, 254, 255t
  inorganic perovskite, 254–256
  layer requirement, 259–262
    absorption layer, 260–261
    hole transport layer, 261
    metal electrode, 261–262
    photoanode/electron transport layer, 259–260
  mesoporous perovskite solar cells, 257–258, 258f
  N-i-p structure approach, 258, 258f
  organometal perovskite, 256
  planar p-i-n approach, 258–259, 258f
  solar energy, 249–250
  specific characterization
    current density–voltage characteristics, 263–264
    time-resolved photoluminescence, 264–265
  synthesis techniques, 252–254
    hybrid solution and vapor method, 253–254
    solution-based methods, 253
    vacuum-based deposition, 253
  working mechanism, 262–263, 263f
Phosphor-converted light-emitting diodes (pc-LEDs), 449–450
Phosphorescence, 409
Phosphoric acid fuel cell (PAFC), 373
Phosphors, 407. *See also specific phosphors*
  applications, 427, 427f
  characterization for
    photoluminescence, 422–425, 422f
    scanning electron microscope (SEM), 418–420, 419f
    transmission electron microscope (TEM), 420–421
    X-ray diffraction (XRD), 418, 418f
  host lattices, 445–446
  impurities, 411
  light-emitting diodes (LEDs), 427–435, 430f–432f
  synthesis and characterization of
    combustion synthesis, 416
    coprecipitation method, 417
    sol-gel synthesis, 416–417
    solid-state diffusion, 415–416
    solvothermal synthesis, 417
  types of
    blue, 412–413
    green, 413–414
    red, 413
    yellow, 414–415
Photoaging, 531
Photoanode materials, 143–148, 144f, 190–191
  carbon-based materials, 148
  1D construction for
    absorbance spectra, 232
    electrospinning method, 229–230
    parameters, 231t
    porous morphology, 229
    refinement analysis, 234t
    TiO₂/PAN composite fibers, 229–230, 230f
    ZnO nanoparticles, 232–233, 234f, 235, 236f
  3D nanostructures, 146
  2D TiO2 nanostructures, 145–146
  electron transport layer, 259–260
  metal oxide/semiconductor materials, 147
  plasmonic-based materials, 148

Photoanode materials (*Continued*)
titanium dioxide-based materials, 143–147
Photocurrent, 186
Photocurrent decay (PCD), 123–124, 123f
Photodegradation, 531–532
Photoluminescence (PL), 77–78, 408, 422f,
429f, 432f, 445–446, 510
blended thin films, 496–499
RE(TTA)$_3$dmphen blended films, 492
spectrofluorophotometer
components, 422–423
optical system of, 423–425, 424f
Photosensitizer, 184
Photovoltage, 187
Photovoltaics cell, 14, 15f, 569–570, 571f
Physical vapor deposition (PVD), 46, 47f
Piezoelectric oscillators, 5
Piezoelectric technologies, 5–6
Planar p-i-n approach, 258–259, 258f
Plasma display panels (PDPs), 472
Plasma-enhanced chemical vapor deposition
(PECVD), 50, 51f, 548
Platinum-based materials, 149
Polarized light microscopy, 77
Pollutants, 593–594
Poly(methyl methacrylate) (PMMA), 318, 324
Poly(vinylidene fluoride) (PDVF), 323t
Polycrystalline silicon
cells, 587
deposition, 57
glass, 55
Polydimethylsiloxane (PDMS), 84–86, 87t
adhesion and crosslinking, 93–94, 94t
antifog analysis, 97–100, 99f
chemicals specifications, 85t
evaporation time of droplets, 99, 100t
field emission scanning electron microscopy
(FESEM), 86–87, 90, 91f
Fourier transform infrared (FTIR), 86–87, 92–93, 93f
interaction, 89f
rainfall impact, 96–97, 97f
water contact angle (WCA) measurement, 88–90, 88f
X-ray diffraction, 94–95, 95t, 98t
Polyethylene glycol (PEG), 416–417
Polyethylene oxide (PEO), 314–315
Polymer electrolyte fuel cell (PEFC), 373
Polymethylmethacrylate (PMMA), 487, 492
Polystyrene (PS), 487, 492
Portable instruments, 363
Portable power packs, 364
Possible solar cell structures, 53, 54f
Potassium niobate (KNbO$_3$), 7
Power efficiency, 21, 22f, 22t

Propylene carbonate (PC), 314–315
Pt-free composites, 156

Q
Quantum dot-sensitized solar cells (QDSSCs), 591
Quantum efficiency (QE), 74–75, 189–190, 526
Quinoline, 508–509
advantages, 507
chromophore, 507–508
metal chelates, 508, 508f
small molecules, 508, 508f

R
Radiative process, 408–409
Radiative relaxation, 78
Radioluminescence, 408
Rare earth-based RE(TTA)$_3$dmphen. *See* Solid-state
lighting (SSL)
Rare-earth elements
dysprosium, 425
europium, 425
properties, 426
solid-state lighting (SSL), 486–487
Rare-earth ions, 445–446
Reactive sputtering, 49
Recombination minimization
bulk passivation, 33
surface passivation, 32
Recombination velocity at the surface, 124
Recycling, 581
advantages, 584, 584f
aluminum, 584
bauxite, 584
electrical and electronic equipment, 586
paper, 584–585
photovoltaic panels, 585
Red phosphors, 413, 445–446
Reflection, 541
RF-5301 PC instrument, 422–423, 422f
Rhodamine B, 424–425

S
Salt, 324–326, 324t
Samarium (Sm)
chemical properties, 487, 487t–488t
physical properties, 487, 487t–488t
Saturation current at the emitter, 124
Scanning electron microscope (SEM), 68, 124,
265–266, 265f, 418–420, 419f
principles of, 419–420, 420f
Scintillators, 473
Sealing material, 185
Second-generation solar cells, 140, 588

INDEX

amorphous silicon thin film, 588
cadmium telluride thin film, 588
copper indium gallium selenide thin film, 588
Seebeck effect, 12, 569
Shockley—Read—Hall (SRH)/multiphonon
    recombination, 78
Silicate-based phosphors, 461, 470t—471t
    applications of
        cathode ray tubes (CRTs), 473
        display devices, 469—472
        field emission display (FED), 472
        light-emitting diodes (LEDs), 469
        plasma display panels (PDPs), 472
        scintillators, 473
        solar cell, 473
    $BaLu_2Si_3O_{10}:Eu^{3+}$ red-emitting phosphors, 465—466,
        466f
    Ce^{3+}/Tb^{3+} codoped $CaLa_4Si_3O_{13}$ phosphors, 464—465,
        465f
    Eu^{2+}-activated barium silicate phosphors, 462—463,
        462f
    Eu^{2+}-doped $Ca_8Sc_2(PO_4)_{6-y}(SiO_4)_{1+y}$ phosphor,
        468—469, 468f
    near-UV-converting cyan-emitting $RbBaScSi_3O_9:Eu^{2+}$
        phosphor, 463—464, 464f
    Tm^{3+} and Dy^{3+} ion-codoped $Ba_3La_6(SiO_4)_6$
        phosphors, 466—468, 467f
Silicon, manufacture technology
    base material, 118
    crystal pulling method, 118—119
    fabrication of material, 118
    fabrication of silicon wafers, 119
    refractioning process, 118
Silicon nitride (SiN), 413, 548
Silicon solar cell technology, 119—121
    antireflection technologies, 121
    auxiliary technologies, 120—121
    oxidation technologies, 120
    P-n junction fabrication process, 119—120
    solar cell metallization, 121
Small medical equipment, 363
SnO_2, 225—227, 225t, 226f, 227t, 228f
Sodium calcium aluminosilicate glass systems,
    389—390
Sodium-ion batteries (NIBs), 594
Solar cell design
    analytical techniques, 34—37
        current—voltage characteristics, 34, 35f
        current—voltage curve under illuminated
            conditions, 35
        dark current characteristics, 35—36
        PCVD analysis technique, 37, 38f
        photocurrent decay (PCD), 37

spectral response solar cell, 36, 37f
basic functions, 29
cell parameters, upper limits of
    efficiency, 29
    fill factor (FF), 28—29
    open-circuit voltage (VOC), 28
    short-circuit current (I_{SC}), 28
high fill factor, 34, 34f
high open-circuit voltage, 33—34
    back surface field (BSF), 33—34
    requirements for, 33—34
high short-circuit current, 30—33
    collection probability, 30
    generation rate, 30
    junction depth, 30
    optical losses minimization, 30—32
    orientation, 30
    recombination minimization, 32—33
    requirements, 30
losses in solar cells, 29
Solar cells, 14, 473, 533f—534f, 544f, 559f—560f
    antireflection coatings (ARCs), 548—550, 549f
    encapsulation, 550—551, 551f
    factors, 526, 531f—532f
        antireflection coatings (ARCs), 531—532
        bandgap, 529
        durability, 532
        $I—V$ characteristics, 527—528, 528f
        nanofilters, 531—532
        photoaging, 531
        photodegradation, 531—532
        power curve $(P—V)$, 528—529, 528f—529f
        preventive stabilizers, 532
        single-junction solar cells, 529, 530f
        solar irradiance, 527—528
        temperature, 528—529
    first-generation, 587—588
    limitations, 556—557
    maintenance, 526
    operation, 526
    performance, 185—190
    quantum efficiency (QE), 526
    second-generation, 588
    solar concentrator, 546—547, 547f—548f
    solar filter, 544—546, 545f—546f
    spectral response (SR), 526, 534, 536f
        defined, 535
        external quantum efficiency (EQE), 536—538, 537f
        internal quantum efficiency (IQE), 538—539, 540f
        light conversion process, 534, 535f
        reflectance and transmittance, 541—543, 542f
        source spectrum, 534—535
        spectral mismatch, 539—540, 541f

614 INDEX

Solar cells (*Continued*)
  spectral splitting, 551–554, 552f–554f
  tandem system, 554–556, 555f–556f
  third-generation, 588–592
Solar cell technology
  commercial Si cell technology
    crystal pulling process, 39
    polycrystalline silicon material manufacturing, 39
  high-efficiency Si solar cells, 40–43
    antireflection technologies, 43
    auxiliary technologies, 42
    commercial Si solar cells, 38–39
    oxidation technologies, 41–42
    p–n Junction production technologies, 41
    refractioning processes, 39
    solar cell metallization technique, 43
  process used for, 40
    sheet materials, 40
    silicon wafer manufacture, 40
    silicon wafers out of polycrystalline material, 40
  solar cell design. *See* Solar cell design
  thin-film solar cell technologies, 43–59
Solar concentrator, 546–547, 547f–548f
Solar devices, 526, 527f
Solar energy, 5–6, 526, 583
Solar filter, 544–546, 545f–546f
Solar irradiance, 527–528
Solar power
  photovoltaic panels, 5–6
  storage, 364
  technology, 214–215
Solar radiation, 214
Solar thermophotovoltaic (STPV) systems, 545–546
Sol-gel method, 63, 456–459, 457t–458t
Sol-gel synthesis, 416–417
Solid-electrolyte interphase (SEI) layer, 297
Solid oxide fuel cell (SOFC)
  components, 377–378, 377f
    requirements, 378–379
  definition, 375
  methodology, 380–381
  operation, 375–377, 376f
  sealants materials
    alkaline-earth aluminosilicate glass sealants, 386–387
    alkaline-earth-based aluminosilicate glass-ceramics, 390
    alkaline-earth-based silicate glasses, 393–394
    alkali oxides, 387
    barium-free glass-ceramic composition, 397
    glasses and glass-ceramics, 381–397, 382t–384t, 385f

    nonalkaline glass-MgO glass-ceramic composites, 397
    requirements, 379–380
    sodium calcium aluminosilicate glass systems, 389–390
    X-ray diffraction, 385–386, 385f
    zirconia, 386
    ZnO, 389
Solid polymer electrolytes (SPEs), 318–319, 319f
Solid-state diffusion method, 62–63, 415–416
Solid-state electrolyte lithium–air battery (LAB), 317–318, 317f
Solid-state lighting (SSL), 482
  advantages, 483, 484f
  blended thin films, 490
    characterization, 492, 492t, 493f
    photoluminescence spectra, 496–499, 496t, 497f–498f, 498t, 500f–501f
    preparation of, 491, 491f
    UV-Vis absorption spectra, 492–495, 494f–495f, 496t
  chemical structure, 489, 489f
  color-rendering index (CRI), 483
  Commission Internationale de l'Eclairage (CIE) coordinates, 483
  correlated color temperature (CCT), 483
  eco-friendly and energy-efficient organic complexes, 483, 484t–486t
  rare-earth elements, 486–487
  reagents and solvents, 488–489
  requisite of, 483
  synthesis procedure, 489, 490f
Solid-state method, 450–451, 451f
Solution-based methods, 253
Solution casting, 318–319
Solvents, 328–329, 328f
Solvothermal synthesis, 417
Sonoluminescence, 408
Spatial and time-resolved photoluminescence, 77–78
Spectral mismatch, 539–540, 541f
Spectral response (SR), 526, 534, 536f
  defined, 535
  external quantum efficiency (EQE), 536–538, 537f
  internal quantum efficiency (IQE), 538–539, 540f
  light conversion process, 534, 535f
  reflectance and transmittance, 541–543, 542f
  solar cells, 122
  source spectrum, 534–535
  spectral mismatch, 539–540, 541f
Spectral sensitivity, 526
Spectral splitting, 551–554, 552f–554f, 558
Spectrofluorophotometer
  components, 422–423

INDEX

615

optical system of, 423–425, 424f
Spray pyrolysis method, 64, 453–454
Sputtering, 48–49, 49f
SSL. *See* Solid-state lighting (SSL)
Structural evaluation
  Fourier transform infrared spectroscopy, 66
  IR spectroscopy, principles of, 66–67
  x-ray diffraction, 67
Substrate (TCO), 183
Sulfur cathode, 361–362
Supercapacitors, 15–16, 16f, 572, 573f
Surface evaluation
  atomic force microscopy (AFM), 70
  scanning electron microscopy (SEM), 68, 69f
  tunneling electron microscopy (TEM), 69
  X-ray photoelectron spectroscopy (XPS), 70–71, 71f
Surveillance/alarm systems, 364
Sustainability, 481–482, 482f, 581, 583–584, 583f
Sylgard resin, 84–86, 87t
  adhesion and crosslinking, 93–94, 94t
  antifog analysis, 97–100, 99f
  chemicals specifications, 85t
  evaporation time of droplets, 99, 100t
  field emission scanning electron microscopy
    (FESEM), 86–87, 90, 91f
  Fourier transform infrared (FTIR), 86–87, 92–93, 93f
  interaction, 89f
  rainfall impact, 96–97, 97f
  water contact angle (WCA) measurement, 88–90, 88f
  X-ray diffraction, 94–95, 95t, 98t

T

Tandem system, 554–556, 555f–556f
Tetraethylene glycol dimethyl ether (TEGDME), 328–329
Thermal energy harvesting, 5–6
Thermoelectrics, 12, 13f, 569, 570f
Thermoluminescence, 408
Thin-film crystalline Si solar cell technologies, 55
Thin-film solar cells, 140
  advantages, 44
    higher energy production, hot conditions, 44
    minimal power loss with better response
      to shading, 44
    spectral advantage, humid environments, 44
  amorphous Si solar cell technologies, 54–55
  atmospheric pressure chemical vapor deposition
    (APCVD), 50, 51f
  chalcopyrite (CIGS) solar cell technology, 55
  chemical vapor deposition (CVD), 47–48
  closed space sublimation, 51, 52f
  common features, 52–53
  evaporation, 48, 48f

gas-flow sputtering, 50
high-target utilization sputtering, 50
hot wire chemical vapor deposition (HWCVD), 51, 52f
ion-assisted deposition (IAD), 50, 52, 53f
ion-beam sputtering (IBS), 49
large-grain thin-film crystalline Si solar cells, 58, 59f
light trapping, 52–53
low-pressure chemical vapor deposition (LPCVD), 50, 50f
materials for, 44–45
  cadmium telluride, 44
  copper indium gallium selenide, 45
  gallium arsenide, 45, 45f
methods of, 46–52, 46f
microcrystalline Si thin-film technology, 55–56
physical vapor deposition (PVD), 46, 47f
plasma-enhanced chemical vapor deposition
  (PECVD), 50, 51f
polycrystalline silicon
  deposition, 57
  glass, 55
possible solar cell structures, 53, 54f
reactive sputtering, 49
sputtering, 48–49, 49f
thin-film crystalline Si solar cell technologies, 55
thin-film epitaxial Si solar cells, 58–59, 59f
thin-film polycrystalline Si solar cell technology, 56–57
transparent conductive oxide, 52–53, 53f
Third-generation solar cells, 588
  dye-sensitized solar cells (DSSCs), 589–591
  organic solar cells (OSCs), 592
  perovskite solar cells (PSCs), 591–592
  quantum dot-sensitized solar cells (QDSSCs), 591
Titanium (IV) oxide (TIO_2), 218–228, 218t–219t, 220f, 222f, 224f
Traditional p-n junction solar cells, 183, 183t
Transition-metal carbides, 157
Transition-metal oxide, 157–158
Transmission electron microscopy (TEM), 266, 266f, 420–421
  construction and operation, 420–421, 421f–422f
Transmittance, 541
Transparent conductive oxide, 52–53, 53f
Transparent encapsulation, 18–19
Triboluminescence, 408
Trisilicates, 446
Tunneling electron microscopy (TEM), 69

U

Ultraviolet (UV) light, 16
Upconversion phosphors, 414, 415f
UPS, 364

616 INDEX

V
Vacuum-based deposition, 253

W
Water contact angle (WCA) measurement, 88—90, 88f
White light-emitting diodes (WLEDs), 445—446
  phosphor, 407
    blue, 412
    green, 414
    red, 413
    yellow, 414
    silicate-based phosphors, 461
Wind power, 583

X
X-ray diffraction (XRD), 94—95, 95t, 98t, 385—386, 385f, 418, 418f
  spectrometer, 265
X-ray photoelectron spectroscopy (XPS), 70

Y
Yellow phosphors, 414—415
  downconversion, 415, 415f
  upconversion, 414, 415f

Z
Zeta instruments, 124—125
  dimensions, 125
  film thickness, 125
  finger contact profiling, 125
  images, 125
  metrology, 125
  texture characterization, 125
Zirconia, 386
ZnO nanoparticles, 218—228, 221t, 222f, 224f, 232—233, 234f, 235, 236f, 389

Printed in the United States
by Baker & Taylor Publisher Services